you ^Can only ~~have to~~ be Cruel to be ~~good~~ Kind.

DESIGN OF MACHINE ELEMENTS

Other Books by **VIRGIL MORING FAIRES**

V. M. Faires	ELEMENTARY THERMODYNAMICS, 3D ED.
V. M. Faires	THERMODYNAMICS, 4TH ED.
V. M. Faires	THERMODYNAMICS OF HEAT POWER, 2D ED.
V. M. Faires, C. M. Simmang, and A. V. Brewer	PROBLEMS ON THERMODYNAMICS, 4TH ED.
V. M. Faires and S. D. Chambers	ANALYTIC MECHANICS, 3D ED.
V. M. Faires and R. M. Wingren	PROBLEMS ON THE DESIGN OF MACHINE ELEMENTS, 4TH ED.

There is no great concurrence between learning and wisdom.

Francis Bacon (1561-1626)

DESIGN OF MACHINE ELEMENTS

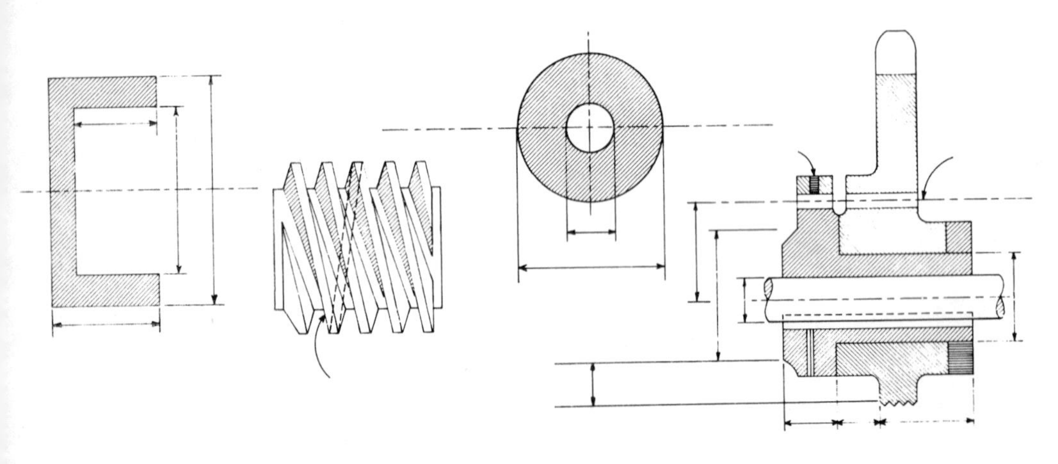

FOURTH EDITION

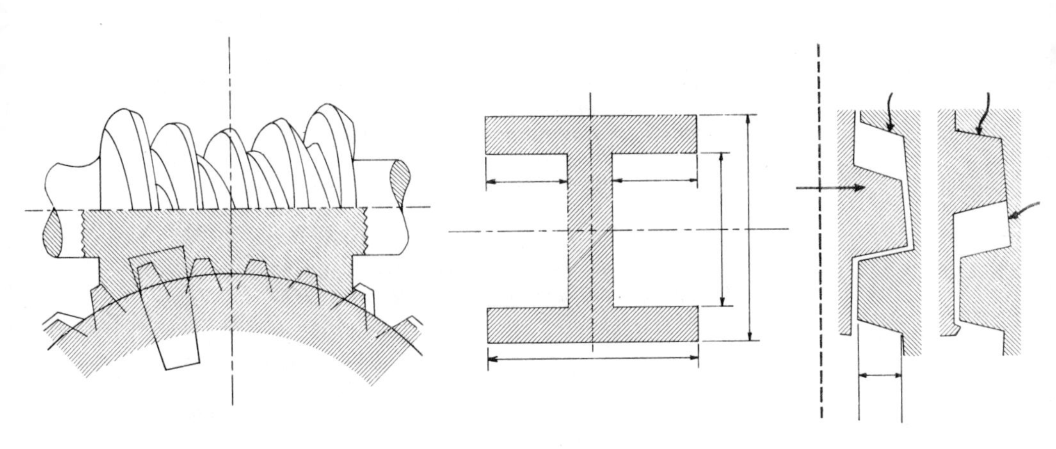

VIRGIL MORING FAIRES

UNITED STATES NAVAL POSTGRADUATE SCHOOL

THE MACMILLAN COMPANY, NEW YORK/COLLIER-MACMILLAN LIMITED, LONDON

PRINTING 20 YEAR 23456

Library of Congress catalog card number: 65–10732
The Macmillan Company
Collier–Macmillan Canada, Ltd., Toronto, Ontario

Printed in The United States of America

PREFACE

Since this book deals with engineering, it necessarily deals with science, because engineering is the *art* of applying the physical *sciences* to the problems of mankind. For the most part, the science involved here is the science of mechanics, and it is assumed that the reader is familiar with the contents of typical undergraduate courses in analytic mechanics and strength of materials. Therefore, derivations common to these courses are not repeated, but there are extensive reminders of limitations and of how to use the resulting equations.

The amount and quality of the presentation of the science of mechanics have been enhanced, but not at the expense of the engineering. On the contrary, a large percentage of the increased number of pages is attributable to the new engineering information included. This increased emphasis on engineering is more than ever desirable as an offset to the apparently inevitable notions of students about the infallibility of science in an engineering situation. The most used parts of the science of mechanics are precise and impeccable, but there are innumerable rough spots in the path of the practitioner. One must have an instrument (or the tools) to practice an art, and the machine elements serve admirably as instruments for the beginner. While it is not necessarily true that good judgments are inevitable when arrived at from good knowledge, it is self evident that good knowledge is a necessary prerequisite for good judgment; and we have tried to emphasize this point throughout by presenting certain engineering knowledge, by directing attention to references on specific points, and by including a long list of references whose titles are suggestive of sources of additional knowledge. Since the point is repeatedly made that good engineering judgment is a consequence of good engineering knowledge, we are hopeful that the student who reads practically the entire text understands that good decisions are not made in a cerebral void, that science must be tempered to the "facts" of life, that he, the student, has just begun to acquire engineering knowledge, and that engineering changes because we learn more but that there is no point in ignoring what we "know" now. We might even be hopeful that some readers become interested enough to want to extend mankind's knowledge in certain areas or to correct what we presently "know." Considering the variability of engineering data, teachers are challenged to impress upon students that this variability does not imply that any old guess is as good as another.

The objective added in the third edition—to wit, to make the book more useful to the practicing engineer—has been so much appreciated by students that the current changes in tone and content are as much due to this objective as to the original pedagogical objective. New procedures, new engineering information, and different points of view are so prevalent throughout that (I am sorry to report) nearly all my solutions to the problems must now be revised. You will observe: another large expansion of the treatment of fatigue, more emphasis on probabilities and the natural variability of data, a more precise treatment of the science of mechanics, frequent references to the effects of residual stresses, a single equation easily adaptable to design by either the maximum shear stress theory or the octahedral shear theory (for a normal and a shear stress), computer solutions of Reynolds' equation for journal bearings, several approaches to gear-tooth design, and more detail on contact stresses, on theories of failure, and on the mechanics of brake shoes.

The textbook writer performs a function similar to that of a reporter. Since there is always so much that is relevant, the writer is responsible for choosing that which is most important to be said. Because opinions as to what should be done in mechanical engineering design are so varied and because machine design, the special province of mechanical engineers, deals with myriads of machines, the choice of material that emphasizes the right fundamentals and that allows a flexibility of approach is perhaps more difficult for machine design texts than for most others.

It is true that for each of the previous revisions the literature has been searched and judged as to its appropriateness for the objectives of this text. For this edition, more than usual time has been spent on this search; but in addition, every sentence has been examined and rejudged, with the result that it would be difficult to find a sentence dealing with engineering identical to one of the previous edition. Although most of the illustrations will be familiar to long-time friends of the book, there have been so many minor improvements in them that all together they make a significant change.

Problems are given in a separate book, *Problems on the Design of Machine Elements, 4th Ed.*, by Faires and Wingren. This arrangement is a convenience, both for the teacher and student. The book of problems includes many tables, charts, and formulas that increase its usefulness; in addition, it contains more problems than would be feasible to present within the covers of this volume.

As always, I shall be most grateful to those who will be so kind as to report errors they have discovered or who have suggestions for improvement.

VIRGIL M. FAIRES

Monterey, California

ACKNOWLEDGMENTS

FIRST EDITION

. . . to Mr. T. M. Durkan of the Gleason Works for . . . suggestions . . . on bevel gears; to Mr. M. D. Hersey . . . for reading the chapter on journal bearings . . . to Mr. A. M. Wahl of the Westinghouse Electric . . . for reviewing the chapter on springs; to Mr. D. T. Hamilton of the Fellows Gear Shaper Company for reading the chapter on spur gears; and to Mr. D. F. Windenburg of the United States Experimental Model Basin for as yet unpublished material on thin shells under external pressure . . . to Professor Earle Buckingham for repeatedly rendering assistance during the preparation of the chapters on gearing and for manuscript material which he has not yet published.

REVISED EDITION

. . . to Professors R. M. Wingren and J. G. H. Thompson, Professors A. H. Burr and M. L. Price . . . to Professor Earle Buckingham . . . to Mr. S. J. Needs . . . journal bearings.

THIRD EDITION

. . . to . . . Professors R. L. Acres of Texas A. & M. College, C. T. Grace of the University of New Mexico, Boynton M. Green of Stanford University, Fred Hirsch of the University of California, L. C. Price of

Michigan State College, and D. K. Wright of Case Institute of Technology. . . . Among others . . . W. W. Austin of North Carolina State College; A. M. Wahl, R. E. Peterson, and John Boyd of the Westinghouse Electric Co.; W. Coleman of the Gleason Works; H. G. Taylor of the Diamond Chain Co.; R. D. Knight of American Steel & Wire; E. N. Swanson of Brown & Sharpe Manufacturing Co.; E. Siroky of the Wagner Electric Corp.; F. A. Votta, Jr., of the Hunter Spring Co.; W. S. Worley of the Gates Rubber Co.; S. J. Needs, Kingsbury Machine Works. . . . to Professor P. B. Leonard of North Carolina State College for his careful work on the line drawings. . . .

FOURTH EDITION

I am grateful to several people for their helpful interest: John Boyd for the solutions of the journal bearing equations, F. A. Votta and W. R. Johnson for information on springs, O. W. Blodgett for material on the design of welds, T. E. Winter and W. D. Cram anent gearing, R. M. Wingren for the many helpful comments in general. There are numerous acknowledgments for specific help, including illustrations, in the book. And most of all, thanks to my wife, Lucile, for her patience, understanding, and valuable help during the preparation of the manuscript.

V.M.F.

SYMBOLS

The symbols agree in general with the recommendations of the American Standards Association. Some exceptions were deemed advisable. In gearing, where symbols had not already been established, the recommendations of the American Gears Manufacturers Association were followed.

a	linear acceleration; a dimension; acoustic speed
A	area; allowance
b	breath; a dimension
B	life of rolling bearings
c	distance from neutral axis to fiber where stress is desired, usually to extreme fiber; bearing clearance
C	center distance; spring index; a number; a constant
C_1, C_2, etc.	constants
D	diameter; D_o, outside diameter; D_i, inside diameter; etc.
e	eccentricity of load; effective error in gear-tooth profiles; efficiency
E	modulus of elasticity in tension
f	coefficient of friction
F	a force; total load; F_1, initial force or force at 1; F_m, mean force; F_A, force applied at point A; etc.
g	local acceleration due to gravity; g_o, standard acceleration of gravity (use 32.2 fps^2)
G	modulus of elasticity in shear or torsion

h	height; a dimension; h_o, minimum film thickness in journal bearing
$\boldsymbol{h}$	heat-transfer coefficient (transmittance)
hp	horsepower
i	interference of metal
I	rectangular or polar moment of inertia
J	polar moment of inertia; geometry factor, bevel gears
k	radius of gyration, $(I/A)^{1/2}$ or $(I/m)^{1/2}$; spring constant, the load-per-unit deflection; conductivity
K	Wahl factor for design; K_c, factor for effect of curvature in springs and curved beams; K_s, factor for shear in spring
K_t	theoretical stress-concentration factor; K_f, fatigue strength reduction factor
K_s, K_m	shaft-design factors from ASME *Code*
K_g, K_w, K_c	wear factors, spur gears, worm gears, cams
KE	kinetic energy
L	length; a dimension
m	mass in slugs (W/g)
m_ω	velocity ratio
M	moment of a force; bending moment; M_v, vertical component of the moment; M_m, mean value of the moment; etc.
n	angular velocity; revolutions per minute; n_s, revolutions or cycles per second; also n_c, number of cycles of fatigue loading
N	design factor or factor of safety; sometimes, load normal to a surface
N_t, N_c, etc.	N with a subscript stands for the quantity of something, as number of teeth or number of threads, number of coils, etc.
p	pressure in pounds per square inch
P	pitch of springs, gear teeth, threads, etc.; P_d, diametral pitch; P_c, circular pitch
q	quantity of fluid; notch-sensitivity index
Q	quantity of heat; sometimes a force, a constant
r	radius
R	reaction or resultant force; radius of the larger of two wheels; ratio; roughness; R_{1v}, vertical component of R_1; R_{1h}, horizontal component of R_1; etc.
R_C	Rockwell C hardness; R_B, Rockwell B hardness, etc.
s	stress; s_a, alternating component of the total stress; s_{as}, alternating component in shear; s_c, compressive stress; s_d, design stress; s_e, equivalent stress; s_{es}, equivalent shear stress; s_f, flexural or bending stress; s_m, mean stress; s_{ms}, mean stress in shear; s'_n, endurance limit; s_n, endurance strength; s_{no}, endurance strength in torsion, load from zero to maximum; s_{ns}, endurance strength in shear,

reversed load; s_s, shear stress; s_t, tensile stress; s_u, ultimate strength; s_{us}, ultimate shear strength; s_{uc}, ultimate compressive strength; s_y, yield strength in tension; s_{ys}, yield strength in shear or torsion; s_1, initial stress or one part of a total stress; s_A, stress at point A; see also σ and τ

S — Sommerfeld number; centrifugal force; separating force; the distance a body moves, displacement; scale

t — thickness; temperature in degrees Fahrenheit

T — torque; tolerance; T_m, mean value; T_a, alternating component

U — work, U_f, work of friction; U_s, work of spring

v — velocity; v_s, velocity in fps; v_m, velocity in fpm

V — volume; shearing force in beam section

w — load-per-unit distance; weight-per-unit distance; mass; weight

W — total weight or load; force

Y — Lewis' factor in gearing

Z — section modulus, I/c; absolute viscosity in centipoises

Z' — section modulus based on polar moment of inertia, J/c

α (alpha) — coefficient of thermal expansion; an angle; angular acceleration

β (beta) — angle of limiting friction; an angle; cam angle

γ (gamma) — pitch angle of bevel gears; shearing-unit strain

δ (delta) — total elongation; total deflection of a beam

ϵ (epsilon) — normal-unit strain; eccentricity ratio

η (eta) — efficiency of riveted or welded joint

θ (theta) — an angle

λ (lambda) — lead angle of worm or screw threads

μ (mu) — Poisson's ratio; absolute viscosity in lb-sec. per sq. in. (reyns)

ν (nu) — kinematic viscosity

π (pi) — 3.1416 . . .

ρ (rho) — density; sometimes variable radius

σ (sigma) — resultant normal stress in combined stresses; standard deviation

Σ (sigma) — shaft angle, bevel and crossed helical gears; summation sign

τ (tau) — resultant shearing stress in combined stresses; time; represents unit of time

ϕ (phi) — angle of twist; pressure angle in gears and cams; frequency in cycles per second or minute

ψ (psi) — helix angle in helical gearing; spiral angle

ω (omega) — angular velocity in radians per unit of time

ABBREVIATIONS

AFBMA	Anti-Friction Bearing Manufacturers Association
AGMA	American Gear Manufacturers Association
AISC	American Institute of Steel Construction
AISI	American Iron and Steel Institute
ALBA	American Leather Belting Association
ASA	American Standards Association
ASLE	American Society of Lubrication Engineers
ASM	American Society for Metals
ASME	American Society of Mechanical Engineers
ASTM	American Society for Testing Materials
AWS	American Welding Society
BHN	Brinell hardness number
CC	counterclockwise
cfm	cubic feet per minute
c.g.	center of gravity
C.I.	cast iron
CL	clockwise
cp	centipoises
cpm	cycles per minute
cps	cycles per second
fpm	feet per minute
fps	feet per second

fps^2	feet per second-second
gpm	gallons per minute
hp	horsepower
ID	inside diameter
ips	inches per second
ips^2	inches per second-second
ksi	kips per square inch
mph	miles per hour
mr	millions of revolutions
OD	outside diameter
OQT	oil quenched and tempered
psi	pounds per square inch
psf	pounds per square foot
QT	quenched and tempered
rpm	revolutions per minute
rps	revolutions per second
SAE	Society of Automotive Engineers
SCF	stress concentration factor
SESA	Society for Experimental Stress Analysis
WQT	water quenched and tempered
YP	yield point
YS	yield strength
μin.	microinch = 10^{-6} in.

SELECTED CHEMICAL SYMBOLS

Al	aluminum	Fe	iron	Sb	antimony
B	boron	Mg	magnesium	Se	selenium
Bi	bismuth	Mn	maganese	Si	silicon
Be	beryllium	Mo	molybdenum	Sn	tin
Cb	columbium	Ni	nickel	Ta	tantalum
Cd	cadmium	O	oxygen	Ti	titanium
Co	cobalt	P	phosphorus	V	vanadium
Cr	chromium	Pb	lead	W	tungsten
Cu	copper	S	sulfur	Zn	zinc

CONTENTS

DESIGN OF MACHINE ELEMENTS

FIGURE 1.2

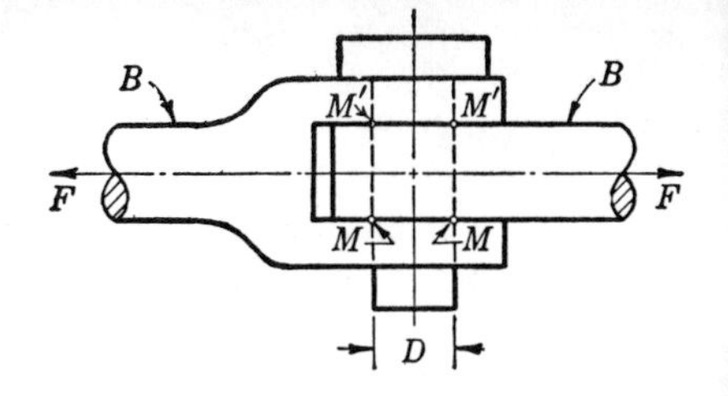

sections is said to be in *shear*, and the stress is called a *shear stress* s_s; its average value is

$$s_s = \frac{F}{A} \text{ psi or ksi,}$$

where, for this particular case, A is twice one sectional area of the pin, because both the areas, at MM and $M'M'$, offer resistance to the shearing action of the load, that is, $s_s = F/(2\pi D^2/4)$ in Fig. 1.2. The pin is said to be in *double shear*. This pin is also subjected to bending and compressive stresses. See the analysis of the pin in the example of § 1.21. Pure shear can be obtained only from torsion (§ 1.13).

1.8 TENSILE STRENGTH AND YIELD STRENGTH.

When a member is subjected to the action of a force, it is deformed, no matter how small the force. A test specimen subjected to increasing stress will undergo increasing strain. Recall, by referring to Fig. 1.3, some of the characteristics of stress-strain curves. By ***strain*** (tensile or compressive), we mean the deformation per unit gage length, in. per in. It is thus a

FIGURE 1.3 Comparison of Stress-strain Diagrams. (Scale of strain beyond the dot is 10 times scale up to the dot.) The modulus lines (see Fig. 1.4) have been drawn to the same scale and show relative values; the remaining part of the curves should be considered as qualitative but typical of certain metals. Some soft copper alloys elongate more than soft steel. See admiralty metal, Table AT 3. As the carbon content increases from some low value, the upper and lower YP merge into one YP (almost horizontal), and then disappears with increasing hardness of the steel. See Fig. 1.4.

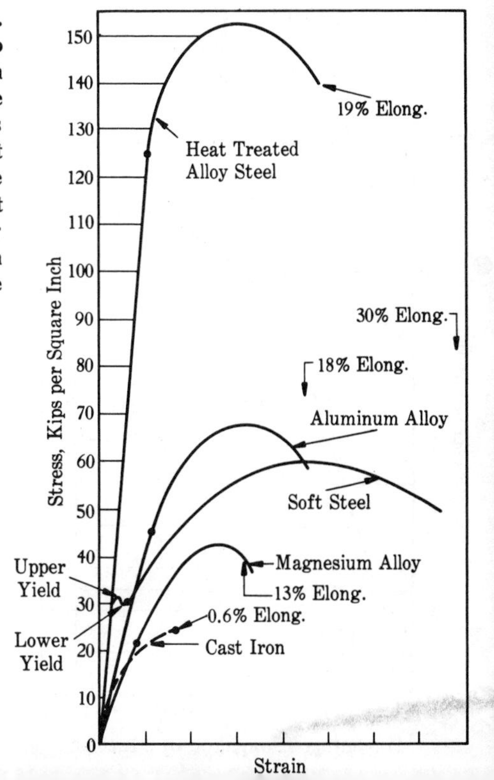

percentage increase (or decrease) in length. If a strain is 0.004 in. in a gage length of 2 in., the unit strain is 0.002 in./in., or 0.2%.

The ***ultimate stress*** s_u or ***tensile strength***, the highest point on the stress-strain curve, Fig. 1.3, is the maximum load divided by the original area before straining occurs. The stress at which a low or medium carbon steel undergoes a marked elongation without an increase in load is called the ***yield point*** *YP*, Fig. 1.3.

Higher carbon steels and nonferrous metals generally do not have a characteristic yield *point*. For these, there is a defined ***yield strength***, which is the stress for a specified deviation from the straight part of the stress-strain curve. The yield strength is determined by drawing an offset line, Fig. 1.4, starting at a certain offset *A*, parallel to the straight part of the s-ϵ curve, and noting where it crosses the stress-strain curve *B*. The ordinate at *B* is the yield strength s_y. (We shall use the symbol s_y, whether or not the material has a characteristic yield *point*.) The amount of the ***offset*** is usually 0.2% (0.002 in./in. of strain) for steel, aluminum, and magnesium alloys. Sometimes, the yield strength is specified for a particular *total extension* (deformation or strain), Fig. 1.4; for example, usually 0.5% ***extension*** for copper base alloys (see Table AT 3).*

1.9 MODULUS OF ELASTICITY.

The ***elastic limit*** is the maximum stress to which a standardized test specimen may be subjected without a permanent deformation. The specimen returns to its original length if subjected to stresses below the elastic limit. The ***proportional limit*** is, for practical engineering purposes, coincident with the elastic limit, but is precisely defined as the stress at which the stress-strain curve deviates

* See Appendix.

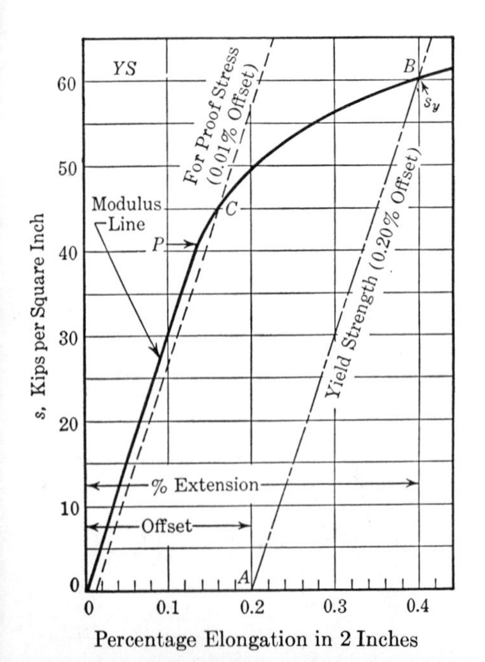

FIGURE 1.4 Yield Strength. At 0.2% offset, $\epsilon = 0.002$; etc. The yield strength at 0.2% offset is found by drawing the line *AB* from the value of 0.2 at *A* parallel to the straight part of the $s-\epsilon$ curve, called the *modulus line;* the intersection of *AB* with the $s-\epsilon$ curve at *B* is taken as the yield strength. *In this particular case,* the yield strength at 0.2% offset and at 0.4% extension is the same. The yield strength of nonferrous alloys is sometimes stated in terms of percentage extension. The so-called *proof stress* is at the intersection of a very small offset line and the $s-\epsilon$ curve, usually an offset of 0.01%, as at *C*. The proof stress is seen to be closer to the elastic limit *P* than the yield strength. Since there are different bases of determining the yield strength, a yield strength should be accompanied by its basis of determination. However, this fact is not always made clear in the literature.

from a straight line. It is indicated approximately by point P in Fig. 1.4. Recall that for some materials, for example, cast iron, Fig. 1.3, little or none of the stress-strain curve is straight.

Below the proportional limit, stress s is proportional to strain ϵ, and the proportionality constant in tension is called the ***modulus of elasticity*** E, the slope (s/ϵ) of the straight part of the stress-strain curve, Fig. 1.4;

$$s = E\epsilon \text{ psi or ksi.} \tag{1.2A}$$

Since the strain ϵ is dimensionless (in./in.), the units of E are the same as those of s. In terms of total deformation δ, where $\delta = \epsilon L$, the normal stress in simple tension or compression is

$$s = \frac{E\delta}{L} \text{ psi} \quad \text{or} \quad \frac{F}{A} = \frac{E\delta}{L} \quad \text{or} \quad \delta = \frac{FL}{AE}, \tag{1.2B}$$

in which L is the total length undergoing a total deformation of δ.

The modulus of elasticity for ordinary steel usually falls between 28 to 31 million pounds per square inch (psi), and most designers use either 29×10^6 or 30×10^6 psi (29,000 or 30,000 ksi). See Table AT 7. Since some alloy steels have values of E below 30 million, it may be desirable to find more exact values where E would be significant in the design.

The modulus of elasticity is a measure of ***stiffness***. For particular values of stress s and E, we see from equation (1.2) that there results a certain strain and, therefore, a certain total deformation of an actual part. It is interesting to note that, since all grades of steel have about the same value of E, there would be little or no reduction in deformation if a high-strength alloy steel were substituted for a low-strength carbon steel. From another point of view, the unit deformation can be reduced only by reducing the stress, or by choosing a material with a higher value of E.

Stiffness (or minimum deformation) is an important criterion in many designs, such as those of machine tools for accurate work (lathes, milling machines, etc.), rotor shafts in motors and generators and turbines.

The modulus of elasticity for most metals in *compression* is usually taken the same as that in tension. There are some exceptions among the nonferrous metals.

1.10 BENDING.

(a). Bending, or flexure, produces two kinds of normal stresses, tension on one side of a neutral plane and compression on the other. From strength of materials, we have

$$s_f = \frac{Mc}{I}; \quad \underset{[\text{BELOW } x\text{–}x, \text{ FIG. } 1.5]}{s_t = \frac{Mc_t}{I}} \quad \text{and} \quad \underset{[\text{ABOVE } x\text{–}x, \text{ FIG. } 1.5]}{s_c = \frac{Mc_c}{I}} \text{ psi or ksi,} \tag{1.3}$$

where: s_t (or s_c) is the stress at some point in the beam (Mc/I is a maximum at the fiber farthest from the neutral plane—at maximum c in., as in Fig. 1.5); M in-lb. or in-kips is the bending moment at the section of the beam that contains said point; c_t is the distance from the neutral plane to a point on the tensile side, and c_c is measured to a point on the compressive side (*if the section is symmetric*, the distance of the *external fibers* in both directions is $c = c_t = c_c$); I in.[4] is the centroidal moment of inertia (that is, I_x, Fig. 1.5) of the section containing said points; the material is homogeneous; the beam is straight in the longitudinal direction (in the unstressed state) and the neutral and centroidal axes coincide; the said point is not located in the vicinity of the point of application of a force or of a discontinuity of section (as at B and P, Fig. 1.5); the loading is static or gradually applied; there are no residual stresses (otherwise $s_f = Mc/I$ is

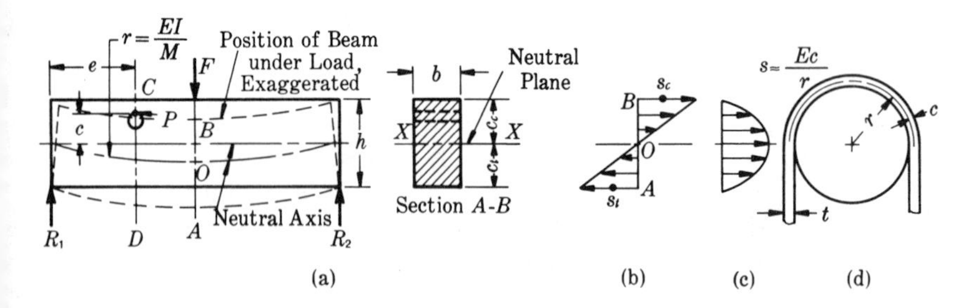

FIGURE 1.5 Stresses from Bending. The figure (b) shows the distribution of tensile and compressive stresses at some section *AB* in (a), as computed from Mc/I. At some section, such as *CD*, other than the section of maximum bending moment, and at some point *P* other than an external fiber, the nominal or computed stress is $s = Mc/I$, where $M = R_1 e$ in this figure, *I* is the centroidal moment of inertia of the section *CD* with respect to the neutral axis, approximately *X-X*, and *c* is the distance from the neutral axis to point *P*. The distribution of the vertical (and horizontal) shear is suggested by (*c*) —and equation (1.6). In (d), *t* must be very much smaller than *r*.

the change of stress due to M); the beam does not twist; flanges (if any, as in an H-beam) do not buckle (§ 1.24); the shear stress (vertical shear), Fig. 1.5(c), is negligible compared with the flexural stress; there is no longitudinal component of the forces on the beam; and *stress remains proportional to strain* (Hooke's law—$s_f <$ proportional limit).

The ratio I/c is called the ***section modulus*** Z ($Z = I/c$) and is quite convenient to use for symmetric sections. Expressions for I and Z for some sections commonly met in machine design are given in Table AT 1.* If the section is not symmetric, the centroid of the section must be located and c_t and c_c computed for use in equation (1.4).

When the metal is ductile, the design stresses in tension and compression are usually taken the same, no matter what shape the section has. When the material has significantly greater compressive strength than

* All table and figure numbers preceded by A are in the appendix at the rear, placed together for convenient reference.

tensile strength, as for example cast iron (Table AT 6), there are two cases:

1. Symmetric sections—use the ultimate flexural stress (bending modulus or modulus of rupture in bending) as a criterion for the design stress (see cast iron in Table AT 6);

2. Unsymmetric sections—use different design stresses in tension s_{dt} and compression s_{dc}, and equations (1.3). For cast iron, the usual procedure is to design for the design tensile stress and then check the compressive stress.

Recall from strength of materials that the radius of curvature r after the straight beam is bent is related to bending moment by

$$(1.4) \qquad \frac{1}{r} = \frac{M}{EI}; \qquad \frac{1}{r} = \frac{s}{cE} \qquad \text{or} \qquad s = \frac{Ec}{r}, \quad \text{[Fig. 1.5(d)]}$$

where the last two forms are obtained by using $M = sI/c$. Both equations (1.3) and (1.4) are virtually true in straight beams if the maximum stress does not exceed the proportional limit, and for (1.4), r should be large compared to c.

(**b**) The shearing stress in a beam, distributed as shown in Fig. 1.5(c) for a rectangular section, is computed from

$$(1.5) \qquad s_s = \frac{VQ}{Ib},$$

where V (lb. or kips) is the shear (from shear diagram, § 1.11) at the section under study, b is given in Fig. 1.6, I is the centroidal moment of inertia as previously defined (I_x in Fig. 1.6), and $Q = \int y\, dA' = \bar{y}A'$, where A' is that area "outside" of a transverse line through the point whose stress is desired, partly shaded in Fig. 1.6. Thus equation (1.6), with the same restrictions as given for (1.3), gives the approximate average stress along a line such as BB, Fig. 1.6. For the rectangular section in Fig. 1.6(a), $Q = gb(h - g)/2$. In the case (b), divide the A' into two rectangular parts A_1 and A_2 and get the sum of the moments of these parts for Q. It is important to observe that if the section is symmetrical, the flexural stress is zero when the vertical shear is a maximum, and in any case, the *shear is zero where the flexural stresses are maximum* (external fibers). Moreover, a metal beam has to be unusually short for the vertical shear to be significant, but it should be looked into for short beams (and wooden beams), and on

FIGURE 1.6 Distribution of Shear Accompanying Bending. Equation (1.6) becomes more inaccurate with an increase of width with respect to the depth; hence the s_s distribution for area A_1 is undependable. Recall that the shear stress on the *exposed* underneath part of the flange (at H) is zero.

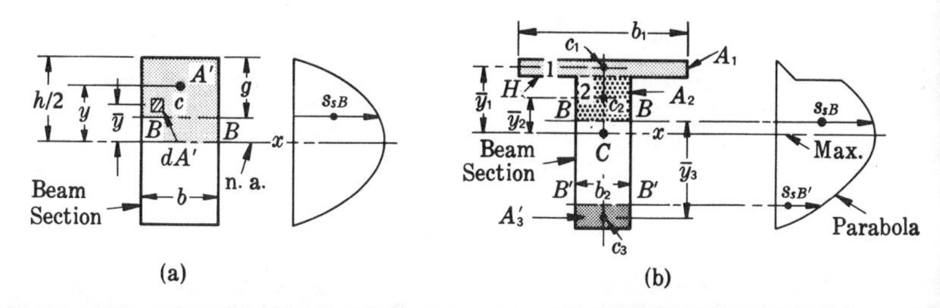

occasion, the combination (Chapter 8) of shear and normal stresses at some interior point of a beam should be checked. It may be significant to note that in a member subjected to torsion and bending, the maximum vertical shear at the neutral plane adds vectorially to the torsional stress, say of the front or back fiber. The shearing stress also affects the amount of deflection of a beam, and again, this effect is significant only for short beams.[1.1]

(c) In a beam where the maximum loading is applied only a few times during its expected lifetime, designing for some inelastic response is permissible. If the material has a distinctive yield point (Fig. 1.3), such as low-carbon structural steel has, the fibers that reach this stress s_y first will remain at more or less constant stress; and while the loading is increased, additional fibers will become stressed s_y, say as shown in Fig. 1.7(b). Here, the beam is stressed to the yield point to a depth *ab* (and *dc*), with elastic action from *b* to *c*, and the beam could be designed for this distribution of stress. However, when design, called *limit design*, is based on plastic action, a common assumption is that the loading is such that

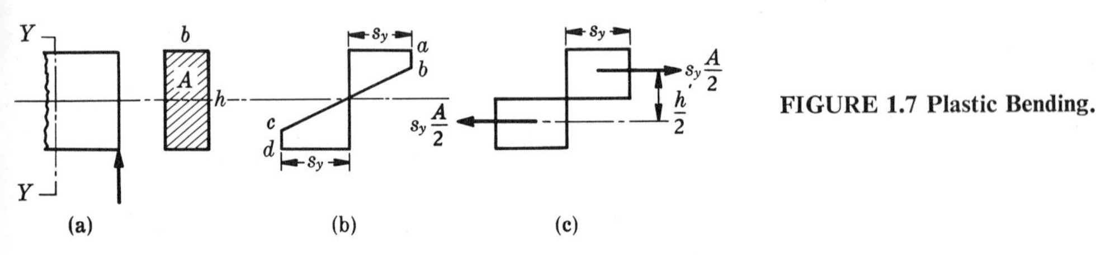

FIGURE 1.7 Plastic Bending.

the material at the neutral plane has just been stressed s_y as in Fig. 1.7(c). In this case, the forces on each half of a rectangular cross section are seen to be $s_yA/2$, which gives a resisting moment of $(s_yA/2)(h/2)$; or with the applied moment M equal to the resisting moment, we have

$$M = s_y \frac{bh}{2} \frac{h}{2} = s_y \frac{bh^2}{4}, \tag{1.6}$$

vs. $M = sZ = sbh^2/6$ for all elastic stresses (Fig. 1.5), where $Z = bh^2/6$ for a rectangular section. On this assumption, the beam can tolerate a maximum bending moment 50% greater than for all elastic action. However, machine parts are so often made of the higher-strength steels that do not exhibit a true yield point (no constant stress with increasing load) and they are too commonly subjected to varying loading for an indefinite time, in which case, *assumption of plastic action as a basis of design is downright dangerous*. For more detail about limit design, see books devoted to structural design and strength of materials.[1.4,1.5]*

* Superscript numbers in brackets designate references in the list of references at the rear; they indicate either that more detail is available in the indicated reference or that the factual information comes from the reference, or both.

1.11 MATHEMATICAL RELATIONS FOR BEAMS; SHEAR CENTER. Let the deflection of a straight beam be represented by y; then for small elastic deflections (the corresponding slopes are small $\tan \theta \approx \theta$), we have

$$\text{(1.7)} \qquad \text{Slope } \theta = \frac{dy}{dx} \text{ radians}$$

$$\text{(1.8)} \qquad \text{Moment } M = EI\frac{d\theta}{dx} = EI\frac{d^2y}{dx^2};$$

$$\text{(1.9)} \qquad \text{Shear } V = \frac{dM}{dx} = EI\frac{d^3y}{dx^3};$$

$$\text{(1.10)} \qquad \text{Loading} = \frac{dV}{dx} = \frac{d^2M}{dx^2} = EI\frac{d^4y}{dx^4}.$$

Thus if the loading (or shear or moment) can be expressed as a function of x, successive integrations eventually yield the deflection y. A rational sign convention is needed in applying these equations: *the shear at a section is positive if the part on the left of the section tends to move "upward"* (*horizontal beam*) *relative to the part on the right; the bending moment and curvature are positive when the "top" of the beam is in compression.* More on obtaining deflections of beams with several loads (graphical integration) is given in Chapter 9. For the more elementary problems that we shall be concerned with at first, deflections of certain common types of beams are given by formula in Table AT 2.

For immediate use, recall that the bending moment at any section of a beam is the sum of the moments, about an axis in the section and through its centroid, of *all* external forces acting to the right of the section (*or* to the left of the section) where the moment is desired. To determine the *maximum* bending moment on a beam, use the principle that the maximum moment will occur where the shearing-force diagram crosses the zero axis. The procedure is first to draw the shear diagram. For many if not most problems, this can be a freehand sketch made from actual values of the shear. The *shear at any section* for coplanar force systems is the algebraic sum of *all* forces perpendicular to the neutral axis on *either* side of the section. Then compute M at *each* section where the shear diagram passes through zero. At one such section the moment will be a maximum. The designer must consider the section of maximum moment without fail, but because of stress raisers elsewhere and for other reasons, he may also need to use moments at other sections. Given the loading as shown in Fig. 1.8, use the principles of analytic mechanics[1.6] and compute the reactions R_1 and R_2. The shear diagram can be sketched, but to determine the exact location of section C of maximum moment in this case, set up

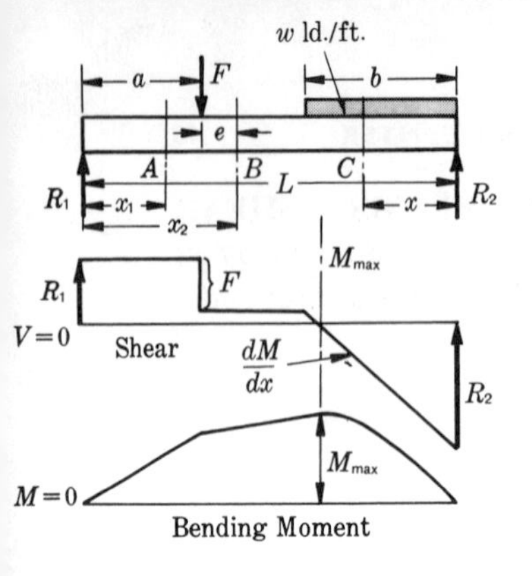

FIGURE 1.8 Shear and Moment Diagrams. At A, $M = R_1x_1$. At B, $M = R_1x_2 - Fe$. While the positive direction is customarily toward the right, it is often convenient to measure from the right and toward the left; thus, at C, $M = R_2x - wx^2/2$.

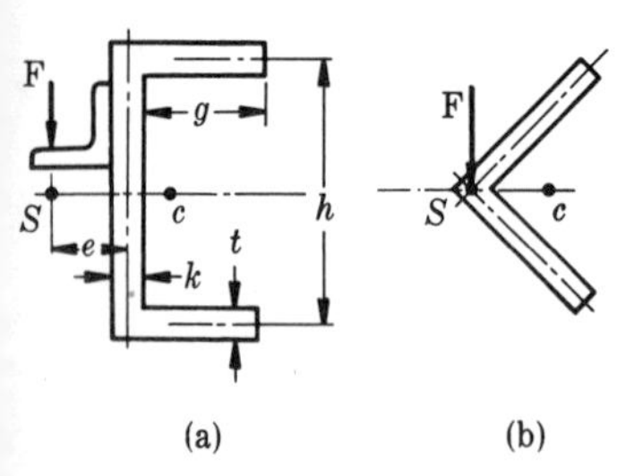

FIGURE 1.9 Shear Center. Beam length is perpendicular to paper.

the equation for shear measured from the right-hand end ($V = R_2 - wx$), equate to zero and solve for x. See the caption to Fig. 1.8 as a reminder of the process of computing moments. More on shear and moment diagrams is given in Chapter 9.

If the beam section is unsymmetric with respect to the plane of the loads, as for the channel loaded as shown in Fig. 1.9(a), one must be sure that the lines of action of the loads are correctly located if twisting is to be avoided. In short, they should pass through the ***shear center***, which is defined by a force analysis of the internal forces,[1.7] and is always on any existing line of symmetry. For an angle section, Fig. 1.9(b), the shear center is at the intersection of the center lines of the legs. For the channel, with dimensions as shown, the shear center S is located by

(a)
$$e = \frac{3g}{6 + kh/(gt)}.$$

1.12 FINDING THE MOMENT OF INERTIA. Centroidal moments of inertia and section moduli may be found in handbooks (and in books on analytic mechanics) for the most common sections, some of which are given in Table AT 1. To save time, memorize those for circles and rectangles. However, for composite areas, that is, areas made up of two or more "fundamental" areas, such as T-sections (two rectangles) and H-sections (three rectangles), it is better to find the desired moment of inertia by using the parallel axis theorem than to copy a formula from a table. The procedure may be in steps as follows:

(a) Locate the centroid of the section. If the section is symmetric about any axis, the centroid lies on that axis. If the section is symmetric about two axes, the centroid lies at the intersection of the two axes. If the section is non-symmetric, divide it into a group of rectangles, triangles, the locations of whose centroids are known, naming them A_1, A_2, A_3, etc. Let A represent the *total* area whose centroid is desired. Then

$$\textbf{(b)}\qquad A\bar{x} = A_1x_1 + A_2x_2 + A_3x_3 + \cdots,$$
$$\text{where} \quad A = A_1 + A_2 + A_3 + \cdots;$$

x_1 is the distance from the centroid of A_1 to any convenient moment axis, x_2 is the distance from the centroid of A_2 to the same moment axis, etc., and $\bar{x}$ is the distance from the chosen moment axis to the *centroid of the composite area*. Solve for $\bar{x}$.

(b) Knowing the location of the centroid of the composite section, find the moment of inertia of each "fundamental" area (A_1, A_2, etc.), about the centroidal axis of the composite section, using the parallel axis theorem.

$$\textbf{(c)}\qquad I_1 = \bar{I}_1 + A_1d_1^2,$$

where I_1 is for area A_1 with respect to the composite centroid, $\bar{I}_1$ is for area A_1 with respect to *its own centroidal* axis, and d_1 is the distance from A_1's centroidal axis to the composite centroid. Apply equation (**c**) to each "fundamental" area and then find

$$\textbf{(d)}\qquad \bar{I} = I_1 + I_2 + I_3 + \cdots,$$

where $\bar{I}$ is the centroidal moment of inertia of the section, the value of $\bar{I}$ in $s = Mc/I$. (Notice also how the value of I_A was obtained for the H-section (5) in Table AT 1.)

1.13 TORSION. (**a**) The only shape of section for which the simple torsion equation,

$$(1.11)\qquad T = \frac{s_sJ}{c} = s_sZ',$$

is strictly applicable is circular (hollow or solid); T (in-lb. or in-kips to accord with the units of s_s) is the applied ***torque*** or twisting moment, s_s (psi or ksi) is a shearing (torsional) stress, which in designing is a design stress; J in.^{4} ($= \pi D^4/32$ for circular area—Table AT 1) is the centroidal polar moment of inertia of the section; c in. is the distance from the neutral axis to the point where the stress s_s is desired (usually the maximum stress on an external fiber); and $Z' = J/c$ in.^{3} ($= \pi D^3/16$ for circular area—point on circumference) is the polar section modulus. The conditions given for equation (1.3) also apply to (1.11) except that the member does twist but does not bend. Equation (1.11) for an external point on a solid circular member becomes

$$\text{(e)} \qquad s_s = \frac{16T}{\pi D^3} \text{ psi} \quad \text{or} \quad \text{ksi}.$$

(**b**) Since plane sections do not remain plane in twisting of noncircular members, equation (1.11) is not exact for these.[1.2,1.7] However, more precise analyses suggest that the use of particular values of Z', as for rectangular and elliptical sections in Table AT 1, give reasonable approximations. The Z' for the rectangular section varies with the ratio of h/b;[1.7] the value in Table AT 1 is good for about $h/b \gtreqless 2.2$. The maximum torsional stress occurs at the center of the long sides. For thin-walled tubes, Fig. 1.10(a), equation (1.11) becomes difficult to evaluate. The

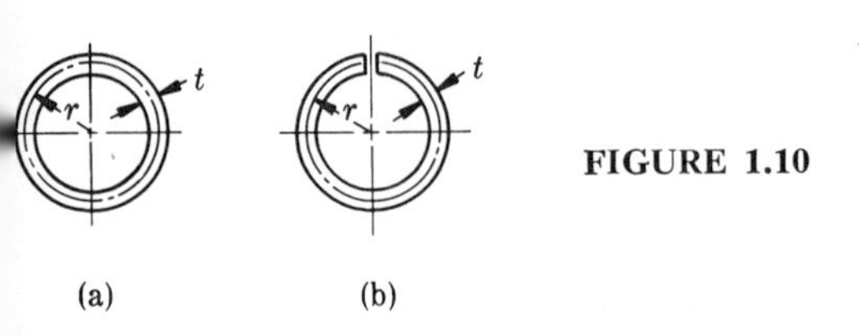

FIGURE 1.10

membrane (soap-film) analogy [or letting $t \ll r$ in (1.11)] results in the approximate equations[1.7]

$$\text{(f)} \qquad s_s = \frac{T}{2\pi r^2 t} \qquad \text{and} \qquad \theta = \frac{T}{G(2\pi r^3 t)},$$

where r = mean radius of tube wall, t = thickness of tube wall, G is the modulus of elasticity in shear (see below) and θ = angle of twist (Fig. 1.12). Equation (**f**) says nothing about the thin wall buckling locally before elastic stresses are exceeded, a type of failure that needs to be checked when a significant torque is applied. The big surprise comes when the thin wall tube is slit along its length, Fig. 1.10(b), and its loss of torsional capacity as compared with the solid tube is noted:[1.7]

$$\text{(g)} \qquad s_s = \frac{3T}{2\pi r t^2} \qquad \text{and} \qquad \theta = \frac{3T}{G(2\pi r t^3)},$$

where the resistance is equivalent to that for a long slender rectangle and the symbols mean the same as before.

(**c**) Recall that shearing stresses on an element appear as indicated in Fig. 1.11. With respect to the base ad, the top of the element is deformed an amount δ_s. The unit deformation or strain is $\delta_s/L = \tan\gamma$; but since

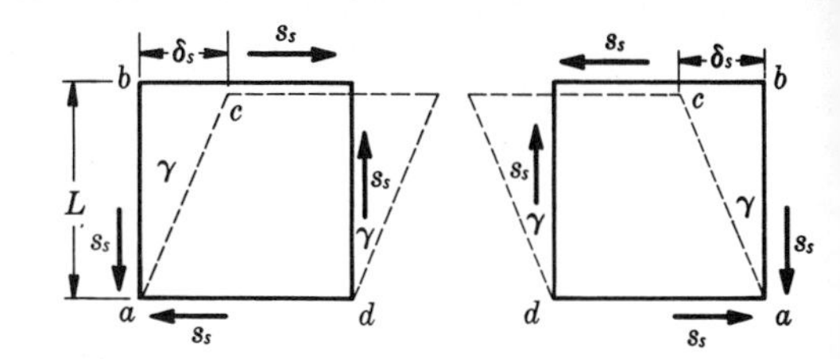

FIGURE 1.11 Shear Strain.

the angle γ is very small, the unit strain is taken as $\gamma = \delta_s/L$. Within the proportional limit, strain is proportional to stress,

$$(1.12) \qquad s_s = G\gamma \qquad \text{or} \qquad \gamma = \frac{s_s}{G},$$

where the proportionality constant G is called the ***modulus of elasticity in shear*** (also *modulus of rigidity* and *transverse modulus*); Fig. 1.12.

(**d**) In many situations, the design of a part is defined by the permissible deformations, rather than by some safe stress. The angle of twist θ of a shaft between two sections M and H, Fig. 1.12, is given by

$$(1.13) \qquad \theta = \frac{TL}{JG} \text{ radians},$$

which is obtained from the arc $A_1A_2 = (D/2)\theta = L\gamma$, with the use of equations (1.11) and (1.12). Notice that, for a shaft with a given twisting moment on it, θ, the angle that one section M twists through with respect to another section H, depends on the distance L in. between sections; J in.^{4} is the polar moment of inertia; $T = FD$ (in Fig. 1.12) in-lb. or in-kips; for steel, G is usually taken as 11.5×10^6 psi (11,500 ksi), sometimes 12×10^6 psi. Equation (1.13) applies to hollow or solid round shafts [but for thin walls, see equation (**f**)]. Pulleys and gears (and keyways) on shafts affect the angle of twist θ, but common practice often ignores these effects, taking the length L between two gears, for instance, as the center distance. Of course, when deflections must be computed accurately, such effects cannot be disregarded. If a member is subjected to different torques, it is true that the torque is constant between sections where twisting loads are applied, in which case, equation (1.13) applies between these sections. If the member changes diameter, (1.13) is applied for lengths of the same diameter only.

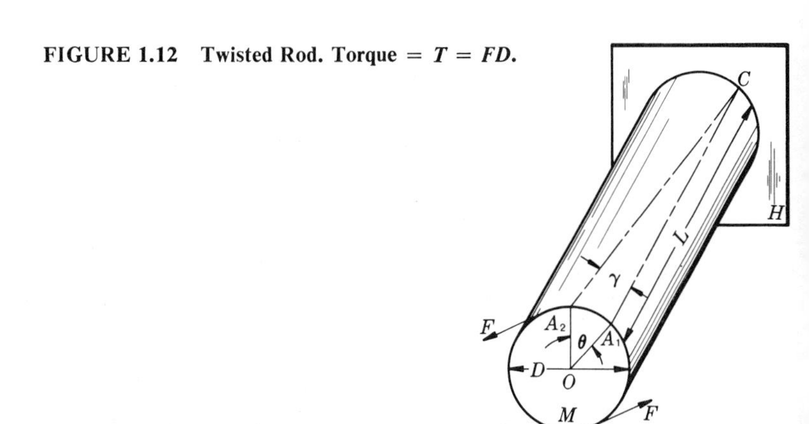

FIGURE 1.12 Twisted Rod. Torque = $T = FD$.

(e) The theoretical relation between the tensile and shear modulii (E and G) is given by (when the deformation angle = tangent of angle)

$$G = \frac{E}{2(1 + \mu)}, \tag{1.14}$$

where μ is Poisson's ratio. See Tables AT 3, AT 6, and AT 7 for values; $\mu = 0.3$ is often used for steel.

1.14 TORQUE. The horsepower equation is used so frequently that we had better review its origins briefly. Assume a force F, Fig. 1.13, acting at the circumference of a circle of radius r inches. (The size of pulleys, gears, etc., is always given as the *diameter* in inches.) The work done by this force moving once around the circle is $F(2\pi r/12)$ ft-lb. If the force

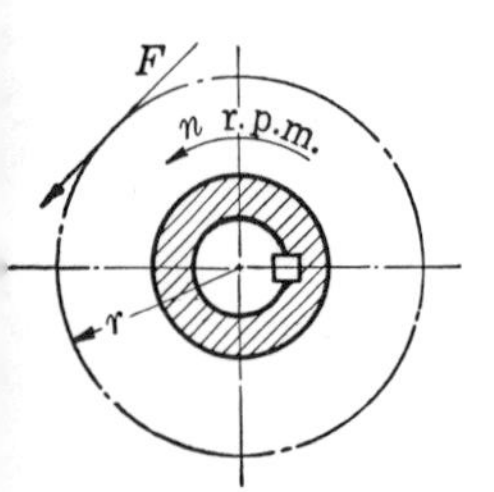

FIGURE 1.13

moves around the circle n times, the total work is $F(2\pi r/12)n$ ft-lb.; if n represents revolutions per *minute*, then $F(2\pi r/12)n$ is work per minute, ft-lb./min. But 33,000 ft-lb./min. is equivalent to one horsepower; hence

$$\text{hp} = \frac{(F)(2\pi r)(n)}{(12)(33{,}000)} = \frac{Frn}{63{,}000} = \frac{Tn}{63{,}000}, \qquad [r \text{ in.}] \tag{1.15}$$

where 63,000 is approximately equal to (12)(33,000)/(2π). Referring to Fig. 1.13, we see that Fr is the turning moment or torque T in-lb.

Another useful form of the horsepower equation is obtained by letting $v_m = (2\pi r/12)n$ fpm in equation (1.15), where v_m fpm (or v_s fps) is the speed with which F "moves through a distance." Then

$$\text{hp} = \frac{Fv_m}{33{,}000} = \frac{Fv_s}{550}. \tag{1.15A}$$

If we use T ft-lb., instead of T in-lb., hp = $Tn/33{,}000$. For T in-kips, hp = $Tn/63$; etc.

1.15 STRENGTH OF MATERIALS. The strength of a material is its capacity to resist the action of applied forces. Unfortunately, the strength of a material cannot be represented by a single number, because its ability to resist the action of loads and forces depends upon the nature of those loads and forces, the kinds of stresses induced, and other circumstances.

If a member is stressed beyond its elastic limit, the permanent deformation it receives may render it unfit for further service. Thus, the elastic limit is one significant criterion of strength. Instead of the elastic limit, however, we invariably use the yield strength, a stress that is generally not too far from the elastic limit and much easier to determine experimentally. (§ 1.8 and Fig. 1.4.)

The ultimate stress is also an important criterion of strength because a part that has ruptured has almost certainly lost its usefulness. There are other criteria which we shall take up later—for example, ability of a material to absorb energy without failure, endurance strength, buckling strength, creep strength, and excessive deflection.

All criteria of strength are modified in some way in order to obtain a design criterion. In its simplest form, the design criterion is a ***design stress***, or *working stress*, which may also be called a *safe stress* or *allowable stress*. The stress used in *design* must be a *safe* one to use for computations if failure is not to occur, and such a stress is said to be *allowable*.

1.16 FACTOR OF SAFETY—DESIGN FACTOR.

Ordinarily, the ***factor of safety*** is a number that is divided into a criterion of strength in order to obtain a design criterion. In the literal meaning of the words, factor of safety would indicate by what factor the design is safe, but as actually used this is not true. Since its meaning does not accord with the true meaning of the words, the number would better be called the ***design factor***.* For the time being we shall use the design factor N or factor of safety to define a design stress s_d; thus, for the ultimate-stress s_u and yield-stress s_y criteria, we have

$$\text{(h)} \qquad s_d = \frac{s_u}{N} \qquad \text{and} \qquad s_d = \frac{s_y}{N}.$$

Inasmuch as the stress used in design is the significant number and because for a particular design procedure, the best design stress is a particular number, the values of N in the foregoing equations must be different. Thus the design factor depends upon the criterion used for design. Suppose $s_u = 80$ ksi, $s_y = 50$ ksi, and that a good $s_d = 20$ ksi. Then

$$N_u = \frac{80}{20} = 4, \text{ the design factor based on ultimate strength and}$$

* We shall tend to favor *design factor* and *design stress*, but the most common usage is *factor of safety* and *working stress*.

TABLE 1.1 FACTORS OF SAFETY (DESIGN FACTORS)

The factors of safety marked with * are primarily for beginners' use, although they are traditional values. They should not be used when a detailed accounting is made of the variable loading, stress concentrations, etc., Chapter 4. Acceptable for use with typical strengths.

KIND OF LOAD	STEEL, DUCTILE METALS		CAST IRON, BRITTLE METALS	TIMBER
	Based on Ultimate Strength	*Based on Yield Strength*	*Based on Ultimate Strength*	
Dead load, $N =$	3–4	1.5–2	5–6	7
Repeated, one direction, gradual (mild shock),* $N =$	6	3	7–8	10
Repeated, reversed, gradual (mild shock),* $N =$	8	4	10–12	15
Shock,* $N =$	10–15	5–7	15–20	20

$$N_y = \frac{50}{20} = 2.5, \text{ the design factor based on yield strength}$$

If you are finding or stating a factor of safety N, state also its basis; as, "factor of safety based on yield strength" or "based on ultimate strength." Perhaps the more basic definition of factor of safety is

$$\text{(1.16)} \qquad \text{Factor of safety} = \frac{\text{loading that would cause failure}}{\text{actual loading on part}},$$

at least if a single load is involved. Also this definition is used when the stress does not vary linearly with force, as in some column formulas (Chapter 7).

Table 1.1 gives some rule-of-thumb values for your guidance. In recent years there has been a trend toward using the yield strength as the preferred criterion for getting a design stress—very desirable *for dead loads.* This practice is based on the logic that failure occurs when a body ceases to perform its allotted function and that most machine elements will not perform properly after they have received a permanent deformation. However, the practice is questionable in machine design *where loads vary*, because the endurance strength of steels is nearly proportional to the ultimate strength but not to yield strength. If the loads vary in a definable manner, the methods of design developed in Chapter 4 should be used. Otherwise, for variable loads, it is better to use $s_d = s_u/N$ with a suitable design factor. An unnecessarily large degree of safety means unnecessarily high cost. A stress that is computed from a stress equation, such as $s = F/A$ or $s = Mc/I$, is best called a ***computed*** or ***nominal stress***.

1.17 VARIABILITY OF STRENGTH OF MATERIALS AND THE DESIGN STRESS. Whatever design criterion is used, we must accept the fact that it is not a simple precise number. Although the properties of materials given in Tables AT 3–AT 11 look uncompromisingly fixed, they are either simply typical values or minimum values such as would be found in specifications. For example, Fig. 1.14 shows the variation in ultimate strength of 1011 test coupons taken from hot-rolled, SAE-1020 structural shapes. The variability of test specimens from one heat would be less than the variability of specimens taken from several heats and several furnaces.

FIGURE 1.14 Strength of SAE 1020 Structural Steel. The test coupons, taken from the finished *rolled product*, were $1\frac{1}{2}$ in. wide, 9 in. long, and varied in thickness from $\frac{1}{4}$ to $\frac{1}{2}$ in. The diagram of rectangles is called a *histogram*. The height of each rectangle represents the number of pieces (or percentage of the total) testing within the particular range; for example, 108 pieces (10.7%) showed an ultimate strength of between 61 and 62 ksi, step *C*. The dotted curve, called a *normal curve*, is an idealized distribution of the ultimate strengths as obtained from these data. It is the curve that would be obtained from an infinite number of tests plotted on infinitesimal steps, the samples having been produced in a constant environment (if some factor varies, it varies the same way "constantly"). As the statistician puts it, all tested samples are taken from the same *universe*. The significance of the 3σ (three-sigma) limits, *A* and *B*, of the normal curve is that it is very unlikely that values will fall outside of these limits, which are 55.8 ksi and 71.5 ksi, which are computed by statistical methods. Some of the product falls above the higher 3σ limit at *B*, which suggests that something happened during the manufacture that tended to increase the strength of the material. (Maybe the operators were "making sure" that the material would meet a minimum ultimate strength test?) The variability of the strength of *heat-treated parts* could be kept much less than in this illustration by adjusting the tempering temperature from heat to heat.

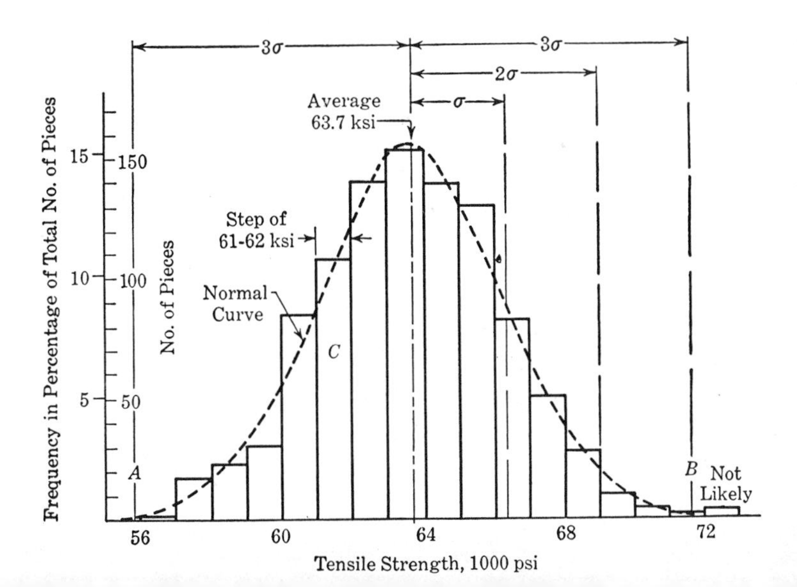

Knowledge of the variability of strength affects the designer's choice of a design factor. If the probable minimum strength is used to determine the design stress, a lower design factor (factor of safety) may logically be used than if the average or mean strength were used—that is, to obtain the same design stress. For a factor of safety of 5, the design stress from the approximate average stress from Fig. 1.14 would be 63.7/5 = 12.74 ksi. This design stress corresponds to a factor of safety of 55.8/12.74 = 4.48, compared with the probable minimum strength of 55.8 ksi from Fig. 1.14. Compared with the probable maximum strength, the factor of safety is 71.8/12.74 = 5.76. The difference between the maximum and minimum values is 28.6% of 4.48, the minimum value.

The variability of heat-treated steel in normally controlled processes should be materially less than in the foregoing example. Strength values for another group of specimens from *various heats* but with each specimen given the *same heat treatment* varied 13% from the minimum strength, compared to 28.6% in the previous example. The variability might have been less than 13% if the tempering temperature (heat treatment) had been varied from heat to heat in accordance with the particular characteristics of a heat.

Usually, the design factor is such as to allow for this variability if average or typical strength values are used. When a material specification includes the minimum tensile strength, it is likely that the manufacturer will control his processes so that there is little probability of a lower tensile strength. He would hesitate to guarantee a higher strength than that at A for the case shown in Fig. 1.14. This situation accounts for *typical* and average mechanical properties being better than the specifications. The properties given in Tables AT 3–AT 11 are not minimum properties unless a note so states. Where the extra expense may be justified, the purchaser may obtain closer limits than are customary in standard specifications.

All mechanical properties of a particular steel vary in a pattern similar to the histogram shown in Fig. 1.14: yield strength, endurance strength, hardness, elongation, etc. If the designer wishes to make some allowance for this variability, he may somewhat conservatively estimate the minimum strengths (s_u, s_y, s_n) of metals that are supposedly alike but are from untested universes, or from different universes as in Fig. 1.14, as about 10% less than "typical" or average values, and the minimum strength of carefully controlled processes, as heat-treating processes, as about 5% less than average. Reference *(2.1)* shows the distribution of some mechanical properties for a few metals. The composition also varies in a similar statistical manner: the percentages of carbon, sulfur, silicon, and of other alloying elements. Moreover, the same generalities concerning variability apply to every material.

The materials whose properties are given in the Tables AT 3–AT 11 and the charts of Figs. AF 1, AF 2, and AF 3 in the Appendix have been carefully selected to present an informative variety, and the tables are therefore worth some study, especially by way of comparisons. However, since the sources of these data are varied and numerous, caution should be exercised in drawing firm conclusions. The materials chosen for presentation are not expected to be either better or worse on average than many other similar ones not mentioned here. Do not specify materials by the specification numbers given without referring to the source. Frequently there are classifications within an ASTM specification number, for instance. For additional information about a particular material, refer to specialized books and standard specifications. See Refs. *(2.1–2.21)* at the back of the book.

1.18 MORE ON FACTOR OF SAFETY AND DESIGN STRESS.

In a general way, the greater the uncertainties, the greater the design factor, and the lower the design stress, should be. For this reason, factor of safety has been called "factor of ignorance." When the loads and their manner of variation are known accurately, when the properties of materials are carefully controlled within *known* narrow limits, when the maximum stresses can be computed with confidence, and when the proper criterion (yield strength, endurance strength, etc.) is used, the design factor of safety may be relatively low (perhaps as low as 1.2). Uncertainties and other factors that affect the magnitude of the design stress are as follows.

(a) Material. See § 1.17. As regards strength, it would be best to choose design stresses with knowledge of the variability as depicted in Fig. 1.14, but such complete information is not generally available. Ignorance of the limits of the properties leads to conservative design stresses. The possibility of internal flaws, more likely in improperly controlled casting procedures, suggests higher N's. It has been traditional to use a higher design factor for cast metal (especially cast iron, which is more brittle) than for wrought metals, but if the processes are well controlled and stress concentrations are accounted for (Chapter 4), even for dead loads, there needs be little difference. However, brittle materials are definitely less suited for impulsive loading than are ductile materials. If the material is bought on specifications and suitably inspected, its properties are reliably known.

(b) Size Effect. Tables quoting strength values are generally based on "standard" sized specimens, commonly about $\frac{1}{2}$-in. diameter (sometimes 1 in.) for the ultimate tensile stress and about $\frac{1}{4}$-in. diameter for fatigue strength. However, it is well known that failure of large parts occurs at lower computed stresses (F/A, Mc/I, Tc/J) than of small parts. See Tables AT 8 and AT 9. For dead loading, the loss is not significant up to say a 2-in. dimension (as the thickness of a flange—the flange width would not be involved), but a 10-in. axle would require a higher apparent factor of safety, or else experimental information must be obtained, which is expensive for large members. In Chapter 4 we begin making allowance for size effect above $\frac{1}{2}$-in. size.

(c) Loading. Deciding upon the service loads for a machine is frequently practically impossible. (Think of an automobile going over rough roads.) But some sort of estimates are essential, and in some cases, the results are reasonably close to the truth, and can be stated as falling between certain limits with a high probability of being right. For many elements where experience is available, design is based solely on experience. When the nature of the loading is known in some detail and is accounted for in the design, a lower design factor is appropriate as compared with a case (such as those in Table 1.1) that covers a considerable ignorance. For static loads and ductile materials, the yield strength is the best criterion if a significant permanent deformation would destroy the part's usefulness (ultimate strength if only rupture destroys usefulness). For varying loading, the endurance strength is the best criterion.

(d) Computed Stress. The computed average stress for a member subjected to simple tension can be viewed with considerable confidence. However, the uncertainty increases substantially for parts of complicated shapes. In some cases of nonuniform stresses, good theoretical equations, perhaps with practical modifying constants, may be available; in other situations, suitable theoretical equations may be all but impossible to obtain. Experimental stress determinations on a finished part give information that does not exist when design begins and will show whether or not the design is acceptable (see Fig. 4.12). Another frequent complication is the *residual stresses* left in the unloaded part by some manufacturing process (say, heat treatment), which may or may not be beneficial under operating conditions. The pattern of residual stresses may not be permanent and properties may change in service.

(e) Environment. Some working environments introduce a considerable uncertainty. For example, salt water, a corrosive atmosphere, etc., may result not only in a cracking of stressed material, but also in an actual disappearance of material, and the roughened surface is detrimental to fatigue life. If nuclear bombardment is involved, the properties of materials are changed—for some materials, in an unknown manner. At very low temperatures (§ 2.22) and at very high temperatures (§ 2.21), metal properties change significantly.

(f) Inspection. The thoroughness of inspection and the strictness of the specifications are factors in the decision of the magnitude of the design factor. If a rational statistical approach that insures accurate knowledge of the final product is used, some uncertainty is removed and a lower design factor may be used.

(g) The Chance of an Accidental Load. A part should be amply strong to withstand an accidental blow that, for example, may occur in moving the machine about. The machine should be able to take without serious damage some overload arising from unexpected causes.

(h) The Danger to Life or Property. Higher factors of safety are desirable if life or valuable property would be endangered in case of failure. The factor of safety of a key that keys a pulley to a shaft should be less than that for the shaft because its failure might then save the failure of more expensive parts of the machine.

(i) The Price Class in Which the Machine Is to Sell. Sometimes, cheaper machines have a lower factor of safety in order to reduce the cost of materials and manufacture.

In the final analysis, *the choice of the design factor N rests with the judgment of the designer*, which in turn depends upon his experience. In many instances, design is necessarily an experimental procedure because so little is known about the actual maximum loads. A part of a certain size and material is tried. If it does not fail, it may be replaced with a smaller one or one made of less expensive material. If it does fail, a larger part, or a stronger material, or a change in shape may be tried. Ultimately, the factor of safety to use in this particular instance may be determined from experience. As you proceed with this study, observe how often the method of determining the design stress is suggested for particular machine elements; then remember that these design stresses are based on experience and are subject to such revision as experience may justify.*

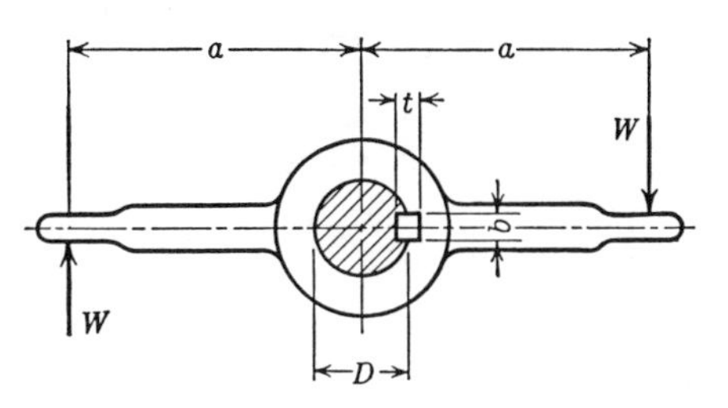

FIGURE 1.15 Torque on Circular Member.

1.19 EXAMPLE—DESIGN FOR TORSION. Let the shaft in Fig. 1.15 be subjected to a twisting moment by loads $W = 100$ lb. at $a = 20$ in. on each side. The material is C1030, as rolled, and a design factor of 4 based on yield strength should be sufficient (to cover the loss of strength from cutting the keyway, too). (a) Determine the shaft diameter D. (b) If the torsional deflection is to be limited to 0.2 deg./ft. of length, will the diameter found be satisfactory? (c) What is the torsional stress at a point $\frac{1}{2}$ in. from the axis of the shaft found in (a)?

Solution. (a) First find the design stress. From Table AT 8 for 1030 as rolled, we find $s_y = 51$ ksi (tension). Table AT 7 suggests using $s_{ys} = 0.6s_y = (0.6)(51) = 30.6$ ksi. For a design factor of 4, the design stress is

$$s_s = \frac{30.6}{4} = 7.65 \text{ ksi.}$$

The applied moment is $T = W(2a) = (100)(2)(20) = 4000$ in-lb. Using equation (e) for the maximum stress condition, we find

$$D^3 = \frac{16T}{s_s\pi} = \frac{(16)(4000)}{7650\pi} = 2.67,$$

$$D = 1.39 \text{ in.} \qquad \text{Use } 1\tfrac{3}{8} \text{ in.}$$

* One might say facetiously that the best design stress is the one that results in the right answer—the right answer as conclusively proved by experience.

Observe in Table AT 8 that there is a minor decrease in the yield strength of a section of this size as compared to that of a $\frac{1}{2}$-in. specimen.

(b) For $J = \pi D^4/32 = \pi(1.375)^4/32 = 0.358$ in.^{4} and $L = 1$ ft. $= 12$ in., from equation (1.13),

$$\theta = \frac{TL}{JG} = \frac{(4000)(57.3)(12)}{(11.5 \times 10^6)(0.358)} = 0.668^\circ,$$

which compared with the permissible value of 0.2° shows that the shaft must be made larger if stiffness is the significant criterion. [Solve for D from (1.13) with $\theta = 0.2/57.3$.]

(c) For the shaft in (a) with $c = 0.5$ in.,

$$s_s = \frac{Tc}{J} = \frac{(4000)(0.5)}{0.358} = 5600 \text{ psi}.$$

1.20 SAFE COMPRESSIVE STRESS. For highly ductile materials, a definite ultimate compressive stress, that is, a stress above which the body *breaks*, is impossible to determine because such materials simply flatten out under a compressive load without showing a fracture. For the purpose of deciding upon design stresses, the ultimate compressive strength of this type of material is taken equal to the ultimate tensile strength. When tested in compression, ductile materials usually exhibit *approximately the same* characteristics up to the yield strength as they do when tested in tension.

An illustration of a situation where experience suggests an upward modification of the design compressive stress is when the maximum stress exists on the surface and decreases inward with distance from the surface.

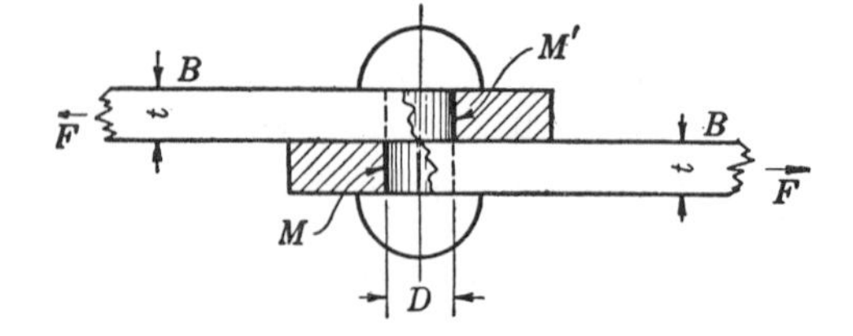

FIGURE 1.16 Compressive Stress on Cylindrical Surface. Loaded as shown, the rivet is also subjected to shear and bending.

In Fig. 1.16 the fibers on the *surface* of the *hole* and those on the *surface* of the *rivet* are pressed together but the compressive stress due to F does not pervade the whole rivet nor extend far into the plates. If the compressive stress exists mainly on the *surface* of a body, a higher design stress may be used than when the compressive stress pervades the entire body, as in Fig. 1.1(b). This surface type of compressive stress is sometimes called a ***bearing stress***. The designer must decide how much higher design stress is permissible. As one example, the ASME *Code* for riveted joints permits the design surface compressive stress to be about 60% higher than the design tensile stress (where $N = 5$ on s_u). If you should use this idea, be sure to use a percentage increase less than 60% if $N < 5$ on the ultimate

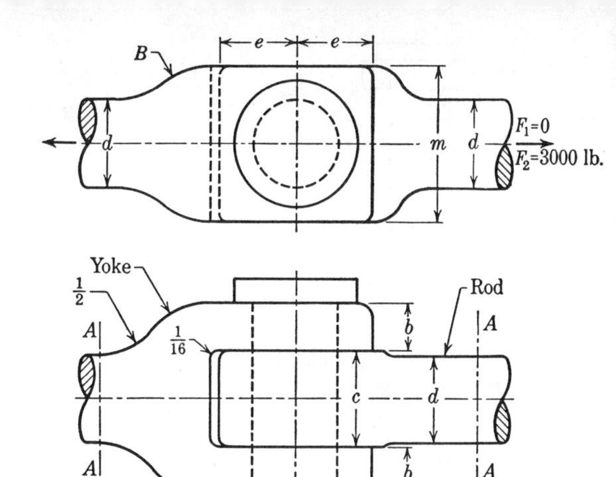

FIGURE 1.17 Rod and Yoke Connection.

stress, and in any case, apply it only to ductile materials. Moreover, the resulting design stress in bearing should generally be safely below the yield strength.

Recall that projected area tD is used for computing the surface compressive stress in Fig. 1.16, the nominal value being $s_c = F/A = F/(tD)$ psi.

1.21 EXAMPLE—STRESS ANALYSIS.

The beginner's first difficulty is in sorting out various areas subjected to different kinds of stresses; that is, recognizing bending, shear, tensile, or compressive stresses and the areas on which they act when they exist in a single unit to be designed. The thought process involved in this act is called stress analysis, which is best illustrated by an example.

Design a yoke connection, similar to the one shown in Fig. 1.17, to withstand a load of $F = 3000$ lb. repeated in one direction, if the material is AISI C1022, as rolled.

Solution. Let the first step in a *design* problem be to decide upon the design stresses. Since the load is repeated, choose a design factor for use with *ultimate* stresses. From Table 1.1 for ductile material, use $N = 6$. The various ultimate stresses are found in Table AT 7 for the 1022, as rolled (look them up now). The corresponding design stresses are:

$$s_t = \frac{72}{6} = 12 \text{ ksi}, \qquad s_c = \frac{72}{6} = 12 \text{ ksi}, \qquad s_s = \frac{54}{6} = 9 \text{ ksi}.$$

The most obvious stress is tension across the circular section A–A, Fig. 1.17. From $F = sA$, we get

$$3000 = (12{,}000)\left(\frac{\pi d^2}{4}\right)$$

$$d = \left[\frac{(3000)(4)}{12{,}000\pi}\right]^{1/2} = 0.565 \text{ in.}; \qquad \text{standard } \tfrac{9}{16} \text{ in.}$$

A "standard" fraction will be chosen for each dimension (see § 1.22).

Determining the dimensions a, b, c, e, and m is not so simple. The best thing to do in most cases where several design equations are involved is first to set up

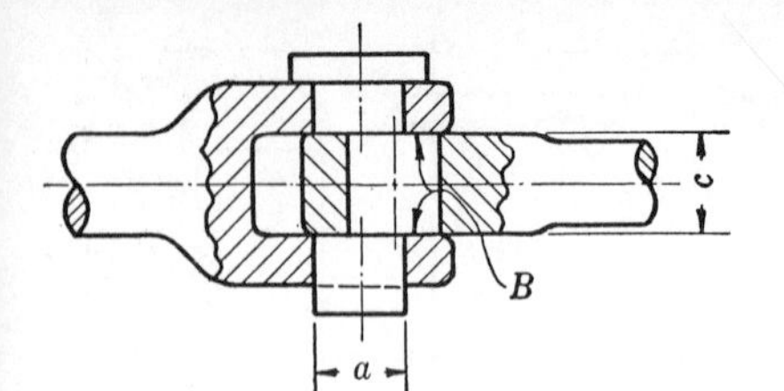

FIGURE 1.18 Shear of Pin.

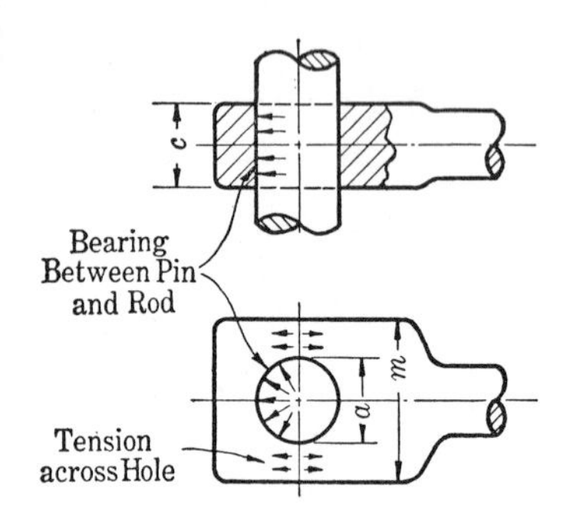

FIGURE 1.19 Compression Between Pin and Rod, and Tension Across the Hole in Rod.

all the strength equations representing the various possible methods of failure so that they may be studied. We shall follow this plan, being sure that the areas are given in terms of the dimensions shown in Fig. 1.17.

The *pin* may fail by ***shearing*** across the section B, Fig. 1.18. One sectional area of the pin is $\pi a^2/4$, and the total resisting area is twice this. Thus

$$\text{(i)} \qquad F = sA = s_s(2)\frac{\pi a^2}{4} = \frac{s_s \pi a^2}{2}$$

The maximum shearing stress computed as a vertical shear in a beam [§ 1.10(b)] will be 33% larger than the average stress as given by (i). Therefore, if shear of the pin governs its size, this difference should be considered. The *compressive stress* s_c between the *pin* and the *rod* may be excessive. For a projected area ac, Fig. 1.19, we get

$$\text{(j)} \qquad F = sA = s_c(ac).$$

The compressive stress between the pin and the *yoke* may be excessive. For one side of the yoke, the projected area is ba (see Fig. 1.20) and for two sides, $2ba$. This gives

$$\text{(k)} \qquad F = sA = s_c(2ba).$$

The rod or yoke may fail in *tension* across the hole for the pin, the section of minimum area. See Figs. 1.19 and 1.20. We have

$$\text{(l)} \qquad F = sA = \underset{\text{[ROD]}}{s_t(m - a)c} = \underset{\text{[YOKE]}}{s_t(m - a)2b}.$$

The pin is not only subjected to shear across the sections B, Fig. 1.18, but also to bending. There are several assumptions that might be made, some of which are shown in Fig. 1.21. This decision as to what idealization to use for design purposes is an elementary illustration of the kind that the designer is continually making. The caption to Fig. 1.21 suggests some possible logic. The best answer would be found by comparing computed results according to one or more ideal models with experimental measurements. On this basis, for working fits (Chapter 3), the assumption in Fig. 1.21(e) agrees best with actual measurements,[1.8] except for relatively high loading when the points of support move

FIGURE 1.20 Compression Between Pin and Yoke, and Tension Across the Hole.

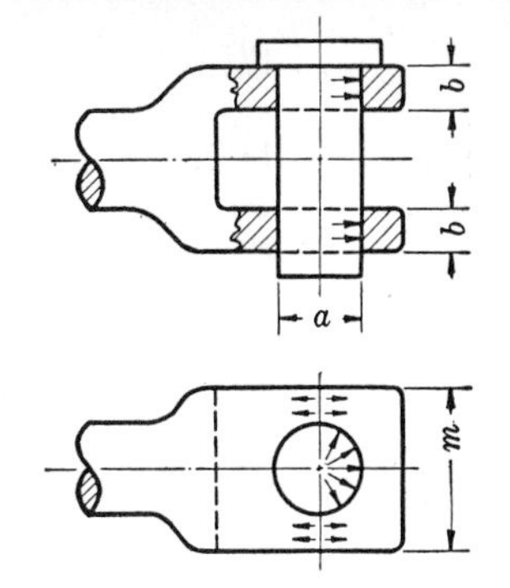

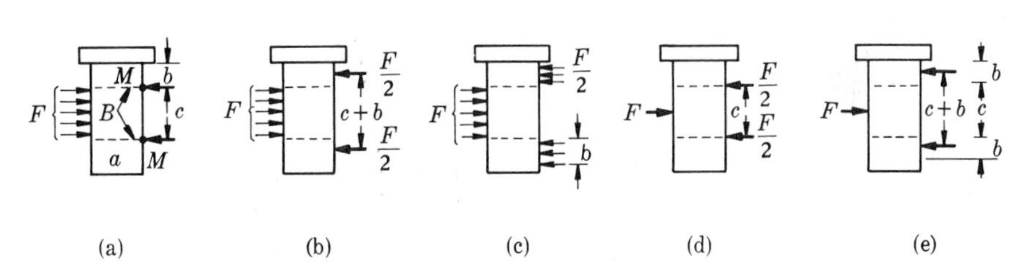

Figure 1.21 Pin as a Beam. (a) Loose fits in yoke arms: assuming that the pin bends enough that it rests at M on the inside edge of the holes in the yoke; uniform load; max. $M = Fc/8$. (b) Loose fits in yoke, but support assumed at midpoint of b; uniform load; max. $M = F(b+c)/8$, which is greater than in (a) and therefore more conservative. (c) Tight fits in all parts: almost a fixed end beam with uniform load, $M = Fc/12$; or a uniformly loaded cantilever beam, $M = Fb/2$ (which is larger than $Fc/12$ for normal proportions). (d) Loads not uniformly distributed, $M = Fc/4$. (e) Loads not uniformly distributed, $M = F(b+c)/4$.

toward each other approaching Fig. 1.21(d). Moreover, the situation in Fig. 1.21(e) results in the largest maximum bending moment and is therefore the most conservative shown and therefore it would perhaps be a normal choice in the absence of experimental verification. Thus for the pin in bending,

$$\text{(m)} \qquad M = \frac{F(b+c)}{4} = s_f \frac{\pi a^3}{32} \qquad \text{or} \qquad F = \frac{s_f \pi a^3}{8(b+c)}$$

One other type of failure that might be checked would be the pin tearing out the end of the rod or yoke as suggested by Fig. 1.22. There is a shear on length e and a depth c on both sides of the pin;

$$\text{(n)} \qquad F = sA = \underset{\text{[ROD]}}{s_s(2ce)} = \underset{\text{[YOKE]}}{s_s(2)(2be)}.$$

By a comparison of equations (j) and (k), we see that if $2b = c$, the yoke will be safe if the rod is safe; hence, let $c = 2b$ in all calculations. There are still several decisions to be made. One could *assume* a ratio of say a/c, solve the various

FIGURE 1.22 Pin Shearing Out End of the Rod.

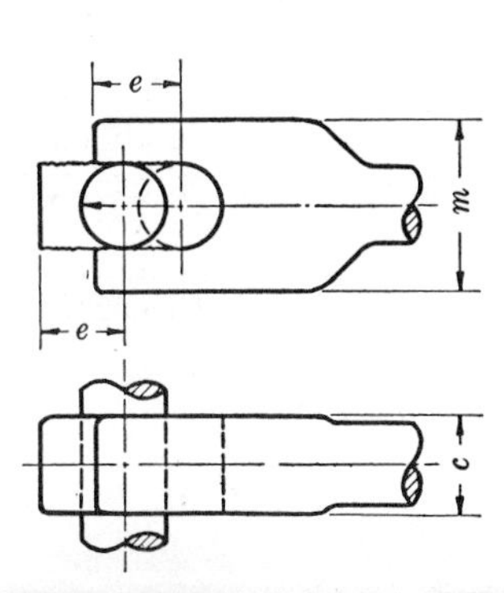

equations and use the largest values of a and c obtained. We might think in terms of a more optimum design and get proportions that would make the pin equally strong in compression and bending or equally strong in shear and bending. Try equal compression and bending by equating the F's from equations (**j**) and (**m**).

$$\textbf{(o)} \qquad F = s_c ac = \frac{s_f \pi a^3}{8(b + c)} \qquad \text{or} \qquad a = 1.955c$$

for $b + c = 1.5c$, $s_c = s_f$. (NOTE. For a fixed connection with no relative motion, one could safely assume the bearing stress $s_c \approx 1.6 s_f$ in accordance with § 1.20, but the resulting proportions would be even worse; of course, one does not know this at the start.) Solving for c from (**j**), we have

$$\textbf{(p)} \qquad F = s_c 1.955c^2 \qquad \text{or} \qquad c = \left[\frac{3}{(12)(1.955)}\right]^{1/2} = 0.358 \text{ in.}$$

This dimension is less than $d = \frac{9}{16}$, and while it would give optimum proportions from the standpoint of utilizing the capabilities of the material, most such connections have $c \gtreqless d$. (See the design for relative motion below.) Assume then that $c = \frac{5}{8}$ in., which means that the compressive stress will be lower than the design value. For $b + c = 1.5c = \frac{15}{16}$ in., equation (**m**) gives

$$\textbf{(q)} \qquad a = \left[\frac{8(b + c)F}{s_f \pi}\right]^{1/3} = \left[\frac{(8)(0.9375)(3)}{12\pi}\right]^{1/3} = 0.842: \qquad \text{Use } \tfrac{7}{8} \text{ in.}$$

Check the pin in shear, equation (**i**);

$$\textbf{(r)} \qquad a = \left(\frac{2F}{s_s \pi}\right)^{1/2} = \left[\frac{(2)(3)}{9\pi}\right]^{1/2} = 0.461 \text{ in.,}$$

which, being less than that required for bending, means that the pin is safe in shear. (IMPORTANT NOTE. If the load were a dead load, the shear computation could be allowed to govern, because a consideration of actual rupture suggests that failure would not occur until the pin actually sheared. However, for a repeated load, failure may well occur by fatigue in bending (Chapter 4) and therefore cannot be ignored in this problem where the load is repeated.)

The dimension m can now be computed from (**l**);

$$\textbf{(s)} \qquad m = \frac{F}{s_t c} + a = \frac{3}{(12)(0.625)} + 0.875 = 1.275.$$

This designer is worried about the small margin around the hole*; use $m = 1\frac{1}{2}$ in. The computed value of e from (**n**) is, for $c = \frac{5}{8}$,

$$\textbf{(t)} \qquad e = \frac{F}{2s_s c} = \frac{3}{(2)(9)(0.625)} = 0.266 \text{ in.,}$$

which, being less than the radius of the hole, is unreasonably small; that is, this is an instance where the area needed is too small to be physically possible. Besides the assumption that such shear starts at the center line of the hole is overly optimistic. The dimension m and the margin distance e are roughly analogous to a hub diameter (for gear or pulley), which is generally empirically chosen (for steel) from $1.25D$ to $1.8D$, where D is the hole diameter. Considering e from this

* The stress concentration effect at the hole may be surprisingly large; see Fig. AF 6 and ref. *(4.62)*.

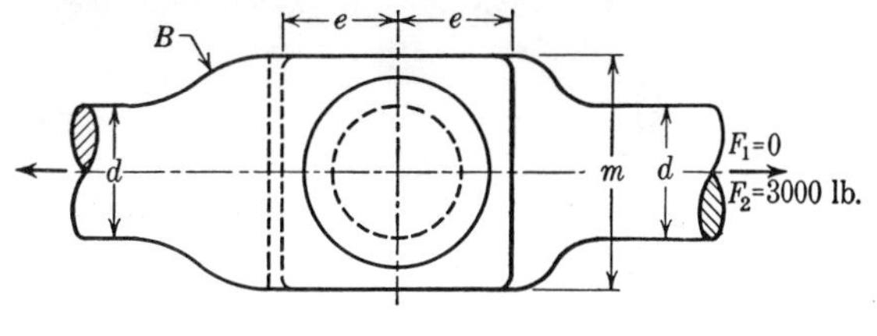

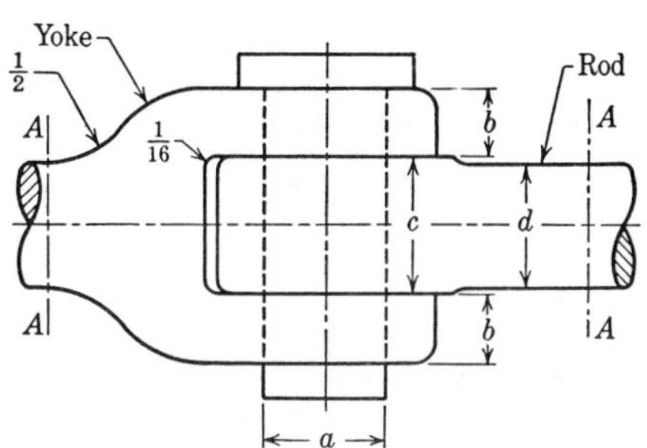

FIGURE 1.17 (Repeated).

point of view, let it be $e = m/2 = \frac{3}{4}$ in. Summarizing, we have the dimensions:

$$d = \tfrac{9}{16} \text{ in.}, \quad a = \tfrac{7}{8} \text{ in.}, \quad c = \tfrac{5}{8} \text{ in.}, \quad b = \tfrac{5}{16} \text{ in.}, \quad e = \tfrac{3}{4} \text{ in.}, \quad m = 1\tfrac{1}{2} \text{ in.}$$

The next thing to do would be to sketch this design to scale, to check the over-all appearance and proportions. If changes are made that cause an increase in stress, check the stress in the affected parts. Provide generous fillets and curves at adjoining sections, as at B, Fig. 1.17, in order to keep the stress concentrations at a minimum (Chapter 4). Specify all other dimensions; help on tolerances, allowances, and surface finish, is given in Chapter 3.

Solution for Relative Motion. If the ends are round with a diameter m, Fig. 1.17, significant relative motion is possible. If relative motion occurs, the permissible compressive stress must be very much smaller than before in order to avoid excessive wear on the surfaces. If the motion is small, meaning that the heating from friction introduces no difficulties, the allowable stress may be, say, 1000 to 4000 psi or more. Experience in a particular machine is needed to define the proper value. The wear rate, which also is a function of a surface hardness that has not been specified, increases with the pressure, and the life expectancy desired would govern. We shall let the design value of $s_c = 4000$ psi $= 4$ ksi and present the following solution after the above pattern but without the discussion. From equation (**o**),

$$s_c ac = 4ac = \frac{12\pi a^3}{12c} \qquad \text{or} \qquad a = 1.13c$$

$$F = s_c ac \qquad \text{or} \qquad c = \left[\frac{3}{(4)(1.13)}\right]^{1/2} = 0.814; \qquad \text{Use } \tfrac{13}{16} \text{ in.}$$

$$a = \left[\frac{8(b + c)F}{s_f \pi}\right]^{1/3} = \left[\frac{(8)(1.5 \times 0.8125)(3)}{12\pi}\right]^{1/3} = 0.918; \quad \text{Use } \tfrac{7}{8} \text{ in.}$$

$$m = \frac{F}{s_t c} + a = \frac{3}{(12)(0.8125)} + 0.875 = 1.183 \text{ in.}$$

As before, use $m = 1\frac{1}{2}$ in.; the rounded end takes care of e. There is no need to repeat the shear-of-pin calculation. Other suggestions are the same as before. Summarizing,

$$d = \tfrac{9}{16} \text{ in.}, \quad a = \tfrac{7}{8} \text{ in.}, \quad c = \tfrac{13}{16} \text{ in.}, \quad b = \tfrac{7}{16} \text{ in.}, \quad m = 1\tfrac{1}{2} \text{ in.}$$

The maximum deflection of the second pin for the loading assumed in Fig. 1.21(e), by formula taken from Table AT 2, is

$$y = \frac{FL^3}{48EI} = \frac{(3)(1.25)^3(64)}{(48)(3 \times 10^4)(\pi)(0.875)^4} = 1.42 \times 10^{-4} \text{ in.},$$

where $L = b + c = \frac{7}{16} + \frac{13}{16} = 1.25$, $I = \pi a^4/64$, $E = 3 \times 10^4$ ksi. The next question is of course what deflection is permissible. Such values for shafting are typically expressed in terms of deflection per inch of length between supporting points, $y/(b + c)$ in this problem. A value of 0.002 in./ft. or $0.002/12 = 0.000167$ in./in. is relatively strict. This compares to $0.000142/1.219 = 0.0001165$ in./in. for our pin. Therefore, even without prior experience in this particular case, it is fairly safe to assume that the deflection is not excessive.

By way of résumé, the general plan of attack is to write down all the strength equations applying to various areas of the part in terms of letters for dimensions and then to study the equations for a logical method of solution, not overlooking a chance for optimum proportions (the ideal design, like the old one-horse shay, is one in which every area fails simultaneously, but not until after the machine has served its purpose). Compute the dimensions and study them for their practicality. Adjust as deemed desirable and check questionable stress areas. The procedure from here depends on the end use and may involve building the item and testing it, perhaps more than once.

1.22 PREFERRED SIZES (STANDARD FRACTIONS).

For many machine elements, there are standardized sizes, as bolts, keys, I-beams, which means that such sizes are more readily available on the market and are also cheaper. The designer always uses standard items and standard proportions unless he feels strongly that some custom design is desirable. We shall give information on standards as we proceed. If there are no such standard sizes, let the preferred dimensions[3.13] vary by

$\frac{1}{64}$ between $\frac{1}{64}$–$\frac{1}{32}$; $\frac{1}{32}$ between $\frac{1}{32}$–$\frac{3}{16}$;
$\frac{1}{16}$ between $\frac{3}{16}$–$\frac{7}{8}$; $\frac{1}{8}$ between $\frac{7}{8}$–3;
$\frac{1}{4}$ between 3–6; $\frac{1}{2}$ above 6.

In choosing a standard size from a decimal number, avoid a smaller size if the change is more than some 4–5%; take the next larger. Also observe that now and then a larger dimension results in a weakening, for example, a longer beam.

1.23 COURTESY IN THE WRITTEN FORM OF CALCULATIONS.

Before going too far, we should suggest that you present your problem work in a form that will be pleasing and quickly understood by those concerned. Neatness and completeness of calculations are necessities, either in the school room or *in the engineering office*. In both cases, the design calculations are likely to be checked by another person. The time taken by the original designer to make his work legible and complete in every detail will consistently be less than the time spent by a checker (or the boss!) in trying to comprehend the meaning of semi-illegible figures. However, it is not only a matter of the economical use of time and of having your work so that you can quickly get the most out of it in reviewing for quizzes, it is also a question of *common courtesy*, a concern for others. The following suggestions for the form of calculations should be followed except where they may conflict with special directions of your instructor or of your supervisor.

(a) Name the calculation with a phrase or sentence, for example, *Diameter of Pin for Shear.*

(b) Give the equation to be used in terms of defined symbols: $F = s_s \pi a^2/2$. Where appropriate, the symbols should be defined by a sketch. *Be liberal with sketches.* It is wise to identify an equation taken from a book by page number and equation number (if any). If appropriate, state the assumptions on which the equation is based (as the pin in bending above).

(c) Substitute numerals for known symbols in the *same order* in which they appear in the equation. Be sure the units are consistent. Highlight the design stresses, making it clear how they are obtained.

(d) Mathematically simplify the equation by solving for the unknown. Many equations are so simple in form that this step is unnecessary. A good rule for school purposes is to have in the final calculations all steps in the solution except those that you can easily make mentally.

(e) Write down slide-rule results, *with units*. Use separate scratch sheet for cancellations and juggling for slide-rule purposes.

(f) Highlight your answer or result—by heavy underlining, placing it in a marginal column, etc.

(g) If you are computing a dimension, do not leave it in decimal form; select a standard dimension or part.

(h) give pertinent conclusions, if any, derived from your computations.

(i) Reflect on your calculations and try to be sure that you have thought of contingencies and alternatives. Compared to the other parts of the problem, *does the answer look reasonable?* Have you located points of maximum stress? Have you allowed for the fact that a computed answer may be a bad answer, practically? Put on record in your calculations significant reasons for designing the way you did.

(j) In general, avoid overcrowding and the use of small script and letters. Work toward the desired result by using a series of fundamental equations rather than by substituting into a more complicated derived equation, except for repeated calculations.

(k) Separate the various computations clearly by drawing a straight line after each entirely across the page.

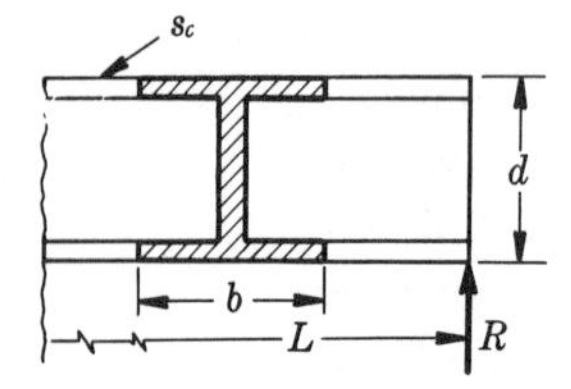

FIGURE 1.23

1.24 BUCKLING OF BEAM FLANGE. Some structural shapes, I-beams, etc., are designated by the nominal depth and the weight per foot. For example, 14WF34 is a wide-flange beam with a depth $d \approx 14$ in. and it weighs 34 lb. per foot of length. If the unbraced span L, Fig. 1.23, is too great, the flange that is in compression may buckle, even with an otherwise safe stress level. This contingency is covered by building codes for structures; for example, the AISC gives the permissible flange compressive stress ($s_c = Mc/I$) as[1.1]

$$s_c = \frac{22.5}{1 + (L/b)^2/1800} \text{ ksi} \qquad \left[15 < \frac{L}{b} < 40\right] \tag{1.17}$$

where, Fig. 1.23, b = flange width and the maximum allowed $L/b = 40$. For $L/b < 15$, the permissible $s_c = 20$ ksi.

1.25 THIN-WALL PRESSURE VESSEL. A thin-wall pressure vessel is one whose plate thickness is small compared to the diameter of the vessel. The pressure p psi gage, Fig. 1.24, acts in a radial direction. Summing the vertical components $\int p \, dA \sin \theta$, we find the total force tending to rupture the vessel on a diametral plane, say section AA, as $p(D_iL)$—the pressure times the *projected* area. Assuming uniform stress across thickness t and $A = 2tL$, we have

$$\text{(u)} \qquad pD_iL = s_t(2tL) \qquad \text{or} \qquad t = \frac{pD_i}{2s_t} \qquad \text{and} \qquad s_t = \frac{pD_i}{2t}.$$

When these vessels are made with a welded (Chapter 19) or riveted joint,

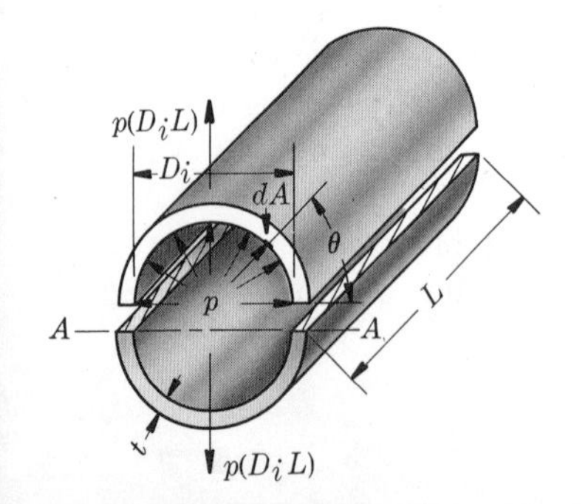

FIGURE 1.24 Separation of Shell along Longitudinal Section.

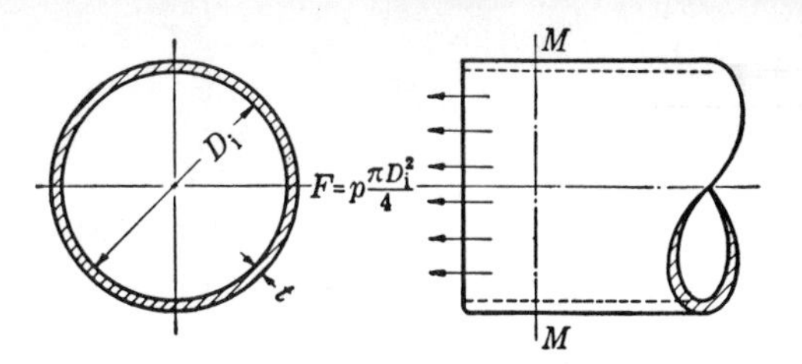

FIGURE 1.25 Stress in Diametral Plane.

as they usually are, the weakness of the joint as compared with the parent plate is cared for by a joint efficiency η, defined as

$$\text{(v)} \qquad \eta = \frac{\text{Minimum strength of joint}}{\text{Strength of solid plate}},$$

which would preferably be called the ***relative strength***. Introducing η, we find the required thickness of shell from

$$(1.18) \qquad t = \frac{pD_i}{2s_t\eta}.$$

The stress in a diametral section is different. The force is $p\pi D_i^2/4$ and the area, Fig. 1.25, is closely $\pi D_i t$; substituting into $F = sA$, we have

$$\text{(w)} \qquad \frac{p\pi D_i^2}{4} = s_t \pi D_i t \qquad \text{or} \qquad s_t = \frac{pD_i}{4t},$$

which, compared with equation (**u**), shows that the nominal stress on a transverse section is half that on a longitudinal section. These two stresses are principal stresses (Chapter 8); the one on the longitudinal section is the maximum principal stress and is the basis of design. In most if not all states, it is required that pressure vessels for land installations be designed to meet code specifications,[1.9,1.10,1.11] where the factor of safety on specified material strengths is taken as 5. For thick-wall vessels, see § 8.26.

1.26 EXAMPLE—TITANIUM VESSEL. Helium is used to provide pressure on the fuel and lox (liquid oxygen) in rocket motors. Let helium be stored in a cylindrical vessel, 20 × 24 in. (the first number is always the diameter), at 4000 psia. The welded joints are assumed to have an efficiency of 85% (Chapter 19) and for the purpose of saving weight, the vessel is to be made of annealed titanium alloy, Ti 6 Al 4 V. For normal temperature, what thickness of plate is needed with a design factor of 1.4 on the yield strength?

Solution. In Table AT 3, we find $s_y = 130$ ksi. (At low temperature, the yield strength is higher, § 2.22). From equation (1.18),

$$t = \frac{pD_i}{2s_t\eta} = \frac{(4)(20)}{(2)(130/1.4)(0.85)} = 0.507 \text{ in.},$$

use $\frac{1}{2}$ in. NOTE. The low factor of safety proves satisfactory here because the vessels are proof tested and closely inspected besides. If failure involves human life, the factor of safety may be raised to about 2. It is assumed that support brackets and fittings do not result in a significant increase in stress. Weight is too important a factor here for designers to be free and easy about the degree of safety and the codes do not apply. See § 2.19.

1.27 CONTACT STRESSES. Stress equations that are usually omitted in the undergraduate course on strength of materials because of the complications and because of the time needed for solutions are those giving the so-called contact stresses. The most frequently met cases in engineering are the equivalent of two cylinders in contact along an element, Fig. 1.26, and a sphere in a groove (ball bearing). The case of the cylinders is relatively uncomplicated, it will serve our purposes, and it can be applied to gear teeth, cam and follower, roller bearings, chain power drives, and to other elements. Theoretically, two perfect cylinders with axes parallel make contact along a line, but when a load F is applied, deformation occurs and the line becomes a finite area, say w by b as shown

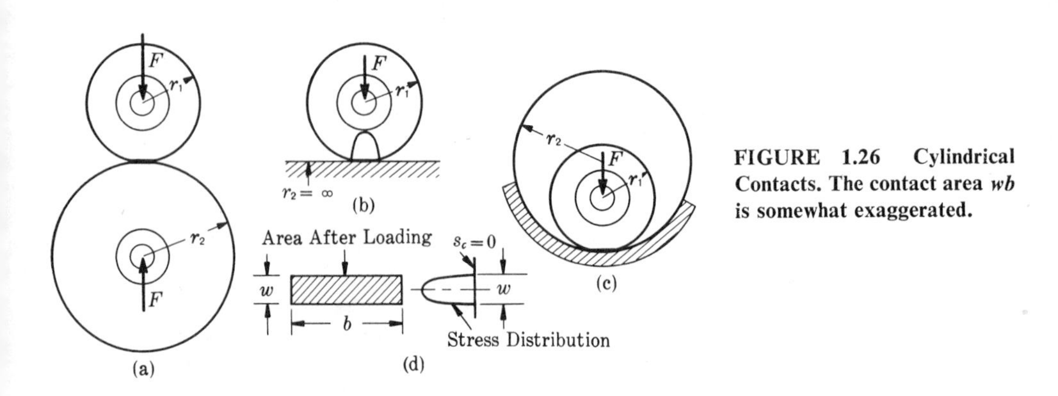

FIGURE 1.26 Cylindrical Contacts. The contact area *wb* is somewhat exaggerated.

in Fig. 1.26(d). The original derivations were by H. Hertz (the corresponding stress is therefore often called a Hertz stress) on the assumption that the stress distribution normal to the surface across the short dimension is elliptical, Fig. 1.26(b) and (d). The equation is

$$(1.19) \qquad s_c^2 = \frac{F(1/r_1 + 1/r_2)}{\pi b[(1 - \mu_1^2)/E_1 + (1 - \mu_2^2)/E_2]} \text{ (psi)}^2 \text{ or (ksi)}^2,$$

where s_c is the maximum principal stress; r_1 in. = radius of smaller cylinder, r_2 = radius of larger cylinder [r_2 is *negative* when it wraps about smaller as in Fig. 1.26(c)], b = length of cylinder in contact, E_1, E_2 (psi or ksi to match F lb. or kips) are the modulii of elasticity for the materials, and μ_1, μ_2 are Poisson's ratios. The usual assumptions apply: elastic action, homogeneous and isotropic materials. Poisson's ratio does not vary greatly with different metals (but see values given in Tables AT 3 and AT 7), and it is not always accurately known. Assuming a single value of $\mu = 0.3$, equation (1.19) reduces to

$$(\text{x}) \qquad s_{c\,max} = \left[\frac{0.35F(1/r_1 + 1/r_2)}{b(1/E_1 + 1/E_2)}\right]^{1/2} \text{ psi or ksi.}$$

Since the stress is compressive, it is often given the negative sign from the square-root operation. The maximum shearing stress is a consequence of the three principal stresses, all compressive, and is

$$(1.20) \qquad s_{s\,\max} = 0.3 s_{c\,\max} \text{ at } z = 0.393w \text{ in.}$$

below the surface, where

$$(y) \qquad w = \frac{4 s_{c\,\max}}{1/r_1 + 1/r_2}\left(\frac{1-\mu_1^2}{E_1} + \frac{1-\mu_2^2}{E_2}\right).$$

The maximum shear stress is often taken as the significant stress, because it is thought that this stress is effective in promoting the flaking of metal particles from the surface (the visible evidence of surface fatigue).

A sphere in contact with another surface has theoretical point contact and therefore, for a particular force, the maximum stress is much larger than for cylinders. While the equations for two spheres in contact are easy to use, this case is not too often found in engineering. For detail pertaining to various shapes of surfaces in contact, see Seely,[1.7] where there are charts to expedite solutions.

When there is relative motion, especially sliding, the stress situation is more complicated,[1.7] and if the coefficient of friction is large (large frictional force), the maximum stresses are significantly greater than given by the foregoing equations. In any case, there seems to be a good correlation between the wear life of a pair of lubricated surfaces (rolling and/or sliding on one another) and the maximum stress. When the stress is above the endurance-limit stress of the surface, an increase of stress greatly shortens the life. Buckingham,[14.3] experimenting with two case-hardened rollers, found a life of 10^6 cycles for a load of 12.89 kips (computed $s_{c\,\max} = 362$ ksi) and a life of 10^8 cycles for a load of 8.27 kips (computed $s_{c\,\max} = 295$ ksi), the life changing by a factor of 100 while the load decreased about one-third. However, it is an observed fact that the computed contact stress may be allowed to be very much higher than that due to tension, bending, etc., if a reasonable finite life is expected.

1.28 STATICALLY INDETERMINATE PROBLEMS. Problems are said to be statically indeterminate when the load on the various parts cannot be found by the principles of statics alone ($\Sigma F = 0$, $\Sigma M = 0$). For example, if a beam with parallel loads has two suppports, the reactions at the supports can be found by statics, but if there are three points of support, an additional unknown is introduced and an additional condition must be found. Where the problem is solvable by rational methods, the extra conditions usually relate to deformations (deflections). These problems exist in wide variety, but generally some idealization can be made that leads to a solution.

For example, suppose a link A, Fig. 1.27, supports a load W and is in turn supported by slender members 1 and 2 as well as the pin B. If A has

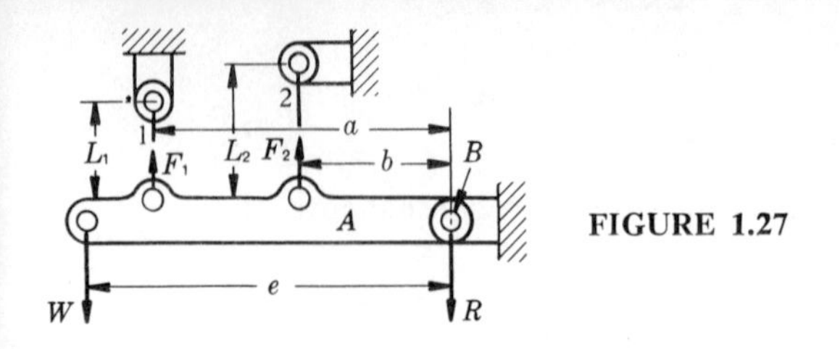

FIGURE 1.27

relatively large stiffness against bending in the plane of the paper, the deformations of 1 and 2 can be considered to be the only deformations. The deflection of the pins must also be relatively negligible. This simplifies the problem considerably.

Thinking of A as a free body, we see that, knowing W and all the dimensions, there are three unknowns, F_1, F_2, and R, but for a parallel force system, there are only two independent conditions. First, take moments about B;

$$\text{(z)} \qquad \sum M_B = Fe - F_1 a - F_2 b = 0,$$

in which F_1 and F_2 are unknown. If A is rigid, the deflection δ of the members 1 and 2 are proportional to their distances a and b from the pivot B. We can write this proportionality and then use $\delta = eL$, $\epsilon = s/L$, and $F = sA$ [see equations (1.1), (1.2), and (1.3)] to find

$$\frac{a}{b} = \frac{\delta_1}{\delta_2} = \frac{\epsilon_1 L_1}{\epsilon_2 L_2} = \frac{s_1 L_1 E_2}{E_1 s_2 L_2}; \qquad \text{or} \qquad s_2 = \frac{s_1 L_1 E_2 b}{E_1 L_2 a};$$

$$F_1 = A_1 s_1; \qquad F_2 = A_2 s_2 = \frac{A_2 L_1 E_2 b}{E_1 L_2 a} s_1.$$

These values of F_1 and F_2 into equation (**z**) give

$$Fe - A_1 s_1 a - \frac{A_2 L_1 E_2 b}{E_1 L_2 a} s_1 b = 0.$$

If the dimensions of the members 1 and 2 are known, the only unknown here is s_1. Solve for s_1; then for F_1; then for F_2; then another static equilibrium equation will give R and all the forces are known. If the problem is a design problem, one first decides which member 1 or 2 has the maximum stress. If this is s_1, then s_1 can be a safe design stress. Perhaps for convenience in manufacture, the rods 1 and 2 have the same cross section, $A_1 = A_2 = A$; if so, solve for A. Let the reader assume that $A_1 = A_2 = A$, $E_1 = E_2 = E$, and show that it would be more economical of material to reverse the positions of 1 and 2; that is, the shorter member should be closer to B.

1.29 THERMAL STRESSES. If a structure or machine element undergoes temperature changes, it expands or contracts, the amount in simple situations being defined by the coefficient of thermal expansion α; the total linear deformation $\delta = \epsilon L = \alpha(\Delta t)L$ in. However, α is not constant and, for practical purposes, it varies with the temperature. The values given in the Tables AT 3–AT 6 are for ambient temperatures, but other values are available in the literature. If an unstressed body is heated so that temperatures are different in its different parts, there will exist thermal stress gradients (as well as temperature gradients), which, in many cases, have led to service failures. If a body is restrained from deforming, even though its temperature is uniform, it will be stressed proportionally to the strain ($s = E\epsilon$ within the elastic limit), which is the difference between the unrestrained dimension and the same dimension under strain. Situations of thermal stresses are typically special problems whose solutions depend upon the circumstances, the ingenuity of designer, and sometimes experimental stress analysis. A theoretical approach soon becomes involved, but it should be used where appropriate.[1.12] In general, a designer as best he can avoids mounting members so that temperature changes are met with restraints that tend to prevent a change of dimensions (thus he uses expansion joints, sliding supports, etc.); but restraints are often lived with and, frequently, it is impossible to avoid differences in temperatures in a particular part.

1.30 FOR STUDENTS ONLY. Education without misrepresentation is impossible. There are so many things that need to be said at once to the beginner that many statements must be simplifications in order to be intelligible. For pedagogical reasons, safe design procedures are often given in this book—usually in too much detail. Although it is a convenience to a teacher to be confronted with some uniformity of approach for grading purposes, there are likely to be other "correct" points of view. Your teacher or supervisor may ask for another. After a few years of experience, your conclusions will rest more and more on your own background, but it is hoped that you will always be in search of better design approaches. Since there is no single correct answer to a design problem unless the procedure is specified completely, including the design factor and the material, your instructor is more interested in how you attack a problem and in the decisions you make than in the result. Therefore, *work for a good solution*, not so much for an answer.

In any subject, there is a certain amount of new language to be learned. Since difficulty with a new subject is often synonymous with an ignorance of the language of the subject, pay close attention to new words, making a real effort to master them. Practice using them and make the subject easy. In studying the examples, work them out yourself, looking up all values referred to. Comprehending study prior to working a problem is truly a time-saver, student practice to the contrary notwithstanding. Those tables in the Appendix of immediate interest can be marked with a paper clip for easy location.

This text should be looked upon as a transition step toward real engineering work. The attitudes that you might begin to acquire have been expressed in *Report on Engineering Design*,* from which the following is quoted.

"(1). Willingness to proceed in the face of incomplete and often contradictory data and incomplete knowledge of the problem.

"(2). Recognition of the necessity of developing and using engineering judgment.

"(3). Questioning attitude toward every piece of information, every specification, every method, every result.

"(4). Recognition of experiment as the ultimate arbiter.

"(5). Willingness to assume final responsibility for a useful result."

Engineering is the *art of applying the physical sciences* to the solution of the problems of mankind. If, after completion of your study of this book, you feel somewhat knowledgeable yet uncertain, our aim will have been at least partially accomplished. The art is *never* perfected. Moreover, since uncertainty is the father of progress, only the ignorant can afford to be certain. As you will see, machine design *is* engineering.

* *Journal of Engineering Education*, Vol. 51, No. 8, p. 650.

2. MATERIALS AND THEIR PROPERTIES

2.1 INTRODUCTION. This chapter will serve as a handy but abbreviated reference and it should be studied with this purpose in mind. We hope that most of the topics have been studied elsewhere in more detail. Since the number of available materials and the quantity of information on their properties is staggering, one must search for more complete knowledge when designing "for real." The references given will be helpful. There are many materials of interest to one designer or another that are not mentioned, for example: glass, asbestos, wood, concrete, cork, and of course a number of "plastics."

In making a choice of material, past experience is a good guide, so good in fact that engineers too frequently have overlooked possibilities in new materials. The best material is the one that will serve the desired purpose at the lowest cost for the manufacture and operating maintenance of the finished part. This best material is not always easy to find. It may involve trial and error. Sometimes the choice of material involves an intensive and expensive search by a group of engineers and scientists. Since the material often "makes or breaks" the machine, a sound decision, which calls for a wide background of knowledge, is important.

2.2 DEFINITIONS. See also § 2.3. For easy reference, we shall define briefly some of the terms (not elsewhere defined in this book) with which the reader should be familiar.[2.1,2.2,2.8]

Age hardening (precipitation hardening) occurs in some metals, notably certain stainless steel, aluminum, and copper alloys, at ambient temperature after solution heat treatment, the process being one of a constituent precipitating from solid solution. Where used, the consequences include increased strength and hardness, decreased ductility. Aging at moderately elevated temperature expedites the process and is called *artificial aging.*

Alloy is a substance with metallic properties, composed of two or more elements of which at least one is a metal.

Alloying elements in steel are usually considered to be the metallic elements added for the purpose of modifying the properties.

Anisotropy is the characteristic of exhibiting different properties when tested in different directions (as tensile strength "with the grain" or "across the grain").

Brittleness is a tendency to fracture without appreciable deformation. See *ductility*.

Charpy test is one in which a specimen, supported at both ends as a simple beam, is broken by the impact of a falling pendulum. The energy absorbed in breaking the specimen is a measure of the impact strength of the metal. See *Izod test*.

Cold shortness is brittleness of metals at ordinary or low temperatures.

Cold working is the process of deforming a metal plastically at a temperature below the recrystallization temperature and at a rate to produce strain hardening. Cold-drawn steel is frequently used because it increases strength and machinability, and improves surface finish. It reduces ductility. Commercial amounts of cold working of steel are of the order of 10–20%.

Damping Capacity is the ability of a material to absorb or damp vibrations, which is a process of absorbing kinetic energy of vibration owing to hysteresis. The absorbed energy is eventually dissipated to the surroundings as heat. At a particular stress level, cast iron is a much better damping material than steel.

Decarburization is a loss of carbon from the surface of steel, occurring during hot rolling, forging, and heat treating, when the surrounding medium reacts with the carbon (as oxygen and carbon combining).

Ductility is that property that permits permanent deformation before fracture in tension. There is no absolute measure of ductility, but the ***percentage elongation*** and the ***percentage reduction of area*** are used as indices; the higher these indices, the more ductile the material is said to be. Ductility is the opposite of brittleness, but there is no sharp division line. For purposes of definition, it is frequently assumed that

Ductile material → Elongation greater than 5% in 2-in. gage,
Brittle material → Elongation less than 5% in 2-in. gage.

Ductility is frequently a valuable property because, by virture of it, a member may take an occasional exceptionally-high load without breaking.

Elasticity is the ability of a material to be deformed and to return to the original shape. Stress is proportional to strain only during an elastic deformation (see *proportional limit*, p. 8).

Embrittlement involves the loss of ductility because of a physical or chemical change of the material.

Free carbon is that part of the carbon content of steel or iron that is in the form of graphite or temper carbon.

Hard drawn is a temper produced in a wire, rod, or tube by cold drawing. See *temper* and §§ 2.16, 2.17.

Homogeneous materials (have homogeneity) have the same structure at all points. (Steel consists of randomly oriented iron crystals of different sizes, with other matter in between and is thus not homogeneous.

Isotropic materials have the same properties in all directions. (Wood has a grain; rolled steel is not isotropic.)

Izod test is a test in which a specimen, supported at one end as a cantilever beam, is broken by the impact of a falling pendulum. The energy absorbed in breaking the specimen is a measure of the impact strength. Impact values in the tables should be considered more qualitative than quantitative because the actual variation of samples from the same universe is quite wide (see Fig. 1.14). See *Charpy test.*

Killed steel is steel that has been deoxidized with a strong deoxidizing agent, such as silicon or aluminum, in order to eliminate a reaction between the carbon and oxygen during solidification. Ingots of killed steel are sounder, containing fewer gas holes, and more homogeneous than non-killed or *rimmed steel*; these are desirable characteristics for forgings and heavy rolled sections.

Machinability is a somewhat indefinite property that refers to the relative ease with which a material can be cut. In the case of steels, cold-drawn AISI B1112 being cut with a high-speed tool-steel tool and with a proper cutting oil is usually taken as 100%. Free-cutting brass is a reference for copper alloys. Such data as in Table AT 7 are roughly relative at best, since the actual conditions of operation vary widely. There are significant production variables, such as the sharpness and shape of the cutting tool, exact nature of the material, the cutting lubricant, and the use of carbide tools.

Malleability is a material's susceptibility to extreme deformation in rolling or hammering. The more malleable the metal, the thinner the sheet into which it can be formed (usually cold). Gold and aluminum are quite malleable.

Mechanical properties are those that have to do with stress and strain: ultimate strength and percentage elongation, for example. See *physical properties.*

Percentage elongation is the extension in the vicinity of the fracture of a tensile specimen, expressed as a percentage of the original gage length, as 20% in 2 in.

Percentage reduction of area is the smallest area at the point of rupture of a tensile specimen divided by the original area.

Physical properties exclude mechanical properties, and are other physical properties such as density, conductivity, coefficient of thermal expansion. See *mechanical properties*. *Chemical* properties include corrosion resistance.

Plasticity is the ability of a metal to be deformed considerably without rupture. In a plastic deformation, the material does not return to its original shape. See *elasticity*.

Poisson's ratio is the ratio of the lateral strain (contraction) to the longitudinal strain (extension) when the element is loaded with a longitudinal tensile force.

Precipitation heat treatment brings about the precipitation of a constituent from a supersaturated solid solution by holding the body at an elevated temperature, also called *artificial aging*. In some alloys, precipitation may also occur at ambient temperatures, a process called *aging*.

Proof stress is that stress which causes a specified permanent deformation of a material, usually 0.01% or less. See *yield strength*, § 1.8.

Red shortness is a brittleness in steel when it is red hot.

Relaxation, associated with creep, is the decreasing stress at a constant strain; important for metals in high-temperature service.

Residual stresses are those not due to applied loads or temperature gradients; they exist for various reasons, as unequal cooling rates, cold working, etc.

Rimmed steel is incompletely deoxidized steel. Ingots of this steel have a surface layer quite free of slag inclusions and gas pockets, which results in the optimum surface on rolled sheets.

Solution heat treatment is the process of holding an alloy at a suitably high temperature long enough to permit one or more constituents to pass into solid solution and then cooling fast enough to hold the constituents as a supersaturated solution. (Precipitation may occur with time.)

Stiffness is the ability to resist deformation. It is measured by the modulus of elasticity in the elastic range; the higher the modulus, the stiffer is the material.

Strain hardening is increasing the hardness and strength by plastic deformation at temperatures lower than the recrystallization range. See *temper*.

Temper is a condition produced in a non-ferrous metal by mechanical or thermal treatment: for example, annealed temper (soft), hard temper, spring temper. See §§ 2.16 and 2.17.

Toughness is the capacity of material to withstand a shock load without breaking. The impact strength (see Charpy and Izod tests), though not an absolute measure, evaluates toughness. Formerly, the energy required to pull a standard tensile specimen in two was taken as the toughness, but this quantity is not representative because of the effect of the cold working of the specimen during the slow-speed test.

Transverse strength refers to the results of a transverse bend test, the specimen being mounted as a simple beam; also called *rupture modulus*. It is frequently applied to brittle materials, especially cast iron.

Work hardening is the same as strain hardening.

Wrought steel is steel that has been hammered, rolled, or drawn in the process of manufacture; it may be plain carbon or alloy steel.

2.3 HEAT-TREATMENT TERMS.[2.3,2.6,2.8]

Heat treatment is an operation or combination of operations involving the heating and cooling of metal or an alloy in the solid state for the purpose of altering the properties of the material. A few of the most common terms have meanings as given below. See § 2.8 for case-hardening processes.

Aging (and ***age hardening***) is a change in a metal by which its structure recovers from an unstable or metastable condition that has been produced by quenching or cold working. The change in structure, which proceeds as a function of time and temperature, consists in precipitation often submicroscopic. The result is a change of mechanical and physical properties, a process that may be accelerated by using a temperature slightly higher than room temperature.

Annealing, a comprehensive term, is a heating and slow cooling of a solid metal, usually done to soften it. Other purposes of annealing include those of altering the mechanical and physical properties, producing a particular microstructure, removing internal stresses (stress relieving), and removing gases. See *normalizing* below.

Critical range has the same meaning as *transformation range* (below).

Drawing is often used to mean *tempering*, but this usage conflicts with the meaning of the drawing of a material through a die (§ 2.9) and is to be avoided.

Graphitizing, and *annealing* process, causes the combined carbon to transform wholly or in part into graphitic or free carbon; it is applied to cast iron, sometimes to high-carbon steel.

Hardening is the heating of certain steels above the transformation range and then quenching, for the purpose of increasing the hardness. In the general case, hardening is any process of increasing the hardness of a metal. See § 2.9.

Malleablizing is an annealing process whereby combined carbon in white cast iron is transformed wholly or in part to temper carbon. Temper carbon is free (graphitic) carbon in the form of rounded nodules, characteristic forms in graphitizing and malleablizing. See § 2.12 concerning malleable iron.

Normalizing is the heating of an iron-base alloy to some 100°F above the transformation range with subsequent cooling to below that range in still air at room temperature. The purpose is to produce a uniform structure.

Spheroidizing is any heating and cooling of steel that produces a rounded or globular form of carbide. Typically, it is a prolonged heating at a temperature slightly below the transformation range, usually followed by slow cooling; or, for small objects of high-carbon steel, it may be prolonged heating alternately within and slightly below the transformation range.

Stress relieving (thermal) is the heating of a metal body to a suitable temperature (generally just below the transformation range for steel, say 1100–1200°F) and holding it at that temperature for a suitable time (1 to 3 hours for steel) for the purpose of reducing internal residual stresses. The internal stresses may be present because the body has been cast, quenched, normalized, machined, cold worked, or welded.

Tempering is a reheating of hardened or normalized steel to a temperature below the transformation range, followed by any desired rate of cooling. Quenched steel is tempered in order to reduce internal stresses, to restore a certain amount of ductility, and to improve toughness. The time and temperature of tempering are selected in order to give the steel the desired properties. See Figs. AF 1, AF 2, AF 3, Appendix.

If, for example, a particular steel with a particular yield strength is desired, do not specify the tempering temperature. This temperature can be varied slightly to produce closely the desired mechanical property.

Transformation range for ferrous metals is the temperature interval during which austenite is formed during heating; it is also the temperature interval during which austenite disappears during cooling. Thus, there are two ranges; these may overlap but never coincide. The range on heating is higher than that on cooling.

2.4 HARDNESS.

The hardness of a material is a measure of its resistance to indentation, and is one of the most significant properties because, properly interpreted, it says much about the condition of the metal. The most common instruments used to determine hardness are the Brinell, Rockwell, Vickers, and Shore scleroscope. Increasing hardness numbers indicate increasing hardness.

The Brinell hardness number (BHN) is determined by a standard pressure (3000 kg. standard, 500 kg. for soft metals) applied to a 10-mm. ball which presses for 10 sec. or more on the surface of the material being tested. The load in kilograms *divided by* the area of the surface of the indentation in square millimeters is the BHN. This hardness number is closely related to the *ultimate tensile stress of steel* as follows:

$$s_u \approx (500)(\text{BHN})\text{ psi} \quad \text{or} \quad (0.5)(\text{BHN})\text{ ksi.} \tag{2.1}$$

[FOR STEEL WHEN 200 < BHN < 400]

The probable range of values is $(470)(\text{BHN}) < s_u < (530)(\text{BHN})$. Use this approximation only when reliable test data are not available, and *do not apply it to any other metal.*

The Rockwell tester, faster than the Brinell and widely used commercially, utilizes several different indenters and, in effect, measures the *depth* of the penetration by the indenter. Each indenter has an identifying symbol, as follows (always specify the Rockwell scale):

Rockwell B (R_B), $\frac{1}{16}$-in. ball, 100-kg. load, for medium soft metals, as for many copper alloys and soft steel.

Rockwell C (R_C), diamond indenter, 150-kg. load, for hard metals, as hard steel.

Rockwell A (R_A), diamond indenter, 60-kg. load, for extremely hard metals, such as tungsten carbide.

Rockwell D (R_D), diamond indenter, 100-kg. load, sometimes used for case-hardened metal.

Rockwell E (R_E), $\frac{1}{8}$-in. ball, 100-kg. load, for soft metals, such as bearing metals and magnesium; F, ($\frac{1}{16}$-in. ball); H, K, L, M, P, R, S, V (with different sizes of balls), are all used for soft materials as a substitute for the E scale; scale G ($\frac{1}{16}$-in. ball) for phosphor bronze.

The Rockwell superficial tester, a different machine, is used for a piece of material too thin for the standard tester. The scales for this tester are N (for C-hard materials) and T (for B-hard materials). The N-scale indenter is a diamond, and the load may be 15, 30, or 45 kg., designated thus: 15-N, 30-N, 45-N. The T-scale indenter is a $\frac{1}{16}$-in. ball with loads as before and designated thus: 15-T, 30-T, and 45-T.

The Vickers tester has a square-base, diamond pyramid indenter, and the Vickers number is the load in kilograms divided by the impressed area in square millimeters.

The Shore scleroscope number is obtained by letting a freely falling hammer with a diamond point strike the object to be tested and measuring the height of rebound. This height is the Shore number; the higher the rebound, the harder is the material. The Shore machine can be used on large parts and is often used as a quick inspection aid, but it is less accurate than the other tests. Conversions from one hardness to another may be made in Fig. AF 4, which gives the approximate relationships for steel. The term *file hard*, often found in the literature, should be a hardness of perhaps 600 Brinell.

2.5 AISI AND SAE SPECIFICATION NUMBERS.

There are numerous "standard" materials specifications.[2.2,2.3,2.6] Many large consuming organizations and nearly all producers have some standards of their own. The armed forces have numerous ones. However, the principal agencies whose specifications are most widely used are the American Society for Testing Materials (ASTM), the Society of Automotive Engineers (SAE), and the American Iron and Steel Institute (AISI). The SAE and AISI specification numbers are alike for steel except that the AISI uses prefixes B, C, D, and E to indicate the method of manufacturing the carbon grades; see Table 2.1.

In a general way for steel, the first digit (or the first two digits) of the number represents a type of steel, for example: 1XXX is a plain carbon steel, 11XX is a plain carbon steel with greater sulfur content for free-cutting, 2XXX is nickel steel. The last two digits in four-digit numbers invariably give the approximate or average carbon content in "***points***" or hundredths of per cent. For example, an SAE 1030 or an AISI C1030 has about 0.30% carbon, spoken of as 30 points of carbon (nominal range is 0.28–0.34%). Or in 8620, the average carbon content is close to 0.20% (range of 0.18–0.23%).

TABLE 2.1 SYSTEM OF SPECIFICATION NUMBERS FOR STEEL—AISI AND SAE

In the AISI system, prefixes have the following meanings: B, acid bessemer steel; C, basic, open-hearth carbon steel; D, acid open-hearth carbon steel; E, electric-furnace steel (usually alloy). Letters B or L in the middle of the number indicate that boron or lead, respectively, has been added; as 94 B 40 and 11 L 41 (§ 2.6). An H at the end indicates that material can be bought on hardenability specification, as 9840H (§2.7).

STEEL	SAE	STEEL	SAE
Plain carbon	10XX	Molybdenum-chromium-nickel	47XX
Free cutting	11XX	Molybdenum-nickel	48XX
Manganese	13XX	Chromium	5XXX
Boron	14XX	heat and corrosion resistant	514XX
Nickel	2XXX		515XX
Nickel-chromium . .	3XXX	Chromium-vanadium . .	6XXX
heat and corrosion resistant	303XX	Nickel-chromium-molybdenum	8XXX
Molybdenum	4XXX	Silicon-manganese	92XX
Molybdebum-chromium	41XX	Nickel-chromium-molybdenum (except 92XX)	9XXX
Molybdenum-chromium-nickel	43XX		
Molybdenum-nickel . .	46XX		

Examination of Table AT 7, for example, shows that in general the strength of steel increases with carbon content, while the ductility decreases. A brief suggestion of typical uses of plain carbon wrought steel is as follows.

Carbon, 10–20 points, 10XX group. Used for tubing, forgings, pressed-steel parts, screws, rivets, and for carburized case-hardened parts.

Carbon, 10–20 points, 11XX group. Due to higher sulfur content in certain grades, it is free-cutting and good for use in automatic screw machines for miscellaneous parts, including screws; it also may be carburized. For case hardening, the open-hearth steels, identified by the symbol C in the AISI number, are to be preferred. Higher carbon-content steels in the 11XX group, as 1141, contain more manganese and are heat treatable for improved mechanical properties. See Table AT 9. These steels are not usually welded.

Carbon, 20–30 points. General purpose grades, used for forged and machined parts; screws; also for boiler plate and structural steel.

Carbon, 30–55 points. With 0.40–0.50% C, frequently used for miscellaneous forged and machined parts; shafts. Frequently heat treated for improved mechanical properties. Cold finished for shafting and similar parts.

Carbon, 60–95 points. May be hardened to a good cutting edge, especially in the higher ranges of carbon; therefore, used for tools. Also for springs. High strength, low ductility. Nearly always heat treated, say, to a Brinell hardness of 375 or higher.

2.6 ALLOY STEEL. Wrought alloy steel is steel that contains significant quantitites of recognized alloying metals, the most common being aluminum, chromium, cobalt, copper, manganese, molybdenum, nickel, phosphorus, silicon, titanium, tungsten, and vanadium. Alloys are used to improve the hardenability of steel (§ 2.7), to reduce distortion from heat treatment, to increase toughness, ductility, and tensile strength, and to improve low-temperature or high-temperature properties. See remarks on alloys below. With alloys, steel may be heat treated to the desired hardness with less drastic quenching and therefore with less trouble from distortion and cracking. For small parts, a relatively small amount of alloy is needed in order for the part to respond in depth to heat treatment. Larger parts should have greater amounts of the alloying elements for hardenability purposes. Alloy steels may be classified as:

(a) Low-alloy structural steels (not heat treated). These steels ($s_y \geqq 50$ ksi as rolled) were developed for structural uses where light weight is important (but not extremely so as in aeronautics), such as in the transportation industry, but they are also used in other structures. Phosphorous (0.03–0.15%) is an effective strengthener, as is nickel (0.5–2%) and copper (0.2–1.25%). Copper also imparts resistance to atmospheric corrosion. The carbon is typically some 0.15–0.20%; but more may be used on occasion. Other alloys are manganese, silicon, chromium, and molybdenum, but not necessarily all at once. They weld easily and do not air-harden.

(b) Low-carbon alloy steels (0.10–0.25%C), AISI steels, used chiefly for carburizing.

(c) Medium-carbon alloy steels (0.25–0.50% C), usually quenched and tempered to hardnesses between 250 and 400 Brinell.

(d) High-carbon alloy steels (0.50–0.70% C or more), ordinarily heat treated to hardnesses between 375 and 500 Brinell, for use as springs, wear resisting parts, etc.

(e) High-alloy steels, such as stainless steels.

A few brief remarks about the principal alloying elements will suggest other functions of alloys (*chemical symbol* in parentheses).

Aluminum (Al) is an efficient deoxidizer, an alloy in nitriding steels (nitralloys), and it promotes fine grain size.

Boron (B) in very small amounts (0.001% or less) is an economical hardenability agent in low- or medium-carbon deoxydized steels. It has no effect on tensile strength.

Chromium (Cr) improves hardenability economically, resistance to corrosion (with other alloys), strength at high temperatures, and wearing properties (high carbon).

Cobalt (Co) improves red hardness.

Columbium (Cb) is often used to "stabilize" stainless steel (that is, it preempts the carbon and forestalls the formation of undesired carbides).

Copper (Cu) improves steel's resistance to atmospheric corrosion; up to 4% Cu, it increases the fluidity of the melt; it improves tensile strength and the yield ratio in the normalized condition. Yield ratio $= s_y/s_u$. With more than 0.75% Cu, steels can be precipitation hardened.

Lead (Pb) improves machinability, but affects different alloys differently.

Manganese (Mn) improves strength and increases hardenability moderately, counteracts brittleness from sulfur. Present in all steels, manganese becomes an alloying element when it's amount exceeds about 0.6%, as in the 13XX steels. Medium-carbon manganese steels are subject to embrittlement at temperatures above 600°F. *Austenitic manganese steel* (not 13XX) typically contains 1.2% C and 12–13% Mn and responds to work hardening most readily.

Molybdenum (Mo) increases hardenability markedly and economically (when Mo > Cr), tends to counteract temper brittleness, improving creep strength and red hardness; it improves wear by forming abrasion-resistant particles. It is the most effective alloy for improving strength at high temperatures. (Alloys of molybdenum—not molybdenum alloys of steel—are proving to be very effective at temperatures above about 1500°F. See § 2.21.)

Nickel (Ni) strengthens unquenched and annealed steels, toughens steel (especially at low temperatures), and simplifies heat treatment by lessening distortion. It is the most effective element for reducing the brittleness of steel at very low temperature; see § 2.22. It is one of the principal alloys for stainless steel (§ 2.15).

Phosphorus (P) increases hardenability, strengthens low-carbon steels, improves machinability of free-cutting steels, and improves resistance to corrosion.

Selenium (Se) improves machinability of stainless steel; also added to leaded resulfurized carbon steels for the same purpose.

Silicon (Si) strengthens low-alloy steels and improves resistance to high-temperature oxidation; it is a good general-purpose deoxidizer and promotes fine grain.

Tantalum (Ta) is a stabilizer (see Columbium).

Titanium (Ti) is used for deoxidation and for stabilizing austenitic stainless steels (preventing intergranular corrosion and embrittlement); it increases the hardness and strength of low-carbon steel and improves creep strength.

Tungsten (W) increases hardenability markedly in small amounts and improves hardness and strength at high temperature. An expensive alloy, it is used only where a particular advantage results, as in high-speed tool steel in which it forms a hard, abrasion-resisting carbide.

Vanadium (V) promotes fine-grain structure, improves the ratio of endurance strength to ultimate strength of medium-carbon steels (average about 0.57), increases hardenability strongly when dissolved, and results in retention of strength and hardness at high temperature; it is the most effective element in retarding softening during tempering.

Since alloy steels are more costly than plain carbon steels, an alloy should not be employed unless its use yields some advantage. By properly balancing the alloy and carbon contents, one may obtain a particular strength with higher ductility or a particular desired ductility with higher strength than is possible without alloys. Since alloys generally improve the mechanical properties, their strength/weight ratio is higher, and therefore their use may result in smaller parts which partly offsets the increased cost per pound. It is important to note that the modulus of elasticity E (and G) is virtually the same for alloy steels as for carbon steels and that therefore *if rigidity is the basis of design, there is no advantage in using alloy steel*—an alloy steel deflects the same amount per unit stress as does a carbon steel.

If an alloy steel is used, it should, in general, be heat treated in order to obtain the best properties for the purpose. See tables in the Appendix. Illustrative of the uses of alloy steels, we have:

AISI 2330: bolts, studs, tubing subjected to torsional stresses.

AISI 2340: quenched and tempered shafting, connecting rods, very highly stressed bolts, forgings.

AISI 2350: high-capacity gears, shafts, heavy duty machine parts.

AISI 3130: shafts, bolts, steering knuckles.

AISI 3140: aircraft- and truck-engine crankshafts, oil-well tool joints, spline shafts, axles, earth moving equipment.

AISI 3150: wear-resisting parts in excavating and farm machinery, gears, forgings.

AISI 3240: shafts, highly stressed pins and keys, gears.

AISI 3300 series: for heavy parts requiring deep penetration of the heat treatment (hardenability) and high fatigue strength per unit of weight.

AISI 4063: leaf and coil springs.

AISI 4130, 4140: automotive connecting rods and axles, aircraft parts and tubing.

AISI 4340: crankshafts, axles, gears, landing gear parts; perhaps the best general purpose AISI steel.

AISI 4640: gears, splined shafts, hand tools, miscellaneous heavy duty machine parts.

AISI 8630: connecting rods, bolts, shapes; air hardens after welding.

AISI 8640, 8740: gears, propeller shafts, knuckles, shapes.

Alloys with 10–20 points carbon are widely carburized (§ 2.8) in producing pins, bolts, gears (teeth), shafts (at wearing surfaces), cams, and worm threads.

2.7 HARDENABILITY.

Hardenability is the capacity of steel to through-harden when cooled from above its transformation range. It is determined from a standard 1-in. round specimen, Fig. 2.1, with the test, called a Jominy test, conducted according to a standard procedure. Flats

FIGURE 2.1 End-Quench Hardenability Test. The specimen is heated to the proper quenching temperature, placed in a fixture as shown in this illustration, after which a spray of water strikes the lower end. The hardest part of the specimen will be the end exposed to the jet of water, the most drastic quench. Hardness will decrease as the distance from the sprayed end is increased, because the rapidity of cooling decreases to practically that of air cooling at the far end. Good hardenability is especially important when the entire section is subjected to high stresses; it is less important when the high stresses are on or near the surface. (Courtesy U.S. Steel Corp., Pittsburgh).

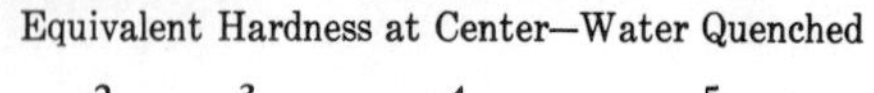

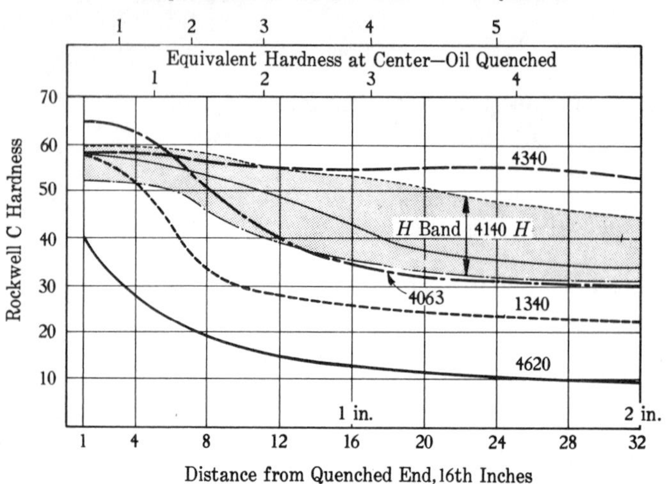

FIGURE 2.2 Hardenability Curves. The materials shown were selected to show differences. Notice that the hardness of AISI 4340 holds up well with depth, that the higher carbon 4063 has the hardest surface, that the same carbon steels, 4340 and 1340, have practically the same surface hardness (at $\frac{1}{16}$ in.), and that the low carbon 4620 (actual 0.17% C) has lower hardness all the way. The individual lines are typical actual test values. The H band shown for 4140 is suitable for specifications. The top coordinates are defined by this example: a 2 in., oil-quenched piece of 4063 has a hardness at its center of $R_c = 42$. (Courtesy Bethlehem Steel Co., Pittsburgh).

are ground longitudinally on opposite sides of the cooled specimen, after which the Rockwell C hardness is determined at each $\frac{1}{16}$-in. interval from the quenched end. The hardness of a quenched surface is largely dependent on its carbon content, while the depth hardness, or *hardenability*, depends on the amount of carbon, the alloy content, and the grain size.

The most effective alloys in improving hardenability are boron, vanadium, manganese, chromium, molybdenum, phosphorus, tungsten, and to a lesser extent, nickel and silicon. Smaller amounts of several elements increase hardenability more than does a larger amount of one alloy.

Typical hardenability curves are shown in Fig. 2.2, where it is seen that the hardness at a certain point on the test specimen corresponds to the hardness at the center of a certain size rod. This relationship is worked out experimentally for a particular kind of quench and steel. Although it is easy to harden a light piece all the way through, the material of a heavy piece must be chosen with forethought if through hardness (or nearly through hardness) is desired. Through hardness is not always wanted. A softer core may be an asset.

Some steels, frequently designated by H as a suffix to the AISI number (Table 2.1), may be bought on the basis of hardenability by specifying, for example, the limits of hardness of a Jominy specimen at a certain point along its length. Suppose that a hardness of $R_C = 40$ is desired at a certain depth and that this corresponds to $R_C = 45$ at $\frac{12}{16}$ in. in the Jominy test; the H band of Fig. 2.2 shows that AISI 4140 would be satisfactory in this respect.

2.8 CASE HARDENING. Case hardening of iron base alloys is a process of surface hardening whereby the surface or ***case*** is substantially harder than the ***core*** or inside metal. Case hardening is done by carburizing, cyaniding, nitriding, carbonitriding, induction hardening, and flame hardening.[2.1,2.3,2.6,2.8]

The purpose of case hardening is usually to provide a hard, wear-resistant surface while retaining great toughness in the core. It is also used for increasing the endurance strength of certain machine parts and for other purposes where high strength and hardness on the surface are advantageous.

(a) Carburizing. Carburizing is a process of adding carbon to the surface of steel by exposing it to hot carbonaceous solids, liquids, or gases—above the transformation temperature. Quenching, and usually tempering at 300–450°F for the purpose of relieving residual stresses induced by quenching, follow carburizing. The common methods of adding carbon are ***pack (or box) carburizing*** and ***gas carburizing.*** In pack carburizing, the part is heated in contact with solid carburizing compounds of various constituents, including charcoal, burned bone, charred leather, tar, and barium, sodium, and calcium carbonates, especially barium carbonate and charcoal. The depth of the case and the rapidity of the process depend in part on the soaking temperature, which is of the order of 1650–1750°F.

In ***gas carburizing,*** which has been developed to an efficient and economic procedure, especially for large quantities, the part is heated in carburizing gases, such as methane, ethane, propane, and CO. The temperatures of operation and case thicknesses obtained are much the same as those in pack carburizing. After 4 hr. at 1700°F, the thickness of the case should range between 0.04 to 0.05 in. In ***liquid carburizing*** the part is immersed in a molten salt bath that imparts a case similar to that obtained with gas or pack carburizing except that the case is thinner, usually not in excess of about 0.025 in.

For heavy duty, as in some gear teeth, a case thickness of 0.06 to 0.09 in. may be desired. It would seem that a safe design value of the surface hardness of carburized steel would be about 600 BHN. The hardness should generally fall between the limits

$$55 < R_C < 65 \qquad \text{or} \qquad 560 < \text{BHN} < 730.$$

Carburizing steels are low carbon steels, say 0.15–0.25% carbon.

(b) Cyaniding. As in liquid carburizing, cyaniding is accomplished by immersing the part in a hot (about 1550°F) liquid salt bath, sodium cyanide (NaCN) being a common medium in both processes. The difference in the processes lies largely in the use of a catalyst in liquid carburizing that results in a more rapid penetration of carbon and a relatively small amount of nitrogen in the case. Thus, the so-called cyanided case has more nitrogen, which is also a hardening agent. Whereas the thickness of the case of liquid-carburized parts may be somewhat greater than 0.02 in., the cyanided case is seldom thicker than 0.010 in. Low- and medium-carbon steels are usually used for cyaniding, and the case hardness may be of the order of that obtained by carburizing.

(c) Nitriding. In surface hardening by nitriding, the *machined and heat-treated part* is placed in a nitrogenous environment, commonly ammonia gas, at temperatures much lower than those used in the previously described processes, say 1000°F or somewhat less. Since a nitrided part does not need to be quenched rapidly, this process *avoids the distortion* that accompanies quenching, sometimes a big advantage, especially for complicated shapes. The hardening is the result of a reaction of the nitrogen, dissociated from the ammonia, with the alloying elements in the steel to form nitrides. For maximum hardness of the case, special steels, called nitra-alloys and containing aluminum as an alloy, are used; yet other steels, notably AISI 4340, are frequently nitrided. The case hardness of AISI 4340, tempered at 1025°F, and nitrided at 975°F for about 40 hr., may be over 600 Vickers (560 BHN) and the case depth some 0.025–0.030 in. In other situations, the nitriding time may be upwards of 90 hr., which together with the control problem, accounts in part for its high cost.

The carbon content of the nitralloys falls within the approximate range of 0.20 to 0.40% carbon. The case hardness of Nitralloy N, for example, nitrided at 975°F for 48 hr. should be above Vickers 900 (equivalent Brinell of about 780), or better than $R_C = 67$. The case thickness may be from 0.010 to 0.020 in. The case is quite strong for a body subjected to bending or tension, so that failures that result from repeated stresses usually originate in the transition region between the case and core.

(d) Carbonitriding is a process of case hardening steel by the simultaneous absorption of carbon and nitrogen from a surrounding hot gaseous atmosphere, followed by either quenching or slow cooling, as required. It is used for both batch and continuous processes. With a sufficient percentage (up to 15%) of ammonia in the carburizing gas, carbon steel parts may emerge file-hard without quenching. (See *nitriding*.) Very small amounts of ammonia (less than 1%) in the carburizing gas are sufficient to permit attainment of maximum hardness with an oil quench. The use of small quantities of ammonia in combination with quenching is cheaper than using a larger amount of ammonia. In continuous furnaces at 1500–1550°F, the case depth may range from 0.003 to 0.010 in., depending on time and temperature.[2.1] This process is used as a low-cost substitute for cyaniding and produces a product of good quality. Some hard cases produced by this method have been reported to wear many times longer than the best cyanided or carburized cases previously used.

(e) Induction Hardening. Induction hardening consists of heating a thin surface layer, preferably of annealed or normalized steel, above the transformation range by electrical induction and then cooling, as required, in water, oil, air, or gas. Since this process rapidly heats a thin layer of the surface, leaving the core relatively cool, the process is widely used for surface hardening of steels whose carbon content is in the range of 0.35 to 0.55%, the carbon content with which steel readily responds to heat treatment. The depth of the hard case may be regulated so that, for example, the Rockwell hardness is C 50 or greater to depths of 0.02 to 0.17 in. The surface hardness may be of the order of $R_C = 50$ to 55 or higher (BHN = 500 seems to be a reasonable design value); the core is of the order of $R_C =$ 30 to 35. See Fig. 2.3. After quench hardening of the heated surface, the part is preferably tempered at some 400–450°F. Surface hardening through induction heating is used also for cast iron and malleable iron. Since

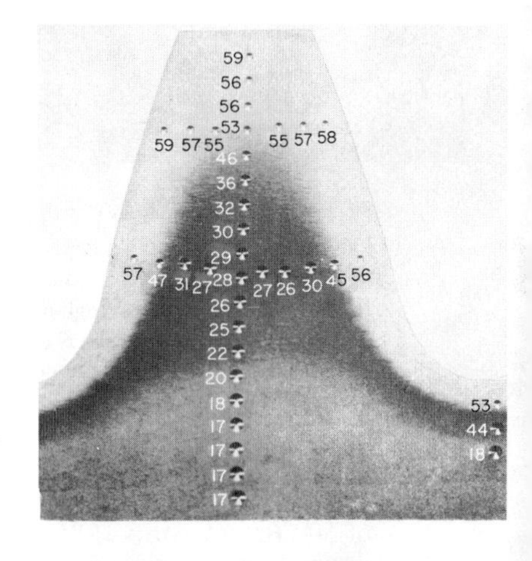

FIGURE 2.3 Hardness of a Gear Tooth, Induction Hardened. Observe the hardness close to the surface and near the middle of the base of the tooth (where the lower hardness indicates a tough core). Hardness readings are Rockwell C. (Courtesy Ohio Crankshaft Co., Cleveland).

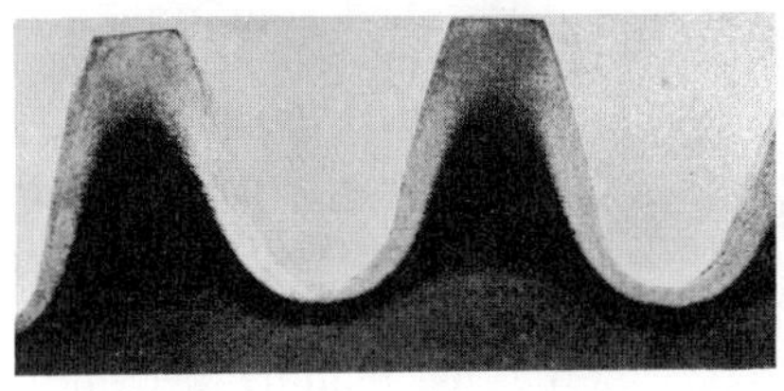

FIGURE 2.4 Induction-hardened Gear Teeth. If gear teeth are heated in an induction hardening machine, the heating follows the contour of the teeth as suggested by this illustration and Fig. 2.3. The lighter areas were rapidly heated, after which the gear was quenched. Besides being advantageous for contour hardening, as in gear teeth and cams, this process is well adapted to some selective hardening jobs, as hardening journal surfaces while leaving the rest of a shaft unaffected.

induction heating follows well the contour of the part being heated, it is appropriate for cams, gears (Figs. 2.3 and 2.4), and other irregular surfaces to be hardened.

(f) Flame Hardening. Flame hardening, like induction hardening, is a process of heating the surface of an iron-base alloy, which is preferably annealed or normalized, and then quenching it. Typically, neutral acetylene flames are played upon the surface to be hardened, followed closely by jets of water for cooling. While the process can be carried on manually, more consistent results are obtained by using especially designed machines. This method is applicable to the same metals as is induction hardening (say 0.45% C for steel), and while it is used for both small and large parts, it has particular advantages for very large parts where selected surfaces

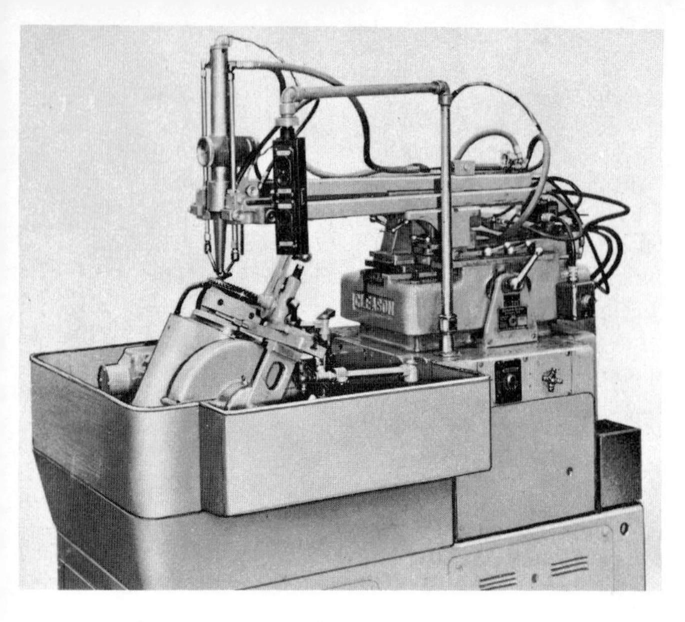

FIGURE 2.5 Machine for Flame-hardening Gear Teeth. An example of a special flame-hardening machine. This machine flame-hardens the surfaces of bevel-gear teeth on gears up to 24-in. The hardening operation is electroniclly controll ed and automatic; adaptable to spur, helical, and herringbone gears, and to straight, spiral and Zerol bevel gears. Flame hardening in general, which is especially well adapted to hardening selected surfaces on large parts, may also be manually controlled. (Courtesy Gleason Works, Rochester, N.Y.)

are hardened—such as the ways of a lathe or the surface of the teeth on large gears (Fig. 2.5). The resulting surface hardness should be of the order of 500 Brinell for 0.45% C steel, and the distortion may be negligible under controlled conditions.

2.9 WORK HARDENING. Work hardening is the result of a metal being stressed at some point into its plastic range, usually ordinary temperatures (certainly below recrystallization temperature); metal cold worked in this manner becomes stronger and more brittle. "Cold-finished" material* has had its cross section significantly reduced by cold rolling (usually used for flat products) or cold drawing through a die (usually employed to produce cold-finished rods). By 10% cold work, for example, is meant that the cross-sectional area is reduced 10% during the process. Typically for steel, the dimensional reduction per pass is $\frac{1}{32}$ or $\frac{1}{16}$ in., with a total reduction of 20% to 12% or less. A 12% reduction of steel (whose $s_u < 110$ ksi) results in an increase of about 20% in ultimate strength (closely a straight-line variation to this point), an increase of about 70% in yield strength (values to this point are all greater than the straight-line variation), and a decrease of about 35% in percentage elongation.[2.1] The toughness also decreases, and machineability increases. Cold drawing improves

* *Cold finished* is a term also applied to rounds that are turned or ground, but our use of the term will be as defined.

FIGURE 2.6 **Cold-Drawn AISI 1117. Scatter of yield strengths of 1 in. round bars, from 25 heats and 2 vendors. After Ref. (2.1).**

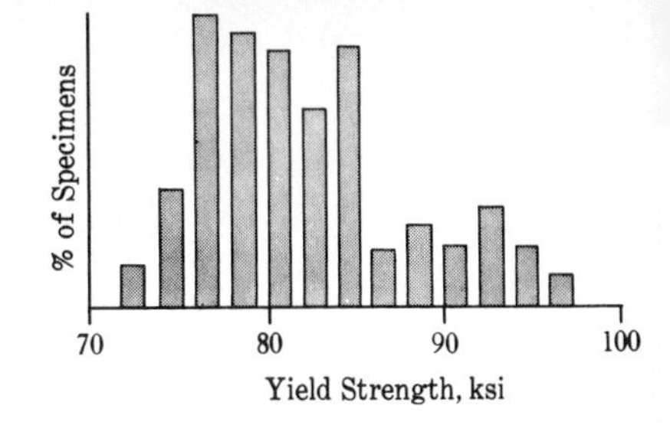

the strength of carbon steel enough that it is sometimes used instead of a more expensive heat treatment. However, the natural spread of mechanical properties is likely to be greater than for heat treated parts, suggesting caution. See Fig. 2.6. Cold drawing leaves residual tensile stresses that may be deleterious to fatigue strength; see § 4.30.

Steel is also drawn at elevated temperatures, with an even greater increase in strength (up to a material temperature of about 1000°F during drawing) as compared to cold-drawn material, stress relieved at the same temperature. Maximum strength for AISI 1144 bars, 1-in. diameter, is obtained when the material is drawn at about 600°F (a proprietary process* called elevated-temperature drawing, ETD) and no stress relieving is needed[2.1]—but a higher drawing temperature may be needed for improved ductility. The benefits of this method of finishing will in some cases eliminate heat treating operations and avoid the need for alloy. For comparison purposes (see Tables AT 7 and AT 10), some properties of AISI 1144 processed in this manner are: s_u = 140 ksi min. (150 ksi typical), s_y = 125 ksi min., BHN = 280, elongation ≈ 10%; and it has a good machinability rating of 80.

Additional remarks concerning cold working of other metals are found below. The cold working of surfaces (plastic deformation limited to a *thin surface layer*) by peening and rolling is taken up in Chapter 4.

2.10 WROUGHT IRON.

Wrought iron is made by burning the carbon from molten iron and then putting the product through hammering and rolling operations. The product contains some 1–3% slag and less than 0.1% carbon. The material is very soft and ductile and is easily forge welded. It is used principally for rivets, welded steam and water pipes, and for general forging purposes. Its most advantageous properties are its ductility and resistance to corrosion as compared to steel.

2.11 CAST IRON.

Cast iron in a general sense includes white cast iron, malleable iron, and nodular cast iron, but when *cast iron* is used without a qualifying adjective, gray cast iron, spoken of as ***gray iron*** is meant. In general, gray iron contains so much carbon (2.6–3.6% usually) that it is not malleable at any temperature. In gray iron, the excess carbon is uncombined, and a fracture is gray.

* La Salle Steel Co.

The ASTM, in specification A 48–46, has classified gray iron according to minimum tensile strength.[2.8] See Table AT 6. Thus, a "30 gray iron" will have a standard test strength of not less than 30,000 psi (30 ksi). Strength is increased by reducing the carbon content (about 3.7% for class 20, 2.8% for class 60), and the higher-strength cast irons contain typical alloys.[2.1,2.14,2.15]

Cast iron is sometimes given a heat treatment, but ordinarily it is cheaper to improve strength and other properties by reducing silicon and/or carbon or by increasing alloy content than by heat treating. The histogram of Fig. 2.7 shows a typical distribution of tensile strength as obtained from say 60 tests. Rigidly inspected, this lot would classify as class 35.

Gray iron has excellent wearing properties that are improved by certain alloys and by heat treatment (including flame and induction hardening).

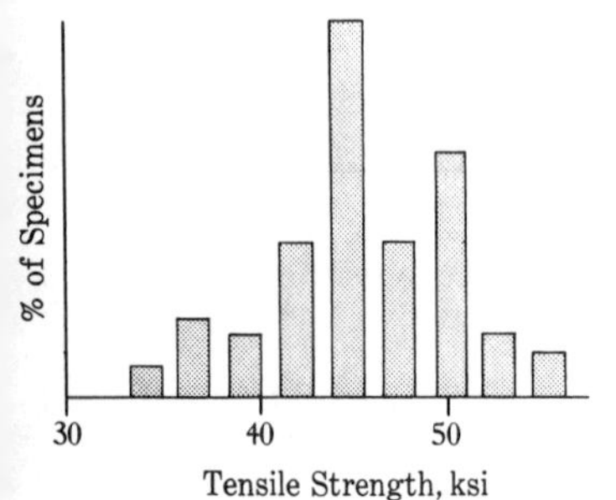

FIGURE 2.7 Tensile Strength of Cast Iron.

Thus, it is widely used for cylinder blocks, brake drums, gears, machine tool ways, and in general where there is metal-to-metal contact and relative motion. The lower-strength varieties and all grades in the annealed state are easily machined. Gray iron is more resistant to many kinds of corrosion than are ordinary or low-alloy steels.[2.1] It has much greater damping capacity (for vibrations) than steel, which suggests its use under some vibrating conditions. It is successfully used in crankshafts, as in automobile engines. See Fig. 2.14. Being the cheapest of metals, ordinary (low-strength) gray iron is the most widely used of all cast metals. Its principal disadvantages are its brittleness and lack of toughness, but these properties are often not significant. Cost tends upward with specification numbers, especially above ASTM 35; the cost of melted metal for good castings from class 60 iron runs about 2.3 times that for class 30 iron,[2.1] assuming large production; and there are other sources of increased cost.

Cast iron loses strength with increase of the minimum section dimension. A $\frac{7}{8}$-in., class-20 bar has a tensile strength of 28 ksi; a 4-in. bar has $s_u \approx 11$ ksi, no longer class 20. For a 1-in., class-60 bar, $s_u \approx 65$ ksi; for a 4-in. bar, $s_u \approx 44$ ksi.[2.1] Hooke's law is not too well approximated by cast iron, so that one must be careful of the values of the modulus of elasticity E used when the operating stress is high. See Fig. 1.3. Typical quoted values of E would be the slope of a straight line from the origin of the s-ϵ diagram to a point on the curve at $s_u/4$.

In ***white cast iron*** (the fracture is "white"), most of the carbon is combined chemically with the iron, and as a result, the metal is very hard. If the combined carbon is as much as 1.5%, the cast iron may be difficult or impossible to machine. When an extremely hard surface is desired, white cast iron, called ***chilled iron,*** is intentionally produced by using an iron plate in the mold to cause rapid cooling of the surface. The rapid cooling does not allow time enough for the carbon to be released as free carbon. Such a surface may be finished only by grinding and is suitable for car wheels, rolls, etc.

2.12 MALLEABLE IRON.

Malleable iron is heat-treated white cast iron.[2.1,2.17] The white cast iron is obtained not by chilling, as mentioned above, but by using the proper composition in the melt. The heat treatment of the white cast iron, in which substantially all of the carbon is combined in the form of iron carbide, is an annealing, called ***malleablizing,*** during which the white iron changes to ferrite and free (or temper) carbon.

Malleable iron produces strong, ductile, and easily machined castings at low cost in quantity. Best results are obtained on relatively thin sections. See Fig. 2.15. If the part is thicker than about 3 in., there will be difficulty in producing a white-iron casting devoid of uncombined primary graphite. Since it is necessary that all the carbon in the original casting be combined carbon for the best results, malleable castings are generally produced in sections of from $\frac{1}{8}$ to 2 in. thick. See grades 32510 and 35018 in Table AT 6. Induction hardening, as for gear teeth, may produce a surface hardness of $R_C \approx 55$.

A special form, pearlitic malleable iron differs from normal malleable iron in that there is a significant amount of combined carbon in the finished product. The difference is achieved by adding alloys, changing the heat treatment, or both.

2.13 NODULAR CAST IRON.

Nodular cast iron, also called ***ductile iron,*** has the castability (for complex forms), machinability, and wearability of gray iron, but higher strength and ductility. It thus has good toughness (shock resistance). Instead of flakes as in gray iron, the graphite has a spheroidal shape, obtained by the addition of certain alloys, usually both magnesium (a few hundredths of 1%) and cerium (a few thousandths of 1%), both because they are helpful to each other. See Table AT 6, but there are other standard specifications.[2.1]

Nodular iron is used for a wide variety of items, including casings, crankshafts, hubs, rolls, forming dies. It has a good resistance to thermal shock, and its "growth" at high temperature is less than that of gray iron. The as-cast 80-60-03, Table AT 6, is pearlitic and harder than the annealed 60-45-10, which is ferritic, and therefore the pearlitic has better wear properties. In general, nodular iron has wear properties comparable to those of gray iron of the same hardness, which is good. Nodular iron (pearlitic) responds well to flame or induction hardening, perhaps with $R_C > 55$, but residual casting stresses should be relieved before treatment to avoid possible cracks, and after treatment, a stress relief at 300–400°F may be desirable.

2.14 CAST STEEL. The combination of highest strength and highest ductility in a cast ferrous metal is obtained in cast steel.[2.1,2.16] When steel castings are heat treated, the carbon content generally falls within the range of 0.25 to 0.50%. However, many steel castings with carbon less than 0.20% are used in the as-cast condition.

Steel castings may be plain carbon or alloy steels. Among the ASTM grades of plain carbon steel castings in Table AT 6 are (ASTM A27-58): 60-30, 65-35, and 70-36, where the first number, say 60, represents the *minimum* tensile strength and the second number, say 30, represents the *minimum* yield strength at 0.2% offset, each in ksi. There are several high-strength cast steels in Table AT 6 (ASTM A148-58) that are often referred to in a similar manner, as 80-35, where the numbers have the meanings explained above. To obtain the high strength, the foundryman regulates the carbon and alloy content (manganese, silicon, and other alloys) in such a way as to produce the mechanical properties stated in the specifications. A minimum heat treatment is annealing or normalizing. Alloy cast steels respond to heat treatment in much the same way as similar wrought alloy steels do. Not all grades of cast steel are adapted to welding, but usually there is no trouble in this respect. Special alloys for corrosion resistance, heat resistance, and for other particular purposes are common.

2.15 STAINLESS STEEL. Table AT 4 gives the mechanical properties of some of the "standard" stainless steel specifications. Stainless steel is relatively expensive, but where the environment is significantly corrosive or at high or quite low temperatures, it provides an economical answer for many problems. There are three classes: austenitic steels (200 and 300 series—that include 3.5 to 22% nickel for its stabilizing of austenite), martensitic steels (usually with no nickel, but some types have 2.5% maximum), and ferritic steels (no nickel) that do not harden by quenching and tempering. All classes contain chromium (4 to 26%). Methods of hardening are:

(a) Cold working, § 2.9, which is the usual way of hardening the austenitic types because of their potent response to this treatment. Quenching these steels from about 1850°F avoids austenite transformation and leaves the steel soft, an effect opposite to that for ordinary steel. Cold-worked austenitic steels are classified according to temper as $\frac{1}{4}$ hard (for which minimum $s_u = 125$, $s_y = 75$ ksi), $\frac{1}{2}$ hard (minimums, $s_u = 150$, $s_y = 110$ ksi), $\frac{3}{4}$ hard (minimums, $s_u = 175$, $s_y = 135$ ksi), full hard (minimums, $s_u = 185$, $s_y = 140$ ksi). Austenitic stainless is the best stainless steel for high-temperature service; its corrosion resistance is better in the annealed than in the work-hardened state. Uses of some types include: 301—hardware, utensils, structural; 302—strong at elevated temperature, decorative and corrosive applications, widely used; 303—free machining 18-8; 321—carbon stabilized by titanium, which makes it good for welding without subsequent annealing; 347—columbium (Cb) and tantalum (Ta) stabilized for welding without annealing.

(b) Age hardening, usually termed ***precipitation hardening*** with reference to stainless, which occurs because of the precipitation of a constituent from a supersaturated solid solution. It is generally conducted at elevated temperatures in order to increase the rate of precipitation. The stainless steels 302, 303, 304 and 316 are subject to precipitation hardening at 800–1650°F. The 17-7PH (17% Cr, 7% Ni), Table AT 4, which contains considerably less carbon, manganese, and silicon than 301, is precipitation hardened when mill annealed at 1950°F, air cooled and reheated to 1400°F for 90 min., air cooled and, to accelerate precipitation of carbides, reheated to 1050°F for 90 min. Carbide precipitation is a migration of carbon to the grain boundaries where it combines with the chromium to form chromium carbide. The depletion of the chromium, the element that makes the steel stainless, adjacent to the grain boundaries results in the boundary material of some grades being susceptible to highly corrosive media. Corrosion of this sort is called intergranular corrosion. The lower the carbon content, the less carbide precipitation there is; hence, these special grades of stainless steels are made with carbon content less than 0.08% (versus a normal 0.15–0.25%), to be used, for example, where a body is to be welded without subsequent annealing—as in the case of a body in the field or too large for available annealing furnaces. Other grades of austenitic stainless steels have the stabilizing elements columbium plus (maybe) tantalum or titanium added. These stabilizers combine with the carbon, thus preventing it from combining with the chromium. The 347 grade is stabilized with columbium and, moreover, contains less than 0.08% carbon. Another treatment for precipitation hardening stainless steels involves cooling to some −100°F to transform the austenite to martensite; annealing at this stage results in a very high yield strength. Some types of austenitic manganese steels are capable of being precipitation hardened, as well as many nonferrous alloys.

(c) **Quenching and tempering,** as for usual steels, except that transformations are so slow that the quenching is by air cooling for maximum hardness. This is the common way of hardening the martensitic types (although they respond in varying degrees to cold working). The type 410 is perhaps the least expensive, and its response to heat treatment is less than that of the 431 and some other types. The 403 is used for forged turbine blades; 410 is a general purpose type (screws, rods, shafts, rivets, pistons, knife blades) and can be successfully cold formed; 431 is a high strength stainless.

The ferritic stainless steels do not harden significantly by heat treatment, nor do they work harden excessively. Cold working these steels raises their yield strength 30% or more, but the increase of ultimate strength is much less. Type 430 is widely used in the annealed state for auto trim, restaurant fixtures, heat exchanger flues, chemical equipment, etc.

2.16 COPPER ALLOYS. Since copper is one of the oldest known metals, it has been the base of many alloys, as well as being used in a relatively pure form.[2.1,2.9] Originally, the words brass and bronze, which have been used for hundreds of years, had fairly distinct meanings: brass being an alloy of copper and zinc; bronze being an alloy of copper and tin. However, the names have become so confused that the only safe way to know an alloy is to know its actual composition. Frequently copper-zinc alloys contain some tin or copper-tin alloys contain zinc. Moreover, many other alloys for copper have come into use, notably aluminum, silicon, beryllium, and cadmium. The tendency is to call the newer copper alloys *bronze*, even though there is little or no tin present, as in aluminum bronze and silicon bronze. Manganese *bronze* is really a high-strength *brass*, the improved mechanical properties being obtained by including small amounts of aluminum, iron, manganese, and tin. Phosphor bronze *is* a bronze, but the finished product may have only a trace of the phosphor that was added primarily to deoxidize the melt, a treatment that improves the mechanical properties. See Table AT 3 for a few common copper alloys; the compositions given are approximate.

Copper and its alloys have characteristics that determine the advisability of their use; among these may be mentioned: electrical and thermal conductivity, resistance to corrosion, malleability and formability, ductility, strength, excellent machinability (especially with lead added), nonmagnetic, pleasing finish, ease of being plated, and castability (sand castings, permanent-mold castings, die castings, and others).

Quenching does not harden copper alloys (except beryllium copper), so the usual manner of increasing strength and hardness is by cold work. Strip and sheet are cold worked by rolling; rods, bars, shapes are cold worked by drawing through a die. The percentage reduction of area by cold work determines the temper; the softest grade is annealed. Reductions for other tempers are approximately as follows:[2.1]

TEMPER	STRIP	WIRE	TEMPER	STRIP	WIRE
¼ hard	10.9%	20.7%	Hard	37.1%	60.5%
½ hard	20.7%	37.1%	Extra hard	50 %	75 %
¾ hard	29.4%	50 %	Spring hard	60.5%	84.4%
			Extra spring hard	68.7%	90.2%

Copper alloys high in zinc (over say 35%; e.g., naval brass) cannot take the extreme cold working because they become excessively brittle. A few uses of some of the copper alloys follow.

Admiralty metal: condenser and other heat-exchanger tubes and plates.

Aluminum bronze: corrosion-resistant parts; marine pumps, shafts, valves; parts where high strength, toughness, wearability, low coefficient of friction, and damping are important, as some bearings, gears, worm wheels, cam rollers; also a decorative metal, as in statues and costume jewelry.

Beryllium copper (also called beryllium bronze): parts where high formability, high *yield*, *fatigue*, and *creep* strengths, and also good corrosion resistance are advantageous; springs, bolts and screws, firing pins, dies, surgical instruments, spark resistant tools. It is solution annealed by holding at 1450°F for about 1 hr. per in. or fraction of inch; quench rapidly in water.[2.1] It is precipitation hardened by starting with solution-annealed metal and then for say ½ hard, heat for 2 hr. at 600°F. The properties of beryllium bronze are quite attractive, but it is some 5 times as expensive as brasses.[2.1]

Cartridge brass: electrical parts, automotive radiator cores, pins, rivets, springs, ammunition components, tubes.

Manganese bronze: clutch disks, pump rods, shafts, valve stems, welding rod.

Naval brass: condenser plates, marine hardware, propeller shafts, piston rods, valve stems, welding rod, balls, nuts, bolts, rivets.

Phosphor bronze: bellows, diaphragms, clutch disks, cotter pins, lock washers, bushings, springs, wire, welding rod, chemical hardware, wire brushes.

Silicon bronze: hydraulic pressure lines, hardware, bolts, nuts, rivets, screws, electrical conduits, heat-exchanger tubes, welding rod.

Yellow brass: electrical fixtures, plumbing, wire, pins, rivets, screws, springs, architectural grillwork, radiator cores.

2.17 ALUMINUM ALLOYS.

The lighter alloys are especially adapted for use where it is desired to reduce the inertia forces of moving parts and where, in general, reduced weight is an inherent advantage, as in airplane construction and in some parts of trucks, trains, and other vehicles. Other characteristics of aluminum alloys that suggest their use include: high electrical and thermal conductivity; resistance to some corrosive effects (imparted by a film of oxide that forms on the surface); ease of casting, working (hot and cold), and joining by most manufacturing methods; and high mechanical properties in certain alloys. Various rolled shapes are obtainable, including structural shapes.[2.1,2.12,2.13]

Temper designations have been standardized and are briefly defined as follows. The softest (annealed) temper is designated by 0, as 2014-0. The symbol F designates as-fabricated, as 360-F, 3003-F. The symbol H designates strain (work) hardening and is followed by two digits, the second (last) digit indicates the amount of cold work, 8 being the maximum practical temper; H14 is strain hardened only, half hard (4 is half of 8). The symbol H2x indicates strain hardening plus partial annealing (to reduce hardness a little); the symbol H3x indicates strain hardening and then stabilized (used when magnesium is present, and consists of low-temperature heating to hasten a transformation that takes much longer at ambient temperature); thus H34 is half hard and stabilized. The symbol T indicates other treatment: T2, annealed cast products; T3, solution heat treated, cold worked, naturally aged to a stable state; T4, solution heat treated and naturally aged to a stable state; T5, artificially aged only; T6, solution heat treated and artificially aged; T7, solution heat treated and stabilized; T8, solution heat treated, cold worked, artificially aged; T9, solution heat treated, artificially aged, then cold worked. Typical uses for some aluminum alloys are as follows (the first digit indicates by code the principal alloying element—given in parentheses after the identification number below).

3003 (Mn):—formerly 3S: vessels, tanks, tubing, cooking utensils, chemical equipment; in general, where good formability, weldability, and resistance to corrosion are desired.

2014 (Cu)—formerly 14S: aircraft fittings, truck frames, heavy duty forgings; in general, for high strength, high hardness, good formability, low weight.

2024 (Cu)—formerly 24S: aircraft structures, truck wheels, screw machine products, rivets, hardware, miscellaneous structures.

6061 (Mg and Si)—formerly 61S: aircraft landing mats, canoes, furniture, marine applications, piping, welded parts, transportation equipment; in general, good strength, formability, weldability, and resistance to corrosion.

7075 (Zn)—formerly 75S: structural applications, aircraft structures in particular.

360: thin-wall and intricate castings; has excellent castability and resistance to corrosion.

355: fuel pump bodies, air-compressor pistons, liquid-cooled cylinder heads, crankcases, and various housings; good castability, weldability, and pressure tightness.

Aluminum takes a wide variety of pleasing finishes and colors, including anodized surfaces that additionally protect the base metal from corrosion. Aluminum bonded to steel (steel ingot or slab) and subsequently rolled comes out as steel with an integrated layer of aluminum, called aluminized steel; also obtained by hot dipping.

2.18 MAGNESIUM ALLOYS. Since magnesium alloys are about two-thirds as heavy as aluminum, lightness is one of the most significant characteristics of this metal. The relative weights of steel, aluminum, and magnesium, in that order are: 1, 0.35, 0.23. Other characteristics of magnesium include: nonsparking and nonmagnetic, good machinability, and low modulus of elasticity. Because of its light weight, it is often found in portable devices, pneumatic tools, sewing machines, typewriters; as parts where light weight is important, as in aircraft for blowers, nose pieces, housings, wheels, levers, brackets, etc.; and in accelerating parts where it is desired to reduce the inertia forces. Magnesium is cast as sand castings,[2.1,2.22] permanent mold and die castings; it is a good material for extrusion.

The system of designating magnesium alloys is defined briefly as follows. The first two letters indicate the principal alloying elements (as AZ = aluminum and zinc); the second part indicates the rounded-off amounts of these elements (as AZ61 $\approx$ 6% Al and 1% Zn); the third part is a letter designating the order in which the compositions became standard (as AZ61A is the first of this principal composition to be standardized); the last part indicates the temper, for which the symbols defined for aluminum apply (as AZ61A-T4 means that the alloy has been solution heat treated). There are already a large number of more or less standard alloys. For the ones in Table AT 3,

AZ61A: good for extrusions and press forgings (forging is generally done at slow speed).

AZ80A: also used for forgings and extrusions; can be heat treated.

Two characteristics should be noted particularly: magnesium alloys are highly notch sensitive (important when a part changes section and is subjected to a varying load) and, contrary to usual expectation, the compressive yield strength of the wrought form is less than the tensile yield strength, each taken at an offset of 0.2%. The following comparisons of the strength/mass and stiffness/mass ratios will prove interesting; ρ lb./in.^{3} is the density. In drawing conclusions from these index numbers, keep in mind the particular alloys being compared.

STRENGTH AND STIFFNESS COMPARISONS

MATERIAL	$\frac{s_u\text{(ksi)}}{\rho}$	$\frac{E \times 10^{-6}}{\rho}$	MATERIAL	$\frac{s_u\text{(ksi)}}{\rho}$	$\frac{E \times 10^{-6}}{\rho}$
Magnesium (AZ61 A-F extruded)	693	100	Gray iron, ASTM 40 ..	150	61
Aluminum (2024-T4) ..	694	108	Stainless 303, annealed	314	98
Steel C1020, annealed ..	200	105	Aluminum bronze (B148, cast) ..	290	62
			Yellow brass, ½ hard ..	200	49
Steel 9255, OQT 1000 ..	635	105	Titanium (B265, hard.)	1060	94

2.19 TITANIUM. Since titanium is expensive, it is not used except where its properties are important—in particular in extreme-temperature

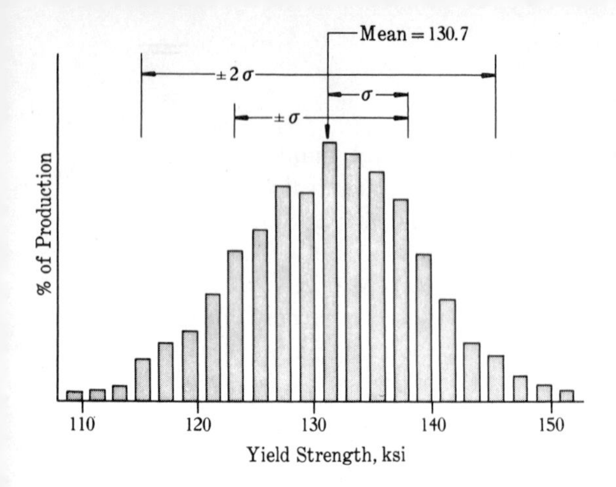

FIGURE 2.8 Yield Strength of Ti 6 Al 4 V, annealed. After Ref. (2.1)

situations where strength is needed, especially for aeronautical purposes. It is also gradually finding other uses; for example, for tubes in a condenser handling 60% nitric acid. Where stainless steel needed replacement every 6–8 months, titanium paid for itself in the first year and is expected to last for 10 years. The 6 Al 4 V alloy given in Table AT 3 has high strength to above 600°F and is used for aircraft turbine blades and disks, air-frame fittings, sheet metal, and extrusions. Figure 2.8 shows a typical distribution of yield strengths of annealed Ti 6 Al 4 V with the $\pm\ 2\sigma$ spread indicated (see § 3.9).

2.20 LEAD, TIN, AND MISCELLANEOUS ALLOYS. There is a host of other alloys, but we shall comment on only a few.

The babbitt B23-46T, grade 8, in Table AT 3 has a lead base and is a suitable bearing material for light and moderate service in various machines.

The tin babbitt B23-49, grade 1, has a tin base and is a general purpose bearing material, also used for die-castings. Tin base alloys are particularly easy to bond, have excellent anti-seizure qualities, and they resist corrosion better than lead base babbitts.

Hastelloy B is an expensive alloy of nickel, molybdenum, and iron (5%) that is very useful in the chemical industry because it resists certain corrosion admirably—for example, hydrochloric acid up to its boiling point, phosphoric acid, sulfuric acid up to 50% concentration, cuprous chloride, and other corroding substances. Since it holds its strength well at high temperature (e.g., 60 ksi at 1500°F), it classifies as a superalloy (§ 2.21). There are several other grades of hastelloy, some containing considerable chromium.

Monel is primarily an alloy of nickel and copper (67 Ni, 30 Cu), the different "kinds" of monel having small amounts of other alloys. The K monel is used where a combination of high strength and good corrosion resistance is needed. It is also used for nonmagnetic aircraft parts such as pump rods, springs, valve stems, shafts. Despite monel's resistance to corrosion, its fatigue strength in adverse environments is sharply reduced, to about 20 ksi at 10^8 cycles in fresh water.

The zinc alloy in Table AT 3, which has the trade name Zemak-5, may be used for either die castings or sand castings for such articles as automotive parts, building hardware, padlocks, toys, and novelties.

2.21 HIGH-TEMPERATURE SERVICE. Ordinary steel begins to lose strength (and elasticity) significantly at about 600–700°F, which was formerly an approximate widespread boundary condition. However, modern requirements for much higher operating temperatures in petroleum refining, chemical processing industries, steam power plants, gas turbines, and now rocket engines and objects travelling at supersonic speeds in the atmosphere, lead to the development of materials that retain significant strength at higher temperature levels (and also that have good corrosion resistance). The most advanced alloys in this respect are called *superalloys* or *superstrength alloys*; these alloys are some combination of nickel, cobalt, chromium, iron, molybdenum, tungsten, columbium, titanium, and aluminum, but never containing all of these. With the advent of space exploration, the search for and development of superalloys has been intensified and new knowledge is available every day.

Any material begins to lose strength rapidly at some temperature; as the temperature increases, the deformations cease to be elastic and become more and more plastic. When plastic deformations are involved, the criterion for design at a particular operating temperature is the ***creep strength*** or the ***rupture stress*** at a specified length of time. Creep results in a permanent deformation, and for a given material and stress, this deformation in a particular time is greater for a higher than for a lower temperature. When constant dimensions must be maintained, measurable creep cannot of course be tolerated; but there are many cases where small permanent changes have negligible effect. *Creep strengths* at a particular temperature are variously defined, and much of the data is so new that these strengths are not always presented in a consistent and orderly manner. A common definition of creep strength is that stress that produces a creep rate of 0.0001% per hr. This is mathematically the same as 1% in 10,000 hr., but it does not mean that the material can maintain this rate for 10,000 hr. without rupture unless it is known from other data that it can. Because of the time it takes to conduct an experiment (100,000 hr. = 11.5 yr.), test data are most frequently found for 1000 hr. and less. If one must design for 10–20 yr. or longer life with a material for which test data at said life is unavailable, the only thing to do is to extrapolate courageously.*

The dangers of extrapolation are shown by the typical behavior curve of Fig. 2.9(a). Upon application of a creep producing load (some materials creep measurably at ordinary and low temperatures—as you know from

* Do not confuse a *courageous extrapolation*, which is done with as much knowledge as is available (unless you have this knowledge now, you will not have time to obtain it in this course), with a *foolhardy extrapolation*, which may be a thoughtless jump to a conclusion.

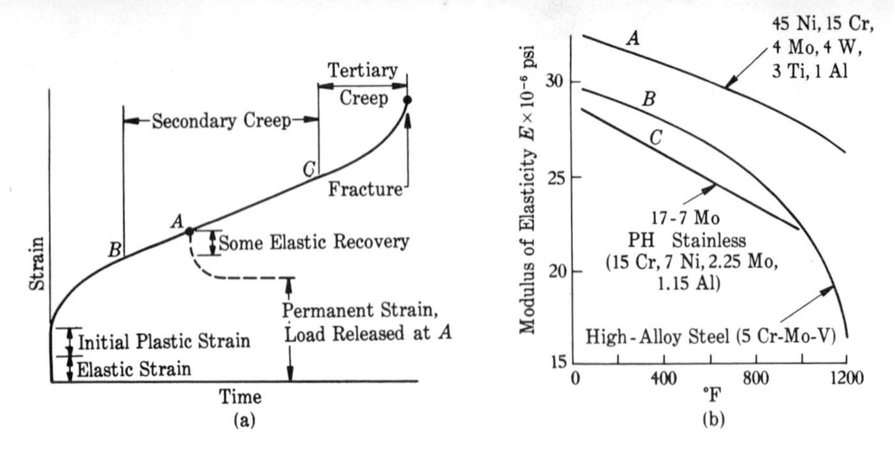

FIGURE 2.9 Strain-time Curve Illustrating General Characteristics of Creep and Decrease of E with Temperature. The steels in (b) are for higher-temperature service; A is a nickel-base alloy.

personal experience with wax), there is, as Finnie and Heller[2.29] point out, first an initial strain, Fig. 2.9(a), partly plastic and partly elastic; then the initial rate of increase of strain is generally high for a while, after which there may be a period during which the creep proceeds at a more-or-less constant rate, called *secondary creep*. If the load is high enough and if the time is long enough, there is a *tertiary creep* at an accelerating rate leading eventually to fracture. Thus, a steady rate of creep as between B and C is not necessarily a safe basis of design. The time of life must be included. Not only the strength falls off, but the modulus of elasticity decreases as the temperature goes up, as seen in Fig. 2.9(b).

A proper design stress, as usual, depends on the circumstances. The life of a missile may be measured in minutes; so, accumulating experimental data to fit this case is not too time consuming. On the other hand, a suitable life of a steam power plant would be nearer 200,000 hr., in which event, some extrapolation of data will be necessary. (There are of course some long-time data available.) Where relatively frequent inspection is certain, one can take bigger chances. Except where life or valuable property is threatened, the basis of decision as to what material to use is largely an economic one, and costs seem amazingly high to one accustomed only to ordinary steel; for example, the superalloy M 252 (Ni, Cr, Co, Mo, Ti, etc.) costs about 170 times more than carbon steel plate. Frequent complicating factors include repeated loading, thermal shock (quick temperature changes), cycling thermal stresses (especially detrimental for brittle materials). For their suggestive value: the ASME Code allows a design stress where the creep is 0.01% per 1000 hr. in unfired pressure vessels and 80% of this in fired pressure vessels; furnace tubes in the oil industry have been designed for the stress at 1% creep in 10,000 hr.; military jet engines for 1000 hr. of life; stationary gas turbine plants, 100,000 hr. Fairly long life should be obtained at a particular operating temperature if the design stress is two-thirds of the 1% in 10,000 hr. creep. Some typical creep data are pictured in Fig. 2.10. Since so many of the superalloys are patented, the designer should be in touch with the manufacturers for the best material and its best treatment. Because so much specialized knowledge is required for intelligent design for creep conditions, we shall leave it largely for another study.

2.22 LOW-TEMPERATURE PROPERTIES.

Modern science and engineering are going to extremes in both directions of temperature, what with the very low temperatures involved in cryogenic studies, in shipping and using liquified gases, etc. Although the standard test tensile strength, hardness, and modulus of elasticity of steels and other metals increases as the temperature decreases, ductility and toughness decrease. Over the years, there have been many mysterious (at the time) brittle failures of tanks, bridges, and other structures, but the significance of the loss of toughness, as measured by an Izod or Charpy test, did not become general engineering knowledge until World War II experiences of vessels cracking in two. Sometimes the weather was heavy, sometimes not, but invariably the air and water were cold (say less than 40°F). Investigation showed that the low-carbon steel had a *transition temperature* (also true of low alloy steels), below which it failed by a (tensile) brittle failure (typically sudden and complete) and above which the failure is ductile (typically a shear rupture). Usually, the fracture starts at a point of high stress, a point of

FIGURE 2.10 Creep Strength and Rupture Stress. Solid curves give creep strength for a rate of 0.1% in 1000 hr. at each temperature. Dotted curves give the stress at rupture at each temperature. Casting alloy HT is a nominal 35% Ni, 15% Cr alloy of iron; Incoloy is principally a Ni (32%)-Cr(20.5%) alloy of iron; Ni-Resist D2 is a ductile iron casting, 20% Ni, 2% Cr, treated with Mg to produce the spheroidal graphite; Greek Ascoloy, a steel alloy with 13% Cr, 2% Ni, 3% W. Data from Refs. (2.1, 2.3, 2.31).

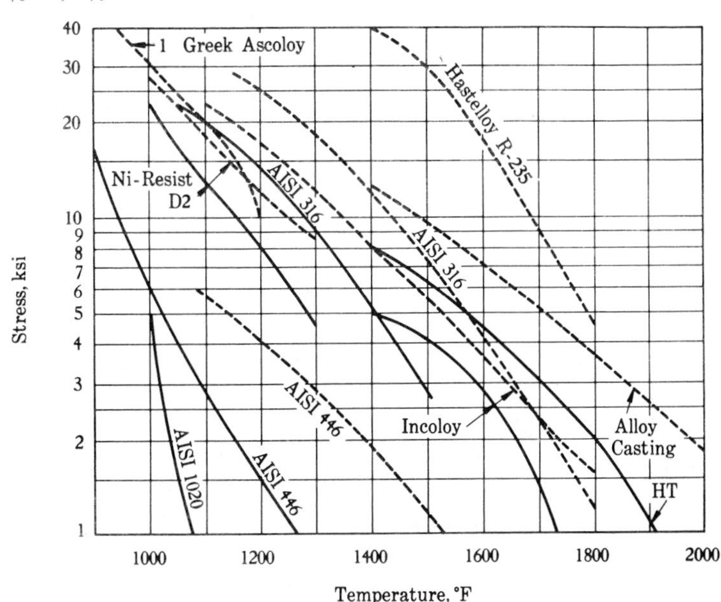

stress concentration (Chapter 4). The fracture for complete brittle failure of steel has a bright faceted appearance; the ductile fracture has a silky fibrous appearance. The transition, except in pure iron, does not occur instantaneously at a particular temperature, but rather over a temperature range within which the failures are partly brittle and partly ductile (Fig. 2.11). In general, it is dangerous to use a material below its transition temperature because it has lost so much of its capacity to absorb energy without rupture. A few materials show increasing impact strength with decreasing temperature, notably nickel, copper, and aluminum. Nickel is the most effective alloy in increasing toughness of steel. Peening (§ 4.28) lowers the transition temperature for the steels.

The impact test, Charpy or Izod, is currently most used as a measure of the low-temperature suitability of steel, but we have no way of using the data quantitatively in design. Comparing the two curves given for the 9% Ni steel in Fig. 2.11, we observe that the V-notch Charpy test may indicate the transition more distinctly than the keyhole Charpy. You realize of course that the measured energies from a group of test specimens do not fall on a single curve as shown in Fig. 2.11, but within some band. These curves are therefore merely typical and should not be used in actual design unless it is known that they apply. Prior treatment significantly affects the transition band and the Charpy energy (different types of specimens do not give correlated quantitative data). The 9% Ni alloy and stainless steel 304L shown have been used successfully for storing and transporting liquid N_2 (about −320°F); aluminum and titanium alloys (to about −300°F) are used for similar service.

It is not uncommon for the specifications to state a minimum impact strength (as 15 ft-lb., Charpy keyhole) at a particular temperature, but experience suggests that this sort of specification does not necessarily eliminate brittle failure. However, when the magnitude of the energy is

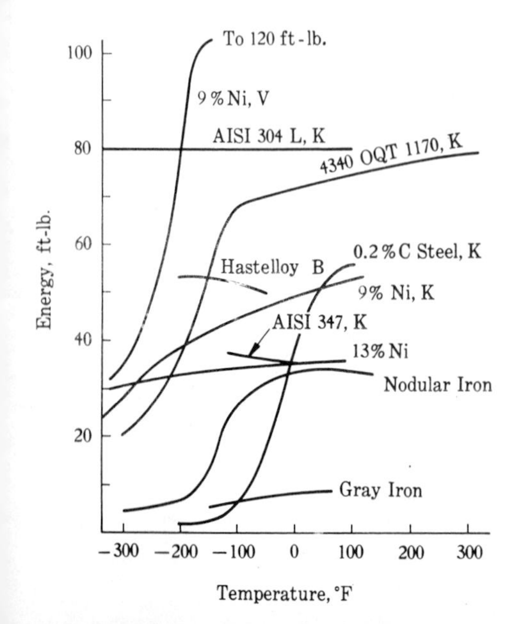

FIGURE 2.11 Low-Temperature Toughness. The symbol V indicates a V-notch specimen; the symbol K, a keyhole notch. The 9% Ni contains about 0.1% C; it is double normalized and stress relieved; now available in rolled structural shapes. The stainless steel 304L (the same as 304 except 0.03% C vs. 0.08% C) does not have a horizontal curve for the V-notch specimen. The AISI 347 is cold drawn, ½ hard. The 0.2% steel was normalized. The nodular iron shown has higher energy properties than many others. The gray iron or cast iron (C.I.) would qualify as approximately ASTM 30.

related to a material whose properties are well known, the purpose may be well served. The exploration into the very-low-temperature region is so recent that engineers are still looking for suitably simple ways of specifying the desired material properties. In general, one does not use a material better than required for the service; for example, 2.5% Ni steel is good to −75°F, 3.5% Ni steel to −150°F, 5% Ni steel to −200°F, and 9% Ni steel to −320°F.

In general, alloys with high-nickel content retain toughness to quite low temperatures. Nickel alloy benefits cast steel similarly, especially if it is quenched and tempered. Stainless steel with nickel (as the 18-8 variety) has good low-temperature properties, but the non-nickel bearing stainless is not so good. Copper and copper alloys are also used at temperatures below −150°F.

Work hardening steels (§ 2.9) at low temperatures often produces economic benefits. For example, 301 stainless steel vessels have had their ultimate tensile strength increased to 260 ksi (contrast with Table AT 4) by stretching them 13% with cold nitrogen inside at temperatures to −300°F.

2.23 PLASTICS. Plastics have come into use not only as decorative and unstressed parts but also as load-carrying members.[2.19,2.20,2.21,2.27,2.30] They may be divided into two main classes—thermosetting, which undergo chemical change and harden on being heated, usually under pressure; and thermoplastic, which soften as the temperature rises and remain soft in the heated state. The thermoplastics may be reshaped on heating, whereas the thermosetting variety cannot be.

The phenolics, Table AT 5, constituting one of the largest, least expensive, and most useful groups, are available in forms for molding and casting and in laminated forms. They are made from formaldehyde and phenolic bodies such as phenols (carbolic acid), cresols, or cresylic acid. Laminated phenolic materials machine readily, have good wearing properties, and are used for gears, gaskets, seals, compressor-valve plates, tubing, bearings, insulators, etc. Grade X has a paper base, comes in sheet, tube, and molded forms, and is used primarily for mechanical applications. Grade XX has a paper base, comes in sheet, tube, rod, and molded forms; it has good machinability, greater moisture resistance than Grade X, and is suitable for electrical (insulating) applications. Grade C has a fabric base, comes in sheets, tubes, and rods, and is used for gears. Grade A has an asbestos base, comes in sheets, tubes, and rods; it is more resistant to heat than are those previously mentioned. There are several other groups of thermosetting plastics, including the urea-formaldehyde, the melamine-formaldehyde, and the polyester plastics.

Many thermoplastics are widely used in unloaded forms, such as handles, knobs, containers, grills, covers. Polyethylene, a relatively inexpensive thermoplastic, is used widely for bottles and other containers. Nylon (polyamide), while relatively expensive, has load carrying capacity up to 200–250°F; it has a low coefficient of friction, 0.05 to 0.20[2.27] and good wear resistance—hence, its use for bearings, dry or lubricated; it is machined or molded. Nylon is used for gears (quietness) and balls (valves, lightly loaded bearings for quietness). A trade name for nylon resins for molding is Zytel.®

Teflon® (polytetrafluoroethylene—also called tetrafluoroethylene TFE) has uses similar to those of nylon, but is more expensive and would therefore be used in more exacting situations; it is not molded in the conventional manner, but more as in powder metallurgy; it makes a tough bearing with a dry coefficient of friction of 0.1–0.2[2.27]. Teflon, a very inert plastic, solves many obdurate problems of gaskets and seals because of its ability to tolerate a continuous temperature of 500°F, higher for short periods, and because it retains some flexibility to a low temperature of −300°F. A new relative of the original Teflon, fluorinated ethylene-propylene, designated FEP, has a distinct advantage in that it can be processed in a more conventional manner. Engineering data on these Teflons are found in Ref. (*2.33*, *2.35*). Thin films of Teflon provide excellent corrosion resistance and make the surface nonsticking (as in cooking utensils and many industrial uses). One of its recent uses is for piston rings in air compressors, eliminating the need of oil lubricant, the accompanying danger of explosion, and in some cases, product contamination by lubricant.

There are a number of rubbers or elastomers (buna-S, butyl, neoprene, buna-N, natural rubber) of engineering interest, but space does not permit a discussion here. All problems involving plastics and rubbers should be solved in the presence of more information than contained herein, which is intended to be suggestive only. A "newer and better" plastic may be available tomorrow.

The mechanical properties of these materials not only have a normal manufacturing variation, but they also vary with temperature and moisture content; there is dimensional change with time; the modulus of elasticity is not a constant; many will creep measurably under load at ambient temperature. Such characteristics, if not accounted for in the design, may lead to an unnecessary failure.

2.24 DESIGN HINTS. The basic dimensions of a part with known loads on it are computed by the designer, but these dimensions and the details of other dimensions depend to some degree on the method of manufacture. Is the part machined from milled or extruded stock; is it forged or pressed; is it cast; is it welded; is it sintered?

If it is to be forged,[2.24] it may be a smith forging done with flat dies (manually or by machine); it may be a drop forging, Fig. 2.12, in which the part is formed between shaped dies attached to the anvil cap and the ram; it may be an upset forging, Fig. 2.13, done in a machine that can form a solid piece into a desired shape by pushing the piece into dies; or it may be a press forging, which squeezes the metal into shape between dies. Drop

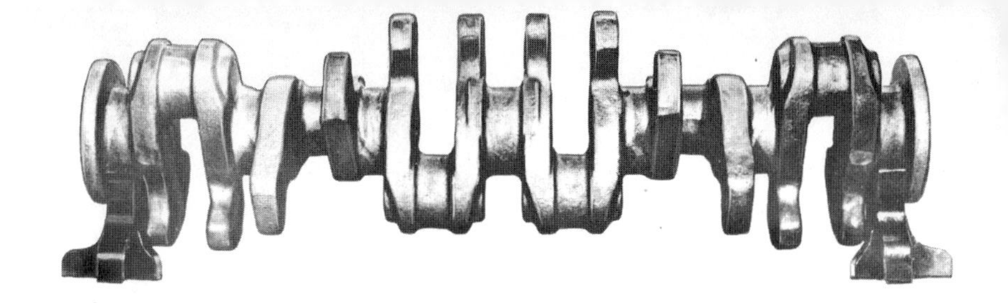

FIGURE 2.12 Forged Crankshaft. A drop forging, made in impression dies, of a large internal-combustion-engine crankshaft, ready for machining. (Courtesy Drop Forging Association, Cleveland.)

FIGURE 2.13 Upset Forging. The original bar is shown in (a); (b) and (c) are the two steps in the forging; (d) is the finished machined part. (Courtesy M. H. Harper Co., Morton Grove, Ill.)

forging is a hammering process, while upset and press forging are done by exerting pressures adequate to cause plastic metal flow into the required shapes. In forgings shaped in dies, several steps are required to produce the final form; for instance, the forging of a 3-ft. connecting rod for a Diesel engine is done in seven steps, including the final step of trimming off the flash. To facilitate removing the forging from the die, the sides of the die are tapered. This taper is called the draft and is about 7° for outside, and 10° for inside, surfaces, although more or less draft may be used. Fillets and radii joining different parts of a forging should be as large as possible (see Chapter 4), because small fillets increase the tendency toward forging defects, such as unfilled structure, and they increase die wear. Corner radii should also be as large as possible. Keep rib heights as small as possible and rib sections as thick as possible because of the difficulty of forcing the plastic metal into thin, deep pockets. Metal may be distributed in a forging in a manner that will take advantage of the "grain flow," inasmuch as wrought metals are stronger under impact in the direction of the "grain" (direction of the rolling) than in the transverse direction. This factor may need to be considered in the design phase of the job.

If the part is to be cast, there are many alternative materials some one of which may or may not be clearly indicated by the nature of the part. There are cast iron (Fig. 2.14), malleable iron (Fig. 2.15), cast steel (Fig. 2.17), nodular iron, nonferrous metals; any one of which may be sand castings, permanent mold castings, die castings (Fig. 2.18), centrifugal castings, or precision investment castings (a lost-wax process).

Perhaps the most important policy in the design of castings is to *make the various sections of the casting as nearly as possible the same thickness.*

FIGURE 2.14 (*Above*) Cast-iron Crankshaft. A machined cast-iron crankshaft, 11 ft. 4 in. long, 1750 lb., for a 5-cylinder, 2000-hp. Diesel engine. Cast of iron from an electric furnace, the shaft was annealed at a low temperature before machining in order to relieve stresses and avoid distortion after machining, and to improve the mechanical properties. Test coupons from such shafts suggest average properties as follows: $s_u = 68$ ksi, E = 23,000 ksi, G = 9200 ksi, BHN = 300. This company casts Diesel crankshafts up to 22 ft. long. (Courtesy Pacific Car and Foundry Co., Renton, Wash.)

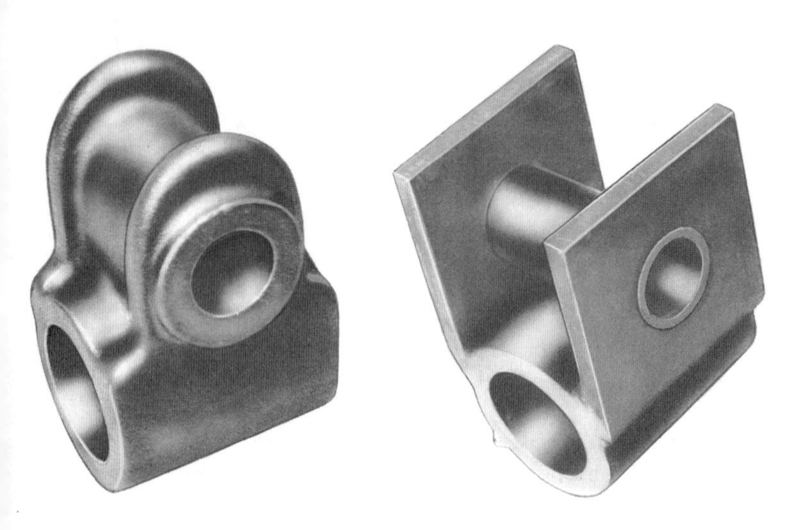

FIGURE 2.15 (*Above left*) Malleable Iron Swivel Hitch. Typical of many small parts suitably made of malleable iron. (Courtesy Eastern Malleable Iron Co., Wilmington, Del.)

FIGURE 2.16 (*Above right*) Welded Swivel Hitch. When compared with Fig. 2.15, this one illustrates the different ways in which a particular part may be made. (Courtesy Eastern Malleable Iron Co., Wilmington, Del.)

FIGURE 2.17 (*Below left*) Cast Steel Gear Box. Made of ASTM A 27-50, T, class 70-36. An example of a large complicated casting, weight about 1900 lb. (Courtesy Steel Founders' Society, Cleveland.)

FIGURE 2.18 (*Below right*) Die Casting—Carburetor. Excellent example of complex coring and shapes obtainable with die castings. A four-jet carburetor (two carburetors in one), zinc bowl. Sharp corners, inside and outside, should be avoided in die castings. (Courtesy General Motors).

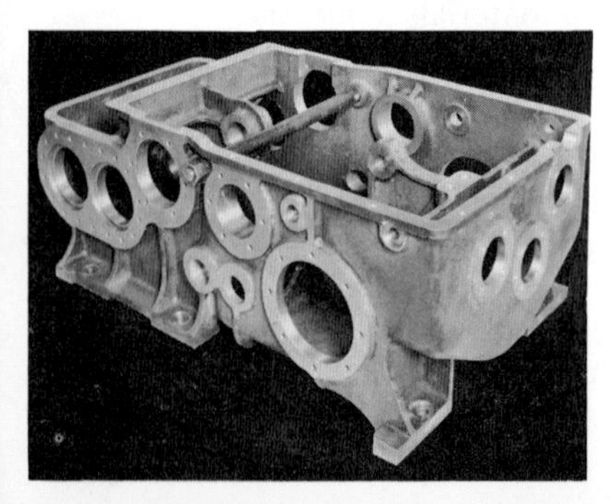

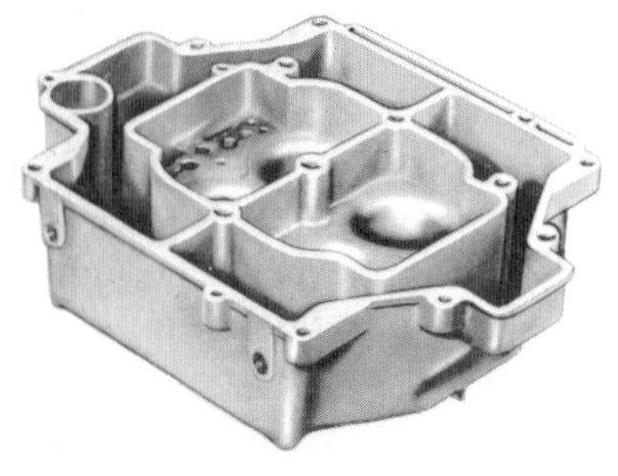

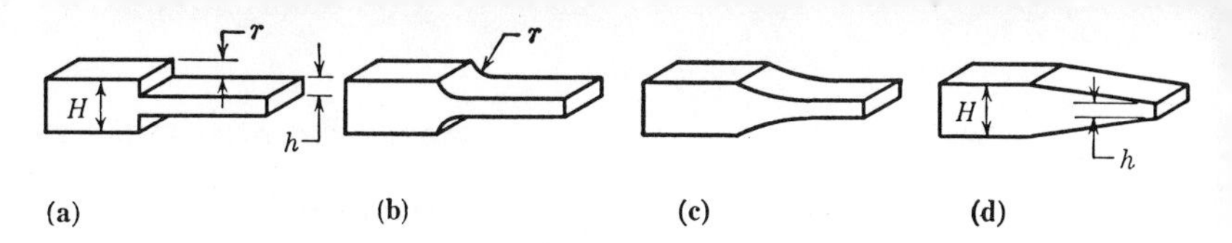

FIGURE 2.19 Joining Section Changes—Castings. Unless an important useful purpose is served (unlikely), changing sections with sharp corners as in (a) should be avoided in all design no matter what the manufacturing method may be. A radius as suggested by the proportions in (b) is acceptable; the proportions in (c) are good; and the plan in (d) is better yet. It happens that these changes from (a) to (d) are progressively better for a part made in any manner when the load is repeated (Chapter 4). The best alternative for castings is not to change the section thickness at all (not always possible to abide by this). (After *Steel Castings Handbook*[2.16]).

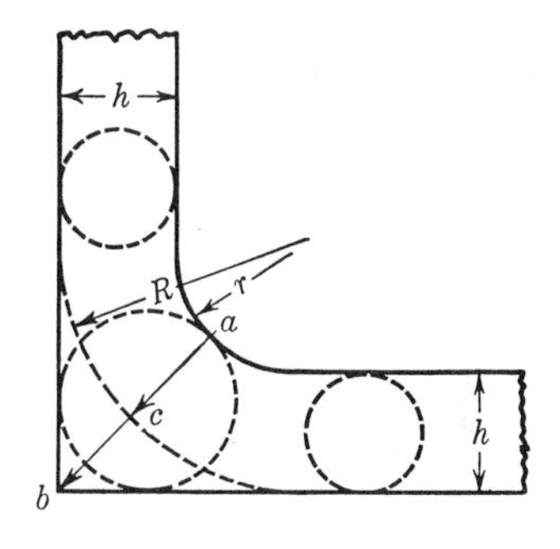

FIGURE 2.20 Hot Spot at Joined Sections. The extra mass of metal at the corner, dimension *ab*, means slower cooling, possibly resulting in shrinkage cavities or cracking. Instead of the corner at *b*, the outside radius *R* is preferably such that the thickness *ac* to the dotted curve is somewhat less than *h* (not always feasible on a stress basis). The radius *r* should be about $r = h$ or more. Another solution may be to core a hole near the center of mass at the corner. (After *Steel Castings Handbook*[2.16]).

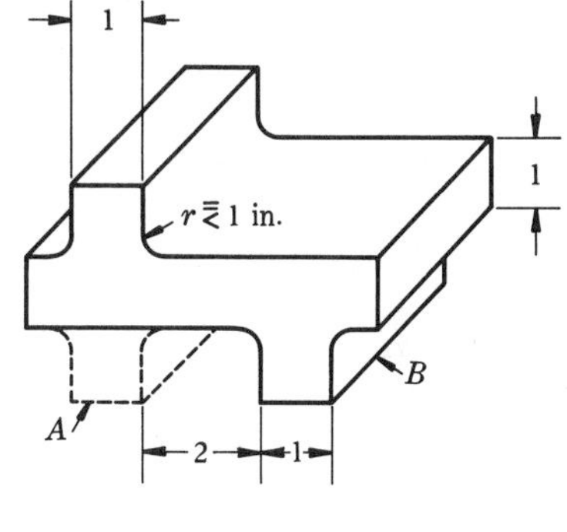

FIGURE 2.21 Casting a "Cross" Section. If sections cross, as shown dotted at *A*, a serious hot spot will occur at the junction. If one part of the "cross" is offset, as at *B*, the foundryman can use external chills to prevent cavities. The minimum recommended offset is suggested by the dimensions shown. The fillet radius *r* should be between $\frac{1}{2}$ and 1 in., regardless of size of the "cross". (After *Steel Castings Handbook*[2.16]).

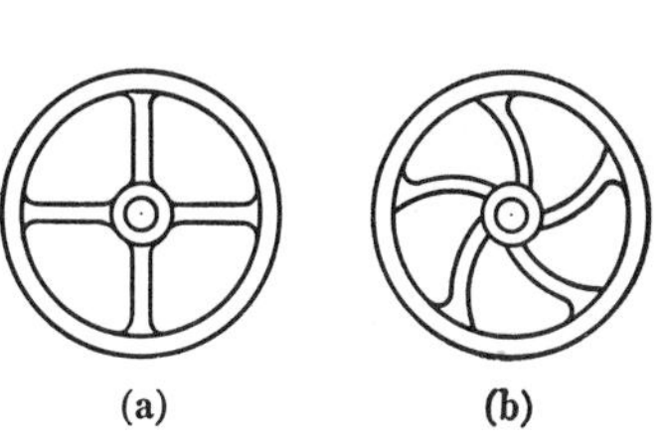

FIGURE 2.22 Wave Construction. The design at (a) resulted in cracked spokes. Curved spokes as in either the plan view (b) or end view (c) will flex under the cooling stresses, probably eliminating distortion and cracking. (From *Fundamentals of Steel Casting Design* by G. W. Briggs).

The tendency toward hot spots, and therefore internal stresses due to unequal cooling rates, is thus reduced. The general principle is to avoid extra mass concentrations. This can sometimes be done by using cores. See also Figs. 2.19, 2.20, 2.21, and 2.22, and Ref. (*2.16*) for more information.

In one situation or plant, a certain part may be fabricated by welding, Fig. 2.16, whereas in another situation or plant, a casting, Fig. 2.15, may be a better answer. "Better" generally has to do with cost if the service conditions are met satisfactorily, the better one being the cheapest. The unit cost will be much affected by the quantity. For one quantity, a certain manufacturing process may be cheaper, but another process may be cheaper for another quantity. Of course, the process may be determined almost entirely by the part—it is difficult to think of a cheaper way to produce a complicated part such as a carburetor, Fig. 2.18, if many are to be made, than by die casting.

2.25 MISCELLANEOUS. Complicated shapes and parts can be made from sintered metals (powder metallurgy). Their properties depend in part on the metals used, which include iron, copper, lead, tin, silver, tungsten, molybdenum (and often carbon) in various combinations. The metals are in powder form, and mixed in the desired proportions (as 88% Cu, 10% Sn, plus iron and graphite for bearings); the bonding of the adjacent surfaces is accomplished by prolonged heating (sintering) at a temperature below the melting point. An additional process sometimes used, called *infiltration*, is to melt a layer of another metal, such as lead or copper, into the pores of the sintered material. Since the manufacture of such parts is highly specialized, a manufacturer should be consulted. Some standard items, as bearings, can be bought off the shelf.

One never knows when a startling new engineering material will be announced. An example is Pyroceram, a group of glass products of the Corning Glass Works that have unusually good mechanical properties at high temperature. This material is already in use in a number of consumer products and it has many industrial uses. A recent announcement describes *maraging nickel steels**, which attain yield strengths of as much as 250 ksi without quenching. There are compositions of 18% Ni, 20% Ni, or 25% Ni, together with several other alloys, all with low carbon (0.03%). The 18% Ni is treated as follows: transformation to martensite occurs below 310°F, but the low carbon content results in soft and tough metal; transformed metal is held at 900°F for 3 hr. (maraging); it may be annealed and cold worked before maraging for improved properties. Typical properties: $R_C = 50$, $s_y = 250$, $s_u = 260$, $E = 27 \times 10^3$ ksi, $\epsilon = 11\%$, Charpy = 16 ft-lb. at −420°F. Compare with 17-7 PH, § 2.15(b). Another development, recently announced by Du Pont, is a nickel alloy (about 98% Ni, 2% thorium oxide), called TD nickel, with unusually good creep strength; retains useful strength to above 1800°F.

2.26 CLOSURE. There are a number of properties of engineering materials that have not been mentioned at all in this chapter. Some of these

* International Nickel Co.

properties will be taken up later—in particular, the endurance strength (Chapter 4). No doubt, you have already noted that a particular material has many different "strengths"; it will serve you well to master the various "strength" concepts as you come to them. We must proceed with our study assuming that many environmental factors involving corrosion, embrittlement, etc. will not be overlooked; for example, hydrogen embrittlement, sulfide cracking, stress-corrosion cracking, nitrogen embrittlement.

There are many decisions that a designer must make, but perhaps the two crucial ones are the choice of material and the choice of design stresses. There are always choices and this is where the art (engineering) of design enters. After the decisions are made, many people (and computers) will be able to follow an outlined routine using the mechanics-of-materials equations (science) to produce certain answers. It is suggested therefore that in all your problem work and in text examples and discussions you think critically of the design stresses, whether they are specified or not, and of the material, which is frequently specified. Since the engineer must first of all be able to solve the science side of design, our principal attention is directed to stress analysis and design stresses, with concurrent discussion of engineering factors.

3. TOLERANCES AND ALLOWANCES

3.1 INTRODUCTION. It is easy for the designer to specify a certain dimension, say $2\frac{1}{2}$ in., but for the workman or machine to turn it out exactly 2.50000 in. is quite another matter. In general, the closer the dimension must be to precisely 2.5 in., which would be called the ***nominal size***, the more costly will be the manufacture (Fig. 3.2, p. 84).

If the designer does not know something about the *natural* variability of a process, he is not in a position to specify the best tolerance. We have seen something of the ways in which the strength of a material of the same specification varies. Similarly, a 2.5-in. part leaves the machine with a dimension greater or less than 2.5 in. by some amount which may be anything from less than a millionth of an inch to more than several hundredths.

How far off will it be? This question can be answered statistically, if enough information about the process is known. Without statistics, a satisfying answer cannot be given. For this reason, machine designers need to interest themselves in statistical analysis applied to manufacturing (§§ 3.9–3.13).

3.2. TOLERANCE. First, let us define ***tolerance,*** which is *the stated permissible variation of the size of a dimension*. If a particular dimension is to *fit* inside of, or outside of, another dimension, as a pin in a hole, this permissible variation or tolerance is important. If no fit is involved, as in the external diameter of some rods—for example, dimension d, Fig. 1.17—it would be unwise, unnecessary, and uneconomical to insist that the final size be very close to the nominal size. Thus, if there is no fit, a liberal tolerance of the order of ± 0.010 in. or $\pm \frac{1}{64}$ in. in machining work could be permitted. Such tolerances are frequently cared for by a blanket note on the drawing; for example, "Tolerances not specified to be ± 0.010 in." See § 3.3 *et seq.* for information on *fitting dimensions.* The tolerance may be:

(a) *Bilateral,* where the size of a part is permitted to be *either* larger or smaller than the given dimension, or

(b) *Unilateral,* where the size of a part may be larger only, or smaller only, than the given dimension. For example,

$$\frac{9}{16} \pm \frac{1}{32}, \qquad 1.062 \pm 0.010; \qquad 1.062^{+0.010}_{-0.000}, \qquad 1.060^{+0.000}_{-0.010}$$

[BILATERAL TOLERANCES] [UNILATERAL TOLERANCES]

Unilateral tolerances are generally used on those dimensions involved in a fit such as a pin in a hole. See Fig. 3.1.

3.3. ALLOWANCE. For fitting dimensions (a fit of mating parts is involved), the tolerance, which depends in part upon service needs, should be chosen with some knowledge of the natural spread of the processes that produce the mating dimensions. If a pin is to turn freely in a hole, the pin must be somewhat smaller than the hole. In manufacture for interchangeable assembly, the difference in size is cared for by the ***allowance,*** which, for running fits, is the *minimum specified difference* between the dimensions of the pin (male part) and the hole (female part). It is not the same in general as the actual minimum difference.

When the pin is larger than the hole, the allowance is as before the difference in sizes for the tightest fit, a difference also called the ***interference of metal*** i (or *negative allowance*).

3.4. FITS. There are several systems of tolerances and allowances. Many companies have drawn up standards to be used in their own practice. In this country, the American Standards Association, ASA Standard B4.1-1955,[3.1] has classified running and sliding *fits* as follows:

RC 1; close-sliding fits. For accurate location of fitting parts without perceptible play.

RC 2; sliding fits. These fits move and turn easily but are not intended to run freely. With a small temperature change in the larger sizes, this fit may seize.

RC 3; precision-running fits. For precision assemblies operating at low speed, light load, and small temperature change.

RC 4; close-running fits. For cases of accurate location and minimum play, but for moderate speed, journal pressure, and temperature rise.

RC 5 and RC 6; medium-running fits. Suitable for higher running speed and heavy journal pressure.

RC 7; free-running fits. Suitable for large temperature variations and where wider tolerances are permissible.

RC 8 and RC 9; loose-running fits. For use with commercial cold-rolled shafts and tubing.

The same standard B 4.1-1955 also gives tables of limit dimensions for clearance location fits, transitional location fits (accuracy of location is important but can tolerate either small interference or small clearance), interference location fits (accuracy of location of prime importance), and other information; those concerned with actual design should have a copy of the standard.

The ASA fits are based on the ***basic hole system***; the nominal size is the minimum hole diameter. In the basic hole system, the tolerance on the hole is always positive (with zero negative tolerance), measured from the ***basic size***, and the tolerance on the shaft is all negative for working fits. The basic hole system has an advantage over the basic shaft system in that the hole produced by a standard reamer may be made the minimum hole size. Also, there is the matter of inspecting with go and no-go gages. Generally, the mating parts are not so sensitive to clearance but that a plant can adhere to some preferred tolerances on the basic hole system, thereby materially reducing the number of gages needed for inspection. In short, the engineer should not scatter tolerances for a particular size of hole all over the spectrum, but should conform to company standards as far as possible. Depart from such choices only for a confirmed engineering reason.

Table 3.1 (pp. 82, 83), from the ASA Standard B 4.1-1955, defines the details of the various classes of fits. The columns headed "Hole" are the tolerance limits on the hole; the columns headed "Shaft" give the same for the shaft (for example, a 0.1-in. shaft of class RC 1 has a maximum diameter of $0.1 - 0.0001 = 0.0999$ in., a minimum diameter of $0.1 - 0.00025 = 0.09975$ in.). The allowance, as defined above, is the numerically smaller number in the "Shaft" column (for example, the allowance for a 0.1-in. nominal diameter, RC 1 fit, is 0.0001 in.). There is no rule that says the values in the standard must be used, but most of the quoted values are part of an ABC (American, British, Canadian) agreement. Think of a standard as what appeared to be a good engineering guide at the time it was formulated.

The standard B 4.1-1955 gives a list of over 40 preferred tolerances and allowances; where it is apropos, we may consider the following ones as our "company preferred" values:

0.0001	0.0006	0.0016	0.0060	0.016
0.0002	0.0008	0.0020	0.0080	0.020
0.0003	0.0010	0.0025	0.010	0.025
0.0004	0.0012	0.0040	0.012	0.030

"Practical" tolerances, appropriate to particular manufacturing methods and materials, are scattered through the literature, where one must go for detail [e.g., Refs. *(2.1, 2.14, 2.15, 2.16, 2.17, 2.20, 2.23, 3.11,*

TABLE 3.1 RUNNING AND SLIDING FITS

Extracted with permission (ASME) from Table 1. ASA B 4.1-1955. The numbers given are the standard limits in thousandths of an inch; multiply by 10^{-3}. Limits for hole and shaft are applied algebraically to the basic size to obtain the limits of size of the parts. (The Standard goes to 200-in. size.)

NOMINAL SIZE RANGE, INCHES Over To	CLASS RC 1		CLASS RC 2		CLASS RC 3		CLASS RC 4	
	Hole	*Shaft*	*Hole*	*Shaft*	*Hole*	*Shaft*	*Hole*	*Shaft*
0.04– 0.12	+0.2 0	−0.1 −0.25	+0.25 0	−0.1 −0.3	+0.25 0	−0.3 −0.55	+0.4 0	−0.3 −0.7
0.12– 0.24	+0.2 0	−0.15 −0.3	+0.3 0	−0.15 −0.35	+0.3 0	−0.4 −0.7	+0.5 0	−0.4 −0.9
0.24– 0.40	+0.25 0	−0.2 −0.35	+0.4 0	−0.2 −0.45	+0.4 0	−0.5 −0.9	+0.6 0	−0.5 −1.1
0.40– 0.71	+0.3 0	−0.25 −0.45	+0.4 0	−0.25 −0.55	+0.4 0	−0.6 −1.0	+0.7 0	−0.6 −1.3
0.71– 1.19	+0.4 0	−0.3 −0.55	+0.5 0	−0.3 −0.7	+0.5 0	−0.8 −1.3	+0.8 0	−0.8 −1.6
1.19– 1.97	+0.4 0	−0.4 −0.7	+0.6 0	−0.4 −0.8	+0.6 0	−1.0 −1.6	+1.0 0	−1.0 −2.0
1.97– 3.15	+0.5 0	−0.4 −0.7	+0.7 0	−0.4 −0.9	+0.7 0	−1.2 −1.9	+1.2 0	−1.2 −2.4
3.15– 4.73	+0.6 0	−0.5 −0.9	+0.9 0	−0.5 −1.1	+0.9 0	−1.4 −2.3	+1.4 0	−1.4 −2.8
4.73– 7.09	+0.7 0	−0.6 −1.1	+1.0 0	−0.6 −1.3	+1.0 0	−1.6 −2.6	+1.6 0	−1.6 −3.2
7.09– 9.85	+0.8 0	−0.6 −1.2	+1.2 0	−0.6 −1.4	+1.2 0	−2.0 −3.2	+1.8 0	−2.0 −3.8
9.85–12.41	+0.9 0	−0.8 −1.4	+1.2 0	−0.8 −1.7	+1.2 0	−2.5 −3.7	+2.0 0	−2.5 −4.5

3.12)]. See also Table 3.4 and Fig. 3.9. Many mass-produced items, bolts, screws, cold-rolled forms, structural steel shapes, tubing, keys, pulleys, bushings, gears, etc. fall within certain specified tolerances.

3.5 EXAMPLE. In the example of § 1.21, we assumed a "working fit" for the $\frac{7}{8}$-in. pin in its mating holes. How much larger than the pin must the hole be? Assuming that the connection of § 1.21 is not a fine piece of machinery requiring a "just-so" fit, we easily decide that the tolerances can be wide with no harm done, say RC 9. From Table 3.1: hole tolerance = 0.005, shaft tolerance = 0.0035, allowance = 0.007 in. To show that it can be done, let us decide that there is nothing gained in an allowance as large as 0.007, and make it 0.004 in.; also, let us decide to stay with the "company's" preferred values stated above (nothing

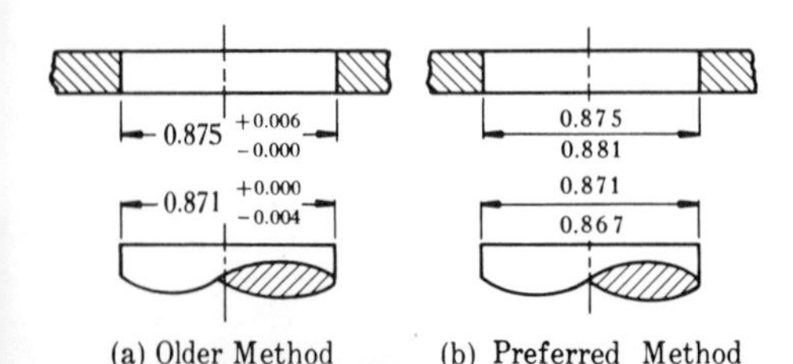

FIGURE 3.1 Methods of Dimensioning. The method in (b) shows the *limits*, the extreme permissible dimensions. The top limit in the dimension should be the one the workman reaches first—smaller one for holes, larger one for shafts.

CLASS RC 5		CLASS RC 6		CLASS RC 7		CLASS RC 8		CLASS RC 9	
Hole	*Shaft*	*Hole*	*Shaft*	*Hole*	*Shaft*	*Hole*	*Shaft*	*Hole*	*Shaft*
+0.4 0	−0.6 −1.0	+0.6 0	−0.6 −1.2	+1.0 0	−1.0 −1.6	+1.6 0	−2.5 −3.5	+2.5 0	−4.0 −5.6
+0.5 0	−0.8 −1.3	+0.7 0	−0.8 −1.5	+1.2 0	−1.2 −1.9	+1.8 0	−2.8 −4.0	+3.0 0	−4.5 −6.0
+0.6 0	−1.0 −1.6	+0.9 0	−1.0 −1.9	+1.4 0	−1.6 −2.5	+2.2 0	−3.0 −4.4	+3.5 0	−5.0 −7.2
+0.7 0	−1.2 −1.9	+1.0 0	−1.2 −2.2	+1.6 0	−2.0 −3.0	+2.8 0	−3.5 −5.1	+4.0 0	−6.0 −8.8
+0.8 0	−1.6 −2.4	+1.2 0	−1.6 −2.8	+2.0 0	−2.5 −3.7	+3.5 0	−4.5 −6.5	+5.0 0	−7.0 −10.5
+1.0 0	−2.0 −3.0	+1.6 0	−2.0 −3.6	+2.5 0	−3.0 −4.6	+4.0 0	−5.0 −7.5	+6.0 0	−8.0 −12.0
+1.2 0	−2.5 −3.7	+1.8 0	−2.5 −4.3	+3.0 0	−4.0 −5.8	+4.5 0	−6.0 −9.0	+7.0 0	−9.0 −13.5
+1.4 0	−3.0 −4.4	+2.2 0	−3.0 −5.2	+3.5 0	−5.0 −7.2	+5.0 0	−7.0 −10.5	+9.0 0	−10.0 −15.0
+1.6 0	−3.5 −5.1	+2.5 0	−3.5 −6.0	+4.0 0	−6.0 −8.5	+6.0 0	−8.0 −12.0	+10.0 0	−12.0 −18.0
+1.8 0	−4.0 −5.8	+2.8 0	−4.0 −6.8	+4.5 0	−7.0 −9.8	+7.0 0	−10.0 −14.5	+12.0 0	−15.0 −22.0
+2.0 0	−5.0 −7.0	+3.0 0	−5.0 −8.0	+5.0 0	−8.0 −11.0	+8.0 0	−12.0 −17.0	+12.0 0	−18.0 −26.0

demanding about this design) and use a hole tolerance of 0.006, a shaft tolerance of 0.004 in. Then the dimensions are (basic hole, nominal size = 0.875):

$$\text{Hole, } 0.875 \begin{array}{l} +\,0.006; \\ -\,0.000 \end{array} \text{ shaft, } 0.871 \begin{array}{l} +\,0.000 \text{ in.,} \\ -\,0.004 \end{array}$$

and the drawing is dimensioned as shown in Fig. 3.1. On the other hand, if the standard is followed, for which there may be good reasons, the limit dimensions are: hole, 0.875 to 0.880 in.; pin, 0.868 to 0.8645 in.

3.6. INTERCHANGEABILITY.[3.1,3.13] Allowances and tolerances are a practical necessity when many mating parts are to be produced. If it were intended to make only one of the yoke connections of § 1.21, good results would be obtained in a small plant by describing the fit to the workman and leaving the rest to his judgment. But if many parts are to be made, interchangeable manufacture, the basic ingredient of our mass-production technique, is economically essential.

For a completely interchangeable system of manufacture, a workman with a box each of the yoke ends, rod ends, and pins may expect any pin to fit satisfactorily with any rod or yoke. Where engineering calls for closer fits than can be obtained economically with a completely interchangeable plan, selective interchangeability is practiced. In this instance, the manufactured parts are classified according to two or more size groups. Then, if there are two size groups, the larger pins would be assembled with the larger holes, smaller pins with smaller holes, etc. This procedure may produce closer fits more economically than if the same fits were obtained by reducing the allowances and tolerances, even though 100% inspection is required. This is because the cost of obtaining small tolerances may mount rapidly. See Fig. 3.2. Also, it will be cheaper on occasion to manufacture some scrap than to take the steps necessary to produce all parts within tolerance (§§ 3.9–3.12).

When selective interchangeability is used, the problem of servicing a customer who needs a replacement part should not be overlooked. An example will illustrate.[3.2] Suppose that the desired clearance between a 1-in. shaft and bearing is between 0.002 in. and 0.005 in., giving a desired variation of 0.003 in. Suppose that the best tolerances that the machines will hold (determined by experience) are:

On the shaft, 0.002 in., On the bearing, 0.003 in.,

which add up to 0.005 in., whereas the desired variation is 0.003 in. A solution to the difficulty would be to hold the bearings to the limits 1.000 and 1.003 in. and divide them into three groups with limits as follows

(A) 1.000 to 1.001; (B) 1.001 to 1.002; (C) 1.002 to 1.003.

Also make the shafts to the limits 0.999 to 0.997 in. and divide them into two groups with the limits:

(1) 0.997 to 0.998; (2) 0.998 to 0.999.

Group (1) shafts may be assembled with bearings of either group (A) or (B), and group (2) shafts with either (B) or (C) bearings, with the clearances in every instance being as desired. Check it yourself. (But this does not mean that the actual clearances as randomly assembled will be as stated—see § 3.12.) Whenever a customer needs another bearing, he will be sent bearing (B) which correctly fits either shaft. This plan does not solve all problems, but the idea is suggestive.

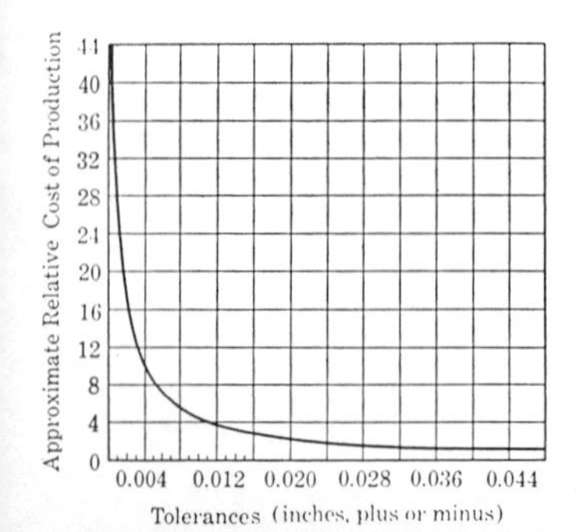

FIGURE 3.2 Tolerance vs. Relative Cost. This curve, derived from a variety of data, gives a good idea of the general trend of costs for increasing or decreasing tolerances. Costs may fall well off of this curve on the favorable side because of some factor such as smart design with manufacturing in mind, the choice of an especially appropriate machine, an efficient quality control system, or quantity production with special machines. (From Bolz, R. W., *Production Processes,* Penton Publishing Co.).

TABLE 3.2 FORCE AND SHRINK FITS (INTERFERENCE FITS)

Extracted with permission (ASME) from Table 5, ASA B 4.1-1955. The numbers given are the standard limits in thousandths of an inch. Limits for hole and shaft are applied algebraically to the basic size to obtain the limits of size for the parts.

NOMINAL SIZE RANGE, INCHES *Over* *To*	CLASS FN 1		CLASS FN 2		CLASS FN 3		CLASS FN 4		CLASS FN 5	
	Hole	*Shaft*	*Hole*	*Shaft*	*Hole*	*Shaft*	*Hole*	*Shaft*	*Hole*	*Shaft*
0.04– 0.12	+0.25 −0	+0.5 +0.3	+0.4 −0	+0.85 +0.6			+0.4 −0	+ 0.95 + 0.7	+0.4 −0	+ 1.3 + 0.9
0.12– 0.24	+0.3 −0	+0.6 +0.4	+0.5 −0	+1.0 +0.7			+0.5 −0	+ 1.2 + 0.9	+0.5 −0	+ 1.7 + 1.2
0.24– 0.40	+0.4 −0	+0.75 +0.5	+0.6 −0	+1.4 +1.0			+0.6 −0	+ 1.6 + 1.2	+0.6 −0	+ 2.0 + 1.4
0.40– 0.56	+0.4 −0	+0.8 +0.5	+0.7 −0	+1.6 +1.2			+0.7 −0	+ 1.8 + 1.4	+0.7 −0	+ 2.3 + 1.6
0.56– 0.71	+0.4 −0	+0.9 +0.6	+0.7 −0	+1.6 +1.2			+0.7 −0	+ 1.8 + 1.4	+0.7 −0	+ 2.5 + 1.8
0.71– 0.95	+0.5 −0	+1.1 +0.7	+0.8 −0	+1.9 +1.4			+0.8 −0	+ 2.1 + 1.6	+0.8 −0	+ 3.0 + 2.2
0.95– 1.19	+0.5 −0	+1.2 +0.8	+0.8 −0	+1.9 +1.4	+0.8 −0	+ 2.1 + 1.6	+0.8 −0	+ 2.3 + 1.8	+0.8 −0	+ 3.3 + 2.5
1.19– 1.58	+0.6 −0	+1.3 +0.9	+1.0 −0	+2.4 +1.8	+1.0 −0	+ 2.6 + 2.0	+1.0 −0	+ 3.1 + 2.5	+1.0 −0	+ 4.0 + 3.0
1.58– 1.97	+0.6 −0	+1.4 +1.0	+1.0 −0	+2.4 +1.8	+1.0 −0	+ 2.8 + 2.2	+1.0 −0	+ 3.4 + 2.8	+1.0 −0	+ 5.0 + 4.0
1.97– 2.56	+0.7 −0	+1.8 +1.3	+1.2 −0	+2.7 +2.0	+1.2 −0	+ 3.2 + 2.5	+1.2 −0	+ 4.2 + 3.5	+1.2 −0	+ 6.2 + 5.0
2.56– 3.15	+0.7 −0	+1.9 +1.4	+1.2 −0	+2.9 +2.2	+1.2 −0	+ 3.7 + 3.0	+1.2 −0	+ 4.7 + 4.0	+1.2 −0	+ 7.2 + 6.0
3.15– 3.94	+0.9 −0	+2.4 +1.8	+1.4 −0	+3.7 +2.8	+1.4 −0	+ 4.4 + 3.5	+1.4 −0	+ 5.9 + 5.0	+1.4 −0	+ 8.4 + 7.0
3.94– 4.73	+0.9 −0	+2.6 +2.0	+1.4 −0	+3.9 +3.0	+1.4 −0	+ 4.9 + 4.0	+1.4 −0	+ 6.9 + 6.0	+1.4 −0	+ 9.4 + 8.0
4.73– 5.52	+1.0 −0	+2.9 +2.2	+1.6 −0	+4.5 +3.5	+1.6 −0	+ 6.0 + 5.0	+1.6 −0	+ 8.0 + 7.0	+1.6 −0	+11.6 +10.0
5.52– 6.30	+1.0 −0	+3.2 +2.5	+1.6 −0	+5.0 +4.0	+1.6 −0	+ 6.0 + 5.0	+1.6 −0	+ 8.0 + 7.0	+1.6 −0	+13.6 +12.0
6.30– 7.09	+1.0 −0	+3.5 +2.8	+1.6 −0	+5.5 +4.5	+1.6 −0	+ 7.0 + 6.0	+1.6 −0	+ 9.0 + 8.0	+1.6 −0	+13.6 +12.0
7.09– 7.88	+1.2 −0	+3.8 +3.0	+1.8 −0	+6.2 +5.0	+1.8 −0	+ 8.2 + 7.0	+1.8 −0	+10.2 + 9.0	+1.8 −0	+15.8 +14.0
7.88– 8.86	+1.2 −0	+4.3 +3.5	+1.8 −0	+6.2 +5.0	+1.8 −0	+ 8.2 + 7.0	+1.8 −0	+11.2 +10.0	+1.8 −0	+17.8 +16.0
8.86– 9.85	+1.2 −0	+4.3 +3.5	+1.8 −0	+7.2 +6.0	+1.8 −0	+ 9.2 + 8.0	+1.8 −0	+13.2 +12.0	+1.8 −0	+17.8 +16.0
9.85–11.03	+1.2 −0	+4.9 +4.0	+2.0 −0	+7.2 +6.0	+2.0 −0	+10.2 + 9.0	+2.0 −0	+13.2 +12.0	+2.0 −0	+20.0 +18.0
11.03–12.41	+1.2 −0	+4.9 +4.0	+2.0 −0	+8.2 +7.0	+2.0 −0	+10.2 + 9.0	+2.0 −0	+15.2 +14.0	+2.0 −0	+22.0 +20.0

3.7 PRESS AND SHRINK FITS. When the hole is smaller than the shaft, it will take force or pressure to put the cold parts together. When this occurs, the allowance is said to be negative and is termed the ***interference of metal.*** The ASA Standard B 4.1-1955 gives details for five classes of interference fits, from which Table 3.2 is taken: FN 1, for light drive fits, thin sections, long fits, cast-iron external member; FN 2, for medium drive fits, ordinary steel parts, shrink fits on light sections, the tightest fit it is advisable to use with a high-grade cast-iron external member; FN 3, for heavy drive fits, heavy steel parts, shrink fits in medium sections; FN 4 and FN 5, for force fits when the parts can safely withstand high stress. Shrink fits (heating the hub or cooling the shaft or both) can be used where pressing the fit is impractical. The mating parts may be sorted into size groups so that the amount of the interference of metal does not vary greatly, obtaining a *selected average interference* i of metal.

3.8 STRESSES DUE TO THE INTERFERENCE OF METAL. The tight fits may produce large bursting pressures on the part with the hole. The stresses in the hub can be estimated with reasonable accuracy from the thick-cylinder equations (§§ 8.26, 8.27). For a quick check on the safe side, one can assume that the shaft is rigid and all the deformation occurs in the hub; then $s = E\epsilon = E\delta/L$. The length L is the hole circumference, $L = \pi D$, and δ is the difference: circumference of the shaft, $\pi(D + i)$, minus the circumference of the hole, πD; or $\delta = \pi i$. Hence, $\delta/L = \pi i/(\pi D) = i/D$; and the corresponding stress is $s = E\delta/L = Ei/D$. For a cast-iron hub on a solid steel shaft, use $E = 10{,}430$ ksi. But see § 8.27.

3.9 NATURAL SPREAD OF DIMENSIONS.[3.14,3.15] If tolerances set by the designer are to be respected, the process of manufacture should be such that the tolerances can be met, or the designer should have incontrovertible evidence that his tolerances are essential. In one sense, not the designer, but the material, the machine and tools, and the workmen determine the tolerance. To gain a better understanding of this statement, consider briefly a technique of the quality-control engineer—statistical analysis. Statistical control of quality during manufacture is primarily aimed at forestalling the manufacture of scrap, but there results a wealth of useful information for the designer. It is not enough for the designer to know that the size of a part varies as it comes from a manufacturing process, but he also needs to know something of how it varies and what the ***probable limits*** of size are.

A theoretical aspect of the answer to this problem is embodied in the ***normal curve,*** Fig. 3.3, which can be described in terms of the ***standard deviation*** σ (sigma). The curve extends to plus and minus infinity, but

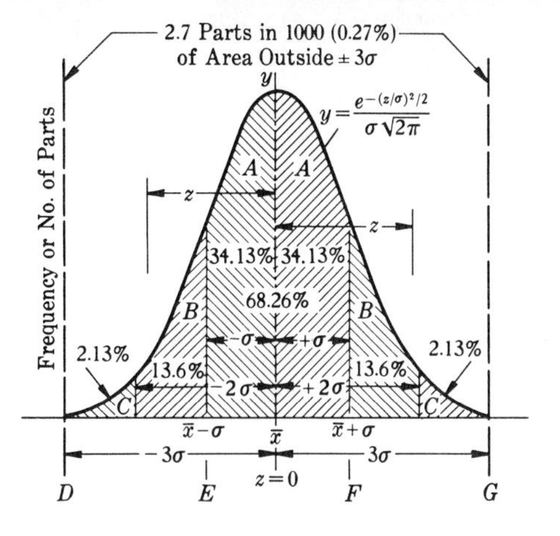

FIGURE 3.3 Normal Curve. Also called Gaussian distribution. Total area is 100%; $\bar{x}$ = average. For example, area A is 34.13% of the total area under the curve. Area $2A$, between $\pm\sigma$, is 68.26%; etc. These figures come from the mathematics of the curve and apply to all normal curves.

99.73% of the area "under" it lies between $\pm 3\sigma$. Suppose a nominal $\frac{1}{4}$-in. drill is used to drill a large number of holes under a constant set of chance causes (the drill is maintained at the same sharpness, the machine adjustments do not vary, etc.). All the holes will not be $\frac{1}{4}$ in. in diameter. Of a large number of drilled holes, there will be some of approximately average size, the highest point of the normal curve, Fig. 3.3; some will be larger and some smaller than the average. If the number of measured holes is large enough, the distribution of sizes of all holes produced will be much as suggested by the normal curve, as shown by actual checks of production. The total production is, in statisticians' terms, the ***population*** or ***universe*** from which the samples are taken. (Not all industrial measurements have a normal distribution, but usually the assumption of normal distribution does not result in a significant error.) The area under the curve, when it is fitted to the graphic distribution of a particular production, represents closely the percentage production between certain "extreme" dimensions. For example, between $+\sigma$ and $-\sigma$, where σ is measured from the average, there is over 68% of the total area; and we can logically and reasonably expect that about 68% of the product from a certain process will fall between $\bar{x} + \sigma$ and $\bar{x} - \sigma$, where $\bar{x}$ = the average or mean value of the process. (The standard deviation σ must be determined from the output of the process itself. See Fig. 3.5.) Since only 0.27% of the area lies outside of the $\pm 3\sigma$ limits, it is very unlikely, 3 chances in 1000, that any of the production falls outside of the $\pm 3\sigma$ limits. On this account, we often refer to the 3σ limits as being the ***natural spread*** (NS) of the process, though other values are sometimes used, such as $\pm 2.5\sigma$. The significance of the natural spread (sometimes called *natural tolerance*) is that *if tolerance limits are set narrower than the natural spread, the manufacture of scrap is inevitable*.

Thus, if the designer specifies a total tolerance of 0.002 in., corresponding say to EF, Fig. 3.3, and if the natural spread of the process is 0.006 in. and corresponding to DG, it would be expected that about 32% of production would not pass inspection. Therefore, unless there is an otherwise *imperative* reason, always specify tolerances greater than the natural spread, at least

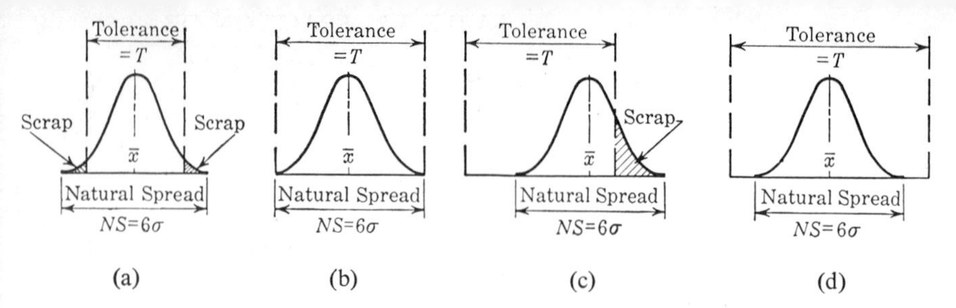

FIGURE 3.4 Tolerance and Natural Spread.
(a) $T < NS$. Scrap is sure to be manufactured — the minimum amount being when the average dimension of the process $\bar{x}$ falls at the midpoint of the tolerance limits. The percentage of out-of-tolerance parts made is represented by the shaded areas.
(b) $T = NS$. No scrap is manufactured in a process in control — provided the process is kept precisely centered; i.e., $\bar{x}$ falls at the midpoint of the tolerance limits. But, see (c).
(c) $T = NS$. It is difficult to center a process exactly and impossible to keep it there indefinitely; hence; when $T = NS$, some scrap is almost sure to follow. The amount is represented by the shaded area, the whole area being virtually 100%.
(d) $T > NS$. The proper relation. A good objective is $T = 1.3NS$, which allows the process to be off center somewhat without the tolerance limits being exceeded.

a third greater,[3.3] unless you are willing to pay the price of the scrap or unless you are willing for the shop to ignore the specified tolerances. Study Fig. 3.4.

3.10 EXAMPLE—ANALYSIS OF ACTUAL PRODUCTION. The data for Fig. 3.5 were taken from actual production. The histogram, a graphical device of tabulation by classes, shows the number of pieces falling into each *cell*. Cell *d* in (a), for example, shows that 18 pieces measured between 0.2503 and 0.2505 in. The facts that only one piece lies in cell *h*, one in cell *i*, and, especially, none in *g*, suggest that there was some abnormal factor that caused these "unusually" large holes. A statistical analysis confirms this observation, since cells *h* and *i* are outside of the 3σ limits when *all* pieces are included in the computations. However, the value of sigma used in drawing the normal curve in Fig. 3.5(b), shown dotted in (a), was obtained by omitting the cells *h* and *i*. Thus, the 3σ limits of this normal curve should predict the natural spread of the process when normal factors only are present. The quality-control engineer calls an abnormal factor an "assignable cause," since a complete knowledge of the events leading to a part falling outside of the 3σ limits enables him to assign its nonconformity to a particular "cause."

Statistical analysis is the only rational tool that can be used to decide with practical certainty whether or not to expect pieces normally to fall into cells *h* and *i*. In this case, it tells us that it is unlikely that pieces larger than 0.2510 or smaller than 0.2498 in. will be made, provided the process is kept in control. If holes outside of these limits are made, there is some unusual condition—an assignable cause, such as worker's carelessness or inexperience, dull tools, poor machine setup, etc.

Actually, 50 measurements, as in Fig. 3.5, are not enough to give an accurate measure of the natural spread of the process; hence, these first computed limits should be considered as tentative. A recalculation could be made after measuring 100 pieces and another still later. The larger the number of pieces measured, the better will be the definition of the process. However, measurement of three hundred pieces will define the limits of the controlled process quite accurately. Judging

TABLE 3.3 FRACTION OF AREA UNDER NORMAL CURVE

From $-\infty$ to z measured from zero at $\bar{x}$

z/σ	A	z/σ	A	z/σ	A	z/σ	A
0.00	0.5000	−0.8	0.2119	−1.6	0.0548	−2.4	0.0082
−0.1	0.4602	−0.9	0.1841	−1.7	0.0446	−2.5	0.0062
−0.2	0.4207	−1.0	0.1587	−1.8	0.0359	−2.6	0.0047
−0.3	0.3821	−1.1	0.1357	−1.9	0.0287	−2.7	0.0035
−0.4	0.3446	−1.2	0.1151	−2.0	0.0228	−2.8	0.0026
−0.5	0.3085	−1.3	0.0968	−2.1	0.0179	−2.9	0.0019
−0.6	0.2743	−1.4	0.0808	−2.2	0.0139	−3.0	0.00135
−0.7	0.2420	−1.5	0.0668	−2.3	0.0107		

from the appearance of Fig. 3.5(a), it would not be surprising to find that the natural spread of the process is somewhat larger than that computed (the normal curve), but this conclusion could be justified only by additional production and new computations of the natural spread.

3.11 STANDARD DEVIATION AND AREA UNDER THE NORMAL CURVE.

The standard deviation of a group of measurements x taken from a particular population is given by [3.15]

(a) $$\sigma = \left[\frac{\Sigma (x - \bar{x})^2}{N}\right]^{1/2},$$

where $\bar{x}$ is the arithmetic average $(\Sigma x/N)$ and N is the total number of measurements. Tables giving the areas under normal curves are available in handbooks and statistics books. (Notice that σ is a root mean square (rms) deviation—§ 3.14.) Table 3.3 is a very abbreviated version, suitable for our purposes. To make the tables universal, the total area is taken as unity; the table value is that fraction of the total area as measured from $-\infty$ to a point located by z/σ, where z, the deviation from the mean, is measured from zero at $\bar{x}$, Fig. 3.3. Since the curve is symmetric, areas for only half the curve are sufficient. To illustrate the use of Table 3.3, suppose

FIGURE 3.5 Distribution of Size of $\frac{1}{4}$ in. Reamed Holes. Refer to a book on statistical control [3.17] for the method of computing σ, which was done in (b) from the same data as plotted in the histogram. Note that the spread between the largest and smallest holes is 0.0016 in. (0.2514 − 0.2498), but that if the process is kept in control, the probable spread is 0.0012 in. (0.2510 − 0.2498).

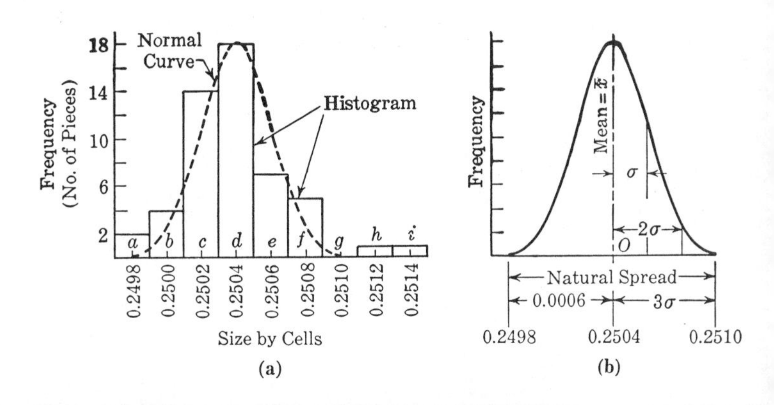

the lower tolerance limit on the hole of Fig. 3.5 is 0.2500. What proportion of the production will be less than 0.250 in. (scrap)? The mean $\bar{x} = 0.2504$; the deviation from the mean is $z = x - \bar{x} = 0.2500 - 0.2504 = -0.0004$; $\sigma = 0.0006/3 = 0.0002$; hence $z/\sigma = -0.0004/0.0002 = -2$; from Table 3.3, read 0.0228 corresponding to $z/\sigma = -2$, which means that about 2.28% of the production should normally be smaller than 0.2500 in. If the top tolerance limit is 0.2508, there would be here an equal loss—holes too large (by symmetry).

Observe that while the preferred dimensioning practice is to give limit dimensions, one works from the mean dimension in statistical studies.

3.12 STATISTICAL DISTRIBUTIONS OF FITS. Given a production of two mating parts, one might suppose that the allowance is actually the closest fit that will be obtained during the assembly of these parts. No doubt, the allowance is sometimes mistakenly chosen on this assumption. However, such a fit under controlled manufacturing conditions is quite unlikely, and it is sometimes important for the designer to know this.

For example, suppose the shaft tolerance has been set at 0.003 in., the bore tolerance at 0.004 in., and the allowance at 0.001 in., as shown in Fig. 3.6(a). *If* the tolerances are a third or more greater than the natural spread of the processes and *if* each process is centered with respect to its specified average dimension, the distribution of sizes of the shafts may be represented by the normal curve *S*, and of the bore by the normal curve *B*, Fig. 3.6. If a shaft is picked at random, its most likely size is 1.0000 in., the size at the maximum ordinate on the normal curve *S*. Similarly, the most likely size of bore is 1.0045 in., which gives the most likely difference (or clearance) as 1.0045 − 1.0000 = 0.0045 in. This 0.0045 dimension is the most frequent difference and is therefore that found at the highest point of the normal curve *D*, which shows the distribution of the clearances. Statistical theory defines the standard deviation σ_D of the difference (or sum) of two independent variables as the square root of the sum of the

FIGURE 3.6 Distribution of Clearance in Fits. If the processes should move off center, as they will, the effect on the normal curve of the differences in (b) is only to shift its mean from 0.0045 in. to some other value. For example, if *S* moves leftward 0.0005 in. so that *A* is at *A′* (shown dotted) and if *B* moves rightward 0.0005 in. so that *F* is at *F′*, the mean difference would be increased by 0.001 in. to 0.0055 in. the maximum fit would become 0.0073 in. (point *H′*), and the minimum fit 0.0037 in.

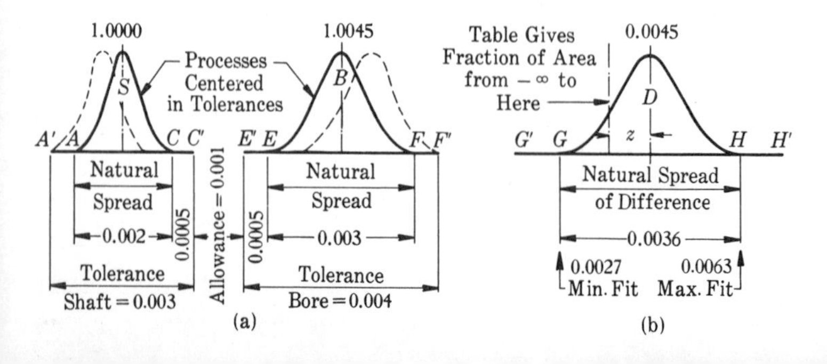

squares of the standard deviations, σ_1 and σ_2, of the variables; in equation form,

(b) $$\sigma_D = (\sigma_1{}^2 + \sigma_2{}^2)^{1/2},$$

where $6\sigma_D$ is the natural spread of the differences, $6\sigma_1$ is the natural spread of the shaft dimensions, and $6\sigma_2$ is the natural spread of the bore dimensions. Substituting $\sigma_1 = 0.002/6$ and $\sigma_2 = 0.003/6$ into equation **(b)**, we find $\sigma_D = 0.0006$ (to four decimal places). Adding $3\sigma_D = 0.0018$ to the known mean of 0.0045, we get the maximum likely clearance of 0.0063 in. for the situation as defined. Likewise, by subtraction, we find the minimum clearance is likely to be 0.0027 in. These limits are much larger than the 0.001-in. allowance.

Even though the curves S and B are off center toward each other but still within their respective tolerance ranges, the probable minimum fit will be greater than the allowance. Thus, let S move toward the right so that point C is at C' and let B move toward the left so that E is at E', a total movement toward each other of 0.001 in. The standard deviation of the differences does not change, but the mean difference will now be 0.001 in. less ($0.0045 - 0.001 = 0.0035$), and the minimum clearance or fit will be $0.0027 - 0.001 = 0.0017$ in. (versus an allowance of 0.001 in.). It is seen that it is possible to obtain working fits on an interchangeable basis even though the allowance is zero. There are occasions where the designer can take advantage of this fact by decreasing the allowance and increasing the tolerance with a saving in costs and scrap.

Similarly, if several parts are assembled together externally, one next to the other, the standard deviation of the dimension of the (sum of the) assembled parts is given by

(c) $$\sigma = (\sigma_1{}^2 + \sigma_2{}^2 + \sigma_3{}^2 + \cdots)^{1/2},$$

where σ_1, σ_2, σ_3, etc., are the standard deviations of the dimensions of the respective parts. Thus, *if* the tolerances are proportional to the standard deviations, the overall tolerance may be

(d) $$T = (T_1{}^2 + T_2{}^2 + T_3{}^2 + \cdots)^{1/2},$$

and not the sum of the individual tolerances, as often assumed. Even though tolerances are *not* proportional to standard deviations, the conclusion in equation **(d)** has a general validity; the exact effect on actual dimensions can be determined only from statistical analysis based on adequate data. To illustrate the idea, assume that the processes involved in manufacturing parts 1, 2, and 3, Fig. 3.7, are centered in the tolerance range and that tolerances are equal to the natural spread of the processes (the ideal situation). Suppose that the desired overall tolerance is $T = 0.018$ in. $= 6\sigma$.

FIGURE 3.7.

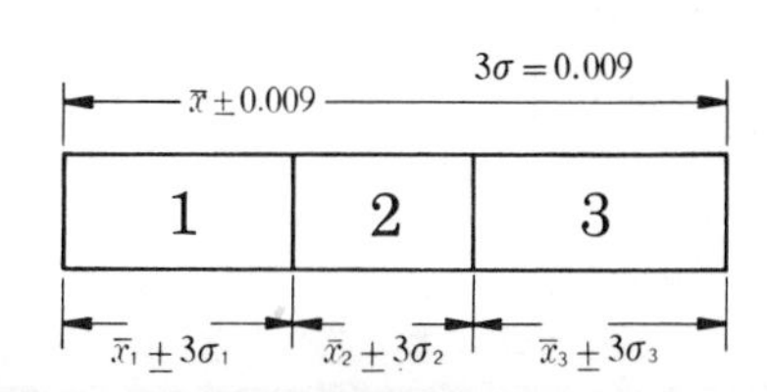

What tolerances must be applied to the individual parts? Let us assume that $T_1 = T_2 = T_3$ and $\sigma_1 = \sigma_2 = \sigma_3$. On the basis of simple arithmetic, it seems logical to divide the total tolerance by 3 and make each part with a tolerance of 0.006 or ± 0.003 in. But the laws of probability show that the

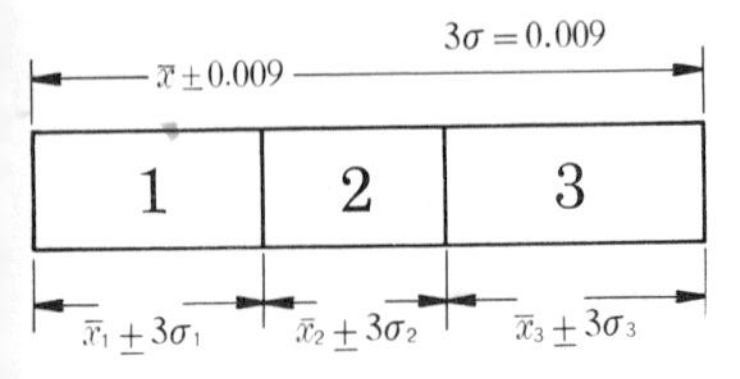

FIGURE 3.7 (Repeated).

individual tolerances can be made significantly larger. Using equation **(c)** with $\sigma = 0.018/6 = 0.003$ and $\sigma_1 = \sigma_2 = \sigma_3$, we find

$$\sigma = 0.003 = \sqrt{3\sigma_1^2} \qquad \text{or} \qquad \sigma_1 = 0.00173 \text{ in.}$$

This corresponds to a tolerance of $T_1 = 6\sigma_1 = 0.0104$ in., which, if used instead of 0.006 in., may mean a considerable reduction in cost. If for some reason, the tolerance on one part, say 1, should be more or less than on another, decide upon a ratio of tolerances, reduce the radical in equation **(c)** or **(d)** to one unknown, and solve as described.

Since a little knowledge is dangerous, one would acquire more knowledge of the statistical approach before use in actual engineering problems. In general, our conclusions are valid only when the production processes are "in control."

3.13 TOLERANCES ON LOCATING HOLES. It often happens that two (or more) parts are to be assembled with matching holes for bolts or screws where accuracy is important. If the holes are close to a matching location and somewhat undersized, the parts may be put together at assembly and the holes finished to size, a procedure that automatically obtains a good match. Frequently the most economical, this procedure is common. However, if assembly on an interchangeable basis is to be made, the various tolerances need to allow for this fact and they must be practical.

Suppose we wish to locate a pair of holes on each of two articles which are to match at assembly. Suppose that one pair is at the minimum spacing $L - T/2$ and that the other pair is at the maximum spacing $L + T/2$,

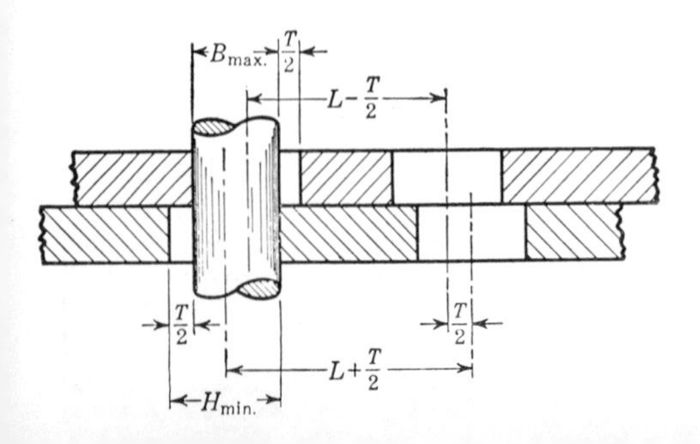

FIGURE 3.8 Tolerance on Hole Location. The spacing of the holes is $L \pm T/2$. The difference between $L + T/2$ and $L - T/2$ is much exaggerated in the illustration.

Fig. 3.8. Considering the foregoing discussion of statistical aspects, we recognize that this particular combination of parts is very unlikely in a random assembly operation, so unlikely for a manufacturing process in control that it almost surely will not happen. However, by using this combination, our conclusions will be on the "safe side." Making the match even more unlikely, we shall assume the worst geometric situation, whose parts have the minimum size of hole and the maximum diameter of bolt permitted by the tolerance. Recall the definition of allowance A as the difference between the minimum hole H_{min} and the maximum "shaft" B_{max}, Fig. 3.8; that is,

(e) $$A = H_{min} - B_{max}.$$

But we also see from Fig. 3.8 that[3.4]

(f) $$H_{min} - B_{max} = \frac{T}{2},$$

where T is the tolerance on the hole spacing L. From **(e)** and **(f)**,

(g) $$\frac{T}{2} = A \qquad \text{or} \qquad T = 2A;$$

that is, the tolerance T on the hole spacing should be twice the allowance A for hole and bolt.

A similar geometric study[3.4] shows that if there are more than two holes, the tolerance on the spacing comes out

(h) $$T = \frac{A}{\sqrt{2}} \approx 0.7A.$$

In this case, tolerances in two directions are necessary, and they are preferably the same.

Where there are more than two holes located with reference to one another, there should be a "master hole" (or reference surfaces) from which all others are located. It is cheaper to locate two holes to, for instance, ± 0.002 in. with respect to one another and the other holes to ± 0.010 in. than to hold all holes to a close tolerance, and it is usually just as satisfactory. Bilateral tolerances, $L \pm T/2$, are preferable for hole locations.

3.14 TOLERANCE AND SURFACE FINISH. There is necessarily a close relationship between the smoothness of a surface and how fine the tolerances can be made. It would be ridiculous to specify a tolerance of a few ten thousandths of an inch for a surface that has irregularities of several thousandths of an inch. As would be expected, the smoother the finish, the more expensive—generally. The standard B 4.1-1955 gives the standard tolerances shown in Table 3.4 and suggests the methods of finishing the surface appropriate to the grade as given in the heading of the table. The

TABLE 3.4 STANDARD TOLERANCES (from ASA B 4.1-1955)

(This is only part of the tabulation in the Standard.) Tolerances are in thousandths of an inch; for use with standard gages. Includes values agreed to by ABC (American, British, Canadian). Not part of the standard but given as a guide: Grades 4, 5—lapping and honing; Grades 5, 6, 7—cylindrical grinding, surface grinding, diamond turning and boring, broaching; Grades 6–10 incl.—reaming; Grades 7–13—turning; Grades 8–13—boring; Grades 10–13—milling, planing, shaping, drilling. See also Fig. 3.9.

NOMINAL SIZE RANGE INCHES	GRADE									
Over *To*	4	5	6	7	8	9	10	11	12	13
0.04– 0.12	0.15	0.20	0.25	0.4	0.6	1.0	1.6	2.5	4	6
0.12– 0.24	0.15	0.20	0.3	0.5	0.7	1.2	1.8	3.0	5	7
0.24– 0.40	0.15	0.25	0.4	0.6	0.9	1.4	2.2	3.5	6	9
0.40– 0.71	0.2	0.3	0.4	0.7	1.0	1.6	2.8	4.0	7	10
0.71– 1.19	0.25	0.4	0.5	0.8	1.2	2.0	3.5	5.0	8	12
1.19– 1.97	0.3	0.4	0.6	1.0	1.6	2.5	4.0	6	10	16
1.97– 3.15	0.3	0.5	0.7	1.2	1.8	3.0	4.5	7	12	18
3.15– 4.73	0.4	0.6	0.9	1.4	2.2	3.5	5.0	9	14	22
4.73– 7.09	0.5	0.7	1.0	1.6	2.5	4.0	6	10	16	25
7.09– 9.85	0.6	0.8	1.2	1.8	2.8	4.5	7	12	18	28
9.85–12.41	0.6	0.9	1.2	2.0	3.0	5.0	8	12	20	30
12.41–15.75	0.7	1.0	1.4	2.2	3.5	6	9	14	22	35
15.75–19.69	0.8	1.0	1.6	2.5	4	6	10	16	25	40

valuable information given in Fig. 3.9, while it does not agree in all particulars with the typical roughnesses mentioned in ASA 46.1-1955,[3.9] is worthy of detailed study. It stands to reason that the material itself must be nearly flawless in order to obtain uniform highly-smooth surfaces.

Roughness is the relatively finely spaced irregularities of the surface, Fig. 3.10. ***Waviness*** is the irregularities or departures from the nominal surface of greater spacing than roughness. ***Lay*** is the direction of the predominant surface pattern, Fig. 3.10, usually determined by the production method, as tool marks. The numbers that specify the roughness R are either:

1. The average arithmetical deviation from the mean line, in which case the area above the mean, Fig. 3.10(b), is the same as the area below. With none of the y coordinates taken as negative, the roughness is

$$R = \frac{1}{L}\int_0^L y\,dx. \tag{i}$$

Most instruments now in use give this value.[3.17]

2. The root mean square (rms) average, for which

$$R = \left[\frac{1}{L}\int_0^L y^2\,dx\right]^{1/2} \tag{j}$$

The units are microinches or μin., which is one millionth of an inch (10^{-6} in.) The arithmetical mean is theoretically 11% less than the rms;

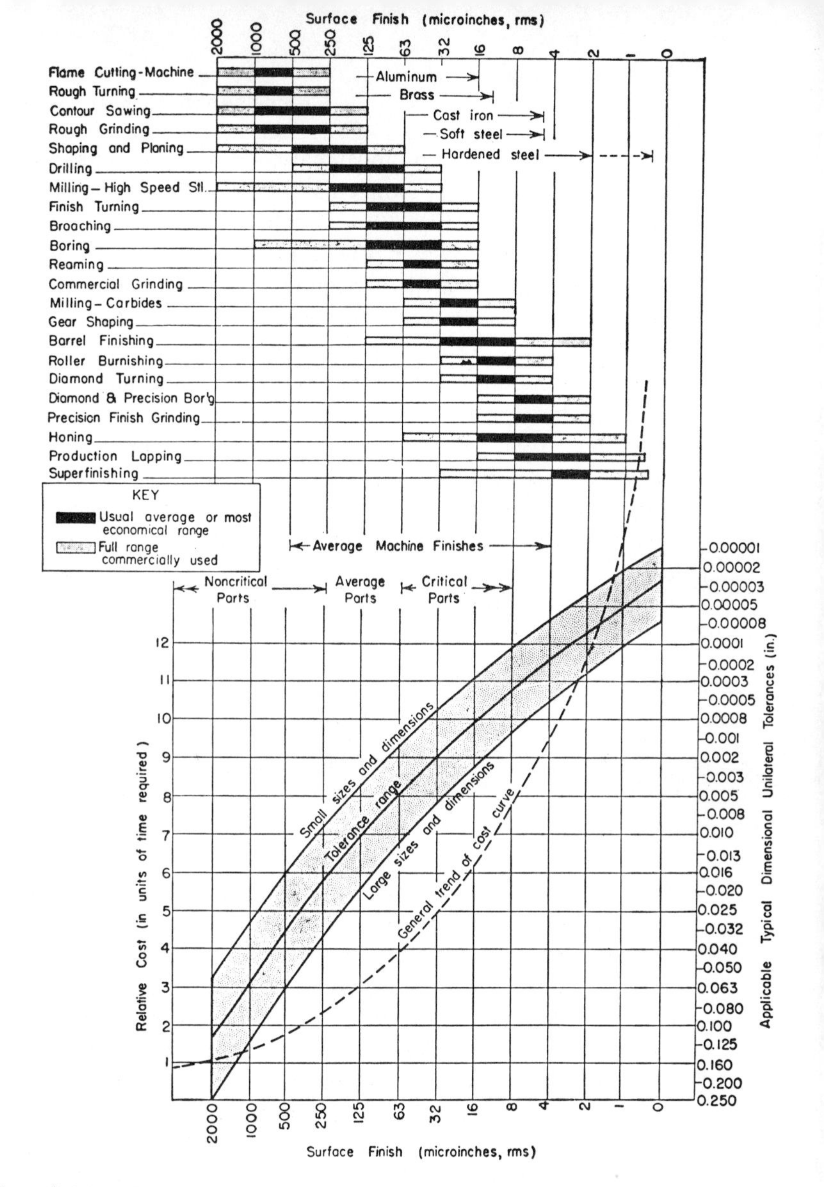

FIGURE 3.9 Surface Finish Related to Process and Tolerance. The tolerances shown at the right on the lower part are indicative and should not be used unless they are known to apply to the manufacturing process. For a roughness of 63, the tolerance may be between about 0.010 and 0.001 in., dependent to some extent on size. Enter this figure and check this range for yourself. (From Bolz, R. W., *Production Processes,* Penton Pub. Co.).

FIGURE 3.10 Roughness and Waviness. The lay shown in (a) is perpendicular to the line *AB* that would represent the nominal surface.

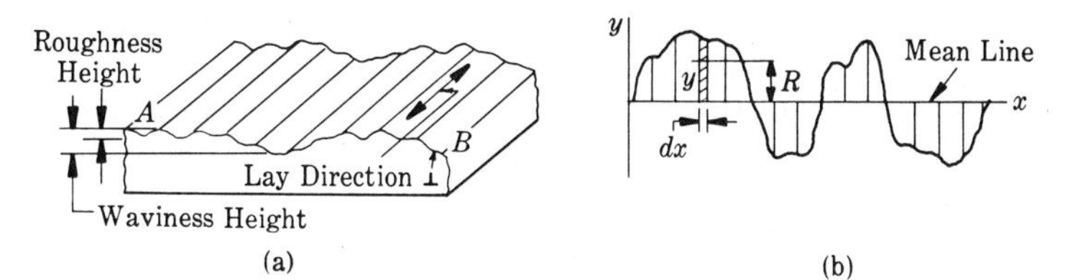

this is not sufficient in most cases to warrant changing the roughness specifications in a shift from rms to arithmetic. Much of the literature deals with the rms, as in Fig. 3.9 and below.

The symbol shown in Fig. 3.11 is used on drawings to designate the desired maximum surface roughness.[3.9] The values preferred for specification purposes are those numbers on the abscissa of Fig. 3.9 and in the summary below; that is, specify a maximum of 32 or 63 or 125, etc., or a *range* of acceptable values. Standard reference surfaces for these numbers are available. For the ordinary run of shop work, a fingernail test has proved good, but visual tests are deceptive. The fingernail test is a comparison of a given surface with graded and standardized roughness specimens,[3.5] obtained by rubbing the fingernail across the surfaces and matching the feel. Specify the waviness when it is significant. The symbols used for lay are: =, parallel to the surface boundary line; ⊥, perpendicular to the surface boundary line; ×, angular in both directions; M, multidirectional;

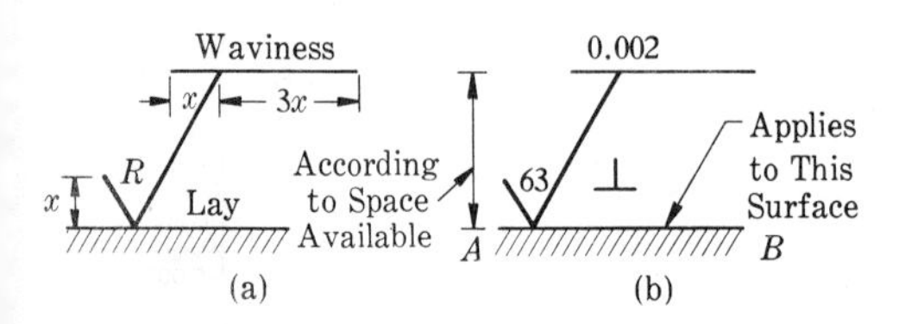

FIGURE 3.11 Symbol for Specifying the Surface.

C, circular; R, radial. The symbol in Fig. 3.11(b) is interpreted as a surface of maximum roughness of 63 μin., a maximum waviness of 0.002 in., and the lay perpendicular to the boundary *AB*.

Remember that the ideal surface finish is the roughest one that will do the job with satisfaction to all concerned. It should be observed that the surface finish for parts that have been forged, cast in permanent molds or dies, rolled, cold-drawn, or extruded, will depend to some extent on the surface finish of the die; for example, cold-rolled steel may have a finish in the range 63–250; extrusions, 32–125. The summary below,[3.7, 3.8, 3.10] will be helpful.

500 or greater rms: heavy cuts and coarse feeds; sand castings; hot-rolled steel.

250 rms: shows definite tool marks from rapid feed; forgings; very coarse grinding; suitable for machined external surfaces in general; chased threads.

63–125 rms: high-grade machine work; smooth machine finish obtained by a high-speed, fine-feed, light-cut, sharp-tool pass; coarsest finished to be used for bearing surfaces when the motion is slow and the load light; surfaces for soft gaskets; forgings; permanent mold castings; investment castings; die or tap-cut threads; datum surfaces for tolerances over 0.001 in.; outside gear diameter; mating surfaces of brackets, pads, housing fits, bases, etc.

32–63 rms: finest machined finish, obtained with carbide or diamond tool; medium grind, reamed, broached, burnished finish, die castings; teeth of ratchets and pawls; gear teeth for ordinary service; datum surfaces for tolerances under

0.001 in.; milled threads; pressed fits; keys and keyways; rolled surfaces; cams and followers; surfaces for copper gaskets; ordinary journal bearings (nearer 32 rms); mating sliding surfaces; worm gears.

16–32 rms: specified when finish is of primary importance; fine cylindrical grinding; smooth reaming; coarse hone or lap; automotive valve stems; pistons; cylinder bores; ground screw threads; gear teeth for severe service; precision journal bearings; shaft seals; precision-mating sliding surfaces; cold drawn surfaces; splined shaft; brake drums and other friction surfaces.

8–16 rms: used only when coarser finishes are inadequate; finest cylindrical grinding; microhoning, honing, lapping, buffing; automotive cylinder bores; seats for antifriction bearing races; crankpins; valve seats; valve stems; rolled threads (this is a low-cost production method); precision rolling surfaces; cam faces.

2–8 rms: produced by honing, lapping, superfinishing, very fine buffing and polishing; generally expensive; piston pins; pressure lubricated bearings; precision tools.

1 rms: gages, micrometer anvils, mirrors.

3.15 CLOSURE. The designer should remember that tolerance is the *stated* permissible variation of a dimension and that the relation between this stated variation and the actual variation is sometimes coincidental and often not as the designer hopes. Tolerances should be realistic, as well as in accord with good engineering practice. Realism in tolerances will follow from a knowledge of what the manufacturing department can and cannot do. The "natural spread" is sometimes called the "natural tolerance," but this latter term should be avoided because the word *tolerance* has a meaning that results in confusion to those not familiar with the idea.

Special considerations are involved for interference fits of materials that are stressed plastically. Quantitative data for some plastics commonly used in machines are found in Ref. *(3.18)*. Tolerances concerning out-of-roundness, straightness of holes, rods, etc. are important but beyond the scope of this book.

4. VARIABLE LOADS AND STRESS CONCENTRATIONS

4.1 INTRODUCTION. It has been said that 80% of the failures of machine parts have been fatigue failures. Whatever the true percentage, it is a large one, which suggests that designing machine elements should always be done from the viewpoint of the possibility of a fatigue failure. Even where the pattern of variability of the loading cannot be predicted and where the magnitudes of the forces are unknown, which is so often true, consideration of the principles of design for fatigue should be rewarding.

This point of view gives the designer considerably more to think about, as we shall see, and often challenges him to extra effort in trying to define the loading more completely. This chapter involves problems without combinations of different kinds of stresses, so that the reader may acquire a working knowledge of the design attitudes and of the language of metal fatigue before facing involved stress analysis (as in Chapter 8). Although we shall present a fairly straightforward procedure for design, this is an area about which we are still fairly ignorant. What knowledge we have is largely empirical, but the literature on the subject has grown to formidable proportions within a single generation, a commentary on its importance. If you practice mechanical engineering design, this is an area for further extensive study.

4.2 MECHANISM OF FATIGUE. On a macroscopic scale, fatigue failure starts at some point (because the repeated stress there exceeds the endurance strength of the material) in the form of a minute crack that gradually spreads with repetitions of the excessive stress until the resisting area becomes so small that complete fracture suddenly occurs, probably without warning and perhaps by now even with a small applied load. The fracture for quite ductile materials is without significant plastic action, Fig. 4.1; hence, such fractures are often referred to as ***brittle fractures,*** or *brittle failures* (§ 2.22). The suddenly fractured surface has a lustrous crystalline appearance characteristic of all brittle failures. Since fatigue failures are also the consequence of a spreading crack, they are even more aptly called ***progressive fractures.***

While there are several descriptions of the fatigue phenomenon in the literature,[4.1,4.10] we can indulge only the briefest one here, which may be

FIGURE 4.1 Shaft after Fatigue Failure. The slots in the cylindrical surface are ends of keyways, where the fatigue failure started. Failure gradually progressed from both keyways toward center. The surface that failed early was subsequently rubbed nearly smooth under load. The final break is at the rough ridge, about on a diametral line. Observe nearly identical lines of progressive failure leading from both keyways toward the center. (Courtesy Joseph T. Ryerson & Son, Chicago).

just as well because knowledge of the subject is incomplete. Fatigue failure is believed to start at random points as slip on shear planes of crystals that happen to be oriented so that this can happen. Imperfections in the crystal and other imperfections, such as the penetration of grain boundaries by oxides, help to start such failures, which are soon microscopic. Crystal slip continues with repeating stresses until visible cracking occurs. Although shear produces the crystalline slip, the crack will spread in the direction of a plane on which there is a tensile stress, if there is a tensile stress. Almen[4.64] shows illustrations of cracks that are in the direction of a plane on which there is compression from an external load, but the cracking is attributed to the residual tensile stress (§ 4.23) known to exist. Because the cracks were subjected to compression by the external load, they did not propagate to failure. On the other hand, once a crack exists in a plane subjected to tension, the high stress concentration at the ends of the crack (§ 4.24) encourages its rapid spread. The crack from a load producing a primary shear stress tends to follow the plane of the principal tension. The crack from fatigue failure under a repeated compressive stress follows approximately the maximum shear direction (so fatigue failure by compression only is a special case). Briefly compare the mechanism described above with that described for creep failure (§ 2.21), which follows from a deterioration and movement at the grain boundaries.

In actual machine parts, the cracking usually starts at a *discontinuity*, a fillet, a scratch, an inclusion or hole inside the member, a keyway (Fig. 4.1), etc. This is because discontinuities have the effect of increasing the stress locally (in the vicinity of the discontinuity), as described in more detail later. The fatigue failure of a smooth, polished, rotating-beam specimen *without internal flaws* starts at the outside surface not only because the maximum stress is there but perhaps because the outermost crystals, not being strengthened by other crystals on all sides, are more prone on a statistical basis to be the first to slip or shear.

4.3 ENDURANCE LIMIT, ENDURANCE STRENGTH.

Instead of using the yield stress or ultimate stress as a basis of finding a design stress for a part subjected to a varying load, we should now use some sort of endurance strength. When one speaks of the ultimate stress, it is presumed that it is for a standard 0.505-in. specimen unless otherwise specified. It is correspondingly convenient for endurance strengths to refer to standard specimens. We shall use ***endurance limit*** and ***fatigue limit*** s_n' to mean *the maximum reversed stress that may be repeated an indefinite* number of times on a polished, standard (nominal 0.2 to 0.4 in. diameter) specimen in rotating bending without causing failure.* There are other sizes of "standard" specimens and other ways of testing for fatigue strength (as torsional or axial loading). Not every material exhibits an endurance *limit*, particularly true for many nonferrous metals. See the curve for aluminum alloy 2017, Fig. 4.3. We shall use ***endurance strength*** and ***fatigue strength*** s_n in a general sense for the strength of actual parts, specimens with notches, etc., and for materials without a particular *limit*. Thus, because of its general application, s_n will be the symbol found in most equations, and the reader should learn when to use an endurance limit s_n'. For materials that do not have a limit, the number of cycles for which the given strength applies should be stated. For example, see Table AT 10.

4.4 CHARTING ENDURANCE STRENGTH.

Finding the endurance strength is time consuming and expensive. The quickest way is to

* It is only fair to warn that some of the literature does not abide by this definition of *endurance limit*, but it is a convenience to have such a reference.

use a machine such that the specimen can be subjected to a constant bending moment, then rotate the specimen. In 360° of rotation, a point on the specimen undergoes a complete cycle of stress from maximum tension to maximum compression, back to maximum tension. Typically, fatigue data are plotted on log-log or semilog paper, stress vs. number of cycles to failure. If done in the approved manner,[4.56] the solid line of Fig. 4.2(a) represents the *median* strength as obtained from a number of tests. The median is the middle-value of a group of measurements (or if the sample size is even, it is the average of the two middle values), which is not necessarily the same as the mean or arithmetic average; the median and mean values usually approach each other as the number of measured pieces (from the same population) increases. Thus, the median curve represents the stress state at or below which 50% of the specimens failed and at which 50% survived; it can be designated as the 50% survival level. For example, if the stress is about 104 ksi for the polished (upper) specimens, we expect that 50% would survive more than 10^5 cycles of stress. However, we are not certain of this and the probability of our being exactly right on the 50% depends on the number of specimens tested. If 100 specimens were tested at each of several stress levels, the corresponding median line would be so close to the true value that the probability of 50% survival at a particular point would be quite high. On the other hand, if there are 4 pieces tested at each stress level, a more likely number, we are much less sure that the true median has been found, and the probability of a particular point representing the true 50% survival is lower. For the same reasons, there is a corresponding uncertainty in the distribution of say fatigue strengths at any particular number of cycles. The degree of certainty in this regard is called the ***confidence level***, which is easiest defined by examples. It happens that the confidence level for Fig. 4.2(a) with 4 specimens tested at a given stress level is 50%; this means that if we pick up 100 groups of

FIGURE 4.2 Endurance Strength Bands. Rotating beam tests.[4.35] The specimens in (a) had a nominal diameter of 0.2 in. The tests, represented by the dots, were for carefully manufactured specimens, which accounts in part for the small scatter. With sufficient test points, the bands typically narrow as shown, the boundaries tending to intersect in the vicinity of 10^3 cycles at s_u for ductile steel.[4.64] In (b): the distribution of failures against $\log n_c$; at a certain stress level; after Dolan.[4.1]

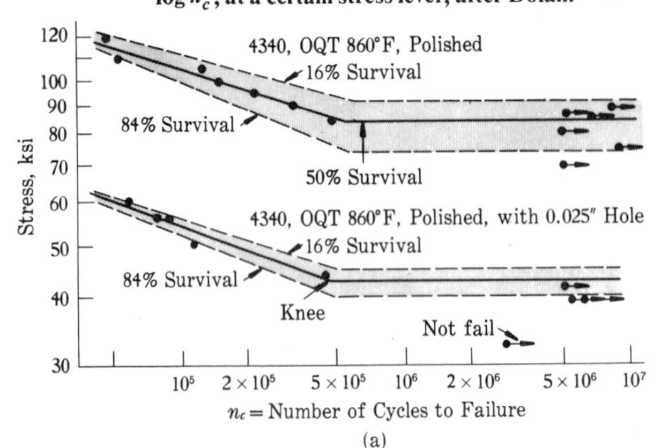

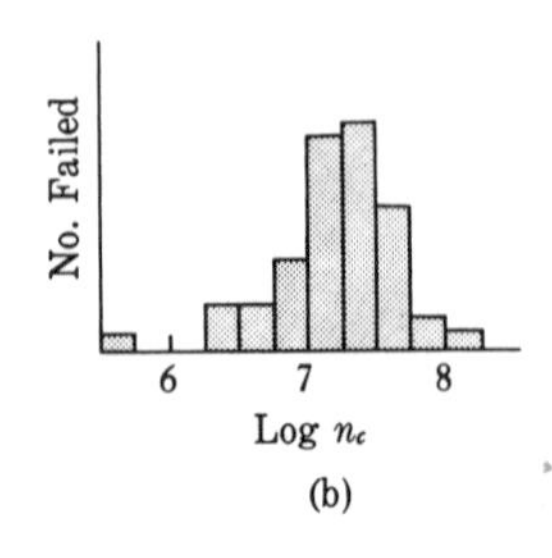

pieces and for each group say "Half of this group will fail at stress s by the time the median curve is reached," we will be right 50% of the time. To get a higher probability of being right, the sample size would have to be increased. If the sample size had been 10, we could make the same statement with the probability of being right 99.9% of the time.

Not all specialists in this area agree that the normal curve (Fig. 3.3) is the best distribution form to use, but it is seen in Fig. 4.2(b) that the distribution of the logarithm of the number of cycles is reasonably approximated by a normal curve. Hence, we may, with appropriate approximation, think of the failing stress distribution at any number of cycles to be distributed normally, with a certain standard deviation σ. As previously pointed out, one would expect that almost the entire population would fall within $\pm 3\sigma$ from the mean. From Table 3.3, we find [84% survival: $(1 - 0.84)/2 = 0.08$] for 0.08 that $x/\sigma = 1.4$; that is, the 84% level in Fig. 4.2 represents about 1.4σ from the median (strictly from the mean)—with a 50% confidence level. This is to say that the *median* of future groups from the same universe should fall above the 84% survival line 84% of the time. If one must reduce the odds of failure, a lower survival line must be the basis of design (or the design factor must cover the gap). If the distance is doubled, this would be about 2.8σ and about 99.5% do not fail ($0.995 \approx 1 - 2 \times 0.0026$, from Table 3.3). Details of a more complete handling of the statistical aspects are given in working form in Ref. *(4.56)*.

Good distributions for indefinite life are not easily obtained for materials with a fatigue *limit*, because so many of the specimens in this vicinity run out without failing. Hence, extrapolations are made. Unquestionably, we should know something of the spread of endurance strengths as a means of being sure that a design is not being made within the failure range, but to get these data with a high confidence level is quite expensive, and one seldom finds reports that have included the necessary statistical analysis. Additional information may become available in the future and designers can be guided accordingly, but in the meantime, we shall accord with current custom and discuss the design from the standpoint of average or median strengths (the literature seldom distinguishes). Stulen *et al.*[4.26, No. 13] report standard deviations σ for long-life fatigue strengths of AISI 4340 varying from 4.9 to 7.8% of s_n (8% looks safe); for titanium 6 Al 4 V, $\sigma = 0.064s_n$; for aluminum 7076-T61, $\sigma = 0.06s_n$; for beryllium copper, $\sigma = 0.075s_n$; for Al-Ni-Bronze, 5 Ni 10 Al, $\sigma = 0.094s_n$. Thus the lower limit of fatigue strength for most metals might be estimated as $3\sigma = 3 \times 8 = 24\%$ below the average s_n. Steels less carefully manufactured may well have greater dispersion and σ. Also, working against safety is the probability of a greater spread (larger σ) in the strength of the finished part because of less uniformity in geometry and composition. In design, it should be advisable to be as sure as possible that the operating stress does not exceed $(1 - 0.24)s_n = 0.76s_n$; that is, that the factor of

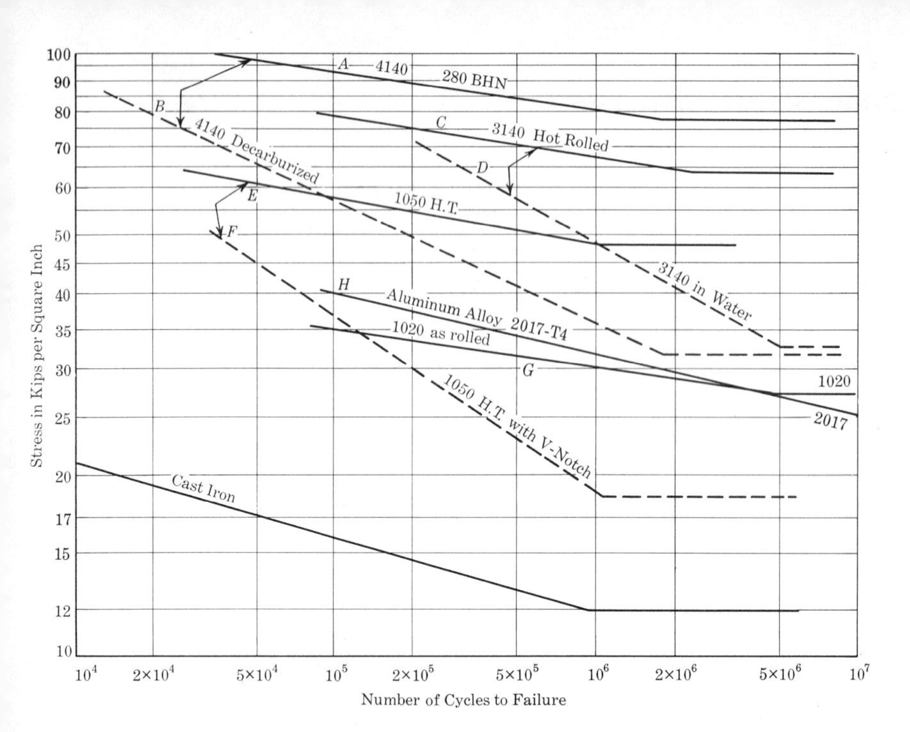

FIGURE 4.3 Typical sn_c Curves. The solid curves are typical 50% probability curves for standard polished specimens in rotating-beam tests; the dotted curves are for polished specimens but with some deviation—a notch, in water, and decarburized. The horizontal portions of the curves represent the endurance limits. Note the absence of a horizontal part for the aluminum alloy 2017–T4, typical of nonferrous alloys.

Observe that SAE 1050 notched is weaker than as rolled 1020; and that the sloping parts of the dotted curves tend to be steeper than the corresponding solid curves; that is, that the effect of stress raisers is less for finite life (§ 4.16) than for indefinite life. Curves *A* and *B* for SAE 4140, OQT to a hardness of 280 Brinell; *C* and *D* for 3140 hot-rolled material; *E* and *F* for 1050, quenched and tempered at 1200°F. Despite the fact that the decarburized layer in curve *B* was thin, it materially weakened the specimen. It is not shown above, but processes that leave a residual compressive stress on the surface, such as peening, § 4.28, result in a smaller slope (as compared to polished specimens) for stresses greater than s_n'. For a particular material, these sloping lines tend to intersect at a point defined approximately by 10^3 to 10^4 cycles and $0.9s_u$ to s_u stress.

safety used with average endurance strengths covers this eventuality after various stress raisers explained below have been accounted for.

Figure 4.3 shows the typical stress curves for a number of different materials and configurations, and it should be studied. Observe that fatigue limits are well below elastic limits.

4.5 HOW STRESSES VARY. Stresses may very well vary in a quite irregular and perhaps unpredictable manner, as the stresses in the structure of an airplane passing through a storm, Fig. 4.4(a). An impulsive load may set up a vibration, Fig. 4.4(b), that gradually damps out; and vibrational stresses may be and often are imposed upon some irregular spectrum as in Fig. 4.4(a). In some cases, the designer can estimate the number of times that some maximum loading will be imposed on a part during its lifetime, and choose design stresses for this many repetitions, § 4.16. Whatever the pattern of variation, given enough experience (measurements), an idealized model of the load variation can be constructed as a basis of design. The most common models are sinusoidal and are given in Fig. 4.5. There will be a maximum and a minimum stress, an average or mean stress s_m, and a variable or alternating component of the stress s_a. If both kinds of normal stresses, tension and compression, are involved, we must use algebraic signs, negative for compression. In Fig. 4.5, we see that the alternating component is in each instance that stress which when added to (or subtracted from) the mean stress s_m results in the maximum (or minimum) stress. The average or mean stress s_m and the alternating component s_a are

$$\text{(a)} \qquad s_m = \frac{s_{\max} + s_{\min}}{2} \qquad \text{and} \qquad s_a = \frac{s_{\max} - s_{\min}}{2},$$

where a *compressive stress is a negative number*. For a complete reversal, Fig. 4.5(a), $s_m = 0$; that is, $s_{\min} = -s_{\max}$ and $s_a = s_{\max}$. In every case,

$$\text{(b)} \qquad s_{\max} = s_m + s_a.$$

A parameter used to locate the curves of Fig. 4.5 is a *stress ratio R* defined as

$$R = \frac{s_{\min}}{s_{\max}},$$

with stresses used algebraically; $R = -1$ for completely reversed stress, Fig. 4.5(a).

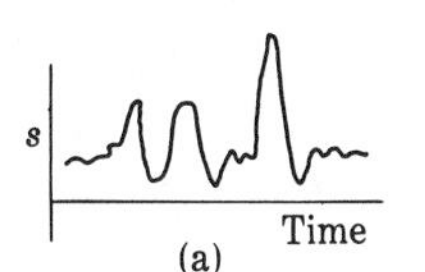

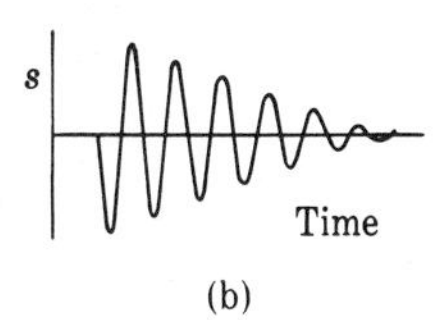

FIGURE 4.4 Spectrum of Stress Variation.

FIGURE 4.5 Sinusoidal Stress Variations. (a) Represents completely reversed stress; rotating beam with constant moment; $R = -1$. (b) A repeated stress has either zero minimum stress as shown ($R = 0$) or zero maximum stress (all compression $R = \infty$).

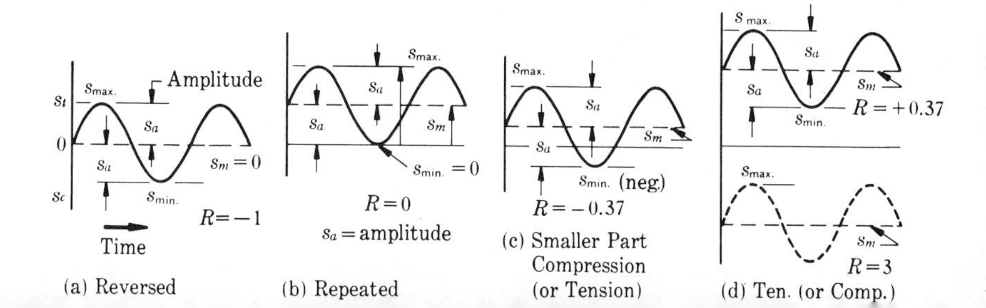

4.6 REPRESENTATION OF THE ENDURANCE STRENGTH IN TERMS OF ALTERNATING STRESS. The type of diagram in Fig. 4.6, with variations, is frequently used as an aid for design. Its construction is indicated by the symbols s_n', s_y, s_u, etc. Straight lines are often used, as HAQ, but if test values are known, a curve such as HPQ can be constructed. The lines HQ and JQ are thought of as the limiting (maximum and minimum) stresses for indefinite life (say 50% confidence). For finite life, as 10^5 cycles, some line VQ (with a matching one for a lower boundary) would apply. With a factor of safety N, the maximum and minimum stresses would fall on the heavy dotted lines. Note that the maximum stress is limited to the yield strength, so that the boundary becomes XYZ. Conventionally, the static yield strength is used, because its values are available, but the yield strength under short-duration loads is greater than for gradually applied loads.

As you may have inferred, we have no theory that says how the mean and variable-stress components are related. Hence, the empirical approach.

FIGURE 4.6 Goodman Type of Diagram. The influence of the mean stress is clearly shown by this diagram. For example, if the mean stress is zero, failure occurs with a maximum stress slightly over s_n' at H. At any mean stress s_m at M, the limiting range of stress is AB, a safe range is CD. Compressive ranges K defined by this diagram are overly conservative; so it is not uncommon to let the permissible range of stress TW at $R = -1$ remain constant as $|s_c|$ increases (but for indefinite life, it is probably best for the maximum $|s_c| \gtreqless s_y$). Tests show that the mean compressive stress can be increased substantially with no decrease in the amplitude of the variable component and without decreasing the factor of safety (see Fig. 4.8). See Ref. (4.65) for Goodman diagrams drawn to scale for numerous materials, including ones for torsion.

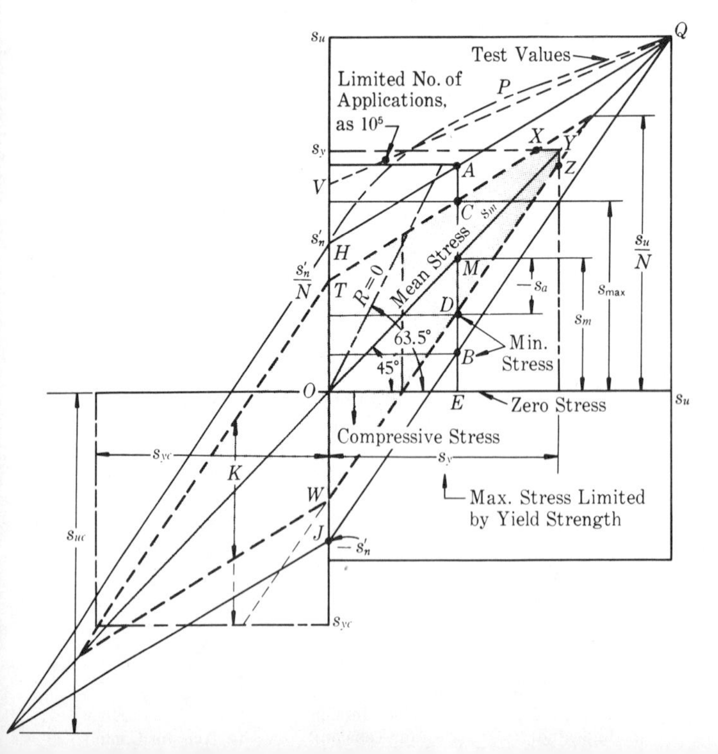

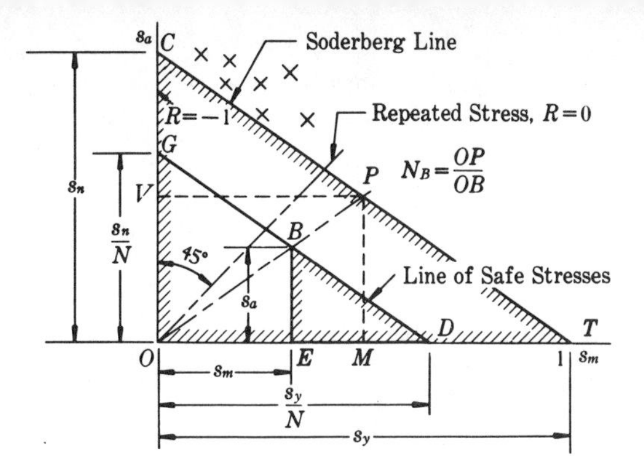

FIGURE 4.7 Soderberg Line. Recommended for ductile materials only.

We shall use a diagram, Fig. 4.7, in which the ordinate is the alternating stress and the abscissa the mean stress. Draw a line through the endurance limit (or strength) at C and yield strength at T. This line is called the Soderberg line,[4.10,4.13] and points on it are assumed to represent a state of stress that puts the part on the point of failing after an indefinite number of alternations of s_a. For example at P, a variable stress OV on a mean stress OM is the limiting condition. Since most actual points of failure of polished steel specimens fall outside of this line, as suggested by the points shown, the Soderberg line is a conservative basis of design. However, a factor of safety is necessary; divide it into s_y and s_n to get points D and G; and the line DG then represents a locus of points that in turn represents safe conditions. The combination s_m and s_a at B is for a design factor of N (also $N = OP/OB$). The equation for line DG is easily obtained by setting up proportions for the similar triangles BED and COT; its useful form is obtained by solving for $1/N$:

(c)
$$\frac{s_a}{s_n} = \frac{s_y/N - s_m}{s_y},$$

(4.1)
$$\frac{1}{N} = \frac{s_m}{s_y} + \frac{s_a}{s_n}.$$

There are several other possible criteria for design, including the modified Goodman line and the Gerber line of Fig. 4.8. The working curve for the Goodman line is BD, and by analogy with equation (4.1), its equation is

(4.2)
$$\frac{1}{N} = \frac{s_m}{s_u} + \frac{s_a}{s_n},$$

often used for brittle materials, such as cast iron. The Gerber line is a parabola with apex at C; the corresponding failing curve is

(d)
$$1 = \left(\frac{s_m}{s_u}\right)^2 + \frac{s_a}{s_n}.$$

The design curve is obtained by dividing s_u and s_n by N thus obtaining a quadratic for N. The German practice is to use this equation, which in truth

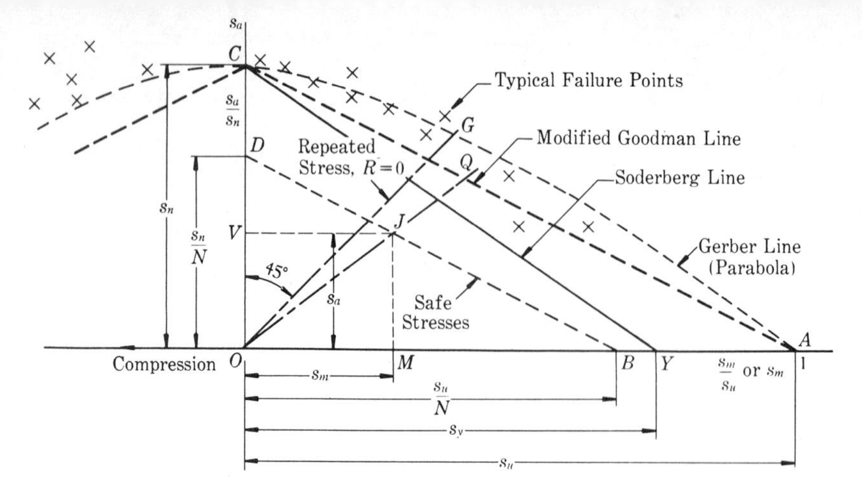

FIGURE 4.8 Comparison of Variable-stress Criteria. The Goodman line is used for brittle materials, preferably with a higher design factor than would be used for ductile materials. Many designers use it also for ductile materials. In this type of diagram, the coordinates are frequently made dimensionless by dividing s_a by s_n and s_m by s_u (or s_y). This change makes the value of the end points A and C unity.

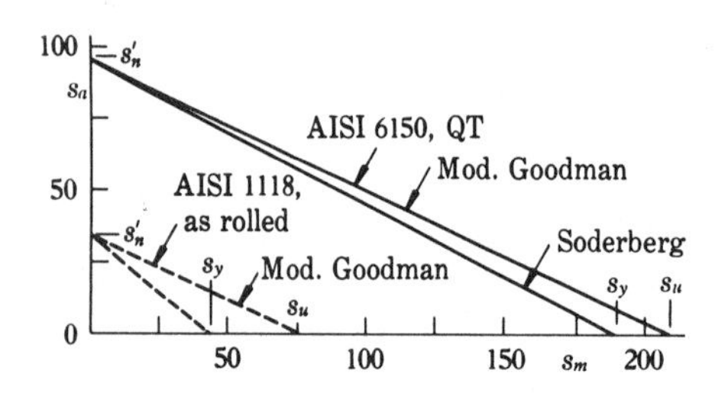

FIGURE 4.9 Soderberg vs. Goodman Criteria.

does follow better the trend of actual failure points. It may be worthwhile to observe from Fig. 4.8 that the difference between the Goodman line CA and the Gerber line is about a maximum at OG and below, where OG represents a repeated stress, $R = 0$, Fig. 4.5. Thus, if CGA is a better approximation of failure, one may wish to use **(d)** in design or give special attention (which we do in § 6.6) to situations of approximately $R = 0$. As shown in Fig. 4.9, the difference between the Goodman and Soderberg lines decreases to a negligible amount for high-strength, heat-treated steels where the yield strength is close to the ultimate.

If the foregoing equations were developed for a varying shear (torsional) stress, the various lines apply and the equation analogous to (4.1), involving yield strength, is

(e)
$$\frac{1}{N} = \frac{s_{ms}}{s_{ys}} + \frac{s_{as}}{s_{ns}},$$

where s_{ns} is the shearing endurance strength, s_{ys} the yield strength in shear, s_{ms} the mean stress, and s_{as} the alternating or variable stress in shear.

The same equations are often applied when the predominant stress is compression, but as suggested by the points on the compressive side of the s_a axis in Fig. 4.8, the results are ultraconservative. Compressive stress inhibits the propagation of a crack.

In solving any of the foregoing equations, s_m and s_a are computed as ***nominal stresses*** s_o, which are those corresponding to F/A, Mc/I, and Tc/J; that is (if there is no rotation),

(f) $$s_m = \frac{F_m}{A} \quad \text{or} \quad s_m = \frac{M_m c}{I} \quad \text{or} \quad s_m = \frac{T_m c}{J},$$

(g) $$s_a = \frac{F_a}{A} \quad \text{or} \quad s_a = \frac{M_a c}{I} \quad \text{or} \quad s_a = \frac{T_a c}{J}$$

Special handling of Mc/I for a rotating member is required because the stress undergoes cycles even though the moment is constant.

4.7 ESTIMATIONS OF ENDURANCE STRENGTHS.

A number of endurances limits and strengths for rotating beam tests are given in Tables AT 3, AT 4, AT 6, and AT 10. Since, however, the possible variations in composition and heat treatments are infinite, a number of empirical formulas have been suggested,[4.24] all of which are of limited application. For wrought steel in its more commonly met commercial forms, it is often assumed that the average endurance limit for an average s_u (50% survival) is

(h) $$s_n' \approx 0.5 s_u, \qquad s_n' \approx (250)(\text{BHN})\ \text{psi}, \qquad s_n' \approx (0.25)(\text{BHN})\ \text{ksi},$$

[WROUGHT STEEL, BHN < 400; GOOD DUCTILITY]

which should be limited to a maximum Brinell of about 400. There are many exceptions that suggest caution. Cazaud[4.24] quotes values for steel showing s_n/s_u ratios, often called the *endurance ratio*, from 0.23 to 0.65. The lowest values for ordinary steels go with untempered martensite, a structure left by rapid cooling of medium- to high-carbon steel, as in quenching (but the usual heat treatment includes tempering after quenching, which results in more typical values of the endurance ratio). If the microstructure is pearlite or austenite, a working value of endurance ratio would better be 0.4 or less (check some of the austenitic stainless steels in Table AT 4); but in the plain carbon, ferrite structure (for example, mild and soft steel, quite ductile), s_n'/s_u may be greater than 0.6. In general, for heat treated steels, the ratio s_n'/s_u tends to decrease with decrease of tempering temperature, and it drops substantially when BHN > 400, as suggested by Fig. 4.10.

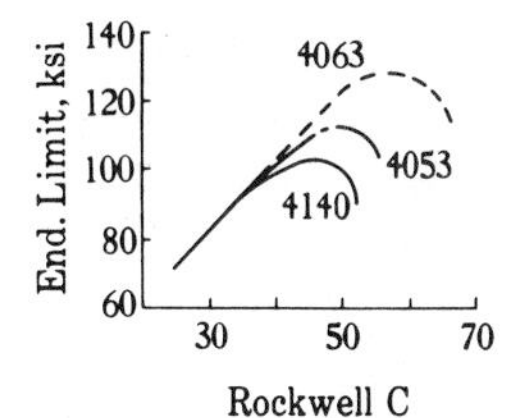

FIGURE 4.10 The static s_u continues to increase, but s_n' turns down.

Probably a better general-purpose estimate of s_n' for *high-strength alloy steel*, which is used in Germany, as reported by Lessells,[4.33] is

(i) $$s_n' = as_y + bs_u,$$

where a is a function of s_y and is obtained by a straight-line interpolation between the following values: $a = 0.2$ for $s_y = 85.3$ ksi, $a = 0.4$ for $s_y = 190$ ksi. Similarly, b is a function of s_u, and its value is obtained by a straight line interpolation between: $b = 0.45$ for $s_u = 85.3$ ksi, $b = 0$ for $s_u = 199$ ksi. Where the difference between the values from **(h)** and **(i)** for alloy steels is significant, the value from **(i)** is probably the better.

For ***cast steel***, it is safer to use $s_n' \approx 0.4s_u$ if test values are not available; for cast iron, use $s_n' \approx 0.35s_u$; for nodular iron, use $s_n' \approx 0.4s_u$ (but 0.33 for normalized). The nonferrous metals either do not have a *limit* or the variation of s_n/s_u for different alloys is so great that arbitrary values are seriously in error. The endurance strengths of aluminum alloys at 10^8 cycles change little with large increases in tensile strength. The copper alloys best suited for fatigue, in order of fatigue strength, are: beryllium copper, phosphor bronze D, C, and A, nickel silver B, and silicon bronze A.[2.1] See Ref. *(4.31)* for a large number of numerical values of fatigue strength.

The endurance strength of steel specimens subjected only to ***axial loading*** (no bending) is usually not as high by test as the endurance limit. This may be because of the difficulty of applying axial loads with no eccentricity, because the material is not homogeneous, and because the stress is never truly uniformly distributed. Also, the presence of a stress gradient in bending and its absence in pure axial tension may be a factor; moreover, since an axial load subjects a greater volume of material to the maximum stress than a bending load, the probability of a high stress at a flaw or of slippage beginning at a weak crystal is greater. The literature reveals a range of from some 0.6 to over 1 for the ratio of axial fatigue to rotating beam endurance strengths. The lower values are more typical of plain carbon steels, and the higher values of alloy steels. But the evidence is inconclusive. In the absence of test values, use

(j) $$s_{n\,max} = (0.8)(s_n \text{ from standard specimen, rotating beam})$$
[REVERSED AXIAL LOADING]

The ratio for Al 2014-T6, Table AT 10, is seen to be $15/18 = 0.83$. Other tests for aluminum alloys[4.62] show a higher ratio (0.82–1.06). In situations where the procedure is justified, axial fatigue testing simulating the particular problem should be conducted. Only in a few cases is s_y (0.2% offset) $< s_n'$; see AISI 321 annealed in Table AT 4. If this should be true, one would wish to have the maximum stress in a satisfactory relation to s_y (see Fig. 4.6); refer to § 4.33.

The octahedral shear theory (§ 8.12) predicts that the shearing strength is about 58% of the tensile strength (elastic action). We recall that static tests of yield and elastic strengths are in good agreement with this factor, and endurance tests similarly show that the torsional endurance strength of steels ranges from about 0.5 to 0.6 of s'_n, with 0.58 being a good typical value. In the absence of test values, consider that the *endurance strength* of a standard steel specimen in *reversed torsion* is

(k) $$s_{ns} = 0.6s'_n = (0.6)(\text{endurance strength, rotating beam}).$$
[TORSION]

But be careful. For the cast irons, the range is closer to $0.8s'_n < s_{ns} < s'_n$; for copper, $0.4s'_n < s_{ns} < 0.56s'_n$.

4.8 STRESS RAISERS. Any discontinuity or change of section, such as scratches, holes, notches, bends, or grooves, is a "stress raiser." It will result in a *concentration of stress* or a localized stress, that is greater than the average or nominal stress. In some situations the theoretical value

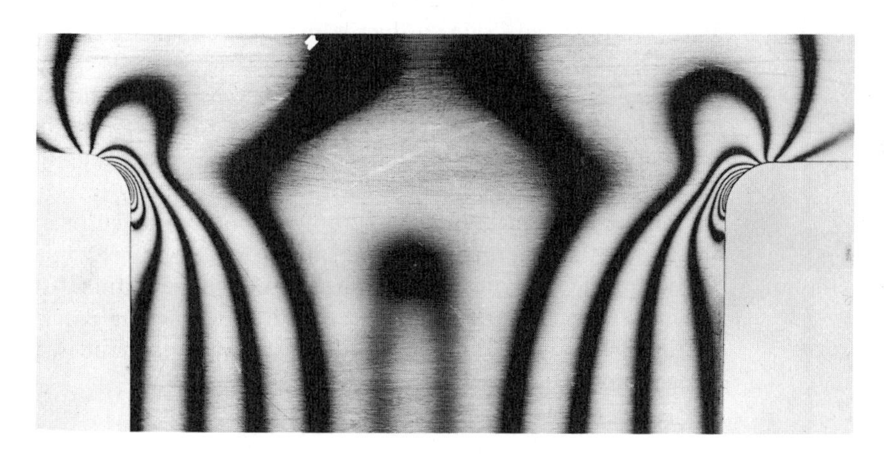

FIGURE 4.11 Stress Concentration at Re-entrant Corners. The stress distribution as indicated by the photoelastic method. The crowding together of the fringe lines at the fillets shows a high stress concentration. The larger the fillet, the less the theoretical concentration. But see § 4.10. In terms of the dimensions in Fig. AF 9, $h/d = 2.97$ and $r/d = 0.0792$ in this picture (not all of h is shown). Observe the parallel fringe lines away from the vicinity of the discontinuity. (Courtesy M. M. Leven, Westinghouse Research Laboratories).

of a stress concentration can be computed by elasticity theory, or it can be determined by various experimental techniques. Among these techniques is the photoelastic method, which uses transparent models of various plastics. A monochromatic light passing through a *loaded* model in a photoelastic instrument emerges as black and white lines, as suggested by Fig. 4.11. The dark lines are called fringes, and the magnitude of the stress at any point is a function of the number of fringes. Many of the available theoretical stress concentration factors (§ 4.9) have been found by

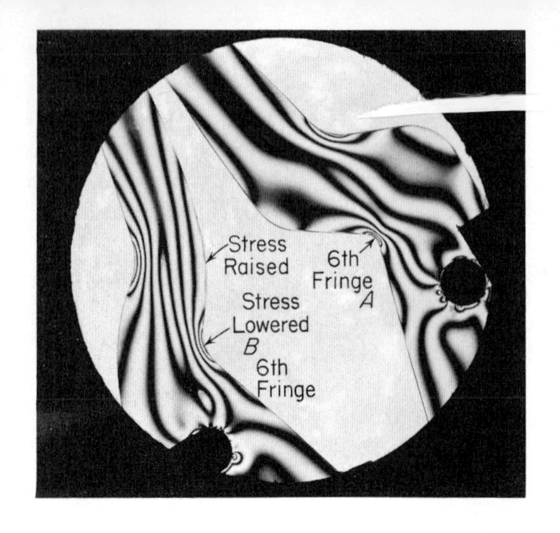

FIGURE 4.12 Improving Design by P elastic Study. Especially in very irregu and in indeterminate structures, phot studies of models of machine elements points of stress concentration and the cor ponding magnitude of stress; thus they are very helpful to the designer. In this illustration, the maximum theoretical stress at *A* in the original design was reduced by one-third at *B* in the revised design, and the weight of the part was reduced 300 lb.—less material yet greater fatigue strength. (Courtesy Chapman Laboratories, West Chester, Pa.).

photo-elastic analysis, and the method is now commonly used to aid in the design of parts that are difficult to analyze mathematically, as illustrated in Fig. 4.12.

Stress concentrations are significant for ductile materials only when the loads are repeated. Since a ductile material under a steady load yields at points of high stress concentration if the stress exceeds the yield strength, there occurs a redistribution of stress and no perceptible damage to the part as a whole. The yielding is local (confined to a very small area). If the load is repeated, however, the stress at the concentration points may exceed the endurance strength, and if so, the part eventually fails by fatigue.

Stress raisers include holes, grooves, tool marks, surface roughness of any sort as that from corrosion or pitting, keyways, welds, clamps (equivalent to a sudden change of section—introducing additional stresses on the surface), wrench marks, accidental bruises, quenching cracks, blow holes and inclusions in castings, small fillets instead of large ones, point where a thread ends.[4.3]

4.9 THEORETICAL STRESS-CONCENTRATION FACTORS.

The degree of stress concentration is usually indicated by a ***stress concentration factor***. If the factor is obtained from a theoretical analysis or a photoelastic study (photoelastic results are close to theoretical ones), it is called a theoretical stress concentration factor K_t, which is the ratio of the theoretical maximum local stress divided by a nominal computed stress s_o. Thus the maximum theoretical stress due to some discontinuity in the section is $s_{max} = K_t s_o$ for a reversed stress ($R = -1$), where s_o represents the nominal stress *computed* from $s_o = F/A$, $s_o = Mc/I$, or $s_o = Tc/J$.

Inasmuch as changes of sections and discontinuities in general can occur in an infinite variety of ways, it is not always possible to find a stress concentration factor that applies to the case at hand. Thus, the exercise of judgment with considerable estimation is often essential. It is also frequently necessary to let experiment proceed with design, with determinations of stresses experimentally either in an actual model or, for example, as in Fig. 4.12. Figures AF 5–AF 14, inclusive and Tables AT 12, AT 13 in the appendix, give a good selection from available data. See also Table AT 18 for curved beams and Fig. AF 15 for curved torsion members. An excellent compendium of stress concentration factors is given by Peterson;[4.21] others are in the literature.[2.1,4.2,4.9,4.15,4.18,etc.] In choosing a value of K_t, *always note the method of computing the nominal stress* s_o; a particular value of K_t refers to a certain resisting area, which is usually the minimum. For example, in Fig. AF 13, I/c and J/c make an allowance for the hole. Also observe that K_t depends on the loading; it is not the same for torsion as for bending.

4.10 NOTCH SENSITIVITY. The quantitative effect of a particular discontinuity in "raising" the stress is different for different materials. Some materials are more notch sensitive than others. To account for these different responses, we use a sensitivity index q, called the ***notch sensitivity*** of a material, Fig. AF 7, defined by

$$q = \frac{K_f - 1}{K_t - 1}, \qquad \text{or} \qquad K_f = 1 + q(K_t - 1), \tag{4.3}$$

where K_f = (actual $s_{\max})/s_o$ is the estimated ***fatigue-strength reduction factor*** (or *fatigue notch factor*).* Let the subscript f suggest fatigue. *For steels*, fatigue tests on the coarser-grain steels (normalized or annealed) show low values of the notch sensitivity; fine-grain steels (quenched and tempered) have high values of q. While there is no sharp dividing line between coarse and fine grains, 200 BHN may be used if the simplification of two classes is desired (see below). Lipson and Juvinall[4.65] state that K_f is never larger than 4 and seldom exceeds 3.

Peterson discusses notch sensitivity at some length in Ref. *(4.1)* and derives various equations based on theoretical and test considerations. He recommends the use of

$$q = \frac{1}{1 + a/r}, \tag{l}$$

where r is the radius of curvature (Fig. AF 7) and typical values of a for normal stresses, with the corresponding (s_u) in parentheses, are: *Q*&T steels, $a = 0.0025$ (122 ksi); annealed or normalized steel, $a = 0.01$ (63 ksi); which give the equations of the curves in Fig. AF 7. For other values

* Peterson[4.21] defines several K factors: theoretical factor, combined factor accounting for failure theory used, shear-stress factor, fatigue notch factor, shear-fatigue notch factor. Available evidence suggests that q for shear, say q_s, is higher than that for normal stress. Those who wish to make a distinction may use a for shear in equation (l) as 0.6 times the values given for normal stress;[4.1] $a_s = 0.6a$.

... $= 0.005$; $s_u = 180$, $a = 0.0005$; ...ws a straight line relation between s_u and $\log a$. Other values ...uminum alloy 2024-T4 bars, $a = 0.008$, and for sheet, $a = 0.05$; aluminum alloy 7075-T6 bars, $a = 0.003$, for sheet, $a = 0.02$.[4.57] Since there is considerable scatter in test results, one cannot be arbitrary about the accuracy of q; some indicated values for fine-grain alloy steels are $q > 1$. In view of the scatter and their relatively high notch sensitivity, many designers use $K_f = K_t$ ($q = 1$) for aluminum ($n_c > 10^5$), magnesium, and titanium alloys. The endurance ratio s_u/s_n for polished titanium alloy specimens is usually greater than 0.55.[4.62]

Cast iron, with its flakes of graphite, is effectively saturated with "stress raisers," so that the addition of another discontinuity seems to have little effect on its fatigue strength;[4.62] that is, $q \approx 0$, at least for small sizes and radii. Like steel, its notch sensitivity tends to increase as the radius of a fillet or groove increases and as the size increases, with K_f approaching K_t (full notch sensitivity). Even for small parts with sharp radii, the designer may wish to be conservative, using $q \approx 0.2$. Certain evidence[2.1] for class 45 cast iron shows that *when the stress is computed for the net section*, a severe stress raiser ($K_t = 2.25$) had little effect on the static tensile strength and reduced the fatigue limit by only about 15%. (Incidental intelligence: this cast iron has about the same fatigue strength up to some 800°F.) Thus, a general observation[2.1] is that at least for the lower-strength cast irons, say less than class 45, cast iron is not very notch sensitive. Designers tend to avoid brittle materials for fatigue loading, especially when impulsive loading is possible, but sometimes there is no desired economic alternative; moreover, ductility *per se* is not too important a property for fatigue resistance.[4.64] More homogeneous brittle materials, such as glass, exhibit full notch sensitivity under static loads, as well as variable, in which case K_t is applied to both the mean and alternating components of the stress. Recall the long-known technique of scratching glass to cause it to break along a certain line.

Explanations of why K_f is sometimes very much less than K_t are: (1) highly local yielding redistributes the stress so that the theoretical peak stresses are not reached; (2) the stress gradient, which increases as the notch radius decreases, relative to the grain size is related to this phenomenon. As a generalization, the radii of notches, fillets, etc. should be as large as possible; although q approaches unity and $K_f \to K_t$, K_t decreases as the radius increases.

For ductile materials, the fatigue-strength-reduction factor (FSRF) is applied only to the variable component s_a, so that we think of the maximum stress as

$$s_{\max} = s_m + K_f s_a.$$

This $s_{\max}$ on the Goodman-type diagram, Fig. 4.6, is represented by EC (as a safe value); thus, pictorially, MD in magnitude represents $K_f s_a$.

4.11 EFFECT OF SURFACE CONDITION ON ENDURANCE STRENGTH. The factors for some unpolished surfaces are given by Karpov's curves,[4.4] Fig. AF 5, Appendix, which show that the effect on fatigue strength may be quite large. These curves include the notch-sensitivity effect, and they thus provide visual evidence that as tensile strength and hardness increase, the fatigue strength does not increase in proportion; for example, an as-forged forging of $s_u = 110$ ksi is little stronger than the same forging of 50-ksi steel when the loads vary. See also Fig. 4.10. These factors can be treated as fatigue-strength reduction factors, right-hand scale, and used as K_f's; or the percentage number on the left-hand scale can be multiplied by the endurance limit to get a corrected endurance strength (Fig. AF 5).

Decarburized surfaces result in lower endurance strengths also, the amount depending on how much decarburization has occurred (see also §§ 4.23, 4.28).

4.12 ENDURANCE STRENGTH AS AFFECTED BY SIZE. The evidence is conflicting, but most of it shows that the endurance strength per unit area tends to decrease as size increases.[4.7,4.8,4.9,4.23,4.33] For example, wire has a higher strength than say a standard 0.3-in. specimen. The reasons are not fully understood. There is the statistical aspect; with more volume of stressed material and more surface, the probability increases of a "weak" point developing. The larger the specimen, perhaps the less uniform are the properties. The material may have different properties because of different rates of cooling, for example. There may have been unfavorable residual stresses. Whatever the reasons, we find the following in the literature. An $11\frac{1}{2}$-in. axle of SAE 1045, with $s'_n = 55$ ksi, had an endurance strength of 11 ksi. An alloy-steel crankshaft with $s_u = 120$ ksi, showed endurance strengths in *torsion* as follows:[4.3] 0.55 in., 36 ksi; 1.18 in., 31 ksi; 1.77 in., 26.5 ksi. A 7-in. rotating beam, with $s'_n = 33$ ksi, had an endurance strength 17.5 ksi. On the other hand, a 1-in. rotating beam, with $s'_n = 32$ for a 0.273-in. specimen, had $s_n = 33.8$. A 9.75-in. shaft ($s_u = 62$ ksi) with a fillet radius $r = 0.305$ in. had a torsional s_{ns} of 13 ksi at 8×10^6 cycles; for $r = 1.5$ in., $s_{ns} = 19$ ksi at 11.5×10^6 cycles. Limited evidence suggests that for *axial loading*, the size effect is *not* significant up to 1.3-in. diameter.[4.33]

The better part of valor, when the significant dimension is greater than about $\frac{1}{2}$ in. for cyclic bending or torsional stresses, would seem to be to reduce the fatigue limit by about 15% for sizes up to 2 in. (arbitrary point)—and, conservatively, for axial loading. Thus[4.2]

(m) Between $\frac{1}{2}$-2-in. size, let $s_n = 0.85\,s'_n$ and $s_{ns} = 0.85 s'_{ns}$,

[BENDING AND TORSION]

unless test data are at hand. Since large-scale testing is costly, appropriate

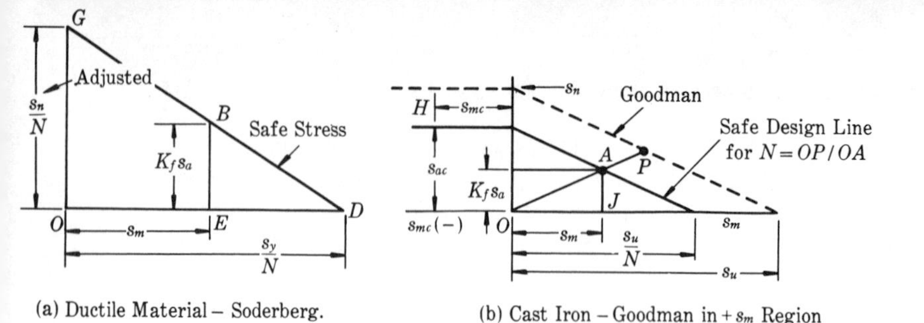

(a) Ductile Material – Soderberg.

(b) Cast Iron – Goodman in $+s_m$ Region

FIGURE 4.13 Working Diagrams for Ductile Materials and Cast Iron. In (a) and (b), s_n should be adjusted for the various factors that affect fatigue strength; surface factor may be included in deciding upon s_n or counted with K_f. The plan in (b) for cast iron [2.1] suggests that when the mean stress is negative, the alternating component does not need to be diminished with decreasing s_m. To some extent, this plan may be permissible for steel, but confirming data are not plentiful.

data are scarce for larger sizes, and one needs to use the best engineering judgment (which will be based on the best engineering *knowledge*).

A size effect, not so pronounced, has been observed also in *static* tensile tests, Table AT 8; hence, appropriate adjustment of the yield stress s_y is advisable unless one is confident that the design factor N cares for this eventuality.

4.13 VARIABLE-STRESS EQUATION WITH K_f. Since the presence of a stress concentration leaves the static strength of ductile materials (based on the minimum section) unchanged, we do not apply K_f to the mean stress. Thus, equations (4.2) and (e) become

$$\frac{1}{N} = \frac{s_m}{s_y} + \frac{K_f s_a}{s_n}, \qquad (4.4) \quad \text{[Normal]}$$

$$\frac{1}{N} = \frac{s_{ms}}{s_{ys}} + \frac{K_{fs} s_{as}}{s_{ns}} \qquad (4.5) \quad \text{[Shear]}$$

where K_f and K_{fs} are determined from equation (4.3); s_u and s_{us} ($s_{us} = 0.75–0.8 s_u$ for steels) may be substituted for s_y and s_{ys} by those who prefer the modified Goodman line as a criterion. Equation (4.4) is represented by the straight line GD, Fig. 4.13(a). In application, the s_n used is the one that includes any weakening or strengthening surface effects, of which some have been mentioned and more are discussed later. A design stress pattern for cast iron is shown in Fig. 4.13(b).

4.14 EXAMPLE—PISTON ROD. A piston rod is to be subjected to a maximum reversed load of 31,416 lb. It is to be made of AISI 8742, OQT 1200°F, machined to 63–125 rms (§ 3.14). What should be its diameter for a design factor of safety of $N = 1.75$ if there are no stress-raising discontinuities and no "column action"?

Solution. From Table AT 9, for AISI 8742, QT 1200°F, we find $s_u = 130$ ksi. Use the approximation $s_n' \approx 0.5s_u = (0.5)(130) = 65$ ksi. For $s_u = 130$ and from Fig. AF 5, we get the factor 0.81 for a machined surface; since the load is axial, we apply the 0.8 factor (§ 4.7). The corrected endurance strength is then $(65)(0.81)(0.8) = 42.1 \text{ ksi} = s_n$ in equation (4.1). Since $s_m = 0$ for a reversed load, equation (4.1) reduces to $1/N = s_a/s_n$ or $s_a = s_n/N$. For the data given, we have a design stress of 42.1/1.75, and

$$s_a = \frac{42.1}{1.75} = \frac{F_a}{A} = \frac{31.416}{\pi D^2/4} \text{ ksi},$$

or $D = 1.29$ in.; use $1\frac{1}{4}$ in. (p. 32). Some available tests for a 1.25 dimension suggest that there is little size effect in the case of axial loads (§ 4.12). (NOTE. Since it is likely that the method of attaching the ends of the piston rod to the piston and crosshead will introduce an additional stress concentration, the answer obtained is too low.)

4.15 EXAMPLE—VARYING TORQUE. A hot-rolled rod of AISI C1035 is to be subjected to a cyclic torque varying from +3000 in-lb. to −1000 in-lb. The crank is attached to the rod via a sled-runner keyway; let $N = 1.8$. What should be the diameter of the rod?

Solution. From Table AT 7 for C 1035, as rolled, we find $s_u = 85$ ksi and $s_y = 55$ ksi. Direct data on shearing strengths are not available; so they are estimated. In accordance with a note in Table AT 7,

$$s_{ys} \approx 0.6s_y = (0.6)(55) = 33 \text{ ksi}.$$

The endurance limit is $s_n' \approx s_u/2 = 85/2 = 42.5$ ksi; by § 4.7, for a polished specimen, $s_{ns} \approx 0.6s_n' = (0.6)(42.5) = 25.5$ ksi. This value is now reduced by the surface factor (60% from Fig. AF 5 for $s_u = 85$) and the size effect (85%), provided that the required size is over $\frac{1}{2}$ in. Thus s_{ns} in equation (4.5) is

$$s_{ns} = (25.5)(0.6)(0.85) = 13 \text{ ksi}.$$

From Table AT 13, we get $K_{fs} = 1.3$. To use equation (4.5), we need the nominal mean and alternating stresses, which are found from the mean T_m and alternating T_a components of the torque. Paraphrasing equations (a), § 4.5, we have

$$T_m = \frac{T_{max} + T_{min}}{2} = \frac{3000 + (-1000)}{2} = 1000 \text{ in-lb.}$$

$$T_a = \frac{T_{max} - T_{min}}{2} = \frac{3000 - (-1000)}{2} = 2000 \text{ in-lb.}$$

The corresponding nominal stresses are then

$$s_m = \frac{Tc}{J} = \frac{(1000)(16)}{\pi D^3} \text{ psi} = \frac{16}{\pi D^3} \text{ ksi}, \quad \text{and} \quad s_{as} = \frac{(2000)(16)}{\pi D^3} \text{ psi} = \frac{32}{\pi D^3} \text{ ksi},$$

where $J/c = Z' = \pi D^3/16$. Substitute the various values into equation (4.5);

$$\frac{1}{N} = \frac{s_{ms}}{s_{ys}} + \frac{K_{fs}s_{as}}{s_{ns}}$$

$$\frac{1}{1.8} = \frac{16}{(\pi D^3)(33)} + \frac{(1.3)(32)}{(\pi D^3)(13)}$$

from which $D = 1.28$ in.; use $1\frac{1}{4}$ in.

4.16 ENDURANCE STRENGTH FOR A FINITE LIFE. The endurance limit is that stress below which the specimen will have indefinite life. We have observed, Fig. 4.3, that an endurance *limit* has not been found for certain non-ferrous metals. For such materials, the design may be made on the basis of some estimated cycles of stress to occur during the desired life of the part. Similarly, there are many cases where the number of repetitions of the maximum stress in the lifetime of steel parts is relatively small. If, for example, it were known that the number of applications of the maximum stress would be 2×10^5 and that other repeated stresses would be lower than the endurance limit, we could use the endurance strength corresponding to 200,000 cycles obtained from fatigue-strength curves such as those in Fig. 4.3, or otherwise from estimation. The use of the endurance *limit* in designing for finite life, while on the safe side, may be inadvisable or unnecessarily expensive—certainly if weight is important or if very many parts are to be made.

In Fig. 4.3 we observe a certain parallelism of the curves for the steels without stress raisers of any kind. (CAUTION: There are steels for which the slope of the sn_c curve is quite different from any average value that might be chosen.) Lipson and Noll[0.2,4.2] found that in a large percentage of cases, the inclined part of the line (with ground, machined, hot rolled, forged, etc., surfaces) was fairly well approximated by straight lines on log-log paper going from an ordinate of $0.9s_u$ at 10^3 cycles to the endurance

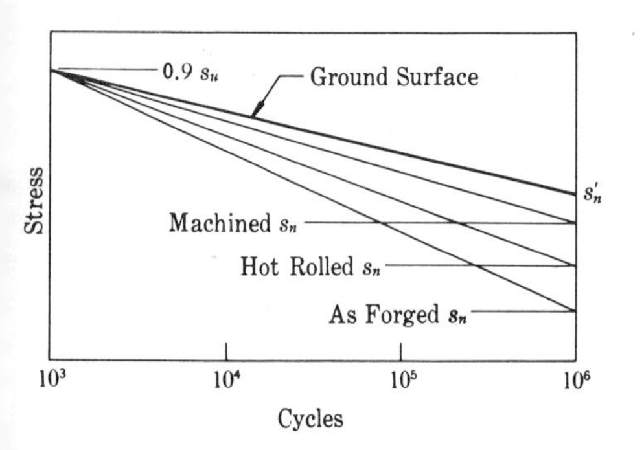

FIGURE 4.14 Endurance Strength of Steel for Finite Life. After Ref. (4.2).

strength at 10^6 cycles, Fig. 4.14. The equation of such a line to an endurance strength s_{n10^6}, that is, at 10^6 cycles, is

(n) $$\log s_n = \log \frac{0.81 s_u^2}{s'_n} - \frac{1}{3} \log \frac{0.9 s_u}{s'_n} \log n_c,$$

which, if $s_{n10^6} = s'_n = s_u/2$, reduces to

(o) $$s_n \approx s'_n (10^6/n_c)^{0.085},$$

[CERTAIN STEELS, NO STRESS RAISER]

where $n_c < 10^6$ is the number of cycles for which s_n is desired, log is to the base 10, and the other symbols have the usual meanings. The significance of the 10^6 is that the knee of the sn_c curve is so often in the vicinity of 10^6 cycles, Fig. 4.3. Factors for surface condition and other discontinuities should be applied to the result from **(n)** if s_{n10^6} is taken as s'_n for polished specimens. It would be nice if equations **(n)** and **(o)** were always reliable, but unfortunately the knee does not always occur at 10^6 cycles and the actual curve does not always intercept the ordinate at $0.9s_u$. Some evidence suggests that the knee moves to the right as specimen size increases; for example, to 10^7 cycles for a 7-in. size.[4.22] Another relation that fits some data for polished steel specimens is

(p) $$s_n = s'_n\left(\frac{10^6}{n_c}\right)^{0,09}$$

and this one, also based on the knee being at 10^6 cycles; it gives results similar to those from **(o)**.

As instances of finite rather than infinite life requirement, J. O. Almen[4.12] points out that the number of cycles at maximum loading is 100,000 for automobile differential gears and automobile suspension springs, and 500,000 for clutch springs.

The fatigue strength reduction factors are less for finite life than for indefinite life, perhaps because local yielding redistributes the stress and also work hardens, and thus strengthens, the local material. Observe the steeper slopes of the curves with stress raisers in Fig. 4.3. At the extreme of one gradual application of the load, the stress raiser has no significant effect on the ultimate stress; in between, its effect would be expected to be somewhere between 1 and K_f for indefinite life. As a matter of fact, the curves for a heat-treated 1050 steel, with and without a notch, intersect at about 10^4 cycles, at which point $K_f = 1$; other such pairs of curves intersect in the vicinity of 10^3 cycles. An experimental but satisfactory relation for *steel* is[4.2]

(q) $$K_{fl} = \frac{n^{(\log K_f)/3}}{10^{\log K_f}} = \frac{n^{(\log K_f)/3}}{K_f},$$

where K_{fl} is the strength reduction factor for a limited life of $n < 10^6$ cycles, and the other symbols are as defined above. See also Peterson.[4.21] Above 500,000 cycles, the point would be near or in the probable range of the endurance limit, so one might as well base the design on the endurance limit. A part designed for a finite life may turn out to have indefinite life because the variation of properties resulted in a certain part or parts operating below their individual endurance strengths; but there exists the economic advantage of the design having been made with a higher design stress. Interest in designing for limited life has been increasing and considerable help can now be found in the literature.[4.41,4.46,4.47,4.59]

4.17 EXAMPLE—LIMITED LIFE. A forged crank, Fig. 4.15, has a constant thickness of 2 in., a length of 25 in., and a depth of 6 in. at a section *B–B* where it was found necessary to place two $\frac{1}{4}$-in. holes, each $2\frac{1}{2}$ in. from the center line of the link. It is expected that there will not be more than 100,000 cycles of stress during the lifetime of the link, which is made of AISI C 1035. What is the safe reversed load F, Fig. 4.15, for a design factor of safety of 1.7?

Solution. As usual, first decide upon the stress situation. From Table AT 10 for 1035 in air, we find

$$s_n' = 40.6 \text{ ksi}, \quad s_n'/s_u = 0.46 \text{ or } s_u = 40.6/0.46 = 88.3 \text{ ksi}, \quad s_y = 58 \text{ ksi}.$$

We next estimate the endurance strength for finite life. From equations (**n**) and (**p**), we have

$$s_n = \text{antilog}\left[\log\frac{(0.81)\,(88.3)^2}{40.6} - \frac{1}{3}\log\frac{(0.9)\,(88.3)}{40.6}\log 10^5\right] = 50.8 \text{ ksi},$$

$$s_n = 40.6\left(\frac{10^6}{10^5}\right)^{0.09} = 50 \text{ ksi}.$$

Use 50, the more conservative value. Also, we elect to use a surface factor for as-forged, as well as the strength reduction factor K_f for the hole (§ 4.22),

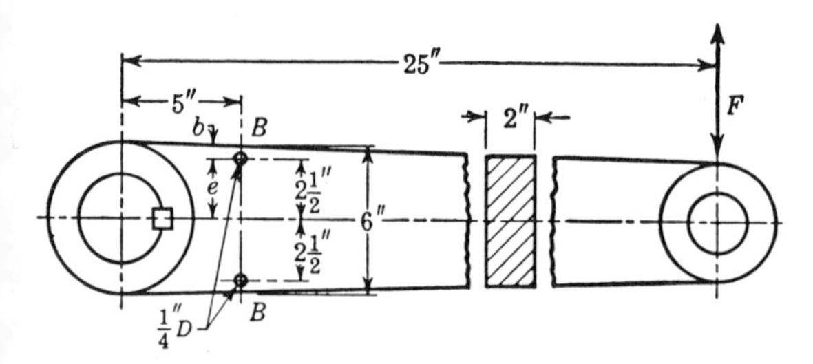

Figure 4.15.

assuming that one stress raiser reinforces the other. For $s_u = 88.3$ ksi, surface factor = 0.45 from Fig. AF 5; for size effect, use 0.85 (§ 4.12); thus, s_n in equation (4.4) is

$$s_n = (50)(0.45)(0.85) = 19.1 \text{ ksi}.$$

To get K_t, use Fig. AF 11; $d/b = 0.25/0.5 = 0.5$; $e/d = 2.5/0.25 = 10$; from which $K_t = 3.5$. Then from Fig. AF 7, $q = 0.92$ for $r = d/2 = \frac{1}{8}$ in. and curve A for coarse-grain, as-rolled steel. Because it is possible, use $q = 0.89$ and get,

$$K_f = 1 + q(K_t - 1) = 1 + 0.89(3.5 - 1) = 3.22.$$

Allow for a limited life of $n = 10^5$ cycles by equation (**q**);

$$K_{fl} = \frac{n^{(\log K_f)/3}}{10^{\log K_f}} = 10^{(2 \log K_f)/3} = 2.18.$$

For the nominal stress $s_o = Mc/I$, Fig. AF 7 shows that I for the net section is used. Since there are two holes, the moment of inertia is closely (notice that this is not the exact value)

$$I = \frac{(h - 2d)^3 t}{12} = \frac{(6 - 0.5)^3(2)}{12} = 28 \text{ in.}^4$$

and the distance to the top edge of the hole, Fig. 4.15, is $c = 2\frac{5}{8} = 2.625$ in.

The moment at section *B-B* is $M = (F)(20) = 20F$. Using $N = 1.7$, $s_n = 19.1$ ksi and $s_m = 0$ in (4.4), we have

$$\frac{1}{N} = \frac{K_f s_a}{s_n} = \frac{1}{1.7} = \frac{(2.18)(20F)(2.625)}{(19.1)(28)}$$

from which $F = 2.75$ kips or 2750 lb., a safe reversed load. The endurance strength of this link can be materially increased (say doubled) with no change of nominal size by certain changes of the surface, as by machining off all the de-carburized layer, or by cold working the edges of the holes, and least expensive, by shot peening the as-forged surface (§ 4.28).

4.18 EXAMPLE. Letting all the data of the previous example be the same except that F varies from $+3000$ to -1000 lb.; to find the corresponding factor of safety.

Solution. Paraphrasing equation (**a**) we find the mean and variable forces to be

$$F_m = \frac{3000 + (-1000)}{2} = 1000 \text{ lb.}, \qquad F_v = \frac{3000 - (-1000)}{2} = 2000 \text{ lb.}$$

then, with $s = 20Fc/I$,

$$s_m = \frac{(20)(1000)(2.625)}{28} = 1880 \text{ psi}, \quad K_{fl} s_a = \frac{(2.18)(20)(2000)(2.625)}{28} = 8200 \text{ psi},$$

where the values of c, I, and K_{fl} have been taken from the previous example. Using equation (4.4), we have ($s_y = 58$ ksi)

$$\frac{1}{N} = \frac{s_m}{s_y} + \frac{K_f s_a}{s_n} = \frac{1.88}{58} + \frac{8.2}{19.1} = 0.462$$

or $N = 2.16$, a larger factor of safety than in the previous example even though the maximum load is greater in this one.

4.19 EQUIVALENT STRESS. It is often convenient to have an *equivalent stress*, one that when divided into say s_y (or s_u or s_n') would indicate the degree of safety. Also the equivalent stress idea is useful when dealing with variable combined stresses. To obtain such a stress, one needs only to multiply equation (4.4) through by s_y and consider $s_y/N = s_e$ as the equivalent stress:

$$s_e = \frac{s_y}{N} = s_m + \left(\frac{s_y}{s_n}\right) K_f s_a. \tag{4.6}$$

The analogous equation for shear is

$$s_{es} = \frac{s_{ys}}{N} = s_{ms} + \left(\frac{s_{ys}}{s_{ns}}\right) K_{fs} s_{as}. \tag{r}$$

Since $s_{ys} \approx (0.5 - 0.6)s_y$ and $s_{ns} \approx (0.5 - 0.6)s_n$, then $s_{ys}/s_{ns} \approx s_y/s_n$, some values of which are given in Table AT 10. Analogously, multiplying equation (4.4) through by s_n gives an equivalent stress based on s_n; $s_e = s_n/N = s_m s_n/s_y + K_f s_a$, different from that in (4.6).

4.20 DESIGN FACTOR FOR VARIABLE LOADING. Design stresses based on experience with over-all safety factors, as in Chapter 1, are invaluable criteria, but they are not always the most economical. When such factors are used with the yield or ultimate strengths and live loads, they are necessarily large in order to cover large ignorance. However, when we delve into the various factors that affect strength and life, as in this chapter, and make proper allowances, a much smaller design factor is permissible, as low as about 1.4, although if it is this low based on the *median* fatigue strength and if all other operating conditions have been well evaluated, the actual factor of safety of the part may be close to unity because of the natural spread of properties (§ 4.4). If the uncertainties for a particular design are so great that (or if experience suggests that) $N > 3.5$ to 4 is advisable, a more intensive search for more knowledge is in order, especially for important designs and large production. Some factor of safety greater than unity must be provided in order not to have a failure during the required life, because it is necessary to allow not only for variations in properties, but also for variations in manufacturing processes, deflections in housings, warpage from heat treatment, etc., for the inadequacies of theoretical stress equations, and for unpredictable departures from the ideal model. There are additional factors for the designer to consider, but first a résumé of the discussion to this point.

4.21 RÉSUMÉ OF DESIGN FOR VARIABLE STRESSES. Observe that there are two kinds of strength reduction factors, those that are used to reduce the endurance limit to a realistic endurance strength and those (K_f) that are multiplied into a nominal s_a to get an estimate of an actual maximum stress. Many of the surface factors are in a sense stress concentration factors, but there are a number of surface treatments (decarburization, peening) that lower or raise the endurance strength without a stress-raising discontinuity being involved. To treat surface factors in the same way at all times, we shall use them as factors to reduce (or raise) endurance stress. This is in accordance with the way the examples are worked, and the adjustments include those in Fig. AF 5 and those for size and axial loading. For fillets, holes, etc., the adjustment is by $K_f > 1$, as explained.

(a) Brittle Materials. We have mentioned evidence (§ 4.10) that indicates that cast iron has a low notch sensitivity index q. However, for brittle materials in general, use the full theoretical K_t on both mean and variable stresses.

(b) Ductile Materials. For dead loads, ignore stress concentrations.

For live loads, proceed as detailed in this chapter. The following check list will be a reminder of the details.

(1) What is the basic endurance strength for the kind of stress involved—reversed bending, torsional, or axial loading?

(2) Has the loss (or gain—see below) of strength because of surface condition been accounted for? It is always on the safe side to include a surface factor, Fig. AF 5, as well as a fatigue strength reduction factor K_f, but see § 4.22.

(3) Has the possible size effect been overlooked?

(4) Are there discontinuities requiring an estimation of stress concentration via a K_t and the sensitivity index q?

(5) Will a finite life satisfy requirements? This affects the value of s_n and K_{fl}.

(6) Has the natural spread of properties been accounted for?

(7) Have the kind and condition of material been considered? A casting may have a concealed blowhole near a stressed surface. Properties are changed by heat treatment. Directional properties may be significant, as from rolling, extruding, forging.

(8) Are the surroundings corrosive? See the correction factor for salt water surroundings in Fig. AF 5.

(9) Are the residual stresses helpful or harmful? This topic and several other pertinent ones are discussed below.

In your problem work, check off stress, surface, size, SCF, and life, as a minimum.

4.22 SUPERPOSED STRESS RAISERS. It often happens that there is a hole or other discontinuity at a machined or other kind of rough surface, and the question arises as to whether both strength reduction factors should be used. One set of torsion tests[4.34] indicates the following for AISI 4140, 350 BHN: (a) when the stress-raising effects are about the same in magnitude, both should be included at full value; (b) when one effect is significantly greater than the other, the contribution of the lesser one is small, perhaps negligible. Another group of bending tests[4.35] on 4340, 410 BHN, also indicated that the net effect of two stress raisers ($K_f = 2.38$) is about 20% less than the product of the individual values for a rough machined surface ($K_f = 1.52$) and a hole ($K_f = 1.96$). Generalizations are not justified on the basis of the evidence, except to say that it is always on the safe side to count both effects full value, but that if it is important and if one factor is larger than the other, some value less than the product can be justified.*

4.23 RESIDUAL STRESSES. In a part at uniform temperature and not acted upon by an external load, any internal stresses that exist are called ***residual stresses***. Since static equilibrium obviously exists, the forces from the residual tensile stresses are balanced by those from the residual

* Another approach is to let the overall $K_t = K_{t1}K_{t2}$; then $K_f = 1 + q(K_t - 1)$.[4.65]

compressive stresses; also the internal shear forces are in equilibrium. These stresses exist because of what has happened in the past history of the part (as welding, rolling, heat treating, shot peening, work hardening, etc.), and they may be beneficial or harmful. If a tensile member is loaded somewhat beyond the yield point (§ 1.8), what actually happens is that the surface material reaches the yield strength s_y before the core does; when the load is removed, the outside layers of material that received a permanent deformation will have residual compressive stresses produced by the residual tensile stresses in the core, which stresses may or may not be helpful, depending on how the part will be stressed when it is in its allotted place. Whatever the source of the residual stresses, they can be eliminated in whole or in part by a suitable tempering or annealing, called ***stress relieving***, which is what should be done if the residual stresses are expected to be harmful. Residual stresses of the order of 32 ksi for as-cast cast-iron wheels have been reported,[2.1] but such high values in this case are not typical of good design and good engineering practice.

More and more, residual stresses are put into a part intentionally for the good they will do. ***Autofrettage*** is a process of *prestressing* or *overstressing* a hollow cylindrical member beyond the elastic range by hydraulic pressure. A long-standing use of autofrettage is as applied to gun barrels (and now to heavy cylinders for many uses). The tangential stress at the inside of a cylinder due to an internal pressure is tension, and this is the

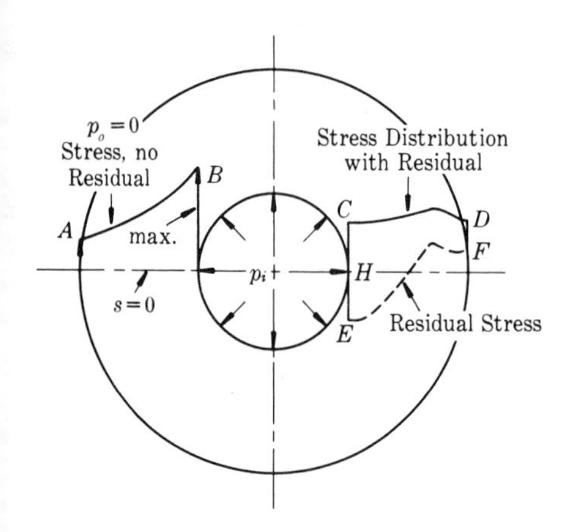

FIGURE 4.16 Effect of Residual Stress.

point of maximum normal stress, as at B, Fig. 4.16. (See § 8.26.) If the cylinder is subjected to a pressure that induces stresses into the plastic range at some distance through the wall thickness from the inside, the inside material takes on a permanent deformation. After the pressure is released, the outer material that did not get much if any permanent deformation, contracts elastically exerting pressure on the material with permanent deformation, so that the internal tangential stresses become compressive, and the residual stress distribution is as shown dotted at *EF*, Fig. 4.16. The deformation may be such as to enlarge the ID permanently about 6%, and the OD about 1%.[4.10] Now when the cylinder is subjected to a pressure, the tangential compressive stress at *E* first begins to decrease, becomes zero, and finally becomes tensile, with some distribution *CD*, Fig. 4.16, where the peak stress is lower than before. In application to guns, the idea was to make a higher internal pressure possible without producing a permanent deformation, that is, without the maximum stress exceeding the approximate elastic limit. The cold working by autofrettage also raises the elastic limit of the material. For vessels that are subjected to repeated pressures, the residual compressive stress on the inside improves fatigue strength, because it is the excessive tensile stresses that produce the fatigue damage.

If a member has a residual compressive stress on its surface, which is subjected to repeated tensile stresses from an external load, its endurance strength may be much higher than when the surface is residual-stress-free. The only trouble is that there are no commercial (economic) means of determining directly what these stresses are and what their quantitative effect on endurance strength will be. However, if the variable working load is always *applied in the same sense*, parts with a stress raiser can be

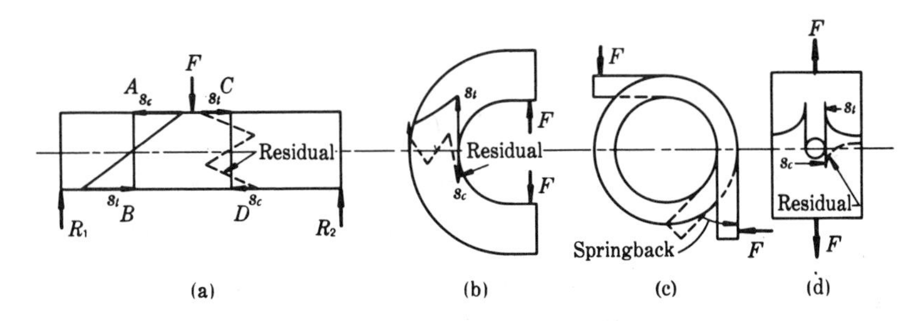

FIGURE 4.17 Overstressing, Presetting (Diagrammatic). Part (a) could represent diagrammatically a leaf in an automotive spring; such springs are commonly prestressed. Part (b), a C-frame, and part (c), a torsion spring, react as described for (a); but in these cases, there is a stress-concentration effect because of the curvature. The net stress for the tensile member in (d) is the algebraic sum of the dotted residual and the stress distribution shown solid; this reduces the magnitude of the peak concentration stress at the edge of the hole.

overstressed in that *same sense*, generally with improvement in endurance strength as suggested by Fig. 4.17. For example, in Fig. 4.17(a), if *F* always acts downward, the Mc/I stress distribution is as at section *AB*, where the line *AB* represents zero stress. If a load large enough to produce plastic yielding is applied and then removed, some pattern of residual stresses as indicated at *CD* remains. The net stress *at a particular point in a particular section* with the working load on is: the Mc/I stress algebraically plus the

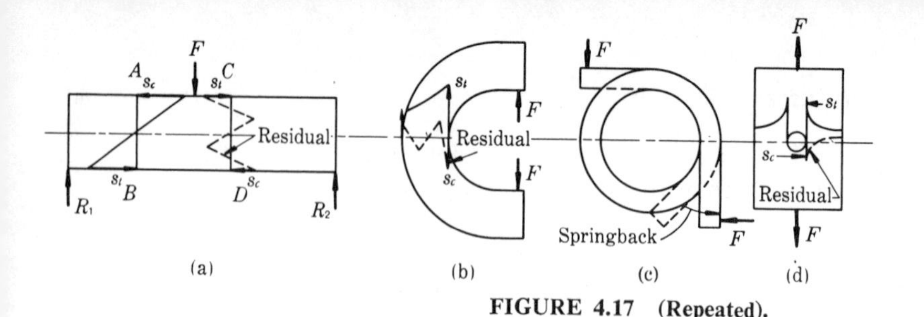

FIGURE 4.17 (Repeated).

residual stress (s_c negative). This is seen to reduce the resultant tensile stress on the bottom fiber and therefore it increases endurance strength. This idea of overstressing may be used as an extra precautionary step where the operation may be close to the endurance strength and when experience (testing or service) indicates an appropriate strength improvement is thereby obtained. If F is a reversing load, Fig. 4.17, *prestressing as shown results in a weaker part*, because the residual s_t and the $s_t = Mc/I$ will add numerically for one direction of F. It is reasonable and logical to presume that if the residual stress at a point is such that the residual stress plus the stress by the external loading exceeds the yield strength of the material, yielding results in some relaxing of the residual stress. Hence, if a part has been overstressed for the purpose of inducing favorable residual stresses and then if some loading (perhaps accidentally) induces stresses that remove part or all of the favorable residuals, the original improvement in strength is thereby lost.

A point that is frequently overlooked is that stresses induced in a part by the assembly operation have effects similar to residual stresses. If any such induced stresses are tensile, fatigue failure is more likely.

Quenching heat treatments leave residual stresses. Quenching in water is a more drastic cooling than quenching in oil; the part warps more and the residual stresses are much larger. The surface of a solid piece in contact with the quenching medium cools first, and as the inside cools and shrinks, residual compressive stresses (tangential and longitudinal) are left in the outer layers. (A hollow tube quenched from the outside generally has residual compressive stresses on the outside and tensile on the inside.) Medium carbon (or better) steels have a much higher residual stress if water quenched than if oil quenched; for example, 0.49% C steel, 2-in. bar, quenched from 1560°F; in water, tangential residual stress is about 70 ksi; in oil, about 35 ksi.[4.10] The size of the part naturally affects the numerical results. The tempering operation reduces these stresses by an amount that depends on the tempering temperature, the length of time held at this temperature, and the rate of final cooling. The foregoing piece cooled in air from a tempering temperature of 1200°F has a tangential residual stress of $s_c = 6$ ksi.

Not only does quenching and tempering tend to leave a favorable residual s_c on the surface, but it also improves the mechanical properties generally. Nevertheless, there are pitfalls in heat treatment, especially for the high-strength steels. If it is done in an oxidizing atmosphere, the surface will be decarburized—one of the weakening factors also in hot-rolled and as-forged parts. Decarburization not only leaves a weaker material but usually a residual tensile stress.[0.2] The effect of decarburization alone, smooth specimens, is illustrated by the values for SAE 2340:[0.2] hardened to $R_C = 28$, it lost 47% of its endurance strength (83 to 44 ksi); hardened to $R_C = 48$, it lost 71% (122 to 35 ksi). Coupled with observable surface defects, the damage is worse. The depth of decarburization seems to be unimportant, as it takes only a weak surface (from any cause) or one with a residual tension to serve as a source of fatigue cracks and lower endurance strength. The loss of fatigue strength for limited life, say 10^5 cycles, is not so great; the loss of strength for 1 cycle (gradually applied) is unnoticeable. Steps taken to combat decarburization include heat treating in a non-oxidizing atmosphere; and for forgings, etc., where such control is not practicable, recarburization, surface rolling, peening, and hardening (§§ 4.28, 4.29); and removing the decarburized layer by machining.

4.24 PLATE WITH ELLIPTICAL HOLE. This case is important because it throws some light on the stress at cracks, which are roughly long, slender ellipses. If the load is perpendicular to the major axis of the ellipse, Fig. 4.18, the worst stress condition exists, and the maximum stress occurs at the ends of the major axis where the theoretical K_t is[4.21]

$$K_t = 1 + \frac{2a}{b}$$

a is half the axis that is at right angles to the load and b is half the other axis. Observe that when the ratio of a/b is very large, tremendous stresses are indicated by K_t; but see § 4.10. This observation accords with the rapid

FIGURE 4.18

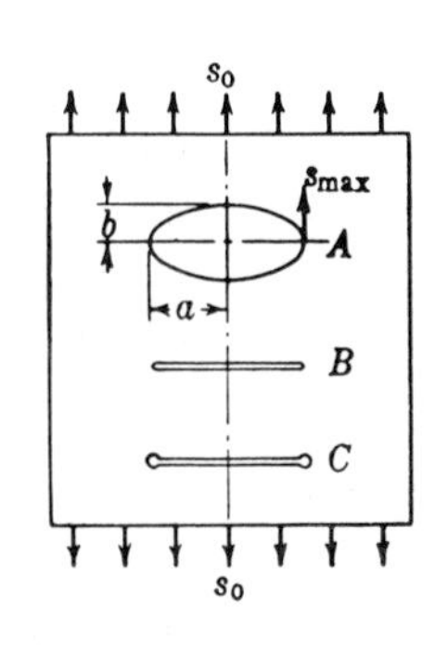

spread of cracks which are at right angles, or nearly so, to the load. If a crack such as that at B, Fig. 4.18, were 100 times its width ($a/b = 100$), $K_t = 201$, suggesting that a relatively small repeated load might cause failure when a flaw equivalent to a crack exists in the material. The spreading of a crack can often be stopped by drilling small holes at each end, as at C, Fig. 4.18. This step materially reduces K_t.

On the other hand, if the ellipse or crack is parallel to the direction of the stress, so that $a < b$, the effect of stress concentration may be negligible.

4.25 BEAM WITH HOLES. The significance of the concentration of stress resulting from a hole in a beam depends upon the location of the hole. If the hole is on the neutral axis and not too large, Fig. 4.19(a), no reduction whatsoever may occur in the fatigue strength of the beam. On the other hand, if the hole is near the external fibers, where the stress is naturally large, a stress much larger than the Mc/I stress at the external fiber may occur at the edge of the hole, Figs. 4.15 and 4.19. Particular care should be taken in locating holes in curved sections of a beam that is to be subjected to repeated loading, for example, at a point where an axle is bent. The neutral axis of a curved beam does not coincide with the geometric axis but is nearer the inside (concave) surface (see § 8.25). Holes located near the geometric axis of a *curved* beam may result in progressive failure.

4.26 CORROSION. Even the normal atmosphere reduces the endurance strength of some materials (as compared to strengths in a vacuum). Whenever the surface operates in a corrosive environment, a dramatic loss of endurance strength follows (Table AT 10); tensile stress promotes the corrosive effect. Some materials are much more resistant to corrosion than others, and resistant coatings may provide considerable improvement. In noncorrosive environments, coatings often lower the endurance strength of a part, but they can help more than harm in other circumstances. Zinc on steel takes the galvanic action, and thus small corroded points where corrosion-fatigue cracks can start do not appear. Other advantageous coating materials, especially for steel, include galvanized, cadmium plated, and enamel.

Corrosion-fatigue cracks generally progress at right angles to the maximum tensile stress, and the combination of corrosion and fatigue loading is more detrimental than either alone. Some copper alloys are subject to corrosion fatigue, others are more resistant. Stainless steels deteriorate in this respect in chlorides, particularly in boiling magnesium chloride. Chromium is more effective than nickel in providing resistance to corrosion fatigue, but all types of stainless are likely to become pitted in sea water. Compressive residual surface stresses inhibit the start of corrosion fatigue cracks.[4.1] The endurance strength reduction factor K_f as predicted by equation (4.3) is not so large in the presence of corrosive action because of the overwhelming effect of corrosion. All in all, materials for parts being designed for corrosive surroundings must be chosen with care. Almen[4.64]

FIGURE 4.19 Stresses at Holes in Beams.

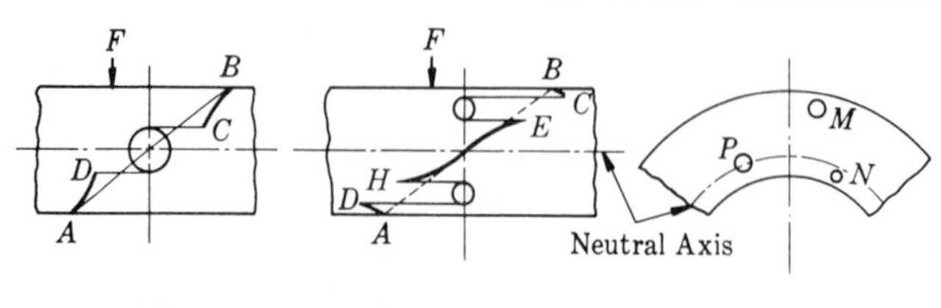

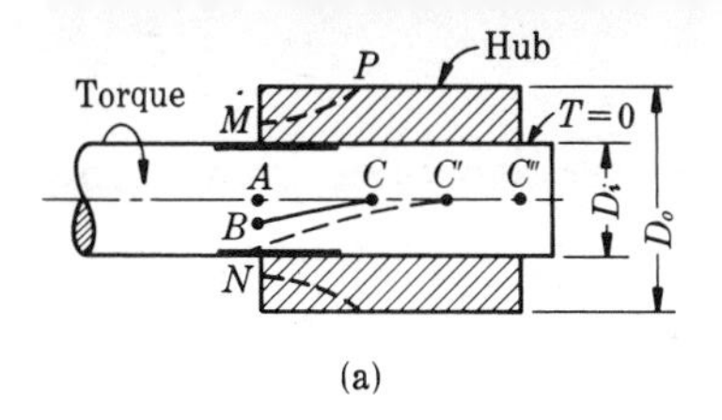

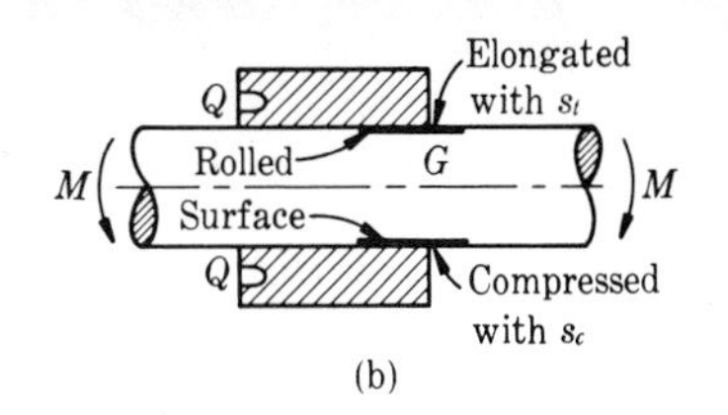

FIGURE 4.20 Fretting Action in Force Fit. In (a), a varying torque causes the amount of twist to vary, and an element AC (unstressed) may take a position BC (much exaggerated) under load. The distance A to C for a given fit, through which slipping occurs, depends on the amount of torque. If the torque is large enough and if the shaft does not fail, the shaft will turn in the hub, point C moving through C' to C'' and beyond. Maximum slipping in any case occurs at one end of the fit, MN here. If the hub is made less rigid at this end by, say, cutting away as indicated by the dotted lines at M and N as MP, then the hub may move more with the twisting shaft, thereby receive the torque T with less relative movement and less fretting. If the member has a varying bending moment on it, or if it is a rotating shaft with a constant moment, as in (b), then a point just inside the hub will have a varying stress which is the same as saying a varying deformation, and the change of deformation results in a slight relative motion and possible fretting. The heavy lines at the entrance to the fits represent a surface area that should be surface rolled for improved strength (§ 4.28). If the shaft is subjected to a bending moment or a varying torque at both ends of the fit, surface roll both ends. Grooving the hub as at Q in (b) also improves the fatigue strength.[4.31]

cites an interesting case; a brass cup formed in a forging die of course contained complex residual stresses. Placed in an ammonia atmosphere, it broke into several parts without external load within 2.5 hr. because of tensile residuals; a similar part peened all over was intact after 100 hr. in gaseous NH_3.

4.27 FRETTING CORROSION. When two touching surfaces have a high contact pressure and when these surfaces have minute relative motion (as in some force fits (§ 3.8), bolted and riveted connections, leaf springs, bearing races, splined connections, etc.), a phenomenon called *fretting corrosion* or just ***fretting*** often occurs. Even with the smoothest surfaces, a microscopic chafing causes small particles of metal to tear loose and oxidize. (Fretting occurs also in the presence of an inert gas—Fenner.[4.28]) In the case of steel, the oxidized particles form a reddish brown powder (if dry), which, once the process has started, acts abrasively and hastens the deterioration. Typical events at an interference fit are suggested by Fig 4.20, but vibration of assemblies with joined parts is a common source of fretting. The damage is greatest with dry surfaces, but no known lubricant stops the action entirely. The damage increases with the load, the slip, and with the total number of oscillations.[4.28] The magnitude of the effective fatigue-strength reduction factor cannot be accurately predicted, but a study of the literature will give one some ideas; for example: on a 7-in. as-forged shaft, $K_f = 3$; the fatigue strength of an SAE 1025 steel was reduced from 40 ksi to 20 ksi by fretting;[0.2] cold-worked aluminum alloy with an endurance strength of 20 ksi at 5×10^8 cycles, had $s_n \approx 7$ ksi with fretting.[4.31] With fretting corrosion, even for steel, there may be no distinct endurance *limit*. The difference between fatigue strengths with and without fretting is negligible when the life is low, say less than 10^4 cycles.

It is not certain whether or not fretting can be entirely suppressed. Lower pressures and reduction of relative motion are obvious things to consider, but these factors often cannot be changed significantly. The most effective help is obtained from a residual compressive stress on the surface most prone to fretting failure. See § 4.28.

Surface rolling is a means frequently used to prolong fatigue life. Steel axles of SAE 1045 had a fatigue strength of about 14 ksi with fretting corrosion (40×10^6 cycles), and about 33 ksi after surface rolling.[0.2] Unrolled magnesium had $s_n = 7$ ksi at 5×10^6 cycles versus 21 ksi when rolled (24 ksi when regular—"regular" not defined).[0.2] Certain heat treatments leave favorable residual stresses. Horger [4.28] shows that when the material is slowly cooled (in furnace) from a temperature of 1160°F, most of the helpful residual stresses from quenching having been relieved; but if water quenched from 1160°F, the residual compressive stresses are large enough to increase strength markedly. Citing numbers, we have

(1) For 9.5-in. OD, 0.51% C steel shafts; normalized and tempered at 1160°F and furnace cooled; 85×10^6 cycles; $s_u = 91.2$ ksi; an endurance strength of 11 ksi (versus $s_n' \approx s_u/2 = 45.6$).

(2) For same size, same material and heat treatment, except that a final treatment is a water quench from 1160°F, fixing favorable residuals; endurance strength $>$ 19 ksi, more than a 70% increase.

(3) For 9.5-in. OD, 3-in. ID, 0.51% C steel shafts; quenched from 1550°F, tempered at 1000°F; 85×10^6 cycles; $s_u = 125$ ksi; an endurance strength of 12.5 ksi (versus $s_n' \approx s_u/2 = 62.5$); residual stresses all small.

(4) For same size, same material and heat treatment, except that a final treatment is a water quench from 1000°F, fixing favorable residuals (63 ksi on outside); endurance strength of 18 ksi, a 44% increase.

(5) For the shafts in (3), except that the tempering temperature was 750°F (residual $s_c = 50$ ksi), $s_n = 22$ ksi or better, showing that a tempering temperature of 750°F does not remove the favorable residual stresses.

Other means of leaving residual compressive stresses are mentioned below. Repeated stressing in the vicinity of the endurance strength may either change the pattern of the residual stresses or, if the specimen was originally stress free, it may induce residual stresses. The suggested explanation is that where the high stresses occur, there will be plastic flow (local yielding). Fuchs[4.1] states that the yield strength for repeated loads is below the static value. See Rosenthal,[4.1] Horger,[0.2] and Sigwart.[4.28] With tests on three other alloy steels (various combinations of vanadium, molybdenum, nickel, and chromium), Horger's results[4.28] suggest that such steels do not have greater endurance strength or longer life if fretting occurs. For example, a Ni–Cr–Mo alloy shaft as in (3) above, with $s_u = 123$ ksi, had $s_n = 9.5$ ksi with fretting.

There has been some success with coatings on the surfaces subjected to fretting, notably molybdenum disulfide MoS_2. For dry steel on dry steel, fretting appeared in less than 100 cycles; with the surfaces coated with a mixture of MoS_2 and grease, fretting did not appear until more than 1.5×10^6 cycles; and if a mixture of MoS_2 and corn syrup is rubbed on and baked until dry, it was over 9.8×10^6 cycles before fretting.[4.31] Anodizing aluminum surfaces inhibits fretting.

4.28 SHOT PEENING AND SURFACE ROLLING.

Both of these processes stress the surface, on quite small areas at a time, beyond the tensile yield strength, producing a local permanent deformation. On the spring-back, the underneath adjacent fibers that had no plastic flow tend to resume their original dimensions and in doing so, produce residual compressive stresses in the plastically deformed surface. Both processes are cold-working processes and in general increase the local mechanical properties. Thus, any resulting increase in endurance strength can be attributed partly to higher s_u, but the principal benefit comes from the residual compressive stress.

In shot peening, a rain of metallic shot (say chilled cast-iron pellets) impinges at high speed, on the surface, perhaps on selected areas as on a fillet. The plastically deformed portion extends inward from several thousandths of an inch to a few hundredths, and the amount of the cold working, as would be expected, depends principally on the plastic work done by the pellets, which in turn is dependent on the size and speed of the pellets ($mv^2/2$) and the total number of impacts. This is not to say that improvement continues with continued energy input, because there is a counteracting effect in the damage that bombardment does to the surface. Besides leaving the indentations, which are stress raisers to a degree (surface roughness of the order of 140 μin. (65–200 μin.), arithmetic average—see Fig. 3.9), the impacts may cause cracks in the surface, one or more of which becomes the source of early fatigue failure. The full improvement to be obtained in fatigue strength is soon gained (say, in 2 min.), and "deep" peening is not needed except for a reason, as in overcoming the effects of decarburized surfaces. For each material and its state, there is some optimum combination (with the present state of knowledge, we do not always know what the optimum is) of shot size and speed, and duration of peening, so that the engineer cannot simply specify "peened surfaces" and be confident of the consequences, unless he knows that the production department has the know-how for the particular part. Moreover, after peening, there is no commercial inspection of the actual part less expensive than sample testing to assure that the peening is proper. To test the peening operation, one side of a thin test strip of steel, $R_C = 47$, is subjected to the same stream of shot as is the part, the strip takes a curvature, peened side convex, and

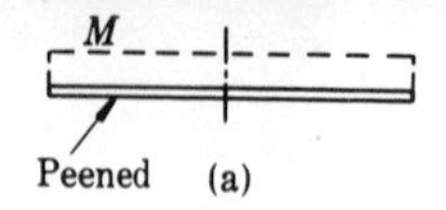

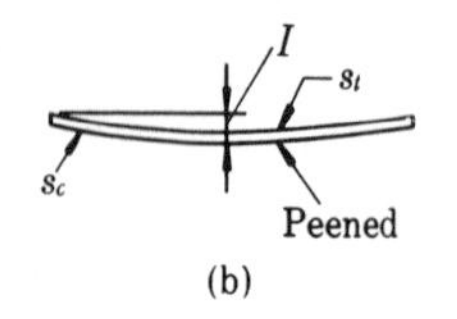

FIGURE 4.21 During peening, the strip is firmly attached to a heavy backing M, suggested by the dotted outline in (a), The residual s_c results in the strip bending as in (b) when released from M.

the amount of the arc height I, Fig. 4.21, is a measure of the peening, called the *Almen intensity*. Because of the annealing effect, the improvement due to any kind of cold working begins to disappear when steel is heated above 500°F; for aluminum, above 250°F.

Typical distributions of the residual stresses in a high-strength steel are shown in Fig. 4.22; SAE 5147, $R_C = 48$. Notice the large increase of compressive residual stress over that left by the heat treatment, and the fact that the *maximum* compressive stress occurs a little below the surface. This means that the dimples produced by peening may be machined or polished off to the extent of about 0.002 in. (if it is worth the cost),[4.58] without losing the strengthening effect of these high residuals. Evidently the penetration of the effect varies somehow with the size of shot, and this may be significant. If the penetration is not far enough, the fatigue failure may start at some point below the surface, as 0.006 in. below for curve B, Fig. 4.22. Since fatigue failure is prone to start on the surface, or close to it (§ 4.2), with a suitable surface stress, failure beginning much below the surface is unlikely unless there is some stress raiser (inclusion) present (but see nitriding in § 4.29). If the curves of Fig. 4.22 are carried deeper into the interior, balancing tensile stresses would be shown because the internal *forces* are in equilibrium. The effectiveness of shot peening improves as the hardness of the steel increases. However, shot peening a decarburized surface brings about a good, sometimes spectacular, increase in strength provided the peening effect extends through the decarburized zone (which is thereby limited to about 0.04-in. thickness) where the material is much weaker. Peening may be better than recarburizing the surface, and tests

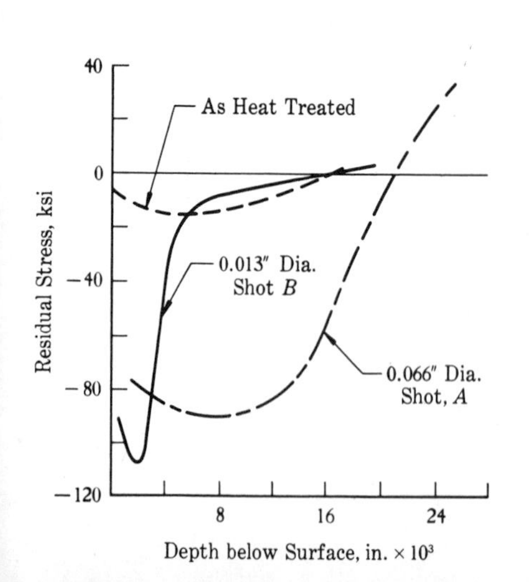

FIGURE 4.22 Residual Stresses from Shot Peening. Steel, SAE 5147; $R_c = 48$. (After Mattson, R. L.[4.28])

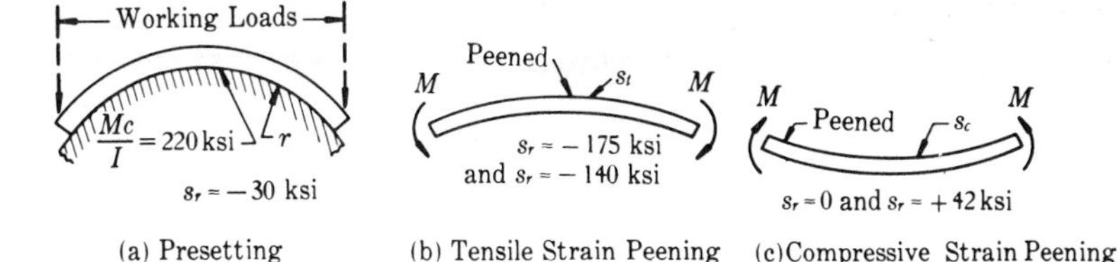

FIGURE 4.23 See Table 4.2 (a) Presetting or overstressing. Held for 30 sec. In (b) and (c), stresses s_t and s_c by M are elastic; s_r = residual stress near surface that is placed in tension by the working loads, top side in illustrations.

show that it is more effective than smooth-finishing forgings.[4.64] Since peening is a low-cost means of increasing fatigue strength, it is often used on forgings, heat-treated items, springs, and hot-formed parts in general, that naturally suffer some decarburization; it is used sometimes instead of machining or grinding as a means of improving strength. Where the cost is justified, decarburized surfaces are recarburized, and peened too. Troubles from service failures have been cured by adding peening to the regular production processes. Mattson[4.28] reports that the shot used on hardened steel does not need to be harder than the steel; that on hardened steel, small shot with some minimum intensity will produce the optimum strengthening; that for softer steel, larger shot and severe treatment tends to give optimum strengthening, perhaps because of greater penetration and greater response to work hardening.

Mattson and Roberts[4.37] report on an interesting series of tests. The specimens were of spring steel, SAE 5160, machined to 0.192 by 1.5-in. cross section, heat treated to Rockwell C 48 $\pm$ 2. All specimens were preset (prestressed), a common practice for leaf springs in the automotive industry, which consists of bending the spring leaf cold to some radius r, Fig. 4.23(a), that takes the external fibers into the plastic range, and thereby leaves residual stresses in the unloaded piece (§ 4.23). All testing was by repeated bending, $R = 0$. The peening was done in such a way as to obtain five different levels of residual stress, in addition to that in the preset specimen. If, for example, the piece is subjected to a constant moment M while it is being peened on the surface in tension, Fig. 4.23(b), the compressive residual stresses due to peening on this surface when the part is unloaded are augmented by the elastic spring back.

In one experiment, the unit tensile strain during peening was + 0.006, ($s = \epsilon E = 180$ ksi); the residual stress reached about -175 ksi, vs. about -128 ksi for conventional peening (piece unloaded). By peening the side with the compressive stress, the spring back may leave a residual tensile stress there, as seen in Table 4.2 (p. 134), which gives a summary of a series of comparative tests. The effect of the residual stresses is so convincing that one feels sure that if there were an easy nondestructive way to determine these stresses, many unexpected troubles might find a ready explanation. The authors state that the improvement in endurance strength is almost entirely due to the residual stresses, not to work hardening.

We find the following in the literature. Spring steel, 0.77% C, 0.67%

Mn, 0.28% Ni, 0.22% Cr, OQT 752°F:[4.28] original machined finish, $s_n = 39$ ksi; with surface polished to 7 μin. finish, 0.0025 in. removed, $s_n = 49$ ksi. SAE 1020:[0.2] as rolled, $s_n = 28$ ksi; polished, $s_n = 35$ ksi; as rolled and shot peened, $s_n = 37$ ksi. Carburized Ni–Cr–Mo steel carburized and heat treated:[0.2] surface as received, $s_n = 58$ ksi; polished surface, $s_n = 69$ ksi; shot-peened surface, $s_n = 71$ ksi; surface shot peened and then honed, $s_n = 74$ ksi. For highly stressed parts that are expected to have a finite life, peening can be expected to prolong life significantly. For example, an unpeened spring wire failed in 105,000 cycles at a stress of $\pm$ 100 ksi; peened, it did not fail in 10^7 cycles. Polishing of the peened surface should further increase life; at 135 ksi, a peened steel spring had a life of 60,000 cycles, and peened and polished, 10^6 cycles.

Surface rolling* is a process that cold works a limited amount of material, thus giving it higher strength, and it leaves a surface compressive stress—as does peening—a stress of the same order of magnitude, except that the compressive residual stress usually goes deeper, perhaps as much as $\frac{1}{2}$ in. at times.[0.2] In general, it is more expensive than peening, but is

TABLE 4.2 EFFECT OF RESIDUAL STRESSES[4.37]

CONDITION OF SPECIMEN (ORIGINALLY $R_C = 48$)	APPROX. SURFACE RESIDUAL STRESS, s_r ksi	ENDURANCE STRENGTH, ksi
$\epsilon = -0.006$ (peened on compression side)	+42	55
$\epsilon = -0.003$ (peened on compression side)	0	78
Heat treated only ($R_C = 48$)	0	88
Preset only	−30	128
Shot peened while unstressed	−128	140
$\epsilon = +0.003$ (peened on tension side)	−140	176
$\epsilon = +0.006$ (peened on tension side)	−175	194

convenient and appropriate for local treatments of round members, as at interference fits (Fig. 4.20), fillets, and grooves. Horger[0.2] reports the following: for a normalized and tempered, 9.5-in., SAE 1050 steel, the permissible design bending stress at the fit is 11 ksi without rolling and 22 ksi or higher with surface rolling,[0.2] a 100% improvement; rolling fillets increases the fatigue strength by 30 to 68%.

Since polishing leaves a residual compressive stress that may be upwards of 15 ksi[4.64]—most beneficial when polished in the direction of the load—a polished tensile member gains little strength by cold working the surface. Approximately, a polished surface may be some 10% stronger in fatigue than a stress-free member.

* Not to be confused with *cold-rolled steel*, which generally connotes a relatively large change of dimensions (as in cold drawn).

If the designer is dealing with a shop knowledgeable in these processes it would certainly seem safe to assume in design at least a 25% improvement in fatigue strength as a result of peening or surface rolling under circumstances where it is known to be beneficial.

4.29 HEAT TREATMENTS FOR IMPROVING ENDURANCE STRENGTH. Besides quenching and tempering for through hardness, which increases strength, there are several surface hardening processes that significantly improve endurance strength and wear resistance. All of the processes below not only make the surface material stronger (s_u), but they leave residual tangential and longitudinal compressive stresses in the surface layers. See § 2.8.

(a) Flame Hardening. Flame hardening is widely used for local heat treatments, such as fillets, bearing surfaces, gear teeth. Naturally, when the surface becomes hot, it first expands, then loses strength and yields; the hot layer transforms to martensite on sudden quenching. The naturally greater volume of the martensite results in residual compressive stresses. (If cooled slowly, the martensite would not be formed and the residual stresses would be tensile.) SAE 1045 steel, $\frac{7}{16}$-in. specimens showed the following strengths;[0.2] untreated, $s_n = 18$ ksi; OQT 400°F, 27 ksi; 0.001-in. fillets flame strengthened, $s_n = 32$ ksi (failure not at fillet); fillets and entire reduced section flame strengthened, $s_n = 51$ ksi. Lessells[4.25] found: 1-in. alloy steel bars, 60° V-groove with bottom radius of $\frac{5}{16}$ in., endurance strength doubled by flame hardening, from 40 to 80 ksi, in reversed bending. At a press fit, a 9.5-in. flame-hardened shaft had 63% greater endurance strength than when normalized and tempered, but surface rolling produced better than 100% increase (85×10^6 cycles). Fretting may start early for flame-hardened fits, but the large residual compressive stresses retard the process. There are few quantitative data. Certainly the flame should not be oxidizing.

(b) Induction Hardening. When flame and induction hardening are handled properly, the quality and properties after treatment are much the same. Induction hardening is likely to be economical only on a production basis, because of the need of special machines, and is well suited to hardening cylindrical surfaces, especially bearing surfaces on crankshafts, cam surfaces, gear teeth, etc. A cylindrical specimen with a transverse oil hole had $s_n = 10.2$ ksi not induction hardened and $s_n = 7.7$ ksi induction hardened,[4.10] which emphasizes the fact that some fine-sounding processes cannot be used blindly. The trouble is a residual tensile stress inside the hole, not too far below the surface, which combined with a repeatedly applied tensile stress exceeded the endurance strength. This same piece induction hardened and shot peened had $s_n = 10.9$ ksi, which is a considerable improvement.

The depth of case influences the magnitude of the residual stresses and presumably the endurance strength. As the depth of a hardened layer increases, the residual stress at the surface increases to a peak and then turns down. If the part is afterwards tempered, the fatigue strength decreases—from say 90 to 81 ksi for a tempering temperature of 300°F; from 90 to 61 ksi for 480°F. In fact, for some steels, 480°F may bring about a substantial relief of stresses, from 74 to 19 ksi in one case.[4.10] Thus, if residual stresses are important, so is the tempering temperature. Depths of hardening as much as 0.15 in. or more may be advantageous.

(c) Carburizing. Absorption of carbon into the case increases its volume, the case material is transformed to martensite, and the core contracts last on cooling, thereby inducing high residual compressive stresses on the surface. Also, the increased carbon content of the case results in its mechanical properties being increased. There seems to be some question as to which factor is most significant in accounting for the improved endurance strength. It is not certain what the stress distribution is in the vicinity of a discontinuity, and since the stress state is surely triaxial, there may be residual tensile stresses. However, if the stress concentration on the surface is localized within the carburized material, the high strength of the high-carbon steel is sufficient to result in a large improvement in fatigue strength. Alloy steels with their better hardenability can be made hard enough by oil quenching and hence they distort less than plain carbon steels (water quenched), so they tend to be favored for carburizing, certainly where no finishing operation occurs after treatment. Some tests indicate a very large increase in strength: a $\frac{5}{16}$-in. AISI 2317, in rotary bending, 0.05-in. case, showed $s_n = 48$ ksi normalized, and $s_n = 120$ ksi carburized, water quenched and tempered; the same as above except that the material was 2513, $s_n = 54$ and $s_n = 123$ ksi.[4.25]

Some less optimistic results are: 0.3-in. diameter, 0.03-in. case, 0.2% C steel, rotary bending, carburizing raised s_n from 33 to 45 ksi;[4.5] bars with a radial hole, untreated, rotary bending, $s_n = 48$ ksi vs. 62.3 ksi after carburizing on surface and in hole; same bars, same treatment, had an increase in reversed torsional endurance strength from 17 to 41 ksi.[4.25] Directly comparative data are scarce; through hardened 4140 shafts failed in 10^5 to 4×10^5 cycles, whereas 4320 shafts carburized with a case of 0.04–0.05 in. failed in 4×10^5 to 8×10^5 cycles, at the same stress level.[2.1] A shaft with a hole drilled *after* carburizing had $s_n = 29.9$ ksi, but drilled *before* carburizing, $s_n = 62.6$ ksi.

The effect of carburizing depends to some extent on the case thickness. If the case is very thin, failure often starts near the junction of case and core. Some tests show fatigue life increasing with case thickness up to 0.08 in.;[4.31] other tests on fatigue bending strength of gear teeth showed the strength increasing to a case depth of 0.008 in., then decreasing gradually as the case depth increased to 0.06 in.[4.28] Considering the information available, it would appear that there are no comprehensive guide lines for design stresses; each design is a special case. Inasmuch as carburizing is so frequently used for the principal purpose of obtaining a good wearing surface, the strength often is secondary; in many of these situations, the part has excess strength. Selected surfaces can be carburized, by putting various coatings over the parts not to be carburized, for example, brushed on copper sulfate, or copper plating.

(d) Nitriding. This produces results similar to those from carburizing, but the residual stresses are higher and the percentage increase in strength is generally greater. Also nitriding has an advantage of much reduced or negligible distortion because severe quenching after the process is not required. However, it is not effective on decarburized surfaces. One series of tests suggests that nitriding increases the endurance limit of 0.3-in. diameter pieces, 0.037-in. case, by about 20–25 ksi;[0.2] other results are as follows:

SPECIMEN	UNNITRIDED, s_n ksi	NITRIDED, s_n ksi
Bar without notch	45	90
With semicircle notch	25	87
With V notch	24	80
Bar (1-in.) without stress raiser	67.5	71
With fillet	32	67.5

When stressed in the vicinity of the endurance strength for indefinite life and failure occurs, the fatigue crack of a nitrided part typically starts in the core just inside the case material, where the residual stress is tension; when highly stressed, the failure starts at the outer surface.[4.10] Failures originating in the core also occur when the other surface treatments are used, especially if the case is unusually thin. As usual, $K_f \to 1$ as the number of cycles to failure decreases (stress increases). Notice the effectiveness of nitriding notches for Nitralloy 135, Table AT 10. If a thin strip is nitrided throughout, it will increase in length about 2% for the Nitralloys (about 6% for 4340) because of the diffusion of additional matter. When a thin layer on a heavier part is so treated, the change in length is prevented by the heavy core, resulting in compressive stresses in the case.

4.30 MISCELLANEOUS SURFACE EFFECTS. The effects of *plating*, as with copper, nickel, chromium, cadmium, tin, vary considerably, but usually the fatigue strength of a part is reduced by plating. The plating process can be adjusted so that the deposited metal has a residual compressive stress,[4.1,4.64] in which case, the fatigue strength may be little affected. Ordinarily, the plating process is such that the deposited metal has residual tensile stresses, which means that the surface is weak in fatigue. If the deposited metal cracks through to the base metal, the tension in the deposited metal tends to put the base metal in tension at the point of the crack, which is also serving as a stress raiser. Other contributing factors include penetration of hydrogen into the steel (hydrogen embrittlement) and the fact that the deposited metal may be weaker than the base metal. Peening may be used to offset the loss of strength. For example, a steel with $s'_n = 46$ ksi had an $s_n = 19$ ksi when nickel plated and $s_n = 55$ ksi when plated and then peened; when chromium plated, $s_n = 38$ ksi, and when peened and then chromium plated, $s_n = 51$ ksi. While the evidence may not be conclusive, peening before plating tends to retain the original fatigue strength; peening after plating tends to result in greater strength than the original base metal. Surface rolling produces effects of the same order as peening. If operation is in a corrosive environment, proper plating that protects the underlying steel will maintain s_n the same as the plated part in air: but if cracks in the plating permit the corrosive medium to reach the steel, s_n and life will be much reduced. Reference *(0.2)* has a collection of quantitative fatigue data on various metal coatings.

Cold drawing and *cold rolling* (§ 2.9) result in all the material being plastically compressed. When the load returns to zero, the material expands. However, after the stress in the outer fibers reaches zero, the internal fibers are still in compression; hence, with expansion continuing until the internal forces are in equilibrium, the outer fibers end up in tension (in both longitudinal and tangential directions) and some inner ones in compression. The residual tensile stress, sometimes of considerable magnitude (60–120 ksi), would be expected to affect unfavorably the endurance strength. One set of tests on 1.5-in. bars[2.1] shows tangential and longitudinal stresses at the outermost fiber of about 48 ksi tension, and at the central fiber, compressive stresses of 45 and 80 ksi, respectively (1045, cold drawn 20%). Thus, any improvement in fatigue strength would be due to the higher mechanical properties from work hardening. Stress relieving to remove the tensile residuals would also remove some or all of the work-hardening effects; if heated above the recrystallization temperature, all effects of cold working are generally lost. Shot peening cold drawn surfaces can change the residual stress to compression and significantly improve its endurance, a frequent practice with coil springs.

Evidence[4.84] suggests that *grinding* may leave surprisingly large residual tensile stresses in the surface of an unstressed part and otherwise injure it (grinding cracks), the residual tension resulting from the high temperature induced in a thin surface layer. Carburized and nitrided steels may lose endurance strength to some 35% of the unground value[0.2] with improper grinding methods, and only a few data show any improvement in strength after grinding. For case-hardened steels, the loss of strength is due to the substitution of a layer with residual tension for a layer that had high residual compression. Soft wheels and light cuts are relatively less damaging, but more expensive in removing metal than rougher cuts. There is the possibility of some alloys with certain heat treatments being resistant to grinding damage,[0.2] and peening and tumbling tends to restore the original

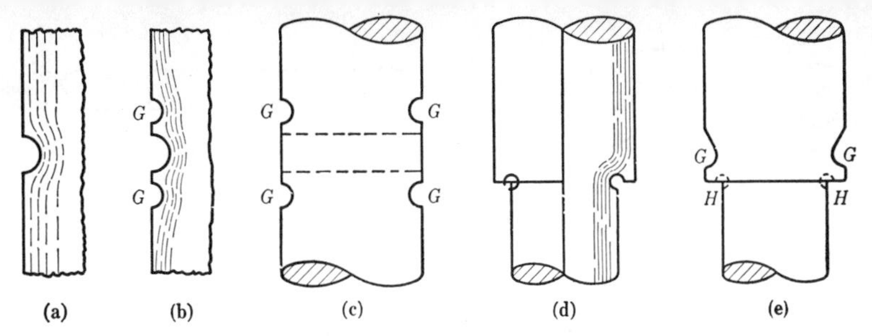

FIGURE 4.24 Relieving Grooves Reduce Stress. When there is a single groove such as at (a), the stress concentration is greater than when it is flanked by two, preferably smaller, grooves *G*, as at (b). The relieving grooves *G* in (c) reduce the concentration about the radial hole. These grooves are more effective if they are pressed in. Undercutting a sharp internal fillet as in (d) or providing relieving grooves *G* in (e), or both, are effective. Undercutting the fillet leaves a face for locating a bearing or other element. (Battelle Memorial Institute[4.3]).

strength. Peening raised the endurance strength of a rough ground flat steel bar, reversed bending, from 42 to 82 ksi.[4.1]

In most cases, *machining* leaves residual tensile stresses in the surface layers; exceptions include austenitic manganese steel and cast iron.

Unexpectedly good results with carbon steel have been obtained by ***shallow quenching,*** which means that the steel has low hardenability (§ 2.7). For example, water-quenched 1046 steel will have a surface hardness approaching 600 BHN, but because of poor hardenability, the hardness falls off to about 280 BHN within 0.25 in. of the surface.[4.1] This treatment produces a residual compressive stress at the surface (the hardened surface material tends to occupy more volume) and residual tension on the inside, a pattern that results in a large increase in fatigue strength for members in bending and torsion. Excellent results have been obtained in this manner for automotive axles and for other heavy duty situations.

4.31 MITIGATING STRESS CONCENTRATIONS. The designer may specify a certain fillet radius, but the shop may make it smaller; or the shop may inadvertently leave a stress raiser that the designer has not counted on, such as clamp marks or the merest change in diameter at the junction of two machining operations. Such things have to be watched. However, the designer too may be responsible for the presence of an unnecessary stress raiser. All designs should be looked at with the question "can this stress concentration point be eliminated?" in mind. If elimination is impracticable, then consider what can be done to reduce its effect. We have already indicated many basic considerations and we shall mention specific steps in connection with some machine elements. In the meantime, a few pointers are suggested by Figs. 4.24–4.27. In Fig. 4.26(a), typical

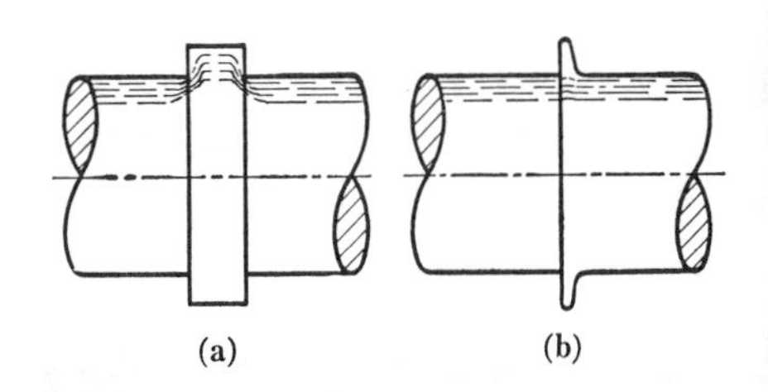

FIGURE 4.25 Effect of Collars. Narrow collars reduce stress concentration. (Battelle Memorial Institute[4.3]).

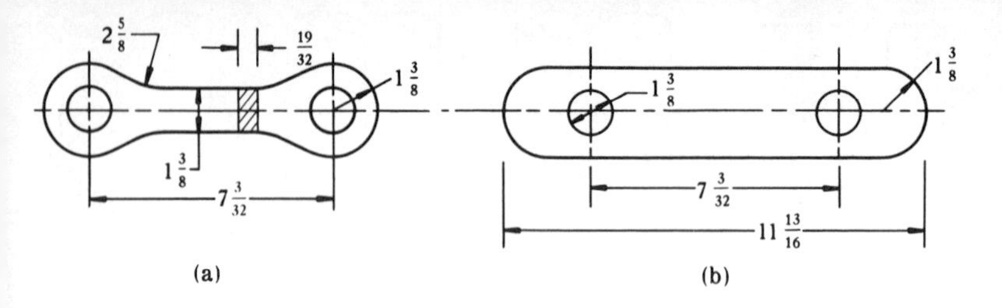

FIGURE 4.26 Rolling Bores.

proportions for simple links, with the holes not rolled, $s_n = 12.7$ ksi; with the holes rolled, $s_n = 12.1$ ksi; surprisingly, no improvement. In Fig. 4.26(b), with the holes not rolled, $s_n = 17.6$ ksi; with rolled holes, $s_n = 26$ ksi. In (a), the collar raised on the edge of the bore by the rolling was not removed; in (b), it was filled off.[0.2] See Fig. AF 8 for a sample value of K_t for a bar being loaded by a pin in the hole. Reference *(4.62)* contains the most complete collection of K_t values covering various configurations of the same basic design. Cold pressing grooves about holes, as in Figs. 4.24(c) and 17.24, improves fatigue strength.

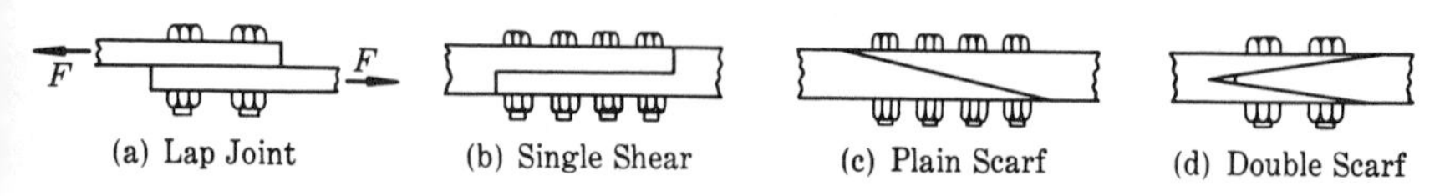

FIGURE 4.27 Improving Fatigue Strength by Design. After Ref. (4.30). The load on (b), (c), and (d) varied from 10.7 to 21.3 kips; material, aluminum alloy 7075-T6. Notice that all surfaces under pressure may have fretting corrosion. Tapers are relatively more expensive to machine. (a) No comparative values. (b) Failed in 42,000 cycles. (c) Failed in 210,800 cycles. (d) Failed in 26,914,000 cycles. Spaulding[4.1] reports stress concentration factors, based on the *gross* section, of 13 for (b), 4.1 to 8.5 for (c), 3.2 for (d). For repeated loading, it is evident that (b) should be avoided. A similar design, except a tongue and groove construction with double shear of the bolts, had $K_t = 4.1$, a sizable saving.

4.32 TEMPERATURE EFFECTS. As temperatures go below normal atmospheric temperature, the fatigue strength tends to increase; for example, copper has $s_n = 14$ ksi at 75°F and $s_n = 30$ ksi at -425°F, both at 10^6 cycles;[4.28] normalized SAE 2330 has an $s_n' = 59$ ksi at 75°F, and $s_n' = 110$ ksi at -320°F. In general, all steels show a similar increase in endurance strength, but those without nickel alloy lose nearly all of their toughness (§ 2.22), and they become more notch sensitive. There is no relation between impact strength and fatigue strength.

With increasing temperature, the effect is generally the reverse, except that plain carbon and very low-alloy steels increase in ultimate and endurance strengths from 70°F to about 600°F or more, then strength decreases rapidly. AISI 4340, unnotched, with $s_u = 160$ ksi at 70°F, has endurance strengths for reversed loads with temperature as follows:[2.1]

70°F, 70 ksi; 600°F, 63 ksi; 800°F 60 ksi; 1000°F, 40 ksi.

The fatigue strength decreases at a slower rate with increasing temperature than the ultimate strength, and most of the failure points at high temperature on an $s_a s_m$ diagram fall *outside* of a parabola (Gerber curve, Fig. 4.8) through s_n for reversed stress and the rupture stress with $s_a = 0$. See Fig. 4.28. This is to say that equation (**d**), § 4.6, is a reasonable design basis, that for a given mean stress, the material can withstand a much higher alternating stress before rupture than that predicted by the Goodman (or Soderberg) line.[4.28] Thus, if the temperature is such that creep is involved, a *conservative* design basis would be the straight line from the static stress for a specified safe creep strain (instead of from s_u/N) to s_n/N, where s_n is the endurance strength in reversed loading at the operating temperature (and theoretically at the frequency of the actual loading). As is true of creep rupture strength, the fatigue strength at a particular frequency, is a function of time; that is, for example, the fatigue strength for 10^5 hr. is higher than that for 10^6 hr. In a general way, with increased temperature, notch sensitivity decreases and the effect of shot peening, so beneficial at room temperature, decreases. Any one starting design work for high temperature situations will find the summary of the state of the art by Allen and Forrest[4.28] helpful; considerable data on high temperature properties are in Ref. *(2.1)*. Notch effects at high temperature are not consistent with those at room temperature.[4.50]

When a hot part is cooled suddenly by quenching, there is momentarily a high temperature gradient that induces a stress gradient. Some metal parts under certain conditions crack as a result; this phenomenon may be called a ***thermal-shock failure.*** If the change is not so severe, repetitions of temperature and stress gradients in metals may be enough to cause eventual failure, a process referred to as ***thermal fatigue.*** It is reported by Allen and Forrest[4.28] that the parameters $s_{n800}/E\alpha$ and $ks_{n800}/E\alpha$ indicate the order of merit of heat-resisting alloys according to their ability to withstand thermal fatigue; s_{n800} = endurance strength at 800°C, k = the thermal conductivity, α = the coefficient of thermal expansion, E = modulus of elasticity. To a limited extent, local heating has been used to induce favorable residual stresses. When a transformation does not occur during cooling, the part that cools first is in compression.

FIGURE 4.28 Curves of Mean Stress vs. Alternating Stress, from Tests. Adapted from Ref. (2.1). The A286 alloy consists of about 55% Fe, 15% Cr, 26% Ni, plus small amounts of several other metals; this metal was stressed by a combination of axial and bending stress. The loading on the stainless 403 was axial. Each curve represents failure by rupture in 500 hr. at the temperature indicated.

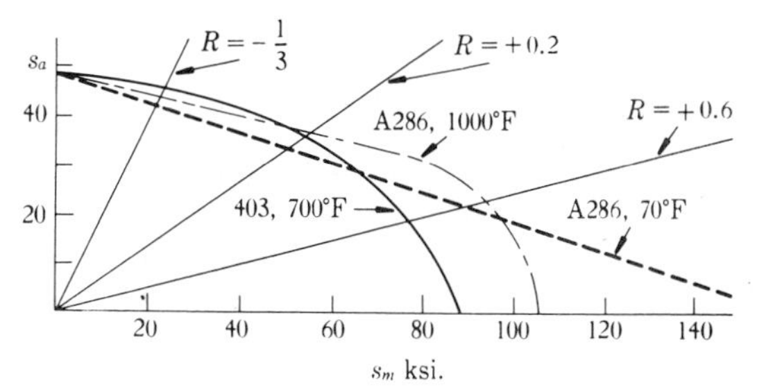

4.33 MORE ON FATIGUE STRENGTH. If one is designing for variable loading and using a material for which no appropriate fatigue data are available, one must purely gamble on the consequences or run tests to acquire information. For some machines, like airplanes, uninformed design is never appropriate. Materials sometimes exhibit unexpected peculiarities that actually seem illogical. For example, in some axial fatigue tests with $R = 0$ (tension) on 17-7 PH and A-286 (a heat and corrosion-resistant steel), the K_f value exceeded K_t when the number of cycles was 10^6 or more.[4.33] Whether the explanation involves the resultant effect of the biaxial and triaxial stresses that exist at the bottom of the notch, as compared to the simple uniform stress of an unnotched specimen, the residual tensile stresses, or whatever, the fact is pertinent to the design of a part whose life is expected to be 10^6 cycles or more.

In some cases, the actual repeated peak stress may be permitted to exceed the yield strength, when the number of lifetime cycles is less than some number, say 10^4 (but the fatigue strength for said number of cycles must not be exceeded), and when the possible change of dimensions (which may be unnoticeable when the highest stress is highly local) is permissible. When the rate at which the load is applied exceeds some 500 cpm, s_{max} may exceed s_y without actual yielding.[4.65] Perhaps the normal operation is such that most of the time the peak stress does not exceed the endurance strength; but if the stress does exceed the endurance limit, the piece is damaged. The damage apparently accumulates and, if expected, must be allowed for.[4.44, 4.46]

Some materials exhibit a marked change with the orientation of the fibers; for example, 4340 showed an endurance strength transverse to the fiber length of 70–75% of that when the longitudinal dimension of the specimen accords with the longitudinal direction of rolling.[4.33] Unless specified otherwise, quoted values of s_n are expected to be for longitudinal specimens.

The rotating *bending*-fatigue strength for only a few cycles, say less than 1000, is greater than the ultimate strength; for 4340, the endurance strength as computed from Mc/I is approximately $1.6s_u$ for a small number of cycles.[4.33] Observe that the stress induced is in the plastic range, that Mc/I is therefore not the true stress, and that the stress distribution is probably some modification of that shown in Fig. 1.7.

In some tests, fatigue strength is increased by repeated loads just below the normal fatigue limit, followed by small step by step increases of the loading, a process called ***coaxing***. A test on Armco iron ($s_n = 26.2$ ksi) was: an initial run of 10^7 cycles at 26 ksi, with about a 2% increase in stress every 10^7 cycles; the stress at failure after nearly 13×10^7 cycles was 30% greater than s_n. Improving fatigue strength by understressing close to the endurance limit or by coaxing is not a commercial way.

For ordinary rates of applying repeated loads, steel shows no significant change of endurance limit, up to say 8000 cps. For very slow rates, as 10–100 cpm, and for very high rates, variations have been found.[4.28]

Also aluminum, copper, lead, and other nonferrous metals are found to have endurance strengths that change with the frequency. Space does not permit elaboration, but if one is involved with unusual frequencies, the matter should be investigated.

4.34 IMPACT. Suddenly applied loads, called *impulsive loads*, produce such complex responses that any procedure of applicable design involves considerable uncertainty. Two bodies are involved, a striking body or load that produces the effect and the struck body, or one being investigated, that has a *response*; keep them mentally distinct. The loading is considered to be impact or shock when the time taken for the response to reach a maximum is less than the lowest natural period of vibration of the struck body. If the impact is repeated, some kind of impact fatigue strength would be involved. If the rate of application of a varying load is increased, the frequency of the applied load will become greater than the frequency response, and the case passes into the range covered by the definition of impact loading. Thus, we see that the two cases, fatigue and impact, merge into each other at their extremes; but in their usual ranges, they are handled quite differently. The usual method of handling impact problems is by the laws of the conservation of energy and the conservation of momentum. We have every confidence in the law of conservation of energy; as it applies to this situation,

$$\text{(4.7)} \qquad \begin{bmatrix}\text{Energy given up} \\ \text{by loading body}\end{bmatrix} = \begin{bmatrix}\text{Energy absorbed by} \\ \text{the loaded body}\end{bmatrix},$$

provided no significant energy is exchanged with the surroundings. However, there are some practical difficulties in calculating all of the energies involved, and the ideal models that permit computations are oversimplified. After a part has been designed and a sample made, experimental methods that measure[4.42] velocity, acceleration, time, and of course stress, can be used to check; appropriate modification and redesign follow.

The duration of an impact may be very short, even a fraction of a thousandths of a second, and impact induces vibrations that affect the amount of the induced stress.[4.40,4.41] One helpful phenomenon is that the yield strength (and ultimate too) increases considerably as the rate at which the load is applied increases. When the unit strain of a certain mild steel being tested is increased at the rate of 10^{-3} per sec, the vicinity of the conventional testing rate, the $s_y = 31$ ksi; when the rate is 10^3, $s_y = 79$ ksi;[0.2] this is closely where elastic action ceased. Another conservative factor in design is that points of support are usually assumed to be rigid, whereas all actually deform to some extent, the consequence being that the actual stress is lower than the computed value. We shall assume that all deformations are elastic in accordance with Hooke's law, force is proportional to deformation.

4.35 ELASTIC ENERGY. If an elastic body, say a spring, is deformed an amount δ under a force F that has increased gradually from zero, the spring's response is also F, and the average force is $F/2$. (See Fig. 6.7.) The work done on the spring (and the energy stored in the spring) is $U = (F/2)\delta$, or

$$U = \frac{F\delta}{2} = \frac{k\delta^2}{2} \quad \text{(in-lb. or ft-lb.)}, \tag{4.8}$$

where δ is the total elastic deflection at the point of application of the force F and $k = F/\delta$ lb./in. (or lb./ft., depending on the units of δ, but in conjunction with s psi, inch units are more convenient); k is a common parameter called the ***spring scale*** or ***spring constant.*** Observe that (4.8) applies to any kind of elastic member, when $F \propto \delta$. If the member is a beam, one thus must have the deflection of the beam at the point of application of F; see Table AT 2.

Recall also from mechanics that work done by a constant torque T is $T\theta$; and if the torque varies linearly from 0 to T, the work is $(T/2)\,\theta$ which is representative of the elastic energy stored in a cylindrical bar as torque on it is gradually increased to T.

$$U = \frac{T\theta}{2} \quad \text{in-lb. or ft-lb.}, \tag{4.9}$$

where θ is the angular deformation of the bar. If a force F is applied on a crank at a radius of r, $T = Fr$ and $U = Fr\theta/2$ is the elastic energy stored in the crank and bar (θ including deformation of crank). If the crank is almost rigid, the bar stores nearly all of the energy. If a prismatical member is in uniform tension or compression, the amount of energy per unit volume stored at any elastic stress s is the area under the stress-strain diagram (Fig. 1.3) up to that stress. This is a triangular area, $(\frac{1}{2})\epsilon s = s^2/(2E)$, which, if multiplied by the volume AL, takes the form of (4.8) above. The energy-absorbing capacity of a material is related to its ductility, its strength, and its strain-hardening characteristics.

4.36 BAR AXIALLY LOADED. Consider Fig. 4.29; assume that: (a) the support at G is rigid (no deformation), (b) the weight W and surface M of the stop are rigid (these assumptions mean that the bar takes all the deformation), (c) the weight has been brought to rest (maximum deformation), (d) the stress s pervades uniformly the entire rod (it actually moves in waves—see below), (e) the system of weight and rod is a constant-energy system (actually some energy is dissipated on impact, which first goes to raise the temperature of the parts and then heat is dissipated to the surroundings—if everything started out at ambient temperature; also some of the energy in the rod is vibrational energy which means less elastic energy), and (f) the mass of the weight is large compared to the mass of the rod (which makes the vibrational energy negligible in amount). Let W, Fig. 4.29, fall freely from rest through a distance h, strike the stop M on the rod, and come to rest after the rod has stretched an amount δ. Assuming

FIGURE 4.29 Impulsive Loading.

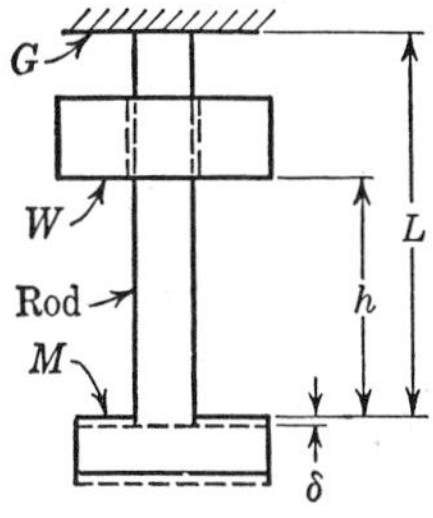

that the loss of potential energy $W(h + \delta)$ is all converted to elastic energy, we have

$$W(h+\delta) = \frac{F\delta}{2} = \frac{sA\delta}{2} = \frac{k\delta^2}{2} = \frac{s^2AL}{2E}, \tag{4.10}$$

$$\text{[A]} \qquad \text{[B]} \qquad \text{[C]} \qquad \text{[D]} \qquad \text{[E]}$$

where the symbols are as defined in the previous article. From the equality of terms [A] and [D], solve for δ and find

$$\delta = \frac{W}{k} + \frac{W}{k}\left(1 + \frac{2hk}{W}\right)^{1/2}, \tag{t}$$

in which δ can be replaced by $\delta = \epsilon L = sL/E$, $W/k = \delta_{st}$ is the deflection of the rod under a static load W, called the *static deflection*, and h can be replaced by its equivalent kinetic energy $v^2/2g_o$, where v is the speed of W at the instant it contacts M (or from another point of view, one can use the kinematic relation $v^2 = 2g_oh$). Conveniently, $g_o = 386$ ips^2, v ips, E psi, k lb/in., s psi, etc. Useful expressions for the scale k that may be used in (**t**) (tensile or compressive member) are

$$k = \frac{F}{\delta} = \frac{sA}{\epsilon L} = \frac{AE}{L}. \tag{u}$$

If we solve for s from terms [A] and [C] in equation (4.10), we find

$$s = \frac{2W}{A}\left(\frac{h}{\delta} + 1\right), \tag{v}$$

where we see that for particular values of A, W, and h, the stress can be reduced only by increasing $\delta = sL/E$, which may be done by increasing L for a given material, or changing material to one with a lower E. Most materials with E lower than that for steel are likely to have lower yield and

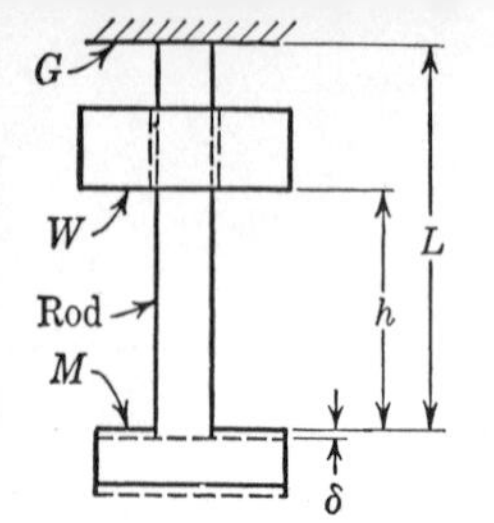

FIGURE 4.29 (Repeated).

endurance strengths too, so a change of material from steel is not too often the solution of a tough impact problem. See also § 4.39.

If we solve for s from terms [A] and [E] in equation (4.10), we get

$$\textbf{(w)} \qquad s = \frac{W}{A} + \frac{W}{A}\left(1 + \frac{2hEA}{LW}\right)^{1/2},$$

where W/A is the stress induced by a static load W, sometimes called the static stress s_{st}; also $WL/(EA) = \delta_{st}$, the static deflection. As we see, a large number of equations differing in details can be obtained by virtue of the interrelations of laws and properties. For this reason, the student is advised not to use equations (**t**) or (**w**), for example, in his problem work, but to work from the more basic numbered forms; the practice in using the basic forms is too valuable to be lost by substituting numbers in an equation.

4.37 EXAMPLE. What should be the diameter of a steel rod 6 in. long if it is to resist the impact of a weight of $W = 500$ lb. dropped through a distance of 2 in.? The maximum computed stress is to be 20 ksi.

Solution. From the given maximum stress, compute the corresponding maximum deflection;

$$\delta = \frac{sL}{E} = \frac{(20{,}000)(6)}{30 \times 10^6} = 0.004 \text{ in.}$$

From equation (4.10), terms [C] and [A], solve for A and find

$$A = \frac{\pi D^2}{4} = \frac{2W(h + \delta)}{s\delta} = \frac{(2)(500)(2.004)}{(20{,}000)(0.004)},$$

from which $D = 5.65$ in.; use $5\frac{3}{4}$ in. Thus, a startlingly large dimension results from rather guileless data.

4.38 SUDDENLY APPLIED LOAD—ZERO VELOCITY OF IMPACT.

If the weight W is held in contact with but not pressing on the stop, Fig. 4.29, and is then released so that the load goes suddenly from zero to W, the distance dropped $h = 0$ and equation (**v**) yields

$$s = \frac{2W}{A}.$$

Suddenly released, the weight vibrates as it would on a spring until the vibration damps out and the stress becomes the static stress W/A. Thus, load W suddenly applied induces a response or stress twice that for a static load W. If the load is applied gradually rather than instantaneously, the maximum stress will be less than twice the corresponding static stress. In the limit, the load can be let down so gradually that the stress never exceeds the static stress.

4.39 TENSION MEMBER WITH TWO OR MORE AREAS. When a member with an axial load has two or more different sectional areas, the deformation of each constant section is determined, and the total deformation δ is the sum of the values for each section. Suppose a member, say a bolt, has an area A_1 of the unthreaded part and a smaller area A_2 at the threaded part. The nominal stresses due to some load $F = s_1A_1 = s_2A_2$ will be inversely proportional to the areas, $s_1/s_2 = A_2/A_1$. Also, since $\epsilon = s/E$,

$$\text{(x)} \qquad \frac{\epsilon_1}{\epsilon_2} = \frac{s_1}{s_2} = \frac{A_2}{A_1}. \qquad \text{[Constant } E\text{]}$$

Suppose it is desired to find the stress s_2 in the smaller section of a two-section rod, Fig. 4.30, subjected to the impact of a falling weight W. The total deformation is

$$\text{(y)} \qquad \delta = \epsilon_1 L_1 + \epsilon_2 L_2 = \frac{s_1L_1 + s_2L_2}{E}.$$

Substitute $s_1 = A_2s_2/A_1$ from (x) and find

$$\text{(z)} \qquad \delta = \frac{A_2s_2L_1/A_1 + s_2L_2}{E} = \frac{s_2}{E}\left(\frac{A_2}{A_1}L_1 + L_2\right).$$

Taking advantage of the convenience of the spring constant $k = F/\delta$, we

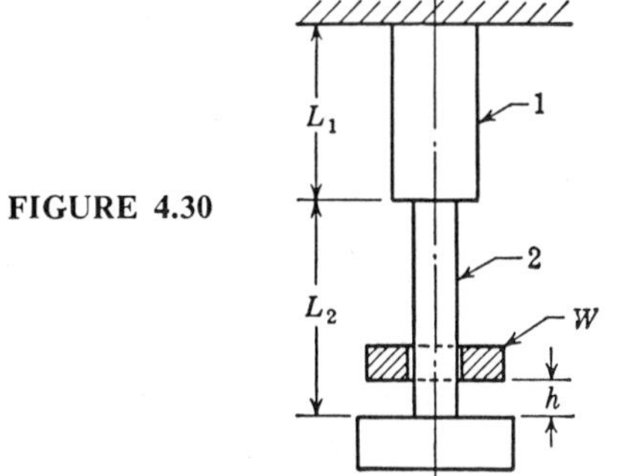

FIGURE 4.30

shall find the equivalent or over-all spring constant k' for the tensile (or compressive) member with two sectional areas; $k' = F/\delta$, where $\delta = \delta_1 +$

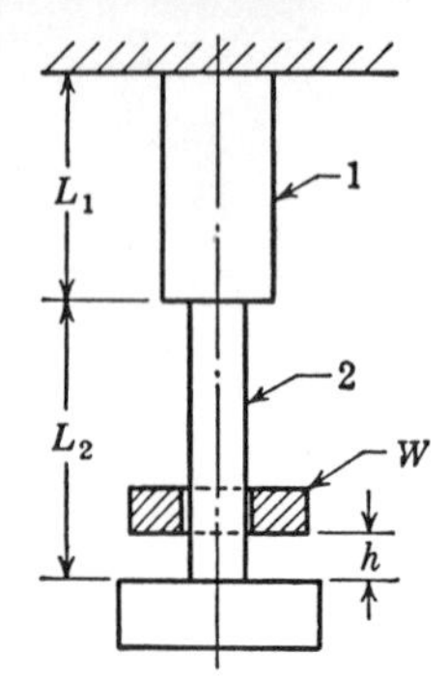

FIGURE 4.30 (Repeated).

δ_2 is the total deflection, equation (**z**), and where $F = s_1A_1 = s_2A_2$ is the response load in the member. Thus,

$$(4.11) \qquad k' = \frac{F}{\delta} = \frac{1}{\dfrac{\delta_1}{F} + \dfrac{\delta_2}{F}} = \frac{1}{\dfrac{1}{k_1} + \dfrac{1}{k_2}} = \frac{k_1k_2}{k_1 + k_2},$$

where k_1 and k_2 are the spring constants for the parts 1 and 2 respectively, Fig. 4.30. This rod is equivalent to *two springs in series*. In general, $k' = 1/\Sigma(1/k)$, where $\Sigma(1/k)$ is the sum of the reciprocals of the individual k's. Now, the energy absorbed by the rod is $U = k'\delta^2/2$, equation (4.8), and is equal to the energy given up by the falling weight, $W(h + \delta)$. This equality results in equation (**t**) except that k' is in place of k. By using the value of δ in terms of s_2 from equation (**z**), we may find the maximum stress s_2. This is a devious but illuminating route. Observe that making part 1, Fig. 4.30, larger, the deformation is decreased and therefore *the part is weakened for handling energy loads.*

4.40 DESIGN FOR IMPACT LOADS. Faced with a shock load, one's first thought should be, "Can it be eliminated?" If not, the second is, "Can it be reduced?" as by shock absorbers on automobiles and aircraft landing gear or by other means. If it must be lived with, the design criterion that should be kept in mind is that the material must have an appropriate ability to absorb energy without failure, which, as we have seen, is related to the amount of deformation. For example, suppose a bolt is subjected to an impact load. The stress on the smaller area at the root of the thread is the larger and critical stress. The unthreaded shank has a lower stress ($A_1 > A_2$), and for this reason its unit strain ϵ_1 is less than ϵ_2 for the smaller area. If the unthreaded shank is turned down to, or somewhat less than, the root diameter of the thread, or if a hole is drilled in the unthreaded shank, Fig. 4.31, the bolt is strengthened for impact loads, because with a higher stress in the unthreaded part, the deformation there is greater.

The method illustrated in Fig. 4.31(a) is cheaper, but the bolt is weaker for an external bending or twisting load. The idea of Fig. 4.31(b) involves difficulties in manufacturing in long bolts, but there is negligible weakening

FIGURE 4.31 Bolts Suitable for Absorbing Energy.

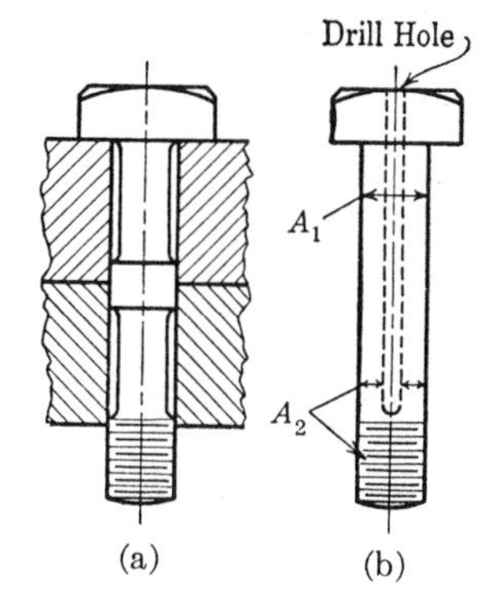

for bending or twisting. The modifications of Fig. 4.31 will not change significantly the static strength; hence, there is no reason to use such bolts except for live loads.

One group of tests indicate that the capacity of the bolt to absorb energy continues to increase as the diameter of the body of the bolt is turned to less than the diameter at the root of the threads—at least to where the body area is 57% of root area.[0.2] But see Chapter 5.

4.41 MASS OF ROD NOT NEGLIGIBLE. If the end of a rod is struck with a hammer, a compression wave is set up that travels along the rod at the acoustic speed v_a in the rod. From physics, recall that $v_a = (E/\rho)^{1/2}$, in which a consistent system of units* must be used; that is, if as before the pound is used for force and inches for length (E psi), then mass and density expressed in pounds must be converted as follows: lb/(386 ips^2), where $g_o = 386$ ips^2, the standard acceleration of gravity. Another phenomenon of this hammer blow is that the material under the hammer is actually moved, given a speed, which is taken as the common speed v_c of the hammer and adjacent material at the end of the deformation period. This first layer of material at speed v_c imparts speed to the next layer, and so on. Thus, in Fig. 4.32, the first layer moves with speed v_c, and the layer at any other point C later moves with speed v. If it is assumed[1.2] that this speed varies as the distance from point D, for instance, we can estimate the kinetic energy, using $v/x = v_c/L$. The element dx at BC, Fig. 4.32, has a mass of (density *times* volume), $dW_b = \rho A\, dx/g_o$, where ρ lb/ft.^{3} divided by 32.2 is $\rho/32.2$ slugs/ft.^{3}; its kinetic energy is $mv^2/2 = \rho A v^2 dx/(2g_o)$. (Consistent inch units are just as satisfactory here.)

* A consistent system of units is defined by saying that it is one in which the proportionality constant in Newton's second law is unity; that is, it is defined by $F = ma$. If the force is in pounds, time in seconds, and length in feet, then from $m = F/a$, the units of mass are $m \to$ lb/(ft/sec.^{2}) = lb-sec^2/ft., which unit is called a *slug*. If the mass is in w pounds, the conversion to the foregoing consistent system is w lb/g_o = (w lb.)/(32.2 fps^2) or the conversion factor is 32.2 lb/slug. There is no generally recognized name for the mass unit in a consistent pound-second-inch system (lb-sec.^{2}/in.), as above. Suggestion: *psin*, a contraction of pounds, seconds, inch; conversion constant is then $g_o = 386$ lb/psin.

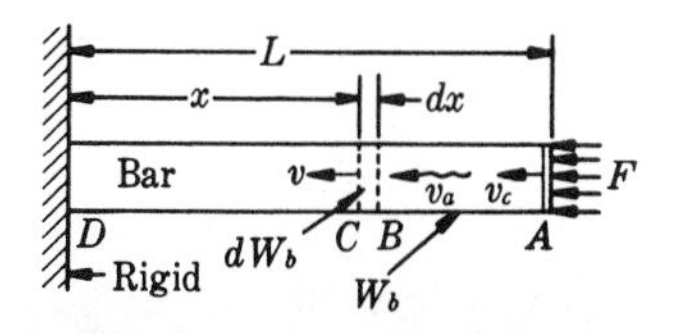

FIGURE 4.32 Kinetic Energy of Struck Bar. The compression (stress) wave with speed v_a is reflected and rereflected from the ends of the rod until it dies out.

This expression can be integrated over the entire volume by substituting $v = xv_c/L$;

$$\text{(a)} \quad KE_b = \frac{\rho A v_c^2}{2g_o L^2}\int_0^L x^2\,dx = \frac{\rho A v_c^2 L}{6g_o} = \frac{W_b}{3}\frac{v_c^2}{2g_o} \quad \text{ft-lb., or in-lb.,*}$$

from which we see that the kinetic energy of the bar KE_b is as though one-third of the bar were given the translational speed v_c. This portion of the whole can be called the effective mass W_e; $W_e = W_b/3$ for the bar with the axial impact loading.

The law of conservation of momentum applies; momentum of striking load W is equal to the momentum of load and rod the instant after contact. Equating momentums, we have

$$\text{(b)} \qquad Wv_w = (W + W_e)v_c \qquad \text{or} \qquad v_c = \frac{v_w}{1 + W_e/W},$$

where the masses are conveniently measured in pounds (conversion constant cancels) and W_e is the equivalent mass. Thus, the kinetic energy of the bar in terms of W and v_w is found by substituting the value of v_c in **(b)** into **(a)**;

$$\text{(c)} \qquad KE_b = \frac{W_e v_w^2}{2g_o(1 + W_e/W)^2} \quad \text{ft-lb. or in-lb.}$$

Now let the body W be moving vertically downward and apply the conservation-of-energy law to the situation at contact (W and equivalent W_e move with v_c) in the form: KE of W and W_e plus loss of potential energy of W is equal to the elastic energy. Then using $v_w^2 = 2g_o h$ or $h = v_w^2/(2g_o)$, v_c from **(b)**, and $W_e = W_b/3$ in **(a)**, we have

$$\text{(d)} \qquad \frac{Wv_c^2}{2g_o} + \frac{W_e v_c^2}{2g_o} + W\delta = \frac{k\delta^2}{2},$$

which simplifies to

$$(4.12) \qquad W\left(h\frac{1}{1 + W_e/W} + \delta\right) = \frac{k\delta^2}{2} \quad \text{ft-lb.} \quad \text{or} \quad \text{in-lb.}$$

where $W_e = W_b/3$ *for an axially loaded* bar of mass W_b. Comparing (4.12) with (4.10), we see that the factor $C = 1/(1 + W_e/W)$ applied to h or to $v^2/(2g_o)$ can be thought of as a correction factor to care for the kinetic energy of the mass of the rod. Thus, the equations previously derived for the axially loaded rod can be easily modified by this factor wherever h or $v^2/(2g_o)$ appears, when it is desired to make this allowance. This second analysis has its shortcomings, but it serves the purpose of this text.

4.42 IMPACT BY A BODY MOVING HORIZONTALLY. This case will throw some light on the principle involved—the conservation of

* For in-lb., use v_c ips and $g_o = 386$ ips^2.

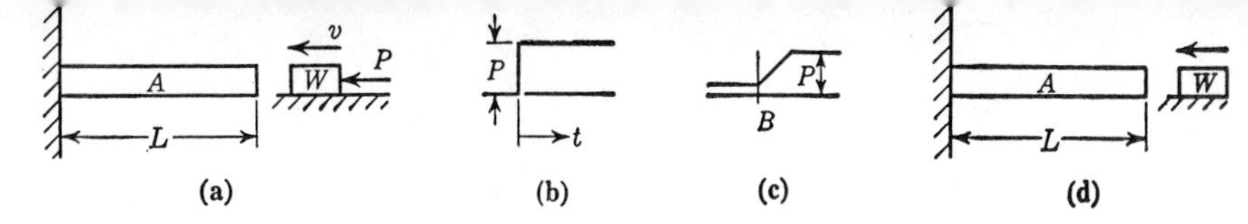

FIGURE 4.33 Horizontal Impact.

energy. Suppose, in Fig. 4.33(a), that a body W acted on by a constant force P strikes the member A at the instant its speed is v and that P continues to act while A and W are in contact. Let P remain constant after contact, Fig. 4.33(b). Then body A must absorb the kinetic energy that W has at the instant of impact plus the work done by P during the deformation of A; thus

Energy absorbed by A = KE of W + work of P

(e)
$$\frac{F\delta}{2} = \frac{Wv^2}{2g_o} + P\delta.$$

If P should be equal to W, this equation would reduce to (4.10) by use of $v^2 = 2gh$. Suggested consistent units: v ips, $g_o = 386$ ips², P lb., F lb., δ in.

If P is not constant, if, for example, it is some small value until contact is made, as at B in Fig. 4.33(c), and then if it increases gradually as suggested, the maximum stress will be less than that obtained from **(e)**.

At the extreme is the case of the body W striking A with no force acting on it, so that the only energy that A must absorb is the kinetic energy of W, $KE = Wv^2/(2g_o)$. The response of A to this impact is some maximum force F at the end of its deformation period and the elastic energy is, as before, $F\delta/2 = sA\delta/2 = s^2AL/(2E)$. Writing an energy balance with the addition of the correction factor of § 4.41, we get

(f)
$$\frac{F\delta}{2} = \frac{s^2AL}{2E} = \frac{Wv^2}{2g_o}\left(\frac{1}{1 + W_e/W}\right)$$

or

(g)
$$s = \left[\frac{Wv^2E}{g_oAL(1 + W_e/W)}\right]^{1/2},$$

where the units must be consistent. Comparing equation **(g)** with **(w)** of § 4.36, for a falling body (where $v^2/(2g_o) = h$), note that **(g)** is a good approximation of **(w)** where $h \gg \delta_{st} = WL/(AE)$.

4.43 ELASTIC IMPACT ON BEAMS.

If the weight that strikes the beam is much heavier than the beam, it is easy to adapt equation (4.10) in the form ($\delta = y$ = symbol for beam deflection)

(4.10)
$$W(h + y) = W\left(\frac{v_w^{\,2}}{2g_o} + y\right) = \frac{ky^2}{2},$$

provided it is assumed further that the deflection curve under impact is the same as under static loading, which is to say that we can use deflection

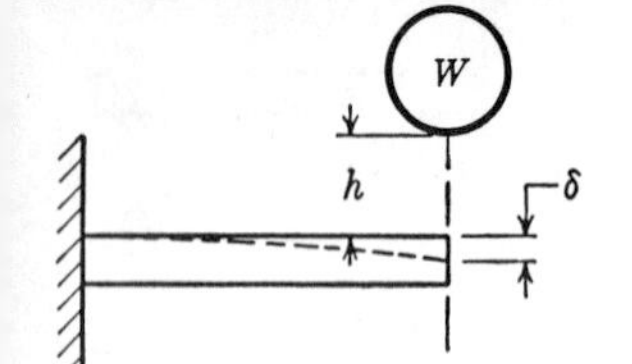

FIGURE 4.34 Impact on Cantilever. Length of beam = L; $\delta = y$.

formulas such as those in Table AT 2; $ky^2/2$ is the elastic energy stored in the beam. For example, for a cantilever beam, Fig. 4.34, we find $y = FL^3/(3EI)$, so that the scale of the beam as a spring, with respect to the point of application of the load, is

(h)
$$k = \frac{F}{y} = \frac{3EI}{L^3} \text{ lb./in.}$$

In one type of problem, one may substitute this value of k into (4.10) and solve for y. Then solve for y_{st} for a static load W and the corresponding stress s_{st}, as usual;

(i)
$$y_{st} = \frac{WL^3}{3EI} \quad \text{and} \quad s_{st} = \frac{Mc}{I} = \frac{WLc}{I}.$$

Finally, if the action is elastic, the stress is proportional to strain; so, the maximum stress under impact is

(j)
$$s = s_{st}\frac{y}{y_{st}}.$$

Other types of beams may be similarly handled.

4.44 EFFECT OF MASS OF BEAM.

If the value of k from **(h)** is substituted into (4.10), one may solve for y, and use $y_{st} = WL^3/(3EI)$; the result is

$$y = y_{st} + y_{st}\left(1 + \frac{2h}{y_{st}}\right)^{1/2} = y_{st} + y_{st}\left(1 + \frac{v^2}{g_o y_{st}}\right)^{1/2}, \tag{4.13}$$

which is seen to be the same as equation **(t)**, § 4.36; $y = \delta$. In this form we can introduce a correction factor on h (or v^2), which in terms of *equivalent* mass W_e is the same as before, namely $1/(1 + W_e/W)$. But the values of W_e are different: for the cantilever beam, Fig. 4.34, $W_e = 33W_b/140$; for a simple beam, loaded at the center, $W_e = 17W_b/35$; where W_b is the mass of the beam.[0.2]

4.45 GENERAL REMARKS ON IMPACT.

It can be said of the foregoing analyses that the deflections are predicted within a reasonable approximation, but the computed maximum stresses may be in serious error. Also, if the rod or beam has a uniform cross-section, the induced stress varies inversely as the volume of material. With respect to a beam of a certain sectional area, for example, if its length is doubled, its elastic capacity to absorb energy is *increased* by the ratio of $\sqrt{2}$, or about 40% ($h \gg y$), whereas its capacity for static loads is *reduced* by half.

The natural frequency of vibration ϕ_n can be expressed in terms of the static deflection y_{st} or δ_{st},

$$\text{(k)} \qquad \phi_n = \frac{1}{2\pi}\left(\frac{g_o}{y_{st}}\right)^{1/2} = 3.13\left(\frac{1}{y_{st}}\right)^{1/2} = 3.13\left(\frac{1}{\delta_{st}}\right)^{1/2},$$

which in turn, can be used to replace y_{st} with ϕ_n in the various foregoing equations.

4.46 CLOSURE. It should be evident by now that to design for varying loads by considering the load static and trying to cover by a large design (ignorance) factor on the yield or ultimate strengths is fraught with danger or conducive to over design. It often happens that a combination of extreme conditions results in failure, even when the designer thought he was being conservative. In spite of the necessarily empirical approach to varying loading, the detailed thinking of and allowing for the various factors that affect the design lead to improvement. Eventually, we may have something in the way of a theory (conceivably related to metallographic structure and microstresses) that provides a more reliable approach, but no suitable one is available now. There is scientific interest in microstresses, getting down to the stress in a single crystal, whereas the stresses we compute are statistical averages, which we call macrostresses. Even for static design, one extrapolates from laboratory data to a finished structure; so live-load design is not too different except in detail. Inevitably in complex systems and machines, there are points of high stresses (weak spots) that the designer cannot possibly foresee; hence, this hazard is always present. There are a number of factors that the designer can misjudge: lead alloy makes for easier machining of steel, but weakens it in fatigue; rolling or peening can be overdone, injuring the material (surface cracks) rather than strengthening it; stress concentrations at points of support, as for a pressure vessel, may be overlooked; distortions from heat treating; unanticipated effects of deflections; etc. In general, a respectable percentage of failures can be eliminated if the engineer follows his design all the way through manufacturing; many failures are attributable to discontinuities (tool marks) left in processing and in wrong fillet radii that are insignificant from the standpoint of manufacture. Be generous with the size of fillets at re-entrant corners, make changes in contiguous dimensions as gradual as possible, let residual stresses be compressive if you can, and as a last exhortation, be wary of the highly simplified generalizations—in this book or elsewhere.

5. SCREW FASTENINGS AND RIVETS

5.1 INTRODUCTION. There are several methods of joining the parts permanently or semipermanently—by welding, by riveting, by the use of screw fastenings, and by numerous special means. The screw fastening is one of the most useful elements of machines. Its design varies from the simple extreme of some casual calculation to dependence on extensive experiments intended to simulate a particular environment.

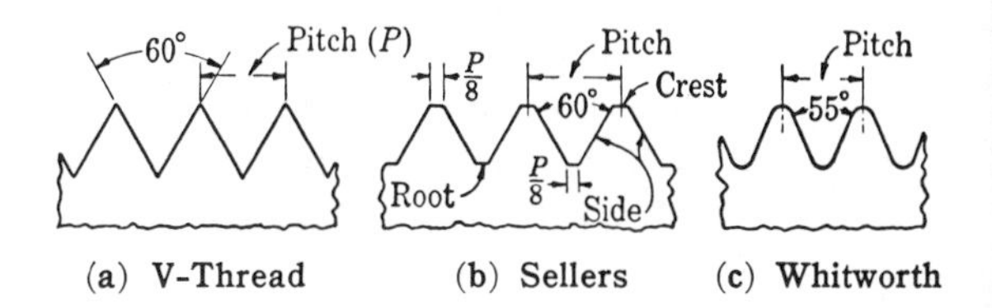

FIGURE 5.1 Thread Forms for Screws.

5.2 KIND OF THREADS. The sharp crest and root of the V-thread, Fig. 5.1, occasionally cut on lathes are undesirable because the thin material is easily injured and because the concentration of stress at the root of the thread is large. William Sellers proposed (1864) the form in Fig. 5.1(b), with flat crest and root, which partially removes the inherent weakness of the V-thread. The Sellers thread was the U.S. standard for many years.

The Whitworth thread (1841), Fig. 5.1 (c), featuring rounded crests and roots, has been the standard in Britain. It has better fatigue strength, because of the rounded root, than the Sellers thread.

The current U.S. standard, shown in Fig. 5.2, is in agreement with the international* Unified Standard. This Standard has the 60° thread angle of the old American standard and the optional rounded root of the British standard for an external thread; the crest may be either flat or rounded, as shown. There are similar choices for the internal thread, Fig. 5.2(b).

5.3 DEFINITIONS. The ***major diameter*** is the diameter of the imaginary cylinder that bounds the crests of an external thread and the roots of an internal thread; it is the *largest* diameter of the screw thread. This dimension has been called the "outside diameter," a confusing term when applied

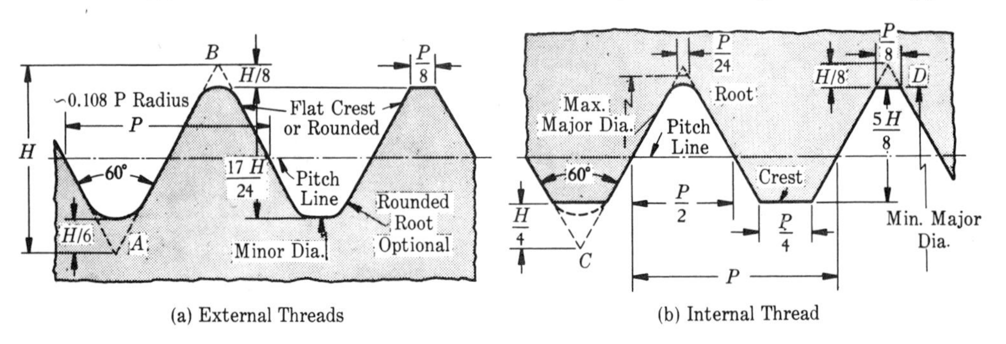

FIGURE 5.2 Unified Screw Threads.[5·1] **At *A* and *B* in (a) and *C* and *D* in (b), see how the thread is derived from the V-thread (height *H*); $H = 0{\cdot}866P$, where P = pitch. Some dimensions and thread variations are shown.**

to threads cut internally. The ***size*** of a screw is its nominal major diameter. The ***minor diameter*** is the diameter of the imaginary cylinder that bounds the roots of an external thread or the crests of an internal thread. This dimension has been commonly called the "root diameter." The ***pitch P*** is the axial distance from a point on a screw thread to a corresponding point on the adjacent thread.

$$\textbf{(a)} \qquad P\text{ (inches)} = \frac{1}{\text{number of threads per inch}}.$$

The ***lead*** is the distance in inches a screw thread (a helix) advances axially in one turn. On a single-thread screw, the lead and pitch are identical; on a double-thread screw, the lead is twice the pitch; on a triple-thread screw, the lead is three times the pitch; etc. See Fig. 8.21.

* U.S.A., Britain, Canada.

5.4 STANDARD THREADS. There are a number of "standard" threads, some quite specialized, concerning which, refer to the standard.[5.1] Tables AT 14 and 5.1 give certain data for some widely used threads.

The ***coarse-thread series*** (designated UNC) is recommended for general use; where jar and vibration are not important factors, where disassembly of parts is frequent, and where tapped holes are in metals other than steel. Always use the coarse thread unless there is a reason for using another.

The ***fine-thread series*** (designated UNF) is frequently used in automotive and aircraft work; especially where jar and vibration (tending to loosen the nut) are present, where a fine adjustment is required with the use of a castle nut, and where tapped holes are in steel (avoid UNF in brittle materials).

The ***extra-fine-thread series*** (designated UNEF) is particularly useful in aeronautical equipment; suitable where thin-walled material is to be threaded, where fine adjustments are required, and where jar and vibration are excessive. Screws with these threads are likely to be made of heat-treated alloy steel.

The ***8-thread series*** (designated 8 UN), Table 5.1, is used on bolts for high-pressure pipe flanges, cylinder-head studs, etc. There are several constant pitch series, for example, 12 UN (see under Fine, Table AT 14), 16 UN, 20 UN. These threads are useful because, in the regular coarse thread, the pitch continues to increase with size and it becomes increasingly difficult to induce the desired initial tension in the bolt, so necessary for tight joints. Even for the 8 UN thread, impact wrenches or very large leverages are needed for tightening to stresses of the order of the yield strength.

5.5 FITS OF THREADS. The same types of fits as defined in § 3.4 are used; for threads, the defined tolerances are designated 1A, 2A, 3A for external threads, 1B, 2B, 3B for internal threads.[5.1] Classes 1A and 1B have the widest tolerances and are used where quick and easy assembly, even with bruised threads, is necessary, as in ordnance.

Classes 2A and 2B, with tolerances suitable for normal production practices, are the most widely used. Clearance for this fit minimizes galling and seizing in highly tightened joints and in high-temperature applications, and it also accommodates plating. Use this class unless there is a reason for using another. Class 2A provides a clearance when mated with any class of internal thread.

Classes 3A and 3B have the closest fit; allowance (but not the clearance—§ 3.12) is zero. Use only to meet exacting requirements. Other fits may be obtained by using internal and external threads from different classes, a 2 with a 3, for example. Interference fits are defined in ASA standard B 1.12.

TABLE 5.1 EXTRA-FINE AND 8-THREAD SERIES

Unified and American Standard. See Subheading on Table AT 14

EXTRA-FINE SERIES (NEF AND UNEF)

Size	*Basic Major Dia. in.*	*Th./In. (tpi)*	*Minor Dia. Ext. Th.*	*Stress Area A_s sq. in.*
12	0.2160	32	0.1777	0.0270
1/4	0.2500	32	0.2117	0.0379
5/16	0.3125	32	0.2742	0.0625
3/8	0.3750	32	0.3367	0.0932
7/16	0.4375	28	0.3937	0.1274
1/2	0.5000	28	0.4562	0.170
9/16	0.5625	24	0.5114	0.214
5/8	0.6250	24	0.5739	0.268
11/16	0.6875	24	0.6364	0.329
3/4	0.7500	20	0.6887	0.386
13/16	0.8125	20	0.7512	0.458
7/8	0.8750	20	0.8137	0.536
15/16	0.9375	20	0.8762	0.620
1	1.0000	20	0.9387	0.711
1 1/16	1.0625	18	0.9943	0.799
1 1/8	1.1250	18	1.0568	0.901
1 3/16	1.1875	18	1.1193	1.009
1 1/4	1.2500	18	1.1818	1.123
1 5/16	1.3125	18	1.2443	1.244
1 3/8	1.3750	18	1.3068	1.370
1 7/16	1.4375	18	1.3693	1.503
1 1/2	1.5000	18	1.4318	1.64
1 9/16	1.5625	18	1.4943	1.79
1 5/8	1.6250	18	1.5568	1.94
1 11/16	1.6875	18	1.6193	2.10

8-THREAD SERIES (8 N AND 8 UN) (PRIMARY SIZES)

Size	*Minor Dia. Ext. Th.*	*Stress Area A_s sq. in.*
1	0.8466	0.606
1 1/8	0.9716	0.790
1 1/4	1.0966	1.000
1 3/8	1.2216	1.233
1 1/2	1.3466	1.492
1 5/8	1.4716	1.78
1 3/4	1.5966	2.08
1 7/8	1.7216	2.41
2	1.8466	2.77
2 1/4	2.0966	3.56
2 1/2	2.3466	4.44
2 3/4	2.5966	5.43
3	2.8466	6.51
3 1/4	3.0966	7.6738
3 1/2	3.3466	8.96
3 3/4	3.5966	10.34
4	3.8466	11.81
4 1/4	4.0966	11.38
4 1/2	4.3466	15.1
4 3/4	4.5966	16.8
5	4.8466	18.7
5 1/4	5.0966	20.7
5 1/2	5.3466	22.7
5 3/4	5.5966	24.9
6	5.8466	27.1

Identification symbols for use on drawings, shop and storeroom, cards etc., should be according to the following examples.

(a) An externally threaded part, 1-in. diameter, unified coarse thread, 8 threads per inch, class 2A tolerance is designated as follows:

1″—8 UNC—2A.

(b) An internal thread, 1-in. diameter, unified fine thread, 12 threads per inch, class 2B tolerance, left-hand thread is designated

1″—12 UNF—2B—LH

5.6 DESIGN OF BOLTS—INITIAL TENSION UNKNOWN. A mechanic with a typical set of wrenches will tighten a small bolt to a higher initial stress than he will a larger one. Partly for this reason, the design stress for bolts and screws should be a function of size when the computations consider only the external load. Some years ago, Seaton and Routhewaite[5.8] proposed that the design stress in terms of the root area A_r of the thread be $s = CA_r^{5/12}$. Considering the approximate nature of the design, we could use an exponent of 1/2 instead of 5/12, use the tensile stress area A_s for convenience (Tables AT 14 and 5.1) instead of the root area A_r, and obtain $s = CA_s^{1/2}$. The value of C depends upon the yield strength of the material and may conservatively be taken as $C = s_y/6$. Thus, the design stress for "well-tightened" bolts or screws may be used as

$$\text{(b)} \qquad \text{Design tensile strength} = s_d = \frac{s_y}{6}(A_s)^{1/2}.$$

Substituting this value of s_d into $F = sA$, we find

$$\text{(5.1)} \qquad F_e = \frac{s_y}{6}(A_s)^{1/2}A_s = \frac{s_y A_s^{3/2}}{6}, \qquad [D < \tfrac{3}{4}\text{ in.}]$$

from which a safe external tensile load F_e may be found, or the needed stress area A_s determined for a particular external load. From a computed area A_s, determine the nominal size of bolt from Table AT 14 or 5.1. The constant 6 in equation (5.1) is *not* a factor of safety. Equation (5.1) may be applied to any thread series, although there are no tests to substantiate its use for other than a coarse thread. AISC[5.34] specifies allowable tensile stresses as follows: $0.4s_y$ in general; for A 307, 14 ksi (see § 5.8); for A 325, 40 ksi; for A 354, 50 ksi; values that serve as a guide for bolt sizes $D \geqq \frac{3}{4}$ in.

5.7 INITIAL TENSION AND TIGHTENING TORQUE.

The stress or load induced by the tightening operation is called the ***initial tension,*** which, with ordinary wrenches, depends upon the workman, how he is feeling, the length of wrench used, as well as the condition of the bolt or screw. Where the amount of initial tension is important, a torque wrench should be used. Even so, there will be a *large variation* of the induced stress because of the way the threads are finished, their lubrication, and the other variables of the application; see Fig. 5.3. The relation between the applied torque T in-lb. and initial tension F_i lb. proposed by Maney[5.10] is

$$\text{(5.2)} \qquad T = CDF_i \text{ in-lb.,} \qquad \begin{cases} \text{As received[5.10]} & C = 0.20 \\ \text{Lubricated[5.14]} & C = 0.15 \end{cases}$$

where D is the nominal bolt size and C, called the *torque coefficient*, is taken as a constant for a particular set of conditions. Equation (5.2) is obtained from a force analysis of the thread (the same as for a worm thread, § 16.8), plus an allowance for friction between face of nut (or head of the

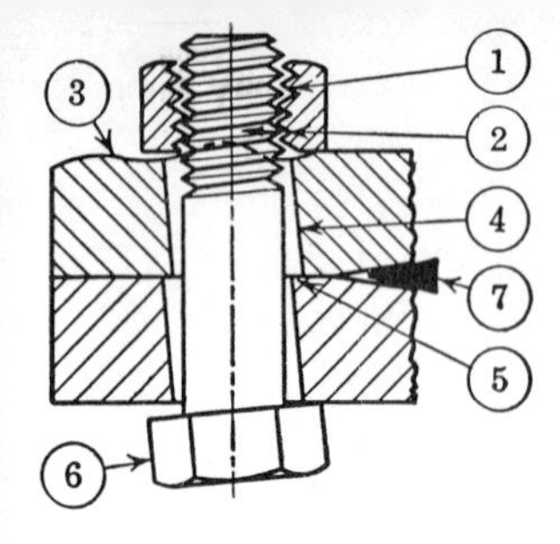

FIGURE 5.3 Difficulties in a Bolted Connection. (1) Load not distributed over all threads. (2) Axis of internal threads not perpendicular to seating face of nut. (3) Surface not flat and perpendicular to axis of bolt. (4) Hole not perpendicular to surface (and parallel to axis). (5) Misaligned holes. (6) Bearing surface on head not perpendicular to axis. (7) Also, the manner of applying the external load may result in bending the bolt. And there is a torsional stress due to tightening. Considering the nature of these faults, we decide that rarely is the load on a bolt purely tensile.

screw if the head is being turned), which is computed as in § 18.10. The value of $C \approx 0.2$ is obtained when the coefficient of friction is taken as $f = 0.15$; this value of C is considered typical and is recommended for use unless another value is known to apply. The "as-received" condition includes the remains of cutting fluid, for example, but no foreign particles. Values of C have been observed in the literature from 0.1 to 0.34 or more. The most unpredictable cases would probably be those where the surfaces have been cleaned and dried. Some data suggest $C \approx 0.14$ if the threads are lubricated with molybdenum disulfide.[5.13] There is some evidence that for UNF threads, C is some 6–10% less than for coarse threads. Plating generally reduces the friction, another situation calling for different values of C. If the actual value of $C = 0.1$ and if the tightening torque used has been computed for $C = 0.2$, the initial tension in the bolt is twice that planned—if the bolt does not yield and break. Thus, if the parts are lubricated, tests may be advisable to avoid overstressing the fasteners. The value of C tends to increase with decrease in size; an average value for $\frac{1}{4}$-UNC-20 is 0.255.[5.14]

Tightening the nut produces a shear stress in the bolt (by the frictional torque);

$$\text{Torque producing shear in bolt} \approx (0.4)(\text{total } T)$$

The torsional stress is generally ignored in design because, first, it is relatively small, and second, it probably mostly disappears as the machine operates (a relaxing of the frictional forces). It is not uncommon for the initial tension to decrease because of seating action on contact surfaces and perhaps because of actual yielding of the materials; also of course if the nut loosens. Where it is important for joints to remain tight, the tightness of the bolts should be checked later.

A torque wrench or other adjustable power tool is the most widely used means of inducing the approximate desired initial tension in smaller bolts. It is easy to tighten smaller bolts, say $\frac{1}{2}$ in. or less for ordinary steel, until they break. For the larger sizes, pneumatic impact wrenches are used, and sometimes the bolts are heated when they are tightened so that the initial tension is greater after they have cooled. The turn-of-the-nut plan is: turn the nut very tight in order to seat the surfaces (a step sometimes omitted), back off nut, turn nut snug (finger snug), then turn it through a precomputed angle to induce a particular strain, which is proportional to the desired stress. (The contact surfaces deform too.) Approved procedures in bolted structures are somewhat different. In another plan, if both ends of the bolts are accessible for the measurement of bolt length, the initial tension may be estimated from the measured total elongation δ. Usually, there is an unknown factor in this procedure, the ***effective strain length*** L_e, which may be defined as that length which when divided into the total elongation gives the unit elongation at some section of the bolt, as at the root section; that is, $\epsilon = \delta/L_e$. Knowing ϵ, we have $s = \epsilon E$. The effective length is closely some function of the ***grip*** and of the proportion of shank that is threaded.

5.8 MATERIALS AND STRENGTH OF THREADED PRODUCTS.

The designer can use any desired material for bolts and screws, but he would use only "ordinary" steel (ASTM A 307 and SAE Grade 1, minimum $s_u = 55$ ksi—equivalent to about 1015, for example), bolts cold headed, without a justifying reason. The SAE and ASTM and several government agencies have standardized specifications for screw-fastening materials. Among the most commonly used steels for bolts are;[2.1] AISI 1013, 1018, 1038, 1041, 1054, 1340, 4037, 4140, 4150, 50B40, 8635, 8735, 4340; but a particular manufacturer's processes may be more readily adapted to one steel than another.

Bolt design is often done with a ***proof load*** or ***proof stress*** s_p. In general, a proof load is some load agreed to by the purchaser and vendor as a capacity to be met. For bolts and screws, there seems to be no unique definition, but the proof stress s_p in the SAE specifications is usually close to 96% of the 0.2% offset yield strength (which may not be the same as the yield strength of a standard tensile specimen). See Table 5.2.

TABLE 5.2
MINIMUM STRENGTHS (ksi) OF BOLTS (SELECTED STANDARD SPECIFICATIONS)

GRADE	SIZES, INCL.	s_p	s_u	s_y
SAE Grade 2	$\frac{1}{4}$–$\frac{3}{4}$	55	69	
	$\frac{7}{8}$–1	52	64	
	$1\frac{1}{8}$–$1\frac{1}{2}$	28	55	
SAE Grade 5, ASTM A325	$\frac{1}{4}$–$\frac{3}{4}$	85	120	88
	$\frac{7}{8}$–1	78	115	81
	$1\frac{1}{8}$–$1\frac{1}{2}$	74	105	77
ASTM 354	BB $\frac{1}{4}$–$2\frac{1}{2}$	80	105	83
	BB $2\frac{1}{2}$–4	75	100	78
	BC $\frac{1}{4}$–$2\frac{1}{2}$	105	125	109
	BC $2\frac{1}{2}$–4	95	115	99
	BD $\frac{1}{4}$–$1\frac{1}{2}$	120	150	125

The ASTM and SAE specifications can be met by many "standard" steels; SAE 1041 QT, for example, can easily meet SAE Grade 5, a good-strength bolt material. In fact, 1041 heat treated to about $s_u = 100$ ksi (BHN ≈ 200) or a little more, will come close to giving the most holding capacity per dollar.[2.1] A low-carbon steel can meet the requirements of Grade 2, Table 5.2, and bolts of this grade are typically cold headed. The medium carbon grades must be hot headed. The resulfurized steels, 11xx series, tend to be damaged by cold heading; so these are used primarily for studs to be machined from bar stock. For ultimate strengths greater than about 100 ksi, medium carbon alloy steels are common; such steels are used to meet the ASTM A 354 specification, Table 5.2. Recall that as size increases, increased hardenability is necessary to maintain the same strength, or else the strength decreases, as allowed by the specification. One large consumer of stud bolts for automotive cylinder heads eliminated assembly-line trouble in the form of stretching and breaking quenched-and-tempered bolts by switching to elevated-temperature-drawn 4140, which came out with properties of $s_u = 150$, $s_y = 130$, $s_p = 120$ ksi, min., without heat treatment; see § 2.9.

If the threads are manufactured by rolling (rolled threads—large deformation of the material), as opposed to cutting, the resulting surface is of the order of 4–32 μin., § 3.14. When the material has $s_u < 80$ ksi, there is little change in fatigue strength as compared with cut threads; but for higher strength steels, rolled threads show a marked improvement in fatigue, as much as 2 or 3 times as great for $s_u > 200$ ksi.[4.62] See Table AT 12 for values of the strength reduction factor K_f, which include the effect of the manufacturing method.

To complicate matters, Sigwart[4.28] found that the residual compressive stress at the root of the rolled thread may not be as great as that at the root of a cut thread, especially if a dull cutter was used. The best improvement was found for ground threads that had been locally rolled at the root (minor plastic deformation—§ 4.28) after heat treatment, an expensive procedure. In any event, if beneficial residual stresses are to remain, the rolling must be done *after* heat treatment, because otherwise the cold-working effects would be lost. As usual, chromium and nickel plating reduce fatigue strength; the effect of cadmium and zinc is smaller. Fatigue failure nearly always occurs at the first thread in the nut, the end of the threaded portion on the shank, or, especially if the threads are rolled, under the head. Thus, if every advantage is being taken to improve fatigue of bolts, the fillet at the head must be considered. It is often rolled. Besides rolling threads after heat treatment, other usual precautions to preserve or improve fatigue strength may be taken, for example, avoid decarburization. Self-aligning nuts, Fig. 5.31, reduce the bending of the bolt mostly due to surfaces not being exactly perpendicular to the bolt's axis, and they prolong fatigue life; an item to consider if there is failure from fatigue and for other reasons.[5.28] Bolts designed for finite fatigue life should be tested to prove the design if failure would lead to dire consequences. One manufacturer claims a repeated-load fatigue strength for his cap screws up to $\frac{1}{2}$-in. diameter of 40 ksi (with threads), stressing them from $s_{min} = 4$ (to 40 ksi), using 8740 and rolled curved roots. For suggestions for geometric improvements of fatigue strength, see Fig. 5.4. With the diameter reduced between "guide surfaces," the capacity to absorb energy is also increased (§ 4.40).

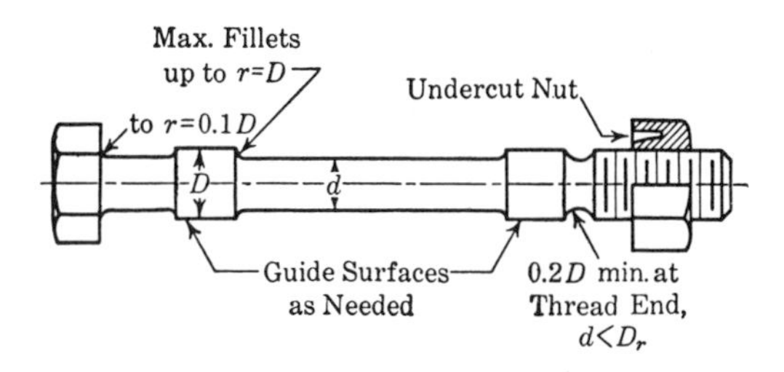

FIGURE 5.4 Bolt for Repeated Loads. The undercut of the nut improves the distribution of the load on the nut threads. The radius of curvature of the groove at the end of the threads should be at least 0.2*D*, preferably 0.5*D*, with a diameter *d* somewhat less than the minor thread diameter D_r. Rather than this groove, the threads could be ended on a gradual taper, instead of an abrupt ending at full depth.

Notice the "undercut nut" of Fig. 5.4. With a regular nut, the first thread may be loaded about 1.8 times the average per thread (some 2.3 for fine threads), whereas the "top" thread takes about half the average ($\frac{1}{3}$ for fine).[0.2] In any case, it is impossible to get a uniform distribution of load on the threads. With the undercut, the reduced rigidity at the bottom of the nut permits the lower part to stretch more thus distributing the load more evenly among the threads. In another design to accomplish the same purpose, the nut is tapered toward the bottom; but of course these special designs are more expensive and would not be used unless needed. In general, the nut material may be somewhat weaker than that of the bolt, but specifications require that the threads not fail before the bolt fails in tension. Thus, better than ordinary, perhaps heat-treated nuts (and hardened washers if the material of the connected parts is not as hard) are required for the higher-strength bolts.

Many screws are made of stainless steel, nonferrous metals, and plastics (for example, Zytel, Teflon,[2.33]) which are used for a reason—corrosion resistance, high or very low temperatures, light weight, electrical conductance, insulation, etc. See Chapter 2. One manufacturer[19.24] indicates that the breaking strength of $\frac{1}{4}$ to $\frac{3}{4}$-in. aluminum bolts (2024-T 4, 6061-T 6) is about 62–64 ksi based on the stress area.

In the design procedure, one might decide upon proof stress. From the proof stress, the next decision is the initial tightening stress. In structures, the tendency is to tighten the bolt to or beyond the yield, $s_i \approx s_y$, in *high-*

strength material (which does not have a distinct yield *point*). Erker[4.28] found that after tightening bolts to just above the yield strength, small plastic deformation in service reduced the clamping stress to an optimum value. On the other hand, a review of the literature suggests that machine designers tend toward an initial stress somewhat less than s_p or s_y; one finds values from $0.75s_y$ (by some automobile manufacturers) to $0.9s_y$ or more. Typical values for bolts to be subjected to a tensile load are

(c) $s_i \approx 0.9s_p$ [WHEN PROOF STRESS AVAILABLE] or $s_i \approx 0.85s_y$, [NO PROOF STRESS]

which values we shall assume are satisfactory, *especially with no gasket*. In the case of a soft steel, a material with a yield point, one must be careful not to tighten to the yield stress because of the relaxation and lower clamping force that follows from the yielding. The higher-strength steels may receive a permanent deformation, but relaxation does not occur at ordinary temperatures.

Having decided upon a suitable s_i, the initial tightening force is $F_i = s_iA_s$; then equation (5.2), with a proper value of C, can be used to estimate the tightening torque (or one can determine the angle through which to turn the nut to produce s_i—§ 5.7).

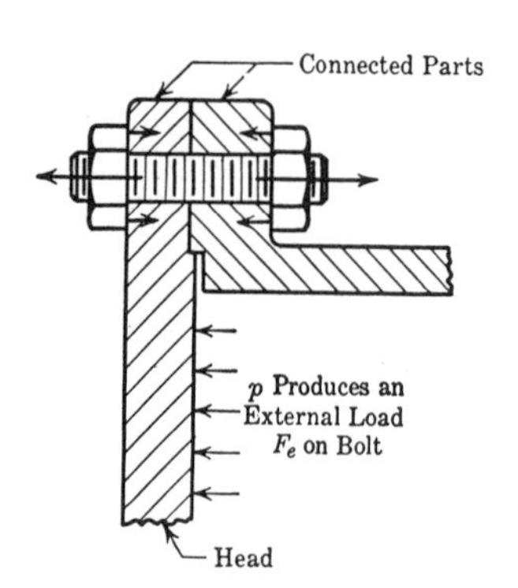

FIGURE 5.5 Head Bolted on Vessel. Internal pressure is *p*.

5.9 ELASTIC ANALYSIS OF BOLTS FOR JOINTS. An oft-heard rule of thumb when the parts to be joined are relatively rigid is *tighten the bolt (or screw) so that the initial tension is greater than the applied external load*, a rule that should yield safe designs if the bolts or screws are known to be tightened to the required initial tension. However, an engineer is more comfortable with an analysis that guides him in the correct direction. First, we shall see what load is required to open a joint, such as the one suggested by Fig. 5.5.

As the nut is tightened, the load on the bolt increases, and the deformation of the bolt increases. Within the elastic range, Hooke's law applies,

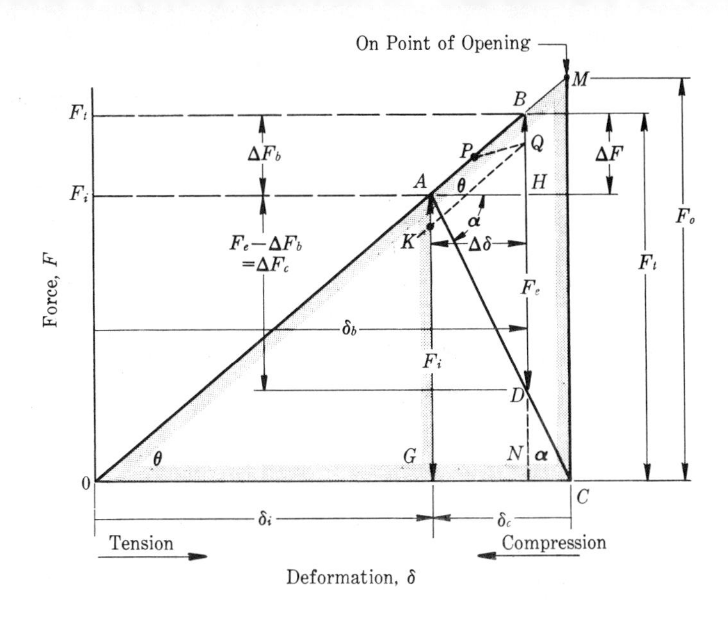

FIGURE 5.6 Forces on a Screw Fastening. The slopes of the F-δ lines are $k_b = F/\delta_b$, $k_c = F/\delta_c$. Suppose, when the external load F_e is applied, that plastic deformation of the bolt occurs, idealized for a good strength material as PQ; F_e is then represented by DQ, the effective F_i is reduced to some value GK because of the resulting permanent deformation.

and the force-deformation curve for the bolt is a straight line, represented by OAM in Fig. 5.6. Also, the connected members deform (in compression), and if they too are elastic, their force-deformation curve is straight, represented by CA in Fig. 5.6.[5.18] The more rigid a member, the steeper is its F-δ curve, because it takes a larger force to produce a particular deformation. Usually the connected members are more rigid than the bolt, as shown in Fig. 5.6 with $\alpha > \theta$. The slope of CDA is negative and represents a compressive deformation.

Suppose we stop tightening at a point A. The load on the bolt and on the connected part is F_i, the initial tightening load. The initial elongation of the bolt is δ_i, and the corresponding compressive deformation of the connected parts is δ_c. To get the external load that would cause a joint such as that in Fig. 5.5 to open, assume that the bolts do not bend, which is equivalent to assuming that the head and flange do not bend, and let an external load F_e be applied. The bolt elongates more $\Delta\delta$, say to B, Fig. 5.6, and the deformation of the connected parts *decreases* the same amount, $\Delta\delta$. The load on the bolt increases an amount ΔF_b; the load on the connected parts decreases a greater amount ΔF_c if they are more rigid. For elastic deformations, the bolt elongation continues along the line OM, and the compressive deformation decreases along AC. The joint will be on the point of opening when the deformation of the connected parts becomes zero, at C, because if the bolt is stretched any further, the connected parts can no longer expand to maintain the surfaces in contact. At the instant marked by C, the total elongation of the bolt is represented by the distance

OC, and the total load on the bolt is $CM = F_o$, the limiting load for opening of the joint, which is also the external load at this limiting condition.

Since the triangles OGA and OCM are similar,

$$\text{(c)} \qquad \frac{F_o}{F_i} = \frac{\delta_i + \delta_c}{\delta_i}, \qquad \text{or} \qquad F_o = F_i\left(\frac{\delta_i + \delta_c}{\delta_i}\right).$$

With the parts acting as springs (Hooke's law), their deformations in terms of their spring constants from $k = F/\delta$, § 4.35, are $\delta_i = F_i/k_b$ for the bolt and $\delta_c = F_i/k_c$ for the connected parts. Using these values in **(c)**, we have

$$\text{(d)} \qquad F_o = F_i\left(\frac{k_b + k_c}{k_c}\right) \qquad \text{or} \qquad F_i = F_o\left(\frac{k_c}{k_b + k_c}\right).$$

In these expressions, F_o is the external load that would place the joint on the point of opening when the bolt had been tightened to a value F_i; or

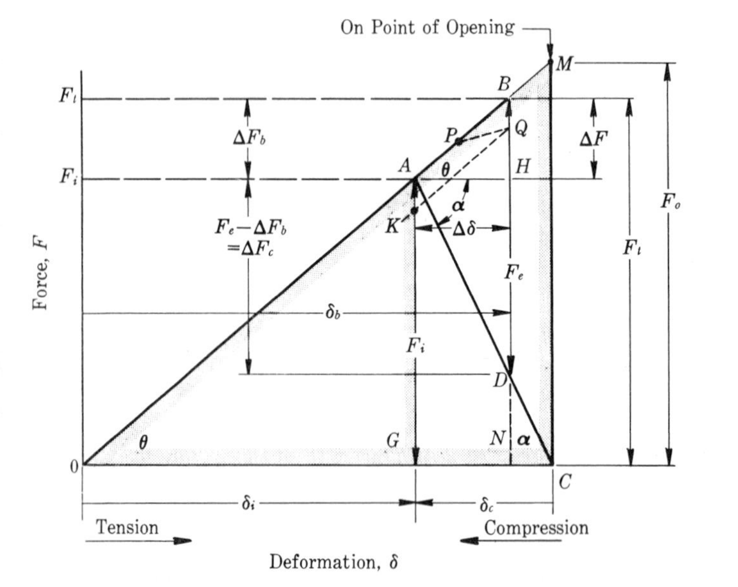

FIGURE 5.6 (Repeated).

if F_o is a known maximum external load, then F_i is the minimum initial tightening load that must be used. Practically, F_i should be greater than this value, 1.2 to 2 times as great when there is no gasket.[5.19] That is, let $F_o = QF_e$ in **(d)**, where F_e is some actual external load, and find

$$(5.3) \qquad F_i = QF_e\left(\frac{k_c}{k_b + k_c}\right) \text{ lb.}$$

Keep in mind that a joint intended to resist leakage may be subjected to a hydrostatic test pressure of 1.5 to 2 times the working pressure. Depending on the values of Q, k_c and k_b, the computed value of F_i from (5.3) may be either smaller or larger than the external load F_e.

In Fig. 5.6, suppose a bolt is tightened to the F_i shown and an external load F_e is applied to the joint, all elastically. The bolt stretches an additional $\Delta\delta$ and the total load F_t on it is marked by point B, with a change of force ΔF_b corresponding to an increase of deformation $\Delta\delta$. The change of load on the connected parts becomes $\Delta F_c = HD = F_e - \Delta F_b$. From $\delta = F/k$, the change in deformation is equal to the change in force divided by the corresponding spring constant; that is,

$$\text{(e)} \qquad \Delta\delta = \frac{\Delta F_c}{k_c} = \frac{F_e - \Delta F_b}{k_c} = \frac{\Delta F_b}{k_b}.$$

Solving for ΔF_b from the last two parts, we get

$$\text{(f)} \qquad \Delta F_b = F_e\left(\frac{k_b}{k_b + k_c}\right).$$

Then the total load on the bolt is, Fig. 5.6,

$$\text{(5.4)} \qquad F_t = F_i + \Delta F_b = F_i + \left(\frac{k_b}{k_b + k_c}\right)F_e.$$

Similarly, the net compressive force on the connected parts is found to be

$$\text{(g)} \qquad F_c = F_i - \left(\frac{k_c}{k_b + k_c}\right)F_e.$$

A negative answer from this equation indicates no force on the connected parts. If the stiffness k_c of the connected parts cannot be determined with assurance, it is always on the safe side to use the term in parentheses as unity; that is, $F_t = F_i + F_e$. See below. When the external load F_e varies, the total load F_t varies in accordance with (5.4); the mean load is $F_m = F_i + \Delta F_b/2$ and the alternating component $F_a = \Delta F_b/2$. The nominal stress for any F is $s_o = F/A_s$.

Examining equation (5.4), we see that if the stiffness of the bolt k_b is very large as compared to k_c, the total load F_t approaches $F_i + F_e$. If k_b is very small as compared to k_c, the term in parentheses becomes small, and the total load approaches F_i. Therefore, the actual load is always *between* the initial tension *and* the sum of the initial tension plus the external load (provided the joint does not open). These remarks are interpreted graphically in Fig. 5.7 (p. 168). Notice that if all parts are elastic, any external load, no matter how small, results in an increased load on the bolt.

The foregoing analyses are inadequate when the bolt is subjected to significant bending moments; hence, when the gasket is inside of the bolt circle, a common arrangement for flat or ring gaskets, Fig. 5.8 (p. 168), the flanges must be heavy enough that bending is small. Also the bolts should be

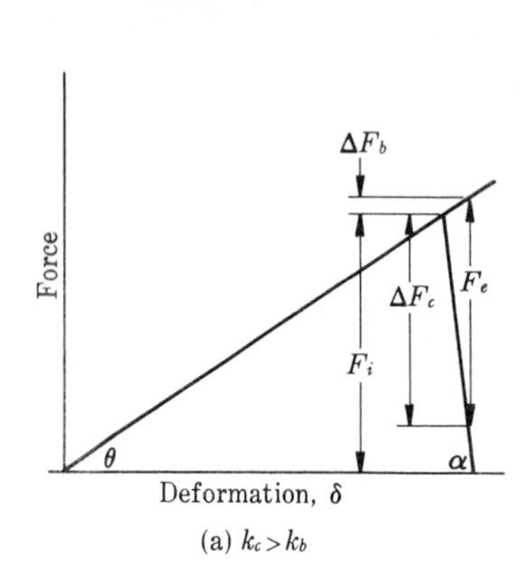

(a) $k_c > k_b$

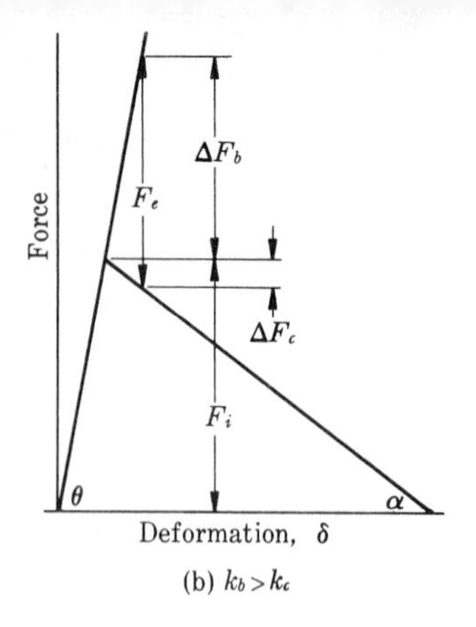

(b) $k_b > k_c$

FIGURE 5.7 Effect of Relative Stiffness of Bolt and Connected Parts. These figures are drawn for the same external load F_e. Notice that when the connected parts are much stiffer than the bolt, in (a), the load F_e does not cause much change ΔF in the load on the bolt. But if the bolt should be much stiffer than the parts, in (b), a very large part of the external load F_e is added to the initial load. The use of a gasket changes the relationship between F_e and ΔF_b in the direction shown in (b).

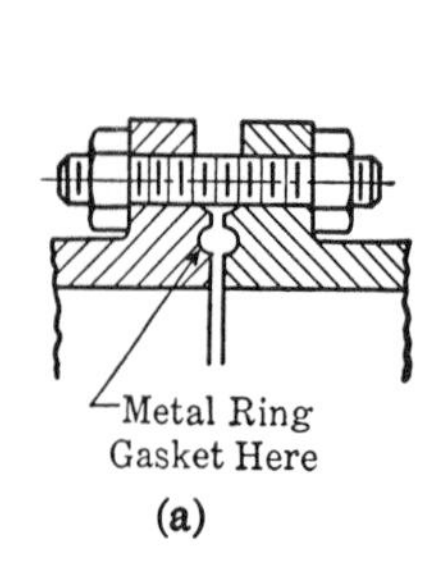

(a)

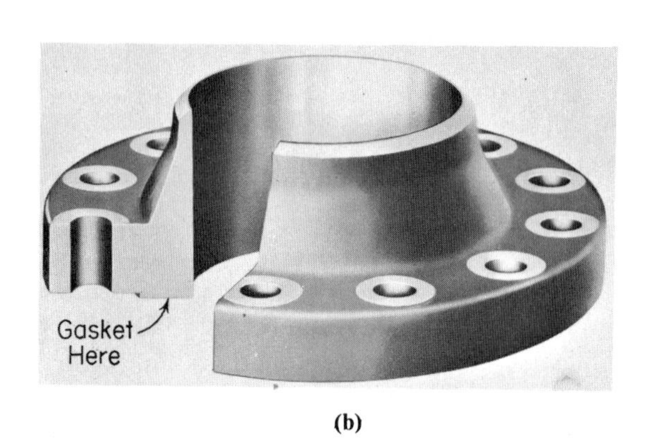

(b)

FIGURE 5.8 Flanges. (Courtesy Taylor Forge and Pipe Works, Chicago).

close enough together that virtually uniform pressure is induced between the faces.[5.22] From one code, we conclude that the spacing or ***pitch*** P of bolts should be: $P \leqq 7D$ for fluid pressures less than 50 psi; $P \approx 3.5D$ for fluid pressures of about 200 psi, where D is the size of bolt.[5.26] Observe that the foregoing analysis ignores the deflection of the bolt in the nut and the deflections of the threads, all of which would have an effect on k.

5.10 ELASTIC CONSTANTS AND GASKETS FOR CONNECTED PARTS. The elastic constant k_b is determined from equation **(u)**, § 4.36; that is $k_b = A_b E_b / L_b$; or if two diameters are involved on the bolt, use the equivalent spring constant from equation (4.11) of § 4.39: $k_b = k'$. The subscript b is a reminder that each variable applies to the *bolt*.

For the connected parts, the same principle holds, but for parts of indefinite extent (area), their deformation at some distance from the bolt is less than that in the immediate vicinity of the bolt. In this event, the usual procedure is to assume some equivalent area of the connected parts A_c, and use $k_c = A_cE_c/L_c$. One such estimation is[5.19]

$$A_c = \frac{\pi D_e^2}{4} - \frac{\pi D^2}{4}, \tag{g}$$

where D is the nominal diameter of the bolt hole, D_e is an "equivalent" diameter of the plate area considered to be in compression; take it as

$$\begin{aligned} D_e &= (\text{Nut or head width across flats}) + \frac{h}{2} \\ &= (\text{Dimension } A, \text{ for example, Table AT 14}) + \frac{h}{2}, \end{aligned} \tag{h}$$

where h is the ***grip*** of the bolt, which is the total thickness of the plates being clamped. If it is clear exactly what area is in compression, do not use **(g)** and **(h)**. See the example of § 5.12.

If the connected members are composed of two or more kinds of material—for example, a gasket between connected parts—the spring constant for the connection is (see equation (4.11), § 4.39)

$$\frac{1}{k_c} = \frac{1}{k_1} + \frac{1}{k_2} + \frac{1}{k_3} \cdots,$$

where k_1, k_2, k_3 are the spring constants for the individual components being clamped; $k_1 = A_1E_1/L_1$, etc.

If needs can be met economically without a gasket, this is the best solution. Next best is the use of as thin a gasket as possible. However, there must be enough thickness to let the gasket material flow into the roughnesses of the flange surfaces and to take care of some slight unparallelism of the mating surfaces. The amount of pressure on the gasket material, called the *flange pressure*, to cause it to respond in this manner is different for each material. For nonmetallic gaskets, it has been found that they should have a certain minimum amount of compression,[5.26] as 62% for a certain cork gasket; this compression corresponds to a certain required minimum flange pressure. Then the bolts must be designed and so tightened as to meet those requirements. Smoley[5.25] recommends the use of an "apparent" flange pressure p_g, which is said to be an upper probability limit such that if the initial bolt tension F_i is computed from this pressure and then if the bolts are tightened with a torque $T = 0.2DF_i$, equation (5.2), the required minimum flange pressure is assured. Certain apparent pressures are: for X, a cork-and-rubber mixture, $p_g = 1500$ psi; for Y, a cork composition,

p_g = 2700 psi; for Z, an asbestos base, p_g = 3300 psi. For a first approximation after a suitable gasket material has been selected, one might proceed as follows: estimate the area of gasket; multiply this by its minimum apparent flange pressure to get the total apparent load on the gasket; decide upon a suitable number of bolts and find the load per bolt F_b; the size of bolt is determined from $A_s = F_b/s$, where $s = 0.75s_y$, say. When the apparent flange pressure with its built-in margin of safety is being used, it would appear reasonable to assume that $F_i = F_b$ and compute bolt torque from $T = CDF_i$. The bolts must also be capable of carrying the external loading. There are many special considerations beyond the scope of this text.[5.26] Metal gaskets (aluminum, copper, monel, and others), with and without sandwich filler, are frequently used, and necessarily so at temperatures above some 850°F.

The modulus of elasticity for the nonmetallic materials is ordinarily not constant over a wide range of stress; but if it is desired to check by the principles of § 5.9, some typical values of E and gasket thickness h are:[5.23] rubber gasket, E = 7500 psi, h = 3/16 in.; vellumoid gasket, E = 19,000 psi, h = 1/16 in.; copper-clad asbestos gasket, E = 20,000 psi, h = 1/8 in.

5.11 EXAMPLE—STUD BOLTS FOR COMPRESSOR HEAD. The head of a 10 × 12-in. air compressor is to be held on by 10 stud bolts; the maximum (repeated) internal pressure is 140 psi. The bolts are to be made of cold-rolled C1118 with threads cut on their entire length. (a) Determine the size of bolt predicted by equation (5.1), which assumes a well-tightened joint. (b) Let a 0.02-in. thick gasket, designated X in § 5.10, be used for which the "apparent" flange pressure is p_g = 1.5 ksi; assume a gasket area of A_g = 70 in.² and a modulus of elasticity of E = 19 ksi. (The E values are so seldom used by the industry that they are hard to find.) Determine the initial tension, the bolt torque to be used, and the ratio of s_i/s_y to obtain the desired flange pressure. (c) The thickness of the steel cylinder head at the bolt holes is 1 in. Estimate the factor of safety by the Soderberg criterion for the bolt size obtained in (a).

Solution. (a) The total load F_t on the cylinder head is

$$F_t = p\frac{\pi D^2}{4} = \frac{(140)(\pi)(100)}{4} = 11{,}000 \text{ lb.}$$

The average external load on each of 10 bolts is 11/10 = 1.1 kips. From Table AT 7 for C 1118 cold finished, we have

$$s_u = 80 \text{ ksi} \qquad s_y = 75 \text{ ksi} \qquad \text{BHN} = 180.$$

Equation (5.1) now yields

$$A_s = \left(\frac{6F_e}{s_y}\right)^{2/3} = \left(\frac{6 \times 1.1}{75}\right)^{2/3} = 0.198 \text{ in}^2,$$

from which, we select the next larger size from Table AT 14; to wit, $D = \frac{5}{8}$ in.–11 UNC, A_s = 0.226 in.²

(b) The computed values of the initial tension and bolt torque that provide the required compression of the gasket are

$$F_b = F_i = \frac{s_g A_g}{N_b} = \frac{(1.5)(70)}{10} = 10.5 \text{ kips.}$$

$$T = 0.2DF_i = (0.2)(0.625)(10.5) = 1.31 \text{ in-kips.}$$

Notice that the initial tightening load needed for the gasket (10.5) is almost ten times the external load (1.1), but relatively higher fluid pressures would change this relation materially. For $s_i = 10.5/0.226 = 46.5$ ksi, we find $s_i/s_y = 46.5/75 = 0.62$, which, compared to the $0.75s_y$ informally suggested in § 5.10 in connection with gaskets, implies that if one designed with this latter approach, a smaller bolt may possibly be used. [After the computations of part (c) have been made, observe that the initial bolt tension required to compress the gasket material is much larger than the initial tension computed from equation (5.3).]

(c) For a grip of $1 + 0.02 = 1.02$ in. and the distance $A = 15/16$ in. from Table AT 14, the equivalent diameter and area from (**h**) and (g) are

$$D_e = 0.9375 + \frac{1.02}{2} = 1.4475 \text{ in.}; \qquad A_c = \frac{\pi}{4}(1.4475^2 - 0.625^2) = 1.34 \text{ in}^2.$$

Use $k = AE/L$ and $1/k_c = 1/k_1 + 1/k_2$ to get k_c.

$$\frac{1}{k_c} = \frac{1}{1.34 \times 3 \times 10^4} + \frac{0.02}{1.34 \times 19},$$

from which $k_c = 1230$ k/in. Observe the large effect of the gasket on k_c.

$$k_b = \frac{A_b E_b}{L_b} = \frac{(0.226)(3 \times 10^4)}{1.02} = 6650 \text{ kips/in.},$$

where the stress area A_s is assumed to be reasonable for this computation. From equation (**f**),

$$\Delta F_b = F_e \frac{k_b}{k_b + k_c} = 1.1\left(\frac{6650}{6650 + 1230}\right) = 0.93 \text{ kips.}$$

For a mean load $F_m = F_i + \Delta F_b/2 = 10.5 + 0.47 = 10.97$ kips, and $F_a = \Delta F_b/2 = 0.47$ kips, and for $s_n = s_u/2 = 40$ ksi and $K_f = 1.8$, Table AT 12, we have

$$\frac{1}{N} = \frac{s_m}{s_y} + \frac{K_f s_a}{s_n} = \frac{10.97}{(0.226)(75)} + \frac{(1.8)(0.47)}{(0.226)(40)},$$

or $N = 1.35$. This design seems to be satisfactory, but it does not necessarily follow that equation (5.1) gives the best size in every instance. There are alternative procedures for design and various codes that apply to pressure vessels and pipe joints. Finally, it is entirely possible that for these relatively small alternating stresses s_a, the Gerber equation (§ 4.6) gives a better estimate of the factor of safety (and a somewhat higher one here). In any case, computations for the next smaller size should be made and studied prior to a final decision.

5.12 EXAMPLE—STIFF JOINT. A bolt of as-rolled C 1118 as shown in Fig. 5.9 is to be subjected to an external load F_e varying from 0 to 1650 lb. It connects the parts C which are made of aluminum 2024-T4, 2-in. total thickness, and with a diameter $2D$ of twice the bolt size D. What should be the diameter of the bolt for a design factor of 2 based on the Soderberg line?

Solution. First, decide about the stresses. From Table AT 7, find $s_u = 75$ ksi, $s_y = 46$ ksi, and $E = 3 \times 10^4$ ksi. Using $s_n \approx s_u/2$, the factor 0.8 for the axial

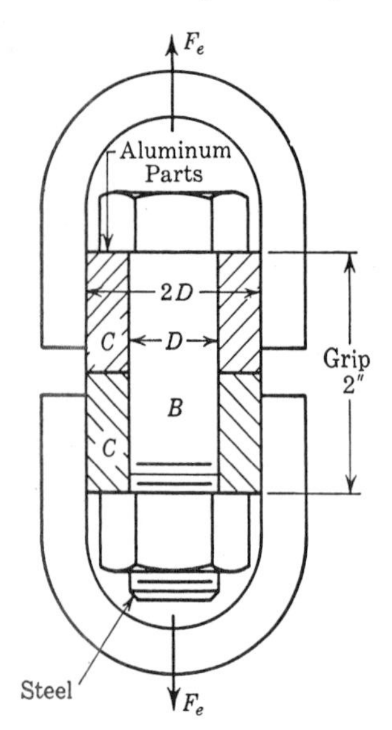

FIGURE 5.9

loading (the bearing surfaces of nuts and heads are almost sure not to be exactly normal to the bolt's axis), and 0.85 to be on the safe side for size, we get

$$s_n = \left(\frac{75}{2}\right)(0.8)(0.85) = 25.5 \text{ ksi}.$$

Since the load varies, find the mean and alternating components, for which are needed the stiffness constants k_b and k_c. Let $A_b = \pi D^2/4$; then $A_c = 4A_b - A_b = 3A_b$.

$$k_b = \frac{A_b E_b}{L_b} = \frac{A_b (3 \times 10^4)}{2} = 15 \times 10^3 A_b \text{ kips/in.},$$

where $L = 2$ in., given. This computation assumes that effective length is the same as the grip and that the threads extend very little beyond the nut. From Table AT 3, for 2024-T4 aluminum, $E_c = 10.6 \times 10^3$ ksi. The effective area in compression in this case should be closely the total area of the connected parts, $A_c = 3A_b$. Therefore,

$$k_c = \frac{A_c E_c}{L_c} = \frac{(3A_b)(10.6 \times 10^3)}{2} = 15.9 \times 10^3 A_b \text{ kips/in.}$$

Thus, with $Q = 1.5$, the initial tension from (5.3) is (kips cancel)

$$F_i = QF_e \left(\frac{k_c}{k_b + k_c}\right) = (1.5)(1650) \frac{15.9 \times 10^3 A_b}{10^3 A_b (15 + 15.9)} = 1272 \text{ lb}.$$

We shall assume that the factory *and* the mechanic in the field use a torque wrench or other method in order to obtain an initial tension reasonably close to this value. From equation (f),

$$\Delta F_b = \left(\frac{k_b}{k_b + k_c}\right) F_e = \frac{(15 \times 10^3 A_b)1650}{10^3 A_b(15 + 15.9)} = 800 \text{ lb.}$$

The alternating force is $F_a = \Delta F_b/2 = 400$ lb.; $F_m = F_i + 400 = 1672$ lb. The corresponding stresses are

$$s_m = \frac{F_m}{A_s} = \frac{1672}{A_s} \text{ psi} \qquad \text{and} \qquad s_a = \frac{F_a}{A_s} = \frac{400}{A_s} \text{ psi.}$$

The stress concentration factor for cut threads (UNC) is $K_f = 1.8$ from Table AT 12. For a design factor of $N = 2$ in equation (4.4), we have

$$\frac{1}{N} = \frac{s_m}{s_y} + \frac{K_f s_a}{s_n} = \frac{1}{2} = \frac{1.672}{46 A_s} + \frac{(1.8)(0.4)}{25.5 A_s},$$

or $A_s = 0.129$ in.² (the force unit is in kips to keep the numbers small). From Table AT 14, we get $D = \frac{1}{2}$ in. (corresponding to $A_s = 0.1419$). Since the answer is borderline regarding use of a size factor, one should now recompute without it and see if a $\frac{7}{16}$ bolt can be used. The load and material are not the same as in the previous article, but some comparisons can be made, especially as related to the stiffness k_c of the connected parts. Some light would be shed upon the situation if you would solve this same problem with steel connected parts instead of aluminum, and it would take a very few minutes; the answer is $\frac{9}{16}$ UNC.

5.13 TYPES OF BOLTS AND SCREWS.

Basically, a bolt is a screw fastening with a nut on it; a screw is one that has no nut and turns into a threaded hole. Some bolt heads and nuts are faced; that is, they have a washer face, Fig. 5.10. A large variety of bolts and screws, standard, near standard, and special, is available; hence, the brief remarks below and the illustrations are suggestive only.

A ***machine bolt***, an old name for an unfinished *through bolt*, comes with a square (Fig. 5.11), hexagonal, or round head and with coarse or fine threads.

A ***coupling bolt*** is finished all over, usually having coarse threads. An ***automobile bolt***, Fig. 5.12, also finished all over, has fine UNF threads, is usually made of heat treated steel, and frequently has a castle nut (Fig. 5.10). A ***cap screw***. Figs. 5.17 and 5.18, also falls into this same class, being finished all over. Cap screws come in a variety of heads: hexagonal, fillister, button, flat head (Fig. 5.28), and hollow heads such as shown in Fig. 5.18.

A ***carriage bolt***. Fig. 5.13, is distinguished by a short portion of the shank underneath the head being square or finned or ribbed. Originally intended for use with wood, this bolt is found useful where, for example, the square part fits into a square hole and prevents the screw from turning while the nut is tightened.

There are many variations of ***eye bolts***, Fig. 5.14, which provide a place for a hook for lifting parts; they have other uses, also.

Stud bolts, Fig. 5.15, are among the most widely used types. They are threaded on both ends and can be used where a through bolt is impossible; for example, in bolting a head on an engine where the holes in the block are threaded. This practice is especially desirable if the hole is in weak or brittle material because

FIGURE 5.10 Castle Nut. (Courtesy National Acme Co., Cleveland).

FIGURE 5.11 Square-head Machine Bolt. (Courtesy Pheoll Mfg. Co., Chicago).

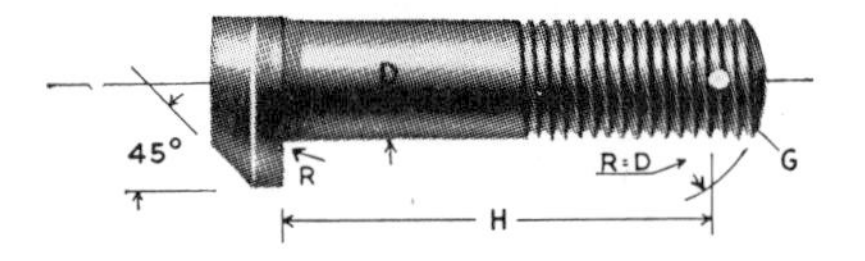

FIGURE 5.12 Connecting-rod Bolt, Fine Threads. (Courtesy Lamson & Sessions Co., Cleveland).

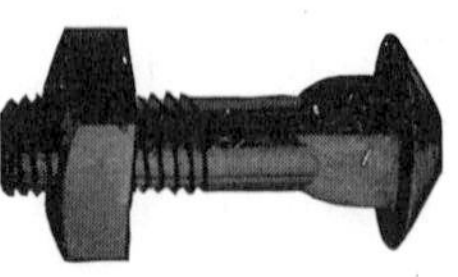

FIGURE 5.13 Carriage Bolt. (Courtesy Link-Belt Co., Chicago).

FIGURE 5.14 Unthreaded Drop-forged Eye Bolt. (Courtesy J. H. Williams & Co., Buffalo).

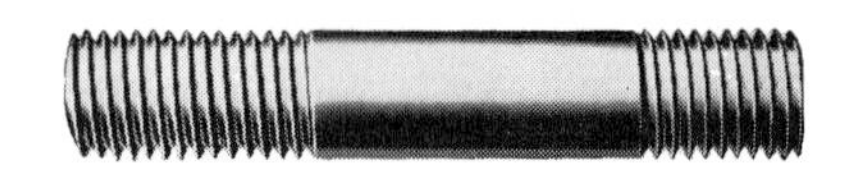

FIGURE 5.15 Stud Bolt—Coarse Threads. (Courtesy National Acme Co., Cleveland).

FIGURE 5.16 U-Bolts. (Courtesy The Bourne-Fuller Co., Cleveland).

FIGURE 5.17 Cap Screw. (Courtesy The National Acme Co., Cleveland).

FIGURE 5.18 Hollow-head Cap Screw. (Courtesy Standard Pressed Steel Co., Jenkintown, Pa.).

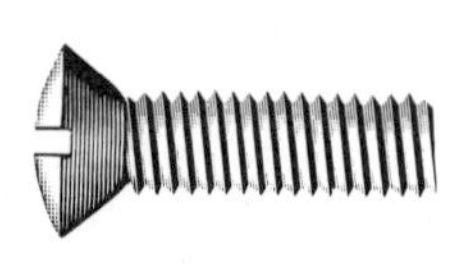

(a) Oval Head.

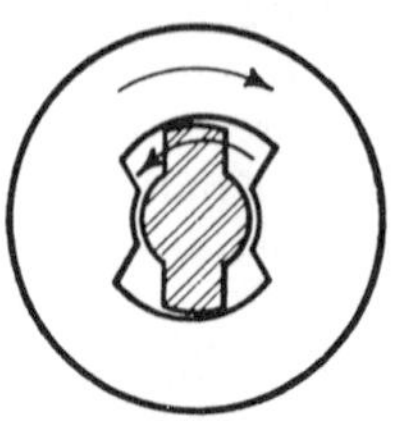

(b) Clutch Head.

FIGURE 5.19 Machine Screw. (Courtesy United Screw and Bolt Corp., Chicago).

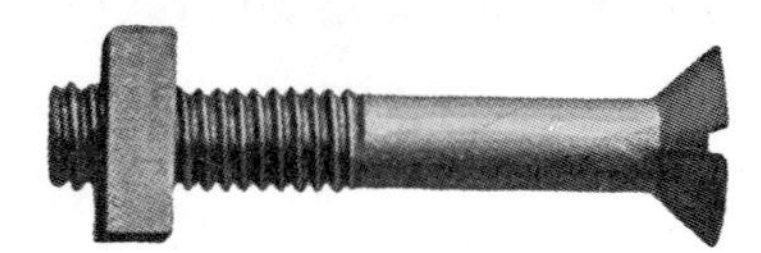

FIGURE 5.20 Stove Bolt.

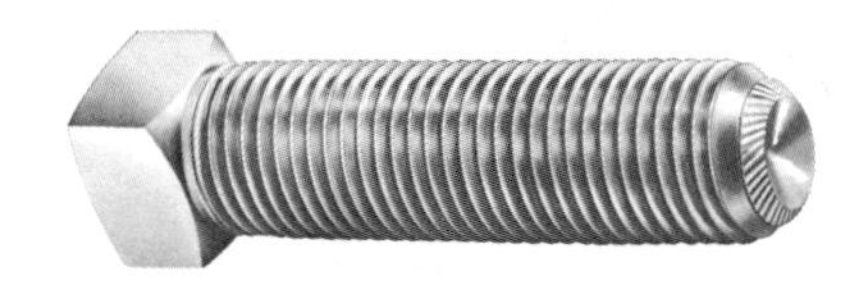

FIGURE 5.21 Square-head Set Screw. (Courtesy Standard Pressed Steel Co., Jenkintown, Pa.).

FIGURE 5.22 Hollow-head Set Screw. (Courtesy Standard Pressed Steel Co., Jenkintown, Pa.).

FIGURE 5.23 Hollow-head Set Screw. (Courtesy The Bristol Co., Waterbury, Conn.).

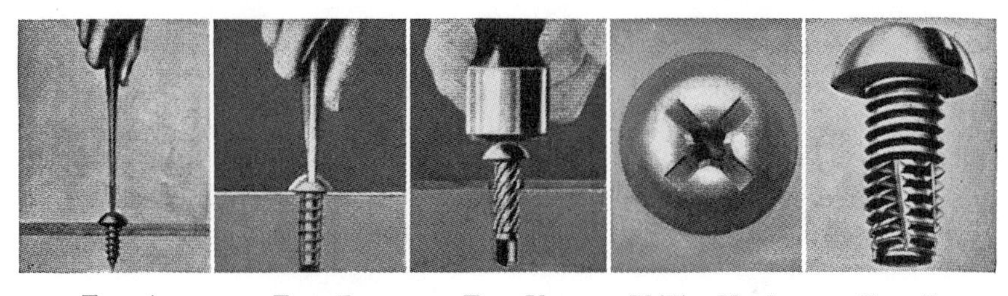

Type A Type Z Type U Phillips Head Type F

FIGURE 5.24 Self-tapping Screws and Phillips Recessed Head. (Courtesy Parker-Kalon Corp., N.Y.).

FIGURE 5.25 Lag Screw. (Courtesy The Bourne-Fuller Co., Cleveland).

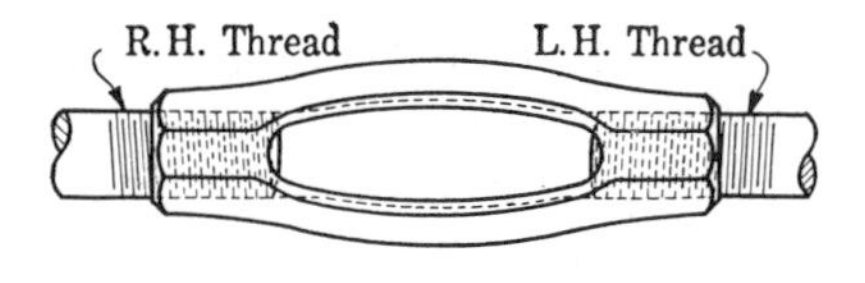

FIGURE 5.26 Turnbuckle.

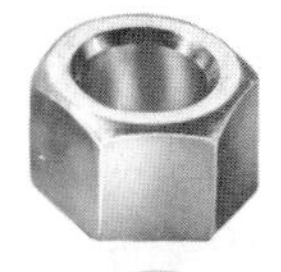
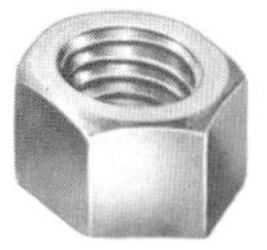

FIGURE 5.27 Forging and Finishing a Nut. (Courtesy H. M. Harper Co., Morton Grove, Ill.).

the studs can be left in place when the head is removed, thus saving wear and tear on the hole. However, there are available hardened steel inserts of various kinds that can be left in the tapped part, including a thin spiral piece, that screws into the tapped hole, covers the weaker threads, and stays there. Stud bolts are also used as through bolts with nuts on both ends.

A ***stove bolt***, Fig. 5.20, is a cheap variety of bolt made in small sizes. ***Machine screws***, Fig. 5.19, are also made in small sizes, the numbered sizes in Table AT 14 and up to $\frac{3}{8}$-in., but they are more accurately made than stove bolts, are finished all over, and have fine or coarse threads, various heads, Fig. 5.28, also the clutch head, Fig. 5.19(b), and the Phillips head (see Fig. 5.24). The length L of a machine screw, Fig. 5.28, is the distance from the end of the screw to the most distant point of contact of the head with the material being clamped.

U-bolts are in the form of a U, Fig. 5.16, and are used as holding clamps, as on an automobile spring. ***Plow bolts*** are widely used on farm machinery. ***Track bolts*** are used in railway track construction.

Self-tapping screws, Fig. 5.24, have proved economical in many assembly operations involving plastics, die castings, and sheet metal (any kind). Types A and Z are applied with a screw driver; type F is made in standard thread pitches, with fluted threads at the start for the purpose of cutting its own threads as it moves into a drilled hole. These screws are hardened, and they are available with the Phillips recessed head, Fig. 5.24. There are many other styles. See Refs. *(5.14)* and *(5.28)*.

A ***lag screw***, Fig. 5.25, a large wood screw, is used to fasten machinery and equipment to a wooden base. On account of its size, it has a square or hexagonal head allowing it to be turned with a wrench. A ***turnbuckle***, Fig. 5.26, is a convenient device used to adjust the length of tie rods, etc. The stages in the forging and finishing of a blanked nut are shown in Fig. 5.27.

5.14 SET SCREWS.

Set screws, Figs. 5.21, 5.22, and 5.23, are used to prevent relative motion between two parts that tend to slide over one another. They are obtainable with several combinations of points and heads, the more common point forms being shown in Fig. 5.29. An unprotected square head should not be used on moving parts, since a workman may catch his clothes on the projecting part and be injured. Slotted set screws for a screw driver or hollow set screws, Figs. 5.22 and 5.23, are preferable for safety. The knurled point of the screw in Fig. 5.22 is designed to resist

FIGURE 5.28 Machine-screw Heads. The dimensions are for drawing purposes only.

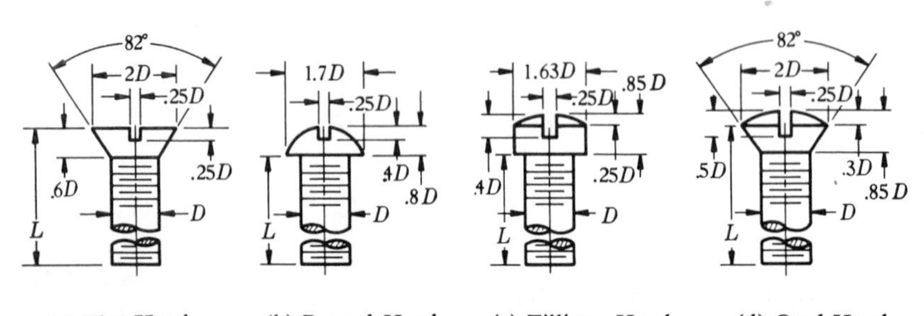

(a) Flat Head (b) Round Head (c) Fillister Head (d) Oval Head

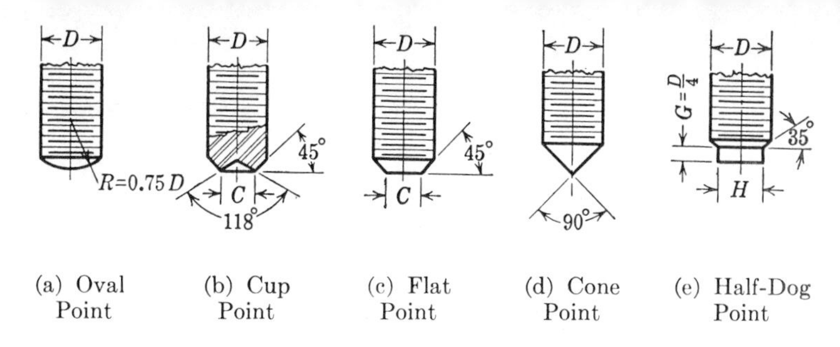

FIGURE 5.29 Set-screw Points. Approximate dimensions; $0.5D < C < 0.6D$; $H > 0.6D$. Short set screws (length = the diameter or less) with cone point should have a cone angle of 118°.

loosening under vibration. Set screws generally have coarse threads and hardened points.

Table 5.3 gives the capacity in pounds of tangential force for a cup-point set screw. This point penetrates somewhat into the shaft material, which should be softer than the screw point by at least 10 points, Rockwell C, or else the capacity is greatly reduced.[5.14] Without spotting holes, the relative capacities for other points, with an index of 1 for a cup point, are: cone point, 1.07; flat or dog point, 0.92; oval point, 0.9. The cone point and dog points are often assembled with the point in a drilled hole, in which case, relative motion is a matter of shearing the point, and Table 5.3 does not apply. Also, if the set screw has a slotted head, the tightening or seating torques indicated are not attained (perhaps only half the table capacity). Lubricating the threads, or plating them, the plating acting as a lubricant, increases capacity because, with less friction in the threads, the normal force will be greater, equation (5.2). Sometimes, trouble with set screws loosening can be remedied by enlarging the hole and using a larger screw. Typically, the size of set screw is about $\frac{1}{4}$ times the shaft diameter. Two set screws on the same side of the shaft, side by side, virtually double the capacity for one screw, but at 180° apart, the increase is only about 30%[5.14] because of the simultaneous loss of friction between shaft

TABLE 5.3 HOLDING CAPACITY OF CUP-POINT SET SCREWS

Approximate "ultimate" values of the tangential holding force at the surface of the shaft when the tightening torque is as given; for cup points with hardness about R_C = 45–50, on a shaft of hardness about R_C = 15. A factor of safety should be applied. (Taken from Standard Pressed Steel Co. pamphlet.)

SCREW SIZE	TIGHTENING TORQUE *in-lb.*	HOLDING FORCE *lb.*	SCREW SIZE	TIGHTENING TORQUE *in-lb.*	HOLDING FORCE *lb.*	SCREW SIZE	TIGHTENING TORQUE *in-lb.*	HOLDING FORCE *lb.*
0	0.5	50	8	20	385	$\frac{1}{2}$	620	3000
1	1.5	65	10	33	540	$\frac{9}{16}$	620	3500
2	1.5	85	$\frac{1}{4}$	87	1000	$\frac{5}{8}$	1225	4000
3	5	120	$\frac{5}{16}$	165	1500	$\frac{3}{4}$	2125	5000
4	5	160	$\frac{3}{8}$	290	2000	$\frac{7}{8}$	5000	6000
5	9	200	$\frac{7}{16}$	430	2500	1	7000	7000
6	9	250						

and hub that exists opposite one screw; this friction thus contributes significantly to holding capacity.

5.15 DEPTH OF TAPPED HOLE AND CLEARANCE AROUND BOLT HEAD AND NUT. The length of thread contact in a tapped hole should be a minimum of about $1.5D$ in. for cast iron and other brittle materials, and about D inches for steel or wrought iron (D = nominal size). If a tapped hole cannot go all the way through the piece, the hole to be tapped should be drilled at least an extra $D/4$ inches deep to allow tool clearance at the bottom. The designer should be sure that heads or nuts to be tightened have sufficient clearance for a wrench and are readily accessible.

5.16 BOLTS AND SCREWS IN SHEAR. Whenever the bolts are to carry a shearing load the holes should be accurately sized and the bolts preferably have a close fit, which necessitates reamed holes and finished bolts. The tightening-up stress is usually neglected in the case of shear if the bolts are larger than $\frac{1}{2}$ in., but if desired, the resultant maximum stresses

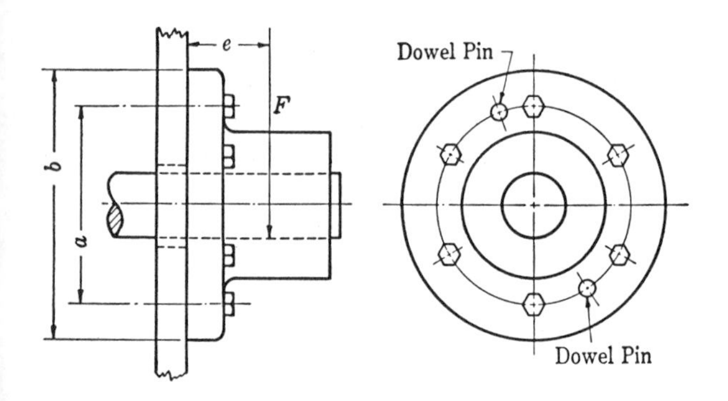

FIGURE 5.30 End Bearing for a Shaft. This figure shows the use of *dowel pins* to carry the shearing load. Dowel pins, which should be on different diameters, are used also to determine and maintain the proper alignment between parts. See §8.11 for the method of determining a tensile load on the bolts in this case.

due to combined shear and tension may be found according to the principles explained in Chapter 8. If through bolts are used, arrange for the shear to be across a major diameter, but the shear may be across the minor diameter, especially for studs and cap screws. If the bolts are well tightened, friction contributes considerably to the holding capacity, so that the actual shear stress is small.

When the location of a bolt is such that it would normally be in shear, it is better practice to use *dowel pins*, Fig. 5.30 to carry the shearing load. With dowels, it is not necessary to take extra precautions to get a close fit for the bolts.

5.17 THREAD-LOCKING DEVICES. Locking nuts and ways of locking nuts have been invented in amazing numbers. Most of them depend on friction to retard or avert the loosening of a nut under vibration. There are other miscellaneous ways, such as nuts held in place by wiring, castellated nuts held in place with cotter pins, Fig. 5.10. It has been found that when the initial tightening load is greater than the external load (close to the proof load), it is helpful in maintaining a tight connection. A few thread locking methods are mentioned below.

Lock washers of the type shown in Fig. 5.32 come in four weights, are heat treated (45-53 Rockwell C), and their purpose is to maintain pressure between the threads of the bolt and nut, thus retarding loosening under vibration.

Lock nuts, to be effective, must have the upper nut, Fig. 5.33, drawn up so tightly that the nuts press on the threads in opposite directions. The "inside" nut (lower nut, Fig. 5.33) may be a *jam nut*, which is about 70% as thick as a regular nut. Lock nuts are often not satisfactory because the "outside" nut is not drawn up as stated above.

In the *Flexloc* locknut, Fig. 5.34, the sections at the top, separated by radial slots, have been permanently deformed inward; they grip the bolt and hold the nut in any position.

In the *Lokut* locknut, Fig. 5.35, the top of the nut has been deformed inward, the idea being that the pressure of the deformed part on the bolt threads will hold the nut in place.

The *Esna insert type*, Fig. 5.36, has a ring insert of fiber or nylon at the *top*; with the nut on the bolt, the insert presses about and against the bolt threads, holding the nut in place at any position.

The *An-cor-lox* nut, Fig. 5.37, has a ring of soft metal, such as soft steel or brass, at the *bottom* of the nut which turns on freely until the nut strikes the part. When the nut is pulled tight, the ring of soft metal flows about the bolt threads, exerting considerable friction. Similar designs, perhaps with nylon, are also used.

The *speed nut*, Fig. 5.38, provides excellent locking properties for light assemblies; when pulled into a flat position, it presses against the root of the screw thread and also against the side of the thread by the nut's "spring" action; the resulting friction holds it. There are innumerable designs of speed nuts, which are widely used.

5.18 DARDELET RIVET-BOLT. The Dardelet rivet-bolt, which has a self-locking thread, Fig. 5.39, is used in place of rivets in assembling steel structures. As suggested by Fig. 5.39, the thread has a wide, tapering root. At the deep end of this root, the tapered crest of the nut thread has clearance, and the nut turns easily on the bolt. When the nut is drawn tight, the two tapered surfaces are pressed together, Fig. 5.39(b), and friction holds the nut in place.

5.19 RIVETS. Most parts that can be bolted together can also be riveted, but of course rivets are not used unless it is expected that the parts

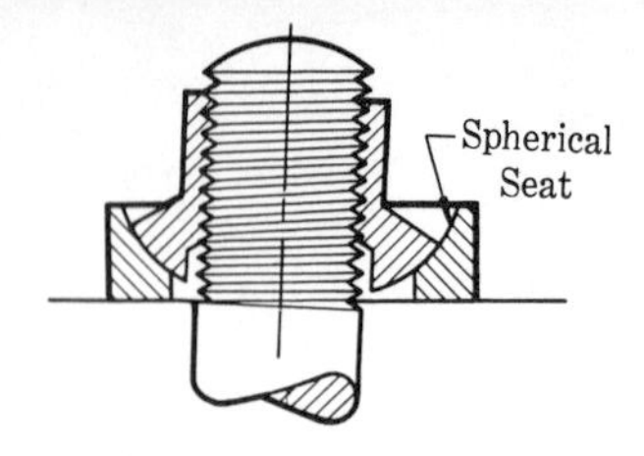

FIGURE 5.31 Self-aligning Nut.

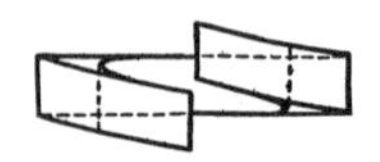

FIGURE 5.32 Spring Lock Washer.

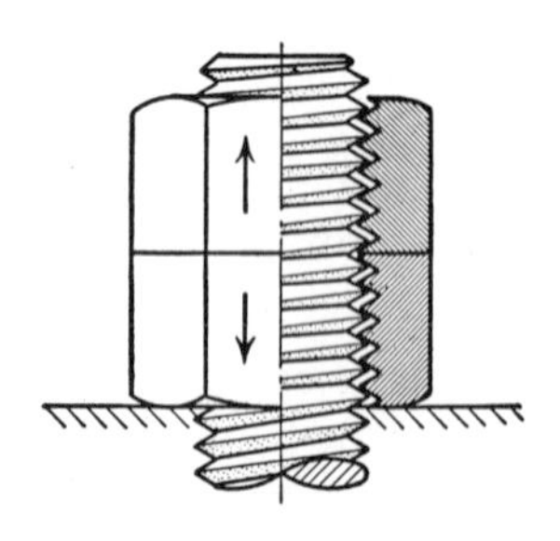

FIGURE 5.33 Lock Nuts. Thread Clearance Exaggerated.

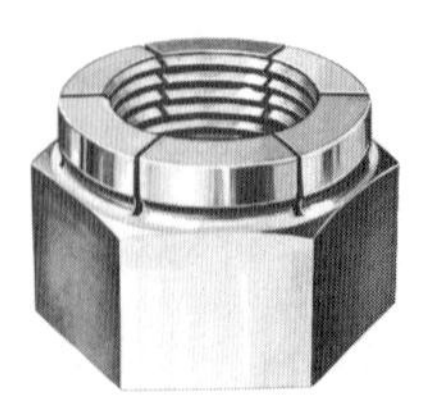

FIGURE 5.34 Flexloc ® Locknut. (Courtesy Standard Pressed Steel Co., Jenkintown, Pa.).

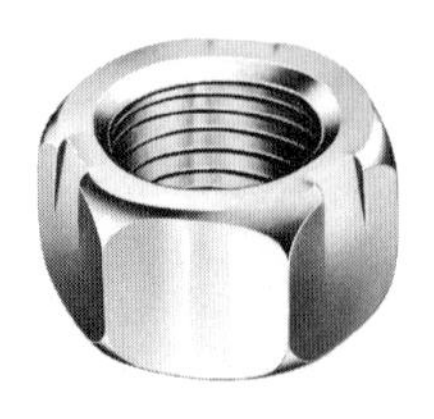

FIGURE 5.35 Lokut ® Locknut. (Courtesy Illinois Tool Works, Chicago).

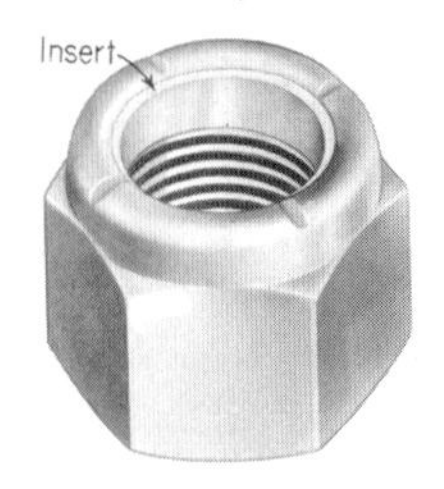

FIGURE 5.36 Esna ® Insert Locknut. (Courtesy Elastic Stop Nut Corp., Union, N.J.).

FIGURE 5.37 An-cor-lox ® Locknut. (Courtesy Schnitzer Alloy Products Co., Elizabeth, N.J.).

FIGURE 5.38 Speed Nut ®. (Courtesy Tinnerman Products, Inc., Cleveland).

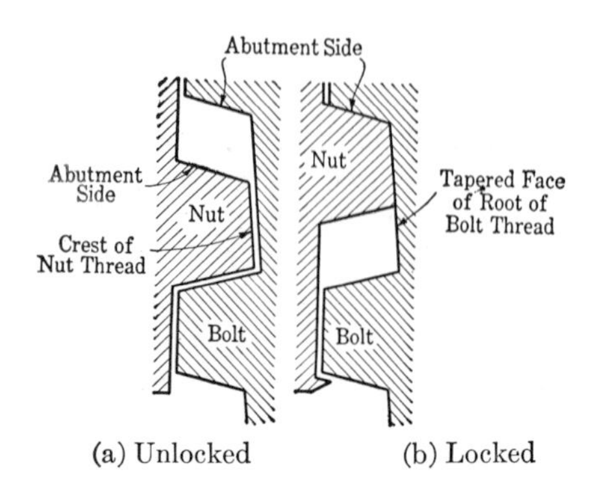

FIGURE 5.39 Dardelet Thread Form.

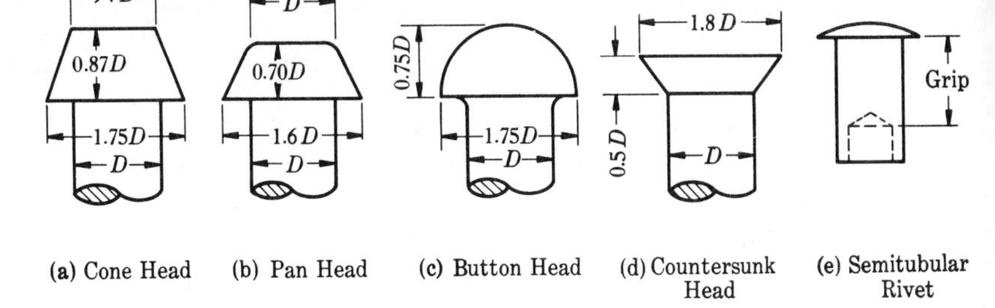

FIGURE 5.40 Rivets with Some Approximate Proportions of Heads. (ASA B 18.4-1950). Tolerances on D are of the order of ± 0.03 in., less for smaller sizes, more for larger ones. In (e), the "heading" is accomplished by bending back the hollow part.

will seldom if ever be disassembled. As is true of threaded parts, there are many styles and types of rivets. The heads in Fig. 5.40(a)–(d) are "regular" styles; these rivets are inserted in the holes and then the opposite ends are headed by machine (hammer or high pressure). The enlarging of a part of the hole for the countersunk head may excessively weaken the connected part; hence this type of head is used only when a flush, or nearly flush, surface is desired. The semitubular rivet of Fig. 5.40(e), drawn enlarged, is an example of small sizes used for thin material, as aluminum sheet. There are a number of patented styles that can be applied and headed from one side only of the connection.[5.12,5.14]

The material of rivets for ordinary steel construction would typically be a soft steel, as 1010, but copper, brass, aluminum, titanium, and others are used for a reason. Because of possible galvanic action, one must be cautious in using different materials together. ASME Boiler Code steel will have a minimum $s_u = 55$ ksi; the design stress for shear is 11 ksi. Aluminum alloys for rivets include 2024 and 6061, Table AT 3; also in tempers 0 and H 13; minimum strengths for rivets will be somewhat less than Table AT 3 values. For some fatigue information for lap connections, see Fig. 4.27, which is as applicable for rivets as for bolts. Reference *(5.30)* gives some design help for bolted and riveted connections, fasteners in repeated shear. When the rivets fill the holes and clamp the parts tightly so as to induce a large frictional force, rivets that are nominally in shear are strong in fatigue because the joint does not slip,[0.2] a condition that many designers would not care to count on. Moreover, where the fasteners are in shear, one has to be on guard not to overlook the possibility of corrosion and fretting. Tight clamping (bolts or rivets) puts the metal adjacent to the hole in compression, which tends to improve the fatigue strength of the connected parts in tension.

The lengthwise shrinkage upon cooling of a hot-headed rivet induces a tensile stress in it that may be close to the yield strength. Nevertheless, tests on $\frac{3}{4}$-in. rivets suggest:[5.29] hot-driven rivets are slightly stronger than the original rod; rivets with long grips are not quite as strong as rivets with short grips (varied from 2 to 6 in.); the ability of a rivet to resist an external tensile load is not reduced by the initial tension, probably because of the high rigidity of the plate as compared to the rivet (§ 5.9). If a riveted connection is such that a repeated external loading tends to put the rivet in both tension and shear, the tensile loads tend to relieve the frictional force between the parts so that the varying shear stress may cause trouble in the joint. At least, this factor must be cared for. Connections involving combined stresses are discussed further in Chapter 8.

5.20 CLOSURE. Variations in the forms of threaded parts and rivets are so numerous that other sources must be consulted for more descriptive detail.

6. SPRINGS

6.1 INTRODUCTION. Springs, which are common and important machine elements, are used for many purposes: to absorb energy or shock loads, as in automobile chassis springs and railroad bumper springs; to act as a source of energy, as in clocks; to produce a pressure or force, as in maintaining pressure between the friction surfaces of clutches and as in keeping a cam follower in contact with the cam; and to absorb vibrations. We shall discuss first, compression coil springs, which may be made of wire with a round, square, or rectangular section.

6.2 STRESS IN ROUND-WIRE HELICAL SPRINGS. Figure 6.1 shows certain forms of helical compression springs, illustrating four

FIGURE 6.1 Compression Springs. Since both ends of a particular spring are usually alike, the variety of ends is for information purposes.

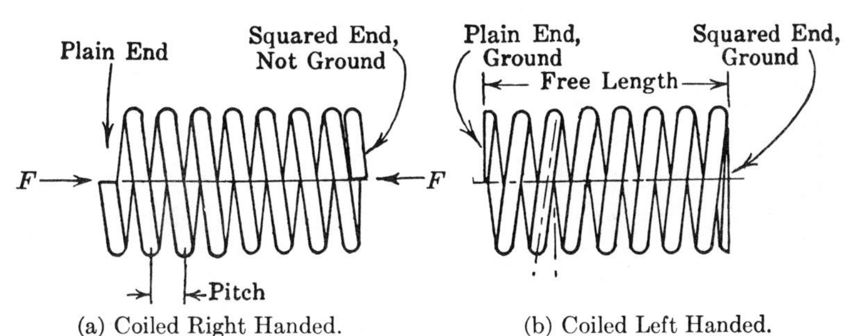

(a) Coiled Right Handed. (b) Coiled Left Handed.

methods of finishing the ends. The discussion of coil springs in this book is limited to those in which the load is collinear with the axis of the spring (no bending of the spring as whole). Hence, for compression springs, we shall have in mind ground ends.

Take a quadrant of a coil as a free body, and imagine some ideal connection to it so that the load F is acting along the axis of the spring, Fig. 6.2. The force system external to the quadrant is then as shown, with the moment $FD_m/2$ being resisted by the torque T at the section. If all the conditions for $T = s_sJ/c$, equation (1.11), § 1.13, were met, the internal resistance would be given by s_sJ/c. A straight *torsion bar*, used as a spring (say, AISI 9260, with 340 < BHN < 450), may be designed to meet the required conditions closely enough for engineering purposes, in which case, the torsional stress in a solid circular member is

$$s_s = \frac{Tc}{J} = \frac{16T}{\pi D_w^3}. \tag{1.11}$$

However, the peak stress in a coil spring is somewhat greater than that obtained from (1.11) because: (1) the member is curved and there is a curvature (stress-concentration) effect K_c on the inside of the coil, Fig. 6.3; (2) there is a transverse shear stress on any section due to F (about $1.23F/A$ max.); (3) there is a compressive stress in the wire arising from the component of F in the direction of the sloping coil; not to mention residual stresses and some bending. Wahl's[6.1] more complete stress analysis yields a satisfactory and practical result for closely coiled springs (within 2% of more exact equations);

$$s_s = K\frac{16T}{\pi D_w^3} = K\frac{8FD_m}{\pi D_w^3}, \tag{6.1}$$

where K, obtained from Fig. AF 15 or from equation (**a**) below, is called

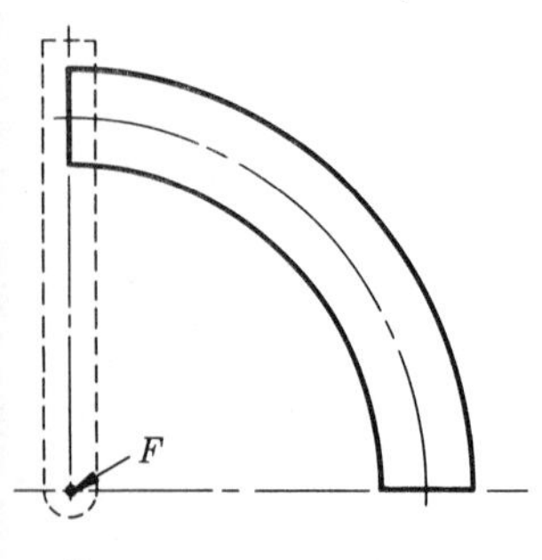

FIGURE 6.2 Quadrant of a Coil.

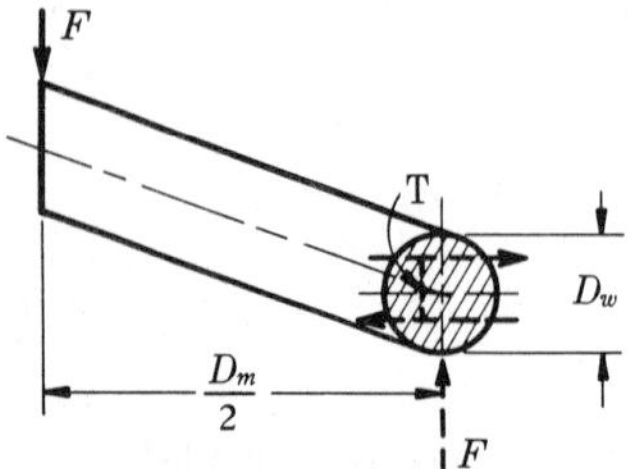

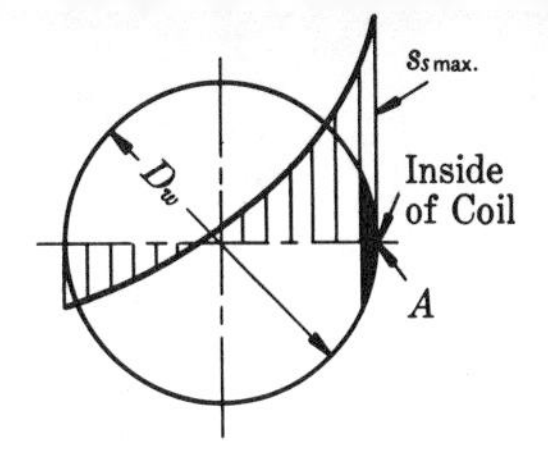

FIGURE 6.3 Shearing Stress in Coiled Wire. This suggests the distribution *after* the uniform or axial shear stress has been added to the torsional stress. Compare with Fig. 1.5, which shows the distribution of stress in a *straight* beam. The highest stresses occur over a small area, as at *A*.

Wahl's factor; D_w = wire size; D_m = mean coil diameter; F = axial load on the spring;

(a) $$K = \frac{4C - 1}{4C - 4} + \frac{0.615}{C},$$

where $C = D_m/D_w$ and is called the ***spring index.*** Observe that the spring index tells of the relative sharpness of curvature of the coil, a *low* index corresponding to *high* sharpness of curvature. The factor K is made up of two factors,

(b) $$K = K_c K_s,$$

where K_c is the correction factor for curvature only, Fig. AF 15, and K_s is the correction factor for the direct shear. For a spring index of $C = 3$, the actual stress in the spring is some 60% ($K = 1.58$ in Fig. AF 15) higher than that indicated by the simple torsion equation (1.11). To minimize the effect of curvature, we see from Fig. AF 15 that values of $C > 5$ are desirable; if $C < 5$, special care in coiling will be necessary to avoid cracking in some wires.

Since equation (6.1) is for closely coiled springs, we must check the pitch (or lead) angle. The distance between corresponding points of adjacent coils is the ***pitch*** P (as in threads). If it is imagined that one full coil is unwrapped into a straight configuration, we note that it advances axially a distance equal to the pitch, Fig. 6.4. As the pitch angle $\lambda = \tan^{-1}P/(\pi D_m)$ exceeds about 12°, equation (6.1) becomes less and less accurate.

6.3 DESIGN STRESSES AND SOLID STRESSES.

As already noted, the "strength" of a metal is a function of size, a variation that becomes significant in spring design. From data available in the literature,[6.1,6.2,6.8] we find that an equation in the form

(c) $$s_s = \frac{Q}{D_w{}^x},$$

seems to approximate various reported strengths, where the constants Q and x depend upon the material. For this reason, Table AT 17 summarizes a considerable amount of such information for wires with $D_w < 0.5$ in. Note that smaller wires are in general stronger than larger ones. Column (3)

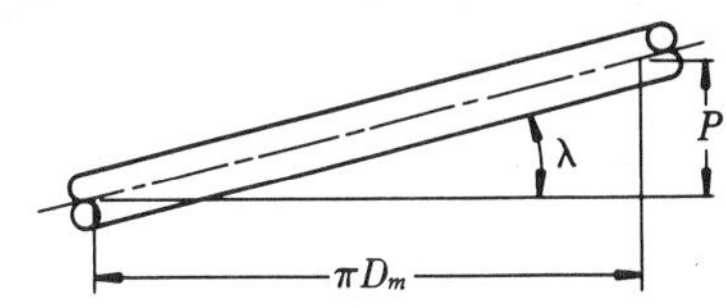

FIGURE 6.4 Pitch Angle.

of Table AT 17 suggests design stresses for the static approach, on which more is found in §§ 6.9 and 6.10.

A compression spring should *not* be compressed solid (with coils touching one another) in operation, because the surface of the coils may be damaged. However, since compression springs are frequently compressed solid in installation or maintenance, they should be designed if possible so that such an occasional deflection will not damage them by inducing a permanent set. Table AT 17 gives values of the maximum "solid stresses," which are in the vicinity of the torsional yield strength. The "solid stress" should be checked for each design. If this requirement is not easy to meet, the manufacturer can help by presetting (§ 6.13).

Larger wires, over about $D_w = \frac{3}{8}$–$\frac{1}{2}$ in., are coiled hot (hot wound) to avoid cracks (and heat treated after coiling). One manufacturer's recommendations[6.2] for the permissible solid stresses in alloy steels (6150, 9260) are approximated by (also use $G = 10{,}500$ ksi)

$$s_s = \frac{117}{D_w^{0.31}} \text{ ksi}, \qquad [D_w > 0.375]$$

when equation (6.1) is used to compute the stress. For SAE 1095, use straight-line interpolations between the following values:

$$D_w = 0.5 \text{ in., } 108 \text{ ksi; } D_w = 1 \text{ in., } 95 \text{ ksi; } D_w = 2 \text{ in., } 86 \text{ ksi.}$$

For extension springs, § 6.21, use 0.8 of these values. A fatigue strength $s_{no} \approx 70$ ksi (§ 6.6) should be attainable for hot-wound springs.

6.4 SCALE OF A SPRING.

The ***scale*** k of a spring is found according to Hooke's law, force per unit of elastic deflection; or the average value is

$$\text{(d)} \qquad k = \frac{F}{\delta} = \frac{\Delta F}{\Delta \delta} \quad \text{(usually lb./in.)}$$

where F is the total force that produces the total deflection δ in the spring, and ΔF is the increase (or decrease) of force corresponding to an increase (or decrease) of the deflection $\Delta\delta$. Other names for the *scale* applied to springs include the ***modulus,*** the ***rate*** (used especially when the scale is not constant), the ***spring constant***, and the ***spring gradient.*** It is often a rather important parameter.

6.5 DEFLECTION OF ROUND-WIRE HELICAL SPRINGS.

The torsional deflection

$$\text{(1.13)} \qquad \theta = \frac{TL}{GJ} \text{ radians,}$$

is applicable in general to round members. If this equation is applied to springs, L in. is the "active" length of wire and is nearly equal to $\pi D_m N_c$ in closely coiled springs, where N_c is the number of active coils in the spring. Substituting into (1.13) the proper values of

$$T = \frac{FD_m}{2}, \qquad L \approx \pi D_m N_c, \qquad \text{and} \qquad J = \frac{\pi D_w^4}{32},$$

we get

$$\text{(e)} \qquad \theta = \frac{(FD_m/2)(\pi D_m N_c)(32)}{G\pi D_w^4} \text{ radians.}$$

The angular deflection in radians times the mean radius of the coil will give the axial deflection of the spring; $\delta = \theta D_m/2$. Therefore, multiplying both sides of equation (e) by $D_m/2$ and simplifying, we find the spring deflection

$$(6.2) \qquad \delta = \frac{\theta D_m}{2} = \frac{8FD_m^3 N_c}{GD_w^4} = \frac{8FC^3 N_c}{GD_w} \text{ in.}$$

Substitute all linear dimensions in inches in this equation. Wahl found that the actual deflection agreed well, within 1 to 2%, with values calculated by this formula, provided that G, Table AT 17, is accurately known and the elastic limit is not exceeded. Note that, given the dimensions and material, the scale, when constant, of a coil spring F/δ can be computed from (6.2).

6.6 ACCOUNTING FOR VARIABLE STRESSES.

Because many springs are subjected to repeated loading, it is logical to use the principles of Chapter 4 in their design, especially when indefinite life is desired. Since springs are rarely subjected to reversing stresses, Wahl proposed a failure line on the s_a-s_m diagram that goes from B, Fig. 6.5, where the mean stress is equal to the variable stress ($R = 0$), to T at the yield strength s_{ys}. See Fig. 4.5(b). The maximum stress at B, Fig. 6.5, is

$$\text{(f)} \qquad s_{s\max} = s_{ms} + s_{as} = 2s_{as} = 2s_{ms} = s_{no},$$

where s_{no} is the shearing endurance strength for a stress from zero to a maximum. From (f), we see that at the failure point B, the mean stress

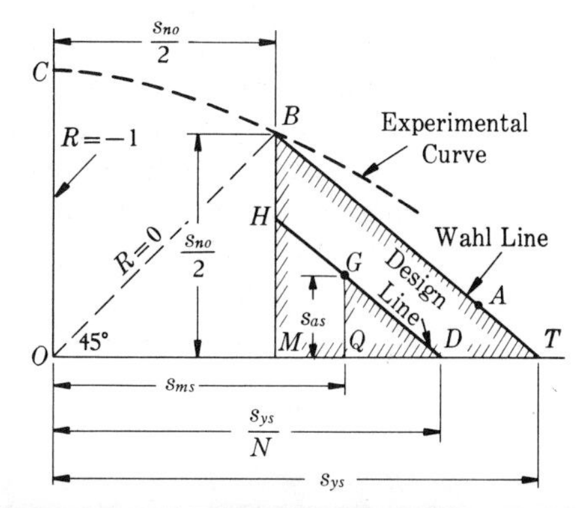

FIGURE 6.5 Variable Stresses in Springs. Point C is at $s_{as} = s_{ns}$ ($s_{ms} = 0$), reversed shear.

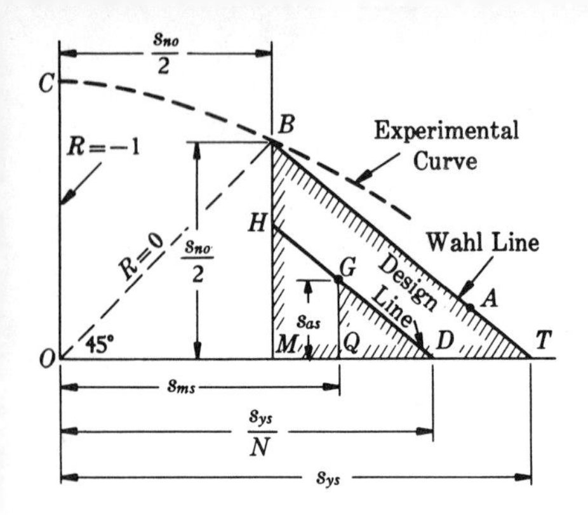

FIGURE 6.5 (Repeated).

$s_{ms} = s_{no}/2$ and the variable stress $s_{as} = s_{no}/2$, as shown in Fig. 6.5. One can see that the line BT could well be a better approximation of test results than one from C to T. To get a design line, divide s_{ys} by a design factor N and lay off $OD = s_{ys}/N$. Through D, draw the line DH parallel to TB. Then, any point G on DH, represents a stress situation for which we say the factor of safety is N. The equation of this design line HD is obtained from similar triangles QGD and MBT;

$$\textbf{(g)} \qquad \frac{s_{ys}/N - s_{ms}}{s_{ys} - s_{no}/2} = \frac{s_{as}}{s_{no}/2},$$

from which

$$(6.3) \qquad \frac{1}{N} = \frac{s_{ms}}{s_{ys}} + \frac{s_{as}}{s_{no}}\left(2 - \frac{s_{no}}{s_{ys}}\right) = \frac{s_{ms} - s_{as}}{s_{ys}} + \frac{2s_{as}}{s_{no}}.$$

The alternating s_{as} and mean s_{ms} stresses are computed from equation (6.1),

$$\textbf{(h)} \qquad s_{as} = \frac{8KF_aD_m}{\pi D_w^{\ 3}} \qquad \text{and} \qquad s_{ms} = \frac{8KF_mD_m}{K_c\pi D_w^{\ 3}},$$

where F_a and F_m are the alternating and mean components of the axial force. The reason for dividing by K_c in the expression for s_{ms} is that the Wahl factor K includes a curvature concentration factor K_c [Fig. AF 15 and equation (**b**)] that experience suggests (Chapter 4) is not needed on the mean stress; therefore, we use only $K_s = K/K_c$. Preferably, s_{no} and s_{ys} are experimental values for the size and kind of spring wire, but data on s_{no} are far from plentiful. The column headed "s_{no}" in Table AT 17 gives endurance strengths (unpeened) in terms of D_w within a maximum difference of some 3% from values by O. G. Meyers,[6.9] who determined the endurance strength for repeated (not reversed) shear by a correlation with reversed bending tests. These results are not always in agreement with actual tests; but shearing endurance tests do not always agree with each other; and besides, one should not conclude that the stresses obtained from the expressions of Table AT 17 are exact. Endurance strength is discussed further in § 6.13.

A definition of yield strength of springs used by the industry is the stress that results in a 2% decrease in the force exerted by the spring at a

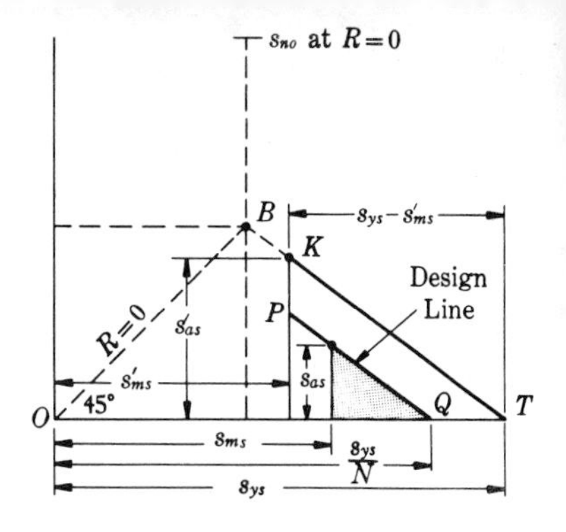

FIGURE 6.6.

particular deflection. In the absence of more specific information the torsional yield strength s_{ys} for no preset (§ 6.13) can be estimated as the values obtained from column (5), Table AT 17.

Do not overlook a check for the "solid stress" as compared to the maximum permissible (yield).

One often finds fatigue strengths given from some $s_{\min}$ to some $s_{\max}$, as from a 10 ksi minimum. (See Fig. 6.9.) If so, a suitable design line for other ranges can be found by using an actual failure point K, Fig. 6.6, whose coordinates are $s'_{ms} = (s_{s\max} + s_{s\min})/2$ and $s'_{as} = (s_{s\max} - s_{s\min})/2$, draw line KT to s_{ys}, draw PQ parallel to KT such that Q is at s_{ys}/N; then write the equation of the line PQ by using similar triangles as before. The equation analogous to (6.3), becomes

$$\frac{1}{N} = \frac{s_{ms}}{s_{ys}} + \frac{s_{as}}{s'_{as}}\left(1 - \frac{s'_{ms}}{s_{ys}}\right). \tag{6.4}$$

6.7 ENERGY ABSORBED BY A SPRING.

If a body with constant k is gradually deflected and Hooke's law is obeyed, the force F required at a particular deflection is directly proportional to the deflection, and the elastic energy is the average force times distance, given by equation (4.8), § 4.35;

$$U_s = \frac{F\delta}{2} = \frac{k\delta^2}{2}, \tag{4.8}$$

where $k = F/\delta = \Delta F/\Delta\delta$. In Fig. 6.7, we see that $F\delta/2$ is represented by the area of a triangle OAD or OBC. Thus, if the force changes from F_1 to F_2 lb. (deflection, δ_1 to δ_2 in.), the work done on the spring between A and B (or by the spring between B and A) is, from Fig. 6.7, p. 190,

$$\textbf{(i)}\quad U_s = \frac{F_2\delta_2}{2} - \frac{F_1\delta_1}{2} = \left(\frac{F_1 + F_2}{2}\right)(\delta_2 - \delta_1) = \frac{k}{2}(\delta_2{}^2 - \delta_1{}^2) \text{ in-lb.}$$

Most computations for elastic energy probably are made with (4.8) or (i). However, for various kinds of springs, the energy can be expressed in terms of certain mechanical and dimensional properties. For a round-wire, coiled spring, the stored energy is obtained by using the value of F from equation

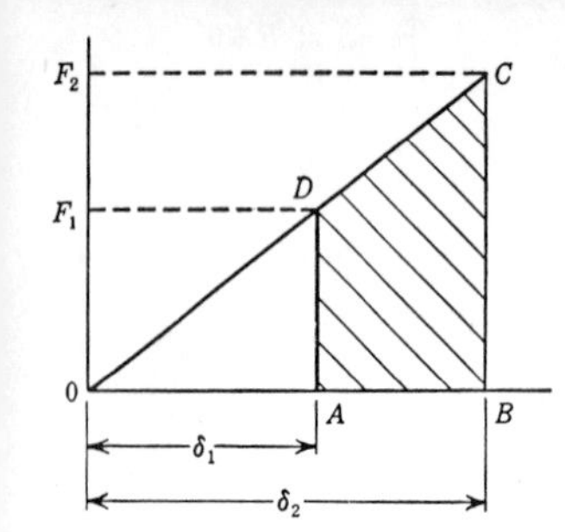

FIGURE 6.7 Work Done on an Elastic Body. The area of the trapezoid $ABCD$ is the average of the bases, $(AD+BC)/2$, times the height, $\delta_2 - \delta_1$.

(6.1) and of δ from equation (6.2) in (4.8) above. From Pappus's theorem, the volume of wire if the spring is closely coiled is, for N_c active coils, $V \approx (\pi D_w^2/4)(\pi D_m)N_c$ in.^{3}; N_c active coils. An exercise for the reader is to find

$$\text{(j)} \qquad U_s = \frac{s_s^2 V}{4K^2 G} \text{ in-lb.,} \qquad \text{or} \qquad V = \frac{4K^2 G U_s}{s_s^2} \text{ cu. in.,}$$

which is the volume of material to store U_s in-lb. with the maximum stress s_s as obtained from (6.1). Similar equations may be obtained similarly for other types of springs. If the spring is rarely to be asked to store U_s in-lb., then the K may be taken as unity on the basis that the peak stress is highly local; of course, K is unity in a straight torsion bar.

Many springs are made for which F is not proportional to δ (§ 6.23), in which case the $F\delta$ curve (or equivalent) is needed in order to evaluate the stored energy. We might note that the energy stored per unit volume when the *stress is uniform* is (§ 4.35): for a normal stress, $s^2/(2E)$; for a shear stress, $s_s^2/(2G)$. The total energy stored is then, say for the normal stress, $U = \int[s^2/(2E)]\,dV$, which could be evaluated when the stress is not uniform. The purpose of showing this relation here is to emphasize that for absorbing the maximum energy with a particular volume of material, the stress should be uniform at its maximum value. Thus, a leaf spring, Fig. 6.22, absorbs much more energy with a given maximum stress than a solid piece of metal of the same shape. Members in bending and torsion inherently have nonuniform stress.

6.8 SOLID HEIGHT AND FREE LENGTH. The ***solid height*** of a coil spring is the over-all length of the spring when it is compressed until all adjacent coils touch. The length of a coil spring under no load is called the ***free length.*** Table AT 16 gives approximate values for different types of ends, Fig. 6.1.

6.9 DESIGN OF HELICAL SPRINGS. The design of springs usually involves a trial-and-error solution. In some cases, sometimes unfortunately, space limitations set limits for certain dimensions; for example, when a spring is to fit into a hole of a certain diameter. In any event, when one or more unknowns must be assumed, the designer should *make several trial designs*, and then choose the one that seems best. If the spring is in a cylindrical hole, a total clearance of $D_w/2$ ($D_w/4$ all around) should be sufficient in general. Special needs may dictate other values.

The simplest procedure is what we may call the "static approach," our Chapter 1 attitude. This is to classify the service as light, medium, or severe (or something in between), and use a design stress from column (3), Table AT 17, in equation (6.1). The other point of view uses the varying load, our Chapter 4 attitude, and considers individually each factor that affects the operation, § 6.6. If a spring is not to operate an indefinite number of times in its lifetime, it is wasteful to design it on such a basis. In these cases, considering the information available, the static approach with light or medium service would seem appropriate. Sharp dividing lines cannot be drawn separating the types of service. Let us say that *light service* means that the load is applied not over 10^4 times; *severe service* corresponds to indefinite fatigue life (10^6 or more cycles); medium service is something in between, such as springs on clutches, brakes, switches. If indefinite life is desired, computations with the load varying would help in arriving at the final configuration; some design information for this approach is found in § 6.6, and its application is shown in § 6.11.

Always, the end use and the quantity to be produced affect the effort put into the design. If weight is a prime consideration or if production is large, it pays to spend more time on the design in search of the optimum answer. The following examples suggest ways in which the information on compression springs can be used in design.

6.10 EXAMPLE—AVERAGE SERVICE.

Design a helical compression spring with squared-and-ground ends to withstand a maximum force of $F = 250$ lb. with a deflection of 1.5 in. The maximum number of applications of F is expected to be about 10^5; say, average service. The spring is to operate over a $1\frac{5}{8}$-in. rod, with a free length of 7 in. if possible. Use hard-drawn wire.

Solution. There are so many unknowns that some assumptions must be made with iterations as necessary. Note that K, Fig. AF 15, does not vary greatly within the normal range of springs; it is thus a good thing to assume; say, $K = 1.3$. Try $D_m = 2$ in. From Table AT 17, $s_{sd} = 0.324 s_u$ for oil tempered, and 0.85 times this expression for hard drawn, or $s_{sd} = (0.85)(0.324)(140)/D_w^{0.19} = 38.55/D_w^{0.19}$ ksi. Equating the design stress to the induced stress, equation (6.1), we have, for $F = 0.25$ kips,

$$s_s = \frac{38.55}{D_w^{0.19}} = \frac{K8FD_m}{\pi D_w^3} = \frac{(1.3)(8)(0.25)(2)}{\pi D_w^3} \text{ ksi} \qquad \text{(TRIAL)}$$

$$D_w^{2.81} = 0.043 \qquad \text{or} \qquad D_w = 0.326 \text{ in.}$$

From Table AT 15, the nearest W & M size is 2-0 or 0.331 in. If the original assumptions are not too far off, this size may do very well; but a *full check must be made*. The inside diameter of the coil is $D_m - D_w = 2 - 0.331 = 1.669$ in.; the clearance between the spring and the 1.625-in. rod is then 0.044 in., which may serve all right since the spring diameter will tend to enlarge as it is compressed. The spring index is $C = 2/0.331 = 6.04$, for which $K = 1.25$ from Fig. AF 15. The computed induced stress is

$$s_s = \frac{(1.25)(8)(0.25)(2)}{\pi(0.331)^3} = 44 \text{ ksi.}$$

compared to the design (permissible) stress of

$$s_{sd} = \frac{38.55}{D_w^{0.19}} = \frac{38.55}{(0.331)^{0.19}} = 47.5 \text{ ksi.}$$

Checking the next smaller wire size, we find that the permissible stress is less than the induced stress and therefore this smaller wire is inadequate. Since $s_{sd} = 47.5 > 44$, the wire size found is satisfactory for strength. The number of coils N_c is obtained from (6.2);

$$N_c = \frac{\delta G D_w}{8FC^3} = \frac{(1.5)(11.5 \times 10^6)(0.331)}{(8)(250)(6.04)^3} = 13 \textit{ active} \text{ coils.}$$

From Table AT 16, the solid height is about

$$\text{SH} = D_w(N_c + 2) = (0.331)(15) = 4.96 \text{ in.}$$

For a spring scale of $k = F/\delta = 250/1.5 = 167$ lb./in., the force to compress the spring to the solid height is $k(7 - 4.96) = (167)(2.04) = 340$ lb. Since the stress is proportional to F, the "solid stress" is obtained by proportion from $s_s = 44$ ksi above;

$$\text{Stress at solid height} = \frac{340}{250}44 = 59.9 \text{ ksi}$$

compared to a permissible solid stress, column (5), Table AT 17, of

$$s_s = \frac{70}{D_w^{0.19}} = \frac{70}{(0.331)^{0.19}} = 86.4 \text{ ksi;}$$

which shows that the spring would not take on a permanent set if it were compressed solid. The equations are inaccurate unless the spring is closely coiled. Checking this, first find the pitch, Table AT 16;

$$P = \frac{7 - 2D_w}{N_c} = \frac{7 - 0.662}{13} = 0.487 \text{ in.}$$

The pitch angle, Fig. 6.4, is

$$\lambda = \tan^{-1}\frac{\text{Pitch}}{\pi D_m} = \tan^{-1}\frac{0.487}{2\pi} = 4.5°,$$

which is well under the maximum of 12° (§ 6.2). The outside diameter of the spring is $D_o = D_m + D_w = 2.331$ in. In practice, several other designs would be made, perhaps with different D_m and/or grade of steel, and a final choice made.

6.11 EXAMPLE—INDEFINITE SERVICE. The load on a compression spring varies from 158 to 316 lb. The mean diameter of the coil is $D_m = 1$ in., and the design factor is to be 1.3 based on the Wahl line. If the material is oil-tempered carbon steel, what size wire should be used?

Solution. There is no easy way to solve for D_w directly. A diameter of wire can be assumed, after which the factor of safety N could be determined from equation (6.3). Depending on how fortunate this assumption is, we might save time by following the procedure of § 6.10 in order to find the order of magnitude of the wire size. A third way is to use the stress values in Table AT 17 and to assume values of K and K_c, so that an equation in terms of D_w can be set up and solved by trial and error. We shall use the third way. Assume $K = 1.48$ and $K_c = 1.29$, corresponding to a spring index of $C = 3.5$, Fig. AF 15. The mean and variable components of the load are

$$F_m = \frac{316 + 158}{2} = 237 \text{ lb.} \qquad \text{and} \qquad F_a = \frac{316 - 158}{2} = 79 \text{ lb.}$$

Then, from equations (**h**), the mean and alternating stresses are

$$s_{ms} = \frac{8KF_mD_m}{K_c\pi D_w^3} = \frac{(8)(1.48)(237)(1)}{(1.29)\pi D_w^3} = \frac{691}{D_w^3} \text{ psi} \qquad \text{or} \qquad \frac{0.691}{D_w^3} \text{ ksi};$$

$$s_{as} = \frac{8KF_aD_m}{\pi D_w^3} = \frac{(8)(1.48)(79)(1)}{\pi D_w^3} = \frac{298}{D_w^3} \text{ psi} \qquad \text{or} \qquad \frac{0.298}{D_w^3} \text{ ksi}.$$

From Table AT 17, use $s_{ys} = 87.5/D_w^{0.19}$ from column (5); try $s_{no} = 30/D_w^{0.34}$ from column (6). In equation (6.3), use $s_{ms} - s_{as} = 0.393/D_w^3$ ksi and find (with kips for force)

$$\frac{1}{N} = \frac{s_{ms} - s_{as}}{s_{ys}} + \frac{2s_{as}}{s_{no}}$$

$$\frac{1}{1.3} = \frac{0.393}{D_w^3(87.5/D_w^{0.19})} + \frac{(2)(0.298)}{D_w^3(30/D_w^{0.34})}$$

$$\frac{1}{1.3} = \frac{0.00449}{D_w^{2.81}} + \frac{0.01985}{D_w^{2.66}}$$

This is a reasonable form for iteration. After several trials with standard wire sizes, we compute $N = 1.37$ from the right-hand side of this equation for $D_w = 0.283$ in., No. 1. This value of N is close to the desired 1.3 and we note that we used the right expression for s_{no} to fit this wire size; *but check with the proper corresponding values of K and K_c*. Thus, $C = D_m/D_w = 1/0.283 = 3.54$; this value is so close to the assumed value of $C = 3.5$ that we can read no significant differences in K and K_c from Fig. AF 15. However, since the computed $N = 1.37 > 1.3$, we try the next lower wire size, W & M No. 2, $D_w = 0.2625$. Using $C = 1/0.2625 = 3.81$, we compute $N = 1.17$, which is too low. Therefore, for strength: $D_w = 0.283$, W & M No. 1, oil-tempered wire.

In order to find the number of coils, solid height, etc., more data are needed. However, the procedure from here would be as before, § 6.10, including the computation of the "solid stress" for comparison with the permissible.

6.12 MATERIALS USED FOR COIL SPRINGS. In general, steel springs are made of relatively high-carbon steel (usually more than 0.5%), heat treated and/or cold worked to a high elastic limit. High elastic limit is important in springs in order to get a large elastic deflection. Coil springs are wound cold in sizes under about $\frac{3}{8}$ to $\frac{1}{2}$ in. and hot in larger sizes. The material may be heat treated (pretempered) before winding (in small sizes) or after winding. When heat-treated wire is coiled cold, it should be stress relieved for bending stresses after coiling by being heated at some 500°F for 15 to 60 min., depending on its size. The wire sizes D_w given below are normal commercially available sizes.[2.1]

Hard-drawn spring wire (ASTM A227) is a low-cost material; sizes, 0.028 to about $\frac{9}{16}$ in.; suitable where service is not severe and dimensional precision is not needed; wound cold; 0.45 to 0.75 %C; not used for indefinite life. The quality of the surface is lower (with, say, hairline seams) than for the other grades (§ 6.13). Cost index = 1.[6.10]

Music wire (ASTM A228) is hard drawn also (80 % reduction), but it is made of high-grade steel; excellent surface, comparable to "valve-spring quality"; 0.7 to 1.0 %C; wound cold; sizes, 0.004 to 0.156 in. It is the best material available in sizes below about $\frac{1}{8}$ in. Cost index 3.5.[6.10]

Oil-tempered spring wire (ASTM A229) is cold drawn to size (reduction 50–70 %) and then hardened and tempered (pretempered); 0.55–0.75 % C; usually wound cold and stress relieved at low temperature, say 450°F; sizes 0.225 to 0.5. Its surface is not the best, but it is significantly better than hard-drawn wire. Cost index 1.5.[6.10]

Valve-spring-quality (VSQ) *carbon steel* (ASTM 230) is the highest quality oil-tempered wire; 0.60–0.75 % C. Since it has an excellent surface, valve-spring quality is the most reliable (with music wire) for fatigue and is thus used for the most severe service; sizes, 0.093 to 0.375 in.

Chromium-vanadium steel (ASTM 231) is oil tempered; 0.45 to 0.55 % C; sizes, 0.28 to 0.375 in. The alloy steels are superior to carbon steels of the same quality above about 250°F. Cost index 4.[6.10] Valve-spring-quality Cr-V (ASTM 232) has the best commercial surface; sizes 0.032 to 0.437 in.

Chromium-silicon (ASTM 401), good quality for impact loads and moderately high temperatures (perhaps to 450°F, depending on the permissible amount of relaxation). Cost index 4.[6.10]

Stainless steel, type 302 (Chromium-nickel, ASTM A313) is corrosion resistant and readily available; sizes, 0.009 to 0.375 in. It is cold drawn and its relaxation at higher temperatures is much less than the grades described above. Cost index 8.5.[6.10] Stainless steel wire with strengths comparable to or better than music wire is available.

Other materials, some not mentioned in this text, are used for coil springs for a reason, as for electrical conductivity; see Table AT 17. Occasionally, there seems to be a reason to use a plastic or glass for springs.

Typical steels used for hot-wound coil springs (§ 6.3) and flat springs include AISI 1095, 50B60, 6150, 8660, 9260, 9850. As we know, the alloys have the better hardenability (§§ 2.6, 2.7). In general, alloy spring steels in small wire sizes are not significantly better or stronger than carbon steels; in the hot-wound larger sizes, alloys with their better hardenability may be more generally advantageous.

6.13 FACTORS AFFECTING FATIGUE STRENGTH OF COIL SPRINGS. If the number of cycles of loading is small (light service), the computed stress in the wire may be relatively high, and normal surface defects may not be significant. When fatigue is involved, the surface condition is primary. Any slight flaw, such as seams, pits, die marks, hardening cracks, inclusions, or an accidentally scratched spot, may result in fatigue failure. For this reason, experimental endurance strengths of wires of a particular size often have a large natural spread. In fact, it is not at all certain that differences found in the fatigue strengths for different wire sizes are significant; that is, fatigue is less dependent on size than on other factors. For unpeened springs of high-quality carbon steel, some engineers let the maximum stress be 90 ksi with a range of 70 ksi; if peened, the maximum stress may be 110 ksi.[6.12]

If a member is subjected to torsion, one principal stress is tensile (Chapter 8). Shot peening, as previously reported, § 4.28, leaves a residual compressive stress which opposes the principal tension. This residual, together with the residual compression "covering" the flaws (especially at the inside of the coil where the stress is a maximum), results in a much improved fatigue strength (peening may be used when $D_w > 0.0625$ in.); examples:[6.1] music wire and Cr-V 6150, $D_w = 0.148$ in., $s_{no} = 70$ ksi when unpeened, 115 ksi peened; stainless 302, $D_w = 0.148$ in., $s_{no} = 45$ ksi when unpeened, 90 ksi peened; phosphor bronze, $D_w = 0.148$ in., the respective values are 15 and 30 ksi. In a torsion bar, the most improvement in fatigue comes from peening the bar while it is overstressed in torsion.[4.64]

Since the force on springs nearly always acts in the same sense, overstressing springs to induce favorable residual stresses, § 4.23, is common practice.[6.14] The process of obtaining the residuals, called ***presetting*** or *setting out*, is to have the coiled spring somewhat longer than desired, then compressing it into the plastic stress range; after which, the spring should

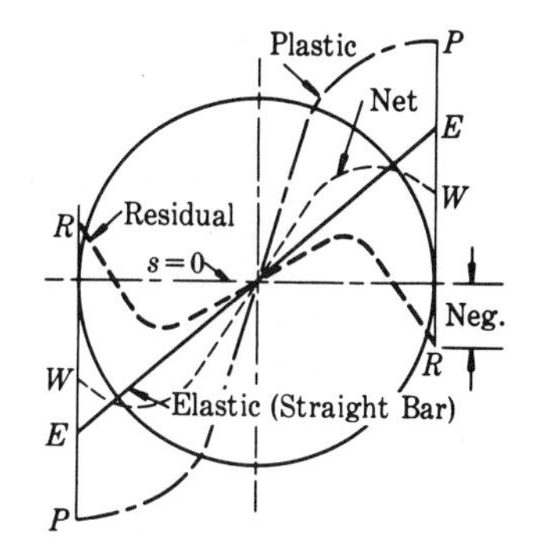

FIGURE 6.8 Effect of Residual Stresses. A torque has induced the plastic stresses *PP* in a straight bar, which, when the torque is released, then has the residual stresses *RR*. Now some applied torque *T* induces the elastic stresses *EE*, $s_s = Tc/J$, but the net stress is obtained by algebraic addition of the residual stresses, to get the distribution *WW*. See Fig. 4.17.

be a suitable length with favorable residual stresses, Fig. 6.8. Reasonable engineering practice with set-out springs is to increase the design stress by an amount up to 50% for *static* loads, but more modestly for fatigue loads;[6.3] for shot-peened coil springs, increase the design stress for fatigue loads by 25%, but make no change for static loads. Above about 500°F for steel, the improvement from peening is rapidly lost. Since presetting of very hard steel ($R_C = 50$) sometimes results in cracks in the direction of the principal tensile stress, peening should be done before presetting in this case; the residual compressive surface stress counteracts the tensile[4.1] and cracking damage is avoided. However, if presetting is not so drastic as to cause cracks, the endurance strength is greater when peening follows presetting than when it precedes.[4.28] The "solid stress" for preset springs may be about $0.60s_u$ without excessive setting. Lastly, presetting may result in some 5–10% reduction in the modulus of rigidity G.[6.1B]

Larger, hot-coiled springs are inevitably decarburized. These springs are then heat treated by quenching and tempering at 400–500°F. If such springs are peened, the effect of peening should reach through the entire decarburized layer; if not, the improvement may be minor or zero because of the existence of heat treatment cracks in the vicinity of the inner boundary of the decarburized zone (Coates and Pope[4.28]). Evidently, the compressive stress from peening about these cracks is sufficient to inhibit their spread. Coates and Pope report fatigue strengths s_a at 10^6 cycles as follows ($D_w = 0.5$ in., $s_m = 56$ ksi, 0.9% C, $R_C \approx 50$, $C = 5.25$, OQT 750°F): as received, $s_a = 11.1$; preset, $s_a = 13.1$; peened and preset, $s_a = 15.9$; preset and peened, $s_a = 19.7$ ksi. Decarburization reduces the effect of peening (lower residuals) in any event because of the lesser response of the material to cold work. If the surface is recarburized, shot peening may result in the failure being initiated at some point below the surface;[4.28] which suggests the possibility of coiling the springs from as-rolled bar, followed by recarburization and heat treatment, stress relieving and peening. Stress relieving can be done at about 400–500°F with no significant loss of the shot peening effect.

If steel is used in corrosive environments, various coatings may control the corrosion. Plating, say cadmium, will provide considerable protection (§ 4.30), but consider the effect on strength. Normal cleaning of high-strength steel in acid, preparatory for plating, is accompanied by atomic hydrogen diffusing into the material. Plating tends to seal the hydrogen in (and some plating, such as chromium, results in considerable hydrogenizing), the consequence being hydrogen embrittlement. A more recent development is a process of "mechanical plating," which is done by wet tumbling of the specially prepared springs in a mixture of metal shot, water, metal powder (say cadmium), and a chemical "promoter."[6.3]

6.14 RELAXATION OF SPRING MATERIALS. The relaxation is sometimes measured by the amount of the loss of force exerted by the spring at a particular deflection, sometimes by the change of deflection for a particular load. Springs relax at temperatures above ambient, the amount being a function of stress, temperature, and material, and the spring is said to *set* when this happens. Spring manufacturers often have enough data to predict the amount, so that it can be allowed for. Also, springs may be preset at a temperature and stress above operating values, which "removes the set." By way of example, the load loss of carbon valve spring steel at a stress of 80 ksi is:[6.2] about 3% at 250°F, 5% at 300°F, 6% at 350°F, 7.5% at 400°F. Normal limiting temperatures of operation are: 400°F for carbon steel, 550°F for 18-8 stainless steel, 700°F for Inconel, 500°F for Monel, 200°F for phosphor bronze. Note that the spring's rate is significantly lower at high temperatures (via decrease of G, E, § 2.21).

6.15 GOODMAN DIAGRAM. A Goodman type of diagram is often used for spring design, an example being Fig. 6.9 for music wire. Because

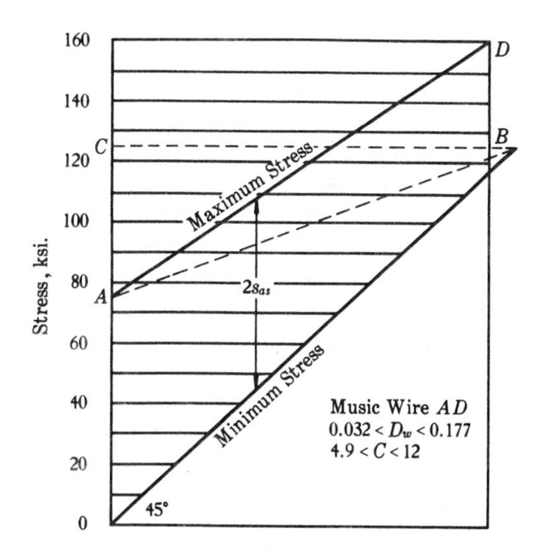

FIGURE 6.9 Goodman Diagram, Music Wire. Tests of coiled springs, unpeened; 185 specimens; stress relieved at 500°F for 1 hr.; ASTM A 228. All test values fell above line *AD*. Knowing either the maximum or minimum stress, the other permissible value is found from the diagram.[2.1]

of the scatter of results, this diagram lets the safe range of stress $2s_{as}$ be the same for all the sizes tested. With sufficient test data, such diagrams may be constructed for a smaller range of wire sizes.

6.16 TOLERANCES. In a particular application, the designer should use close tolerances only where such tolerances are significant. The significant dimension might be the outside coil diameter, the inside coil diameter, or the free length. Perhaps more often, the significant property is the force that a spring exerts at one or more deflections, and all the dimensions, together with the torsional modulus of elasticity, which is not strictly constant, affect the force. In a spring manufactured under controlled conditions, the wire diameter varies between certain natural limits (§§ 1.17 and 3.9); the coil diameter, the precise number of active coils, the torsion

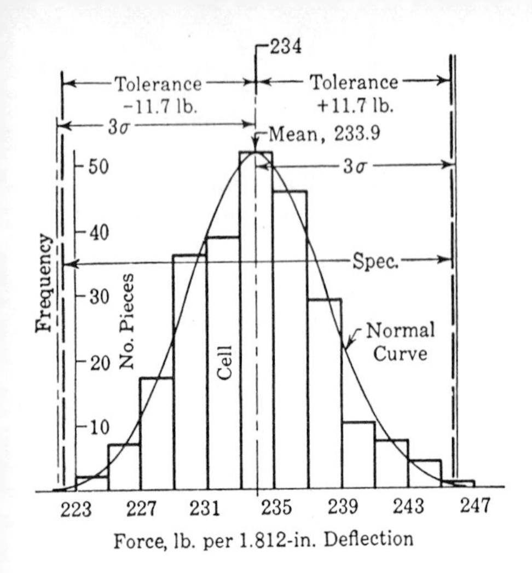

FIGURE 6.10 Distribution of Spring Force. Specifications were 234 lb. $\pm 5\% = 222.3$ to 245.7 lb. at $\delta = 1.812$ in. The value of $\sigma = 4$ lb., as computed from the actual data, giving the natural (6σ) spread from 221.9 to 245.9 lb. on a mean of 233.9. The process is rather well centered. Each cell, which is 2 lb. wide, represents the number of pieces falling within this 2-lb. range.

modulus, and the free length vary similarly. Each of these variables has some effect on the scale of the spring and on the force the spring exerts at a certain deformation. For example, the force at a particular deflection varies as the fourth power of the wire size, equation (6.2); thus a small variation of wire size affects the scale greatly.

Figure 6.10 shows a histogram of the actual distribution of the forces exerted by certain springs produced in a lot size of 250 to a specification of 234 lb. $\pm$ 5% at $\delta = 1.812$ in. Observe, by comparison with the corresponding normal curve for this histogram, that this tolerance is rather close, since the process must be practically centered and always in control to meet the specification. Thus, a tolerance in this instance of something greater than $\pm 5\%$ would probably lower the cost. The *minimum* tolerance on force for hot-coiled as-rolled rod should be about $\pm 10\%$, and for small wire, say under about 0.32 in., the tolerance may need to be larger for economic reasons, because the percentage variation of D_w will be larger.

6.17 SURGE IN SPRINGS. Unless the natural frequency of a spring is quite different from the frequency of the applied load, some resonance may occur, in which case "waves" travel along the length of the spring. These waves are successive compressions and extensions moving from coil to coil, and they may result in deflections of adjacent coils equal to the deflection when the spring is compressed solid. Accordingly, the spring may then be subjected repeatedly to the stress corresponding to the stress for solid compression. Since this stress is often above the endurance strength and since repeated contact of the coils will spoil the surface, surging may be the cause of failures where the calculated stresses for normal loads are apparently safe. The lowest natural frequency (first harmonic) of a spring is given by $\phi = (k/m)^{1/2}/2$ cycles per unit time,[6.1] where k is the spring scale and m is the mass of the active coils, k and m in consistent units. In order to match pounds for force, inches, seconds, the mass must have the units lb-sec.^{2}/in. (from $m = F/a$), which are obtained by dividing pounds of mass by $g_o = 386$ in./sec.² Using the approximate volume of active wire as found in § 6.7, the mass of the active coils is

(k) $$m = \frac{\rho}{386}\frac{\pi D_w^2}{4}\pi D_m N_c \text{ lb-sec.}^2/\text{in.}, \qquad [=\text{psin—footnote, p. 149}]$$

where ρ is in lb/in.³ and the other dimensions in inches. Using the value of k from equation (6.2), the equation for ϕ becomes

(6.5) $$\phi = \tfrac{1}{2}\left(\frac{k}{m}\right)^{1/2} = \tfrac{1}{2}\left(\frac{GD_w^4}{8D_m^3 N_c}\frac{(386)(4)}{\rho\pi^2 D_w^2 D_m N_c}\right)^{1/2}$$

$$= \frac{2.21 D_w}{N_c D_m^2}\left(\frac{G}{\rho}\right)^{1/2} \text{ cps;}$$

G psi, ρ lb/in.³ For steel with $G = 11.5 \times 10^6$ and $\rho = 0.284$ lb/in.³, we get

(l) $$\phi = \frac{14{,}050 D_w}{N_c D_m^2} \text{ cps.} \qquad \text{[STEEL]}$$

If the natural frequency is 12 or more (some say 20) times the rate at which the spring operates, no trouble is anticipated from surging.

6.18 BUCKLING OF COMPRESSION SPRINGS. A compression spring whose free length is more than four times its mean diameter should be checked for buckling. Enter Fig. 6.11 with the ratio (free length)/(mean

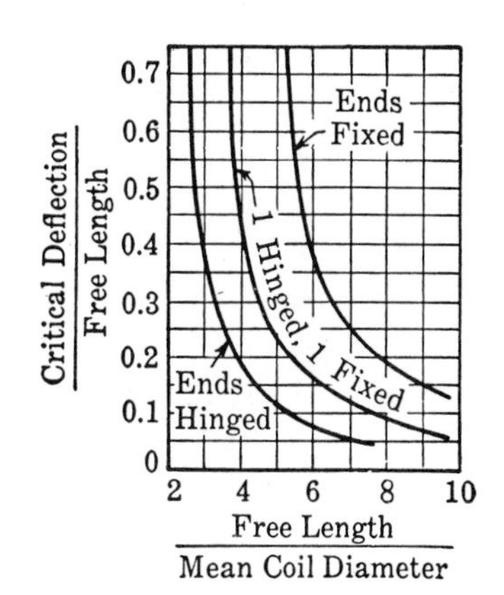

FIGURE 6.11 Buckling Conditions. [6.1, 6.2]

diameter), move vertically to the curve, then horizontally to the ordinate and read the value of (deflection)/(free length). If the deflection per inch of free length in the actual spring is greater than the chart value, Fig. 6.11, buckling with ends supported as stated should be expected. "Ends Fixed" is for the case where the squared and ground ends are on rigid parallel surfaces perpendicular to the spring's axis. The middle curve is for one end on a rigid surface, one end hinged. The bottom curve is for the case where both surfaces on which the ends rest are hinged on a pin. In a common hinge equivalent, the spring rests on a ball, as in a ball valve.

6.19 CONCENTRIC HELICAL SPRINGS. Two concentric (nested) springs may be useful in carrying heavy loads or in eliminating a vibration, such as surge. For a concentric arrangement of springs, there are two conditions that may be, though not necessarily, approximately satisfied: first, the stress in each spring at any deflected position may be the same; and second, the free heights of the springs are usually to be the same. These conditions are approximately obtained in a round-wire spring if both springs have the same index C;

(m) $$C = \frac{D_{m1}}{D_{w1}} = \frac{D_{m2}}{D_{w2}},$$

where subscripts 1 represent one spring and subscripts 2 represent the other. For these conditions, the outer spring takes about $\frac{2}{3}$ the total load when $C \approx 6$, but the proportion is different at other spring indexes. The design procedure may involve extensive iteration to obtain the desired balance of properties; Chandler[6.17] gives considerable help in the way of tabular matter.

6.20 COMPRESSION COILED SPRINGS WITH RECTANGULAR WIRE. Wahl[6.1] gives the following formula for the maximum shearing stress in a spring of rectangular wire when the dimension b of the rectangle is parallel to the axis of the spring (and when $b/t < 3$ if b is the larger dimension):

(n) $$s_s = \frac{K_q F D_m(3b + 1.8t)}{2b^2t^2}; \qquad [b \text{ PARALLEL TO AXIS}]$$

K_q is obtained from Fig. AF 15 for $C = D_m/t$. For a small pitch angle ($<10°$), approximately

(o) $$\delta = \frac{2.45 F D_m^3 N_c}{Gt^3(b - 0.56t)},$$

which will serve also for a spring of square wire ($b = t$). From equation (**n**), the stress in a square-wire spring is

(p) $$s_s = K_q \frac{2.4 F D_m}{b^3},$$

where b is the *average* length of the side of the square section and K_q is found in Fig. AF 15. Note that when the square wire is bent, it bulges on the inside, the section becoming approximately trapezoidal. Wahl gives more accurate means of computation in Ref. *(0.3)*.

6.21 EXTENSION SPRINGS. The foregoing equations for compression springs apply as well to extension springs, except that allowance is made for the initial tension, if any. Extension springs are generally wound with the coils pressing against each other, and the ***initial tension*** is the force on the spring when the coils are on the point of separating. The amount of the initial tension can be regulated to some extent and it varies from manufacturer to manufacturer, but reasonable maximum values of the corresponding stress are as given in the following tabulation[0.3] [as computed from equation (6.1) with $K = 1$]. Lower values should be specified.[6.13]

C 3	4	5	6	7	8	9	10	11	12	15
s_i 24,000	22,500	20,000	18,000	16,250	14,500	13,000	11,600	10,600	9,700	7,000

Hooke's law does not apply until the initial tension F_i is overcome. After the coils are separated, the stress may be computed from (6.1) for the external load F. If the scale of the spring is estimated from (6.2), the load F on the spring is the initial tension *plus* $k\delta$, $F = F_i + k\delta$, where δ is the extension from its unloaded length and $F_i = \pi s_i D_w^3/(8D_m)$. If there is no initial tension, the equations for compression springs apply without modification—provided that the coils are not extended to a pitch angle λ greater than about 12°. Extension springs should be designed to operate with some extension at all times.*

The weak point in an extension spring is likely to be where the coil is bent to make a hook, Fig. 6.12. In case of fatigue loading, the radii used in bending the wire to form the ends, Fig. 6.12(b), should be as large as possible. The maximum stress at a section B, Fig. 6.12(b), may be estimated from $s_s = 8K_cFD_m/(\pi D_w^3)$, where K_c is for $C = 2r_m/D_w$ from Fig. AF 15

* Given an extension spring, to find the initial-tension force F_i: extend it 0.1 in., measure F_1; extend it to $\delta_2 = 0.2$ in., measure F_2; then $F_i = 2F_1 - F_2$. Prove this.

FIGURE 6.12 Extension Spring. Other shapes of loops or hooks are readily available. (Courtesy Associated Spring Corp., Bristol, Conn.).

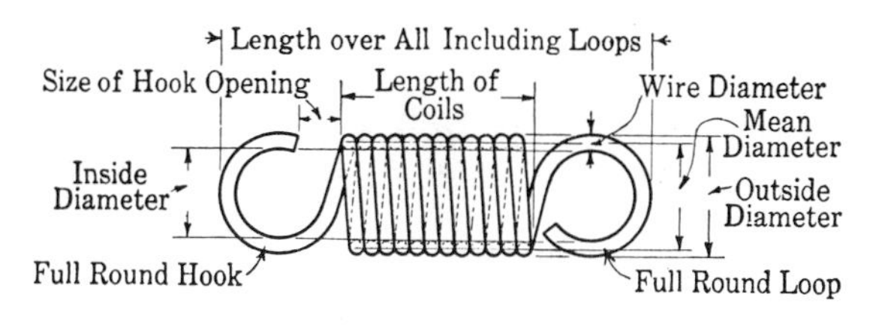

(a)

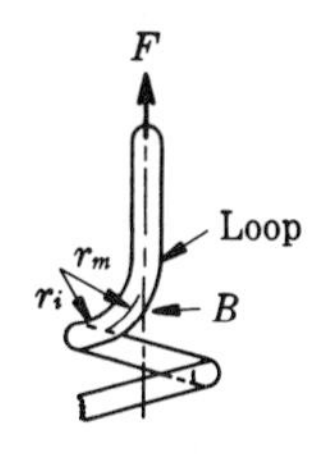

(b)

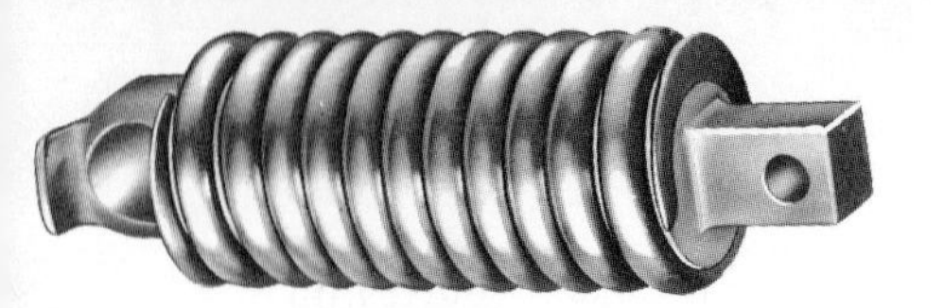

FIGURE 6.13 Threaded Plugs.

(actually, both bending and torsional stresses exist in this vicinity). The flexural stress in the loop itself can be computed using K values for a curved beam (§ 8.25) and a maximum bending moment of $FD_m/2$. Let the design stress for the coils be about 20–25% less than for a compression spring. There are numerous ways of finishing the ends of extension springs, among which is the idea shown in Fig. 6.13.

6.22 TORSION SPRINGS. The preceding discussion of springs assumes in each case that the load is axial, inducing a *torsional* stress in the spring. A torsion spring, Fig. 6.14, is one that resists a moment tending to wind up the spring. Due to the variety of ways in which the spring may be loaded, the calculations considered below are merely suggestive.

If an end coil is attached to a disk or to some member that imposes a pure torsional moment on the spring, the coils will be in pure *bending*. If a force F is applied at the end of a coil, as in Fig. 6.14, then, in order for the maximum stress to approach a bending stress only, a core or arbor, preferably with a diameter greater than $9D_w$, must be used inside the spring in such a position that it bears on the spring at the section B, Fig. 6.14, to

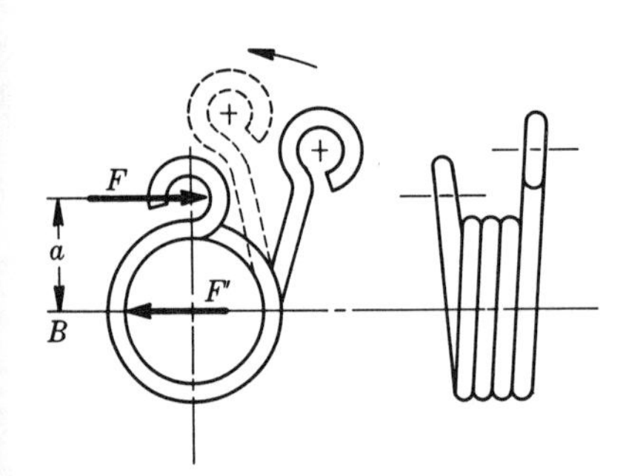

FIGURE 6.14 Torsion Spring. This figure illustrates ends of special design. (Courtesy Associated Spring Corp., Bristol, Conn.).

produce a reaction $F' = F$. In this case, the torque is $T = Fa = M$, the bending moment, where a is the moment arm of the couple F–F'. We are dealing with a curved beam (§ 8.25), in which the curvature acts as a stress concentrator (shifts the neutral axis). Thus, $s = Mc/I$ is modified by a stress factor K_b, obtained for a rectangular section from Fig. AF 15 (or from Table AT 18 for other shapes); thus, the flexural stress is

$$\text{(q)} \qquad s = \frac{K_b Mc}{I};$$

$I/c = \pi D_w^3/32$ for round wire and $I/c = bt^2/6$ for rectangular wire, Fig. 6.15. In general, a force-and-stress analysis is made to fit the situation.

Preferably, the working force winds up the spring, thus taking advantage of the residual stresses. If the working loads act oppositely, the spring should be stress relieved. For repeated loads, watch for sections of stress concentration where the ends are formed.

The angular deflection of a torsion spring subjected to opposing moments $T = M$ at the ends is

$$\text{(r)} \qquad \theta = \frac{ML}{EI} = \frac{M\pi D_m N_c}{EI} \text{ radians,}$$

where the length of active coils $L \approx \pi D_m N_c$; I is the rectangular moment of inertia of the section of the wire taken as a beam. Because of the "winding up" of the spring to a smaller diameter, the actual deflection is

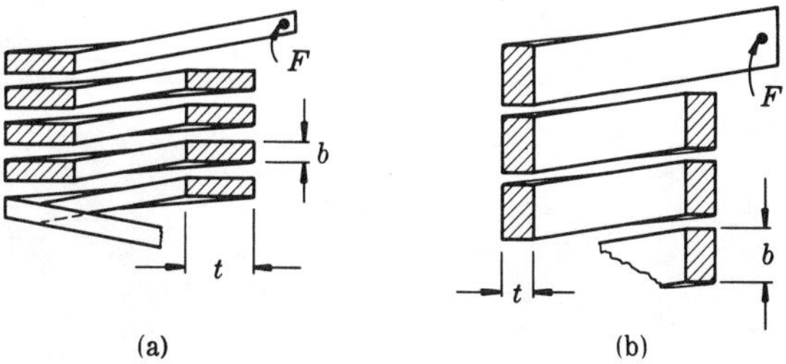

FIGURE 6.15 Torsion Springs. With the t dimension as in (a), the spring is stiffer. $I = bt^3/12$.

somewhat less than this theoretical value.[6.12] In a long torsion spring, θ may be the equivalent of several complete turns. In Fig. 6.14, the deflection of the point of application of F is $a\theta$. The work done on the spring from $F = 0$ to F is $(F/2)a\theta = T\theta/2$.

Some authorities recommend design stresses about 60% greater than those defined for compression springs in column (3), Table AT 17. Basically, with the stress being flexural, something less than the *tensile* yield strength (so that there will be no permanent set under operating conditions) represents a limiting value. Reported values of this strength in wire sizes are not explicit: for oil tempered, use $0.8s_u$; music wire and hard drawn, $0.65s_u$; alloy steels, $0.9s_u$; stainless, A 313, $0.55s_u$ (greater variability than the others); brass, $0.42s_u$; phosphor bronze, monel, and beryllium copper, $0.67s_u$. If the s_u values in Table AT 17 are used to obtain design stresses, note that they are specification *minimums*.

For torsion springs hot-wound from rods, the equations

$$\text{(s)} \qquad \underset{[\text{SAE } 1080]}{s = \frac{120}{D_w^{0.26}}} \quad \text{and} \quad \underset{[6150 \text{ AND } 9260]}{s = \frac{132}{D_w^{0.22}} \text{ ksi}} \qquad [0.5 < D_w < 1.5]$$

give, within 3%, stresses that Associated Spring [6.2] calls "maximum" (say proportional limit) when the surface BHN $\approx$ 437.

6.23 OTHER KINDS OF SPRINGS. As might be surmised from Fig. 6.16, springs are used in a myriad of different forms, and we have space for treatment of only a few. Additional fairly common forms include:

Garter spring, the helical coil is wrapped into a circle forming an annular ring; used in tension to hold oil seals of leather, Teflon, etc., tight against a surface; used in compression as piston-ring expanders.

Motor spring, thin flat strip wound up on itself as a plane spiral, usually anchored at the inside end; an energy source to drive clocks, toys, etc.

Hairspring, wire or strip rolled into a plane spiral, no contact between coils, a sensitive spring used in instruments, watches.

Volute spring, wide-strip material wound in a conical helix (frustum of a cone with long dimension parallel to axis) with strip overlapping; used

FIGURE 6.16 Assorted Springs. (Courtesy Associated Spring Corp. Bristol, Conn.).

where compactness, frictional damping of the turns rubbing on one another, and a spring rate increasing with deflection, are advantageous.

Conical spiral spring, also wound in the form of a conical helix of round wire; not a constant rate spring.

A *belleville spring* is one made in the form of a dished washer, Fig. 6.17, and has been adapted to many uses. The theory[6.18] is too extensive for presentation here, but Wahl has presented charts that reduce the computation time materially.[0.3,6.1B] The load-deflection curves vary widely in shape as the ratio h/t, Fig. 6.17, varies. For $h/t \approx 0.5$, the curve is approximately a straight line up to $\delta \approx t/2$. For $h/t \approx 1.5$, the load is constant for a considerable deflection (zero rate) after a certain initial deflection. These springs are often used in stacks of two or many more. When stacked in series, Fig. 6.17(b), a larger deflection results for a given load (lower rate). A parallel stack, Fig. 6.17(c), supports a larger load for a given deflection.

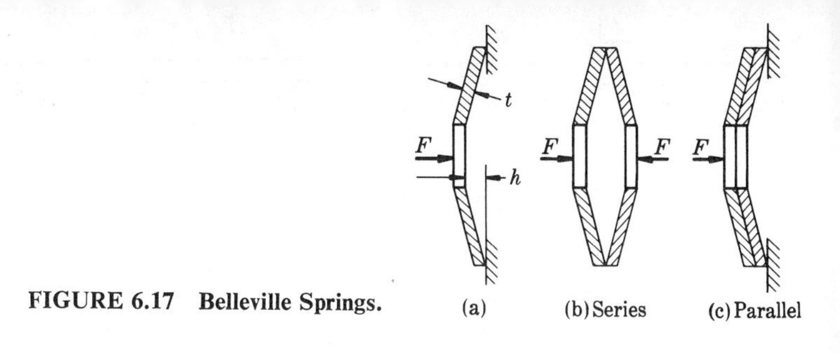

FIGURE 6.17 Belleville Springs.

In either case, a stack of belleville springs can absorb a relatively large amount of energy per unit of volume occupied.

A *Neg'ator spring*, a patented form, exerts virtually a constant force F (zero rate) after a certain initial deflection. The basic idea is that flat strip is made into a plane spiral coil, its natural form without external loading, so that when it is straightened, it exerts a force on account of its tendency

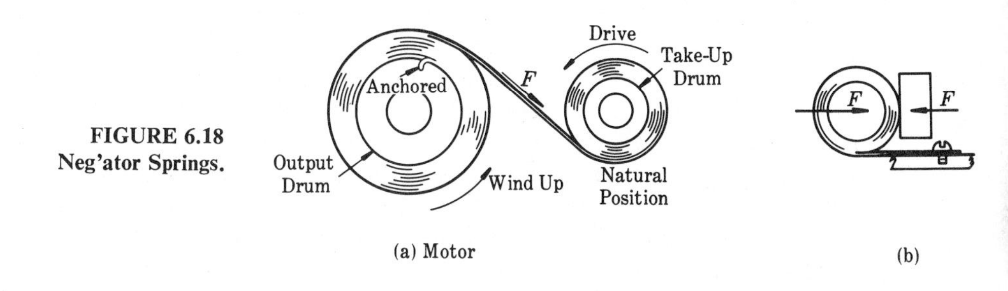

FIGURE 6.18 Neg'ator Springs.

to rewind to its original form. As a motor, Fig. 6.18(a), it exerts nearly a constant torque on the output drum; it is made in many other forms to exert constant force, one of which is suggested by Fig. 6.18(b). Its uses include: exerting constant pressure on commutator brushes; in retriever reels that are used for returning many things to a desired position, as retracting the gasoline hose after the fuel has been delivered. The manufacturer has prepared engineering data for design purposes.[6.15] The same manufacturer makes a constant-force (nearly) compression spring, called a *Flex'ator*, which is basically a tightly coiled helical spring loaded by an eccentric load that bends it. Constant force action can be obtained with ordinary compression and extension coil springs through intermediate cams designed to deliver a constant torque to the shafts on which they are mounted.[6.21]

Rubber springs, and other materials with similar properties, are used in mountings to isolate vibrations (metal springs are too). The modulus of elasticity is not constant and the load-deflection curve depends on the particular compound, but the material has a high capacity per unit volume for storing energy and advantageous damping properties that are absent in metal springs; used in compression or shear. These materials are readily bonded to various metals; and are widely used in connections such as automotive spring shackles and other joints with limited relative motion. (See problems in Slaymaker.[1.15])

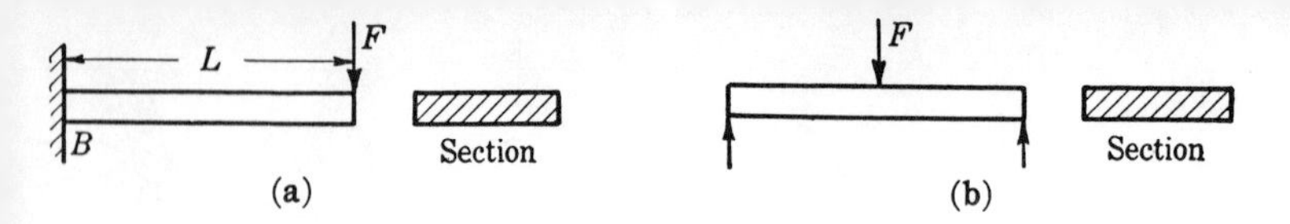

FIGURE 6.19 Flat Springs.

6.24 FLAT SPRINGS. Flat springs may be in the form of a cantilever beam, Fig. 6.19(a), or in the form of a simple beam, Fig. 6.19(b). The stress and deflection of springs like these are computed from formulas given in Table AT 2, just as for any ordinary beam. In the cantilever beam, the maximum stress occurs at the point of support B in Fig. 6.19(a). Since the bending stress for a constant cross section diminishes from B to the point of application of the load F, the section of the beam may be diminished in such a manner that the maximum bending stress on each section is the same. The resulting beam is called a *beam of uniform strength.* For example, a beam tapered in the plan view as shown in Fig. 6.20 has the same bending stress in all sections, and for a particular material, it will support just as large a load with a particular maximum stress as the beam of Fig. 6.19(a), provided the length of the beam and the dimensions at the section B are the same, with a 50% saving in material. Since a shearing stress exists, the beam must not come to a point where the load is applied, because there would be no area to resist the shear. An important difference is that the spring of Fig. 6.20 deflects more under the same load than a beam of constant width b—a distance of $6FL^3/(Ebh^3)$ compared with $4FL^3/(Ebh^3)$. See Tables AT 1 and AT 2. With a larger deflection, the energy that can be absorbed ($F\delta/2$) is greater than that which a beam of constant section can absorb, with the same maximum stress in each beam.

Figure 6.21 represents a simple beam of uniform strength. Similar points of comparison may be made for the two types of simple beams, Figs. 6.19(b) and 6.21, as for the two types of cantilever beams. The nominal stresses in these beams may be computed from the bending moment formula $s = Mc/I$, and the nominal deflections at A of the beams of uniform strength are as shown in Figs. 6.20 and 6.21, provided that the deflection is not large enough to alter the moment arm of F significantly. These springs follow Hooke's law within the elastic limit and the scale is $k = F/\delta$; the maximum stored energy is $F\delta/2$.

6.25 LEAF SPRINGS. If the flat springs of uniform strength just described are divided as indicated by the dotted lines of Fig. 6.22, and assembled as indicated by the solid lines of Fig. 6.22, the same stress and deflection formulas as given in Figs. 6.20 and 6.21 apply, friction between the leaves neglected. (Friction results in the computations being inherently less accurate than those for coil springs.) The result obtained is a leaf spring with all leaves of the same thickness, b in the formula being equal to the sum of the widths of the leaves; that is $b = N_1b'$ where N_1 is the number of leaves.

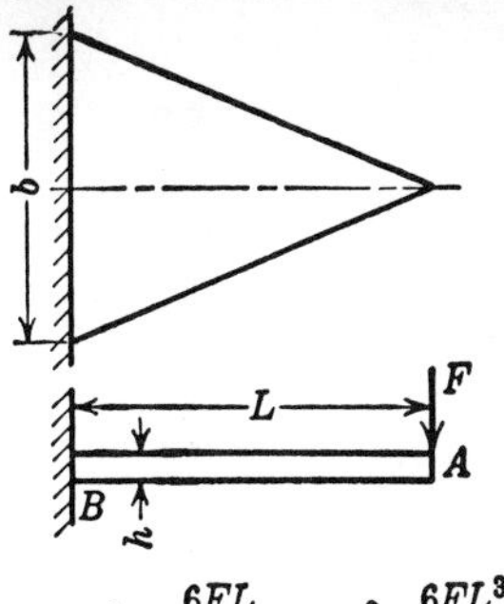

FIGURE 6.20 Cantilever of Uniform Strength.

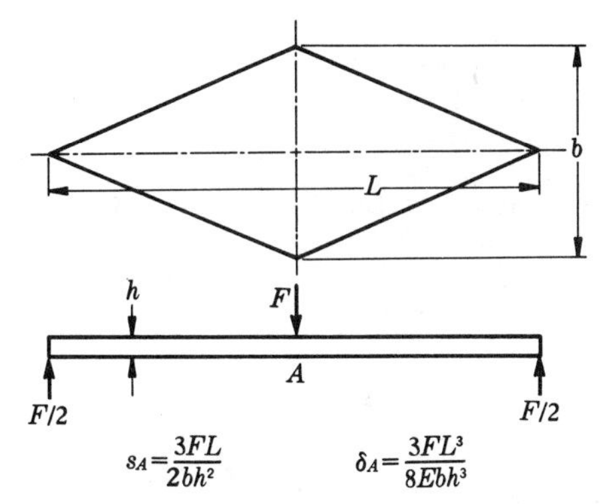

FIGURE 6.21 Simple Beam of Uniform Strength.

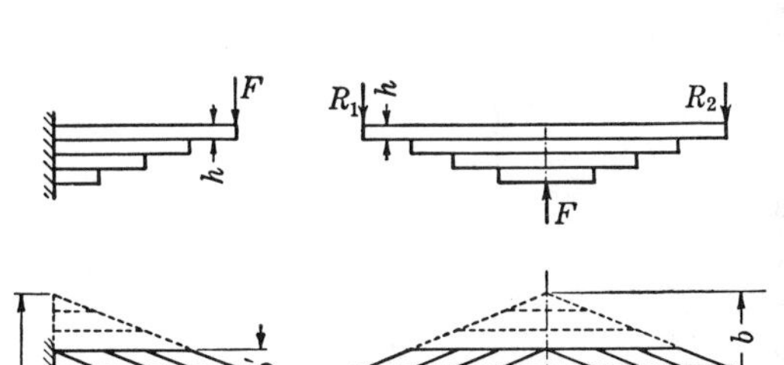

FIGURE 6.22 Deriving Leaf Springs from Beams of Uniform Strength.

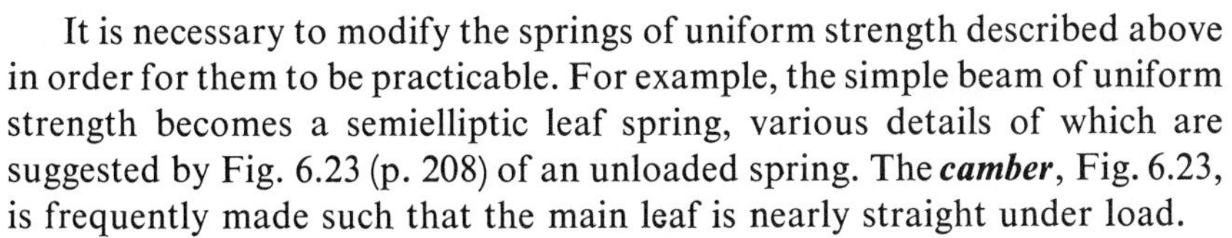

It is necessary to modify the springs of uniform strength described above in order for them to be practicable. For example, the simple beam of uniform strength becomes a semielliptic leaf spring, various details of which are suggested by Fig. 6.23 (p. 208) of an unloaded spring. The ***camber***, Fig. 6.23, is frequently made such that the main leaf is nearly straight under load.

An approximation of actual semielliptic springs which can be presented easily is the trapezoidal spring, Fig. 6.24 (p. 208), for which[6.1]

$$\textbf{(t)} \quad s = \frac{6FL}{bh^2} = \frac{3WL}{bh^2} \quad \text{and} \quad \delta = \frac{K_1 FL^3(1-\mu^2)}{3EI} = \frac{K_1 WL^3(1-\mu^2)}{6EI}$$

where $W = 2F$ is the load at the middle of a simple beam of length $2L$ (F is the load at the end of a cantilever of length L), Fig. 6.24; $b = N_1 b'$,

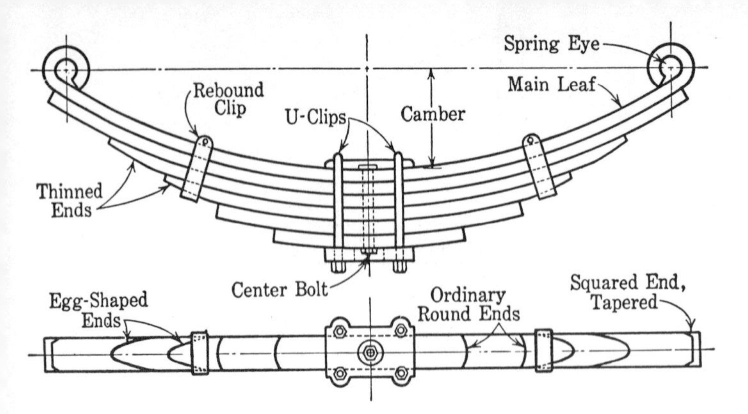

FIGURE 6.23 Leaf Spring. Several styles for finishing the ends of the leaves are illustrated; an unlikely combination in practice. The full-length leaf under the main leaf is often square cut even though the other leaves may have egg-shaped and thinned ends. Beware of the stress concentrations in the vicinity of the clamps (U-clips). Interestingly, Almen[4.64] points out that when the bushing is pressed into the eye as shown, the resulting deformation induces a tensile stress in the top surface of the leaf at this point, where the working stresses are also tensile. On this account, he recommends making the eye by bending, for example, CL on the right instead of CC.

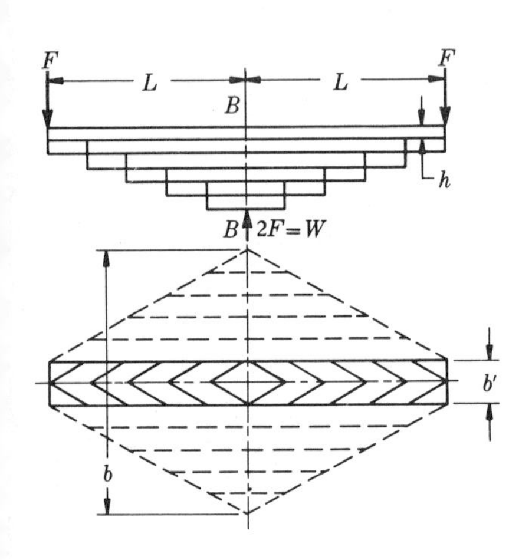

FIGURE 6.24 Trapezoidal Profile into Leaf Spring. The spring may be designed so that b' is twice the width of one leaf, that is, so that there are two full length leaves with squared ends.

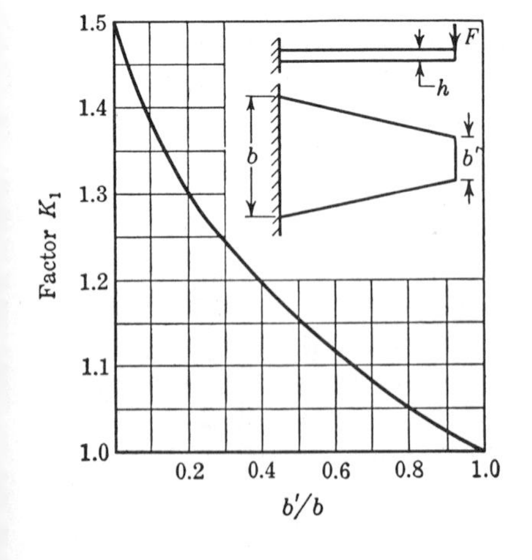

FIGURE 6.25 Correction Factor for Deflection. (From A. M. Wahl, Mechanical Springs[6.1]).

where b' is the width of a leaf and N_1 is the number of leaves; μ is Poisson's ratio and the term $1 - \mu^2$ applied when the spring's width b is large as compared to its thickness h, in which case lateral expansion or contraction of elements near the surface is hindered, resulting in a somewhat stiffer spring than simple flexure theory predicts; I is the moment of inertia of the leaves at section B; K_1 depends on the ratio of b'/b and is taken from Fig. 6.25. In (t), the equation for δ gives the deflection of the end of a cantilever spring (section B, Fig. 6.24, fixed) when F = load on end and L = length of cantilever.

6.26 FATIGUE OF LEAF SPRINGS. Leaf springs often have holes or notches that are points of stress concentration, and the principles of Chapter 4 apply. Even though the clamping action of the center bolt and U-clips reduces the bending stresses at the section of the bolt hole, Fig. 6.23, it would be on the safe side to check this section in accordance with fatigue procedures when the load is repeated. Rubbing of the leaves is conducive to fretting corrosion (§ 4.27). Data are too meager for generalization, but tests of 0.5–0.6% C steel show an actual strength reduction factor of $K_f \approx 1.4$. Sharp edges on the leaves should be avoided in severe situations.

For leaf springs made of as-rolled material, a loss of fatigue strength of the order indicated in Fig. AF 5, should be expected. Typically, the material is later cold rolled, which much improves the surface properties. Heat-treated steel may have a decarburized surface. Chrome-vanadium steel resists decarburization during heat treatment better than silicon-manganese. Thus, suitable surface treatment will greatly improve the fatigue strength of leaf springs. Inducing a residual compressive stress on that surface which operates with a tensile Mc/I stress, either by presetting or peening or both, increases the fatigue strength as usual. Materials used for leaf springs are about the same as for hot-rolled coil springs, principally SAE 1080, 1095, 5155–60, 6150–60, 9250–60. For flat springs in general, bronze, beryllium copper, stainless steel, Inconel, clad stainless and carbon steels are also used for a reason. *Maximum design* fatigue stresses for 10^7 cycles, AISI 1095, flat and leaf springs with $s_{min} = 0$, as a function of thickness are:[2.1] 155 ksi for $t = 0.005$ in.; 140 for 0.010; 130 for 0.020; 125 for 0.030; 120 for 0.040; 105 for 0.060; 100 for 0.090 in.

6.27 GENERAL REMARKS ON LEAF SPRINGS. Prestressing leaves, or flat springs in general, (§ 4.23) in the same direction as the loading leaves a favorable residual stress that increases the safe capacity of the spring (see also §§ 4.26–4.30, inclusive, and other points in Chapter 4). A number of other expedients have been applied to leaf springs.

A common practice is to bend the leaves to different radii of curvature, the radius decreasing on the shorter leaves, Fig. 6.26; the leaves are said to be *nipped* or to have *nip*. Considering the main and second leaves, note that when the leaves are pulled tightly together (by the center bolt), the main leaf is bent opposite to the direction that the working load will bend it. Thus, it is not stressed in the direction of the working load (tension on top, Fig. 6.26) until after the working load bends it past its unloaded

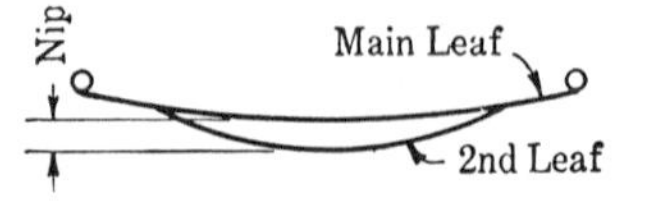

FIGURE 6.26 Spring Leaves of Different Radii.

curvature; consequently, the maximum stress due to purely vertical forces is greater in the second and other leaves than in the main leaf. Therefore the main leaf is left with some capacity for taking nonvertical loading, as it must in springs for vehicles. Nip, which may be applied to the other leaves, also serves to produce forces between the leaves that tend to keep them in contact on rebounds, thereby keeping out dirt.[6.22]

Another idea for improving the load capacity of the main leaf is to make it *thinner* than the others. Using the elastic curve equation of a straight beam $M = EI/r$ and the stress equation $M = sI/c$, we get $s = Ec/r$, where r is the radius to which a straight beam is bent by the moment M that produces a stress s. We see that the stress in a beam bent to a certain radius r is directly proportional to the thickness of the leaf ($2c$). Hence, if one leaf is thinner than the others, it will be stressed less than the others by a particular moment.

A highly destructive action on an automotive spring is the rebound, unless it is held in check, because an unchecked rebound after a bump may bend the leaves until the stress is dangerously high. Thus, shock absorbers act not only to improve the ride but also to prevent spring failures by checking the rebound.

6.28 CLOSURE. The scope of this book limits the coverage of springs to primary considerations of the most common types. In many areas, such as the design of coil and flat springs for precision instruments, for weighing scales, etc., unusual problems call for specialized knowledge that the spring manufacturer may be able to supply. If a spring is to be produced in quantity, it should be tested experimentally to be sure it possesses the required properties.

For a spring subjected to fatigue loading, the most important factor is the surface condition. Where the cost justifies, vacuum melted steels have fewer inclusions and they emerge from the manufacturing processes with an improved surface. As it has been said,[4.3] "An accidental bruise on a highly stressed spring will almost certainly lead to early failure."

7. COLUMNS, CENTRAL LOADS

7.1 INTRODUCTION. A type of failure that we have not yet discussed is one due to an instability, called ***buckling***, though an equation for checking the buckling of wide flanges was given in § 1.24. The buckling to be discussed in this chapter is of centrally-loaded slender members in compression, called columns; a yard stick is a good illustration. The longer and slenderer the column is, the lower the safe stress that it can stand. The slenderness of a column is measured by a ***slenderness ratio*** L/k, where L in. is the length of the column and $k = (I/A)^{1/2}$ in. is the ***radius of gyration*** of a cross-sectional area about a centroidal axis—nearly always the *least* radius of gyration. See Table AT 1, but observe that this table does not give the least radius of gyration in every case.

7.2 EULER'S FORMULA. The Euler analysis applies to a very slender column, and the formula for frictionless rounded ends (no bending moment at the ends), Fig. 7.1(a), is

(a) $$F_c = \frac{\pi^2 EA}{(L/k)^2},$$

where F_c is the axial concentric load, called the ***critical load***, that causes the column to be on the point of buckling, and E is the modulus of elasticity. The units in (**a**) must be consistent, say pounds or kips and inches. Observe that *stress* s is not involved in the failure of a very slender column. If we are to be certain that failure does not occur, the actual load F on a column must be less than F_c; that is, the factor of safety or design factor N *must now be applied to the load* F and is thus defined by

$$\text{(b)} \qquad N = \frac{F_c}{F}.$$

For a particular cross section and length, the load capacity F_c of a column depends only on the modulus of elasticity E. Since there is little variation in E among different grades of steel, there is no advantage in using an expensive high-strength alloy steel instead of structural steel for columns with L/k greater than about 120. See Fig. 7.2.

7.3 EFFECTIVE LENGTH. Euler's equation as written can be applied to a column with ends fixed in any manner if the length is taken as that between sections of zero bending moment; call this length the ***effective length*** L_e. We shall write all our column formulas in terms of L_e. The most common types of columns are as shown in Fig. 7.1. The theoretical values of L_e, followed in parentheses by the AISC[5.34] recommended design values, are: for fixed ends $L_e = L/2$ ($0.65L$); for one end fixed, one end rounded (or guided), $L_e = 0.707L$ ($0.8L$); for one end fixed, one end free, $L_e = 2L$ ($2.1L$). Thus, with a factor of safety N, Euler's equation becomes

$$(7.1) \qquad F_c = NF = \frac{\pi^2 EA}{(L_e/k)^2} \quad \text{or} \quad F = \frac{\pi^2 EA}{N(L_e/k)^2} = \frac{\pi^2 EI}{NL_e^2},$$

[USE FOR STRUCTURAL STEEL WHEN $L_e/k >$ about 120]

where F is some safe central load. A typical design factor for Euler structural columns is $N \approx 3.5$,[1.2] and most designers probably tend to increase N with significant increases in L_e/k. It is important to understand that, if the column remains straight and if the load F is concurrent with the centroidal axis of A, the average stress on a section of the column is $s_c = F/A$ and that local buckling at some point where the stress is considerably lower than $F/A = s_y$ is what will lead to failure.

Theory indicates that an Euler column with fixed ends will support four times as much load ($L_e = L/2$) as a rounded end column, but since the parts to which the ends are connected are not fixed rigidly and because of other departures from ideal, designers rarely if ever use $L_e = L/2$; the extreme limit is more likely about $L_e = L/1.41$, and in machine design, the choice is almost always $L_e = L$ (rounded ends)—except of course when the column has a free end, the weakest type (see reference *7.6*). Use of $L_e = L$ is conservative for pin-ended columns (see § 7.7).

7.4 SHORT COLUMNS. If L_e/k is below a certain value for a particular material, such as the values of L_e/k at points A, B, D in Fig. 7.2, respectively for AISI 8742, 1137, 1015, the Euler formula for F_c gives a stress above the yield strength; that is, below this certain value, the failure may well be a failure of elastic action. Actually, it is probably some combination of buckling and plastic action, and designers usually apply empirical equations in these cases. A popular equation with machine designers is one proposed at the turn of the century by J. B. Johnson (F_c = failing load);

$$(7.2)\qquad F_c = s_y A\left[1 - \frac{s_y(L_e/k)^2}{4\pi^2 E}\right] \quad \text{or} \quad \frac{F}{A} = s_e\left[1 - \frac{s_y(L_e/k)^2}{4\pi^2 E}\right],$$

[APPROPRIATE FOR $30 < L_e/k < 120$, STRUCTURAL STEEL]

where s_e is the *equivalent* stress that indicates the degree of safety for the load F; that is, $N = F_c/F = s_y/s_e$; F/A = actual nominal stress. In design, s_e is a suitable design stress. For these shorter columns and steady central loads, factors of safety between 2 and 3.5 are common. Equation (7.2) is not only called the Johnson formula but also the parabolic formula, because a plot of F/A against L_e/k is a parabola, $s = a - b(L_e/k)^2$. Values of the constants a and b other than those in Johnson's formula are sometimes chosen for particular classes of columns so that the resulting equation reasonably fits available experimental data. The Johnson formula accords fairly well with considerable data for steel columns.

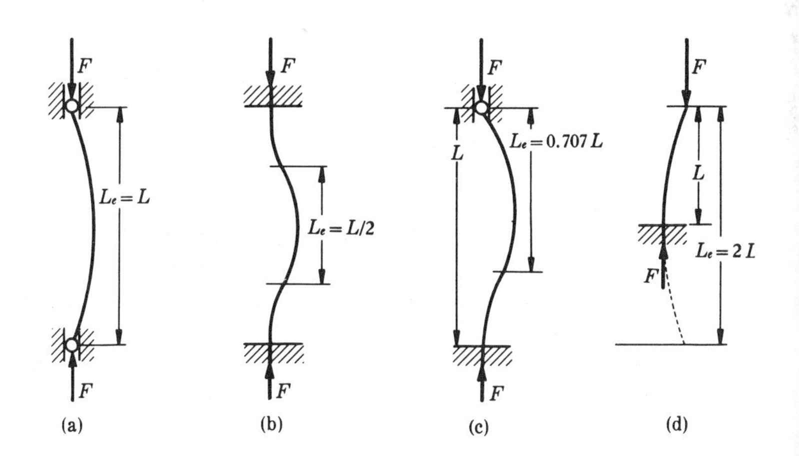

FIGURE 7.1 Types of Columns. (a) Rounded ends; (b) fixed ends; (c) one end fixed, the other rounded; (d) one end fixed, the other free.

7.5 STRAIGHT-LINE FORMULAS. Another common type of column formula, widely used for shorter columns in structures, is the straight-line formula, $s = a - b(L/k)$, where a and b are constants. In some city building codes, Chicago and elsewhere, it becomes, for structural steel,

(c) $$\frac{F}{A} = 16{,}000 - 70\frac{L}{k} = 16{,}000\left(1 - 0.0044\frac{L}{k}\right), \qquad \left[30 < \frac{L_e}{k} < 120\right]$$

where F is the safe or design load for the column.

Cast-iron columns are designed with a straight-line formula (NYC Code),

(d) $$\frac{F}{A} = 9000 - 40\frac{L}{k} = 9000\left(1 - 0.0044\frac{L}{k}\right), \qquad \left[30 < \frac{L_e}{k} < 70\right]$$

where the symbols have the usual meanings. These formulas are not to be applied to a free-end column.

7.6 TRANSITION POINT BETWEEN LONG AND INTERMEDIATE COLUMNS. The intersection point of two column formulas, as the Euler and Johnson, is where they both give the same value of F/A for a particular L_e/k; and if they are tangent, they have a common tangent, $[d(F/A)/d(L_e/k)]_E = [d(F/A)/d(L_e/k)]_J$. The constants in the Johnson equation are such that the curve is tangent to the Euler curve and always at $F_c/A = s_y/2$. Thus, if F_c/A from each formula is equated, we get an equation that gives the corresponding value of L_e/k;

(e) $$\frac{L_e}{k} = \left[\frac{2\pi^2 E}{s_y}\right]^{1/2}.$$

(The student may check the slope for the Euler and Johnson equations at this value of L_e/k and show that they are the same.) The situation with respect to steels 8742, 1137, and 1015, and aluminum 7075-T6 is shown in Fig. 7.2.

In Fig. 7.2, consider for example the curves for 1015. If we had an ideal column (approachable in the laboratory) with $L_e/k \leqq 86$ (point D), say $L_e/k = 70$ at J, the load on it could be increased until $F/A = s_y$ at M and then a further increase of load would result in an *elastic* failure. However, the actual column would be expected to fail more nearly in the vicinity of the stress at K. Notice that if a straight line were drawn between N ($L_e/k = 30$) and T_D ($L_e/k = 120$), it closely approximates the curve for 1015 (straight-line formulas). A similar statement does not hold so well for the 8742, a warning against blind generalizations about straight-line formulas.

Observations from Fig. 7.2 include:

(1) At point N, for $L_e/k = 30$, there is little difference between the load $F = s_y A$ and F_c from Johnson's formula; at the same L_e/k at point P, the difference is more significant.

(2) The tangent point T of the Euler and Johnson curves should not be taken for granted. For example, let a column of 8742, OQT 1200, with

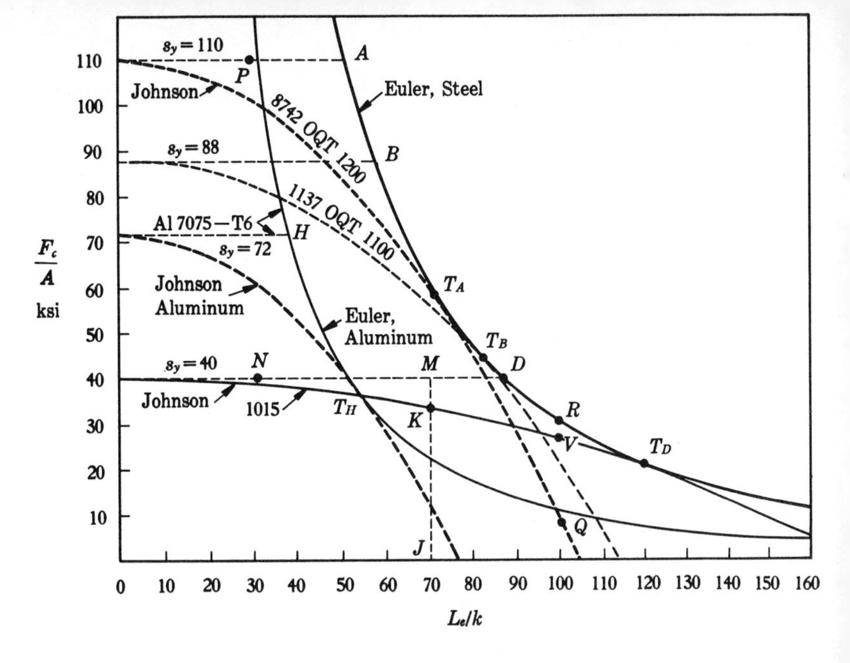

FIGURE 7.2 Johnson and Euler Curves, Different Materials.

$L_e/k = 100$ be considered. If the Johnson formula were used, the failing load at Q would be obtained, whereas the load it would take, at R, is much larger.

(3) There is not a large difference at $L_e/k = 100$ between the strengths of low-carbon steel (at V) and alloy steel (at R), but at L_e/k values somewhat less, the difference is significant.

(4) In the vicinity of the tangent point T, it makes little difference which equation is used.

(5) In general, use the Johnson formula when L_e/k is less than that at the tangent point; use the Euler formula when L_e/k is greater.

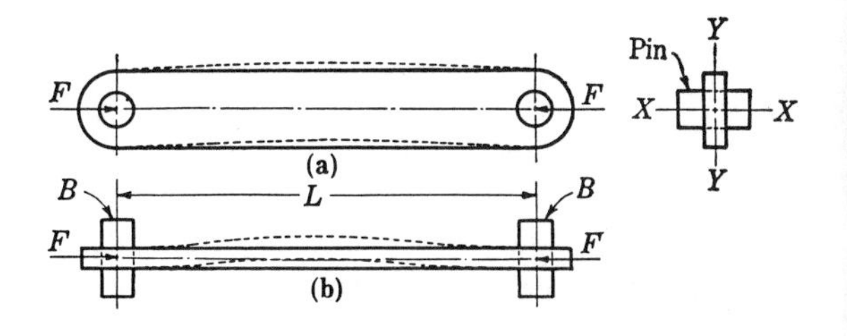

FIGURE 7.3 Buckling of Columns. Assume the pins *B* fixed. Let the member turn on them without clearance, so that there will be no "wobble." Thus, in (a), the column bends as a rounded-end column and, in (b), it bends as a fixed-end column.

7.7 RADIUS OF GYRATION. For rounded-end or fixed-end columns, the radius of gyration would be the one corresponding to the least moment of inertia. To understand the basis of determining the proper radius of gyration, consider Fig. 7.3 and the assumptions in the caption.

The column with pin ends may buckle in one of two directions. It may bend as indicated by the dotted lines in the upper view; or, as shown in the lower view. If buckling in the plane in Fig. 7.3(a) is being considered, the radius of gyration should be with respect to the axis XX; in the plane of Fig. 7.3(b) it should be with respect to the axis YY. As you would guess by now, the fixed-end assumption would not be justified; but there is an additional small restraint with respect to YY. Without tests to substantiate a difference, design for the least k (maximum L_e/k) and $L_e = L$.

7.8 SECANT FORMULA. If one assumes that the load F has an eccentricity e, Fig. 7.4, that the material is elastic, and that the deflection is small, the consequent theoretical equation (see nearly any text on mechanics of materials) is the so-called secant formula. If the limiting (critical) load is that corresponding to s_y, we have

$$\text{(f)} \qquad s_y = \frac{F_c}{A}\left(1 + \frac{ec}{k^2}\sec\frac{L_e}{2k}\sqrt{\frac{F_c}{EA}}\right) = \frac{F_c}{A}\left(1 + \frac{ec}{k^2}\sec\frac{L_e}{2}\sqrt{\frac{F_c}{EI}}\right),$$

where, in the second form we have used $I = Ak^2$. Since s is not directly proportional to F in this case, the factor of safety must be applied only to the force, equation **(b)**. The design equation may then be ($F_c = NF$)

$$(7.3) \qquad s_y = \frac{NF}{A}\left(1 + \frac{ec}{k^2}\sec\frac{L_e}{2}\sqrt{\frac{NF}{EI}}\right),$$

where the symbols have the usual meanings; c is the distance from the centroidal axis to the external fiber; ec/k^2 is called the *eccentricity ratio*.

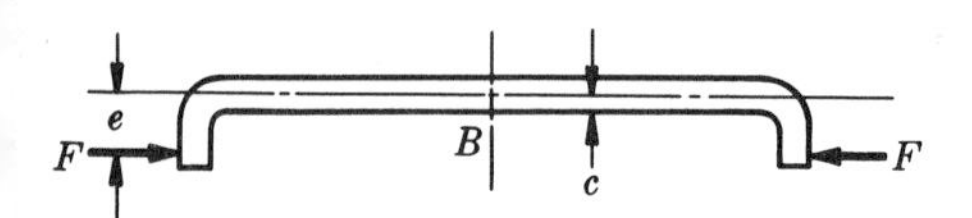

FIGURE 7.4 Effective Eccentricity, Exaggerated.

The secant formula is applied to columns with central loads because in comparing computations from this equation with experimental data, we note that with some "effective eccentricity" the results agree fairly well. The ASCE indicates that the eccentricity ratio $ec/k^2 = 0.25$ is good for structural columns with central load.[1.3] A value of $e \approx L_e/400$ is sometimes recommended.[1.1] But we notice that a suitable eccentricity ratio to fit a given environment is needed. Moreover, the secant formula is awkward to use for design because of the involved way the dimensions of the column enter into the equation. Charts can be constructed that aid in obtaining a solution.[1.1,1.3,1.7,7.1]

7.9 DESIGN OF COLUMNS. At the beginning of the design of a column, it is not known whether the column is slender (Euler) or in the intermediate range. Often too, the section is complicated, for example, an H-section or a built-up column. Thus, we see that the design is frequently one of trial and error. If the section is a simple one (circular, annular, square, or rectangular), k and A may be expressed in terms of a single dimension, which can be solved for. The Johnson formula is more frequently applicable to machines.

If the section is a standard rolled section, a handbook giving the properties (A, I, and k_{min}) of rolled sections can be used to good advantage. A few iterations should locate a satisfactory section.

First, compute $A = F/s_e$, where $s_e = s_y/N$ is the design stress for Johnson's equation, because any section of a particular shape that is subjected to column action must have a larger area than this.

7.10 EXAMPLE. A piston rod is subjected to a maximum compressive load of 31,416 lb. and is 20 in. long. The material is C1040, OQT 1000°F in order to obtain a desired hardness. What would be the diameter for $N = 3$?

Solution. From Fig. AF 1, Appendix, we find $s_y = 71$ ksi. The corresponding design stress is $71/3 = 23.7$ ksi. In Table AT 1, we find $k = D/4$ for a solid round section, which gives $L/k = 20/(D/4) = 80/D$. Since the numerical value of L/k is unknown, we must take a chance on which equation, (7.1) or (7.2), to use. Try (7.2); use force in kips and $L_e = L$.

$$\frac{F}{s_e} = A\left[1 - \frac{s_y(L/k)^2}{4\pi^2 E}\right],$$

$$\frac{31.416}{23.7} = \frac{\pi D^2}{4}\left[1 - \frac{(71)(80^2)}{4\pi^2(3 \times 10^4)D^2}\right],$$

from which $D = 1.44$ in.; use $1\frac{7}{16}$ in. For this diameter, $L/k = 80/1.4375 = 55.6$, which is in the range for the Johnson formula [$L/k \approx 83$ from equation (e)]. Therefore, the result is satisfactory against buckling, and with $N = 3$ should be conservative enough to care for size effects (0.505-in. specimen for Fig. AF 1).

It will be interesting and instructive to refer to the example of § 4.14, p. 117, where we found the diameter of an *alloy-steel* rod for a reversed load of the same magnitude (31,416) as in this problem. The *column* strength of the alloy rod will be only a little larger than that of the C1040 rod in this example; but the *fatigue* strength of the C1040 rod is significantly lower than for the alloy. Therefore, if the load in this example is reversed to approximately the same magnitude, fatigue rather than column action will determine the proper size.

7.11 EQUIVALENT STRESS FOR COLUMNS. Since the stress in an axially loaded column at buckling is less than s_y, it is useful to have an *equivalent stress* s_e that indicates the degree of safety (when compared say to s_y). Such a stress is in the Johnson formula; so solve for s_e from equation (7.2) and get

(g)
$$s_e = \frac{F}{A}\left[\frac{1}{1 - \dfrac{s_y(L_e/k)^2}{4\pi^2 E}}\right] = \alpha\frac{F}{A},$$

where α is equal to the term in the brackets. To introduce an equivalent stress into Euler's equation, let $N = F_c/F = s_y/s_e$ in (7.1) and solve for s_e:

(h)
$$s_e = \frac{F}{A}\left[\frac{s_y(L_e/k)^2}{\pi^2 E}\right] = \alpha\frac{F}{A},$$

where α is equal to the term in the brackets. In the straight-line formula (**c**) for steel, consider the $16{,}000 = s_y/N = s_e$ and solve for s_e:

(i)
$$s_e = \frac{F}{A}\left[\frac{1}{1 - 0.0044L/k}\right] = \alpha\frac{F}{A},$$

where α is equal to the part in the brackets. Thus, we have a simplified point of view; the *equivalent* column stress (but not the actual) being $s_e = \alpha F/A$, where α is greater than unity and is given by the bracketed terms of equations (**g**), (**h**), and (**i**) for the respective situations. If a column is subjected to a combination of stresses, see the next chapter. An equivalent stress from the secant formula is not proposed here.

7.12 OTHER COLUMN FORMULAS. There are several other column formulas designed to cover the above situations.[7.5] The tangent-modulus formula, which the reader probably met in the study of strength of materials, agrees well with experience but is too awkward to use in design unless one is designing expensively and close to the limit. Also one finds formulas to fit a particular shape or a particular material. For example, an equation recommended for magnesium columns is[2.1]

(j)
$$\frac{F_c}{A} = \frac{C}{1 + \dfrac{C(L_e/k)^2}{64.4 \times 10^6}} \text{ psi},$$

where C is a number that depends on the compressive yield strength of the material. For the magnesium alloys in Table AT 3, the values of C are: AZ 91C, $C = 57{,}000$; AZ 61A, $C = 42{,}800$; AZ 80A, $C = 82{,}900$ psi.

A situation that often must be watched in structural columns composed of thin sections, for example, a wide-flange rolled-beam section, and thin-wall tubular columns, is local buckling of the thin metal, mentioned in § 1.24 in connection with beams. See Ref. *(7.5)*.

7.13 CLOSURE. Buckling of columns occurs at a time when they are in unstable equilibrium. The buckling of a flange in a rolled-section beam is of the same family of phenomena. Another collapse type of failure is that of a thin-shell vessel subjected to external pressure (§ 20.2). Residual stresses, such as those left in rolling structural shapes, play a part that has not been fully investigated. In all cases, there is usually a strong element of empiricism in the design; unknowns are answered only by experiment.

8. COMBINED STRESSES

8.1 INTRODUCTION. Until now, we have considered only those cases where the stresses could be considered as simple stresses (F/A, Mc/I, Tc/J). This chapter covers design for some combinations of these simple stresses.

If a normal stress is tensile, we shall give it a positive sign; if compressive, a negative sign, *where convenient*. The reader should not be confused because the use of the negative sign is purely conventional. Thus, if we speak of the minimum stress, the largest negative (compressive) stress, if any, is intended. On the other hand, if the determining stress in design is compression, we may speak of the maximum compressive stress as the maximum stress. The context will make the intention clear. There are so many different kinds of stresses to deal with that symbolization becomes bothersome. Introducing two new stress symbols, we shall use the following:

s, a *normal* stress—tension or compression—computed from F/A, $\alpha F/A$, Mc/I, etc.,
s_s, a shear stress, computed from F/A, Tc/J, etc.,
s_1, s_2, s_3, or s_x, s_y, s_z, etc., various normal stresses,
s_{s1}, s_{s2}, or s_{xy}, s_{xz}, s_{yz}, etc., various shear stresses,
σ, a resultant normal stress due to a combination of the above stresses, either tensile σ_t or compressive σ_c, and
τ, a resultant shearing stress.

8.2 BENDING AND UNIFORM STRESSES. One of the commonest and simplest combinations of stress is a bending stress, Mc/I, and a uniform stress, F/A. For example, when a load is not concentric with a member's centroidal axis, Figs. 8.1 and 8.2, the bodies are said to be ***eccentrically loaded***, the eccentric load inducing the foregoing stresses in the body. Breaking the member at a section CT, Fig. 8.1(b), observe that a moment M and an axial force F must be applied to maintain the severed part in equilibrium. Now introduce equal and opposite forces $F_1 = F$ on the centroidal line in order to replace F by a force F_1 and a couple Fe. Then note that the moment $M = Fe$. The uniform tensile stress at CT is $s_1 = F/A$ and the

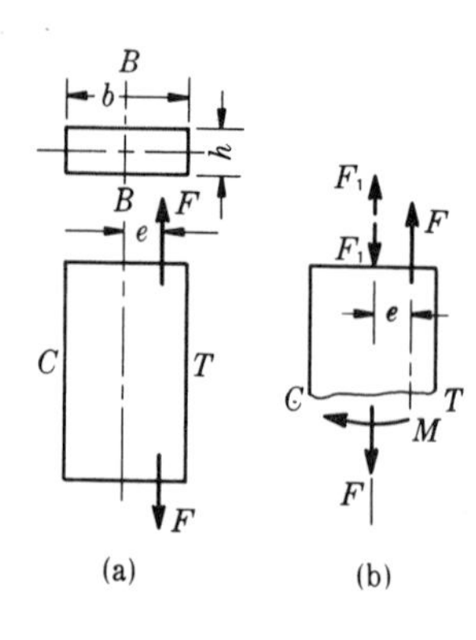

FIGURE 8.1 Prism with Eccentric Tensile Load.

flexural stress is $s_2 = Mc/I = Fec/I$. The bending stress is tensile at T and compressive at C, Fig. 8.1. Thus on the side T, by superposition, stress s_2 adds numerically to s_1; on the side C, s_2 subtracts from s_1; or, with the usual convention of signs, the resultant stress is

$$\sigma = s_1 \pm s_2 = \frac{F}{A} \pm \frac{Mc}{I} = \frac{F}{A} \pm \frac{Fec}{I}, \tag{8.1A}$$

where A is the cross-sectional area, I is the area moment of inertia about the axis BB, Fig. 8.1(a), M is the moment at the section that contains the point at which the stress is desired, c is the distance from the neutral plane to the point where the stress is desired ($c = b/2$ for the maximum). At the point of maximum stress in design, σ would be some safe normal stress. Use the positive sign in (8.1) on the tensile side; on the compressive side C, the stress σ may be either tension ($F/A > |Mc/I|$) or compression ($|Mc/I| > F/A$).

If a bending moment exists at a section where there is a simple compressive stress F/A, the normal stress on the same plane is

$$\sigma = -\frac{F}{A} \pm \frac{Mc}{I} = -\frac{F}{A} \pm \frac{Fec}{I}, \tag{8.1B}$$

where the bending moment may be produced by an eccentric compressive load, Fig. 8.2. If, in design, the member is short (no column action),

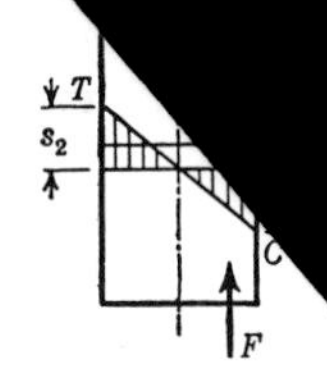

FIGURE 8.2 Prism with Eccentric Compressive Load.

σ would be a suitably safe stress at a point of maximum stress. Since minus signs are somewhat of a nuisance and since this case is so simple that there is never any question, designers frequently use equation (8.1A) for compressive eccentric loads, as well as for tensile.

If a member in compression is long ($L_e/k > 40$) and loaded eccentrically, the secant formula (7.3) is theoretically correct. However, since this equation is difficult to design with, the literature provides a number of alternate procedures, whose results at times vary widely. One method, believed to be reasonably conservative, is to use the equivalent column stress $\alpha F/A$, computed from either equation **(g)** or **(h)**, § 7.11 [and sometimes from the straight-line formula **(i)**], instead of F/A;

(a)
$$\sigma = -\alpha\frac{F}{A} \pm \frac{Fec}{I},$$

where α is introduced to care for column action (§ 7.11) and, in design, σ is an appropriate design stress. When bending occurs in the direction of the maximum resistance, computing α for buckling in the direction of minimum resistance should be on the safe side. The secant formula can be used for checking the design done by equation **(a)**, but if it is important to obtain an optimum design, one must keep in mind that all theoretical equations must be related to the facts of life by experiment, frequently by experiment that simulates some particular actual usage.

8.3 EXAMPLE—DESIGN OF COLUMN WITH ECCENTRIC LOAD. A 10-ft. column is to be designed to support a load of $F = 20$ kips, overhung 15 in. at the free end, Fig. 8.3. It is planned to be made from standard pipe, the material of which is similar to AISI C1015 as rolled. Use a design factor of $N = 2.6$. Specify the pipe size.

Solution. If one were designing many pipe columns, there would no doubt be short cuts at hand (as charts and tables) for at least an initial estimate of size. If only a few are to be designed in the absence of short cuts, the quickest way is

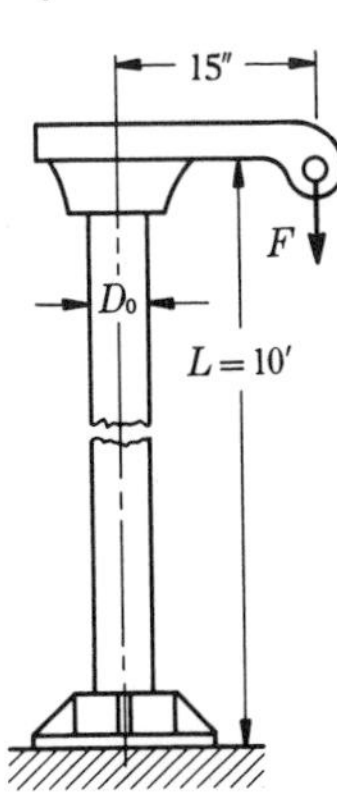

FIGURE 8.3

ply to assume a standard pipe size and compute the corresponding design ...ctor; if the computed N is not in the proper relation to the specified $N = 2.6$, iterate until satisfaction is achieved. We shall use the equivalent stress method, equation (**a**). From Table AT 8, we get $s_y = 45.5$ ksi. From a handbook,[0.5] we find the properties of pipe sections. Assume an 8-in. pipe (nominal size), schedule 60*; then

$$D_o = 8.625 \text{ in.}, D_i = 7.813 \text{ in.}, A_m = 10.48 \text{ in.}^2, k = 2.909 \text{ in.}, Z = 20.58 \text{ in.}^3$$

The equivalent length of a free-end column is $L_e = 2L = 20$ ft. or 240 in., giving $L_e/k = 240/2.909 = 82.5$. From previous discussion, we judge that this value puts this column in the Johnson range, but checking from equation (**e**), § 7.6, to be sure, we get the dividing point as

$$\frac{L_e}{k} = \left[\frac{2\pi^2 E}{s_y}\right]^{1/2} = \left[\frac{(2)(9.86)(3 \times 10^4)}{45.5}\right] = 114;$$

therefore, use Johnson's equation.

$$s_1 = \frac{F/A}{\left[1 - \dfrac{s_y(L_e/k)^2}{4\pi^2 E}\right]} = \frac{20/10.48}{\left[1 - \dfrac{(45.5)(82.5)^2}{(4)(\pi^2)(3 \times 10^4)}\right]} = 2.58 \text{ ksi.}$$

$$\sigma = s_1 + \frac{Fe}{Z} = 2.58 + \frac{(20)(15)}{20.58} = 17.2 \text{ ksi}$$

For this approach $N = s_y/\sigma = 45.5/17.2 = 2.64$. The next smaller size of pipe gives a design factor too low. The answer being so close to the desired value suggests that preliminary iterations had already been done. Thus, an 8-in., schedule 60, pipe is satisfactory. Further study might include checking $N = F_c/F$ by the secant equation and perhaps some experimental stress studies.

FIGURE 8.4 C-Clamp.

8.4 ECCENTRIC LOADING ON AN UNSYMMETRIC SECTION. The principle involved being the same as before, we can explain this case by discussing the stresses in a section of a C-frame. For a T-section C-frame, Fig. 8.4, a load F as shown in Fig. 8.5 induces a uniform tensile stress of

* The schedule symbol designates the weight of the pipe (wall thickness). The outside diameter D_o (= 8.625 in. in this case) remains the same for all schedules, given D.

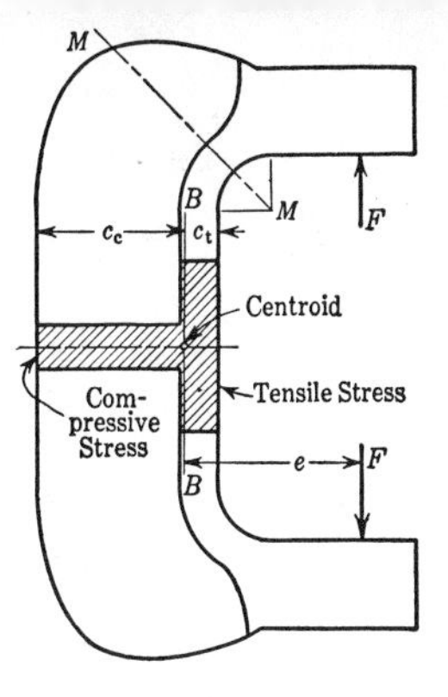

FIGURE 8.5 T-Section C-Frame.

$s_1 = F/A$, a tensile stress at the inside due to bending of $s_2 = Mc/I = Fec_t/I$, and a compressive stress at the outside of $s_c = Mc/I = Fec_c/I$; where I is the moment of inertia of the sectional area about the area's centroidal axis BB, c_t and c_c are as shown in Fig. 8.5, for the maximum stresses due to M. The resultant tensile stress is

(b)
$$\sigma_t = s_1 + s_2 = \frac{F}{A} + \frac{Fec_t}{I}.$$

The resultant compressive stress (if $|Mc/I| > F/A$) is

(c)
$$\sigma_c = s_1 - s_c = \frac{F}{A} - \frac{Fec_c}{I}.$$

This problem is not solely one of getting a safe section, but also of getting *good proportions* in the section. It is frequently possible to save much material, and therefore weight and money, by improving the proportions. If the eccentricity e is large, the strength is more affected by the moment of inertia than by the area. Consequently, less material is needed when the material used is distributed far from the neutral axis. The further away the bulk of the material is from the neutral axis, the greater is the moment of inertia and therefore the resisting moment. However, care must be taken not to make the sections so *thin* that they will be in danger of buckling.

A T-section is particularly appropriate for cast iron. The iron may be concentrated on the tensile side (since cast iron is so much weaker in tension than in compression), and may be spared on the compressive side. The thicknesses of adjoining sections in castings should be not very different from one another; otherwise, unequal rates of solidification will result in significant residual stresses. For steels, which have about the same strength in tension and compression, a box-section, a modified I-section, or an H-section, is better.

In the design of an unsymmetric section, we *might* assume the proportions of a section in terms of one dimension and obtain an equation with only one unknown. However, this method, even for a simple T-section, may become so involved that an error is more probable than not unless it is highly organized (as it would be when such designs are often repeated).

The following procedure with the calculations organized to facilitate checking is recommended.

(a) Make a sketch and place the assumed dimensions on it.
(b) Determine the location of the centroid.
(c) Determine the centroidal moment of inertia.
(d) Determine the eccentricity of the load e.
(e) Solve for the uniform stress s_1.
(f) Solve for the flexural stresses s_t and s_c.
(g) Determine the resultant stresses σ_t and σ_c and compare with the design stresses.
(h) Repeat the preceding calculations until a suitable section is found.

The analyses discussed above apply when the member is *not curved* at the section for which the stress is desired. If the section is curved, as at MM, Fig. 8.5, use the flexural stress as $K_c Mc/I$, where K_c is a curvature factor, § 8.25.

8.5 COPLANAR SHEAR STRESSES IN MORE THAN ONE DIRECTION. Shear stresses in different directions at a point on a plane section of a body may be added vectorially. However, it is probably safer to find the resultant *force* first and then the resultant shear. Consider Fig. 8.6. In the conventional approach, the load W is imagined to have been replaced by a force W through C, the *centroid of the area of the rivets*, and a couple of magnitude Wa. The force W through C is assumed to be resisted equally by the rivets, each rivet being subjected to a shear in the downward direction of $W/N_r = W/4$, where N_r = number of rivets, 4 in this case. Next, it is assumed that the plate is rigid, that all the deformation is taken by the rivets, that the small turning of the plate by the moment Wa occurs about the centroid C, and therefore, with the further assumption that the size and material of the rivets are the same, the deformations, stresses, and forces (F_1, F_2) are each proportional to the distances L_1, L_2 of the rivets from C;

(d) $$\frac{F_1}{F_2} = \frac{L_1}{L_2} \quad \text{or} \quad F_2 = F_1\left(\frac{L_2}{L_1}\right).$$

The forces shown in Fig. 8.6 are those acting *on* the plate. Since the plate is in equilibrium, equate moments about C to zero;

(e) $$Wa - 2F_1L_1 - 2F_2L_2 = 0.$$

Substitute the value of F_2 in (**d**) into (**e**) and solve for F_1;

$$F_1 = \frac{Wa}{2L_1 + (2L_2^2/L_1)} = \frac{WaL_1}{2(L_1^2 + L_2^2)},$$

from which F_1 may be found if the dimensions are known. We now have the forces on the outside rivets, F_1 and $W/4$, acting at right angles to each other, Fig. 8.6(b). The resultant is

(f) $$R = \left[F_1^2 + \left(\frac{W}{4}\right)^2\right]^{1/2},$$ [FIG. 8.6 ONLY]

and $s_s = R/A$, where A is the section area of a rivet. Observe that in this discussion the equations obtained are suitable only for the case analyzed.

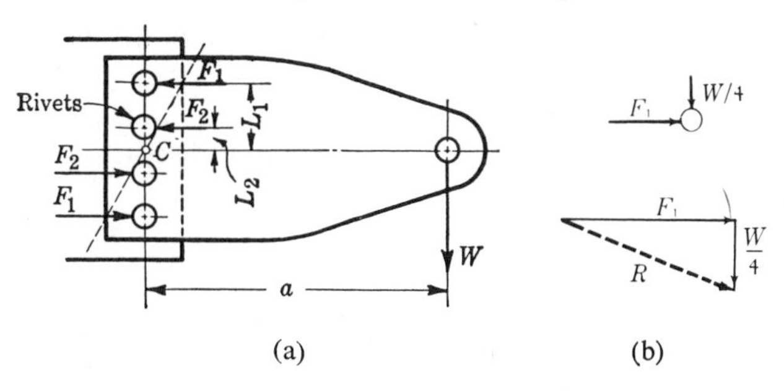

FIGURE 8.6 Shear in Two Directions. If, in this statically indeterminate connection, uniform distribution of W seems too optimistic, probably true with bolts, one could assume the maximum portion on one fastener to be $W/3$ or even $W/2$.

If the deflections are coplanar (no twisting of the plate), this conventional procedure results in conservative results. Other similar connections are analyzed with the same assumptions as outlined.

8.6 COMBINED NORMAL AND SHEAR STRESSES.

Another frequent combination of stresses is a normal stress, tension or compression, and a shear stress, uniform or torsion. Consider the bolt M holding the

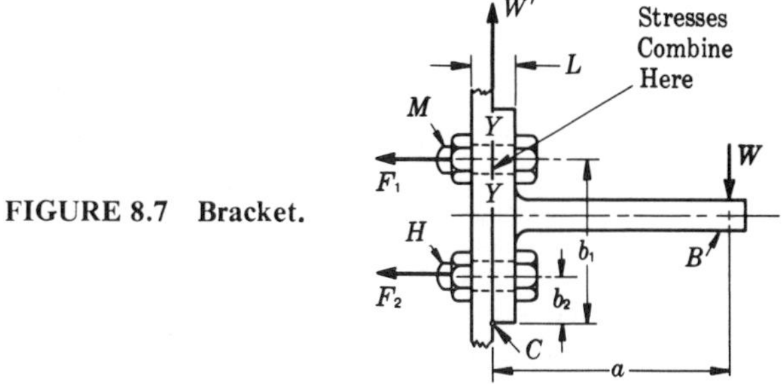

FIGURE 8.7 Bracket.

bracket of Fig. 8.7. In order to approach the design in a rational manner, we must first decide upon an idealized model (more completely defined in § 8.11). Because of the tendency of the load W to tip the bracket about some point C and because of the initial tightening stress, the bolt M is

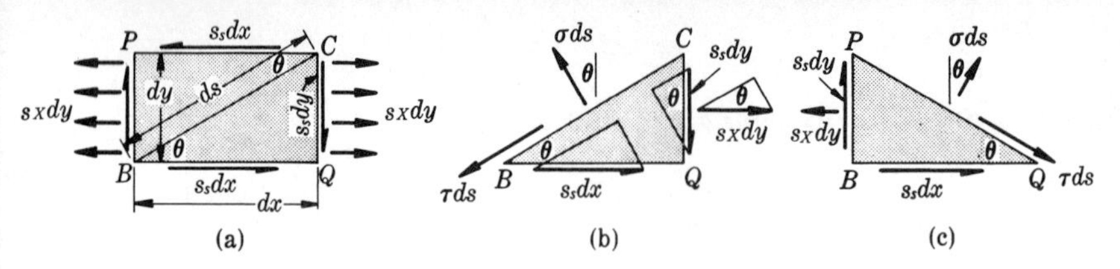

FIGURE 8.8 Force System on Small Prism under Tensile and Shear Stresses. This represents a "point" in a stressed body. It applies no matter how s_t and s_s are induced, say by bending and torsion. The length of the prism normal to the page is unity. Figure (b) shows a diagonal section with forces that produce equilibrium. No stress in a direction perpendicular to the plane of the paper.

subjected to a normal (tensile) stress. In addition, the section of the bolts at the junction of the bracket and the wall is subjected to a shearing stress. These stresses combine to produce a resultant tensile stress greater than the tensile stress on plane *YY* and a resultant shear stress greater than the shear on plane *YY*. The following discussion reviews briefly the theory of the combination.

8.7 PRINCIPAL STRESSES. Use an elementary prism with stresses s_t and s_s, say taken from the bolt *M*, Fig. 8.7, along the section *YY* between the wall and the bracket; width *dx* (along the length of bolt), height *dy*, and depth unity. The forces ($= sA$) acting on the prism are as shown in Fig. 8.8(a). First, set up a general expression for the forces on *any* diagonal plane, *BC*. Remove the upper portion and consider the lower portion as a free body in equilibrium, Fig. 8.8(b). Equate the sum of the forces normal to plane *BC* to zero and solve for σds ($= sA$);

$$\sigma ds = s_X dy \sin\theta + s_s dy \cos\theta + s_s dx \sin\theta.$$

Divide through by *ds*, substitute $dx/ds = \cos\theta$, $dy/ds = \sin\theta$, and find

$$\text{(g)} \qquad \sigma = s_X \sin^2\theta + 2s_s \sin\theta \cos\theta = s_X\left(\frac{1 - \cos 2\theta}{2}\right) + s_s \sin 2\theta.$$

To find the angle θ when the normal stress σ is a maximum or minimum, differentiate (g) with respect to θ and equate $d\sigma/d\theta$ to zero. This gives

$$\text{(h)} \qquad \tan 2\theta = -\frac{2s_s}{s_X},$$

where 2θ is measured from the x direction, positive counterclockwise. Since $\tan 2\theta$ is negative, 2θ is in the second or fourth quadrant and θ is therefore in the first or second quadrant with the two values 90° apart. Using the two values of 2θ, $\tan^{-1}(-2s_s/s_X)$ and $\tan^{-1}[2s_s/(-s_X)]$, in (g) and simplifying, we get

$$\text{(i)} \qquad \sigma = \frac{s_X}{2} \pm \left[s_s^2 + \left(\frac{s_X}{2}\right)^2\right]^{1/2}.$$

At any point in a stressed body, there are three orthogonal planes on which the shear stresses are zero; the normal stresses on these planes are called

principal stresses. For the stress configuration described by Fig. 8.8, two of the principal stresses are given by equation (**i**); the third principal stress is zero. The positive sign in equation (**i**) gives the principal stress that is the maximum (tensile) stress at the point, Fig. 8.8(b). If the radical is greater than $s_X/2$, the negative sign obtained means that the corresponding stress is compressive, and in this case, it is the minimum principal stress. If the negative sign is used and the result from (**i**) is positive, then the minimum principal stress is zero, the third one. (More frequently than otherwise, the negative sign in (**i**) results in a negative stress.)

If the s_X stress in Fig. 8.8 were reversed to compression, the results would be analogous to those discussed except that tensile stresses would become compressive stresses. Therefore, we can generalize equation (**i**) by dropping the subscript X,

$$\sigma = \frac{s}{2} \pm \left[s_s^2 + \left(\frac{s}{2}\right)^2\right]^{1/2}, \tag{8.2}$$

and letting s be the normal stress on a plane where the shearing stress is s_s; s is computed from F/A, $\alpha F/A$, Mc/I, etc., or a combination thereof; s_s from F/A, Tc/J, VQ/Ib; (or other appropriate equations).

If two principal stresses are zero (simple tensile member, for example), the state of stress is said to be ***uniaxial.*** If one principal stress is zero (as in the foregoing discussion), the state of stress is ***biaxial.*** If all principal stresses have finite values (an element on the inside of a pressure vessel), the system is triaxial.

8.8 MAXIMUM SHEAR STRESS. The maximum shearing stress on some diagonal plane, Fig. 8.8, may be found in the same manner as the maximum tensile stress, that is, by summing forces *parallel* to the plane, etc. However, it is worth recalling from the theory of strength of materials that the maximum shearing stress τ is one-half the algebraic difference of the maximum and minimum principal stresses. Using equation (8.2) for the maximum and minimum normal stresses, we get

$$\tau = \tfrac{1}{2}(\sigma_{max} - \sigma_{min}), \tag{8.3}$$

$$\tau = \left[s_s^2 + \left(\frac{s}{2}\right)^2\right]^{1/2}, \tag{8.4}$$

the maximum shearing stress *at some particular point* in a body for the state of stress defined in Fig. 8.8, except that the normal stress may be either tensile or compressive. The maximum shearing stress occurs on a plane at 45° with the plane of the maximum normal stress. Notice that the maximum principal stress (8.2) exceeds (8.4) by the amount $s/2$; that is, $\sigma = s/2 + \tau$.

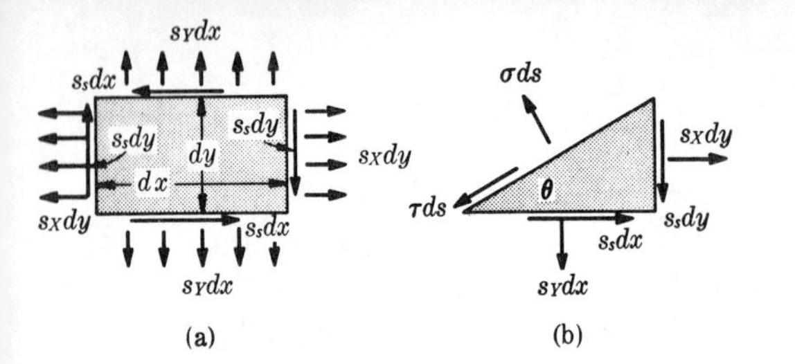

FIGURE 8.9

8.9 ELEMENT SUBJECTED TO TWO NORMAL STRESSES AND SHEAR. The more general case of a plane stress system is an element with two normal stresses s_X and s_Y plus a shearing stress s_s, Fig. 8.9. (In this vicinity, do not confuse the stress in the y direction s_Y with the yield strength s_y.) Making a free body of a triangular portion of the element (of unit depth), Fig. 8.9(b), summing forces and proceeding as explained for Fig. 8.8, we find

$$\sigma = \frac{s_X + s_Y}{2} \pm \left[\left(\frac{s_X - s_Y}{2}\right)^2 + s_s^2\right]^{1/2}, \tag{8.5}$$

$$\tau = \left[\left(\frac{s_X - s_Y}{2}\right)^2 + s_s^2\right]^{1/2}, \tag{8.6}$$

in which s_X and s_Y are algebraic (that is, use a negative sign for compression). The positive sign before the radical in (8.5) gives the maximum principal stress σ_1; the negative sign gives the minimum principal stress σ_2 if the result is negative, otherwise the minimum principal stress is $\sigma_{min} = \sigma_3 = 0$. It is important to note that there are three principal *shear* stresses that can be computed from equation (8.3) by using $\sigma_1 - \sigma_2$, $\sigma_1 - \sigma_3$, and $\sigma_2 - \sigma_3$ (instead of $\sigma_{max} - \sigma_{min}$), the maximum value being as defined by (8.3). Each principal shear stress is on a plane at 45° with the planes of the principal normal stresses from which it is computed. The planes of the principal stresses are as defined by

$$\tan 2\theta = -\frac{2s_s}{s_X - s_Y}, \tag{8.7}$$

where 2θ is positive measured counterclockwise from the x axis.

8.10 MOHR'S CIRCLE. The stresses in *any* direction can be computed from free-body diagrams of an element, after the manner of § 8.7. However, since Mohr's circle provides a simpler approach to other than principal stresses, its properties are briefly reviewed here. Normal stresses are plotted horizontally, shear vertically. Let the stress system be planar and as shown in Fig. 8.10(b), in which it is presumed that the normal (s_X, s_Y both taken as tensile) and shearing (s_s) stresses have been computed from F/A, Mc/I, Tc/J, etc. in accordance with the external loading. Choose a scale and lay out s_X and s_Y from the origin O in Fig. 8.10(a), locating points B and A; tensile stresses are laid out toward the right, compressive

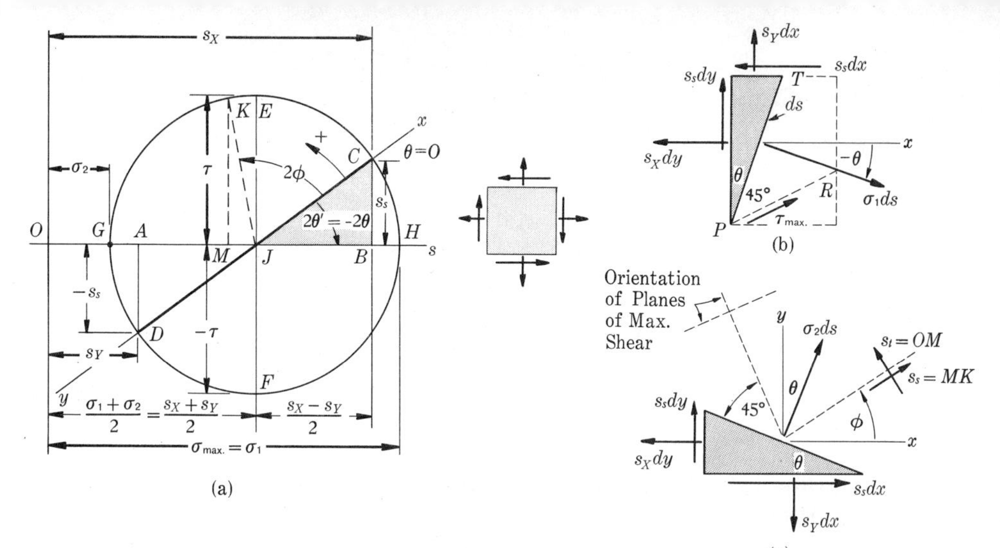

FIGURE 8.10 Mohr's Circle. The principal stress σ_1 is on a plane at $-\theta$ from the x axis, in (b); σ_2 is on a plane at $90-\theta$ with x axis, in (c). Note that if the directions of the normal and shear stresses on elements can be determined by inspection, the planes of maximum and minimum stresses are determined by inspection; in (b), with the shear and normal forces both with tensile components on the diagonal plane *ds*, the stress σ_1 is naturally expected to be larger than σ_2 in (c) where the shear forces, $s_s dy$ and $s_s dx$, tend to result in compression on the diagonal plane *ds*. In (b), the plane *PR* is at 45° with the principal plane *PT*; maximum shear (for σ_1 and σ_2) is on plane *PR*.

stresses toward the left. From B, erect a perpendicular and layout s_s to scale, locating C; point C is a point on Mohr's circle whose center is at J, $(s_X + s_Y)/2$ from O, and whose radius is JC. The sign convention for the shearing stress is: if the $s_s dy$ shear forces form a CL moment, this is positive shear with $s_s = BC$ laid out upward from the end of s_X. With sketches, Fig. 8.10(b) and (c), the planes of maximum σ_1 and minimum σ_2 can be verified; the corresponding planes of maximum shear are at 45° with the planes of σ_1 and σ_2. Draw the circle $CHDG$. Note that in the triangle JCB, the hypotenuse is $[s_s{}^2 + (s_X - s_Y)^2/4]^{1/2}$, which, by comparison with equation (8.6), is seen to represent the principal shearing stress τ. Thus, $JE = \tau_{max}$ and JF represents the shear on a plane at 90° with JE; naturally, $|JE| = |JF|$. Since $JC = JH$ and $OJ = (s_X + s_Y)/2$, the distance OH represents the maximum principal stress σ_1, equation (8.5); by a like logic OG represents the principal stress σ_2. Since σ_2 is positive (tensile), the minimum stress is $\sigma_3 = 0$. Note that on the plane of the principal shear stress τ_{max}, the normal stress is $OJ = (s_X + s_Y)/2$. The x axis, from which 2θ is measured is JC; positive angles measured counterclockwise. Ignoring signs, we see that $\tan 2\theta' = 2s_s/(s_X - s_Y)$, which agrees with equation (8.7). If it were desired to know the stresses on a plane at the *point* being investigated at a counterclockwise angle of ϕ from the x axis, Fig. 8.10(c), lay out angle 2ϕ from the x axis $= JC$, Fig. 8.10(a) and read $\sigma_\phi = OM$ and $\tau_\phi = MK$.

The case of simple tension ($\sigma_2 = \sigma_3 = 0$) is shown in Fig. 8.11(a). It is seen directly from Mohr's circle that $s_{s\,max} = s_t/2$ on a plane at 45° ($2\theta = 90°$) from the x direction. Also from the free body below, the sum

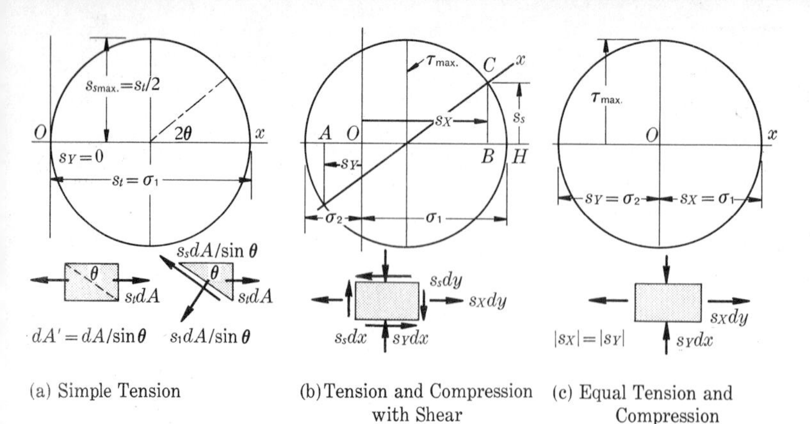

(a) Simple Tension

(b) Tension and Compression with Shear

(c) Equal Tension and Compression

FIGURE 8.11 Mohr's Circles for Various Stress Patterns. The elements $dx\,dy$ have unit depth.

of the forces parallel to the diagonal plane is $s_s dA/\sin\theta - s_t dA\cos\theta = 0$, from which

(j) $$s_s = s_t\cos\theta\sin\theta = \frac{s_t}{2}\sin 2\theta; \qquad s_{s\max} = \frac{s_t}{2},$$

where we have recognized that the maximum value of the sine is unity.

Figure 8.11(b) shows a Mohr's circle for an element subjected to tension s_X, compression s_Y, and shear s_s. In Fig. 8.11(c), Mohr's circle for the special biaxial case of $|s_{Xt}| = |s_{Yc}|$ shows that τ_{max} has the same numerical value; $|\sigma_1| = |\sigma_2| = |\tau_{max}|$. Compare this arrangement with the one of pure torsion in Fig. 8.12 (a)–(d) and note the same numerical equality. The interesting case of two *equal* tensions s_X and s_Y is shown in Fig. 8.12(e) where Mohr's circle becomes a point and the shear stresses are therefore zero ($\sigma_3 = 0$). The idea of Mohr's circle is adaptable for situations of three finite principal stresses.[1.7]

8.11 EXAMPLE—COMBINED TENSION AND SHEAR. A bracket for supporting a shaft bearing, similar to Fig. 8.13(a), is to be riveted to a vertical surface and support a load of $W = 6$ kips at a distance $e = 12$ in. from the wall. As seen, there are two rivets above and one below, at a vertical spacing of $a = 4$ in. The lower rivet is $b_2 = 1.5$ in. from the bottom. Design for the maximum shear stress and compute a suitable diameter of the steel rivets. Only two brackets are to be installed.

Solution. This problem illustrates an application of analytic mechanics to a statically indeterminate situation. (The reader should perceive the principles involved so that he can analyze analogous situations; the equations obtained apply to the configuration of Fig. 8.13.) In indeterminate problems, as in others, one sets up an ideal model for which computations can be made. The principal assumptions here are that the bracket B, Fig. 8.13(b), is rigid and that it tips slightly about the lowest point C when the load W is applied. These assumptions result in the deformations of the rivets, δ_1 and δ_2 being proportional to their distances from the pivot point C. Other assumptions affecting the design are pointed out as we proceed with the solution.

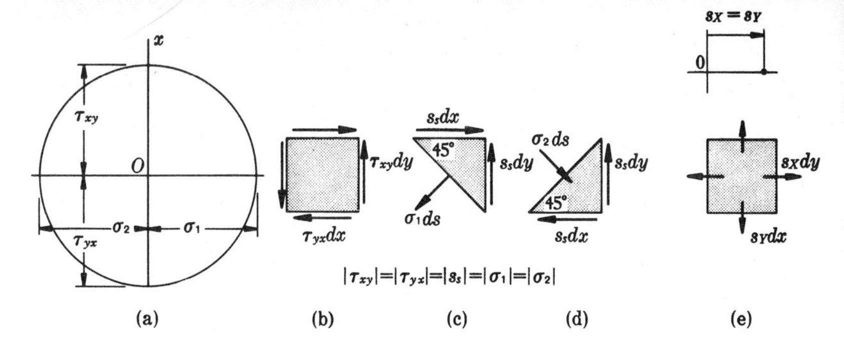

FIGURE 8.12 Mohr's Circle, Pure Torsion and Biaxial Equal Tension. To lay out Mohrs' circle, compute the torsional stress and draw a circle of this radius (to some scale); the horizontal axis through the center of the circle gives the intercepts of principal stresses σ_{1t} and σ_{2c}, numerically equal to $s_{s\max} = \tau_{xy}$. Notice in (e) that the shear stress on any plane perpendicular to the paper is zero when $s_x = s_y$, both tensile.

Rivets, if not otherwise specified, are likely to be equivalent to about AISI 1015. Considering ASME Boiler Code specifications[1.9,1.10] and the strength of 1015, a design stress of $s_{sd} = 8$ ksi should be conservative. (For only two brackets to be made, overconservatism is not bad.) For ease of installing, all rivets should be the same size. Since the upper rivets deform more than the lower one, they are stressed more ($s = E\epsilon = E\delta/L$); hence the design is based on the load F_1 in a *single* upper rivet. The force F_2 is expressed in terms of F_1 by the following logic:

(k)
$$\frac{F_2}{F_1} = \frac{s_2 A_2}{s_1 A_1} = \frac{s_2}{s_1} = \frac{E_2\delta_2/L_2}{E_1\delta_1/L_1} = \frac{\delta_2}{\delta_1} = \frac{b_2}{b_1}$$

or $F_2 = b_2F_1/b_1$. This result is a consequence of certain assumptions: $A_1 = A_2$; $E_1 = E_2$ (rivets of same material); $L_1 = L_2$ (rivets same length); $\delta_2/\delta_1 = b_2/b_1$ from similar triangles, Fig. 8.13. If any of the assumed conditions happen not to be true, one can always proceed through the logic of (**k**) as far as it applies.

(l)
$$F_2 = \frac{b_2F_1}{b_1} = \frac{1.5F_1}{5.5} = 0.273F_1.$$

The forces with the bracket as the free body are as shown in Fig. 8.13(b). Sum moments about C and use $F_2 = 0.273F_1$;

(m)
$$We = 2F_1b_1 + F_2b_2 = 2F_1b_1 + 0.273F_1b_2$$
$$(6)(12) = (2)(5.5)F_1 + (0.273)(1.5)F_1$$

from which $F_1 = 6.31$ kips. We may now solve the stress problem. When the

FIGURE 8.13 Wall Bracket. In (a), the T-slot permits locating, say, a bearing at any distance from the wall. The amount of tipping in (b) is exaggerated.

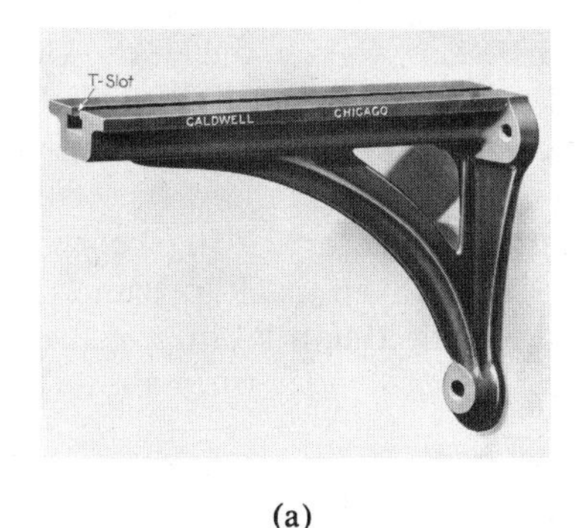

(a)

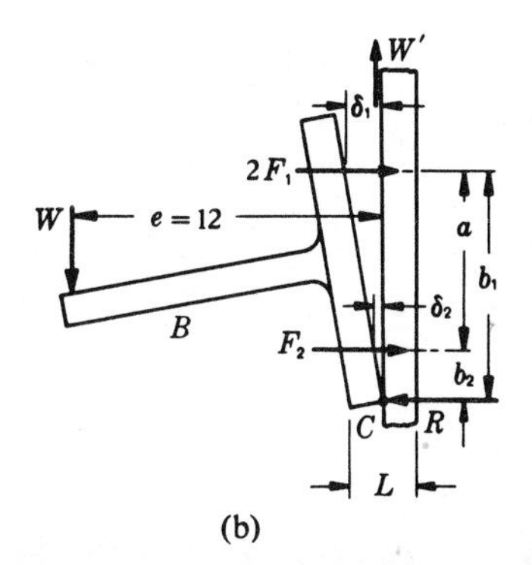

(b)

rivet heads are set, an initial stress is induced in the rivet whether the rivet is headed hot or cold; this stress is ignored or covered by a generous factor of safety in such simple designs. It is also true, as in bolted joints (§ 5.9), that if the connected parts are relatively rigid, very little additional load comes on the rivets when the external load is applied unless the connected parts actually begin to separate. Moreover, there is no practical way of deciding upon the initial stress. Hence, the design assumes that F_1 produces a tensile stress

(n) $$s_t = \frac{F_1}{A} = \frac{(6.31)(4)}{\pi D^2} = \frac{8.04}{D^2} \text{ ksi},$$

and that a uniform shear stress is produced by W of

(o) $$s_s = \frac{W}{3A} = \frac{(6)(4)}{3\pi D^2} = \frac{2.54}{D^2} \text{ ksi},$$

where A = cross-sectional area of a rivet, D = diameter of rivet, $W = W'$ from a vertical sum of the forces, and it is *assumed* that the load is distributed uniformly over the rivets. Substitute the stresses in (n) and (o) into equation (8.4) and get

$$\tau = \left[s_s^2 + \left(\frac{s}{2}\right)^2\right]^{1/2}$$

$$8 = \left[\left(\frac{2.54}{D^2}\right)^2 + \left(\frac{8.04}{2D^2}\right)^2\right]^{1/2} = \frac{4.75}{D^2};$$

from which $D = 0.77$ in.; use $\frac{3}{4}$ in. The foregoing is a conventional approach to this type of problem. The assumptions made tend toward the safe side; critical design may call for more realistic considerations. The principal stress σ_2 is compressive; hence τ is the maximum shear.

NOTE. The AISC[5.34] recommends the use of the interaction principle for bolts and rivets, which for this case is defined by the equation

$$\left(\frac{s_t}{s_d}\right)^2 + \left(\frac{s_s}{s_{sd}}\right)^2 \leqq 1$$

where s_d and s_{sd} are respectively the design stresses for tension and for shear, s_t is the tensile stress computed from the tensile load only, and s_s is the shear stress from the shear component of the load. If the maximum shear stress theory is used (next article), the design $s_d = 2s_{sd} = 16$ ksi in the foregoing example; $s_t = 8.04/D^2$ ksi; $s_s = 2.54/D^2$ ksi. Substituting these values in the above equation, find $D \approx 0.77$ in. as before. But compare the interaction equation carefully and in detail with equation (8.10), § 8.13, below.

8.12 THEORIES OF FAILURE.

Over the years, there have been proposed a number of theories that are designed to predict when a failure of a metal member would occur. Those most frequently used are explained below.

(a) Maximum Principal Stress Theory. This theory is accredited to W. J. M. Rankine (about 1850). In effect, it presumes that when the maximum principal stress exceeds a certain limiting value failure occurs;

this stress is given by equation (8.5) for a biaxial stress system. For *static loading* of *ductile* material, a logical limiting stress is the yield stress s_y; that is, a design stress may be determined from $\sigma_d = s_y/N$. Theoretically, the limit stress is the elastic limit as determined by a simple tensile test (these theories are called theories of elastic failure), but yield stresses are more commonly available. For *static loading* of a *brittle* material (neither a distinct yield point nor elastic limit), as cast iron, the limiting stress is taken as the ultimate stress; $\sigma_d = s_u/N$. For fatigue loading of any material, the limiting stress is logically the endurance strength (of course, with proper allowance for factors affecting this strength—Chapter 4); $\sigma_d = s_n/N$. In this sense, equation (8.5) could be said to represent the normal stress theory of fatigue failure of a member subjected to combined stresses.

Experimental data suggest that the maximum principal stress theory gives good predictions for brittle materials, and it is therefore often so used. It is not recommended for ductile materials. This theory is represented

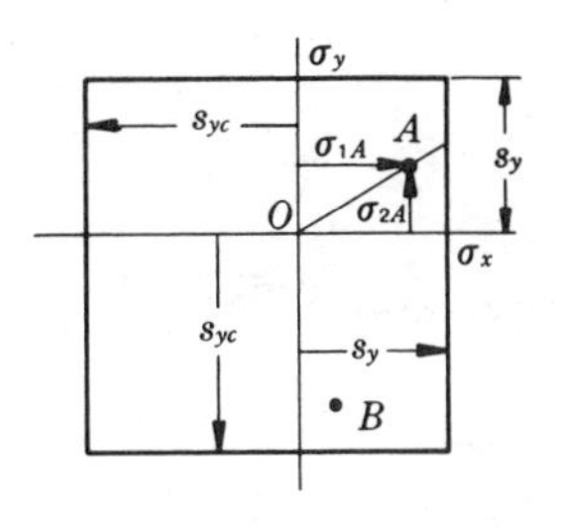

FIGURE 8.14 Boundary for Principal Stress Theory, Static Criteria. Biaxial stresses.

by a rectangle, Fig. 8.14, whose boundaries are defined by the tensile yield s_y and the yield point in compression s_{yc} as shown. When $s_{yc} = s_y$, point O is at the center of a square. The significance is that the boundary represents a failing condition and that inside points, such as A where the principal stresses are σ_{1A} and σ_{2A}, or B, represent a safe state of stress *according to this theory*.

(b) Maximum Shear Stress Theory. This theory is most frequently accredited to J. J. Guest, though others proposed it independently. For the elastic criterion, failure under combined stresses has occurred by this theory when the maximum shear stress, equation (8.6), exceeds the maximum shear stress in a *tensile specimen* when the principal normal stress is the elastic-limit stress. The value of this shear stress is $s_t/2$, as shown in Fig. 8.11(a) and equation (**j**), p. 232. The limiting static shear stress is $s_s = s_y/2$, where s_y = yield strength; and the maximum shear stress, equation (8.3), is given by $(\sigma_{max} - \sigma_{min})/2$; hence the maximum shear stress criterion can also be expressed by $\sigma_{max} - \sigma_{min} = s_y$ (yield). Thus, the design stress in equation (8.6) for static loads is $\tau_d = (s_y/2)/N$ in accordance with this theory (except that yield has been substituted for elastic limit). However, many designers commonly use the *torsional* *yield strength* as the limit stress and find a design stress from $\tau_d = s_{ys}/N$. In a particular application, the significant number is the design stress that is expected to yield suitably safe dimensions as computed from a particular theoretical equation.

The maximum shear stress theory is also used in design for fatigue, in which case the analogous design stress would be $\tau_d = (s_n/2)/N$, where s_n is commonly taken as the fatigue strength in a standard rotating beam specimen, with proper corrections for the factors that affect fatigue strength (Chapter 4); but see §§ 8.13 and 8.15 below. The boundary representation for the maximum shear stress theory is shown in Fig. 8.15 for biaxial stresses. When both principal stresses are positive (or both negative—third quadrant) with $\sigma_3 = 0$, as at A, the principal shear stresses are $(\sigma_1 - \sigma_3)/2 = \sigma_1/2$ and $(\sigma_2 - \sigma_3)/2 = \sigma_2/2$. That is, in the first and third quadrants, Fig. 8.15, the maximum shear stress provides the same limits as the maximum normal stress; say $\sigma_1 = s_y$ when $\sigma_1 > \sigma_2$. When σ_1 is positive, σ_2 negative, and $\sigma_3 = 0$, the maximum shear is $(\sigma_1 - \sigma_2)/2$ and the boundary is given by $\sigma_1 - \sigma_2 = s_y$ or $\sigma_x - \sigma_y = s_y$, where the yield s_y is a constant for a particular material. As seen in Fig. 8.15, this is the equation of the boundary in the fourth quadrant. Thus, a safe state of stress here in accordance with the

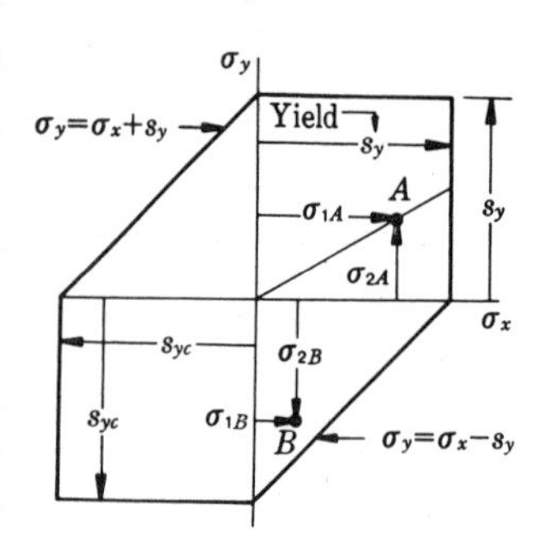

FIGURE 8.15 Boundary for Shear Stress Theory, Static Criteria. Drawn for $|s_{yc}| = |s_y|$. Biaxial stresses.

shear stress theory must be represented by a point, such as B, that is inside of the limit line. The limiting stress for a reversed axial load would be an endurance strength, which can be imagined to be substituted for s_y in Fig. 8.15. The maximum shear stress theory has been the most widely used one for ductile materials.

(c) Octahedral Shear Stress Theory. The theory of mechanics of materials shows that the results from the octahedral shear stress theory and those from the *maximum distortion-energy theory* are the same.[1.7] Hence, the equations given below can be and are referred to by either name. Also, the name von Mises theory, sometimes associated with the names Huber or Hencky,[1.7] as the Hencky–Mises criterion, are used, in honor of the pioneers who developed the theory.

Given a triaxial stress system as shown in Fig. 8.16 with $\sigma_1 > \sigma_2 > \sigma_3$.

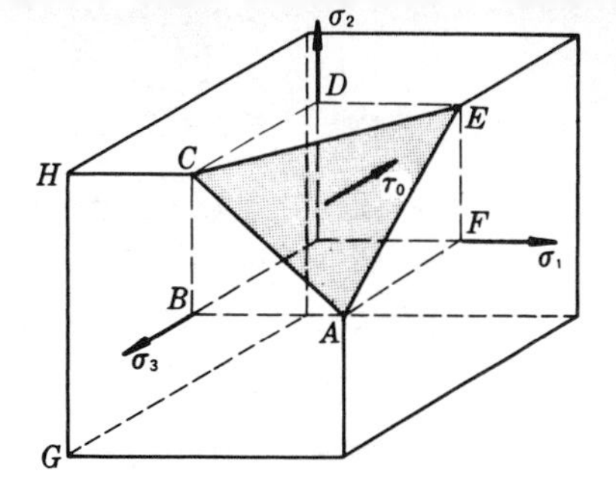

FIGURE 8.16 Octahedral Plane. The octahedral plane is shaded. There is a normal stress σ_o on this plane, not shown.

An octahedral plane may be obtained by passing it through the mid-points of three sides of the cube, as *ACE*. If an element *ABCDEF* is taken as a free body and the forces summed parallel to the octahedral plane, in a manner analogous to procedures previously explained, the shearing stress τ_{ot} for a triaxial system is found to be

$$\textbf{(p)} \qquad \tau_{ot} = \tfrac{1}{3}[(\sigma_1 - \sigma_2)^2 + (\sigma_1 - \sigma_3)^2 + (\sigma_2 - \sigma_3)^2]^{1/2},$$

where the normal stresses are algebraic. The normal stress σ_o on this plane (not shown) is equal to the arithmetic average of the three principal stresses, but we shall not have occasion to use it. Applying equation (**p**) to a uniaxial stress state, $\sigma_2 = \sigma_3 = 0$, we find that $\tau_{ou} = (\sqrt{2}/3)\sigma_1$, which at the point of yielding ($\sigma_1 = s_y$) becomes $\tau'_{ou} = (\sqrt{2}/3)s_y = 0.471s_y$ (axial load) where τ'_{ou} is the limit value (at the point of failure). The octahedral shear theory may now be stated as follows: when the octahedral shear stress τ_o in a body becomes equal to the octahedral shear stress τ'_{ou} existing in a tensile test specimen (uniaxial stress) at the instant that the tensile stress s_t is equal to the elastic limit (use s_y), elastic failure is on the point of occurring. An analogous statement may be made for the fatigue criterion. Equation (**p**) is reduced to that for biaxial stresses τ_{ob} by letting $\sigma_3 = 0$; then squaring as indicated and collecting terms, we find the octahedral shear for a biaxial system as

$$\textbf{(q)} \qquad \tau_{ob} = \frac{\sqrt{2}}{3}(\sigma_1{}^2 + \sigma_2{}^2 - \sigma_1\sigma_2)^{1/2} \qquad \text{[PLANE STRESSES]}$$

Equations (**p**) and (**q**) may be used in design by letting the design value of τ_o be $\tau_{od} = 0.471s_y/N$ for static loading, but more convenient equations can be obtained by setting up an equivalent normal stress σ_e and deriving an equation in terms of conveniently computed stresses, say s_X, s_Y, and s_S. From the tensile test, $\tau_o = (\sqrt{2}/3)s_t$ or $s_t = (3/\sqrt{2})\tau_o$; this value of s_t is the normal stress for simple tension when the octahedral shear is τ_o. Generalizing, call this s_t the equivalent normal stress σ_e, and substitute the biaxial value of $\tau_o = \tau_{ob}$, equation (**q**), into $\sigma_e = (3/\sqrt{2})\tau_o$ and get

$$\textbf{(r)} \qquad \sigma_e = (\sigma_1{}^2 + \sigma_2{}^2 - \sigma_1\sigma_2)^{1/2},$$

for which a static design stress could be $\sigma_e = s_y/N$. It is well to note that σ_e does not designate a particular actual stress, but an equivalent stress that indicates the degree of safety when compared to a failure criterion such as the yield s_y. Substitute into (**r**) the two principal stresses as given by equation (8.5)—use the negative sign for σ_2—and find

$$\sigma_e = (s_X^2 + s_Y^2 - s_X s_Y + 3s_s^2)^{1/2}; \tag{8.8}$$

and for the most frequently encountered state of stress ($s_Y = 0$) of one normal stress and a shear stress, Fig. 8.8,

$$\sigma_e = (s^2 + 3s_s^2)^{1/2}, \tag{8.9}$$

where s, s_X, s_Y, s_s are computed from F/A, Mc/I, Tc/J, etc. When $\sigma_e = s_y$, the octahedral shear stress theory is predicting incipient elastic failure. (It will help the reader to gain familiarity with this material if he will derive (8.8) in detail.)

The relation between the yield (and elastic) strengths in tension s_y and in torsion s_{ys} is obtained by assuming first a state of pure torsion, $\sigma = \sigma_1 = -\sigma_2 = s_s$, Fig. 8.12, in which case, equation (**q**) reduces to ($\sigma_3 = 0$)

$$\tau_{ob} = \sqrt{\frac{2}{3}}\,\sigma = \sqrt{\frac{2}{3}}\,s_s; \qquad \tau'_{ob} = \sqrt{\frac{2}{3}}\,s_{ys},$$

where τ'_{ob} is the limit value when elastic failure is about to occur. The limit value of the octahedral shear in terms of the tensile yield was obtained above for uniaxial stress at $\tau'_{ou} = (\sqrt{2}/3)s_y$. Equating these limit values, $\tau'_{ob} = \tau'_{ou}$, and solving for s_{ys}, we get

$$s_{ys} = \frac{s_y}{\sqrt{3}} = 0.577 s_y, \tag{s}$$

as compared with $s_{ys} = 0.5s_y$ predicted by the maximum shear stress theory. As previously stated, actual test values range widely from below $0.5s_y$ to greater than $0.6s_y$; we have been arbitrarily using $s_{ys} = 0.6s_y$. The fact is that the ratio 0.577 is in better agreement with experiment than 0.5. For this reason and because test points tend to match the boundary defined by the octahedral shear stress theory, there is an increasing tendency toward using this theory. A graphical portrayal of this boundary for biaxial stresses is obtained by letting $\sigma_e = s_y$ in equation (**r**) and squaring both sides. Since the yield s_y is a constant, the resulting equation is that of an ellipse, as seen in Fig. 8.17 (where $\sigma_1 = \sigma_x$, $\sigma_2 = \sigma_y$). The octahedral boundary is everywhere greater than the maximum shear boundary except where one principal stress is zero (also $\sigma_3 = \sigma_z = 0$).

(d) Maximum Strain Theory. This theory, due to St. Venant, is the only other one that we shall mention. At the elastic limit in uniaxial tension, the unit strain is ϵ_e; according to the maximum strain theory, inelastic action begins at a point in a body when the strain at that point begins to

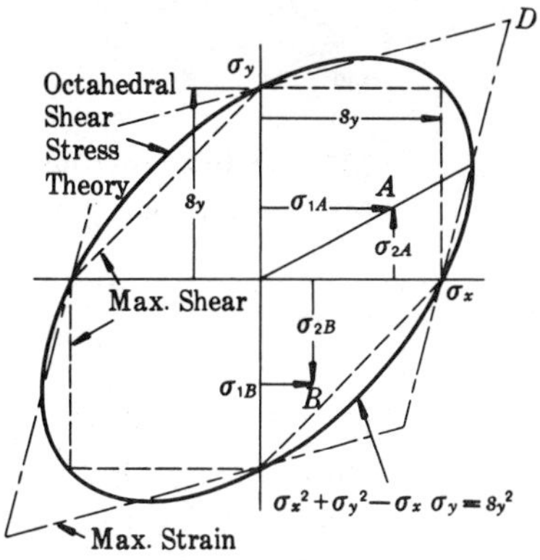

FIGURE 8.17 Boundary for Octahedral Shear Stress Theory. Maximum shear and maximum strain boundaries are shown for comparison.

exceed ϵ_e. The strain is converted to stress by $s = E\epsilon$; and the resulting equation for two normal stresses s_X, s_Y, and a shearing stress s_s is

$$\text{(t)} \qquad \sigma = \frac{1-\mu}{2}(s_X + s_Y) + (1+\mu)\left[\left(\frac{s_X - s_Y}{2}\right)^2 + s_s^2\right]^{1/2},$$

where μ is Poisson's ratio and, in design, $\sigma = s_y/N = \sigma_d$ for static loads. This theory has been widely used for thick cylinders; it fits experimental data on brittle materials better than those on ductile materials. The boundaries are shown by the dot-dash line in Fig. 8.17. Notice that if the other theories are right in the vicinity of $\sigma_1 = \sigma_2$, this theory is relatively dangerous, with point D so far out.

8.13 DESIGN EQUATION FOR MAXIMUM SHEAR AND OCTAHEDRAL SHEAR THEORIES. Consider a biaxial stress state induced by a normal stress s and a shear stress s_s on a particular plane; the maximum shear stress is given by equation (8.4). Elastic failure by the maximum shear stress theory is incipient when the resultant shear $\tau = s_y/2$, as explained above. A safe shear is $\tau = s_y/(2N)$. Substitute $\tau = s_y/(2N)$ in (8.4), divide through by $s_y/2$, and get

$$\text{(u)} \qquad \frac{1}{N} = \left[\left(\frac{s}{s_y}\right)^2 + \left(\frac{s_s}{s_y/2}\right)^2\right]^{1/2} = \left[\left(\frac{s}{s_y}\right)^2 + \left(\frac{s_s}{s_{ys}}\right)^2\right]^{1/2},$$

where s_{sy} is taken equal to $s_y/2$ in accordance with the maximum shear stress theory.

For the same state of stress, the equivalent stress σ_e for the octahedral shear theory is given by equation (8.9); a safe value is $\sigma_e = s_y/N$. Substituting this value into (8.9), dividing through by s_y, we get

$$\text{(v)} \qquad \frac{1}{N} = \left[\left(\frac{s}{s_y}\right)^2 + \left(\frac{s_s}{s_y/\sqrt{3}}\right)^2\right]^{1/2} = \left[\left(\frac{s}{s_y}\right)^2 + \left(\frac{s_s}{s_{ys}}\right)^2\right]^{1/2},$$

where we have used $s_y = \sqrt{3}s_{ys}$ from equation (s), in accordance with the octahedral shear theory. The final form of this equation is seen to be the same as the previous one; for this reason, it will be suitable for design purposes for either theory; to wit,

$$\frac{1}{N} = \left[\left(\frac{s}{s_y}\right)^2 + \left(\frac{s_s}{s_{ys}}\right)^2\right]^{1/2} \tag{8.10}$$

If $s_y/2$ is substituted for s_{ys}, the design would be by the maximum shear stress theory; if $s_{ys} = s_y/\sqrt{3} = 0.577s_y$ is substituted for s_{ys}, the design would be by the octahedral shear theory. If $s_{ys} = 0.6s_y$, a commonly assumed value, the design is not in accordance with either theory; in this case, the equation is an expression of the interaction principle (see note at end of § 8.11).

In equation (8.10), let $N = 1$, let $x = s/s_y$, and $y = s_s/s_{ys}$ and square both sides; the result is seen to be the equation of a circle, Fig. 8.18, a

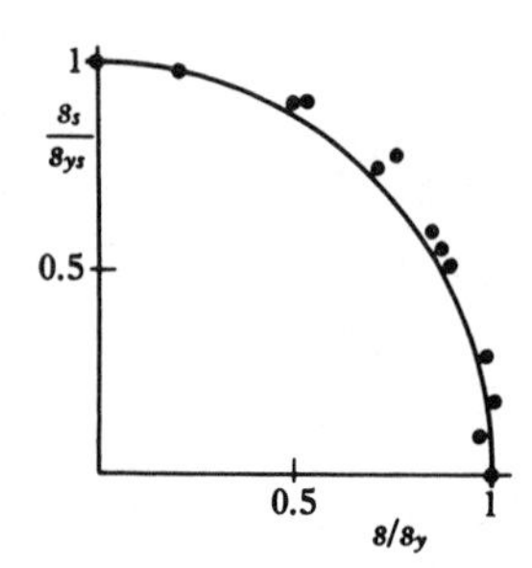

FIGURE 8.18 Points shown are typical.

circle that is the predicted location of combinations of stress that produce elastic failure. (Note that with s and s_s the variables and with s_y and s_{ys} constants, (8.10) is the equation of an ellipse.) There is considerable evidence to support the validity of this conclusion, as suggested by the typical failure points shown. For a design factor of N, the radius of the circle of safe stresses is $1/N$ in Fig. 8.18.

If, at the beginning of this article, we had postulated fatigue failure and used $s_{ns} = s_n/2$ for the maximum shear theory and $s_{ns} = s_n/\sqrt{3} = 0.577s_n$ for the octahedral shear theory,* the resulting equation would become

$$\frac{1}{N} = \left[\left(\frac{s}{s_n}\right)^2 + \left(\frac{s_s}{s_{ns}}\right)^2\right]^{1/2}, \tag{8.11}$$

wherein, for combined stresses, one uses the equivalent stresses based on s_n and s_{ns}; $s_e = s_m s_n/s_y + K_f s_a$ (§ 4.19); surface, size, and other fatigue

* There are some test values showing lower ratios for (torsion fatigue)/(bending fatigue) perhaps lower than 0.52, but other test values reach 0.67 or more for steel; above 0.8 for some cast irons; about 0.63 for aluminum 7075-T6.[4.31] But such data are not too plentiful.

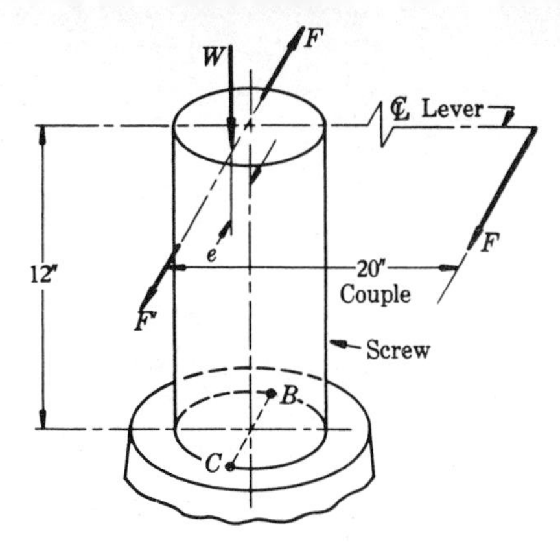

FIGURE 8.19 Force $F' = F$ produces bending moment of 12 F; torque couple FF is $20F$.

factors are applied to the alternating component s_a. Cazaud[4.24] has found that if the shear (torsion) and normal stresses are reversed, periodic, and in phase, this equation agrees well with failure tests ($N = 1$); naturally, corrections are necessary for stress raisers that may exist.

8.14 EXAMPLE—COMBINED BENDING, COMPRESSION, AND TORSION. The jack of Fig. 8.24(a), with a 2-in. Acme thread (Fig. 8.21), is rated at an 8000-lb. load W. Assume that the load can move laterally with the top of the screw, so that a force of $F = 60$ lb., applied at a torque arm of 20 in., produces a bending moment on the screw as a cantilever beam, as well as the torque to raise the load. Also assume that the load is not central, but has an effective eccentricity of $e = \frac{1}{4}$ in. A diagrammatic representation of the force system is given in Fig. 8.19. If the screw is made of C1045 as rolled, is this loading safe? Check by both the maximum shear stress and the octahedral shear stress theory. The maximum lift of the jack is 12 in. (length of the cantilever beam), small enough that α may be taken as 1 (no buckling).

Solution. From Table 8.1, we find the root diameter to be 1.75 in., for a 2-in. screw. Neglecting the strengthening effect of the threads (§ 8.24), we have

$$A_r = \frac{\pi(1.75)^2}{4} = 2.4 \text{ in.}^2; \qquad Z = \frac{\pi D^3}{32} = \frac{\pi(1.75)^3}{32} = 0.525 \text{ in.}^3;$$

$$Z' = 2Z = 1.05 \text{ in.}^3.$$

The screw is subjected to the following stresses: a uniform compressive stress, a bending stress due to the force on the handle, a bending stress due to the eccentricity of the load, and a shearing stress. Compute the normal stresses first.

Uniform stress $$s_1 = \frac{W}{A_r} = \frac{8000}{2.4} = 3330 \text{ psi, on all sections.}$$

Bending by F $$s_2 = \frac{M}{Z} = \frac{(60)(12)}{0.525} = 1370 \text{ psi, at base } (B, C) \text{ only.}$$

Bending by W $$s_3 = \frac{M}{Z} = \frac{(8000)(0.25)}{0.525} = 3810 \text{ psi, on all sections, front and rear fibers.}$$

At some stage of rotation of the screw and handle, all of the normal stresses act together in the same sense at a particular point; for position shown in Fig. 8.19, they are all compressive at point C. If the yield strength in compression is the same as the tensile yield (usual assumption for steel), the check is based on the total stress at C, which is

$$s_c = s_1 + s_2 + s_3 = 3330 + 1370 + 3810 = 8510 \text{ psi.}$$

If the pivot friction is neglected, the torsional stress is

$$s_s = \frac{T}{Z'} = \frac{(60)(20)}{1.05} = 1140 \text{ psi,}$$

which is observed to be relatively small. From Table AT 7, we find $s_y = 59$ ksi. Use $s_{ys} = (0.5)(59) = 29.5$ ksi in equation (8.10) for N based on maximum shear, and use $s_{ys} = (0.577)(59) = 34$ ksi for the octahedral theory.

$$\frac{1}{N^2} = \left(\frac{s}{s_y}\right)^2 + \left(\frac{s_s}{s_{ys}}\right)^2$$

$$= \left(\frac{8.51}{59}\right)^2 + \left(\frac{1.14}{29.5}\right)^2$$

from which, $N = 6.7$ for maximum shear; and

$$\frac{1}{N^2} = \left(\frac{8.51}{59}\right)^2 + \left(\frac{1.14}{34}\right)^2,$$

from which, $N = 6.75$ (octahedral). If the shear had been relatively greater, a greater difference between these answers would have been obtained. Also, note that if s_{ys} had been taken as $0.6s_y$, as you have been doing previously in this text, the result would have changed little. The answers obtained suggest that the jack is conservatively rated. It should be noted, however, that a jack may get some rough handling, that the effective eccentricity of the load may be more than $\frac{1}{4}$ in., and that a workman may use a longer handle and pull harder.

8.15 COMBINING VARIABLE STRESSES. Many machine elements are subjected to a combination of stresses where the axial load varies, the torque varies, or the bending moment varies, or where any two or all vary. Several procedures for combining different kinds of varying stresses have been proposed.[8.12] The one adopted below is logical, though approximate (as are all theories), and it is expected to produce safe designs for ductile metals. Use the equivalent stresses obtained by multiplying equations (4.4), $1/N = s_m/s_y + K_f s_a/s_n$, and (4.5) through by s_n and s_{ns}, respectively. Dropping s_n/N and s_{ns}/N in favor of s_e and s_{es}, we then get

$$\text{(w)} \qquad s_e = \frac{s_n}{s_y} s_m + K_f s_a$$

$$\text{(x)} \qquad s_{es} = \frac{s_{ns}}{s_{ys}} s_{ms} + K_{fs} s_{as}$$

where $s_{ns}/s_{ys} \approx s_n/s_y$,

s_m is a nominal stress obtained from F_m/A, $\alpha F_m/A$, or $M_m c/I$, etc.

If more than one of these stresses is present, the mean stress s_m is determined from the circumstances involved. For example, if a fixed section is subjected to a steady uniform stress F/A and to a variable bending moment, then $s_m = F/A + M_m c/I$ (an *algebraic* sum). If M is constant on a rotating shaft *and* if there is a steady F/A, then $s_m = F/A$ because the mean *bending* stress is zero.

s_a is obtained from F_a/A, $\alpha F_a/A$, or $M_a c/I$, etc. If more than one is present, the value of s_a is determined from the circumstances involved. If the variations of F_a and M_a are "in phase," the two corresponding stresses are added. On a rotating shaft with constant M, the variable bending stress is $s_a = Mc/I$.

s_{ms} is obtained from F_m/A or $T_m c/J$. See comments under s_m above.

s_{as} is obtained from F_a/A or $T_a c/J$. See comments under s_a above.

K_f, K_{fs} are strength-reduction factors owing to stress concentrations.

In the design procedure, the values from the foregoing equations (**w**) and (**x**) for equivalent stresses are substituted into (8.11)*, which is

$$\frac{1}{N} = \left[\left(\frac{s_e}{s_n}\right)^2 + \left(\frac{s_{es}}{s_{ns}}\right)^2\right]^{1/2}, \tag{8.11}$$

good for the case of one varying normal stress and a varying shear stress on the same plane, in phase, Fig. 8.8. Assume that $s_{ns}/s_n = s_{ys}/s_y$; if we take $s_{ys} = s_y/2$, $s_{ns} = 0.5s_n$, we can consider that we are using a maximum-shear-stress theory of fatigue failure, § 8.12; using $s_{ns} = s_n/\sqrt{3} = 0.577s_n$ would correspond to the octahedral shear-stress theory; if $s_{ns} = 0.6s_n$, as frequently assumed, there is no accord with any theory, but the results agree well with tests. The equation resulting from the manipulation of (**w**), (**x**), and (8.11) is

$$\frac{1}{N} = \left[\left(\frac{s_m}{s_y} + \frac{K_f s_a}{s_n}\right)^2 + \left(\frac{s_{ms}}{s_{ys}} + \frac{K_{fs} s_{as}}{s_{ns}}\right)^2\right]^{1/2}, \tag{y}$$

[VARYING NORMAL AND SHEAR STRESSES ON A PLANE AT A POINT]

but the student is urged to solve problems by following a detailed approach to equation (8.11) [or (8.10)], as in the example below, because the logic is lost in substituting numbers into (**y**). For brittle materials, the maximum normal-stress theory is generally better, and it too can be adapted to the equivalent-stress approach. Variations of stresses that are not in phase can be handled with some logic, but this class of problem is beyond the scope of this book except as we may sometimes assume that peak stresses occur simultaneously, an approach on the safe side. The more general case of three finite principal stresses with the same frequency can be set up in terms of principal stresses and the octahedral shear theory[1.14] but justifying experimental data are not plentiful.

* The use of the equivalent stresses from equations (4.6) and (**r**), § 4.19, in equation (8.10) gives the same result. See the discussion on theories below equation (8.10).

8.16 EXAMPLE—COMBINED BENDING AND TORSION, VARYING STRESSES. Figure 8.20 represents a shaft with loads A and C being forces exerted on the shaft by gears keyed on with profile keyways, and B and D being the bearing reactions. For a particular power transmission, the maximum moment occurs at C, which is therefore the section to be investigated since the torque is also transmitted through it. The load on the shaft is variable because the transmitted horsepower varies continuously from 60 hp to 6 hp in half a revolution, and from 6 to 60 in the next half, while the shaft rotates virtually at a

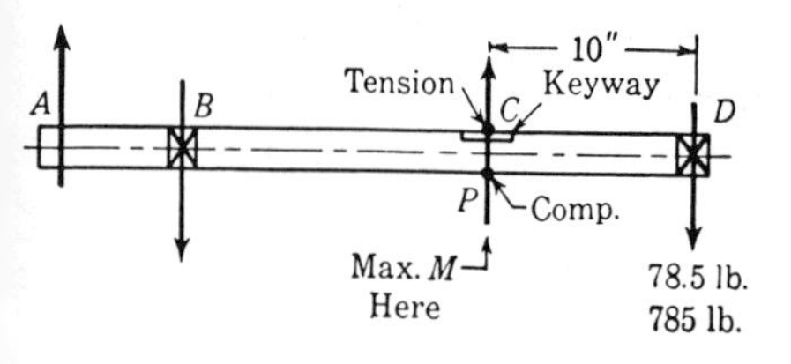

FIGURE 8.20

constant speed of 400 rpm. The force at D varies from 785 to 78.5 lb., changing with the horsepower. For a material of cold-rolled AISI 1137 and a design factor of $N = 1.7$, what should be the diameter of the shaft?

Solution. From an outside source, for 1137, cold drawn, we find

$$s_n = 50 \text{ ksi}, \qquad s_y = 89 \text{ ksi}, \qquad \frac{s_y}{s_n} = 1.78.$$

From Table AT 13, we find the stress concentration factors for a profile keyway as

$$K_f = 2.0 \quad \text{and} \quad K_{fs} = 1.6.$$

The torque on the shaft, from $T = 63{,}000hp/n$, is

$$T_{\max} = \frac{(63{,}000)(60)}{400} = 9450 \text{ in-lb.,} \quad \text{and} \quad T_{\min} = \frac{(63{,}000)(6)}{400} = 945 \text{ in-lb.}$$

From these values, the average torque and the variable component are, respectively,

$$T_m = 5200 \text{ in-lb.} \quad \text{and} \quad T_a = 4250 \text{ in-lb.}$$

$$s_{ms} = \frac{T_m}{Z'} = \frac{5200}{Z'} = \frac{5200}{2Z} \text{ psi} \quad \text{and} \quad s_{as} = \frac{T_a}{Z'} = \frac{4250}{Z'} = \frac{4250}{2Z} \text{ psi,}$$

or $s_{ms} = 2.6/Z$ and $s_{as} = 2.125/Z$ ksi, where $J/c = Z' = 2Z$, because $Z' = \pi D^3/16$ and $Z = \pi D^3/32$, Table AT 1. On the expectation that the size will be greater than 0.5 in., introduce a size factor (SF), § 4.12, applied to the alternating component only; assume that the strength reduction factors K_f, K_{fs} take care of the surface condition, use equation (x) with $s_{ns}/s_{ys} = s_n/s_y = 1/1.78$, and get

$$\textbf{(z)} \qquad s_{es} = \frac{s_{ns}}{s_{ys}} s_m + \frac{K_{fs}s_{as}}{(SF)} = \frac{2.6}{1.78Z} + \frac{(1.6)(2.125)}{0.85Z} = \frac{5.46}{Z} \text{ ksi.}$$

The bending stress would vary even if the bending moment were constant, because the fiber C, Fig. 8.20, which first has a tensile stress on it, has a compressive stress on it 180° later when it has moved to P. Let's say that hp = 60 when the fiber under consideration is at C; then it is 6 hp when the fiber is at P. In the meantime, the force at D has changed from 785 to 78.5 lb., and the moment at C

has changed from 7850 to 785 in-lb. (sum of the moments of the forces to the right of section C). At each position, the tensile stress on the top fiber of the shaft is the same in magnitude as the compressive stress on the bottom fiber, but if we follow the particular fiber at point C as the shaft rotates, we see that the stress variation at that *point* fixed in the shaft is from

$$s_{max} = \frac{M}{Z} = \frac{7850}{Z} \quad \text{to} \quad s_{min} = -\frac{785}{Z}\ \text{psi},$$

[AT POSITION 0] [180° LATER]

where the minimum stress is compression. The mean $(s_{max} + s_{min})/2$ and variable $(s_{max} - s_{min})/2$ components of the bending stress are therefore

$$s_m = \frac{3533}{Z}\ \text{psi} \quad \text{or} \quad \frac{3.53}{Z}\ \text{ksi} \quad \text{and} \quad s_a = \frac{4318}{Z}\ \text{psi} \quad \text{or} \quad \frac{4.32}{Z}\ \text{ksi}.$$

(It is important to note that when the bending moment *on a rotating shaft* varies, one is safer in working directly with stresses than with mean and variable bending moments.) Using the foregoing findings in equation (**w**), we get

$$\textbf{(a)} \qquad s_e = \frac{s_n}{s_y} s_m + \frac{K_f s_a}{(SF)} = \frac{3.53}{1.78Z} + \frac{(2)(4.32)}{0.85Z} = \frac{12.15}{Z}$$

Substitute the values of s_e and s_{es} from (**a**) and (**z**) into equation (8.11), with $s_{ns} = 0.577s_n'$ for the octahedral shear theory, and get

$$\frac{1}{N} = \left[\left(\frac{s_e}{s_n}\right)^2 + \left(\frac{s_{es}}{s_{ns}}\right)^2\right]^{1/2}$$

$$\frac{1}{1.7} = \left[\left(\frac{12.15}{50Z}\right)^2 + \left(\frac{5.46}{(0.577)(50)Z}\right)^2\right]^{1/2} = \frac{1}{50Z}(147.5 + 89.5)^{1/2} = \frac{0.308}{Z},$$

from which $Z = 0.524 = \pi D^3/32$, or $D = 1.748$; use $D = 1\frac{3}{4}$ in. The next chapter contains additional discussion of shaft design.

8.17 ADDITIONAL COMMENTS ON FATIGUE.

Fatigue failure is the initiation of a crack and its propagation. Thus, the logical expectation is that the crack most likely follows a plane of maximum tensile stress. For a member in pure torsion, fatigue cracks typically progress on a plane at 45° with the plane of the maximum shear stress, the plane of the principal tensile stresses (Fig. 8.12). For a member in compression only, cracks will not spread on the plane of the compressive stress, but typically occur in the vicinity of the plane of the theoretical maximum shear, at 45° with the compressive stress. For a member subjected to combined torsion and bending, it has been found for mild steel[8.13] that: if σ_{max}/τ_{max} is distinctly above 1.6, the crack propagates in the normal stress direction; if $\sigma_{max}/\tau_{max} < 1.6$, the crack follows the shear direction; for $\sigma_{max}/\tau_{max} \approx 1.6$, it might go either way. To further complicate the logical processes, Sines[4.1] presents data that indicate that the alternating shear stress produces the fatigue damage. He also shows for *pure shear* that the magnitude of the mean shear s_{sm} has no effect on the magnitude of the alternating shear that produces failure, provided the maximum stress τ_{max} does not exceed the torsional yield strength. (But see Chodorowski[4.28] for contrary data.)

Residual stresses (from the various manufacturing processes—thermal and mechanical) apparently play a more important role in determining fatigue strength than generally conceded. These stresses are triaxial, complicated, and difficult to obtain or to estimate, but the conclusion is that designers should work toward ways and means of including their effect and of controlling them. See Mattson.[4.28]

8.18 POWER SCREWS.

As suggested by the example of 8.14, power screws are often subjected to combinations of stresses; hence, we shall include some engineering information about them at this point. Since they are designed to exert a force with a mechanical advantage, the threads are somewhat different from fastening threads. The common forms are

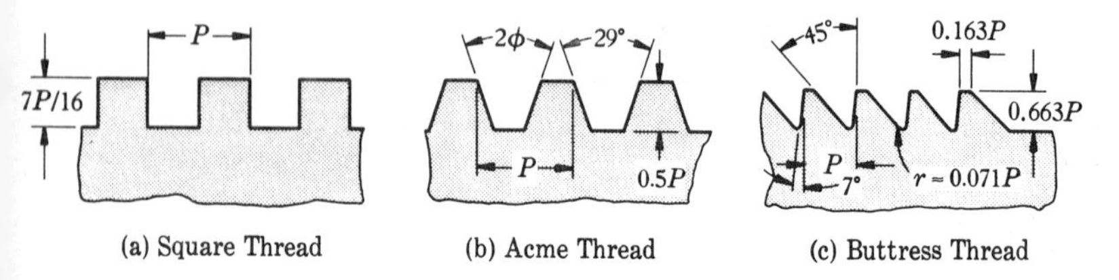

FIGURE 8.21 Power Screw Thread Profiles. (a) Commonly used proportions. In (b), the angle 2ϕ is called the *thread angle*, and ϕ is called the *pressure angle*. (c) Depth of engagement is $0.6P$.

shown in Fig. 8.21, and other information is given in Table 8.1. The *Acme thread*, due to the sloping sides, is theoretically not quite as efficient as the square thread, but it has been shown in practice that the accuracy of manufacture and the condition of the thread surfaces are the important determinants of efficiency. Since the Acme thread may be cut with dies, it is easier and cheaper to manufacture. Furthermore, if a split nut is used, looseness due to wear may be eliminated by "taking up" the nut. The *buttress thread* has virtually the same efficiency as the square thread, but it can transmit power in one direction only.

8.19 PITCH AND LEAD.

The ***axial pitch*** or ***pitch*** P is the distance, measured axially, from a point on one thread to the corresponding point on an adjacent thread, Fig. 8.22

$$P_c = P = \frac{1}{\text{No. of threads per inch}} \text{ in.}$$

The ***lead*** is the distance that a thread advances in one turn; it is the

TABLE 8.1 PROPORTIONS OF POWER THREADS

See Refs. (8.8) and (8.9) for more detail on Acme threads. The stub Acme thread has a height of 0.3*P* instead of 0.5*P*, Fig. 8.21. The minor diameters are given to the nearest thousandth of an inch. The standard[8.21] does not specify a unique number of threads per inch for buttress threads. See Fig. 8.21.

SIZE IN.	SQUARE THREADS		ACME THREADS			BUTTRESS (SUGGESTED)
	Th./in.	*Minor Dia.*	*Th./in.*	*Reg. Minor Dia.*	*Stub Minor Dia.*	*Th./in.*
$\frac{1}{4}$	10	0.163	16	0.188	0.213	
$\frac{5}{16}$			14	0.241	0.270	
$\frac{3}{8}$	8	0.266	12	0.292	0.325	
$\frac{7}{16}$			12	0.354	0.388	
$\frac{1}{2}$	$6\frac{1}{2}$	0.366	10	0.400	0.440	20
$\frac{5}{8}$	$5\frac{1}{2}$	0.466	8	0.500	0.550	20
$\frac{3}{4}$	5	0.575	6	0.583	0.650	16
$\frac{7}{8}$	$4\frac{1}{2}$	0.681	6	0.708	0.775	16
1	4	0.781	5	0.800	0.880	12
$1\frac{1}{8}$			5	0.925	1.005	12
$1\frac{1}{4}$	$3\frac{1}{2}$	1.000	5	1.050	1.130	10
$1\frac{3}{8}$			4	1.125	1.225	10
$1\frac{1}{2}$	3	1.208	4	1.250	1.350	8
$1\frac{3}{4}$	$2\frac{1}{2}$	1.400	4	1.500	1.600	7
2	$2\frac{1}{4}$	1.612	4	1.750	1.850	6
$2\frac{1}{4}$	$2\frac{1}{4}$	1.862	3	1.917	2.050	6
$2\frac{1}{2}$	2	2.063	3	2.167	2.300	5
$2\frac{3}{4}$	2	2.313	3	2.417	2.550	5
3	$1\frac{3}{4}$	2.500	2	2.500	2.700	5
$3\frac{1}{2}$	$1\frac{5}{8}$	2.962	2	3.000	3.200	5
4	$1\frac{1}{2}$	3.418	2	3.500	3.700	4
$4\frac{1}{2}$			2	4.000	4.200	4
5			2	4.500	4.700	4

distance the nut moves along the axis in one turn, Fig. 8.22. A single-threaded screw has a lead equal to the pitch, Fig. 8.22(a). A double-threaded screw has *two starts*, and the lead is twice the pitch. A triple-threaded screw

FIGURE 8.22 Pitch and Lead. The lead angle is λ.

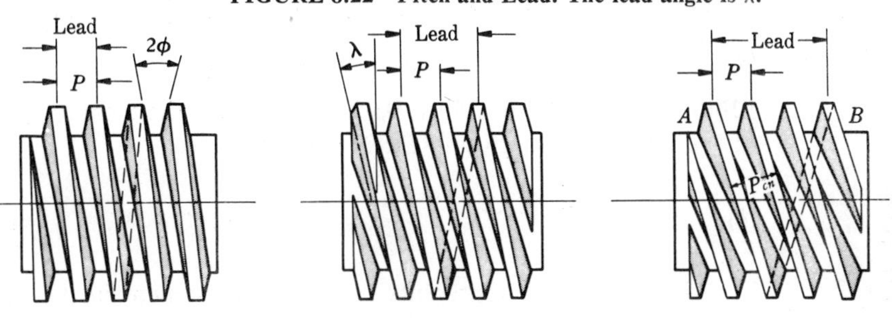

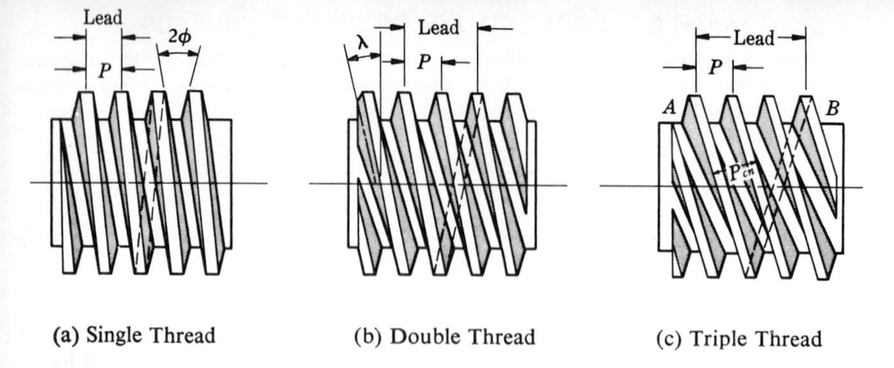

(a) Single Thread (b) Double Thread (c) Triple Thread

FIGURE 8.22 (Repeated).

has *three starts*, and the lead is three times the pitch, as in Fig. 8.22(c); etc. The ***lead angle*** λ is the angle between a tangent to the pitch helix and a plane normal to the axis of the screw, Fig. 8.22(b). If D_m = mean thread diameter [see Fig. 8.23(b)],

(b)
$$\lambda = \tan^{-1}\frac{\text{Lead}}{\pi D_m}.$$

8.20 TORQUE TO TURN SCREW. We may derive an expression for the twisting moment necessary to move an axial load by a screw. Although the load is distributed over several threads, lifting a load by a screw is analogous to moving a block up an inclined plane. Thus the simplest arrangement for a force analysis is in Fig. 8.23(b), which shows the mean line of a thread developed into a plane.

For a square thread, the forces acting on the block, Fig. 8.23(b), are the weight W (axial load), a force Q that pulls the weight up the plane, the limiting frictional force F_f (the block is assumed to be on the point of moving), and the normal plane reaction N; F_f and N are replaced by their resultant R, the total plane reaction. The angle of friction is β (where $\tan\beta = f$, and f is the coefficient of friction); and λ is the lead angle. Sum forces horizontally and vertically:

(c)
$$Q = R\sin(\beta + \lambda).$$

(d)
$$W = R\cos(\beta + \lambda).$$

Dividing Q in **(c)** and W in **(d)** and solving for Q, we find

(e)
$$Q = W\tan(\beta + \lambda).$$

Multiply both sides of **(e)** by $D_m/2$, where D_m is the mean or pitch diameter of the screw, which gives $QD_m/2 = T$, or

(f)
$$T = \frac{QD_m}{2} = \frac{WD_m}{2}\tan(\beta + \lambda). \qquad \text{[SQUARE THREAD]}$$

The right-hand side represents the resisting moment in the threads of the

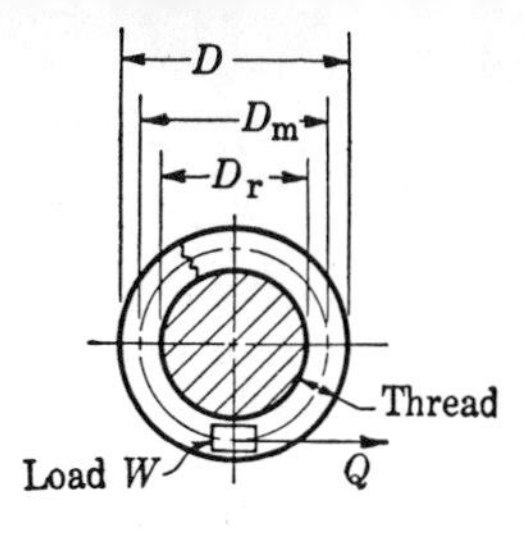

(a) Top View of Thread.

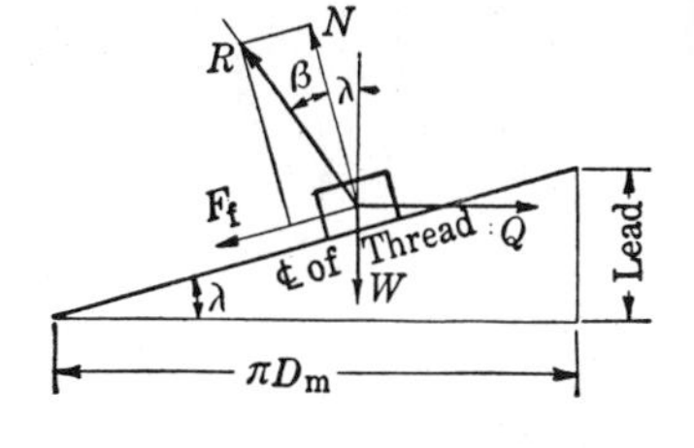

(b) Center Line of Thread Unwrapped from Cylinder.

FIGURE 8.23 Forces on a Body Moving Up an Incline. The force Q is horizontal and normal to the axis of the screw.

screw against turning under a load W. The left-hand side T is the torque applied when the screw is on the point of turning (no other resistance), and Q represents the force that must be exerted at the midpoint of the thread to obtain this torque. However, the external force is applied at the end of a lever, Fig. 8.24(a), or at some equivalent leverage, as in Fig. 8.24(b). Thus $T = QD_m/2 = Fa$, where F is the applied force at some leverage a. Equation **(f)** is sometimes easier to use if the expression $\tan(\beta + \lambda)$ is expanded;

$$\textbf{(g)} \qquad T = \frac{WD_m(\tan\lambda + \tan\beta)}{2(1 - \tan\beta\tan\lambda)} = \frac{WD_m(\tan\lambda + f)}{2(1 - f\tan\lambda)}.$$

The force analysis of an Acme thread is the same as that for a worm. If both sides of equation **(k)**, § 16.8 are multiplied by $D_m/2$ (and let $F_t = W$), we can find the torque to turn the thread *against* the load; in the symbols of this application, it is

$$\textbf{(h)} \qquad T = \frac{WD_m}{2}\left[\frac{\cos\phi\tan\lambda + f}{\cos\phi - f\tan\lambda}\right],$$

where $\phi \approx 14.5^\circ$ for a standard Acme thread, Fig. 8.21(b). Strictly, ϕ should be the pressure angle in a plane normal to the thread, rather than the pressure angle in a diametral plane as shown, the relation being $\tan\phi = \tan 14.5^\circ \cos\lambda$. Since λ is usually small for power screws, $\cos\lambda$ is

FIGURE 8.24 Screw Jacks. In (b), the pivot friction would be small because of the ball bearing. The long nut suggests low wear on threads and stability at maximum extension. Notice the buttress thread and the bevel-gear drive.(Fig. 8.24 (b) courtesy The Duff-Norton Mfg. Co., Pittsburgh).

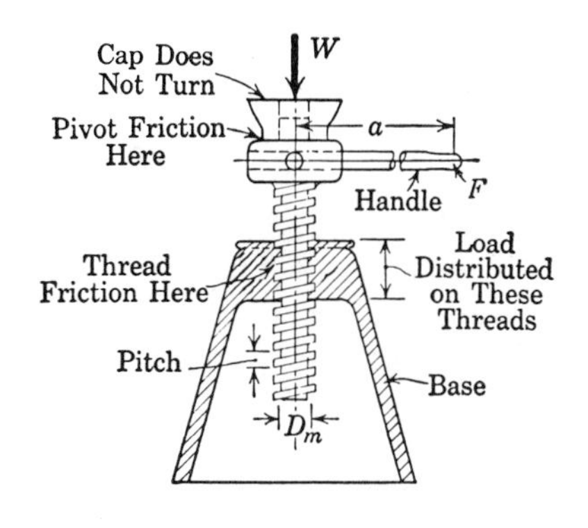

(a)

(b)

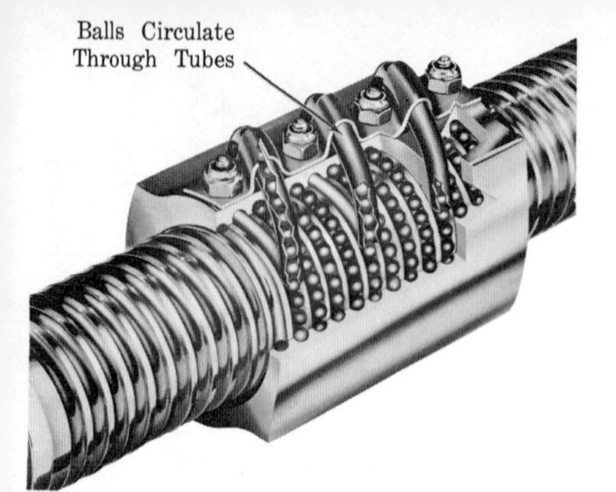

FIGURE 8.25 Ball-Bearing Screw. There is approximate rolling contact between the balls and the grooves in screw and nut. This action is made possible by providing the tubes through which the balls can circulate. (Courtesy General Motors Corp., Saginaw, Mich.).

usually close to unity and the error involved is of a smaller order than that in f. Equation (**h**) applies also to fastening threads where $\phi = 30°$.

If there is friction at a collar or surface other than the threads, the equation for pivot friction may be used, an expression for which is derived in § 18.10, equation (18.2). The total torque to be overcome would be the sum of that at the threads and at the pivot. A ball bearing at the pivot, Fig. 8.24(b), materially reduces the friction. Also where there is a worthwhile advantage, as in automotive steering mechanisms, ball bearings can be used between the nut and screw, Fig. 8.25.

Recall from mechanics[1.6] that the work done by a total torque T_t is $\int T_t \, d\theta$ or $T_t\theta$, when the torque can be considered as constant; θ rad. is the angle through which the screw turns during the application of T_t. Power is $\int T_t \, d\omega$, or $T_t\omega$ for constant torque; hp $= T_t\omega/33{,}000$ for T_t ft-lb. and ω rad./min. Power is consumed at a constant rate if T and ω are constant, otherwise it is an instantaneous value.

8.21 COEFFICIENT OF FRICTION OF POWER SCREWS. If the thread surfaces are smooth and well lubricated, the coefficient of friction may be as low as 0.10, but for materials and workmanship of average quality, Ham and Ryan[8.7] recommend $f = 0.125$. For doubtful workmanship, f may be assumed as 0.15. Increase these values by 30–35% for starting friction.

On the basis of their experiments, Ham and Ryan concluded that the coefficient of friction is practically independent of the axial load; that it undergoes negligible changes due to speed for most speed ranges encountered in practice; that it decreases somewhat with heavier lubricants; that it shows little variation for different combinations of commercial materials, being lowest for steel on bronze; and that the theoretical equations give a good prediction of actual conditions.

8.22 EFFICIENCY OF A SQUARE-THREAD SCREW. Usually, in the design of screws, the aim is to obtain a large mechanical advantage, and since the power transmitted is small, the efficiency is comparatively unimportant. The efficiency of a screw would be 100% if there were no friction. If the friction is zero, f and β are each equal to zero, and equation **(f)** or **(g)** becomes

(i)
$$T' = \frac{WD_m}{2} \tan \lambda,$$

where T' represents the torque (twisting effort needed) to move the load without friction. The efficiency of the screw, which is the effort without friction divided by the effort necessary to turn the screw with friction, is T' in the above equation, divided by T in **(g)**:

(j)
$$e = \frac{T'}{T} = \frac{\tan \lambda \,(1 - f \tan \lambda)}{\tan \lambda + f}.$$

If the efficiency by this equation is plotted against λ with f constant, a curve very similar to that of Fig. 16.6 as given for a worm will be found. Thus an increased efficiency may be obtained by increasing the lead. This variation of efficiency with the lead angle should be kept in mind, and the knowledge applied where practicable. However, increasing the lead decreases the mechanical advantage, and this may be undesirable. Furthermore, a lead angle will be reached where the screw is no longer *self-locking*, a property happily utilized in most screw applications.

8.23 CONDITIONS FOR SELF-LOCKING SCREW.

A self-locking screw is one that requires a positive torque to lower the load, or to loosen the screw if it has been turned tight against a resistance. Figure 8.26 shows the force arrangement on a square thread if the load W is to be lowered (moved down the incline of the thread). Proceeding as we did to get the torque to move the load up the plane, and summing the forces vertically and horizontally, we find

$$\text{(From } \Sigma V) \qquad W = R \cos (\beta - \lambda),$$
$$\text{(From } \Sigma H) \qquad Q = R \sin (\beta - \lambda).$$

Dividing the value of Q by W, solving for Q, and then multiplying each side by $D_m/2$, we get

(k)
$$\frac{QD_m}{2} = T = \frac{WD_m \tan (\beta - \lambda)}{2},$$

which is the turning moment that must be exerted on the screw to *lower* the

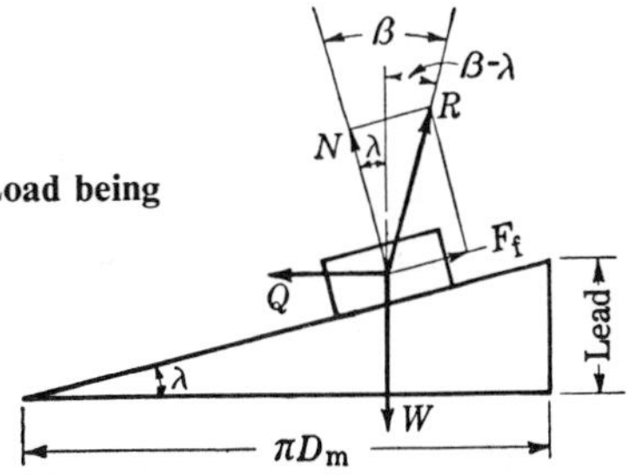

FIGURE 8.26 Forces on a Body Moving Down an Incline. Load being lowered by a square-thread screw.

load. If λ is greater than β, T will be *negative*, which means that *no effort* is required; that is, the load will lower itself. Thus, the condition for self-locking of a square thread is that β must be greater than λ, or that $\tan \beta$ (the coefficient of friction) must be greater than $\tan \lambda$ (the tangent of the lead angle). This same test may be used to approximate the self-locking point for Acme threads, even though the exact equation involves the thread angle; the slope of the thread sides changes the result so little and f varies so much.

Even though a screw is self-locking under static conditions and has a very small lead angle, as in screw fastenings, the load may move down (or a nut loosen) under a vibrating condition.

8.24 DESIGN OF SCREWS. It will be quite conservative to base the design of axially loaded screws on the root area, because the threads provide a definite strengthening. One could use in design some diameter between the minor and major diameters, but if the effort is to get something in the vicinity of the smallest safe size, the design should be substantiated by test. Let the discussion be on the basis of the minor diameter D_r. See the example of § 8.14.

If $L_e/k < 40$, where L_e is the equivalent length, design on the basis of $W = s_c A_r$. If $L_e/k > 40$, use a proper column formula. Be careful of a situation similar to that of an automobile screw jack; for example, if one corner of an automobile is jacked up, the screw would probably not act as a free-end column, but on the other hand, the restraint is not sufficient to classify it as a rounded end.

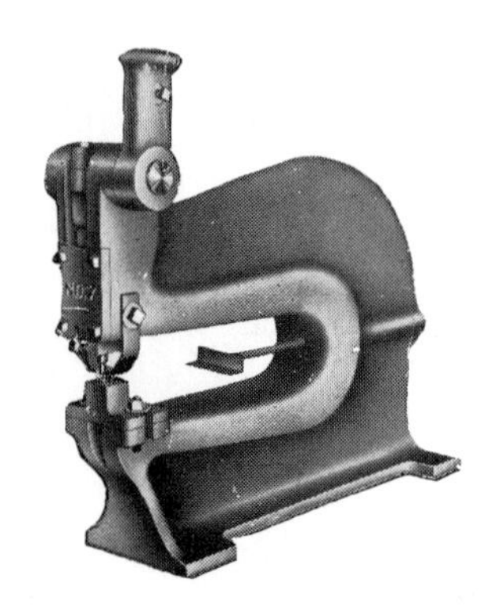

FIGURE 8.27 Hand-lever Punch. (Courtesy Joseph T. Ryerson & Son, Inc., Chicago).

8.25 CURVED BEAMS. The straight-beam formula, $s = Mc/I$, does not apply safely to a curved member subjected to bending, as for example, at the curved section of the frame in Fig. 8.27. In a curved beam, the neutral surface shifts away from the centroidal axis toward the *inside* of the section. In effect, there results a stress concentration on the inside fibers, as suggested

FIGURE 8.28 Photoelastic Study of a Curved Beam. Notice the crowding of lines, indicating a stress concentration effect. (Courtesy T. J. Dolan).

by the crowding of the lines of the photoelastic picture of Fig. 8.28. There are several solutions to this problem; the following is the Winkler–Bach formula.[1.7] The flexural stress at a point in a curved beam is

$$s_f = \frac{M}{Ar}\left[1 + \frac{c}{Z(r + c)}\right], \qquad \text{(8.12)} \quad \text{[BENDING STRESS ONLY]}$$

where M is the bending moment at the section (the moment of the applied force about the centroidal axis) Fe, Fig. 8.29. The bending moment M should be given a *positive* sign when it acts to decrease the radius of curvature, a *negative* sign when it acts to increase this radius.

A is the area of the section in square inches.

r is the distance from the centroidal axis to the center of curvature of the centroidal axis of the unstressed beam.

c is the distance from the centroidal axis to the point at which the stress is desired. It should be given a *positive* sign when measured *away* from the center of curvature, Fig. 8.29, a *negative* sign when measured *toward* the center of curvature.

Z is a property of the section as defined by the relation

$$-ZA = \int \frac{y\,dA}{r + y},$$

y being the variable distance measured in the direction BB from the centroidal axis, Fig. 8.29. See Table AT 18.

Equation (8.12) can be written as

$$s_f = \frac{K_c Mc}{I} \qquad \text{where} \qquad K_c = \frac{\left[1 + \dfrac{c}{Z(r + c)}\right] I}{Arc} \qquad \text{(l)}$$

is a curvature factor related to the radius of curvature r for a particular

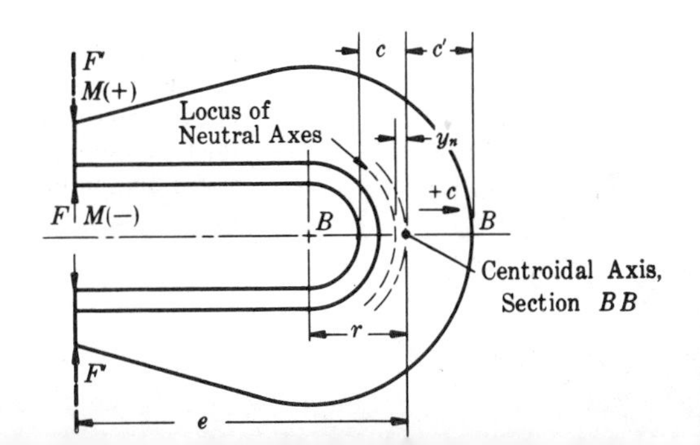

FIGURE 8.29 Curved Beam. The sign of c is negative when substituted in the curved-beam formula, the sign of c', positive. The dotted forces produce a positive bending moment, the solid forces, negative. For the forces F acting, section BB has a uniform tensile stress F/A on it, which must be added to the tensile bending stress given by equation (8.12) in order to obtain the maximum tensile stress.

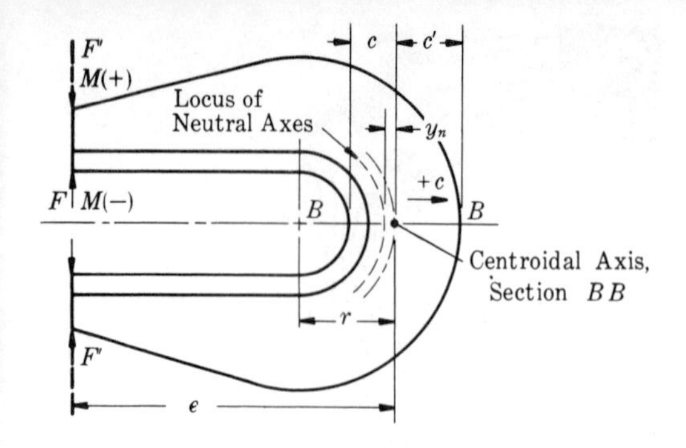

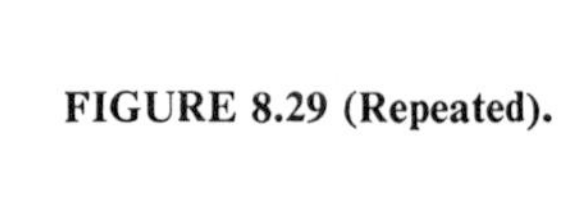
FIGURE 8.29 (Repeated).

point in the section. Some values of K_c are given in Table AT 18 (compare circular-section values with K_c for springs in Fig. AF 15), and also equations for calculating Z for two common sections.

The distance from the centroidal axis to the neutral axis of a curved member in bending only, measured toward the center of curvature, Fig. 8.29, is

(m) $$y_n = \frac{Zr}{Z + 1}.$$

Knowledge of the location of the neutral axis may be important if, for example, a hole must be put into a curved beam subjected to repeated loadings. Their effect as a stress raiser is minimized if the hole is on or near the neutral axis. Holes at some distance from the neutral axis may result in peak stresses higher than those at a smooth external fiber.

If there is a uniform stress F/A on a curved section, the total stress is usually taken as $F/A + K_c Mc/I$, an algebraic sum. The kind of stress obtained from equation (**l**) is easily determined by inspection; but watch the signs for the terms in equation (8.12). Equation (8.12) is somewhat unconservative for curved sections with flanges, as a T or I, because the inevitable distortion of the flange results in stresses somewhat higher than those predicted by (8.12).

8.26 THICK-SHELL CYLINDERS. As pointed out in Chapter 1, the stress in thin-shell cylinders subjected to fluid pressure can be assumed to be uniform. However, when the wall is thick, Fig. 8.30, the tangential stress at the inside surface is much higher than that at the outside surface, and the distribution is no longer approximately uniform. A simple equation, $s_t = p_i r_o / t$, analogous to the thin-shell equation (**u**), § 1.25, except that the outside radius r_o is used instead of r_i, called Barlow's formula, is suitable for an approximation. A thickness too small is dangerous; too large,

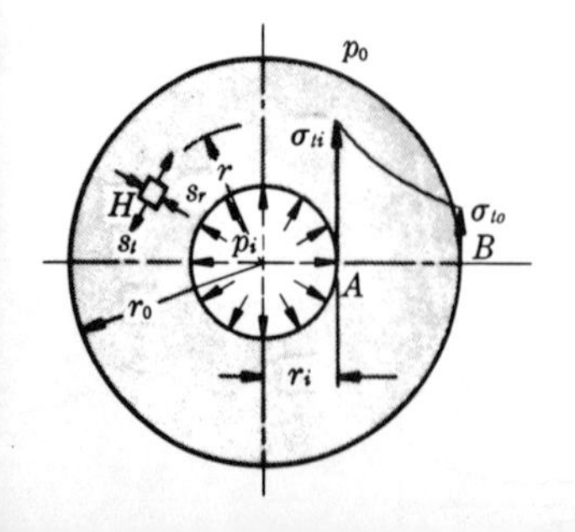

FIGURE 8.30 Stress Distribution in a Thick Cylinder. The manner in which the tangential stress s_t is distributed along a section *AB* for internal pressure only. The maximum stress is much greater than the average. Two of the principal stresses are s_t the tangential stress and s_r the radial stress. At the inside surface, $s_r = p_i$; at the outside surface, $s_r = p_o$.

uneconomical dimensions. A more accurate expression for the tangential stress σ_t, which is a principal stress, is obtained by assuming that the thick shell is composed of a series of thin shells of differential thickness, for each of which the stress is uniform, and that plane transverse sections remain plane (that is, all longitudinal deformations are the same). These assumptions lead to Lamé's formulas[1.7];

$$\sigma_t = \frac{p_i r_i^2 - p_o r_o^2 + r_i^2 r_o^2 (p_i - p_o)/r^2}{r_o^2 - r_i^2} \tag{8.13}$$

where r is the radius to the point where σ_t is desired, as at A, and the other symbols are defined in Fig. 8.30, say, r_o, r_i in., p_o, p_i, σ psi. The radial stress σ_r is given by

$$\sigma_r = \frac{p_i r_i^2 - p_o r_o^2 - r_i^2 r_o^2 (p_i - p_o)/r^2}{r_o^2 - r_i^2}. \tag{8.14}$$

The maximum tangential stress σ_t is seen to occur at the inside where $r = r_i$; let $r = r_i$ and get

$$\sigma_{ti} = \frac{p_i(r_o^2 + r_i^2) - 2p_o r_o^2}{r_o^2 - r_i^2}, \tag{8.15}$$

Let $r = r_o$ and find the tangential stress at the outside as

$$\sigma_{to} = \frac{2p_i r_i^2 - p_o(r_o^2 + r_i^2)}{r_o^2 - r_i^2}. \tag{n}$$

In (8.14), when $r = r_o$, the radial stress $\sigma_r = -p_o$; when $r = r_i$, then $\sigma_r = -p_i$; the maximum numerical value of σ_r is either p_i or p_o, whichever is larger. The radial stress at an internal point is between p_i and p_o. These stresses σ_t and σ_r are principal stresses. For a solid cylinder (shaft), $r_i = 0$.

If the inside pressure p_i is negligible and the external pressure p_o is large (a closed cylinder in deep water), the term with p_i may be safely dropped; σ_t as computed from (8.15) is at the *inside* and algebraic, compressive if negative. Similarly, as in the case of many pressure vessels, if p_o is negligible, as atmospheric pressure when the internal pressure is high, let $p_o = 0$ and equation (8.15) gives the corresponding internal tangential stress σ_t. The third principal stress is the longitudinal stress, equation (**o**) below, which is between the values of σ_t and σ_r. Thus, the maximum shear stress is $(\sigma_t - \sigma_r)/2$, equation (8.3); the maximum shear stress at the inside surface of the cylinder is $(\sigma_t + p_i)/2$, or

$$\tau = \frac{r_o^2(p_i - p_o)}{r_o^2 - r_i^2}, \quad \text{[INSIDE SURFACE]} \tag{8.16}$$

The longitudinal stress σ_l in a closed cylinder is computed on the assumption that it is uniform and that the section in question is not close to the ends; external force on the ends is $p_o \pi r_o^2$, internal force is $p_i \pi r_i^2$; the difference of these forces is resisted by the metal wall, $sA = \sigma_l \pi (r_o^2 - r_i^2)$; thus

$$\text{(o)} \qquad \sigma_l = \frac{p_i r_i^2 - p_o r_o^2}{r_o^2 - r_i^2}. \qquad \text{[CLOSED ENDS]}$$

Equation (8.15) represents the maximum normal stress theory, and it is often used for brittle materials; it is much more conservative for ductile materials than the other stress theories. Equation (8.16) would represent the maximum shear theory as defined when $s_{ys} = s_y/2$, but there are substantiating data for use of the yield stress from a torsion test in the maximum shear equation.[8.16] (The materials of Ref. *(8.15)* had $s_{ys} = 0.55 s_y$, torsion.) The maximum shear stress is the same in closed- and open-end cylinders, for a particular internal pressure, and this stress correlates well with failure conditions, either static or fatigue.[8.16]

Also good correlation is obtained with the octahedral shear stress, which can be obtained from equation (**p**), § 8.12, with the principal stresses as defined above. After considerable algebraic manipulation for internal pressure only,[1.7] the wall thickness by this theory is

$$\text{(8.17)} \qquad t = r_i \left[\left(\frac{1}{1 - \sqrt{3} p_i / s} \right)^{1/2} - 1 \right] \text{ in.,}$$

where s is the circumferential stress, a safe normal stress in design. The cylinder-wall thickness t can be incorporated into the other equations by using $r_o - r = t$.

If the internal pressure is static, or seldom repeated, some vessels might be safe if designed so that the maximum shear stress at the inside does not exceed the shearing yield (or more conservatively, if the tangential stress σ_t does not exceed s_y, but this is not rational design because this theory is not supported by experiment for ductile material). The stress gradient, Fig. 8.30, shows that the stress decreases from the bore outward. If the pressure is repeated a sufficient number of times, fatigue failure becomes the likely method, and in this case, the high tensile stress at the inside becomes significant. To provide a longer life or to reduce the wall thickness for given specifications, autofrettage (§ 4.23) is practiced. If the autofrettage is carried to 100% overstrain, all the material of the wall has been stressed to the yield strength. Assuming plastic action, after the smaller-diameter fibers have reached the yield point, the stress in them remains constant while the stress in the larger-diameter fibers continues to increase to this value as the pressure increases. On this assumption, the internal pressure required for 100% autofrettage [1.7] is

$$\text{(p)} \qquad p_{100} = 2 s_{ys} \ln(r_o/r_i).$$

When the pressure in removed, the residual tangential stress at the inside is a high compression, some 140 ksi in a certain test ($r_o/r_i = 2$) on 4340,[8.16] while the residual tangential stress at the outside is some 90 ksi tension. Thus, the inside fibers will not be stressed in tension until the pressure is more than enough to override the residual compression.* Autofrettage improves the fatigue strength significantly for normal working pressures when the repetitions of load are some 10^4 or more, the improvement being greater the more the number of loading cycles. Because of the stress gradient, the material becomes less efficiently utilized as r_o/r_i increases. Thus, if the r_o/r_i seems excessively large, the use of a stronger material will reduce the ratio, and by using less material, it may be less expensive. From another point of view, the use of autofrettage may permit the use of a less expensive, weaker material. Each situation must be appraised. Indicative of the stress distribution, if $r_o/r_i = 2$ with $p_o = 0$, the inside tangential stress is 2.5 times that at the outside. The difference between these stresses decreases as the wall thickness (and r_o/r_i) decreases, and autofrettage makes possible a more efficient use of the material. But as r_o/r_i approaches one, the benefits of autofrettage approach zero (the material is becoming more nearly uniformly stressed). A jacket shrunk on the working cylinder (§ 8.27) puts the working cylinder in compression, a practice that has been widely used in guns, and provides strength in much the same sense as autofrettage—by inducing compressive stresses in the working cylinder.

8.27 FORCE AND SHRINK FITS. These fits are used for connecting hubs and shafts, sometimes in addition to keys, when an especially rigid connection is desired; see §§ 3.7 and 3.8, and also Fig. 4.20, § 4.27. However, shrink fits become economically advantageous for vessels where, usually, the internal pressure is high, as explained below. Moreover, a steel jacket shrunk onto aluminum, copper, or other metal liner may be a less expensive approach for, say, corrosion resistance.

Consider first the more general case of a hollow cylinder with an internal pressure p_1, with a hoop (or hub) shrunk on, Fig. 8.31 (p. 258); the pressure at the interface is p_i; the external pressure is $p_o = 0$. The equations (8.13)–(8.16) give the stresses as defined for thick cylinders, but p_i must be determined. The pressure p_i at the interface depends on the interference of metal (§ 3.7), the various radii, and the materials of the parts, and it results in an increase of the hoop radius of δ_h in. (diameter increases $2\delta_h$). The radius of the inner member decreases some amount δ_s, its diameter by $2\delta_s$. The interference of metal (§ 3.8) is therefore $i = 2(|\delta_h| + |\delta_s|) = D_s - D_h$, where D_s is the shaft diameter (outside diameter of the inner cylinder) and D_h is the inner hub diameter, both measured in the unstrained state.

Recall from strength of materials that when there are two perpendicular

* In case autofrettage is used, equation (p) may be divided by a factor of safety to get $p_i = 2s_{sd}\ln(r_o/r_i)$ in which p_i is the working pressure and $s_{sd} = s_{ys}/N$ is a design stress. Space is not available for the specialized information needed for this approach.

normal stresses, say s_1 and s_2, the deformation in the direction of s_1 is $\epsilon = s_1/E - \mu s_2/E$, Poisson's ($\mu$) effect considered. If the longitudinal deformation is negligible, frequently assumed even if not strictly true, the unit deformations ϵ_h, ϵ_s in the tangential direction of hoop and shaft are, respectively ($D_s \approx D_h = D_i$ for this purpose),

(q) $$\epsilon_h = \frac{2\delta_h}{D_i} = \frac{\sigma_{th} - \mu_h \sigma_{rh}}{E_h}, \qquad \text{[OUTER]}$$

(r) $$\epsilon_s = \frac{2\delta_s}{D_i} = \frac{\sigma_{ts} - \mu_s \sigma_{rs}}{E_s}, \qquad \text{[INNER]}$$

where μ is Poisson's ratio. The interference of metal is

(s) $$i = 2(|\delta_h| + |\delta_s|) = D_i\left[\frac{\sigma_{th} + \mu_h p_i}{E_h} - \frac{\sigma_{ts} + \mu_s p_i}{E_s}\right],$$

used to compute p_i with a known value of i; where $\sigma_{rh} = \sigma_{rs} = -p_i$ and the stresses are algebraic; E_h and μ_h are for the material of the outer member,

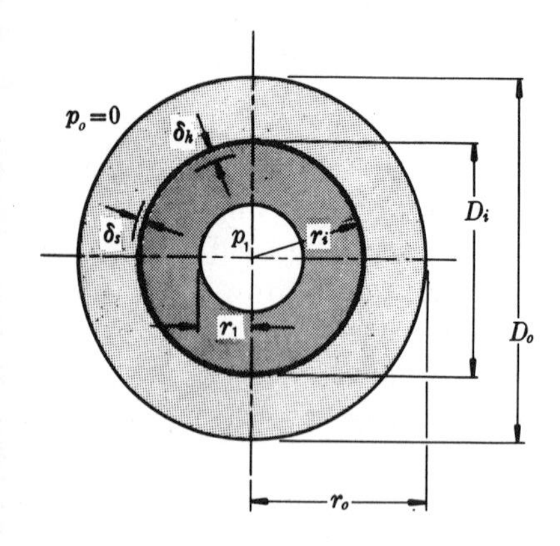

FIGURE 8.31 Hollow Cylinder with Jacket.

E_s, μ_s for the material of the inner member. For the most general case, $\sigma_{th} = \sigma_{ti}$ in equation (8.15), but usually $p_o = 0$; σ_{ts} is σ_{to} obtained from equation (**n**) and is the stress on the inner member at the interface. If the inner member is the liner of a pressure vessel, the internal and external pressures are p_1 and p_i, subscripts defined in Fig. 8.31. In the case of a hub on a shaft, the internal pressure p_1 is likely to be zero; and the most frequent case is a solid shaft ($r_1 = 0$). Adapting equation (s) to a solid shaft and a hub of the *same material*, $E_h = E_s = E$, $\mu_h = \mu_s = \mu$, $r_i/r_o = D_i/D_o$, we get

(t) $$p_i = \frac{Ei}{2D_i}\left[1 - \left(\frac{D_i}{D_o}\right)^2\right], \qquad \text{[SOLID SHAFT]}$$

from which the interface pressure can be computed for a known interference of metal i. Now the tangential stress in the hub is obtained by substituting the value of p_i from (t) into equation (8.15) and letting $p_o = 0$; in equation form, the result is

$$\text{(u)} \qquad \sigma_{th} = \frac{Ei}{2D_i}\left[1 + \left(\frac{D_i}{D_o}\right)^2\right],$$

a tensile stress. If it is desired to have a safe maximum shearing stress in the hub, let $p_o = 0$ in equation (8.16) and substitute p_i from (t);

$$\text{(v)} \qquad \tau = \frac{Ei}{2D_i}, \qquad \text{[SOLID SHAFT OF SAME MATERIAL AS HUB]}$$

Often, the computed stress is allowed to approach closely the yield strength, because the stress decreases away from the bore. The foregoing equations (t), (u), and (v) strictly apply when the shaft is the same length as the hub. The part of the shaft outside of the hub resists the compressive deformation, which resistance results in an increased radial pressure just inside the hub.

If the hub is cast iron and the shaft is steel, the following equations may be obtained by assuming that the modulus of elasticity for steel E is twice that for cast iron E_c, $E = 2E_c = 30 \times 10^6$, and that Poisson's ratio for steel and cast iron are virtually the same[8.18]:

$$\text{(w)} \qquad p_i = \frac{Ei[1 - (D_i/D_o)^2]}{D_i[3 + \mu + (1 - \mu)(D_i/D_o)^2]},$$

$$\sigma_{th} = \frac{Ei[1 + (D_i/D_o)^2]}{D_i[3 + \mu + (1 - \mu)(D_i/D_o)^2]},$$

..ere μ may be taken as $\mu = 0.27$. If the shaft is hollow and if the diameter of the hole does not exceed 25% of the shaft diameter, the error in using the foregoing equations is about 7% or less. However, appropriate allowance for a hollow shaft is made by following the procedure outlined above but not using $r_1 = 0$, Fig. 8.31. Equations for any other combinations of materials can be similarly obtained. See Ref. *(8.20)* for considerable additional detail.

The torque that the fit will transmit and the force needed to make the fit may be estimated after p_i has been calculated. If the length of fit is L and the diameter D_i, the contact area is $\pi D_i L$; the normal force is $p_i(\pi D_i L)$; the frictional force is $fp_i \pi D_i L$; and the moment of the frictional force about the shaft center is $fp_i(\pi D_i L)(D_i/2)$. Thus we find that the torque is

$$\text{(y)} \qquad T = \frac{fp_i \pi D_i^2 L}{2}.$$

The value of $f = 0.1$ is recommended for use here by Baugher.[8.18] For severe duty, it would be advisable to consider $f \approx 0.05$.

The axial force required to press the hub onto the shaft is the product of the area ($\pi D_i L$), the normal pressure p_i, and the coefficient of friction f_1. Expressing the units of this axial force F in tons, we get

$$F = \frac{f_1 p_i \pi D_i L}{2000} \text{ tons.} \tag{z}$$

Since f_1 is quite variable (Baugher[8.18] found variations in f_1 from 0.05 to about 0.3), this formula will give only approximate results.

In practice, the engineer is interested in optimum proportions (especially true in case of a pressure vessel), in which case a decision must be made as to what to optimize. For example, for hub and shaft of the same materials, it may be assumed that the maximum tangential stresses in each part, on the inside of the cylinder in each case, are the same. If the parts are of different materials, these maximum stresses, instead of being made equal, could be, say, in proportion to their yield strengths. Let us assume that the maximum shear stresses are to be made equal for the case of a hollow shaft with no internal pressure ($p_1 = 0$ in Fig. 8.31), with hub and shaft of the same material. Adapting equation (8.16) to the symbols of Fig. 8.31, we get

$$\tau = \underset{\text{[LINER]}}{\frac{r_i^2 p_i}{r_i^2 - r_1^2}} = \underset{\text{[HUB]}}{\frac{r_o^2 p_i}{r_o^2 - r_i^2}},$$

where both shear stresses are taken with the same sign. Solve for r_i and find

$$r_i = \sqrt{r_1 r_o}. \tag{a}$$

These proportions for power transmission through a hub are not necessarily practical. One should not depart from recognized good proportions without knowing that the departure is good engineering. If the system is a pressure vessel, the amount of interference of metal should be optimized. Two different approaches are found in Ref. *(1.7)* and *(8.20)*, too lengthy for detail here. Note that the use of shrink-fit pressure vessels is a means of stressing the material more uniformly, with a resultant saving. Faupel[8.20] gives a curve comparing solid construction with shrink-fit construction; for example, suppose the solid construction requires a wall thickness such that $r_o/r_1 = 3$, corresponding to a rather high internal pressure; for the same pressure, this ratio for shrink-fit construction is about 1.8, indicating a significant saving. Carrying the idea further leads to a laminated construction, a number of relatively thin shrunk-on cylinders, which in the limit of differential thicknesses provides the "most optimum" solution. As the number of laminations are increased, the diminishing returns soon suggest a limit. The greatest percentage improvement comes from the first step, a liner and a single jacket. Do not forget autofrettage, § 8.26.

A hub on a rotating shaft is subjected to centrifugal forces, which should be accounted for when the speed is unusually high.[8.20]

8.28 CLOSURE. This chapter endeavours to relate engineering considerations with some of the more complicated stress states. An understanding of applicable theory should precede the engineering, and for more details on theory consult books on strength of materials. Strength is often affected by anisotropy (§ 2.2) and comprehensive data on this characteristic are seldom available.

It may be helpful to summarize modes of failure: They may occur by: (1) plastic deformations, that ruin the shapes and relations of parts; this is elastic failure; (2) fracture, which is the kind of failure considered for brittle materials and for fatigue of ductile materials; in addition to the external loading, resonant vibrations may result in fracture; (3) creep, which is a subdivision of (1) and is usually associated with high-temperature deformations, although some materials, notably magnesium among engineering materials, may creep excessively at ordinary temperatures; (4) wear, which is due to relative motion of surfaces in contact under pressure; this failure may be totally an abrasive action that removes material or, as in the case of gear teeth, it may be that the contact stresses are so high that a "surface fatigue" occurs; (5) overheating, which is a phenomenon that speeds up failure by other modes; for example, it may result in a lubrication failure with excessive wear, or the overheating may destroy needed mechanical properties, leading to other kinds of failure.

It is appropriate at this time to reread §§ 1.16–1.18 on factor of safety. Observe that, given a particular machine element, one could compute its "factor of safety" by each of the various theories of strength and the answers will be different. Since failure is an engineering phenomenon, it (and the range of the true factor of safety) can be determined only by experiment (past experiment may be used to predict failure); hence it is better to speak of a design factor, rather than a factor of safety, and relate the design factor to the particular way in which it is computed.

9. SHAFT DESIGN

9.1 INTRODUCTION. Although the theory needed for shaft design is covered in Chapter 8, shafts are so ubiquitous that special treatment and more engineering information is worthwhile. We shall also present briefly the former ASME Code on transmission shafting.[9.1]

As the term is usually used, a ***shaft*** is a rotating member transmitting power. Especially as used in the past, an ***axle*** is a stationary member carrying rotating wheels, pulleys, etc. However, common usage favors the word *shaft* whether the member is rotating or not. On the other hand, as a carryover from wagon and buggy days, we speak of the *axle* on an automobile. A ***line shaft***, or main shaft, is one driven by a prime mover; power is taken from it by belts or chains, usually at several points along the shaft. Shafts intermediate between a line shaft and a driven machine are variously called ***countershafts***, ***jackshafts***, or ***headshafts***. Short shafts on machines are often called ***spindles***.

9.2 BENDING FORCES PRODUCED BY BELTS AND CHAINS. The force F in the horsepower equation is the net driving force. In a belt drive, for instance, this net force is $F = F_1 - F_2$, where F_1 is the tension in the belt on the tight side and F_2 is the tension on the slack side (§ 17.2). Since both of these tensions pull on the pulley, the corresponding bending force on the shaft is $F_1 + F_2$. The sum of the tensions is not constant in a drive but depends upon the ratio F_1/F_2, which varies with such factors as the power transmitted, the speed, the initial belt tension (hp = 0); the initial tension to which the belt is subjected depends on other factors in the belt drive (Figs. 17.7, 17.8).

The bending force $F_1 + F_2$ can be computed by assuming F_1/F_2 (see Fig. 17.10) or by assuming C in $F_1 + F_2 = C(F_1 - F_2)$. For a *flat belt*, unless it is significantly overloaded, a value of $C = 2$, which corresponds to $F_1/F_2 = 3$, is reasonable when the belt is mounted with a proper initial tension (see Chapter 17 for more information on belts). If there is reason to be conservative, use a value of C of 2.5 or 3 or more. In the absence of other information, assume the *bending force* of a flat belt to be

$$\text{(a)} \qquad F_1 + F_2 = 2(F_1 - F_2),$$

where $F_1 - F_2 = F$ is usually obtained from the horsepower equation (1.15). As the belt stretches in service, $F_1 + F_2$ decreases for a given power and speed.

The bending force due to V-belts may be taken as $1.5(F_1 - F_2)$; or say $2(F_1 - F_2)$ if there is cause for concern. The bending force exerted by chains and also by gears, is usually taken as the net driving force F; that is, the slack-side tension F_2 in chain drives is taken as zero.

9.3 DESIGN OF SHAFTS FOR STRENGTH. If a shaft carries several gears or pulleys, different sections of the shaft will be subjected to different torques, because the total power delivered to the shaft is taken off piecemeal at various points. Hence one must note the amount of torque on each part of the shaft. Then study the distribution of the bending moment, preferably sketching (freehand is all right) the shear and bending moment diagrams.

From this preliminary examination, which is a problem in mechanics, we note the section where the bending moment is a maximum and the section where the torque is a maximum. If these maximums occur at the same section, the diameter needed for that section is determined—and used for the entire shaft when the diameter is to be constant. If the maximums do not occur at the same section, determine the diameter for the section of maximum torque and also for the section of maximum bending moment, and use the larger value.

The diameter of a shaft is often varied from point to point, sometimes for structural reasons. In this case, check the stress or determine the size needed for each section. The designer makes certain that all sections of the shaft are subjected to safe stresses, taking due note of fillets, holes, keyways, and other stress raisers. An example will serve as a reminder of the mechanics involved.

9.4 EXAMPLE. A 24-in. pulley B receives 30 hp at 360 rpm from below at an angle of 45°, as shown in Fig. 9.1. An 18-in. gear C delivers 40% of the power

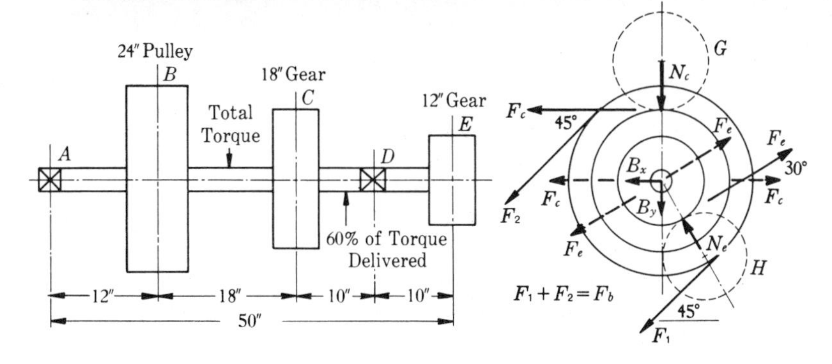

FIGURE 9.1 The gears *G* and *H* "push back" on *C* and *E*, respectively.

horizontally to the right. A 12-in. gear E delivers the remaining power downward toward the left at an angle of 30° below the horizontal. Both gears have 20° involute teeth. The shaft is to be machined from annealed C1137, with profile keyways for each gear and pulley. The load is smooth. (a) Find the diameter of the shaft for $N = 1.8$ when the Soderberg criterion is used for the equivalent stresses. (b) Let there be a step-down in the size of the shaft at the bearing D and compute the diameter from D to E.

Solution. The torques transmitted by the pulley and gears are

$$T_b = \frac{63{,}000 \text{ hp}}{n} = \frac{(63{,}000)(30)}{360} = 5250 \text{ in-lb. (on shaft between } B \text{ and } C).$$

$$T_c = \frac{(63{,}000)(12)}{360} = 2100 \text{ in-lb. delivered.}$$

$$T_e = \frac{(63{,}000)(18)}{360} = 3150 \text{ in-lb. (on shaft between } C \text{ and } E).$$

FIGURE 9.2 Forces on Gear Tooth.

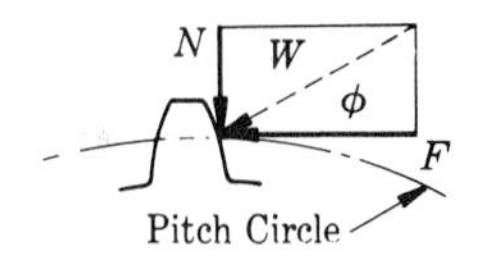

Let the bending force produced by the belt be (r_b = radius of B)

$$F_b = 2(F_1 - F_2) = \frac{2T_b}{r_b} = \frac{(2)(5250)}{12} = 875 \text{ lb.}$$

For the gears, the driving forces are computed as though the contact is always on the pitch circle (r_e = pitch radius of E; etc.)

$$F_e = \frac{T_e}{r_e} = \frac{3150}{6} = 525 \text{ lb.} \qquad \text{and} \qquad F_c = \frac{T_c}{r_c} = \frac{2100}{9} = 233 \text{ lb.}$$

From your study of the kinematics of gearing, recall that the total load W on the gear tooth (ignoring the frictional force) is normal to the tooth surface, with the result that there is a separating force, N, Fig. 9.2, which is seen to be $F \tan \phi$, where F is the computed driving force. (See § 13.9). For $\phi = 20°$, the separating forces for C and E are

$$N_c = F_c \tan 20 = (233)(0.364) = 84.8 \text{ lb.}; \quad N_e = (525)(0.364) = 191 \text{ lb.}$$

If C delivers power to the right, the force F_c on C points leftward, as shown in the end view of Fig. 9.1. Similarly, with *E delivering* power as stated, the force F_e is upward toward the right. By analytic mechanics, those forces acting at some distance from the center of the shaft are replaced by a force through the shaft axis and a couple. Thus, add and subtract forces F_e through the shaft axis as indicated and note that there will now be a CC torsional couple $F_e r_e$, where r_e is the pitch radius of gear E, and a bending force F_e through the center of the shaft parallel to the original F_e. This is the justification of the free bodies to be used later.

At this stage, the easiest way to design a shaft subjected to a three-dimensional force system is to resolve the forces into two perpendicular coplanar systems. The horizontal forces at B, C, and E are

$$B_x = F_b \cos 45 = (875)(0.707) = 619 \text{ lb.}$$
$$C_x = F_c = 233 \text{ lb.}$$
$$E_x = F_e \cos 30 - N_e \cos 60 = (525)(0.866) - (191)(0.5) = 359.1 \text{ lb.}$$

Looking at the end view of Fig. 9.1, imagine the horizontal plane rotated 90° CC, and sketch the forces in this plane as shown in Fig. 9.3. To find the bearing reactions at A and D, take moments about D and about A, and check the results by $\Sigma F_x = 0$. About A,

$$\Sigma M_A = (12)(619) + (30)(233) - (50)(359) + 40D_x = 0,$$

or $D_x = +88.3$ lb. The positive sign for D_x tells that its sense as shown is correct; or, since the clockwise was taken as positive, the positive sign for D_x indicates

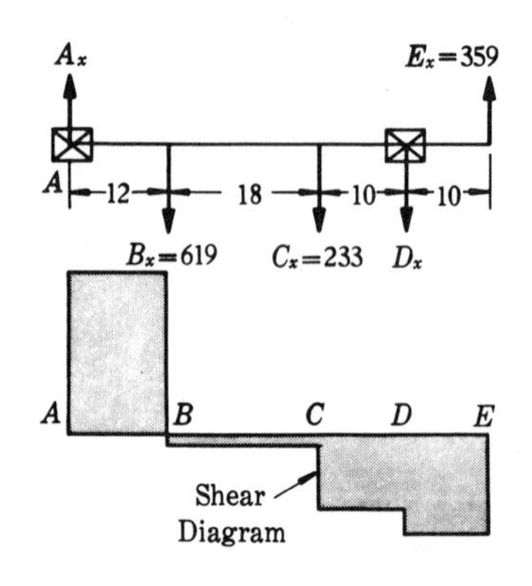

FIGURE 9.3 Section of Maximum Bending Moment, Horizontal Plane.

that D_x should act in such a way as to produce a clockwise moment about A. Summing moments about the right bearing, we have

$$\Sigma M_D = 40A_x - (28)(619) - (10)(233) - (10)(359) = 0$$

from which $A_x = +581.3$ lb. Again A_x was chosen in the correct sense. Sum the forces as a check, $\Sigma F_x = 0$.

Since we are interested in locating the point of maximum stress, we are definitely interested in the sections of maximum bending moment, in each plane.

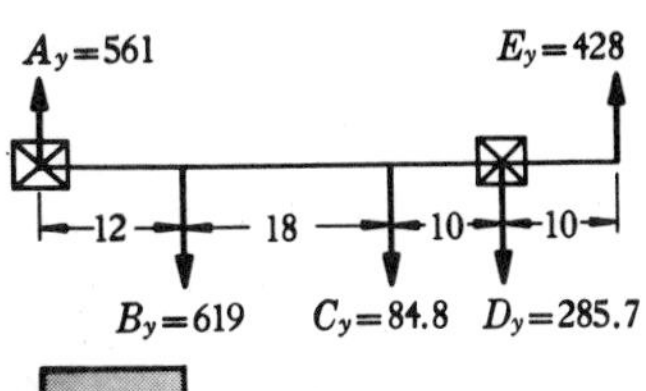

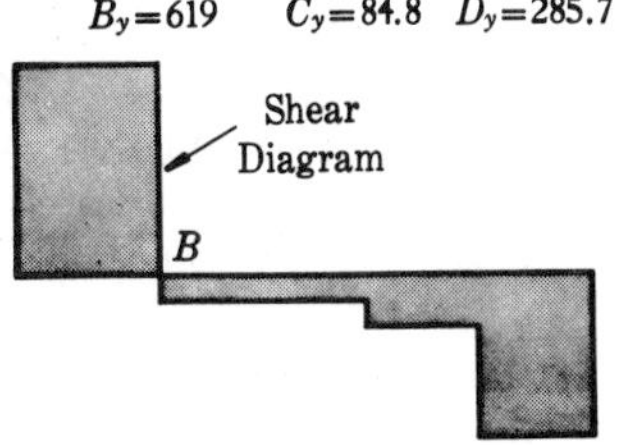

FIGURE 9.4 Section of Maximum Bending Moment, Vertical Plane.

Sketching the shear diagram, Fig. 9.3, note that it crosses at *B*, which is therefore the section of maximum bending moment in the horizontal plane. The sum of the moments to the left of the section gives

$$M_{bx} = (12)(581) = 6972 \text{ in-lb.}$$

The forces in the vertical plane are indicated in Fig. 9.4; the *y* components of the force system of Fig. 9.1 are

$$B_y = F_b \cos 45 = (875)(0.707) = 619 \text{ lb.}$$
$$C_y = N_c = 84.8 \text{ lb.}$$
$$E_y = F_e \sin 30 + N_e \cos 30 = (525)(0.5) + (191)(0.866) = 427.9 \text{ lb.}$$

The values of the bearing reactions A_y and D_y are determined by moments as before, from a free body of the *y* components, Fig. 9.4, and checked by $\Sigma F_y = 0$. The results are as shown. Since the shear diagram crosses zero at section *B*, the maximum moment in the vertical plane is also at *B*, of value

$$M_{by} = (12)(561) = 6732 \text{ in-lb.}$$

(*Note*. The horizontal and vertical shear diagrams do not always "cross" at the same section. Sometimes, one or both cross at more than one section. If the shaft is to be made a constant diameter, each section where the shear diagram crosses zero must be considered to make sure that the size of shaft obtained is safe at all sections. The maximum moment will be where at least one diagram crosses.) Since both shear diagrams cross zero at *B*, this is the section of maximum bending moment, which is the vector sum of the components computed above;

$$M_B = (M_{bx}^2 + M_{by}^2)^{1/2} = (6972^2 + 6732^2)^{1/2} = 9680 \text{ in-lb.}$$

or 9.68 in-kips. We now notice that the maximum bending moment and the maximum torque both occur at *B*, which is therefore the significant section. (NOTE. If the maximum bending moment occurs at a section that is *not* subjected to the maximum torque, some section of maximum torque must be investigated as well as the section of maximum bending moment.)

(a) Designing by variable-stress principles (Chapters 4 and 8), we probably need to include a size factor, say 0.85; there is a keyway at section *B*, for which $K_f = 1.6$ and $K_{fs} = 1.3$ from Table AT 13 (for annealed steel); mechanical properties for AISI 1137 from Table AT 8 are

$$s_u = 85 \text{ ksi}, \quad s_y = 50 \text{ ksi}; \quad \text{use } s_n' = 85/2 = 42.5 \text{ ksi.}$$

Assume $s_{ns} = (0.6)(42.5) = 25.5$ ksi and $s_{ys} = (0.6)(50) = 30$ ksi. (Observe that for these values of s_{ys} and s_{ns} in equation (8.11), the results accord closely but not exactly with the octahedral stress theory.) Since the flexural stress varies through a complete cycle, we need to find the equivalent stress to use in, say, equation (8.11); $s_m = 0$; therefore, $s_a = Mc/I = 9.68/(\pi D^3/32)$ ksi; and from equation (w), § 8.15, we get

$$s_e = \frac{s_n}{s_y} s_m + K_f s_a = 0 + \frac{(1.6)(9.68)(32)}{(0.85)\ (\pi D^3)}.$$

Assume a steady torque; $s_{as} = 0$, and $s_{ms} = Tc/J = (5.250)/(\pi D^3/16)$ ksi; from equation (x), § 8.15, we find

$$s_{es} = \frac{s_{ns}}{s_{ys}} s_{ms} + K_{fs} s_{as} = \frac{(25.5)(5.25)(16)}{(30)(\pi D^3)} + 0.$$

These equivalent stresses are substituted into equation (8.11); with size factor,

$$\frac{1}{N} = \left[\left(\frac{s_e}{s_n}\right)^2 + \left(\frac{s_{es}}{s_{ns}}\right)^2\right]^{1/2}$$

$$= \left[\left(\frac{1.6 \times 9.68 \times 32}{0.85 \times 42.5 \times \pi D^3}\right)^2 + \left(\frac{25.5 \times 5.25 \times 16}{25.5 \times 30\pi D^3}\right)^2\right]^{1/2}$$

$$\frac{1}{1.8} = \frac{16}{\pi D^3}(0.735 + 0.0306)^{1/2},$$

from which $D \approx 2$; use $D = 2$ in. This size shaft should be satisfactory for *strength*, but, especially with meshing gears, it should be determined that the deflections are appropriate (§ 9.11).

(b) Suppose the bearing D is 1 in. wide (axial dimension); then, certainly as a first approximation, the moment at the center of the bearing may be used for design, even though the discontinuity occurs at the left side of D, where there is a fillet whose radius must be assumed; say $r = \frac{1}{16}$ in. Checking in Fig. AF 12, we conclude that an iterative process must be used because r/d and D/d must be known in order to find K_t. Assume $K_f = 2$. The resultant force at E is $F_e = (359^2 + 428^2)^{1/2} = 560$ lb.; the moment at D is $10F_e = 5.6$ in-kips. The torque on the shaft at D is $T_d = 3.15$ in-kips, computed above. As before

$$s_e = K_f s_a = \frac{(2)(5.6)(32)}{(0.85)\ (\pi d^3)}, \qquad s_{es} = \frac{s_{ns}}{s_{ys}} s_{ms} = \frac{(25.5)(3.15)(16)}{(30)\pi d^3}.$$

Substituting into equation (8.11), including the size factor 0.85, we have

$$\frac{1}{N} = \frac{1}{1.8} = \frac{16}{\pi d^3}\left[\left(\frac{2 \times 5.6 \times 2}{0.85 \times 42.5}\right)^2 + \left(\frac{25.5 \times 3.15}{30 \times 25.5}\right)^2\right]^{1/2},$$

from which $d = 1.54$; use $d = 1\frac{1}{2}$ in. (except that if a rolling bearing is to be used, the size would be adjusted to an available bore). In preparation for the next iteration, check the value of K_f assumed above. For $r/d = 0.0625/1.5 = 0.0417$ and $D/d = 2/1.5 = 1.33$, we find, from Fig. AF 12, $K_t = 2.15$. From Fig. AF 7, get $q = 0.86$; thence

$$K_f = 1 + q(K_t - 1) = 1 + (0.86)(1.15) = 1.99;$$

since this is close to the assumed $K_f = 2$, the dimension d found should be safe for strength.

9.5 SHAFT SIZES AND MATERIALS. ***Transmission shafting***, used to transmit power from a prime mover to a machine, comes in the following sizes: [9.5]

$\frac{15}{16}$, $1\frac{3}{16}$, $1\frac{7}{16}$, $1\frac{11}{16}$, $1\frac{15}{16}$, $2\frac{3}{16}$, $2\frac{7}{16}$, $2\frac{15}{16}$, $3\frac{7}{16}$,
$3\frac{15}{16}$, $4\frac{7}{16}$, $4\frac{15}{16}$, $5\frac{7}{16}$, $5\frac{15}{16}$, $6\frac{1}{2}$, 7, $7\frac{1}{2}$, 8.

Commercial bearings and couplings are readily obtainable for most of these sizes. Stock lengths of transmission shafting include 16, 20, or 24 ft.; lengths over 24 ft. can be obtained on order. Typical sizes of ***machinery shafting***, shafts that are an integral part of a machine, are (with tolerances for finished shafting):

By $\frac{1}{16}$-in. increments in this range: $\frac{1}{2}$ to 1 in. with tolerance of -0.002 in.; $1\frac{1}{16}$ to 2 in. with tolerance of -0.003 in.; $2\frac{1}{16}$ to $2\frac{1}{2}$ in. with tolerance of -0.004 in.
By $\frac{1}{8}$-in. increments, $2\frac{5}{8}$ to 4 in. with tolerance of -0.004 in.
By $\frac{1}{4}$-in. increments, $4\frac{1}{4}$ to 6 in. with tolerance of -0.005 in.
By $\frac{1}{4}$-in. increments, $6\frac{1}{4}$ to 8 in. with tolerance of -0.006 in.

Shafts are made in several different ways and of a wide variety of materials. The shafting referred to above is likely to be cold-drawn carbon steel in sizes smaller than $3\frac{3}{4}$ in. Shafts are also cold finished by turning and polishing, sometimes with a grinding operation also included. See §§ 2.9, 4.23, 4.30 for effects of cold working. If the finish meets the requirements, cold-drawn shafting may be the lowest in cost because of its superior mechanical properties. However, because of the residual stresses from cold drawing, the shafts will warp when a keyway is cut, necessitating straightening. Straightening shafts and axles is usually done by cold-bending them opposite to their curvature until there is plastic yielding, which invariably leaves a residual tensile stress somewhere in an outer fiber. A reduction of endurance strength results, which may be aggravated if a tensile residual exists in a fillet or other stress raiser. For example, unstraightened automotive axles had $s_n = 20$ ksi; straightened in normal production, $s_n =$ 13–16 ksi; straightened and peened, $s_n = 43$ ksi.[4.10] Instead of bending the shaft to straightness, it may be straightened by peening with a hammer selectively at the proper points; in one case, this plan left a shaft with unchanged s_n.

Larger shafts are turned and/or polished and ground from hot-rolled material; the turning removes some decarburized steel. Sizes over 5 or 6 in. are usually forged and turned to size. Crankshafts, as for automotive engines, may be forged or cast. They are cast from steel, high grades of cast iron, and nodular iron.

The most commonly stocked material for shafting would be carbon steel with some 30 to 40 points of carbon, perhaps resulfurized (11XX series) for finishing in automatic screw machines. But all kinds of materials, including nonferrous metals and nonmetallics, are used for a reason.

Materials for automotive crankshafts and axles include 1345, 8637, 8650, 3140, 4135, 4150, 5145, 4340. Higher carbon contents are used in the larger sizes for greater hardenability. Heat treatment may be to BHN of 229-269 for crankshafts, to BHN of 300-444 for axles, the higher BHN being used for the larger sizes ($2\frac{1}{2}$ in.). When localized surface hardening is desired, as by flame and induction hardening, low hardenability may be preferable. In general (in all design), no more carbon is used than is needed to obtain the required properties. Sometimes, a reduction in carbon may save trouble with quench cracking. In one instance, a carbon-steel, $2\frac{3}{4}$-in. shaft with a keyway had to be induction hardened at the keyway to obtain sufficient endurance strength. All these shafts with 0.43% C cracked; when the carbon was reduced to 0.37%, there was no further trouble with cracking, and the part had sufficient fatigue strength. If parts of the shaft are to be carburized, lower carbon, as 0.15–0.20%, is used. The possibility of fretting corrosion if forced fits are used on the shaft, the effect of any kind of corrosion, of oil holes, fillets, etc., on the fatigue strength should not be overlooked (§§ 4.21, 4.26, 4.27).

9.6 HOLLOW ROUND AND SQUARE SHAFTS. Hollow round shafts sometimes serve a useful purpose, usually in large sizes, though they are more expensive than solid ones. They have the advantages of being stronger and stiffer, weight for weight, because the outer fibers are more effective in resisting the applied moments, and they respond better to heat treatment because quenching can proceed outward as well as inward. The theory already presented applies exactly to hollow rounds, whose section moduli are given in Table AT 1. The designer commonly assumes a relation between the hole size and the outside diameter. See § 9.13.

So-called square shafts are occasionally needed, but the edges should be, and usually are, generously beveled in order to remove sharp edges which may be the source of failure. The values of Z and Z', from Table AT 1, used for beveled square shafts should give conservative answers. The simple torsion formula is not strictly correct for noncircular members, and the interested reader should look further into the theory for these cases.[1.2,1.7]

9.7 VERTICAL SHEAR. At this time, review § 1.10(b) on vertical shear in a beam. As pointed out there, the shear due to bending is a maximum at the neutral plane where the normal stress is zero, and the magnitude of this shear is relatively small compared with other stresses, except in short beams (a cantilever crank pin may qualify). Its *maximum* value for a solid round member can be computed from equation (1.6), which gives the correct value at the neutral plane;

(b) $$s_s = \frac{16V}{3\pi D^2} = \frac{4V}{3A},$$ [SOLID ROUND ONLY]

where V lb. is the vertical shear at the section under consideration. In a uniformly loaded cantilever beam, the vertical shear may govern when the length is less than about half the diameter. However, if the shaft is subjected to torsion as well as bending, the vertical shear combines with the torsion, say at the neutral plane, so that the consequences may be significant in a somewhat longer shaft, and especially for hollow shafts approaching the tubular.[9.14] It is also true that the deflection (§ 9.9) due to shear becomes significant in short beams (see texts on mechanics of materials), but for most shafts this is a negligible quantity.

9.8 TORSIONAL DEFLECTION. Deflection is another and frequently significant consideration in the design of shafts. Criteria for the limiting torsional deflection vary from 0.08° per foot of length for *machinery shafts* [9.6] to 1° per foot [0.5] or 1° in a length of 20 diameters for transmission shafting. Even short shafts become special problems in rigidity when the load is applied in *impulses*, as on an automobile crankshaft. The impulses produce a torsional vibration, usually compensated by torsional-vibration dampers in an automotive engine. The torsional deflection of a round shaft is given by equation (1.13), $\theta = TL/GJ$ radians, where G psi is the modulus of elasticity in shear, J in.^{4} is the polar moment of inertia of the section, and L in. is the distance from the section where the torque T in-lb. is applied to the section of the resisting torque. Review also § 1.13 now.

9.9 TRANSVERSE DEFLECTIONS. Books on mechanics of materials explain various methods of finding deflections of beams. The basic mathematical approach is to set up an equation for the loading (or bending moment or M/EI) at any section, in accordance with equation (1.10), p. 13, and then make successive integrations, equations (1.9), (1.8), and (1.7) until the deflection y is obtained (Table AT 2).* For one load, or two, in addition to bearing reactions, the method is not too tedious.

* An algebraic method (as opposed to graphical), variously called pointed brackets, curly brackets, singularity functions, that is shorter than any other algebraic approach, is explained in Crandall and Dahl.[1.18]

Another approach is to use the principle of superposition; that is, the deflection at a certain section of a shaft caused by all the loads, $F_1, F_2, \ldots,$ is equal to the *vector sum* at that section of the deflections caused by each of the loads acting alone. After one has learned the method, a graphical integration procedure, described below, has certain advantages whenever there are many loads and changes in the diameter of the shaft.

Data on permissible values of deflections are rare, probably because the range of values would be large and each situation has its own peculiarities. An old rule of thumb *for transmission shafting* is that the deflection should not exceed 0.01 in. per foot of length between supports; although greater stiffness may be desired. Preferably, on transmission shafts, the pulleys and gears should be located close to bearings in order to minimize moments. If journal bearings with "thick film" lubrication (Chapter 11) are used, the deflection across the bearing width should be only a small fraction of the oil-film thickness; if the slope is excessive here, there will be "binding" in the journal. A self-aligning bearing, Fig. 11.13, may eliminate this trouble if the deflection is otherwise acceptable.

On machine tools (lathes, milling machines, etc.), rigidity is a special concern because of its relation to accuracy. If a shaft supports a gear, deflection is more of a consideration than if it carries a V-belt pulley. In general, for machinery shafts, the permissible deflection may be closer to 0.002 in./ft. (rather than 0.01 for transmission shafts). Gleason[15.1] states that at bevel gears, sizes 5 in. to 15 in., the gears "should not lift or depress more than 0.003 in." At a section of a spur-gear mesh (good quality gears), the shafts should have a deflection *relative to each other* of less than 0.005 in. More important perhaps for spur gears is the *relative slope* of the shafts at the mesh; Brown & Sharpe recommends that this slope be limited to 0.0005 in./in.

9.10 GRAPHICAL INTEGRATION. Given a curve defined by $u = f(x)$; then, as seen in Fig. 9.5, a vertical strip of height u and width dx is dA and the entire area under the curve (or curves) is $\int dA = \int u\, dx$, integrated through as many functions as required. By analogy, we have equation (1.8), $d\theta = (M/EI)\, dx$, the integration of which gives θ. Hence, if

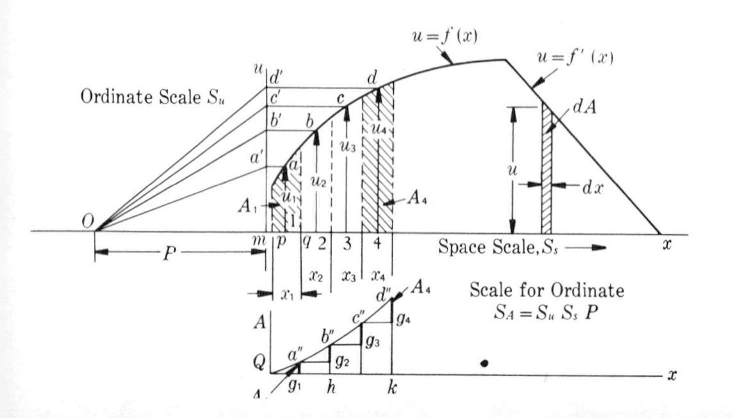

FIGURE 9.5 Graphical Integration.

an M/EI diagram is laid out for a particular beam or shaft (or a moment diagram if I is constant), the $\int(M/EI)\,dx$ between certain limits is the change of slope of the beam between the same limits. Since $\theta = dy/dx$, the next integration $\int\theta\,dx$ gives deflection y.

One method of graphical integration is illustrated in Fig. 9.5.* The curve $abc \cdots$ is known. The area under it is divided into smaller areas A_1, A_2, etc., usually with $x_1 = x_2 = x_3$, etc. for convenience only, so that a straight line is a good approximation of the curve at the top of the areas. Points a, b, c, etc. are then halfway across the areas A_1, A_2, etc. Draw any convenient vertical reference line mu; draw aa', bb', etc.; choose any convenient pole O at a distance P from the reference line mu; draw Oa', Ob', etc.; choose an origin Q for the integration curve; draw Qa'' parallel to Oa', $a''b''$ parallel to Ob', $b''c''$ parallel to Oc', etc., giving the intercepts a'', b'', c'', . . . on the ordinates of the Ax diagram; then g_1a'' represents to scale area A_1, g_2b'' represents A_2, etc. For proof, consider the similar triangles Oma' and Qg_1a''; $g_1a''/Qg_1 = ma'/P$ or

$$\text{(c)} \qquad g_1a'' = \frac{(ma')(Qg_1)}{P} = \frac{(1a)(pq)}{P} = \frac{A_1}{P},$$

where it is seen that $(1a)(pq)$ is equal to A_1 (approximately in this case). Similarly, for triangles Omd' and $c''g_4d''$, note that $g_4d'' = A_4/P$. The conclusion is that the distances g_1a'', $\cdots g_4d''$ represent to scale the areas $A_1, \cdots, A_4$, etc. and that therefore the curve $Qa''b''d''$ represents the integral of the function. The ordinate kd'' represents the total area up to $x = Qk$.

Since the area A_1 in its actual units is the integral at point q (and g_1), the scale to which it is represented by g_1a is obtained by multiplying both sides of (c) by the u scale S_u and the space scale S_s (that is, the x scale); thus

$$\text{(d)} \qquad (g_1a'')(P)(S_u)(S_s) = (1a)(S_u)(pq)(S_s) = A_1.$$

Therefore, the scale of the ordinate of the integral curve is given by

$$\text{(e)} \qquad \text{(Pole distance)(ordinate scale)(abscissa scale)} = PS_uS_s.$$

Reference *(9.2)* gives an analogous numerical procedure for deflections, and numerical help is found in Ref. *(9.3)*.

9.11 EXAMPLE—DEFLECTION OF SHAFTS. For the loading defined in the example of § 9.4, determine the diameter of shaft for a maximum deflection

* The force-and-funicular polygon procedure of making a graphical double integration, described in previous editions of this text, is probably somewhat more accurate when the same scales are used. But the method described in this edition has the advantage of giving the slope, which may be important. Any one frequently concerned with deflections may wish to look into this other method. It can be used to get slopes by starting with the vertical shear diagram.

of 0.003 in. For pedagogical purposes, assume that gear C has a forced fit with the shaft and that the shaft size is to be reduced about 10% at bearing D.

Solution. It is recommended that hubs making forced fits with a shaft be considered as integral with the shaft for the purpose of determining deflections.[9.7] Thus the configuration to be dealt with is shown at the top of Fig. 9.6, laid out to scale axially but not diametrally. Let the hub diameter (§ 13.31) be $1.8D$, Fig. 9.6. One could use graphical integration with the shear diagram, but it will be more accurate and as easy to compute the points on the M/EI diagram (M diagram if I is constant). For the diameter D, $I = \pi D^4/64 = 0.0491 D^4$; at the hub, $I_C = \pi(1.8D)^4/64 = 0.515D^4$; along DE, $I_E = \pi(0.9D)^4/64 = 0.0322D^4$. Now compute the moment at each load and at each change of section size, tabulating the results (below) for ease of use. The results will be little affected if the size change is assumed to occur at the center line of bearing D. A similar approximation would be appropriate at C, but we shall do this one straight for the instructional value. The results of the computations are (E in ksi; let $D = 1$):

	A	B	G_1	G_2	C	H_1	H_2	D_1	D_2
M (in-kip) . . .	0	6.972	6.36	6.36	6.3	5.75	5.75	3.59	3.59
$(M/EI)D^4\ 10^4$.	0	47.2	43.2	4.12	4.07	3.72	39.0	24.4	37.2

Layout the second line to scale, Fig. 9.6, and get the M/EI diagram. If there had been negative moments (which tend to curve the beam concave downward), they would have been plotted below the axis. If D is unknown, as in this example, think of it as unity until the deflection y curve is obtained, then reinstate it (E and I would be handled similarly if the moment M diagram were used). The scale of the M/EI diagram is chosen as 60×10^{-4} in.$^{-1}$. The integration of this curve is carried out in accordance with § 9.10; the procedure is outlined in the caption to Fig. 9.6. The pole O_2 for integrating the slope diagram could be chosen on the zero axis Q (which is not the axis of zero slope) as for the M/EI diagram. Since the area under the θ curve is continuously increasing, the y diagram would then be cocked high on the right. By choosing O_2 randomly as shown, the deflection curve stays closer to a horizontal. Remember what the ordinates to the θ curve mean; they represent the change of slope from the starting point and the change of θ ordinates between stations represents the change of slope between stations, Fig. 9.5. Except by some coincidence, the shaft (beam) will have some slope at the left bearing (support). The loading is such here that the curvature of the shaft never reverses (it probably would if force E_x were reversed).

Having chosen O_2, the integration procedure is completed to obtain the y diagram. The bearings are assumed to be rigid and fixed in space, so that if their positions are spotted on the y curve, a straight line through them is the base from which deflections are measured (vertically—always perpendicular to axis of original beam). Thus the bearings A and D are located at m and n, and a straight line mn through these points is the line of zero deflection (mk is the horizontal line). Since the rays from O_2 have the slopes of the sides of the deflection polygon, draw $m'n'$ from O_2 parallel to mn; this locates an axis $n'x$ of zero slope; where it intersects the θ curve at p, drop a vertical line downward to measure y_{1x}, the maximum deflection between the bearings (which occurs where a line tangent to the elastic curve of the beam is parallel to the line through the points of support). The true elastic curve is of course a smooth curve and would fall inside of the

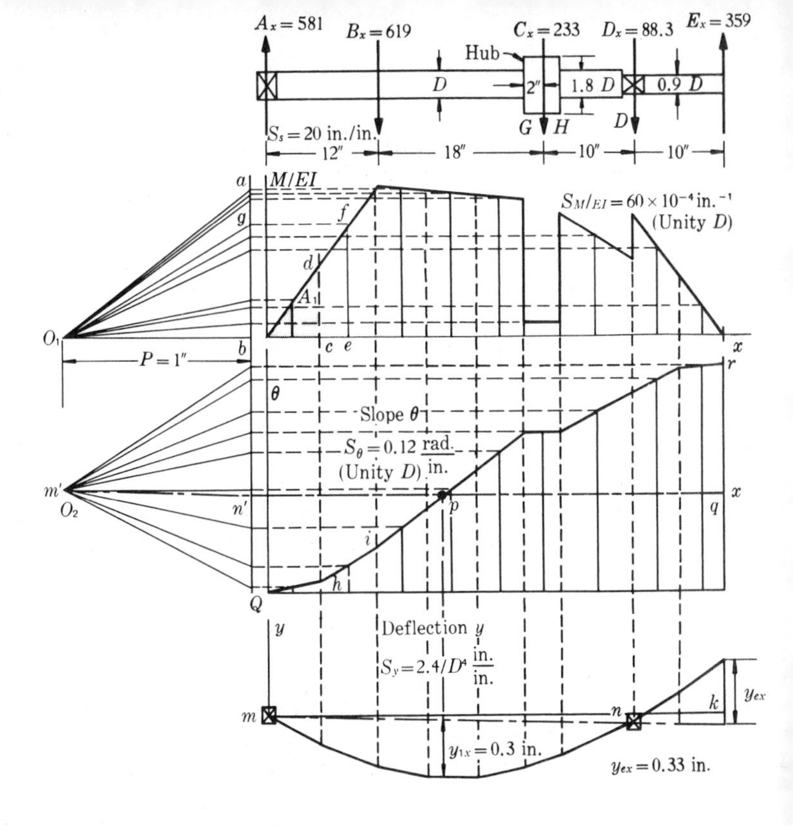

FIGURE 9.6 Slope and Deflection, Horizontal Plane. Having laid out the M/EI diagram to a convenient scale, choose pole O_1 at a convenient pole distance P from some vertical reference line *ab*. Divide the areas under the M/EI diagram into convenient parts; the boundaries here are dotted lines, as *cd*. Draw the mean ordinate of each area, as *ef*. Project the various mean ordinates to the reference line *ab*, as *fg*. Choose a convenient starting point Q for the slope curve and draw lines parallel to the rays $O_1b, \ldots O_1g, \ldots O_1a$, as *hi* is drawn parallel to O_1g. After the slope diagram is completed, choose a convenient pole O_2 for the θ diagram, make the graphical integration to obtain the y curve. Spot the location of the bearings and mark the points, m, n. Draw the straight line mn and measure the appropriate departures of the curve from this line; $y_{ex} = 0.33$ in. and $y_{1x} = 0.3$ in., actual measure. See text for conversion to deflections.

polygon obtained. Thus, the smaller the areas used and the larger the scale, the greater the accuracy of the results. The actual deflection in the horizontal plane is the measured distance y times the scale S_y;

(f) $$y_{1x} = 0.3\left(\frac{2.4}{D^4}\right) \quad \text{and} \quad y_{ex} = 0.33\left(\frac{2.4}{D^4}\right) \text{ in.}$$

The scales are obtained in accordance with equation (e) above. For the θ and y curves:

(g) $$S_\theta = PS_{M/EI}S_s = (1)(60 \times 10^{-4})(20) = 0.12 \text{ rad/in.,} \qquad [D = 1]$$

(h) $$S_y = PS_\theta S_s = (1)\left(\frac{0.12}{D^4}\right)(20) = \frac{2.4}{D^4} \text{ in/in.,}$$

where the diameter D has been reinstated in the latter scale.

The deflection curve for the vertical plane, Fig. 9.7, is found in the same manner, by plotting the following computed values of M/EI:

	A	B	G_1	G_2	C	H_1	H_2	D_1	D_2
M (in-kip) . . .	0	6.73	5.8	5.8	5.7	5.4	5.4	2.86	2.86
$(M/EI)D^4/10^4$.	0	45.7	39.4	3.75	3.69	3.5	37.4	19.1	29.6

The maximum deflections are, between the bearings and at the overhang, respectively

(i) $$y_{1y} = 0.285\frac{2.4}{D^4} \quad \text{and} \quad y_{ey} = 0.29\frac{2.4}{D^4} \text{ in.}$$

If the overhung member at E had been a pulley, the amount of the deflection, within reason, would not be very important, but being a gear, it is important for good conjugate action of the gear teeth, and it is also the largest of all deflections. The resultant deflection is the vector sum of y_x and y_y, or

$$y_{\max} = (0.33^2 + 0.29^2)^{1/2}\frac{2.4}{D^4} = \frac{1.054}{D^4} = 0.003,$$

from which $D = 4.33$, use $4\frac{3}{8}$ in.

With the diameter now known, the slopes at the gear meshes can be checked. At the overhang, the approximate slope is the vector sum of $qr_H = 0.71$ and $qr_V = 0.62$ times the scale, which is $0.12/D^4$. Thus

$$\theta = (0.71^2 + 0.62^2)^{1/2}\frac{0.12}{4.375^4} = 0.000309 \text{ in./in.}$$

compared to a permissible *relative* slope of, say, 0.0005 (§ 9.9). This means that the mating gear should not be sloping very much in a direction to increase the angle between the gear axes.

Comment on Problem. We note that a very much larger shaft is necessary to meet the condition specified for deflection than is needed for the stresses. This will often happen when gears are involved, especially on relatively long shafts. Usually, machine shafts can be made rather compact, with the gears, bearings, etc., crowded much closer together. This plan reduces bending moments as well as deflections. If the dimensions of this problem were those assumed for the first design computations, the next step would be a re-examination of the requirements, hopefully with a much smaller spacing between parts. Additional detail information, some of which is in succeeding chapters, concerning dimensions of the bearings, gears, and pulley are needed for intelligent decisions at this point.

9.12 VIBRATION AND CRITICAL SPEEDS OF SHAFTS.

The center of mass of a symmetric, rotating body does not coincide with its center of rotation. This is because: (1) It is impossible from a practical viewpoint to get the mass uniformly distributed about the geometric center of the body and (2) the shaft on which the body rotates deflects under load, thus moving the center of mass away from the true axis, which passes through the center line of the bearings. Rotation may begin about the geometric axis, but at some speed, the centrifugal force of the displaced

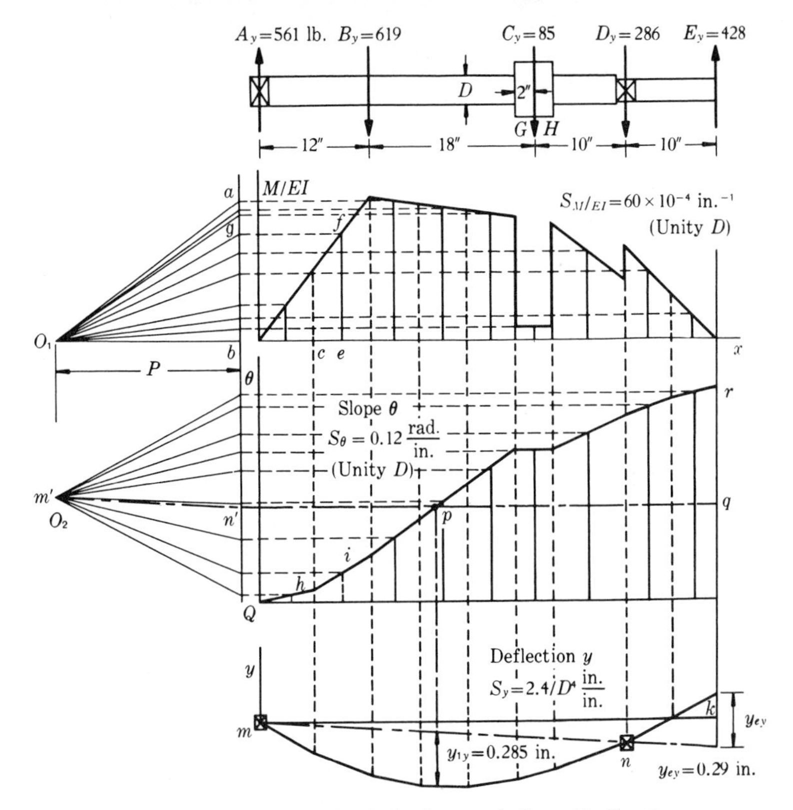

FIGURE 9.7 Slope and Deflection, Vertical Plane. The deflection y_{1y} is theoretically at the same section as y_{1x}.

center of mass will equal the deflecting forces on the shaft; the shaft with its attached bodies will then vibrate violently, since the centrifugal force changes its direction as the shaft turns. This speed is termed the ***critical speed***. Above the critical speed, a smooth-running state of equilibrium is again reached when the body is then rotating virtually about its mass center (centrifugal forces balance). High-speed turbines often operate above the critical speed. Additional critical speeds, harmonics, higher than the first, are successively attained, but the amplitudes of the corresponding vibrations progressively decrease.

Let Fig. 9.8 represent a shaft with any number of loads (three loads are chosen for illustrative purposes) which deflect the shaft to some position shown. Then, the lowest, or fundamental, critical speed n_c is given by[9.7,9.8]

$$n_c = \frac{30}{\pi}\left[\frac{g_o(W_1y_1 + W_2y_2 + W_3y_3)}{W_1y_1^2 + W_2y_2^2 + W_3y_3^2}\right]^{1/2} \text{rpm}.$$

In general the critical speed will be

$$\text{(j)} \qquad n_c = \frac{30}{\pi}\left[\frac{g_o(\Sigma Wy)}{\Sigma Wy^2}\right]^{1/2}, \qquad [g_o = 386 \text{ ips}^2]$$

where ΣWy represents the sum of all the Wy terms and ΣWy^2 represents

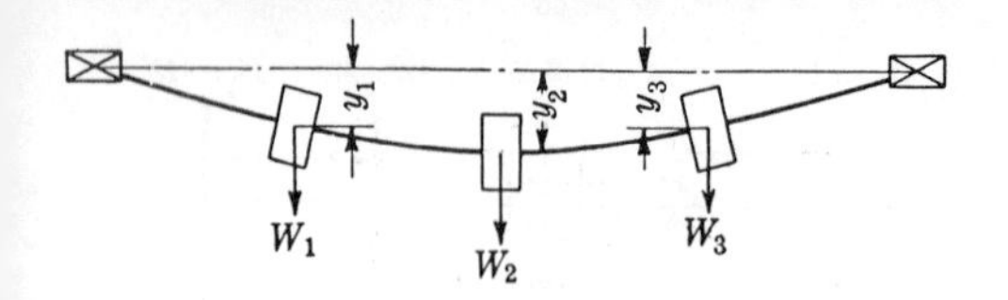

FIGURE 9.8 Shaft in Deflected Position. This is a diagrammatic arrangement showing loads W_1, W_2, and W_3 at points along the shaft where the *static* deflections are respectively y_1, y_3, and y_2.

the sum of all the Wy^2 terms. In both of the foregoing equations, the acceleration of gravity g_o is in inch units to accord with the units of y; that is, $g_o = 386 = (12)(32.2)$ in/sec.^{2}; but n_c rpm. In determining the values of y_1, y_2, y_3, etc., the graphical method of § 9.10 is suitable; the deflections are found for the static loads, *due to weights* of wheels, gears, etc. The static deflections in inches taken from this diagram at the various points of application of the *static* loads W_1, W_2, W_3, etc., are the proper values of the y's. If it is desired to account for the weight of the shaft itself, divide the shaft into parts, compute the weight of each part, consider the weight of each part as a force acting through its center of gravity; and proceed as for any group of concentrated loads. When a large number of loads is involved, tabulate loads, deflections, values of Wy, of Wy^2 to keep the procedure comprehensible.

9.13 SHAFT DESIGN BY CODE. Several years ago, the ASME published the *Code for Design of Transmission Shafting*,[9.1] which was withdrawn a few years ago, but which has been widely used for shaft designs of all kinds. Like codes in general, the results it gives are usually conservative, but there may be situations in which it has failed. The rational approach previously described is recommended, but we shall outline some of the Code provisions briefly.

The design stresses (probably with cold-drawn shafting in mind) are given as

$$\left.\begin{array}{l}\text{Shear}\\ \text{Design}\\ \text{Stress}\end{array}\right\} \begin{array}{c}\tau_d \text{ or } s_{sd} = (0.3)(\text{tensile yield strength})\\ \text{or}\\ \tau_d \text{ or } s_{sd} = (0.18)(\text{tensile ultimate strength})\end{array}$$

whichever is smaller (maximum shear theory). For a shaft in *bending only*,

$$\left.\begin{array}{l}\text{Normal}\\ \text{Design}\\ \text{Stress}\end{array}\right\} \begin{array}{c}\sigma_d \text{ or } s_d = (0.6)(\text{tensile yield strength})\\ \text{or}\\ \sigma_d \text{ or } s_d = (0.36)(\text{tensile ultimate strength}).\end{array}$$

whichever is smaller. Allowance for keyway:

$$(s_d \text{ with keyway}) = (0.75)(s_d \text{ as above without keyway}), \tag{k}$$

where the keyway is at the section being designed. For combined stresses, the design stresses without keyway correspond to a design factor of about 2 on s_y; the Code makes further allowances by use of service factors, called shock and fatigue factors, as follows:

K_s = "numerical combined shock and fatigue factor to be applied in every case to the computed torsional moment or horsepower";

K_m = "numerical combined shock and fatigue factor to be applied in every case to the computed bending moment."

Their values are governed by what the designer thinks of the load and are taken from Table 9.1, where we observe that the minimum value of K_m for a rotating shaft is 1.5, which is intended to care for the reversal of stress during every revolution of the shaft.

For bending or torsion alone, the design equations become

(l) $$s_d = \frac{K_m Mc}{I} = \frac{K_m M}{Z} \quad \text{or} \quad s_d = \frac{32 K_m M}{\pi D^3}$$ [SOLID ROUND]

(m) $$s_{sd} = \frac{K_s Tc}{J} = \frac{K_s T}{Z'} \quad \text{or} \quad s_{sd} = \frac{16 K_s T}{\pi D^3},$$ [SOLID ROUND]

where D is the diameter of a solid circular shaft. For bending, axial load F, and torsion together, the normal stress used is

(n) $$s = \frac{K_m M}{Z} + \alpha \frac{F}{A},$$

where α is defined by equation (**i**), § 7.11, for $L/k < 115$, and by equation (**h**), § 7.11, for $L/k > 115$ when the axial load F is compressive ($L_e = L$); when the

TABLE 9.1 VALUES OF K_s AND K_m

Nature of Loading	K_m	K_s
Stationary shafts (bending stress not reversed):		
Gradually applied	1.0	1.0
Suddenly applied	1.5 to 2.0	1.5 to 2.0
Rotating shafts (bending stress reversed):		
Gradually applied or steady	1.5	1.0
Suddenly applied minor shocks . . .	1.5 to 2.0	1.0 to 1.5
Suddenly applied, heavy shocks	2.0 to 3.0	1.5 to 3.0

axial load is tensile, as on a propeller shaft of an airplane engine, $\alpha = 1$; and the shearing stress used is

(o) $$s_s = \frac{K_s T}{Z'}.$$

In accordance with the maximum shear stress theory, equation (8.4), these values of s and s_s, with Z and Z' for a hollow shaft ($D_i/D = B$ = internal diameter divided by the outside diameter) give

(p) $$D^3 = \frac{16}{\pi \tau_d (1 - B^4)} \left[(K_s T)^2 + \left(K_m M + \frac{\alpha F D (1 + B^2)}{8} \right)^2 \right]^{1/2},$$

where F and/or B may be zero, τ_d is the Code design stress. Whether the Code is used or not, when a number of similar designs are to be made, it pays to derive a formula to fit the situation and reduce it to the simplest form, as in (**p**); but in the learning process, the student should avoid such formulas.

9.14 CLOSURE. Many shafts are supported by three or more bearings, which means that the problem is statically indeterminate. Texts on strength of materials give methods of solving such problems. The design effort should be in keeping with the economics of a given situation. For example, if *one* line shaft supported by three or more bearings is needed, it probably would be cheaper to make conservative assumptions as to moments and design it as though it were determinate. The extra cost of an oversize shaft may be less than the extra cost of an elaborate design analysis.

10. KEYS AND COUPLINGS

10.1 INTRODUCTION. Gears, pulleys, etc., may be attached to shafts by force or shrink fits, by one or more of several kinds of keys, by splined connections, or, occasionally, by some especially devised means. This chapter will cover briefly some conventional design information on keys and describe some typical couplings that are used to connect shafts. Couplings that are readily disconnected are called clutches and are covered in another chapter.

10.2 DESIGN OF FLAT AND SQUARE KEYS. Most keys are so-called ***flat keys*** or ***square keys***. Flat keys are rectangular in section with the smaller dimension placed in a radial direction, Fig. 10.1, and they may or may not be tapered. Square keys have a square section, $b \times b$, Fig. 10.2, and they may or may not be tapered. When one of these keys is in place, the hub pushes on its upper half on one side and the shaft on its lower half on the other side, Fig. 10.1, with the result that a couple acts to tip the key in its seat. The amount that the key actually tips depends somewhat upon

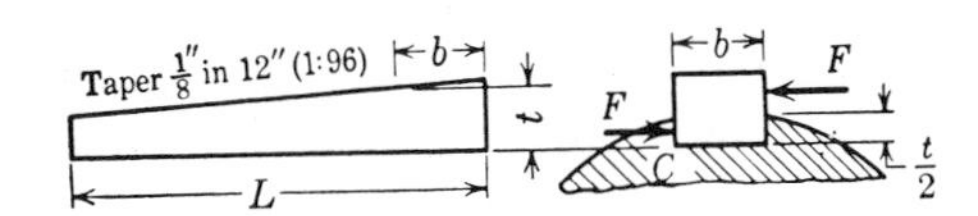

FIGURE 10.1 Flat Key, Tapered. Taper exaggerated. Stock lengths increase by increments of *2b* from a minimum of *4b* to a maximum of *16b*.

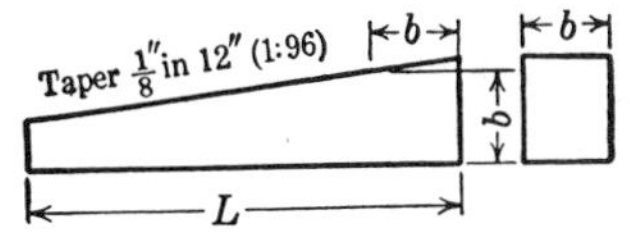

FIGURE 10.2 Square Key, Tapered. Taper exaggerated. Stock lengths increase by increments of *2b* from a minimum of *4b* to a maximum of *16b*.

the fit at the top and bottom, because some of the resisting moment may act on these surfaces.

A conventional stress analysis simplifies the situation considerably, and the strength is usually expressed in terms of the torque that is to be transmitted through the key. For shear, the resisting area is taken as a plane area between the hub and shaft, to wit, $A = bL$, where L in. is the length of key in contact with the hub; the corresponding force is $F_n = s_s A = s_s bL$, Fig. 10.3; with a moment arm of $D/2$, the torque T is $F_n D/2$, or

(a) $$T = \frac{s_s bLD}{2} \text{ in-lb. (or in-kips).} \qquad \text{[SHEAR]}$$

Since one-half of the key is in the hub and one-half in the shaft, as measured at the side of the keyway, the bearing area in each is $(t/2)L$. The force F_n, shown acting at the surface of the shaft, is not the actual force acting on either half of the key. The line of action of the force between the

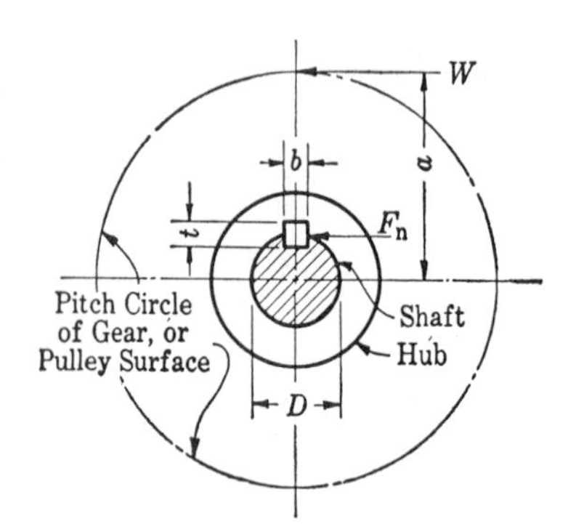

FIGURE 10.3 Loading on Key.

hub and the key would be a little above, and the force between the *shaft* and the key would be a little below, the line of action of F_n. Because of inherent inaccuracies in the analysis and the small difference involved, the moment arm of the force is taken as $D/2$. Thus, $F_n = sA = s_c tL/2$ and the torque capacity is given by $F_n D/2$, or

(b) $$T = \frac{s_c tLD}{4} \text{ in-lb. (or in-kips)} \qquad \text{[COMPRESSION]}$$

In design, the stress in equation (**a**) may be obtained from the shearing yield strength of the key material; the stress in (**b**) is taken according to the weakest of the three parts involved, the shaft, the key, or the hub. It is suggested that the design factor on the yield strength be about 1.5 for a smooth load, about 2 to 2.25 for minor shock loading, and up to 4.5 for severe shock loads, especially when the loading reverses during operation; use $s_y = s_{yc}$ for ductile metals and s_{ys} as before. The usual design procedure would be to find the shaft diameter, choose the cross-sectional dimensions b, t of the flat or square key in accordance with ASA standards from Table AT 19, solve for L from both (**a**) and (**b**), and use the larger value. Although it is not necessary, the length of hub and the length of key are made about the same, but the hub length of course must be at least as much as the needed key length L. Typical hub lengths fall between $1.25D$ and $2.4D$, where D is the shaft diameter.

If the needed key length is greater than about $2D$, consider using two keys, 180° apart, or Kennedy keys, Fig. 10.10. If the load is other than smooth, the keys should fit tightly, either by use of taper keys or by clamping the hub onto shaft and key. A taper key can be driven into place, and out too, facilitating disassembly. The matching taper is in the hub. The stresses induced by driving the taper key into place may be dangerous, but on the other hand, there results a large frictional force between the hub and shaft that aids in transmitting the power. The friction may be high enough that the stresses in the key from the transmitted torque are much below those computed from equations (**a**) and (**b**). A tapered key is tapered throughout its length, but *the* thickness b, t is that measured at a distance b from the large end, Figs. 10.1 and 10.2.

The usual material of keys is cold-finished, low-carbon steel (0.2% C or less), although heat-treated alloy steels are used where needed.

10.3 EXAMPLE—DESIGN OF FLAT KEY. A cast-iron pulley is to be keyed to a $2\frac{1}{2}$-in. shaft, made of 1040, and it is to transmit 100 hp at 200 rpm. A flat key of cold-finished C 1020 is to be used. The drive is expected to be subjected to quite minor vibrations, so that a design factor of 1.75 appears reasonable. Specify a suitable length of key.

Solution. From Table AT 7 for C 1020, we find $s_y = 66$ ksi; thus, since the compressive strength of cast iron and 1040 is greater than that of the key material, the design stresses are

$$s_s = \frac{(0.5)(66)}{1.75} = 18.85 \text{ ksi} \qquad \text{and} \qquad s_c = \frac{66}{1.75} = 37.7 \text{ ksi.}$$

transmitted torque is

$$T = \frac{63{,}000 \text{ hp}}{n} = \frac{(63{,}000)(100)}{200} = 31{,}500 \text{ in-lb. or } 31.5 \text{ in-kips.}$$

From Table AT 19, we choose $b = \frac{5}{8}$ and $t = \frac{7}{16}$ for the $2\frac{1}{2}$-in. shaft. Then, from equations (**a**) and (**b**), we find

$$L = \frac{2T}{s_s bD} = \frac{(2)(31.5)}{(18.85)(0.625)(2.5)} = 2.14 \text{ in. for shear,}$$

$$L = \frac{4T}{s_c tD} = \frac{(4)(31.5)}{(37.7)(0.4375)(2.5)} = 3.06 \text{ in. for compression;}$$

use $L = 3$ in. with $\frac{5}{8} \times \frac{7}{16}$-in. cross section.

10.4 STRESS CONCENTRATION IN KEYWAYS. Cutting a keyway in a shaft reduces its strength and stiffness out of all proportion to the amount of material removed. Shallow keyways reduce the strength less than deep ones. We notice, in Table AT 13, that the sled-runner keyway, cut by a circular milling cutter, has somewhat lower fatigue strength factors K_f than the profile keyway, cut by an end mill whose nominal diameter is the width of keyway. The larger value of K_f is for the ends of the keyway, where the change of section is quite abrupt.

Experiments show, as expected, that fillets at the internal corners along the length improve fatigue strength. *Inco*[10.1] reports impact fatigue test on $\frac{7}{8}$-in. monel shafts with a keyway for a $\frac{3}{16}$-in. square key as follows: average number of blows (all of the same magnitude) to break a shaft without fillets at the root of the keyway, 6722; *with* $\frac{1}{16}$-in. radius fillets, 12,360. While these tests do not give specific information on the comparative stress concentration factors, they suggest that fillets at the root of the keyway may be significantly advantageous if the shaft is subjected to repeated impact loads. The optimum value of the fillet radius appears to be about half the depth of the keyway (about $t/4$). A smaller radius is better than none. The effect may be illustrated by available information for a hollow, heat-treated shaft. Let r = radius of fillet and d = diameter of the concentric hole in the shaft; for a particular keyway size, $K_f = 2.4$ for $r/d = 0.02$ and $K_f = 1.63$ for $r/d = 0.08$, torsion.[4.2]

10.5 OTHER TYPES OF KEYS.

A number of different kinds of keys are in common use, some of which are described below.

Gib-head keys are square or flat and tapered, Fig. 10.4, with a head. The head should not be left exposed so that there is danger that it will catch on workmen's clothing, possibly resulting in serious injury. This key is useful when the small end is inaccessible for tapping it out; the head permits its easy removal.

A ***pin key***, Fig. 10.5 and 10.6 may be either straight or tapered (taper of $\frac{1}{4}$ to $\frac{1}{16}$ in./ft.). It is usually a drive fit. Longitudinal assembly at the end of a shaft, Fig. 10.5, is easy to do, and in this position, it can transmit heavy power; size, about one-fourth of the shaft diameter. In the transverse position, Fig. 10.6, it cannot transmit nearly as much power; and in this

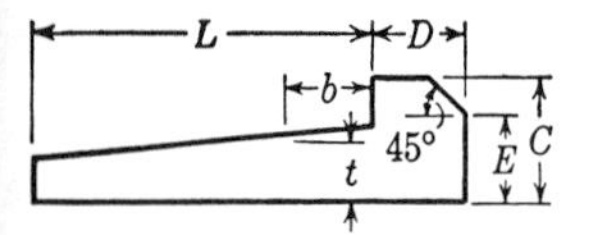

FIGURE 10.4 Gib-head Key.

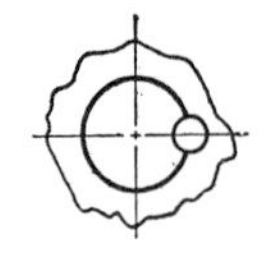

FIGURE 10.5

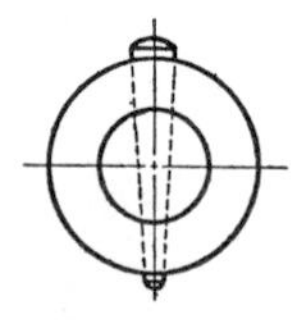

FIGURE 10.6

FIGURE 10.7 Rollpin. (Courtesy Elastic Stop Nut Corp., Union, N.J.).

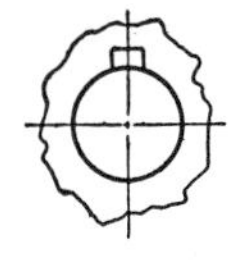

FIGURE 10.8 Saddle Key.

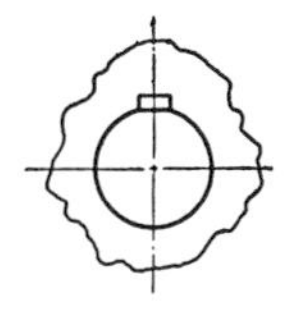

FIGURE 10.9 Flat Saddle Key.

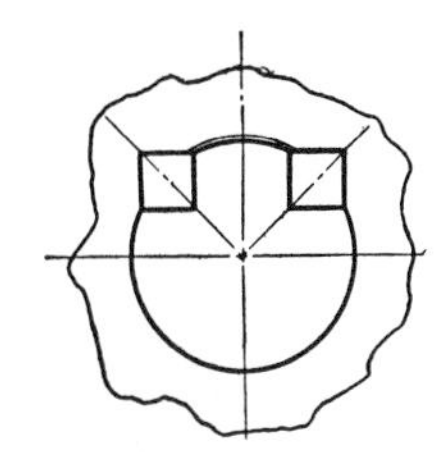

FIGURE 10.10 Kennedy Keys.

configuration, it is sometimes used as a shear pin (§ 10.8). Stock sizes, large end, are available from 0.0625 to 0.7060 in. diameter.

The ***Rollpin***®, Fig. 10.7, one of several patented methods of keying, is driven or pressed into a hole that is small enough to close the slit, assembled in a radial direction as in Fig. 10.6. The chamfered end makes it easy to drive in, and the pressure it exerts holds it in place. Observe that the Rollpin will be effective in a hole made with loose tolerances, a circumstance that is conducive to cost reduction. Made of heat-treated C 1095 (other materials: 420 stainless, beryllium copper), its shear strength is as good as a solid, mild-steel taper pin.

Saddle keys are tapered, suitable for light power, and are either hollow, Fig. 10.8, with a radius or curvature slightly smaller than the shaft radius, or flat, Fig. 10.9, in which case they are assembled on a flat on the shaft. As in Fig. 10.8, they depend entirely on friction to transmit the load; in this type, the hub can be located in any angular position relative to the shaft. A tight fit and the consequent friction also improve the load capacity of the flat saddle key.

Kennedy keys are tapered square keys, with or without gib heads, assembled as shown in Fig. 10.10. These keys are also said to be ***tangential keys***. Rectangular keys assembled with the diagonal dimension virtually in a circumferential direction (Fig. 10.10) are also called ***tangential keys***, and there are other similar ways of assembly. The advantage of this configuration is a significant increase in capacity.

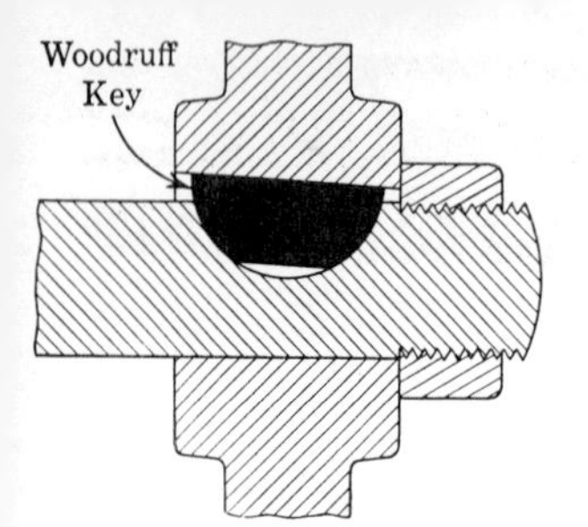

FIGURE 10.11 Woodruff Key. (Courtesy Standard Steel Specialty Co., Beaver Falls, Pa.).

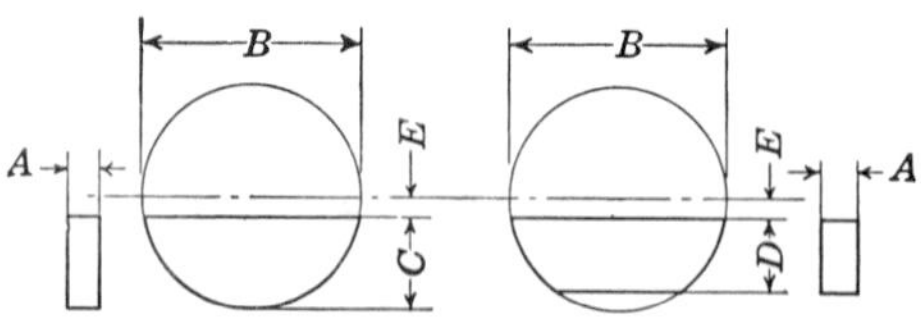

TABLE 10.1 SELECTED WOODRUFF KEY DIMENSIONS

ASA B 17f-1947. Key extends into hub a distance $A/2$. Larger sizes available.

KEY NO.	SUGGESTED SHAFT SIZES	NOMINAL KEY SIZE	HEIGHT OF KEY		DISTANCE BELOW CENTER[1]	SHEARING AREA
	in.	$A \times B$ *in.*	*Max. C*	*Max. D*	*E* in.	*sq. in.*
204	$\frac{5}{16}$–$\frac{3}{8}$	$\frac{1}{16} \times \frac{1}{2}$	0.203	0.194	$\frac{3}{64}$	0.030
305	$\frac{7}{16}$–$\frac{1}{2}$	$\frac{3}{32} \times \frac{5}{8}$	0.250	0.240	$\frac{1}{16}$	0.052
405	$\frac{11}{16}$–$\frac{3}{4}$	$\frac{1}{8} \times \frac{5}{8}$	0.250	0.240	$\frac{1}{16}$	0.072
506	$\frac{13}{16}$–$\frac{15}{16}$	$\frac{5}{32} \times \frac{3}{4}$	0.313	0.303	$\frac{1}{16}$	0.109
507	$\frac{7}{8}$–$\frac{15}{16}$	$\frac{5}{32} \times \frac{7}{8}$	0.375	0.365	$\frac{1}{16}$	0.129
608	1–1$\frac{3}{16}$	$\frac{3}{16} \times 1$	0.438	0.428	$\frac{1}{16}$	0.178
807	1$\frac{1}{4}$–1$\frac{5}{16}$	$\frac{1}{4} \times \frac{7}{8}$	0.375	0.365	$\frac{1}{16}$	0.198
809	1$\frac{1}{4}$–1$\frac{3}{4}$	$\frac{1}{4} \times 1\frac{1}{8}$	0.484	0.475	$\frac{5}{64}$	0.262
810	1$\frac{1}{4}$–1$\frac{3}{4}$	$\frac{1}{4} \times 1\frac{1}{4}$	0.547	0.537	$\frac{5}{64}$	0.296
812	1$\frac{1}{2}$–1$\frac{3}{4}$	$\frac{1}{4} \times 1\frac{1}{2}$	0.641	0.631	$\frac{7}{64}$	0.356
1012	1$\frac{13}{16}$–2$\frac{1}{2}$	$\frac{5}{16} \times 1\frac{1}{2}$	0.641	0.631	$\frac{7}{64}$	0.438
1212	1$\frac{7}{8}$–2$\frac{1}{2}$	$\frac{3}{8} \times 1\frac{1}{2}$	0.641	0.631	$\frac{7}{64}$	0.517

[1] This dimension given to facilitate layout in drafting-room.

The ***Woodruff key***, which fits into a semicylindrical seat in the shaft, Fig. 10.11, is widely used in the automotive and machine tool industries. Since it goes deeper into the shaft, it has less tendency to tip when the load is applied. Note that the construction is such that the key must first be placed in its seat and the hub pressed on. Table 10.1 gives some dimensions of selected sizes. Two grades of material are readily available; SAE 1035 and heat treated alloy steel. For extra strength, two or more keys may be used—usually in a tandem arrangement.

A ***feather key*** is one that allows the hub to move *along* the shaft but *prevents rotation* on the shaft. It is used, for example, to permit moving a gear into or out of engagement with its mate, or to engage and disengage a jaw clutch. The feather key may be attached to the shaft, Fig. 10.12(a), or to the hub, Fig. 10.12(b). A force analysis based on certain assumptions shows that two feather keys 180° apart are preferable to a single key. The axial force required to move a member along a shaft is about half as much when there are two or more feather keys as when there is one.

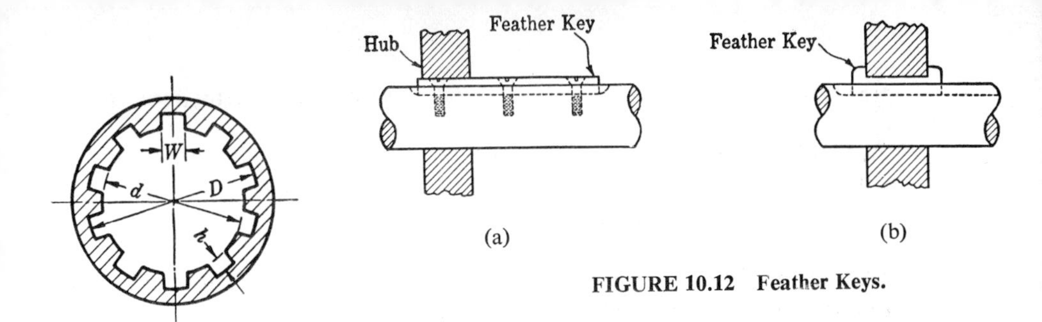

FIGURE 10.12 Feather Keys.

TABLE 10.2 NOMINAL DIMENSIONS OF SPLINES[2.2]

NO. OF SPLINES	STANDARD SIZES, NOMINAL	ALL FITS	PERMANENT FIT		NOT SLIDE UNDER LOAD		SLIDE UNDER LOAD	
		W	h	d	h	d	h	d
4	by $\frac{1}{8}$-in. from $\frac{3}{4}$ in. to $1\frac{3}{4}$; 2, $2\frac{1}{4}$, $2\frac{1}{2}$, 3	$0.241D$	$0.075D$	$0.850D$	$0.125D$	$0.750D$		
6		$0.250D$	$0.050D$	$0.900D$	$0.075D$	$0.850D$	$0.100D$	$0.800D$
10	Same as above, plus by $\frac{1}{2}$ in. from 3 to 6 in.	$0.156D$	$0.045D$	$0.910D$	$0.070D$	$0.860D$	$0.095D$	$0.810D$
16	by $\frac{1}{2}$ in. from 2 to 6 in.	$0.098D$	$0.045D$	$0.910D$	$0.070D$	$0.860D$	$0.095D$	$0.810D$

10.6 SPLINED SHAFTS. Reversing torques and repeatedly applied torques are tough on connections like those previously described. Splined shafts make a stronger connection. Straight-sided splines are widely used in the automotive (Figs. 12.10 and 18.10) and other industries for permanent fits, for fits not intended to slide under load, and for fits that will slide under load. The nominal dimensions of 4-, 6-, 10-, and 16-spline fittings are given in Table 10.2, as taken from the SAE standard,[2.2] from which the details of the various fits may be obtained. The nominal size is the major diameter D, Table 10.2, and this is the maximum size of the fitting, since the tolerances given in the standard are negative and the desired fit is obtained by varying the dimensions on the shaft. In the machine-tool applications, the practice is to vary the hole dimensions in order to obtain the desired fit, because there may be both a press fit and a sliding fit on the same shaft, in which case it is economical to grind the shaft to a uniform size. The machine tool industry, which has need of five classes of fits—free fit, sliding fit, push fit, light drive fit, and press fit[10.4]—prefers the fit on the minor diameter, since the corresponding mating surfaces are the only ones that can be ground economically. The design torque capacity T of a splined connection with axial sliding is based on a side pressure of 1000 psi;[2.2] or

(c) $$T = (sA)r_m = (1000)(hL)(r_m)(N_t) \text{ in-lb.},$$

where L in. is the contact length, r_m in. is the mean radius $(D + d)/4$, and

N_t is the total number of splines. If there is no sliding under load, the side pressure may, of course, be much greater than 1000 psi.

10.7 INVOLUTE SPLINE. Involute splines, Fig. 10.13, are in the form of concentric external and internal gear teeth with a pressure angle $\phi = 30°$ (no undercut with 6 teeth). The splines are specified by the diametral pitch P_d (the standard[10.2] uses two pitches in a ratio, as 3/6, where the numerator is *the* pitch, with the denominator always twice the numerator). Standard diametral pitches are: 1, 2.5, 3, 4, 5, 6, 8, 10, 12, 16, 20, 24, 32, 40, and 48. Many of the rules and much of the nomenclature of gearing apply, some of which may be reviewed in §§ 13.2–13.4. The pitch diameter D, Fig. 10.13, is obtained from the number of splines N_t and the diametral pitch P_d; $D = N_t/P_d$, as in gearing.

The standard[10.2] provides three classes of fits; ***sliding fits*** with clearance on all mating surfaces; ***close fits***, which are close on *either* the major diameter, the minor diameter, or on the sides of the teeth; ***press fits***, which have interference on *either* the major diameter, the minor diameter, or on the sides of the teeth. Since the internal splines are cut with standard size broaches, these various fits are obtained by varying the dimensions of the external splines.

At a splined section, compute the strength of the shaft corresponding to the minor diameter. The length L of nonsliding splines may be computed on the basis of the splines being subjected to the same shearing stress as the shaft when *one-fourth of the splines are in contact* and when the shear is at the pitch diameter D. Thus the shearing area A is $\frac{1}{4}$ times half the circumference $\pi D/2$ times the length L, or $A = \pi DL/8$; the corresponding force is $F = s_s A = s_s \pi DL/8$, and the torque $T = Fr$, for $r = D/2$, is

(d)
$$T = \left(\frac{s_s \pi DL}{8}\right)\left(\frac{D}{2}\right).$$

Equating this torque to the resistance of the shaft, $T = s_s \pi D_r^3/16$, find

$$T = \frac{s_s \pi D_r^3}{16} = \frac{s_s \pi D^2 L}{16},$$

or

$$L = \frac{D_r^3}{D^2}, \qquad \text{[NO SLIDING]}$$

where D_r = root diameter. The compressive stress may be checked by using the projected contact area of a tooth as hL where $h = 0.8/P_d$ = the minimum tooth height in contact; or the area in compression $A_c = hLN_t/4 = 0.8LN_t/(4P_d) = 0.2LD$, at a radius of $D/2$, where N_t is the number of teeth.

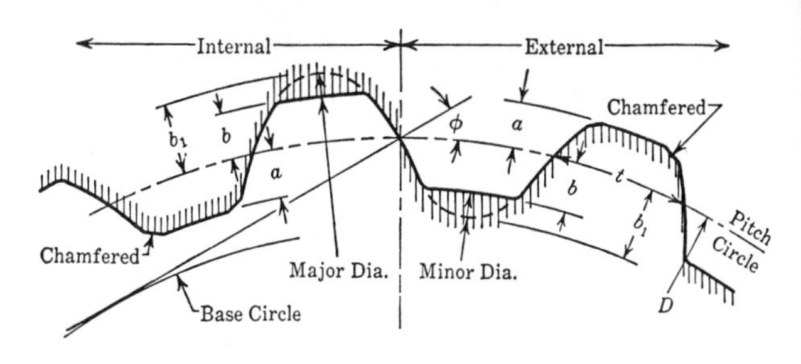

FIGURE 10.13 Involute Spline Profile. See also Fig. 13.1 for additional information on nomenclature. Note that the vertical center line divides internal and external spline forms. The dotted outlines are optional. Certain dimensions are: Pitch diameter, $D = N_t/P_d$; Tooth thickness, $t = \pi/(2P_d)$; Addendum, $a = 0.500/P_d$; Dedendum, $b_1 = 0.900/P_d$ through $P_d = 12$. Dedendum, $b = 0.500/P_d$ on internal spline for major diameter fits, on external spline for minor diameter fits.

FIGURE 10.14 Hob for Splining Shafts. (Courtesy Barber-Colman Co., Rockford, Ill.).

This value of A_c is based on one-fourth of the teeth being under load. However, failure in compression is not likely.

Involute splines have maximum strength at the minor diameter, where needed; they are self-centering and tend to adjust to an even distribution of the load, and when cut by hobbing, Fig. 10.14, they have a smooth surface that does not need grinding for most applications.

Involute serrations, which are used for permanent fits, are similar to involute splines except that the pressure angle is $\phi = 45°$, Fig. 10.13. Standard pitches P_d are: 10, 16, 24, 40, and 48 (the standard[10.3] includes a denominator twice as large, as 10/20). When the fit is on the major or minor diameter, the nominal value of the addendum and of the dedendum is $0.5/P_d$, as in Fig. 10.13. For additional detail, see the standards.

10.8 SHEAR PIN. Shear pins or breaking pins are used as couplings, or in addition to other couplings where, in case of overload, there is danger

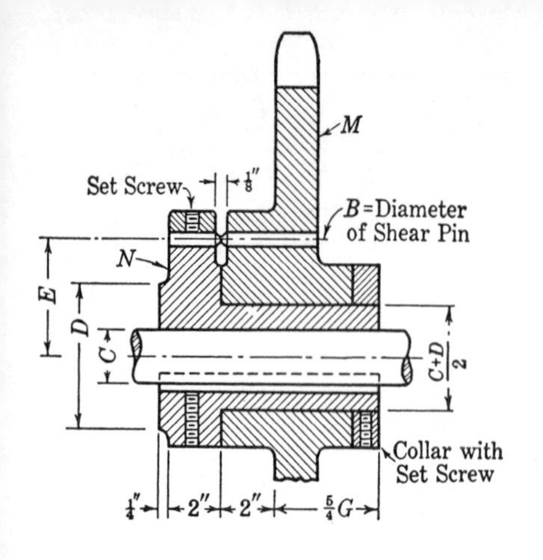

FIGURE 10.15 Shear Pin.

of injury to machines or to material in process. One type, shown in Fig. 10.15, is designed for a roller-chain sprocket wheel. Without the shear pin, the part M would rotate on the part N, which is keyed to the shaft. Thus the drive ceases when the shear pin breaks. The breaking stress in $\frac{1}{8}$-in. to 1-in. pins, sizes manufactured by Link-Belt, is taken as 50,000 psi in shear.

10.9 RIGID COUPLINGS. Shafts are directly connected to one another by couplings; clutches, which are readily disconnected, are discussed in Chapter 18. Rigid couplings (Figs. 10.16, 10.17, 10.18) are used when the shafts are virtually collinear and when they should remain in a fixed angular relation with respect to each other (except for angular deflection). However, nearly true alignment of the axes of the connecting shafts is difficult to achieve and, after it has been accomplished, it is difficult to maintain, because of settling of foundations, unequal deflection of supports, deflection of the shaft under load, temperature changes, wear in bearings, effects of shocks and vibrations. Thus rigid couplings result in stresses of unknown values, which may sometimes lead to failure, and are most successfully used where the shaft is relatively flexible, as a line shaft, and its speed is low.

Each half of the *flanged-face coupling*, Fig. 10.16, usually called a ***flange coupling***, is keyed to a shaft; the faces are finished normal to the axis, the bolts and holes are accurately finished to give a tight fit; standard sizes up to 8-in. bore. The *flanged-compression coupling* (Fig. 10.17) transmits the power via the frictional forces induced by pulling the flanges towards each other over slotted tapered sleeves. This plan eliminates the expense of cutting keyways; standard sizes up to a 3-in. bore. The *ribbed compression coupling* (Fig. 10.18) is a heavy-duty connection; the parts are clamped on a long key that is fitted to both shafts, a plan that helps in aligning. An advantage of this type is that it can be installed and removed without disturbing the shafts; standard sizes up to 7-in. bore.

Stock flanges may be made of cast iron or steel, or of wrought steel; other metals may be used for a reason.

FIGURE 10.16 Flange Coupling.

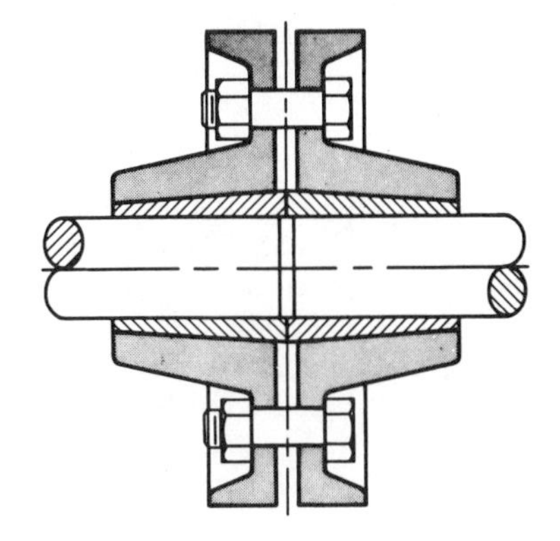

FIGURE 10.17

FIGURE 10.18 Ribbed Compression Coupling. The usual smooth sheet-metal covering improves the appearance and makes it safer. (Courtesy Link-Belt Co., Chicago).

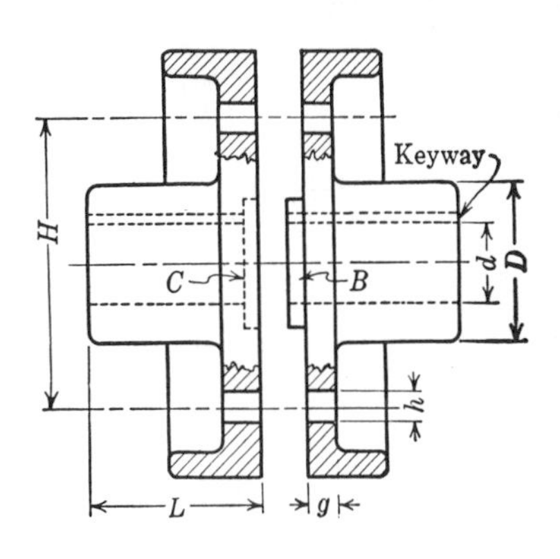

FIGURE 10.19 Flange Coupling. Observe that bolt heads are shielded by the flanges.

10.10 EXAMPLE—FLANGE COUPLING. Couplings are designed by the manufacturer and, ordinarily, the user's function is to choose one that serves his purpose. However, computations will not only provide a lesson in simple stress analysis, but will also indicate by the size of the factor of safety where experience has shown that unaccounted for stresses are involved. A manufacturer's catalog gives the following dimensions in inches for a flange coupling, Fig. 10.19: $d = 3$, $D = 5\frac{3}{8}$, $L = 4\frac{3}{4}$, $h = \frac{3}{4}$, $H = 8\frac{1}{4}$, $g = 1\frac{1}{16}$. Let the shaft be made of cold-finished

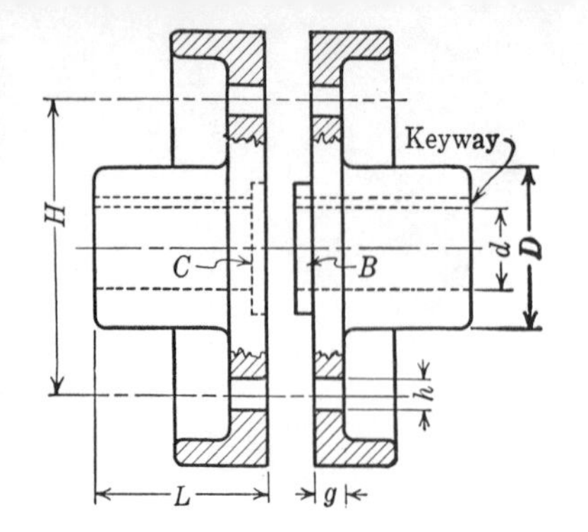

FIGURE 10.19 (Repeated).

C 1035, the bolts and square keys of cold-drawn C 1020, and flanges of as-rolled C 1035. There are $N_b = 4$ bolts equally spaced. Let a design factor of $N = 3.5$ based on the shearing yield strength cover the effect of stress concentration and determine the torque capacity of the shaft in pure torsion. Then, for this torque applied to other parts of the connection, compute the nominal factors of safety based on yield strengths for each conventional method of failure.

Solution. The yield strengths of the materials are (s_y for tension and compression; $s_{ys} = 0.6s_y$):

From Table AT 7, cold-drawn C 1020, $s_y = 66$ ksi; $s_{ys} = 39.6$ ksi;
From Table AT 7, as-rolled C 1035, $s_y = 55$ ksi; $s_{ys} = 33$ ksi;
From Table AT 10, cold-drawn C 1035, $s_y = 78$ ksi, $s_{ys} = 46.8$ ksi.

In computing the nominal stresses, we shall find it convenient to set up the strength equations in terms of the torque, as in the equations (**a**) and (**b**) for keys, because all parts of the connection are subjected to the same torque. For $s_{ys} = 46.8$ ksi and $N = 3.5$, the torque capacity of the shaft is, Fig. 10.19,

$$T = s_s \frac{\pi d^3}{16} = \frac{(46.8)(\pi)(3)^3}{(3.5)(16)} = 71 \text{ in-kips.}$$

The bolts may fail by shearing between the flange faces, where the major diameter of the bolt ($h = \frac{3}{4}$ in.) is assumed to be resisting. For 4 bolts, the area offering resistance is $4A_1 = 4\pi h^2/4 = \pi h^2 = \pi(0.75)^2$; the corresponding resisting force is $s_s A = s_s \pi(0.75)^2$; the moment arm of this resistance is $r = H/2 = 8.25/2 = 4.125$ in., Fig. 10.19. Hence the torque T is

$$T = Fr = s_s \pi(0.75)^2(4.125) = 71 \text{ in-kips,}$$

from which we find $s_s = 9.73$ ksi. The factor of safety is

$$N = \frac{39.6}{9.73} = 4.07, \text{ shear of bolts, friction ignored.}$$

The compression area of one bolt on a flange is hg; for 4 bolts, it is $A = 4hg = (4)(0.75)(1.0625) = 3.1875$; the corresponding resisting force is $s_c A = s_c(3.1875)$, whose moment arm is $r = H/2 = 4.125$. Hence, the torque is (sAr)

$$T = Fr = s_c(3.1875)(4.125) = 71 \text{ in-kips,}$$

from which, $s_c = 5.4$ ksi. With the flange strength governing ($55 < 66$),

$$N = \frac{55}{5.4} = 10.2, \text{ compression of bolts and flange.}$$

The flange may shear at the outside hub diameter. The resisting area is cylindrical, πDg; the resisting force is $s_s\pi Dg = s_s\pi(5.375)(1.0625)$; and, with a moment arm of $r = D/2 = 5.375/2 = 2.6875$, the resisting torque is

$$T = Fr = s_s\pi(5.375)(1.0625)(2.6875) = 71 \text{ in-kips},$$

from which $s_s = 1.47$ ksi, and the nominal factor of safety is

$$N = \frac{33}{1.47} = 22.4, \text{ shear of flange.}$$

There is no danger here of failure by pure shear, even if a fatigue strength reduction factor were included, but this same section may have severe and undefinable bending stresses on it if the flanges are imperfectly aligned, and they surely will be. We might also observe that the bolts will be subjected to some bending, which we hope will be small.

Let the side of a square key be $b = \frac{3}{4}$ in., Table AT 19; let its length be the hub length, $L = 4.75$ in. The computed factors of safety of the key are

$$N = \frac{39.6}{s_s} = \frac{39.6\,bdL}{2T} = \frac{(39.6)(0.75)(3)(4.75)}{(2)(71)} = 2.98 \qquad \text{[SHEAR]}$$

$$N = \frac{55}{s_c} = \frac{55\,tdL}{4T} = \frac{(55)(0.75)(3)(4.75)}{(4)(71)} = 2.07. \qquad \text{[COMPRESSION]}$$

If the shafts cannot be maintained in good alignment and if the loading induces relatively high stresses, eventual fatigue failure of the shaft becomes more likely. The computed factor of safety of the flange at the hub suggests that it could withstand repeated bending if the misalignment is small. The *nearer the flanges are to the bearings*, the smaller will be the deflection of the shaft at that point and the smaller the stresses induced in the flanges by this deflection. However, the slope of the shaft at a bearing may be a source of excessive stressing.

10.11 FLEXIBLE COUPLINGS. Since shaft misalignment is inevitable, rigid couplings often lead to fatigue failures, overheating bearings, and other troubles, which can be avoided by using flexible couplings. A large variety of flexible couplings has been invented and designed, a selection of which are shown in Figs. 10.20–10.27 (pp. 294–295). See Ref. *(10.5)* for additional descriptive material. These couplings care for small amounts of angularity, end play, and axis displacement; they also serve the important functions of absorbing some shock and vibration that may appear on one shaft and of preventing the occurrence of reversed stresses caused by the shafts deflecting at the coupling. With one half keyed to each shaft, the feature of a flexible coupling is an intermediate member that is either flexible or floating (or sometimes both to some extent). Some couplings need lubrication to prevent overheating, some can tolerate more misalignment, intentional or otherwise, than others without ill consequences, but all will benefit by good alignment. That is, if possible, the shafts should be kept carefully aligned; let the coupling care for unintentional (settling of floors) and unavoidable misalignment (due to temperature changes, deflection, bearing wear, etc.). Flexible couplings are not a guarantee against troubles and recommended maintenance procedures should be followed

FIGURE 10.20 Morflex ® Coupling. The intermediate elements are four rubber "biscuits" held in a center piece. Each hub is attached to two opposite biscuits, allowing angularity, which flex torsionally and axially. No metal-to-metal contact. Good absorber of vibrations, including torsional. Other couplings with nonmetallic intermediate elements are available. (Courtesy Morse Chain Co., Detroit).

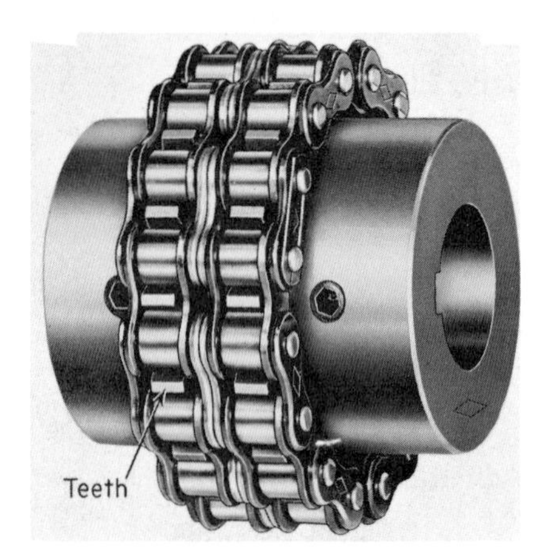

FIGURE 10.21 Roller Chain Coupling. The two opposing hubs are made with integral sprockets over which a double roller chain is fitted. The drive is through the chain. Flexibility is obtained by lateral play in the fit of the chain over the sprocket teeth (no circumferential play). Relatively inexpensive type. Preferably installed with a casing to contain grease lubricant. Similar couplings use silent chain. (Courtesy Diamond Chain Co., Indianapolis).

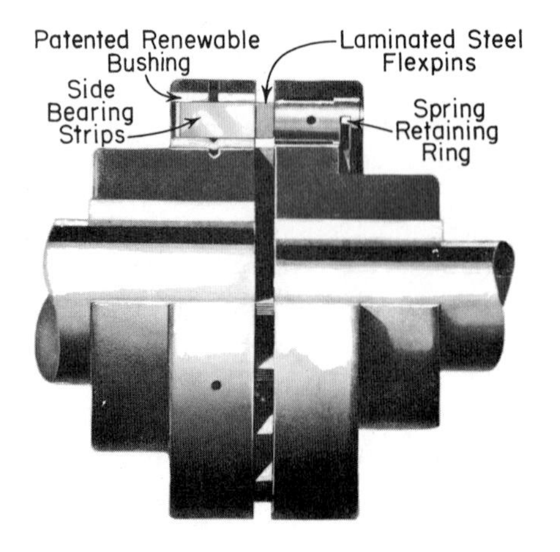

FIGURE 10.22 Flexpin Coupling. The two flanges of the coupling are connected through laminated steel pins that are relatively flexible. In the right flange, the pins are held in place by the spring retaining ring. Freedom of the pins to slide in the left flange permits the coupling to care for some endwise motion, as well as for angular misalignment. (Courtesy Smith & Serrell, Newark, N.J.).

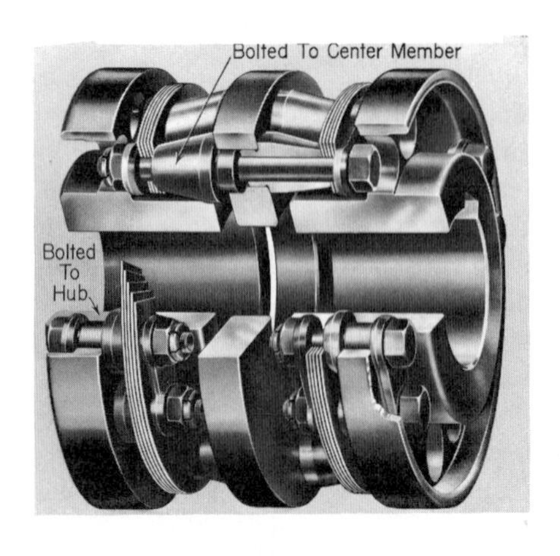

FIGURE 10.23 Flexible Disk Coupling. An all metal coupling with the intermediate flexible elements being thin steel disks. The shaft hubs and center pieces are connected at different points to the disks. Bending of the disks permits misalignment. Two sets of disks permit parallel misalignment. No play or backlash in the connection; no relative motion or lubrication. (Courtesy Thomas Flexible Coupling Co., Warren, Pa.).

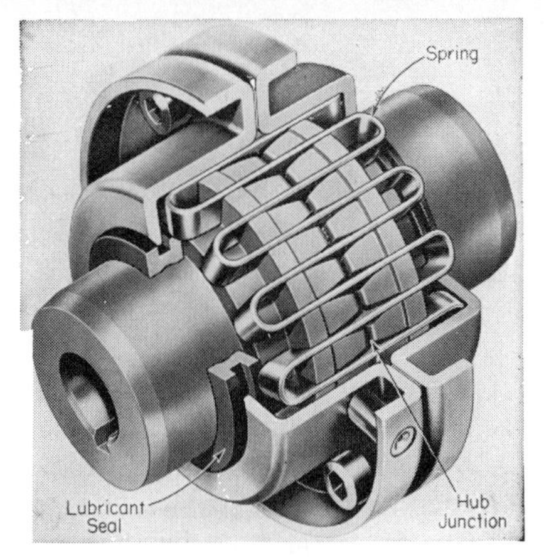

FIGURE 10.24 Flexsteel Coupling. The spring member, which transmits the load, is inserted into grooves in the hubs which widen toward the center. When the load is applied, the spring is bent along the arc of these grooves, whose curvature is such that the spring stresses are safe. (Courtesy The Falk Corp., Milwaukee).

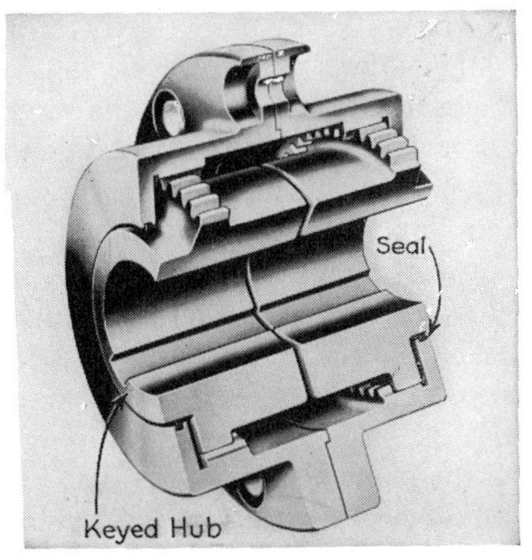

FIGURE 10.25 Gear Type Coupling. The hubs have integral external gear teeth, perhaps crowned (§13.12), that mesh with internal teeth in the casing through 360° as in a splined connection. Flexibility is obtained by play between the teeth. (Courtesy American Flexible Coupling Co., Erie, Pa.).

FIGURE 10.26 Flexible Coupling (Oldham Principle). In the assembled form, the tongues on the central piece engage the grooves on the end pieces which are attached to the shafts to be connected. The action of the parts is similar to that described in Fig. 10.27. (Courtesy W. A. Jones Foundry and Machine Company, Chicago).

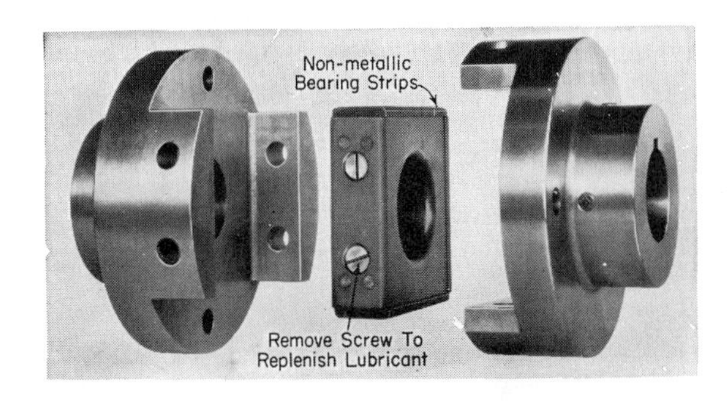

FIGURE 10.27 The American Flexible Coupling. Operates on Oldham principle. In the position shown, the left member can slide up and down, the right member can slide to and fro on the square floating center member. The combined action, when the parts are assembled, produces a flexible connection that will care for misalignment. The center member is hollow, the cavity being filled with lubricant, which reaches the surface of the nonmetallic strips through porous reeds in the block and felt pads in the bearing strip. (Courtesy American Flexible Coupling Co., Erie, Pa.).

Manufacturers' catalogs give rules for selecting a stock coupling for a particular job. If a special design is necessary, their specialized knowledge will be helpful—also see Ref. *(10.7)*.

10.12 UNIVERSAL JOINTS. A universal joint is used to connect shafts whose axes intersect, that is, whose angular misalignment is permanent. Hooke's coupling is shown in Fig. 10.28. We see that if *B*, Fig. 10.28, is held stationary, *C* may be swivelled about either pin *D* or *E*, which property makes the universal-joint action possible. If the connected shafts are at an angle, Hooke's coupling does not transmit a constant velocity ratio, as explained in books on engineering kinematics.[13.2] However, if two universal joints are used with the driving and driven yokes in the same plane, Fig. 10.29, the initial and last shafts, if they are in the same plane and make the same angle with the intermediate part, will turn at a constant velocity ratio of one. Rather than the short coupling of Fig. 10.29, a long drive shaft may generally be safely inserted between the universal joints, as in automotive applications. For a single Hooke's coupling, the shaft angle should not be greater than about 15°; less, if the speed is high.

Some universal joints do transmit a constant velocity ratio, for example the Bendix-Weiss universal joint in Fig. 10.30, where the drive is through steel balls in races that are so shaped that the plane of contact between the balls and races bisects the shaft angle at all times. This position of the plane

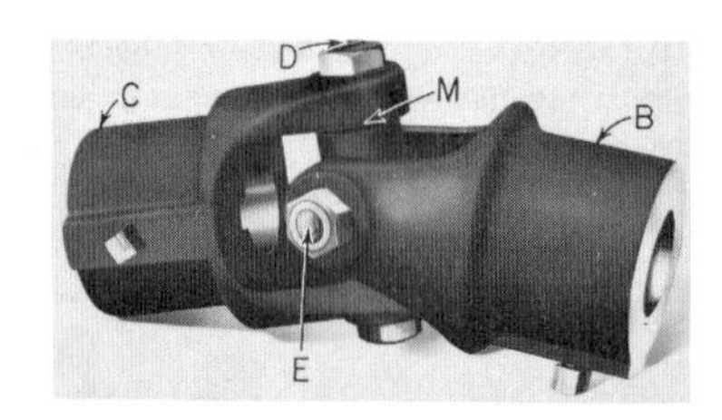

FIGURE 10.28 Hooke's Universal Joint. Observe the added metal over the keyway, and the set screws that hold the key tight in its seat and prevent axial movement of the hubs.

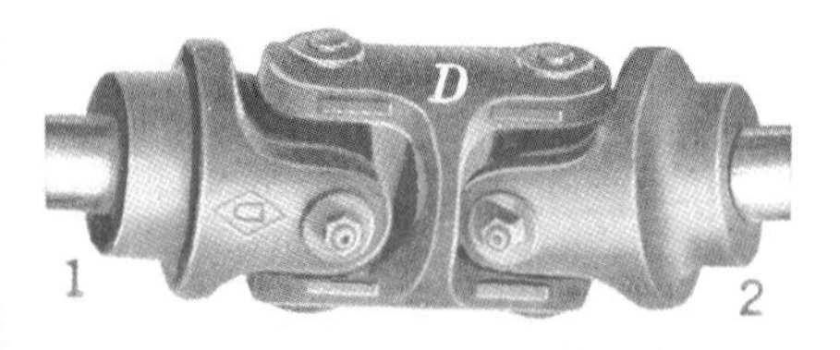

FIGURE 10.29 Double Universal Joint.

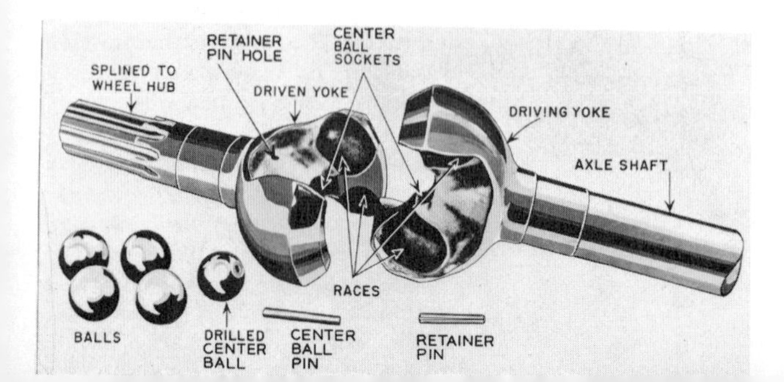

FIGURE 10.30 Constant-velocity Universal Joint. (Courtesy Bendix Aviation Corp., South Bend.).

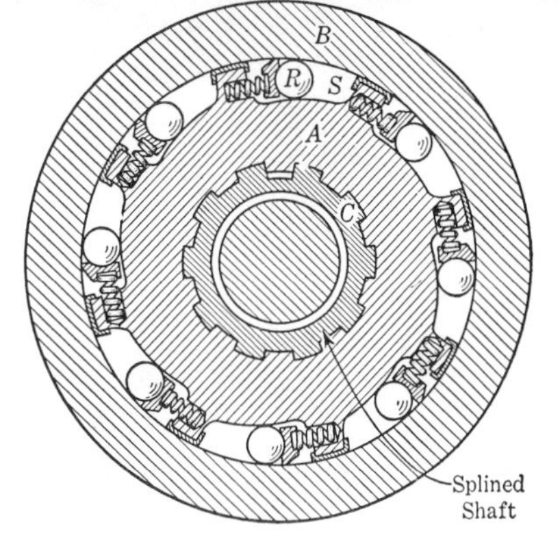

FIGURE 10.31 Overrunning Clutch. If the shaft C *drives* in a counterclockwise direction, the rollers R are edged into the space S between A and B, and B is forced to turn. If C turns clockwise, the rollers are not pressed against B and no drive occurs. The effect is the same if B turns counterclockwise *faster* than C turns counterclockwise.

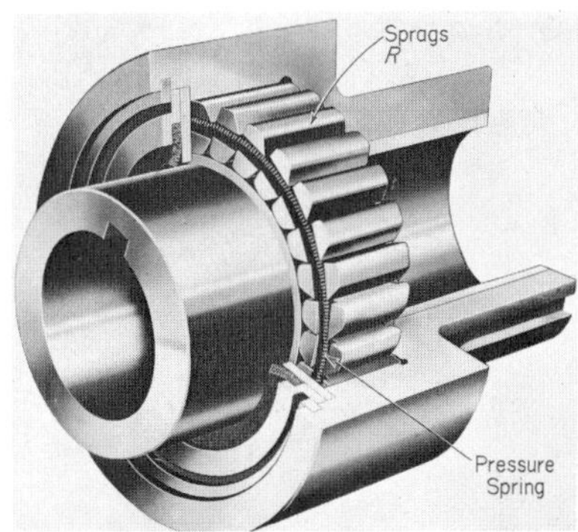

FIGURE 10.32 Sprag-type Overrunning Clutch. The sprags R, which are kept in contact with the members by the pressure spring, wedge tight for one direction of drive and release for the other direction. (Courtesy Formsprag Co., Van Dyke, Mich.).

of contact is the condition for the driven yoke to turn at a constant angular velocity when the driving yoke has a constant speed.

10.13 OVERRUNNING CLUTCH. The overrunning clutch, illustrated in Figs. 10.31 and 10.32, has many uses, among which are: as a "free-wheeling" clutch on automobiles; as a brake on inclined conveyors to lock the conveyor against unintentional backward motion; as a feed mechanism and, in general, in place of a pawl-and-ratchet mechanism; and as a coupling between shafts.

There are variations in the construction of overrunning clutches, but in all cases there are some members, such as R in Figs. 10.31 and 10.32, between the driving and driven members that wedge themselves in one direction and release in the other direction. Figure 10.32 is an example of heavy-duty construction for industrial use.

10.14 CLOSURE. There are many elements of machines whose design and development have become so specialized that usually the best procedure is to choose the item from a catalog in accordance with the manufacturers' instructions. The couplings above are, for the most part, illustrative of this situation. The feature of the strength equations of this chapter is that the strength is expressed as a torque, a form that is often convenient.

11. JOURNAL AND PLANE-SURFACE BEARINGS

11.1 INTRODUCTION. Wherever machine elements move, there must be bearing surfaces, some of which are lubricated easily, some with difficulty, some not at all. When the load is light and the motion slight, one extreme, the designer may be content to specify an oil hole and to depend upon an operator to apply lubricant intermittently; though, this practice is becoming rarer. For an intermediate category of load and speed, the use of a dry lubricant (graphite), grease, porous bearings, synthetic bearing materials, etc. is quite satisfactory. When the load or speed or both are high, as often happens in modern machines, the lubrication of surface bearings, whether by oil, air, or other fluid, must provide a fluid film that keeps relatively moving surfaces from touching.

The theoretical differential equation that interrelates many of the variables in a journal bearing is credited to Osborn Reynolds (1886), who based his work on experimental data obtained by Beauchamp Tower (1883, 1885). A form of this equation restricted to incompressible fluids,[11.6] which is still difficult to solve (x is the direction of motion), is

(a)
$$\frac{\partial}{\partial x}\left(\frac{h^3}{\mu}\frac{\partial p}{\partial x}\right) + \frac{\partial}{\partial z}\left(\frac{h^3}{\mu}\frac{\partial p}{\partial z}\right) = 6v\,\frac{\partial h}{\partial x},$$

where v is the relative velocity of the sliding surfaces. The pressure gradient $\partial p/\partial z$ in the longitudinal direction (perpendicular to motion) is zero in an infinitely long bearing, the kind for which a solution was first made. The first such solutions suitable for engineering design were obtained graphically,[11.9,11.10] followed a few years ago by computer solutions.[11.11] Digital computer solutions have now been obtained for finite-length journal bearings, which allow for end leakage and the pressure gradient $\partial p/\partial z$,[11.6,11.7] and it is from these solutions that we shall make our computations—later.

11.2 TYPES OF JOURNAL BEARINGS. A journal bearing is composed of two principal parts: the ***journal*** (which is the inside cylindrical part, usually a rotating or an oscillating member) and the ***bearing*** or surrounding shell (which may be stationary, as on a line-shaft bearing, or moving, as on a crankpin). One basis of classification depends upon whether the ***bearing surface*** completely surrounds the journal, a type called a ***full bearing***, Fig. 11.5, or only partially surrounds it, a type called a ***partial bearing***, Fig. 11.6. A simple type of partial bearing can be used when the load is carried on the top part of the journal and the journal dips into an oil sump at the bottom. When the line of action of the load (*bearing reaction*) bisects the arc of a partial bearing, the bearing is said to be ***centrally loaded***; when the vector passes to one side of the center, the bearing is ***eccentrically loaded*** (also called an *offset bearing*). While we shall confine our attention to central loading, extensive solutions of equation (**a**) for eccentric loading are found in Pinkus and Sternlicht.[11.6]

Journal bearings may also be classified as ***clearance bearings*** or ***fitted bearings***. In clearance bearings, the diameter of the bearing is larger than the diameter of the journal. The difference in these *diameters* is called the ***diametral clearance*** c_d. The ***radial clearance*** $c_r = c_d/2$ is the difference in the *radii* of the bearing and the journal. The ratio of the diametral clearance to the journal diameter c_d/D, which is the same as c_r/r, where r is the journal radius, is called the ***clearance ratio***. A *fitted bearing* is one in which the radii of the journal and the bearing are the same. Thus, a fitted bearing is necessarily a partial bearing, whereas clearance bearings may be either full or partial bearings. Fitted bearings have a place in the scheme of things but they will not be covered in this work.[11.1,11.9]

11.3 THICK-FILM LUBRICATION. If the film of lubricant between sliding surfaces is thick enough that there is no metal-to-metal contact, lubrication is said to be ***thick film*** or *fluid film*. The rougher the surfaces, the thicker the film required to separate them, Fig. 11.1. If a film gradually gets thinner, first the highest peaks begin to touch; then more and more metal is in contact as the film thins, until friction becomes excessive and a bearing failure soon becomes imminent if this manner of operation was not intended. One of the aims in bearing design is to provide a film thickness whose minimum value h_o is safe.

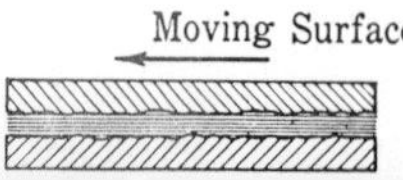

Moving Surface

FIGURE 11.1

11.4 VISCOSITY. When thick-film lubrication exists, *the frictional force that resists relative motion is independent of the nature of the mating surfaces* and is more influenced by *viscosity* than by any other variable. Consider an element of a fluid between two surfaces, Fig. 11.2, one of which M is moving with a constant speed. A layer of fluid adheres to the surface M and moves relative to the next layer, and so on until the layer in contact with the surface N adheres to it and remains stationary. Let the bottom surface of the element E move with speed v, let the upper surface, a distance

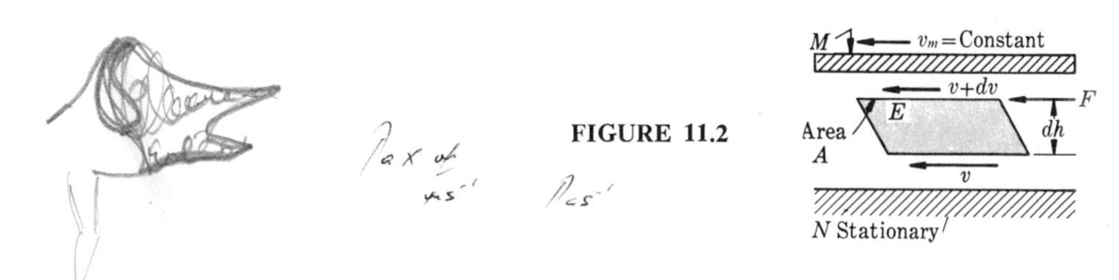

FIGURE 11.2

dh away, move with speed $v + dv$; thus, the difference between these speeds is dv. Among his vast contributions to science, Isaac Newton (1642–1727) enunciated the law that the shearing stress F/A in a fluid is proportional to the velocity gradient dv/dh;

$$\text{(11.1)} \qquad \frac{F}{A} = \mu \frac{dv}{dh} \quad \text{or} \quad F = \frac{\mu A v}{h}, \qquad \text{[NEWTONIAN FLUID]}$$

where A is the fluid area being sheared, and μ is the proportionality constant, called the ***absolute viscosity***, or commonly just ***viscosity***, of the fluid. The second form of equation (11.1) applies where the velocity gradient dv/dh can be taken as a constant. Observe from (11.1) that the frictional force F to shear the fluid increases as either the viscosity μ or dv/dh increases.

Viscosities are commonly given in metric units, poises or centipoises, but the practice in bearing design in this country is to use the ***reyn*** (named after Osborn Reynolds), which is a unit consistent with equation (11.1) in which we use inches, seconds, and pounds for force. Dimensionally and with these units, we have

$$\textbf{(b)} \qquad \mu = \frac{Fh}{Av} \to \frac{(\text{Force})(L)}{(L^2)(L/\tau)} \to \frac{P\tau}{L^2} \to \frac{\text{lb-sec.}}{\text{in.}^2}$$

where P represents the pound unit, τ (tau) the dimension of time, and L length. A poise has the units dyne-sec. per sq. cm. and is equivalent to 100 centipoises. Convenient conversion constants are:

$$6.9 \times 10^6 \frac{\text{centipoises}}{\text{reyn}} \quad \text{and} \quad 6.9 \frac{\text{centipoises}}{\text{microreyn}},$$

where a microreyn is seen to be one-millionth of a reyn. Some viscosities are given in Fig. AF 16. The viscosity varies with pressure,[11.54] but we shall assume that it is constant at some average value.

If the lubricant is not given in Fig. AF 16, it will probably be necessary to convert from Saybolt Universal Viscosity (SUV), which is the commercial viscometer reading used in this country, to absolute viscosity. This conversion is done through another property called the ***kinematic viscosity***, which is the absolute viscosity of the fluid divided by its density, each expressed in the same system of units. The basic dimensions of kinematic viscosity are L^2/τ. Because in the cgs system of units, density ρ is numerically the same as specific gravity SG, it is easier to find the kinematic viscosity ν from the absolute viscosity in centipoises Z;

(c) $$\nu = \frac{Z}{\rho} = \frac{Z}{SG} = 0.22\tau - \frac{180}{\tau} \text{ centistokes,}$$

where τ is the SUV reading in seconds and all properties are for the same temperature t. The specific gravity of a petroleum oil at any temperature t is given approximately by

(d) $$SG_t = SG_{60} - 0.00035(t - 60),$$

where SG_{60} is the specific gravity at 60°F (about 0.89 to 0.93 for petroleum oils).

11.5 PETROFF'S EQUATION.

If a journal is running in a film-lubricated bearing without load (or practically, with a light load and at moderate speed), the journal runs *concentric* with the bearing, Fig. 11.3, and the velocity gradient $dv/dh = v/h$ is constant. Newton's equation defining viscosity can be applied. The area being sheared at the journal is πDL sq. in.; the film thickness is equal to the radial clearance, $h = c_r = c_d/2$; then let the "rubbing" speed (the peripheral speed of the journal) be v_{ips} in./sec. The frictional torque $T_f = Fr$ on the journal, with equation (11.1), becomes

(e) $$T_f = Fr = \frac{\mu A v}{h} r = \frac{\mu \pi D L v_{ips}}{c_d/2} r = \frac{4\mu\pi^2 r^3 L n_s}{c_r} \text{ in-lb.}$$

[PETROFF'S EQUATION]

where the average film thickness $h = c_r$ in a centered journal, $v_{ips} = 2\pi r n_s$, and the units are consistent; μ reyns, D in. = journal diameter, r in. = journal radius, L in. = bearing length (axially), c_r and c_d in., n_s = revolutions per second of the journal (rps). Keep in mind that the unit *second* is involved in reyns. From the torque above, the frictional horsepower fhp is conveniently computed from hp = $Tn/63{,}000$, § 1.14, where n = rpm. For v_m fpm, the frictional work U_f is also $U_f = Fv_m$ ft-lb./min. and fhp = $Fv_m/33{,}000$.

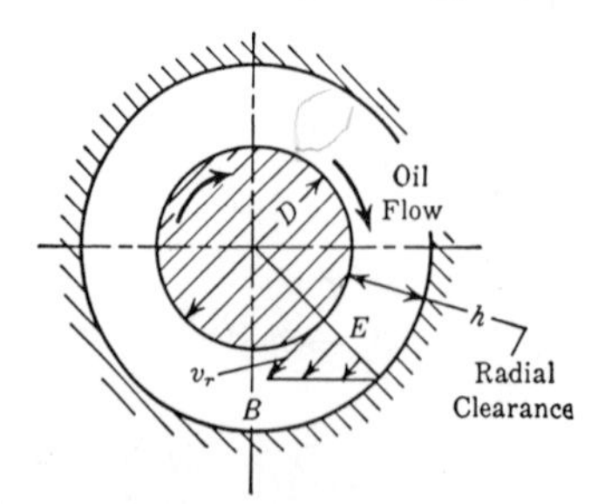

FIGURE 11.3 Concentric Journal.

11.6 HYDRODYNAMIC LUBRICATION. When the surfaces in relative motion are so oriented that their relative motion causes the oil pressure to build up to enough to support the load without metal-to-metal contact, the lubrication is hydrodynamic. The basic requirement for this to occur is that the lubricant enter the bearing via a converging channel, Fig. 11.4. As seen from the curve DEF, the pressure builds up from ambient to a maximum at E, which occurs somewhat closer to section D than to section F. The velocity distribution must be such as to satisfy the continuity of mass law. Note that the velocity gradient is not constant; the fluid is subjected to pressure.

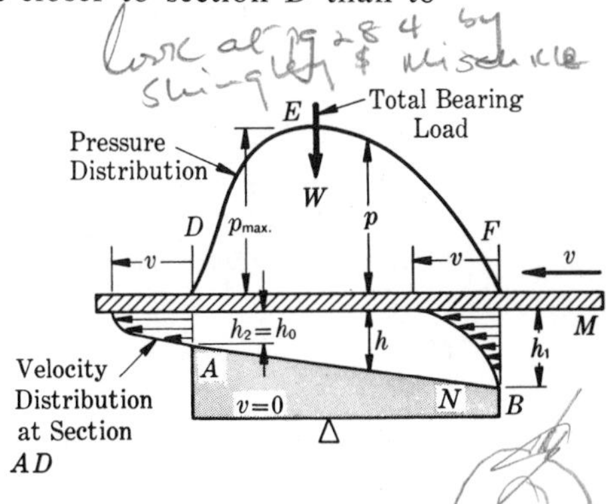

FIGURE 11.4 Hydrodynamic Film. The areas formed by the vectors that represent velocity distribution, typically as shown, are necessarily equal in order to satisfy the conservation of mass law for an incompressible fluid, since the areas are proportional to mass of flow. The maximum film thickness is h_1; the minimum, h_0. The resultant load W does not pass through the point of maximum pressure.

Consider a journal with load W at rest in its bearing in metal-to-metal contact, with the clearance space filled with oil, Fig. 11.5(a). As the journal begins to rotate clockwise, there is first some rubbing of metal on metal, and the journal climbs upward toward the right, Fig. 11.5(b). Since the oil adheres to the journal's surface, an oil film is drawn between the surfaces with the rotation, after which the journal moves to the left of bearing center O'; this is the equilibrium position, Fig. 11.5(c). Notice the wedge-shaped channel required by hydrodynamic theory. The pressure will build up until there is no metal-to-metal contact (if a hydrodynamic bearing has been correctly designed). The minimum thickness of oil film is designated h_o. A little intuitive consideration [especially with equation (11.1) in mind] suggests that (considering one variable at a time): the faster the journal

FIGURE 11.5 Mechanism of Lubrication. The angle of eccentricity ϕ locates h_o.

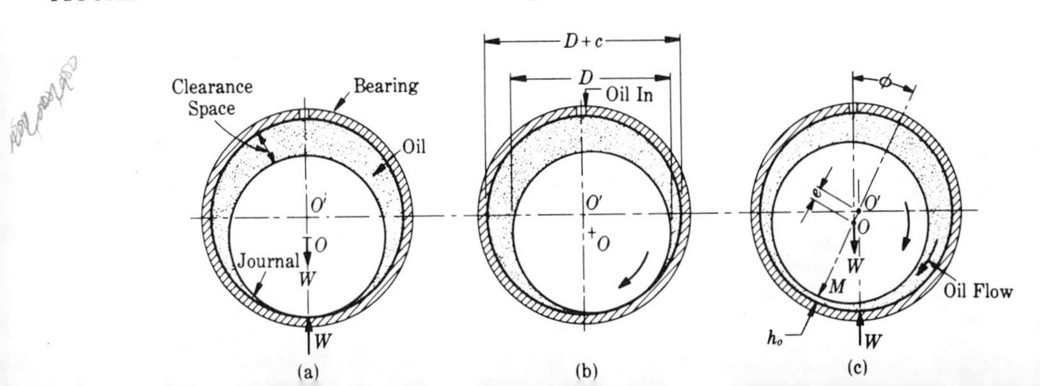

rotates, the greater the amount of oil passing through the load-carrying area and the larger h_o; the greater the viscosity, the greater h_o; the greater the pressure, the smaller h_o.

11.7 GEOMETRIC RELATIONS FOR A CLEARANCE BEARING. The line that passes through the centers of the bearing and the journal is called the ***line of centers***, Fig. 11.6. This line locates the minimum film thickness h_{min}, $= h_o$, provided that the angular length of the bearing is great enough to include the point M. If the bearing extends only as far as some section x, Fig. 11.6, the minimum film thickness h_{min} is at the trailing end of the bearing, and the separation of the circles at M is designated as h_o.

The distance $e = OO'$ between the centers of the journal and the bearing, called the ***eccentricity of the journal***, is

(f) $$e = \frac{c_d}{2} - h_o = c_r - h_o$$

where c_r is the radial clearance in inches. The eccentricity e divided by the radial clearance c_r is called the ***eccentricity ratio***, ϵ (also called *eccentricity factor* or *attitude*); that is,

(11.2) $$\epsilon = \frac{e}{c_r} = \frac{e}{c_d/2} = \frac{c_r - h_o}{c_r} = 1 - \frac{h_o}{c_r} = 1 - \frac{2h_o}{c_d},$$

a geometric relation that must be satisfied for a journal clearance bearing. If two of the quantities are known or assumed, the other must be calculated from equation (11.2).

The film thickness h at any angle θ measured in the direction of rotation from the line of centers is found from Fig. 11.6 as follows:

(g) $$\begin{aligned} h &= ag - ab = aO' + O'g - ab \\ &= e\cos\theta + (r + c_r) - \sqrt{(Ob)^2 + (Oa)^2} \\ &= e\cos\theta + r + c_r - \sqrt{r^2 + e^2\sin^2\theta}. \end{aligned}$$

But since e is quite small compared to r, the term $e^2 \sin^2 \theta$ may be dropped with negligible effect; which gives

(h) $$h = e\cos\theta + c_r = c_r(\epsilon\cos\theta + 1)$$

The load-carrying length of a bearing measured in a *circumferential* direction will be called the ***arc length*** L_A; it is the arc of the journal or the bearing subtended by the angle β, Fig. 11.6,

(i) $$\text{arc length} = L_A = \frac{D}{2}\beta = r\beta,$$

where β is in radians. The length of the bearing in the *axial* direction is called simply the ***length*** L. The angle ϕ, Fig. 11.6,

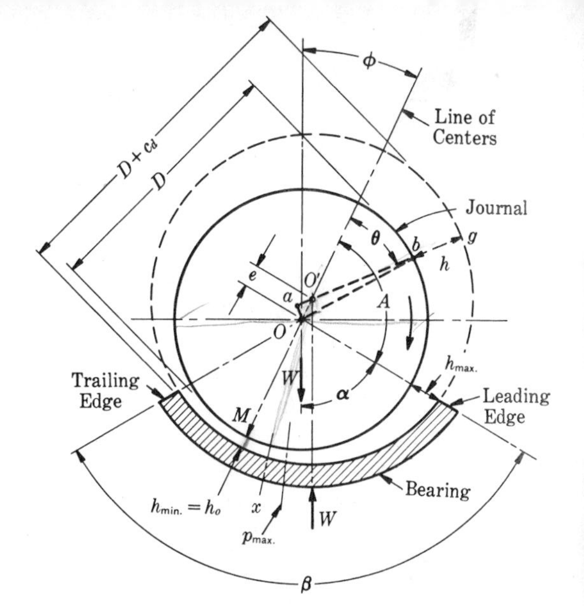

FIGURE 11.6 Central Partial Clearance Bearing, Running Position. Geometric relation for any journal bearing.

called the ***angle of eccentricity***, or *attitude angle*, locates the position of the closest approach h_o of the journal circle to the bearing circle, which is the minimum film thickness if the bearing extends as far as this.

11.8 LOAD CAPACITY AND FRICTION, JOURNAL BEARINGS.

There is a considerable convenience in working with dimensionless parameters in designing bearings. Theory, experiment, and dimensional analyses, when both journal and bearing are smooth right cylinders with axes parallel, lead us to the conclusions that (ϕ used to mean *function of*)

(j) $$\frac{h_o}{c_r} = \phi\left[\frac{\mu n_s}{p}\left(\frac{r}{c_r}\right)^2\right] = \phi(S). \qquad \text{[FIG. AF 17]}$$

Also, the *coefficient of friction* variable $fD/c_d = fr/c_r$ is

(k) $$f\frac{D}{c_d} = f\frac{r}{c_r} = \phi\left[\frac{\mu n_s}{p}\left(\frac{r}{c_r}\right)^2\right] = \phi(S) \qquad \text{[FIG. AF 18]}$$

where n_s rps is the angular speed of the journal, $p = W/(LD) = W/(2rL)$ psi, called the ***unit load*** and *bearing pressure*, which is seen to be the bearing load W divided by the projected area of the journal (use same expression for partial bearings), f is the coefficient of friction (the fluid frictional resistance F divided by the load W; $f = F/W$), μ reyns is the average viscosity, and the other symbols have the usual meanings, except that S stands for the parameter in the brackets, called the ***Sommerfeld number***, or *bearing characteristic number*, dimensionless in a consistent system of units:

(11.3) $$S = \frac{\mu n_s}{p}\left(\frac{r}{c_r}\right)^2 = \frac{\mu n_s}{p}\left(\frac{D}{c_d}\right)^2. \qquad \text{[DIMENSIONLESS]}$$

The relations (**j**) and (**k**) have been presented in graphical form, formerly for an infinitely long bearing, which is one for which the pressure gradient in the axial direction is zero (no end leakage). The procedure then was to correct the results for end (side) leakage by using appropriate factors.[11.15] Now we have solutions of the hydrodynamic equation (**a**) that care for end leakage and the accompanying axial pressure gradient. Figures AF 17 and 18 in the Appendix give the relationships of equations (**j**) and (**k**) for an L/D ratio of unity. Many other charts are given in the original Ref. *(11.7)*, but to save space we have compromised by reproducing Tables AT 20-23. Since various curves that can be plotted from these data are not straight lines, it is best *not* to make linear interpolations in the tables for actual bearings; however, linear interpolation is assumed to be permissible for pedagogical purposes. If the Raimondi and Boyd charts are not available [or others, Ref. *(11.56)* for example], one may interpolate graphically by plotting enough points from the table data to define a particular curve. The use of the various parameters in Tables AT 20-23 will be explained in examples below and by additional discussion later.

For a bearing with the ratio $L/D = 2$, one could interpolate half-way between values for $L/D = 1$ and $L/D = \infty$. For a bearing with $L/D = 1.5$, a rough approximation would be say an interpolation one-fourth of the difference from $L/D = 1$ to $L/D = \infty$. For $L/D > 1$, but not exactly 1.5 or 2, use $L/D = 1$ or with relatively small error, the interpolation for 1.5 or 2, whichever is closest.

11.9 OPTIMUM HYDROMATIC BEARINGS. Of the infinite number of solutions that can be found for a design situation, Kingsbury[11.16] showed that for a particular supporting arc β there is a certain eccentricity ratio ϵ (or certain value of $h_o/c_r = 1 - \epsilon$) that results in a maximum load capacity and another eccentricity ratio that results in a minimum frictional energy loss. The corresponding bearings are called ***optimum bearings***. Especially in bearings subjected to heavy loads or large frictional losses, the designer may very well care to work towards an optimum; the various optimums are defined in the headings of Tables AT 20-23.

11.10 EXAMPLE—FULL BEARING. Returning to the shaft for which computations were made in Chapter 9, design the journal bearing A, the one with the maximum load, § 9.4. From § 9.4, we find $n = 360$ rpm, 30 hp, $A_x = 581$, $A_y = 561$ lb. From § 9.11, a shaft diameter of $4\frac{3}{8}$ in. was found on the basis of deflection. (a) Choose suitable dimensions and find the frictional loss in the bearing. (b) For an average oil temperature in steady state operation of 160°F, choose an oil.

Solution. The resultant bearing load is

$$W = (A_x^2 + A_y^2)^{1/2} = (581^2 + 561^2)^{1/2} = 807 \text{ lb.}$$

There are now a number of decisions to be made. The journal diameter can safely be less than the shaft diameter; also this would provide a shoulder for lengthwise positioning. For the first computations, let $D = 4$ in. Let $L/D = 1$, or $L = 4$ in.; see § 11.17 for a discussion of considerations here. The tolerances and allowances should be commercial if possible. Considering first a class RC 5 running fit, § 3.4, for a 4-in. bearing, we get the following dimensions from Table 3.1:

Hole Limits		Shaft Limits	
	4.0000 in.		$4 - 0.003 = 3.9970$ in.
	4.0014 in.		$4 - 0.0044 = 3.9956$ in.

which, if the manufacturing processes are centered (§ 3.12), give an average clearance of $c_d = 0.0044$ in.; $c_r = 0.0022$ in.; $c_d/D = c_r/r = 0.0011$. Observe that the smaller the clearance, the smaller the permissible slope of the shaft at the bearing (Fig. 11.12); for a closely fitted bearing, this situation should be checked (§ 9.10).

Another important decision is the value of minimum film thickness, which is discussed later in § 11.14; for this design based on the average clearance, use $h_o = 0.001$ in; h_o needs to be less than c_r, a relationship that should be checked for the minimum manufacturing clearance if there is any doubt. Enter Fig. AF 17 [or Table AT 20] with

$$\frac{h_o}{c_d/2} = \frac{h_o}{c_r} = \frac{0.001}{0.0022} = 0.455 \qquad \text{[ENTER CHART]}$$

and find $S = 0.15$ [0.16]. The values in brackets are interpolated in Table AT 20 for comparison. Using this value with the expression for the Sommerfeld number, equation (11.3), ($n_s = 360/60 = 6$ rps, $p = W/A = 807/(4 \times 4) = 50.5$ psi, $c_r/r = 1.1 \times 10^{-3}$) and solving for the viscosity μ, we get

$$\mu = \frac{Sp(c_r/r)^2}{n_s} = \frac{(0.15)(50.5)(1.1)^2(10^{-6})}{6} = 1.53 \times 10^{-6} \text{ reyns},$$

$$Z = (1.53 \times 10^{-6})(6.9 \times 10^6) = 10.55 \text{ centipoises},$$

which is the desired average oil viscosity at the normal, steady-operating 160° temperature. The coefficient of friction variable for the foregoing $S = 0.15$, from Fig. AF 18 [Table AT 20] is 3.8 [3.93] $= fr/c_r$, or

$$f = 3.8\frac{c_r}{r} = (3.8)(0.0011) = 0.00418.$$

The energy loss to friction is

$$U_f = fWv_m = fW(\pi Dn) = (0.00418)(807)\left(\pi \frac{4}{12} \times 360\right) = 1270 \text{ ft-lb/min.},$$

where v_m fpm is the peripheral speed of a point on the surface of the journal to give the energy in ft-lb. This energy must either be radiated and convected from the bearing or the circulating oil must carry it away and dissipate it to the surroundings elsewhere. If calculations are now made for the minimum *likely* clearance (not the same as the allowance of 0.003—§ 3.12), using the same viscosity as found above, it is found that the minimum film h_o is less than the 0.001 in. assumed above, and a decision must then be made as to whether the smaller value is satisfactory. The frictional loss is found not to change much. Hence, the minimum clearance configuration is probably a safe one for operation if everything is as assumed and computed above.

(b) In Fig. AF 16, find the intersection of the line $t = 160$°F and the line of $\mu \times 10^6 = 1.53$ and note that the point is close to the curve for SAE 10 W; use SAE 10 (or 10 W).

11.11 EXAMPLE—OPTIMUM BEARING. What should be the diametral clearance, average oil viscosity, and coefficient of friction, if the bearing in the preceding example is a minimum coefficient-of-friction optimum? What oil would be recommended for the 160°F operating temperature?

Solution. From Table AT 20, we find that $h_o/c_r = 0.3$ for a full bearing, minimum frictional loss, which poses an immediate question because we must change the previous decision on either the film thickness h_o or the clearance c_r. The film thickness decided upon before can be reduced, but if so, extra effort should probably be made to get smooth surfaces and good alignment. In this example, the dilemma is settled by the statement of the problem, which implies that we keep $h_o = 0.001$ in. Therefore, $c_r = h_o/0.3 = 0.00333$ in.; $c_d = 0.0067$ in. (*Ans.*); $c_r/r = 0.00167$; $S = 0.078$ from Fig. AF 17. Solving for the viscosity μ from the Sommerfeld number, we get

$$\mu = \frac{Sp(c_r/r)^2}{n_s} = \frac{(0.078)(50.5)(1.67)^2(10^{-6})}{6} = 1.82 \times 10^{-6} \text{ reyns}$$

Use $fr/c_r = 2.4$ from Fig. AF 18 and get

$$f = (2.4)(0.00167) = 0.004,$$

compared with the previous answer of 0.00418. With the viscosity of 1.82 microreyns and 160°F, we locate a point in Fig. AF 16 as before and find it nearly midway between SAE 20 and SAE 10; the heavier one will cause both the frictional loss and the film thickness to be greater; the lighter one will result in a lower frictional loss and a thinner and less safe film thickness. Assuming that there is an adequate safety margin in the assumed $h_o = 0.001$ in., we recommend SAE 10. Note than an optimum bearing with a lower h_o (same c_r) has a lower frictional loss than an optimum for a thicker film.

11.12 LUBRICANT FLOW THROUGH BEARING. The hydrodynamic action of relatively moving surfaces has been explained (§ 11.6), and Fig. 11.7 shows typical circumferential pressure distributions at different sections of a finite-length bearing. The flow in this direction depends on the pumping action of the journal. For the ideal bearing, it can be computed from the *flow variable* $q/(rc_r n_s L)$, obtainable from Tables AT 20-23; that is,

$$\text{(l)} \qquad \text{Value from table or curves (dimensionless)} = \frac{q}{rc_r n_s L} \qquad \text{[FLOW IN]}$$

where q cu. in./sec. is the needed rate of flow into the leading end of the film to conform to hydrodynamic requirements. The pressure-distribution curves of Fig. 11.7 approximate the results on a $\frac{7}{8} \times 1$-in. full bearing; the maximum pressure occurs in the converging portion of the film; and we note that the pressure distribution is not symmetric about the line of action of the load F. Another point to observe is the small negative pressure on the trailing end of the film, Fig. 11.7. According to theory, for the infinite-length, 360° bearing, there are negative pressures on the non-load-carrying part of the same order of magnitude as the positive pressures; but of course pressures below absolute zero are impossible, and actually the measured

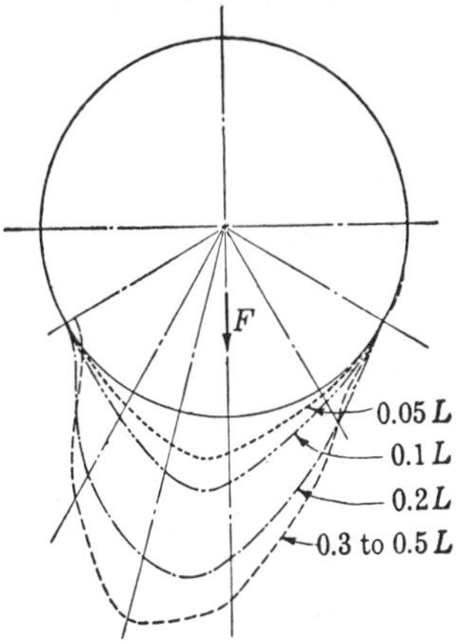

FIGURE 11.7 Circumferential Pressure Distribution.[11,17] The curve labeled 0.05*L* is a record of pressures around the bearing at a distance 0.05*L* from the *end* of the bearing, where *L* is the axial length of the bearing. Thus 0.5*L* represents the distribution of pressure at the center of the bearing.

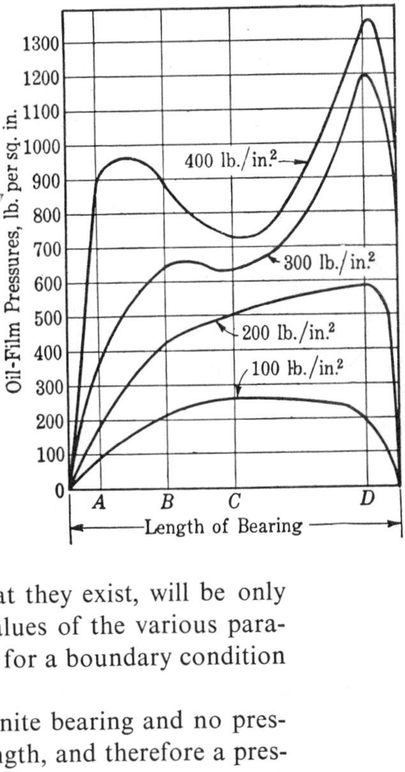

FIGURE 11.8 Longitudinal Pressure Distribution.[11,18] As seen, heavy loads may distort the curves considerably from the ideal parabolic form.

negative pressures, when conditions are such that they exist, will be only a small amount below ambient pressure. The values of the various parameters in Tables AT 20-23 have been determined for a boundary condition of $p = 0$ and no negative pressure.

There is no side flow (end leakage) in an infinite bearing and no pressure gradient, but there is in bearings of finite length, and therefore a pressure gradient. Measured pressure distributions on a 2.5 × 3.875-in. full bearing, when the oil is fed into one end, is shown in Fig. 11.8. The curves would have had better symmetry if the oil feed had been symmetric, a parabolic form in short bearings. For an estimate of the peak pressure in the bearing, see the column p/p_{max} in Tables AT 20-23.

The side leakage is computed from the flow ratio q_s/q, Tables AT 20-23;

(m) $$\text{Value from table or curve (dimensionless)} = \frac{q_s}{q},$$

which, multiplied into q from (l), gives the side flow q_s cu. in./sec., when the oil enters at ambient pressure in the *nonload-carrying region*, as it should in a simple hydrodynamic bearing (Fig. 11.5).

11.13 ENERGY INCREASES OF THE OIL.

The frictional loss in the bearing causes the temperature of the oil to increase Δt_o. Assume that all the frictional loss goes to increasing the lubricant temperature; then the amount of energy into the oil is $wc\Delta t_o$, where $w = \rho q$ lb/sec. is the mass rate of flow when ρ is the density (lb/cu. in.), c is the specific heat of the oil.*

(n) $$\text{Energy into oil} = Q = wc\Delta t_o = \rho qc\Delta t_o \text{ in-lb/sec.}$$

with consistent units. For hydrocarbon oils (from petroleum), the specific gravity will be close to 0.83 giving a density of $\rho \approx 0.03$ lb/cu. in. at usual temperatures; the specific heat is about $c = 0.4$ Btu/lb-°F (or 3734 in-lb/lb-°F). Keeping the units consistent, we have

(o) $$U_f = fW(\pi D'' n_s) = wc\Delta t_o = \rho qc\Delta t_o = (0.03)(3734)q\Delta t_o = 112q\Delta t_o,$$

[PETROLEUM]

where it is assumed that all the oil has a temperature rise Δt_o, the overall average, and where D'' is the journal diameter in inches. Now if all the oil that enters the film wedge leaves the bearing, easy to arrange in a partial bearing, if new oil at a temperature t_i°F is continuously introduced, and if the heat loss to the surroundings per unit of lubricant passing through is negligible, equation (**o**) can be used to compute the average temperature rise of the oil.

The *temperature-rise variable* $\rho c\Delta t_o/p$ in Tables AT 20-23 is somewhat different, as the additional assumption here is that "... the mean temperature of the lubricant leaving the sides of the bearing q_s was equal to the average of the inlet and outlet temperatures [of oil through the bearing]."[11.7] On this basis,

(p) $$\text{Value from table (dimensionless)} = \frac{\rho c\Delta t_o}{p},$$

where $\rho = 0.03$ lb/cu. in. and $c = 3734$ in-lb/lb-°F as given above for consistent units—and where Δt_o is the temperature rise of the circumferential flow $(q - q_s)$ and is larger than the average Δt_o computed from (**o**); therefore, it is a more conservative prediction.

11.14 PERMISSIBLE MINIMUM FILM THICKNESS.

The smallest permissible minimum film thickness is analogous to design stress in that it is a quantity that depends heavily on what experience has shown to be safe. The rougher the surface and the greater the misalignment (and shaft

*Don't confuse specific heat c, without subscript, with c_d or c_r.

deflection) or distortion from thermal gradients, the larger the minimum film thickness needs to be. Some operating conditions are such that the load can be carried only if the finest surfaces are used. In the ordinary commercial situation, thickness is usually such that the smaller particles of foreign matter can pass through without serious surface damage, and of course, it should be enough to care for unpredictable variations in the load (§ 11.32).

Data on design values of h_o are not plentiful, so that in exceptional cases in a particular situation, experiments may have to be conducted to determine safe limits. Karelitz[11.19] suggests $h_o = 0.0001$ in. as a minimum commercial limit, for finely bored small bronze bushings; $h_o = 0.00075$ in. for commercial babbitted bearings. Denison[11.21] suggests $0.0004 < h_o < 0.0006$ in. for 5-in. to 10-in. Diesel-engine bearings, speed 500–1200 rpm. Norton[11.12] proposes $h_o = 0.00025D$ in. as a rough rule, where D is the nominal diameter of the journal. Fuller[11.1] says that for medium-speed (500–1500 rpm) babbitt bearings on electric motors and generators, h_o may be 0.00075 in.; for large shafts at high speed (1500–3600 rpm), babbitt bearings, pressure oil feed, $0.003 < h_o < 0.005$ in.; for automotive and aviation reciprocating engines, bearings with fine surface finish, $0.0001 < h_o < 0.0002$ in.; but a filter to remove solid particles large enough ($\gtreqless 0.0001$ in.) to damage the surface is necessary. The surface for better grades of bearings may be finished to 32 μin. rms or below.[11.23]

11.15 EXAMPLE—PARTIAL BEARING, WITH TEMPERATURE RISE.

A journal ($D = 9$ in.) turning at 1700 rpm with a load $W = 20{,}000$ lb. has a length of 9 in., $L/D = 1$. The bearing is a 120° partial bearing, into which oil enters at an average temperature of 100°F. The diametral clearance is $c_d = 0.006$ in. ($c_r = 0.003$ in., approximately average for RC 5 fit—Table 3.1). Determine (a) a suitable film thickness h_o, (b) the Sommerfeld number and frictional loss, (c) the temperature rise from the temperature rise variable and from equation (**o**), (d) the attitude angle ϕ, (e) the grade of oil to be used.

(**Note.** Principal answers are *chart* values of the variables; answers in brackets [ ], from linear interpolations in the table for the particular h_o/c_r.)

Solution. (a) From § 11.14, $h_o = 0.00025D = (0.00025)(9) = 0.00225$ in. (Norton's rule). Fuller (§ 11.14) suggests 0.003 to 0.004 in. for pressure fed; it has not at this time been determined that the feed must be pressurized. Use $h_o = 0.0025$ in. for the first calculations.

(b) For $h_o = 0.0025$, enter Fig. AF 17 with $h_o/c_r = 0.0025/0.003 = 0.833$ and find $S = 1.2$ [1.39]. With $S = 1.2$, enter Fig. AF 18 and find

$$f\frac{r}{c_r} = 8.7 \; [9.8] \qquad \text{or} \qquad f = (8.7)(0.000667) = 0.0058,$$

where $c_r/r = 0.003/4.5 = 0.000667$. The peripheral speed of the journal is

$$v = \pi D n = \pi\left(\frac{9}{12}\right)(1700) = 4000 \text{ fpm, or } 66.7 \text{ fps.}$$

$$U_f = fWv_s = (0.0058)(20{,}000)(66.7) = 7740 \text{ ft-lb/sec.,}$$

or 92,900 in-lb/sec., or 7740/550 = 14.1 hp.

(c) The temperature-rise variable for $S = 1.2$ [$h_o/c_r = 0.833$] is 37 [41.6] = $\rho c \Delta t_o/p$. With $p = 20{,}000/81 = 247$ psi,

$$\Delta t_o = \frac{37p}{\rho c} = \frac{(37)(247)}{(0.03)(3734)} = 81.6^\circ.$$

The total oil flow should not be less than q from the flow variable, which is $q/(rc_r n_s L) = 3.14$ [3.14], or, for $n_s = 1700/60 = 28.33$ rps,

$$q = 3.14\, rc_r n_s L = (3.14)(4.5)(0.003)(28.33)(9) = 10.8 \text{ cu. in./sec.}$$

Using equation (o), we have the over-all average temperature rise

$$\Delta t_{ov} = \frac{U_f}{112q} = \frac{92{,}900}{(112)(10.8)} = 76.8^\circ,$$

which is fairly close to the previously computed 81.6° because the fractional part q_s leaving the sides is small; $q_s/q = 0.14$ [0.134].

These calculations indicate that if there is a plentiful supply of oil at the leading edge at a temperature of 100°F (given), the outlet temperature of the circumferential flow is about 181.6°F. Although it is not always possible to achieve, a maximum oil temperature below 180°F is preferable (§§ 11.20, 11.28).

(d) The attitude angle ϕ, locating h_o, is 61° by chart, 63° by interpolation on the table. Note that h_o occurs just a little beyond the end of the 120° bearing (the end is at 60° for central loading); therefore actual h_{min} is at the trailing end of the bearing. This bearing has a relatively low eccentricity ratio $\epsilon = 1 - h_o/c_r = 0.167$, and it may well be decided that other proportions shall be tried ($0.3 < \epsilon < 0.7$ would be preferable, other things being equal, which they never are).

(e) The usual assumption is that the average temperature is $(t_i + t_o)/2$, or 146°F in this case. From Sommerfeld number $S = 1.2$, we have

$$\mu = \frac{pS(c_r/r)^2}{n_s} = \frac{(247)(1.2)(0.667)^2(10^{-6})}{28.33} = 4.65 \times 10^{-6} \text{ reyns},$$

or 4.65 microreyns. From Fig. AF 16, the point for 146°F and 4.65 falls between SAE 30 and SAE 40, closer to 30; use SAE 30. The lower viscosity results in a lower frictional loss, so that less heat is generated, and the temperature of the oil film tends downward; this direction of change tends to increase viscosity, with the result that the oil film will not become as thin as one might think.

Since this bearing is operating at other than the best conditions, several other trials should be made before arriving at a decision. Considering the size and loading specified, it is evidently a part of a large and expensive machine. If the clearance is increased, more oil will flow through and the temperature rise will be less. If the oil is supplied under pressure (which is likely for such a large bearing, perhaps pressurized to avoid excessive metal-to-metal friction at start-up), more oil will flow through and the temperatures will again be lower. Lower temperature will permit the use of a lighter oil. The frictional loss fhp = 14.1, while not likely to be a significant percentage of the total power involved, is high enough to make a minimum-friction optimum bearing intriguing enough to investigate. The interested scholar may wish to try out some of these ideas.

11.16 CLEARANCE RATIO.

In the design of an actual bearing of any importance, one should probably investigate the effects of changing some

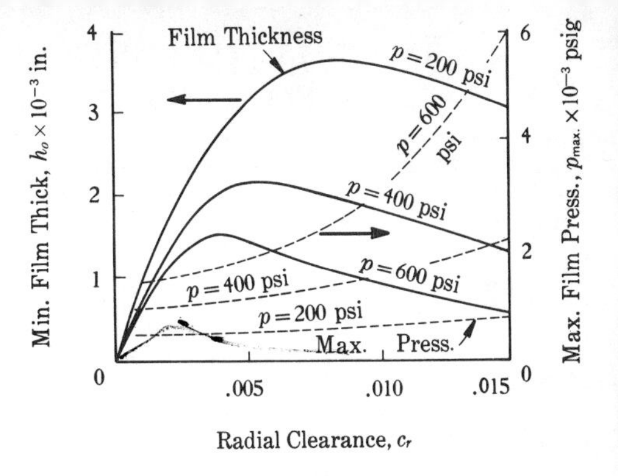

FIGURE 11.9 Clearance vs. h_o and p_{max}. Basic data: $D = 6$ in.; $L = 6$ in.; $n_s = 60$ rps; SAE 20 oil. By permission; after Raimondi and Boyd.[11.7]

parameters over which there is a control, for example, the clearance or clearance ratio. This can be done by assuming different values of, say, c_r and computing various items as suggested by Figs. 11.9 and 11.10. In Fig. 11.9, observe that for a particular load, the minimum film thickness comes to a maximum, but at different values of c_r or c_r/r for different loads. Also, the peak pressure in the film, dotted curves in Fig. 11.9, is sometimes little affected by clearance ($p = 200$), sometimes materially affected (at high loading). For power loss, Fig. 11.10, an increase in clearance sometimes results in a decrease, sometimes an increase; the loss is greater for the smaller films h_o. As would be expected, the amount of lubricant flow, Fig. 11.10, increases with clearance, and decreases as the load increases. The temperature rise as the oil passes through (Fig. 11.10) decreases with a greater flow (and larger c_r) up to a point, but significantly, is little affected after this point is reached. (The curves of Figs. 11.9 and 11.10 are drawn from data obtained for an ideal bearing with the negative pressures, but the general conclusions for this bearing should be valid for the tables and curves given in the Appendix.)

Although the limits of the clearance and clearance ratio are largely determined by manufacturing criteria, one may come closer to some

FIGURE 11.10 Clearance vs. fhp, q, and Film Temperature. Basic data as in Fig. 11.9. (By permission; after Raimondi and Boyd.[11.7])

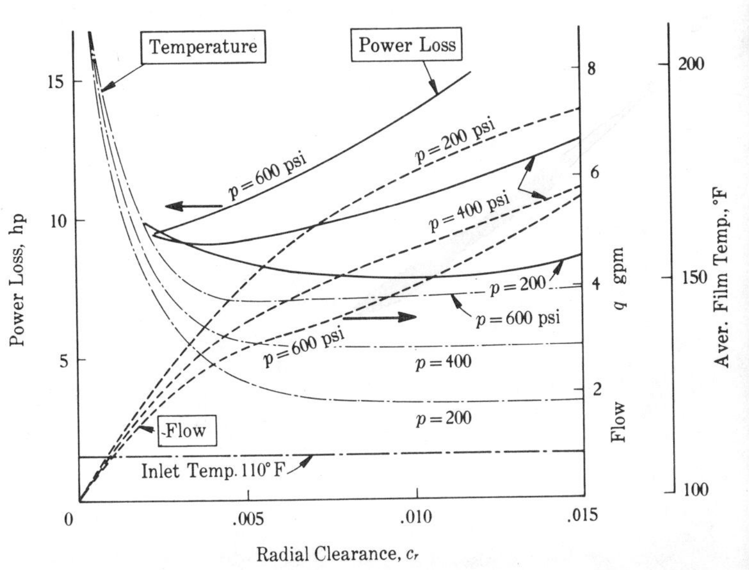

TABLE 11.1 TYPICAL DIAMETRAL CLEARANCES, INCHES[11.24,11.25]

Up to shaft diameter of →	$\frac{1}{2}$	1	2	$3\frac{1}{2}$	$5\frac{1}{2}$
Automotive crankshaft					
Babbitt lined bearing . .			0.0015	0.0025	
Cadmium silver copper .			0.002	0.003	
Copper lead . . .			0.0025	0.0035	
Precision spindle, hardened, ground, lapped into bronze bushing; $v_m < 500$ fpm, $p < 500$ psi; 8–16 μin. rms . .	0.00025 to 0.00075	0.00075 to 0.0015	0.0015 to 0.0025	0.0025 to 0.0035	0.0035 to 0.005
Precision spindle, hardened, ground, lapped into bronze bushing; $v_m > 500$ fpm, $p > 500$ psi; 8–16 μin. rms . .	0.0005 to 0.001	0.001 to 0.002	0.002 to 0.003	0.003 to 0.0045	0.0045 to 0.0065
Electric motor or generator, ground journal in broached bronze or babbitt bushing; 16–32 μin. rms. . . .	0.0005 to 0.0015	0.001 to 0.002	0.0015 to 0.0035	0.002 to 0.004	0.003 to 0.006
General machine practice, turned or cold-rolled journal in reamed bronze or babbitt bushing; 32–63 μin. rms. .	0.002 to 0.004	0.0025 to 0.0045	0.003 to 0.005	0.004 to 0.007	0.005 to 0.008
Rough machine practice, turned or cold-rolled journal in poured babbitt bearing; 63–152 μin. rms.. . .	0.003 to 0.006	0.005 to 0.009	0.008 to 0.012	0.011 to 0.016	0.014 to 0.020

desired optimum condition with little or no extra cost if one knows what optimum he desires. A long-standing rule of thumb is a clearance ratio of 0.001, but evidently this value will be the "best" value only coincidentally. The minimum permissible value of c_r/r, as for h_o, is dependent in some degree on the bearing surfaces. Table 11.1 may be helpful as a guide to practice.

11.17 LENGTH/DIAMETER RATIO. Needs[11.15] concluded some time ago that $L/D \approx 1$, say 0.8 to 1.3, was a good compromise for the general case of hydrodynamic bearings, and nothing has happened to change this conclusion. Thus, ratios of $L/D > 1.5$ are less common today than formerly. However, bearings with $L/D < 1$ are often used for a reason, say, the need for compactness as in a V-8 automotive engine. For a particular diameter and clearance with a certain load and journal speed, it can be said that as L/D decreases: the minimum film thickness decreases, perhaps rapidly below $L/D = 1$, Fig. 11.11; the film temperature increases, perhaps rapidly for L/D below 0.5; the power loss decreases; and the amount

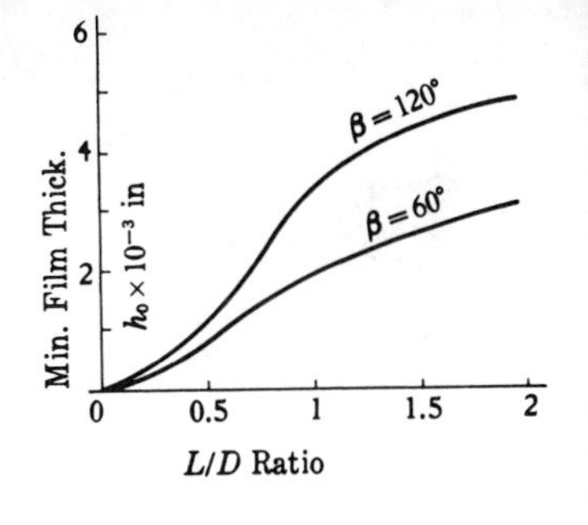

FIGURE 11.11 L/D vs. h_o. Constant clearance ratio. (By permission; after Raimondi and Boyd.[11.7])

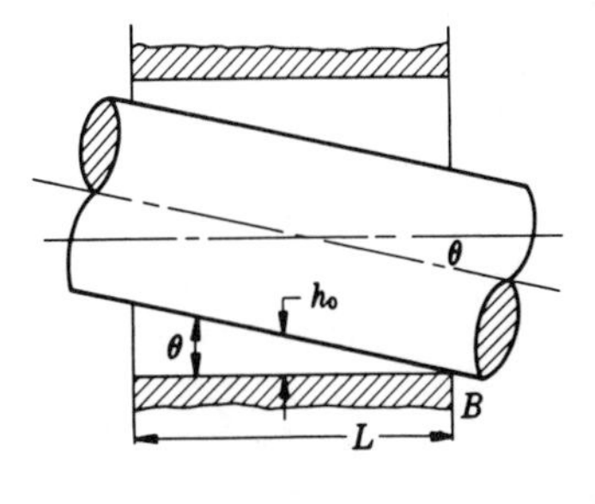

FIGURE 11.12

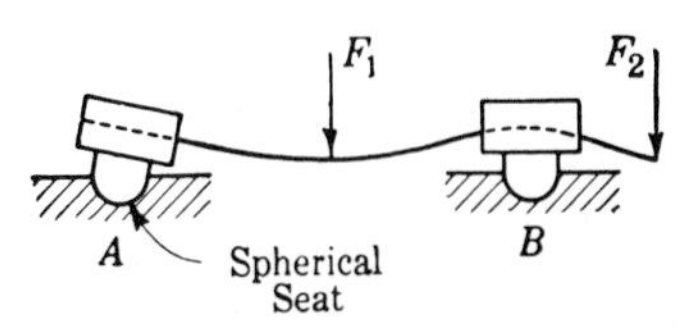

FIGURE 11.13 Self-aligning Bearings.

of oil flow (hydrodynamically pumped) through the bearing decreases. As L/D decreases, the eccentricity ratio increases (with decrease of film thickness) and the approach of the journal surface to the bearing surface soon becomes dangerously close, Fig. 11.11, suggesting a revision in some of the design assumptions. Fuller[11.1] gives the following L/D values as typical of practice: turbogenerators, 0.8–1.5; gasoline and Diesel engines, main and crankpin, 0.3–0.8; generators and motors, 1.2–2.0; shafts with self-aligning bearings, 3–4; machine tools, 2–4; railroad, 1.2–1.8.

Longer bearings (than $L/D \approx 1$) may be used for their help in maintaining alignments (light loads). On the other hand, misalignment may be the cause of bearing failure. In Fig. 11.12, for the indicated average h_o, a journal slope θ at the bearing as shown means the break down of the film at B, metal-to-metal contact, excessive heating that starts locally and spreads until failure occurs. Brown and Sharpe limits the shaft slope at a roller or journal bearing to 0.0005 in./in. ($= \theta$ rad.). A logical conclusion from Fig. 11.12 is that the longer the bearing, the more likely that a given slope θ results in a film breakdown, *but* for a given load, speed, and diameter, the film thickness h_o increases as L increases, Fig. 11.11, and it happens that the maximum misalignment for a particular loading can be tolerated when the $L/D \approx 1$.[11.7] In extreme cases, or wherever it is economically feasible, use self-aligning bearings. At the left-hand bearing, Fig. 11.13, the slope of the bearing adjusts itself well to the slope of the shaft; at an inner bearing, as at B, there could still be trouble if a long bearing were located at such a point as B where the slope is zero.

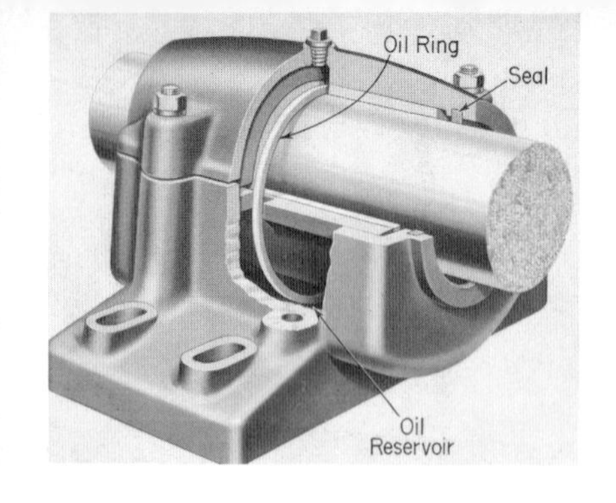

FIGURE 11.14 Ring-oiled Bearing. As the shaft rotates, the ring brings oil to the top of the journal. The lubrication is expected to be effective for a distance of about 4 in. on each side of a ring. Ring diameters are about 1.5 to 2 times the journal diameter. (Courtesy Link-Belt Co., Chicago).

11.18 HEAT DISSIPATED FROM A BEARING. Many bearings are made with an integral supply of lubricant, as for example a ring-oiled, Fig. 11.14, or chain-oiled or collar-oiled, Fig. 11.15, bearing, in which there is a local sump for lubricant storage. This type of bearing can be designed to deliver adequate oil for hydrodynamic needs, but it must be possible for such a bearing to dissipate all the frictional loss to the surroundings without an excessive temperature rise. For example, if a check is made (by the procedure explained below) on the partial bearing of § 11.15, it will be found that a very large temperature rise is necessary to dissipate the large amount of frictional loss. In this case, the normal arrangement is to circulate the oil to an external heat exchanger. Observe in Petroff's equation (**e**) that the

FIGURE 11.15 Collar-oiled Bearing. (Courtesy The Weller Mfg. Co., Chicago).

frictional loss ($Tn/63{,}000$) increases as the cube of the bearing diameter (also as the speed), whereas heat to the surroundings will be more nearly proportional to the first power of D for a particular temperature rise. Hence, as the size increases (implying a larger load too), and as the speed increases with a particular load, a point is reached where natural heat loss, the simplest arrangement, will not be sufficient. In the vicinity of this point, excessive frictional heat can be got rid of by blowing air across the bearing or by circulating a coolant through cooling coils in the bearing (Fig. 11.16).

Consider the case with the oil sump integral with the bearing. It may

FIGURE 11.16 Water-cooled-pillow Block. This style made in sizes from $1\frac{11}{16}$ in. to $3\frac{7}{16}$ in. (Courtesy Dodge Mfg. Corp., Mishawaka, Ind.).

take several hours for the bearing to arrive at steady-state operation. Even then, all parts of the bearing surface are not at the same temperature (the region in which most of the heat originates is in the vicinity of the maximum pressure and minimum film thickness), and conduction will result in a temperature gradient along the shaft. Fuller[11.1] has demonstrated that the loss due to this conduction is small, so that it can be assumed that the frictional energy is radiated and convected from the bearing as heat. Nevertheless, so many uncertainties remain that the computations are approximate; yet, if natural dissipation of heat is to be depended on, an effort must be made to estimate a steady-state operating temperature. Let the heat loss Q be given by*

$$Q = \mathbf{h}_{cr} A_b \Delta t_b \text{ ft-lb/min.,} \tag{11.4}$$

where $\mathbf{h}_{cr}$ is the heat-transfer coefficient for the bearing surface in ft-lb/min-sq. in-°F, A_b sq. in. is the effective area from which heat is lost, and Δt_b°F is the temperature rise of the bearing surface above the ambient temperature. An old rule of thumb[11.12] says to allow a heat loss in still air of 2 Btu/hr-sq.ft-°F, which, converted to the units of $\mathbf{h}_{cr}$ in (11.4), is

$$\mathbf{h}_{cr} = 0.18 \text{ ft-lb/min-sq. in.-°F in still air.}$$

The heat loss to moving air is greater than this, up to ten times greater according to some reports in the literature. Karelitz[11.19] found

$$\mathbf{h}_{cr} = 0.516 \text{ ft-lb/min-sq. in.-°F at 500 fpm}$$

speed of the air. The average temperature rise of the oil Δt_{oa} will be greater than that Δt_b of the bearing surface, how much greater depends on the method of lubrication and the construction of the bearing. The following approximations, adapted from Fuller,[11.1] may be used in connection with equation (11.4) for temperature ranges likely to be reasonable:

Oil-ring bearings, still air, $\Delta t_{oa} = 2\,\Delta t_b$,

Oil-bath bearings, still air, $\Delta t_{oa} = 1.3\,\Delta t_b$,

Waste-packed bearings, still air, $\Delta t_{oa} = 2.5\,\Delta t_b$.

For the ring-oiled bearing in 500 fpm air, Δt_{oa} in terms of Δt_b is some 15–20% greater than in still air.

What area A_b to use is also questionable. For heavy construction, as ring-oiled pedestal bearings, Norton[11.12] recommends $A_b = 25DL$ sq. in., where DL is the projected area of the bearing. For light construction, $A_b = 6DL$ will be safer. Fuller[11.1] recommends an average $A_b = 12.5DL$ for simple pillow-block bearings, and $A_b = 20DL$ sq. in. for pillow blocks with separate shells.

* Don't confuse $\mathbf{h}_{cr}$ with the film thickness h_o; h is commonly used in the sense of (11.4) in heat-transfer circles.

Considering the basic laws of heat transfer, we are reminded of the Stefan-Boltzmann law,[11.27] according to which *radiated* heat is proportional to the absolute temperature to the fourth power. By compromises, and considering that the temperatures do not vary widely, we arrive at

(q) $$\mathbf{h}_r = 0.108 \text{ ft-lb/min-sq. in.-°F}$$

as the unit rate of *radiated* heat (surface coefficient of radiation). Although no directly applicable equations have been developed for the *convected* heat from bearings, the expression for round pipe exposed to a moving external fluid[11.28] is helpful.

(r) $$\frac{\mathbf{h}_c D}{k} = 0.24 \left(\frac{D\rho v}{\mu}\right)^{0.6}$$

where $\mathbf{h}_c$ is the surface coefficient for convection, D is the pipe diameter, k, ρ, and μ apply to the surrounding fluid and are respectively the conductivity, density, and viscosity; the symbol groups are dimensionless and $D\rho v/\mu$ is the Reynolds number. When the surrounding medium is air, this equation reduces to $\mathbf{h}_c = Cv_a^{0.6}/D^{0.4}$, where C is a function of the properties of air and experiment. From meager experimental data, we choose $C = 0.017$; or

(s) $$\mathbf{h}_c = 0.017 \frac{v_a^{0.6}}{D^{0.4}} \text{ ft-lb/min-sq. in.-°F}, \qquad [v_a > 30 \text{ fpm}]$$

the unit rate at which heat is *convected*, where v_a fpm is the air speed and D in. is the nominal bearing diameter. The total coefficient is then $\mathbf{h}_{cr} = \mathbf{h}_c + \mathbf{h}_r$, the value to be used in equation (11.4). If the speed of flow of air across the bearing can be estimated, this detailed procedure may be preferred. Bearings located in the vicinity of flywheels, pulleys, etc., may be assumed to have an air flow at 60 to 100 fpm.

Computing an operating temperature of the bearing is a process of iteration. The basic procedure is to assume an average oil temperature (given the bearing dimensions, oil, etc.), compute the frictional loss U_f, and compute the heat loss Q; if $U_f = Q$, the assumed temperature is assumed to be the operating temperature. If $U_f \neq Q$, try another oil temperature; iterate until the desired agreement is achieved.

11.19 EXAMPLE—STEADY-STATE TEMPERATURE. Return to the full bearing of the example of § 11.10 from a different point of view. Use SAE 10 oil and estimate the average oil temperature during operation for an ambient temperature of 90°F. The data are: $W = 807$ lb., $D \times L = 4 \times 4$ in., $p = 50.5$ psi, $c_r/r = c_d/D = 0.0011$, $n_s = 6$ rps, peripheral speed $v = 377$ fpm. Let the bearing be a pillow-block type of medium mass for which the radiating area is taken as $12.5DL = (12.5)(4)(4) = 200$ sq. in.

Solution. Assume the average oil temperature as 150°F. From Fig. AF 16, find $\mu = 1.82 \times 10^{-6}$ for SAE 10 W. Then

$$S = \frac{\mu n_s}{p}\left(\frac{r}{c_r}\right)^2 = \frac{(1.82)(10^{-6})(6)}{(50.5)(1.1^2)(10^{-6})} = 0.179,$$

from which, $fr/c_r = 4.3$ from Fig. AF 18.

$$f = (4.3)(0.0011) = 0.00473,$$

$$U_f = fWv = (0.00473)(807)(377) = 1440 \text{ ft-lb/min.},$$

the rate at which the lubricant and bearing gain the energy of the frictional loss. To compute the rate of dissipation, assume that the nearby pulley results in an effective air speed over the bearing of $v_a = 80$ fpm, for which

$$\mathbf{h}_c = 0.017\frac{v_a^{0.6}}{D^{0.4}} = \frac{(0.017)(80^{0.6})}{4^{0.4}} = 0.135;$$

$$\mathbf{h}_{cr} = \mathbf{h}_c + \mathbf{h}_r = 0.135 + 0.108 = 0.243;$$

take $\Delta t_{oa} = 2\,\Delta t_b$, or $\Delta t_b = (150 - 90)/2 = 30$.

$$Q = \mathbf{h}_{cr}A_b\Delta t_b = (0.243)(200)(30) = 1460 \text{ ft-lb/min.},$$

compared with $U_f = 1440$ ft-lb/min., the kind of agreement that gladdens the heart of any iterator. If these numbers had been significantly different, another oil temperature would have been assumed and the calculations repeated. The conclusion is that the bearing as designed in § 11.10 will operate in a normal environment without overheating. From Fig. AF 17 and $S = 0.179$, we get $h_o/c_r = 0.5$, or $h_o = 0.5c_r = 0.0011$ in., somewhat larger and a little safer than the $h_o = 0.001$ assumed in § 11.10.

Be sure to distinguish between the Δt_o of §§ 11.13 and 11.15 and Δt_{oa} of this article; Δt_o is the temperature *rise* of the oil as it passes through the bearing from any entering temperature t_i; Δt_{oa} is the *difference* between the average oil temperature in the bearing and the ambient temperature; which may be, and is in this example, quite different from Δt_o. You may compute Δt_o from the temperature rise variable $\rho c\,\Delta t_o/p$.

11.20 OPERATING TEMPERATURES.

Conventional design values of the oil-film temperatures are 140° to 160°F, or less. At higher temperatures, the oil oxidizes, more rapidly above 200°F, and temperatures of the order of 250°F would not be tolerated in industrial equipment unless there is no reasonable alternative. In automotive practice, oil temperatures of some 350°F occur;[11.2] thus, oxidation is a significant cause of deterioration in this application. Some bearing materials will lose considerable fatigue and yield strength at the higher temperatures.

11.21 OIL FLOW FOR PRESSURE FEED.

Introducing the lubricant into the bearing under pressure will serve to increase the flow. The type of bearing in mind here is hydrodynamic with the oil entering the bearing, as usual, in a region of negligible pressure. Since the inlet pressure little affects the amount of the flow through the minimum film area, most of the increased flow becomes end leakage. Even so, the greater flow results in reducing substantially the average temperature of the bearing. Equations for computing the oil flow under pressure are derived in several works[11.2,11.4,11.5,11.13] from a free body of a particle of lubricant and the use of Newton's law of viscous flow, equation (11.1). The amount of flow for a particular clearance, load, and oil depends upon the details of how the oil is handled. For example, an oil hole at the mid-point of the cap may be the whole system, in which case, the axial flow in a 360° bearing as given by Shaw and Macks[11.4] is

$$\text{(t)} \qquad q = \frac{c_r^3 p_i}{3\mu}\left(\tan^{-1}\frac{2\pi r}{L}\right)(1 + 1.5\epsilon^2) \text{ cu. in./sec.,}$$

where the units are consistent (μ reyns), p_i psi is the inlet gage pressure, and the other symbols have the usual meanings (for conversion to gallons: 231 cu. in./gal.). The size of hole, not included in equation (**t**), also affects the rate of flow.[11.24] If, in addition to the hole, there is a longitudinal groove, as in Fig. 11.17, the flow will be two to three times greater than given by (**t**).[11.24]

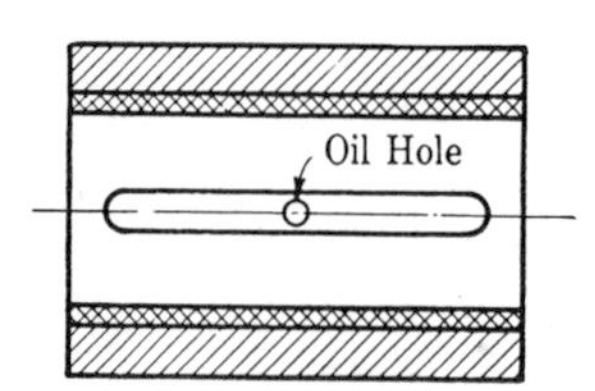

FIGURE 11.17 Grooved Bearing. This is the half of the bearing *not* carrying the load. Such an axial groove is helpful also in distributing the oil in oil-ring bearings.

In general, for hydrodynamic action, *oil grooves are not used in the load-carrying area*, because any marked discontinuity in the surface here causes the film pressure to drop to virtually zero gage. For example, the circumferential groove shown in Fig. 11.18 makes the bearing into two bearings with a pressure distribution something like that shown by the solid curves A, whereas without the groove, the pressure would be more like that shown by the dotted curve B. As you can see, for a given total load with a particular $L \times D$, the peak pressure in the grooved bearing will be greater (h_o less) than that in the ungrooved bearing. Nevertheless, circumferential grooving is common with pressure-fed oil, especially in the bearings of automotive engines, because more oil of a particular viscosity flows through a particular clearance providing greater cooling. The theoretical axial flow through a 360° bearing with a circumferential groove is[11.4]

$$\text{(u)} \qquad q = \frac{2\pi r c_r^3 p_i}{3\mu L}(1 + 1.5\epsilon^2) \text{ cu. in./sec.}$$

where the units are consistent (μ reyns) and have the meanings previously defined; $\epsilon = 1 - h_o/c_r = 1 - 2h_o/c_d$; L is strictly the total axial load-carrying length, Fig. 11.18. To be conservative *for temperature calculations* the designer may wish to assume an actual flow smaller than the amount computed from (**t**) or (**u**), say 70% to 75% of theoretical values [Ref. *(11.24)*]. Also, since it depends on the cube of the radial clearance c_r^3, the amount of

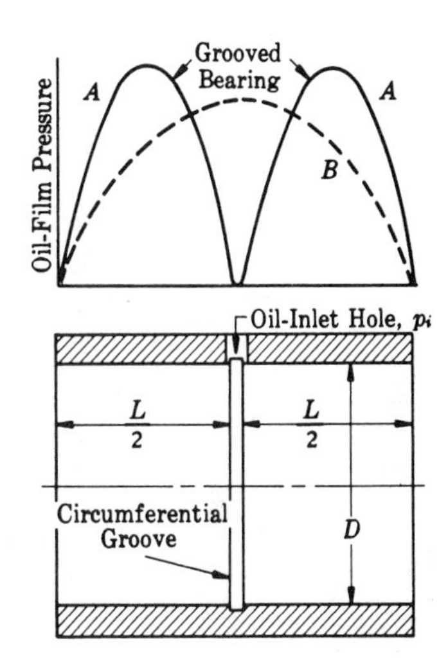

FIGURE 11.18 Circumferential Groove. If solid matter gets into the oil and scores the bearing around the circumference, the distribution of pressure may be affected in a manner similar to that pictured.

flow is critical to variations of this dimension; the minimum flow depends on the minimum actual c_r obtained in interchangeable manufacture.

The design of a bearing with a circumferential groove must be on the basis of a bearing length of $L/2$, Fig. 11.18. Although there is no reason to expect that significant bearing characteristics (h_o, ϵ, f) would be unchanged when the oil supply is pressurized, we can safely decide that oil supply under pressure will not cause the bearing to be less safe in any respect.

11.22 FRICTION LOSS IN CAP. Since the pressure gradients in the unloaded part are trivial, Newton's equation (11.1) for viscosity may be used to estimate the frictional loss in the unloaded part. For a cap arc between 120° and 180°, the average thickness of oil film may be estimated from*

$$\text{(v)} \qquad h_{av} = c_r(1 + 0.74\,\epsilon^2) = c_r\left[1 + 0.74\left(1 - \frac{h_o}{c_r}\right)^2\right] \text{ in.,}$$

which is the value of h to be used in (11.1). To illustrate, let $\epsilon = 0.49$,

* More information on the assumptions made to obtain this equation is given in the previous edition of this text; the path of the center of the journal in the bearing as the load is increased from zero is assumed to be semicircular, as it nearly is, as may be seen for the insert in the lower right-hand corner of Fig. AF 17 for $\beta = 360°$.

$c_r = 0.00392$ in., $\mu = 1.52 \times 10^{-6}$ reyns, peripheral journal speed $v =$ 4000 fpm or 800 ips, $D \times L = 9 \times 10$ in.

$$h_{av} = c_r(1 + 0.74\,\epsilon^2) = 0.00392\,[1 + 0.74\,(0.49)^2] = 0.00462 \text{ in.}$$

The area of lubricant being sheared in a 180° cap is $A = \pi DL/2 = 141.2$ in.^{2} Then from (11.1)

$$F = \frac{\mu A v_{\text{ips}}}{h_{av}} = \frac{(1.52 \times 10^{-6})(141.2)(800)}{0.00462} = 37.2 \text{ lb.}$$

$$U_f = F v_m = (37.2)(4000) = 148{,}800 \text{ ft-lb/min.,}$$

which is equivalent to a loss of 4.51 hp. It may be significant for some situations to note that this loss can be materially reduced by increasing the clearance in the cap, say as suggested by Fig. 11.19, leaving the normal clearance

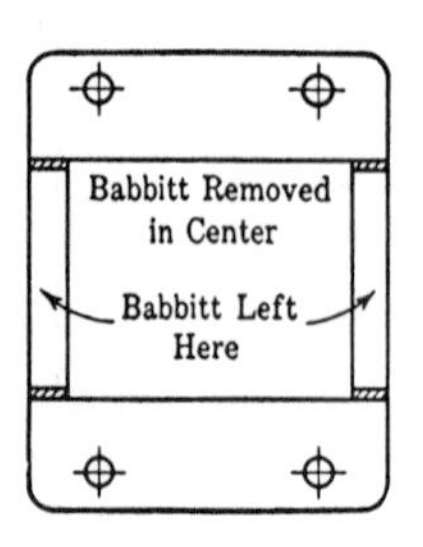

FIGURE 11.19

at each end to avoid excessive axial flow and to take some small unexpected change of load. Suppose the clearance in the cap were increased tenfold to $h_{av} = 0.0462$ in. We see that the fhp is reduced to 0.451 hp.

11.23 SIGNIFICANCE OF *Zn/p*. A few years ago when we were learning the full significance of hydrodynamic action in a bearing, the parameter Zn/p was a popular criterion; Z = viscosity in centipoises, n rpm, $p = W/(DL)$ psi. (Recommended values for different applications are found in the literature.) Values of Zn/p and $\mu n_s/p$ (dimensionless) are different only because of units. Each sleeve bearing has its own (Zn/p)-f curve, but they are all similar to that in Fig. 11.20, which shows what happens to the coefficient of friction f as Zn/p is varied by changes of Z, n, p, any or all. At the higher values of Zn/p, there is hydrodynamic action and thick-film lubrication. As Zn/p decreases from such a state, a point of minimum f is reached at B. If Zn/p is further decreased, the high points on the surfaces begin to touch and f begins to move up rapidly (toward a value of the order of 0.1 for an oil-wet surface). In the transition region BCD, we have boundary lubrication and the relations for hydrodynamic action no longer apply. The location of the lowest point B depends on the surface roughness; for the smoothest surfaces with the bearing carefully run in, which smooths the

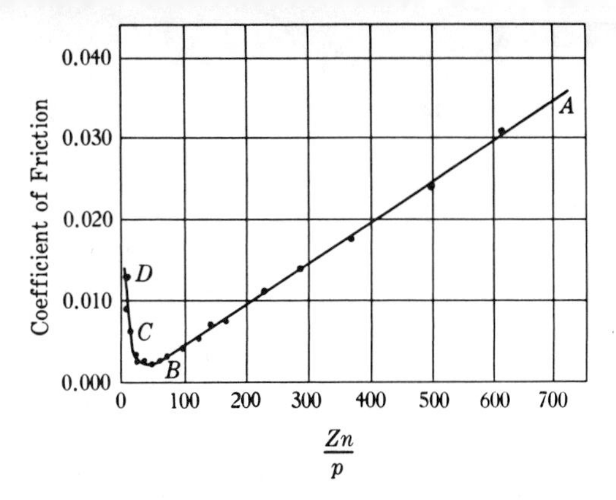

FIGURE 11.20 Values of f versus Zn/p for a Particular Full Bearing.[11.32] With smoother bearing surfaces, the minimum f would occur further to the left.

surfaces more, the minimum f may occur at $Zn/p \leqq 1$ (corresponding to a typical Sommerfeld number of $S \leqq 0.0025$, which is not an ordinary design value). A conclusion repeatedly verified by experiment is that *the smoother the surfaces, the greater the load capacity of the bearing*, in that the lubricant film can be thinner without metal-to-metal contact—a potent argument for breaking-in any new machine carefully with light loading as added insurance against failure.[11.30,11.31] The previous statement must be qualified to the extent that the bearing materials must be strong enough, especially in fatigue if the load is repeated, not to break or deform too much. Only in exceptional circumstances would a design value of $Zn/p < 5$ be used for hydrodynamic bearings; in the general case, $Zn/p \approx 50$ would be more reasonable.

11.24 THIN-FILM LUBRICATION. Many bearings operate as thin-film (boundary) bearings at all times, but of course the capacity of such bearings is not so great as when the surfaces are completely separated by an oil film. Boundary lubrication is also involved in metal cutting, screw threads, meshing gear teeth, piston and cylinder, etc. In bearings, thin-film lubrication may occur because the motion is oscillation, the speed is low, the viscosity is low, the pressure is high, the bearing is subjected to frequent starts and stops, and because of insufficient lubricant. Bearing design for thin films is primarily empirical. Yet in recent years, much has been learned of the phenomena involved; anyone bedeviled by boundary lubrication problems should search the recent literature for help. Space limitations prevent mention of significant detail here.

Polar molecules are typically long-chain, fatty-acid unsymmetrical molecules, one end of which is a nonhydrocarbon group; this end is active and attaches itself tenaciously to many but not all metals, the action being adsorption or chemical reaction. These molecules, as oleic acid and stearic acid, are inherent in vegetable, animal, and marine oils. When they are present in a lubricant, these molecules make a bond with surfaces, extending their long dimension more or less perpendicular to the surface. Several layers of molecules thus form a virtual carpet with reduced coefficient of friction, as compared to a surface wetted with nonpolar (symmetric, as hydrocarbons) molecules.[11.1,11.4] This effect is independent of viscosity and was first called the *oiliness* by Kingsbury, who had noted differences in f with the use of animal fats in thin-film lubrication. Fuller[11.1] quotes some interesting comparisons for boundary lubrication: nickel, a metal to which the polar molecule does not bond showed $f = 0.7$ for a clean surface, $f = 0.3$ with paraffin oil, $f = 0.28$ with oil + 1% lauric acid (not much change in f with the addition of lauric acid); copper, with which the polar molecular reacts strongly, showed $f = 1.4$ (clean), $f = 0.3$ (oil), $f = 0.08$ (oil + 1% lauric acid); aluminum, which reacts somewhat but not avidly with the acid, showed $f = 1.4$ (clean), $f = 0.7$ (oil), $f = 0.3$ (oil + 1% lauric acid); all values at low speed, 0.01 cm/sec. On the basis of surface roughness,[11.1] we find: for 2 μin. rms, $f = 0.128$ (mineral oil), $f = 0.116$ (oil + 2% oleic acid), $f = 0.099$ (100% oleic acid); for 20 μin. rms surface, $f = 0.360$ (oil), $f = 0.249$ (oil + 2% oleic), $f = 0.195$ (100% oleic). From these and other values, we may conclude: the fatty acid results in a significant lowering of the coefficient of friction for the more usual bearing materials, sometimes dramatically; fatty acid is of little help if it does not react with the material; the value of f is also quite dependent on the surface roughness.

The usual assumptions are made as for "dry" friction, to wit, $F = fN$ and F is independent of the area over which the load N acts. About the only general design advice that can be safely given is to be sure that the speed and bearing pressure (say $p \leqq 50$ psi) are low enough that the temperature rise during operation will be conservative. Maximum value of $pv_m = 50{,}000$ psi-fpm (see § 11.27). Rippel[11.24] advises a bearing length given by

$$\text{(w)} \qquad L = \frac{fWn}{15.28\,\Delta t}\ \text{in.,} \qquad [L \leqq 4D]$$

where W lb. = load, n rpm, Δt = temperature rise of the lubricant above ambient, and f = coefficient of friction. Values of f, the most problematic unknown, are likely to be between 0.08 and 0.15, but perhaps as low as 0.02 if there is a partial oil film. Equation (**w**) is based primarily on the assumption that the frictional loss is dissipated to the surroundings. This is an area of design for dependence on experiment or past experience.

11.25 CONSTRUCTION AND LUBRICATION. Only a little space can be devoted to construction details. Figures 11.14, 11.15, and 11.16, with the discussion of the kind of grooving used in hydrodynamic lubrication, have already been mentioned; Fig. 11.21 gives additional information for an oil-ring bearing. Experiments have been run on the amount of oil a ring will deliver to the bearing,[11.19] and such a bearing can be designed so that oil is plentiful. Observe that the lower half of the bearing with an oil ring, perhaps the load-carrying part, can be left without a discontinuity, but that a collar bearing, Fig. 11.15, is necessarily in two parts.

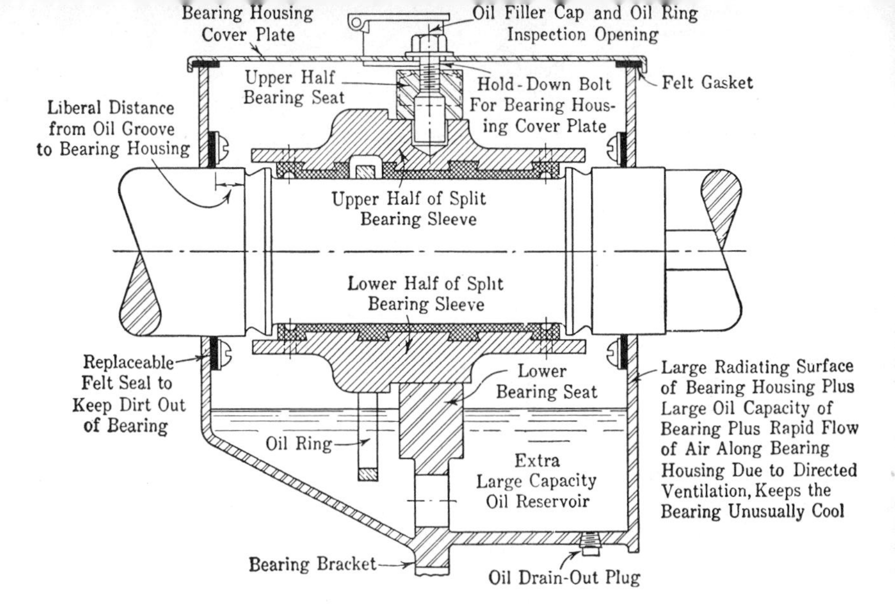

FIGURE 11.21 Construction of Ring-oiled Bearing. (Courtesy Electric Machinery Mfg. Co.).

When the load is on top of a journal, it may be possible for the shaft to dip into a sump of oil, or be in contact with oil-soaked waste as in some railroad wheel bearings. In reciprocating mechanisms, or others with moving links, it may be possible to lubricate bearings by *splash lubrication*, a moving member splashing oil from a sump. Systems that do or might result in boundary lubrication include: oil and grease cups, drop-feed devices, and wick feed (by capillary action).

Since it is insurance against a surface discontinuity in the load-carrying region, the line of action of the resultant load should fall within, say, a 60° angle in the center of one of the bearing halves; pillow blocks with the split on an incline, as in Fig. 11.22, may be helpful in this regard. In small bearings

FIGURE 11.22 Angle Pillow Block. Manufactured with the plane of the split at 30° or 45° with the horizontal. (Courtesy Link-Belt Co. Chicago).

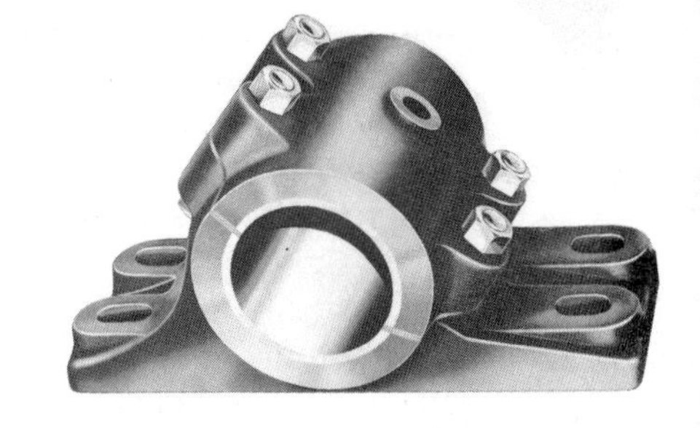

with one-piece bushings pressed into place, allowance should be made for the fact that the size of the bearing will decrease, or the bushing should be finished to size after it is in place. Bearings are made in a myriad of ways,[11.8] but a frequent characteristic, especially in large bearings, is to

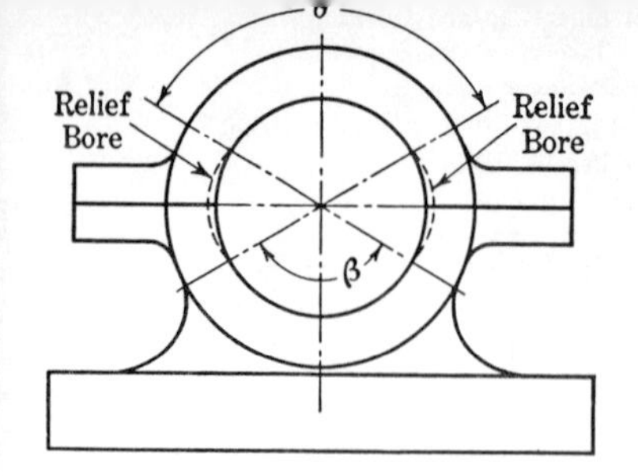

FIGURE 11.23 Bearing Relieved at Split.

V-groove the bearing at the slit, or to counterbore as suggested by Fig. 11.23. This relieved space on the side ahead of the section where the oil reaches the load-carrying surface makes a good place for the entry of lubricant, sometimes with most of it departing at the opposite side. If the bearing is pressure fed, the relieved areas should stop short of the ends of the bearings in order to avoid excessive side flow before the oil reaches the loaded area.

Grooves in the load-carrying area of a bearing are often justified for thin-film lubrication,[11.8,11.24,11.34,11.35] but, except for the circumferential groove, practically never in a hydrodynamic bearing. An interesting development is the availability of rods cased in bronze (that is, the bearing material is *on the journal*); if applicable, this arrangement may result in much longer-lasting bearings because the wear is not concentrated on a small load-carrying area of the softer material.

11.26 MATERIALS FOR BEARINGS. Properties that are considered advantageous for bearing materials include:[11.4,11.8] *conformability* (low modulus of elasticity, which means a larger deformation per unit of load), *compatibility* (which includes antiwelding property with respect to steel and score resistance), *embeddability* (soft enough to allow foreign particles too large to pass through the minimum film to be forced into the surface and thus to be removed from the oil film without scoring and wear), *low shear strength* (capacity to be smoothed easily), *compressive and fatigue strength* (capacity to withstand the maximum hydrodynamic pressure and to resist brittle fracture for repeated loading—at whatever temperature it operates), *corrosion resistance*, *good heat conductance*, nearly the same *coefficient of thermal expansion* as the material of the housing and journal, and, as always, *low cost*.

Most used are babbitts and copper alloys, Table AT 3. The babbitts are either tin-base or lead-base babbitts, depending upon which metal is the principal alloy. Babbitts are relatively weak in all forms, and lose strength rapidly with increasing temperature, with the result that they are used more and more in a thin layer (less than 0.04 in. and down to 0.002 in.) on steel backing. With their low fatigue strength, they are not satisfactory where the load is severe and variable, although the thinnest layers stand up. In 0.016-in. thickness, normal load capacity (thick-film lubrication) is about 1500 psi.[11.25]

The copper alloys used for bearings are mostly bronzes which are much stronger and harder than babbitt. A copper-lead mixture, 25–50% lead, in a layer about 0.03 in. thick, has good fatigue strength; normal load capacity is about 3000 psi. Tin bronzes have a normal load capacity of 5000 psi.[11.25]

Silver bearings for heavy duty are made by depositing a 0.02- to 0.03-in. layer of silver on steel, followed by a lead layer of 0.001 to 0.003 in.; then 4–5% indium is electrolytically deposited and thermally diffused into the layer of lead.

A cast-iron bearing with a hardened-steel journal has been found to be an excellent combination from the standpoint of wear and friction if thin-film lubrication exists. However, cast iron does not possess embeddability and other virtues of a softer material that usually outweigh its low cost.

Aluminum alloys have found favor for bearings in internal combustion engines and in other demanding applications because of good strength, conductivity, corrosion resistance, and low cost; but the mating journal should be hardened. A thin layer of babbitt may be used to improve embeddability.

Elastomeric materials such as rubber, Fig. 11.24, serve excellently with

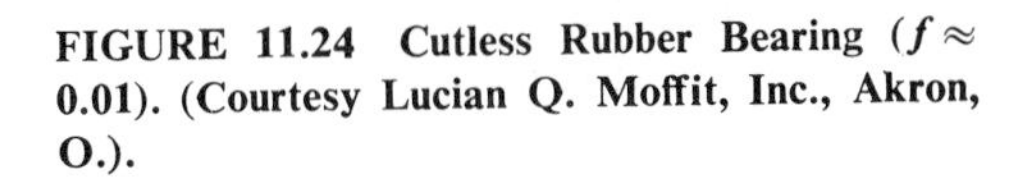

FIGURE 11.24 Cutless Rubber Bearing ($f \approx$ 0.01). (Courtesy Lucian Q. Moffit, Inc., Akron, O.).

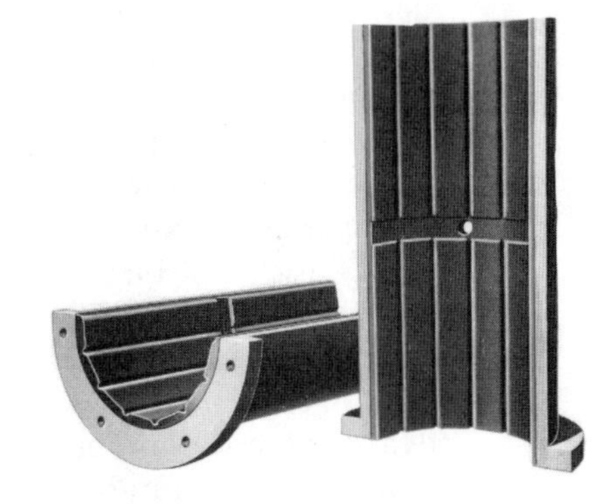

water as the lubricant, and are widely used for shafts of ship propellers, hydraulic turbines, hydraulic dredges, pumps, etc. The soft rubber passes sand and grit without scoring the journal.[11.8,11.36] Numerous other materials are used for bearings with and without oil or water lubrication, including wood (lignum vitae and oil-impregnated oak), plastics (nylon, Teflon, phenolic resins), and ceramics and ceramets (especially in unusually high temperature applications).

11.27 SEMILUBRICATED AND NONLUBRICATED BEARINGS.

There are millions of bearings in machines that are not expected to get much maintenance attention (especially those around the home) and others where maintenance may be impossible, difficult, or costly. In response to these needs, bearings have been developed that carry enough lubricant for a reasonable lifetime of service—or that require no separate lubricant at all. An early development in this direction is the *sintered bearing*, which is made by first compressing to shape a mixture of powdered metals (usually copper and tin or iron and copper), and then sintering at a temperature between the melting points of the two metals. The result is a porous material with voids of 10–35% by volume, which are then impregnated with an oil. The oil comes to the surface when the bearing is subjected to a higher temperature or pressure. Count on boundary lubrication and a relatively high coefficient of friction, say $f = 0.12$ as long as oil is present. Use limit values for temperature °F, bearing pressure p psi, peripheral speed v_m fpm, and pv_m as: 150°F, 4000, 1000, 50,000; Fuller[11.1] recommends design $p = 20$ psi and moderate speeds.

Plastics with and without lubrication are being widely adopted, already in a large variety of forms. Nylon is improved for this usage by adding graphite or molybdenum disulfide (limit $pv_m \approx 3000$). Helpful design information on Zytel (101 Nylon resin):[11.55,11.61] coefficient of friction without lubricant 0.15–0.33, with water 0.14–0.18, with oil 0.09–0.14. Limit values of pv_m for continuous and intermittent use respectively are: dry, 500, 1000; initially oiled, 2000, 2500; water, 2500, 2500; wick lubrication, 50,000, 70,000.

Teflon filled with glass fiber (25%) is dimensionally more stable than other plastics, will withstand a higher temperature, is corrosion resistant, and has a remarkably low coefficient of friction "dry," $0.05 < f < 0.1$ at low speed,[11.8] but more nearly 0.25 for speeds greater than 100 fpm and $p = 50$ psi.[11.41] Limit $pv_m = 10{,}000$ for low-speed ($v_m = 10$) continuous service; 20,000 for intermittent use.[11.37,11.41] Impregnated with molybdenum disulfide, this Teflon mixture has found favor for dry bearing surfaces in space machines. "Plastic" bearings can be designed for hydrodynamic lubrication, but one should keep in mind the dimensional variations with temperature and moisture content.

Carbon-graphite inserts are another type of dry bearing, the mixture of carbon and graphite acting as its own lubricant. They can tolerate temperatures of 750°F; limit $pv_m = 15{,}000$ dry, but not so much for continuous operation. Also, a graphite film some 0.00015 to 0.0005 in. thick, called Electrofilm, can be deposited on bearings, gears, cylinders, splined shafts, etc., to reduce friction and prevent galling. The use of bonded solid lubricants such as molybdenum disulfide in a binder of corn syrup, on bearing surfaces of various metals has proved successful for many inaccessible bearings, where the presence of oil or grease is disadvantageous, in vacuums, at very low temperatures (liquid hydrogen), and at high temperatures[11.40] (but MoS_2 decomposes into abrasive material at some temperature above 700°F).

11.28 LUBRICANTS. Solids, gases, and liquids are used as lubricants, but this article treats briefly of liquid lubricants only. The most important are petroleum oils, which usually contain one or more additives that advantageously change some property. Purposes of additives include: to reduce the rate of oxidation (antioxidants: phosphorous, sulfur, others); to maintain clean IC engine surfaces (detergents) by keeping insoluble particles in suspension; to reduce corrosion (anticorrosives or rust inhibitors), by adding compounds that have polar molecules that attach themselves to the surface (§ 11.24); to increase load capacity for boundary lubrication (extreme pressure (EP) agents), necessary for automotive hypoid differentials, for example; to lower the pour point (pour-point depressants); to improve the viscosity index (VI improvers); to prevent formation of foam (foam inhibitors), helpful where severe churning of oil occurs.[11.5,11.8]

Since petroleum oils are limited to a temperature range of roughly −10°F to 250°F, synthetic liquid lubricants (see solid lubricants above) have been and are being developed. Among the favored ones are the silicones, which have a high viscosity index and have been used to −100°F and to 400°F, or better.[11.5]

11.29 THRUST BEARINGS. Many rotating shafts, some vertical, are subjected to axial forces of significant magnitude, forces that must be balanced at a bearing, called a ***thrust bearing***. The simplest type carries the thrust on parallel surfaces (no wedge shape film), as on the end of the shaft, Fig. 11.25, or on collars, Fig. 11.26. Unless the oil is introduced onto the bearing surfaces under a pressure sufficient to support the load (§ 11.30), boundary lubrication should be assumed ($f = 0.1$ to 0.03), and the bearings used for moderate operating conditions ($50 < p < 200$ psi, $50 < v_m < 200$ fpm).

FIGURE 11.25 Thrust Bearing for Vertical Shafts. (Courtesy Link-Belt Co., Chicago).

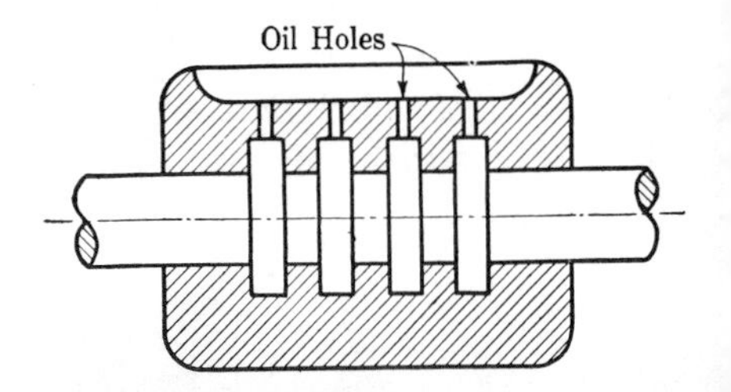

FIGURE 11.26 Sketch of Collar-thrust Bearing.

Hydrodynamic thrust bearings were designed by Albert Kingsbury in this country and A. G. M. Michell in Australia, at about the same time working independently, each being guided by the classical work of Reynolds.[11.42] The basic elements of such bearings are a collar and a group of nonrotating segments, Fig. 11.27. The segments in the actual bearing may be at a fixed angle of attack, they may be pivoted so that they are free to assume any angle of attack, or they may be mounted on flexible supports that permit variation with minor constraint. With oil adhering to the collar,

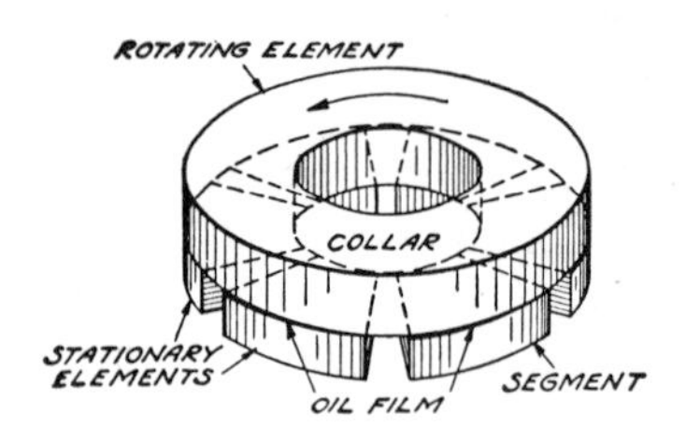

FIGURE 11.27 Basic Elements of the Kingsbury Bearing. Observe that the wedge formation of the oil film insures a complete separation of the metal parts, provided of course that the oil is suitable for the load and the speed. (Courtesy Kingsburg Machine Works, Inc., Philadelphia).

which is attached to the rotating member, and with the converging wedge-shaped film, hydrodynamic pressure is built up on the segments, which support the load. These bearings are designed by the hydrodynamic theory, an application somewhat simpler than the form applied to journal bearings;[11.1] their coefficients of friction are of the same order as those for journal bearings. The load capacity is a function of $\mu n_s/p$. See Ref. *(11.44)* for charts. Fuller[11.1] notes the 96-in. diameter bearing supporting the 2,150,000-lb. load of the rotor of a Grand Coulee Dam hydraulic turbine; $f \approx 0.0009$.

On occasion, large *journals* are supported by pivoted segments, three or more, distributed about the circumference. Such bearings can take a radial load varying in any direction and, moreover, the segments will hold the shaft accurately in a radial position (as for a machine tool spindle).

11.30 HYDROSTATIC LUBRICATION. A loaded journal at rest soon makes surface contact with its bearing. Since this is so, it is evidently wise to start machines under no load or light load to avoid excessive wear of the surfaces on start up, especially for machines frequently started. However, in many cases, as for turbine rotors, this is impossible. Also, there are many other journal bearings that turn so slowly that hydrodynamic action does not build up a separating film. And finally, there are thrust bearings, Figs. 11.25 and 11.26, that are not inherently designed for producing hydrodynamic action. To reduce the large friction that would otherwise exist in these situations, oil is pumped into the load carrying area at sufficient pressure to support the load. The turbine journal may be "floated" in oil at start up, but after it is up to speed, hydrodynamic action may be permitted to take over. In some situations, the oil pumps are kept

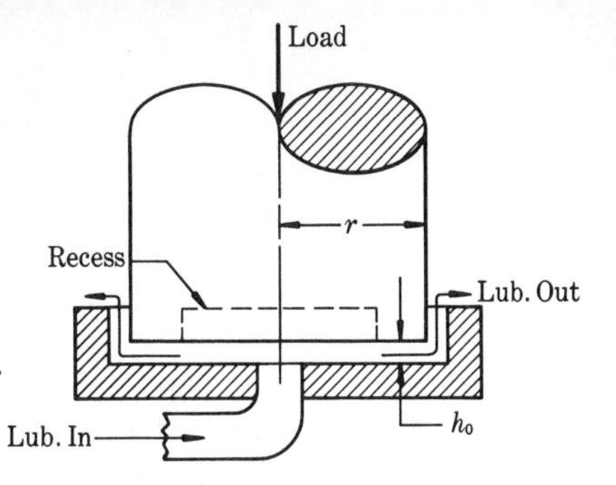

FIGURE 11.28 Hydrostatic Thrust Bearing.

running providing the film; where the film is maintained by oil flow due to external oil pressure, as in the elementary hydrostatic thrust bearing of Fig. 11.28, the lubrication is said to be hydrostatic. Large, heavy telescopes, antennas, etc. are often supported on hydrostatic films with coefficients of friction surprisingly low. Fuller[11.1] calculates a value of $f = 0.00046$ for one case and quotes another of $f = 0.00000075$. These values come about largely because of the low speed, inasmuch as the frictional force to shear a fluid at zero speed is zero. Moreover, if a comparison were to be made on the basis of total loss, the oil pump work is charged against the bearings, which will make the equivalent coefficient of friction perhaps of the same order as those computed for hydrodynamic action. Methods of making theoretical calculations for hydrostatic bearings are found elsewhere.[11.1,11.46,11.65]

11.31 GAS-LUBRICATED BEARINGS.

Gas, especially air, is used as a lubricant and does fine except that its load-carrying capacity is quite small (say 10 psi). The hydrostatic air bearing is used as a thrust bearing, Fig. 11.28; and air bearings are also designed to operate hydrodynamically. The solution of the Reynolds equation that we have previously used is on the basis of an incompressible fluid in laminar flow. Since a gas is compressible, other solutions allowing for varying density are needed and are available (tables and charts in Ref. *11.47*).[11.1] For low Reynolds numbers and light load, the Petroff equation (**e**) gives a reasonable estimate of the frictional moment. Because of the low viscosity of gases, the frictional losses are only a fraction of those for liquid lubricants of any kind, and gas is sometimes appropriate for exceptionally high speeds and light loads, as in gyroscopes, and instruments. Some applications: General Electric reports the development of a small cryogenic system, with a weight of 50 lb., that liquefies helium, whose compressor has been rotated at 350,000 rpm, and in which a small turboalternator rotating at 250,000 rpm expands the helium, both turning on air bearings; Boeing reports a hydrostatic air thrust bearing running at 100,000 rpm; grinder bearings with air hydrostatic lubrication turn up to 100,000 rpm; speeds up to 1,300,000 rpm have been attained in the laboratory.[11.1]

A troublesome phenomenon is likely to be ***whirl*** or ***whip***, which is the orbiting of the journal center with respect to but not necessarily around the bearing center. This is an instability that may manifest itself in a destructive vibration. Journal whirl also occurs in high-speed, lightly loaded, oil-lubricated bearings, and is called *half-frequency whirl.*[11.24]

11.32 DYNAMIC LOADING. Many bearings, as in internal combustion engines, are subjected to high impulsive loading. If the time of the impulse is short, it does not necessarily follow that the film thickness goes to zero, even though the maximum load is theoretically large enough to predict this, simply because it takes time to squeeze the oil from the bearing. The same effect is of course involved in any kind of bearing. Some calculations can be made for this kind of situation,[11.1,11.4,11.24] spoken of as *squeeze-film lubrication.* Not only must there be a copious supply of lubricant, but oil holes or grooves in the area taking the impulse will certainly make surface contact more likely by reducing the attainable squeeze-film pressure.

11.33 CLOSURE. All relatively moving surfaces in contact profit from some lubrication, but the subject is too large to be covered in a survey. In general in this book, we assume that if lubrication is proper and adequate, the design as computed should perform satisfactorily—as in gear-tooth design.

In designing journal bearings of importance by the theory we have outlined, the engineer is likely to make numerous solutions, perhaps in the form of curves similar to those of Figs. 11.9, 11.10, and 11.11, from which he chooses the solution that seems to balance the pros and cons. The application of theory to hydrodynamic bearings for the first time, no matter what traditional practice says, should lead to improvements.

12. BALL AND ROLLER BEARINGS

12.1 INTRODUCTION. Probably the most significant advantage of rolling-contact bearings is that the starting friction is not very much larger than the operating friction (at usual speeds and contrasted with an initial metal-to-metal rubbing of ordinary sliding-contact bearings); that is, the "coefficient of friction" (§ 12.14) varies little with load and speed, except at extreme values. This property makes rolling-contact bearings particularly suited for machines that are started and stopped frequently, especially under load (as axle bearings on railway cars, where they are being slowly adopted). Other characteristics include: they require little lubricant and maintenance; they occupy less axial space but larger diametral space than journal bearings; they are noisier than journal bearings, more expensive; they have a limited life because the raceways are highly stressed repeatedly as the shaft rotates, resulting in eventual fatigue failure; and several types of bearings may take a radial load *and* an axial thrust.

Sliding-contact and rolling-contact bearings have advantages which make one type more suitable in a particular application than the other, and it cannot be said that either one is better except in reference to a given engineering problem. The rolling bearing is a standardized, specialized, and precision device that the machine designer does not design; from a large variety, he selects an appropriate type, style, and size from a catalog. For intelligent selection, he needs to understand the basic considerations affecting capacity and life. Although we speak of "rolling-contact bearings," this is a device for classification inasmuch as in operation both balls and rollers do considerable sliding (sliding and rolling).

12.2 STRESSES DURING ROLLING CONTACT. The stresses involved are quite high because of the small area of contact (a force of 1 lb. pressing a ½-in. ball on a ½-in. ball induces a maximum stress of about 150 ksi[11.4]), and the Hertz equation for two cylindrical surfaces in static contact in § 1.27 is applied to roller bearings. (Review § 1.27 before reading further.) However, at the outset, we should recognize that the sliding (frictional force) results in a tensile stress near the surface, and it is this force that is thought to cause pitting.[4.64] Also, since the contact stresses are greater than the yield strength, rolling induces *residual* compressive stresses in races and balls, which increase with repetitions, but there are balancing residual tensile stresses just below the surface. Moreover, the deformations change significantly the local values of the radii of curvature that appear in the Hertz equation. It follows then that for rolling contact as in these kinds of bearings, the Hertz stress is more of an index than an actual evaluation of the stress. Usually, fatigue cracks clearly start at pits or inclusions; so, it is not unexpected that bearings made from vacuum-melted steel have a longer life.[12.3]

Similar equations for a sphere on variously shaped surfaces have been derived. However it is easy to understand without these equations that one factor affecting the magnitude of the actual stress is what is fetchingly called the *degree of osculation*. The area in contact in Fig. 12.1(a) is greater than, and the peak stress is less than, the corresponding area and stress in

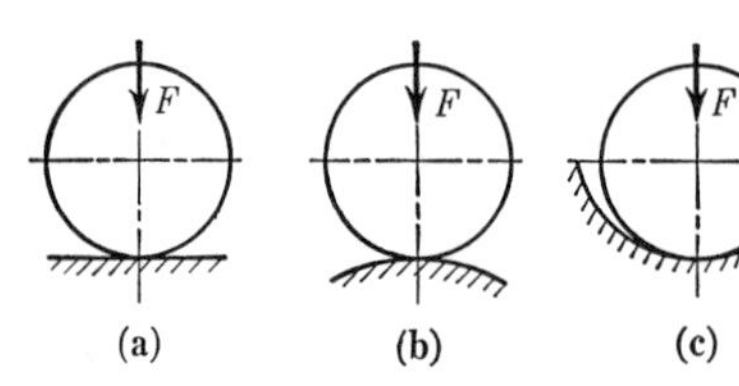

FIGURE 12.1 Degree of Osculation. The greater the osculation, the lower the stress; after some relation of diameters is reached in (c), decreasing the difference between radii of ball and race results in increasing friction.

Fig. 12.1(b). Because the area in Fig. 12.1(c) is greater, the stress is lower than that of either of the other two. Partly for this reason, the ***races*** of ball bearings are curved to wrap about the balls, Fig. 12.2. Even with the most favorable shapes, the peak stress at a particular point in a bearing race (or on a ball or roller) as the rolling element passes across that point under load (even with a "light" load) is quite high. Since such a stress is repeated with shaft rotation, it is natural to expect that eventual failure will be a fatigue failure.[12.22] *And since any normal loading induces stresses above the fatigue strength involved, rolling bearings have a limited life*, limited by the number of repetitions of stress to cause fatigue failure. For example, on average, with a particular load, doubling the speed of rotation reduces

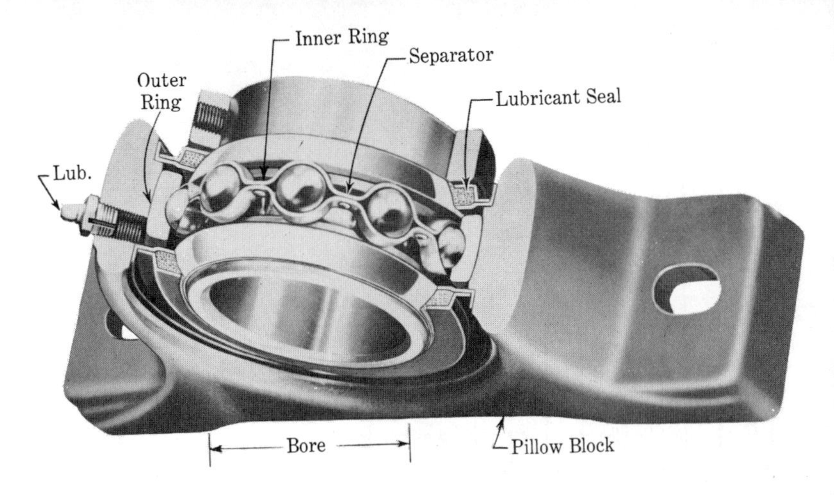

FIGURE 12.2 Single-row, Deep-groove Ball Bearing Mounted in Bearing Block. Since the balls in this type of bearing are loaded by the eccentric displacement of the inner race, the spacing is greater than when a filling notch is used. Compare with Fig. 12.6. (Courtesy Link-Belt Co., Chicago).

the time life of the bearing by one-half. As the speed increases to high values, centrifugal force begins to have an effect; for example, a number 206 ball bearing rotating unloaded at about 20,000 rpm had a life of about 1000 hr.[11.5]

A change of the load has a much more decisive effect on the maximum stress and life. Palmgren[12.1] found experimentally that the life B varies inversely as the k power of the load F; $B \propto 1/F^k$, where values of k from 3 to 4 are found in the literature; the ASA recommends 3 for ball bearings, 10/3 for roller bearings. In equation form,

$$\left(\frac{F_1}{F_2}\right)^k = \frac{F_1{}^k}{F_2{}^k} = \frac{B_2}{B_1} \quad \text{or} \quad \frac{F_1}{F_2} = \left(\frac{B_2}{B_1}\right)^{1/k}. \tag{12.1}$$

Observe that for $k = 3$, halving the load increases the life by 8 times; etc.

Environmental conditions that reduce the fatigue life of metals, corrosion from acid or water for instance, also reduce the life of rolling bearings. The preferred unit for measuring life is millions of revolutions mr, but some catalogs are set up with a certain number of hours at various rotative speeds. Fatigue occurs in the form of spalling or flaking of small particles of material from the surface of a ring or ball (or roller), because of the high shearing stress just below the surfaces in the vicinity of contact (§ 1.27). When a rolling bearing has "failed" in this way, it becomes noisy, thereby giving warning. Continued running may eventually cause a ball or roller to break, the end result of the progression of a fatigue crack.

12.3 STATISTICAL NATURE OF BEARING LIFE. Even for operation under controlled conditions, fatigue tests of bearings which are commercially identical show a range of life as great as 50 to 1. Thus there is no way to predict the life of an individual bearing, and since this is so,

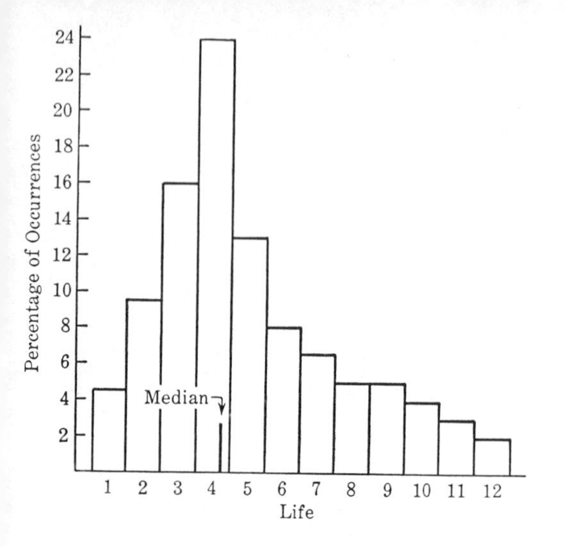

FIGURE 12.3 Histogram Showing Life of Rolling Bearings. A normal curve (Figs. 1.14, 3.3, 6.10 etc.) could be computed to represent these data, but they are evidently skewed. The Weibull distribution is a better representation.

the only logical approach is from statistics and probabilities. The histogram of Fig. 12.3 shows the distribution of the lengths of life, measured by index numbers, as found for a particular series of tests. By way of interpretation, cell 3 shows that about 16% of the population (total number tested) had a life between 2.5 and 3.5 units, where the median life was 4.35 units; the median life is the middlemost value. All the reasons for such a large spread are not clear; actual defects that may result in the very short-lived ones are not evident on examination; perhaps undetectable crystalline flaws, perhaps unusual and unknowable residual stresses.

The ASA[12.8] recommends ratings of rolling bearings on the basis that 90% of a large group of bearings in particular surroundings will survive a specified life; this life may be designated B_{10}, which is then interpreted to mean 10% failures may be expected in B_{10} revolutions (or time). However, one must examine catalogs carefully, because other bases are in use; some catalogs give the rated load for 50% survival B_{50} (median life), which is a number quite different from B_{10}, as predicted by equation (12.1). A commonly assumed relationship that probably accords with different catalog ratings in the USA is

(a) $$\text{Median life} \approx (5)(90\%\ \text{life}); \quad B_{50} \approx 5B_{10}.$$

However, certain extensive tests on deep-groove ball bearings yielded $B_{50} = 4.08B_{10}$,[12.9] which may be used if appropriate. At the other end of the spectrum, one may estimate the *longest life* at 4 to 5 times the median life ($4B_{50}$ to $5B_{50}$). The design life varies widely, depending of course on the machine and service; the values in Table 12.1 will be an aid to judgment. The expression "90% life" is interpreted to mean that the probability of a particular bearing surviving the given time is 90%. To convert to revolutions,

(b) $$\text{Revolutions} = (\text{Hours})(60\ \text{min/hr.})(n\ \text{rpm}).$$

The distribution of Fig. 12.3 is seen not to be symmetric (contrast with Fig. 1.14), so that the equation for the symmetric distribution does not apply so well; [12.11, 12.21] but bearing test data have been analyzed by W. Weibull, who found that the probability P of survival without failure for a particular life B is

$$P = e^{-(B/a)^b} \quad \text{or} \quad \ln P = -\left(\frac{B}{a}\right)^b, \qquad [e=2.7183] \tag{12.2}$$

[WEIBULL DISTRIBUTION]

where ln is to the Naperian base, a and b are taken as constants that can be defined by experiment. We shall use equation (12.2) later to find an equivalent "90% life" when a greater than 90% survival is desired.

12.4 STATIC LOAD CAPACITY. With contact stresses as high as they are, it does not take a large load to produce a permanent deformation. These indentations are called ***brinells*** and the act of indenting is called ***brinelling*** or ***false brinelling***. The objectionable damage is done on the loaded bearing at rest, since during operation the deformations are distributed over all the race and rolling surfaces. The question is, how much brinelling can there be before the bearing is ruined, and there is no simple answer because it depends on the service. A deformation as small as 10 μin. can be detected with an optical flat, but the effect of a deformation in normal operation is not bothersome until the indentation is of the order of 100 μin. (0.0001 in.),[12.8] which is the basis on which the *basic static load rating* F_s, Table 12.3, is calculated—equation (**c**) below. SKF[12.4] suggests that the permanent deformation in the ring and rolling element combined be less than 0.0001 times the diameter of the rolling element. The static capacity F_s, Table 12.3, is a practical measure of the amount of brinelling that can normally be tolerated when the bearing is to rotate. If subsequent noisy operation is permissible, a static load larger than the static capacity may be allowed. Also, where the operating motion is slow, higher static capacities can be used: the bearing loads supporting artillery pieces may be twice the catalog rating; static loads on bearings for aircraft-control pulleys and on sliding-door rollers may be four times their catalog ratings; the static fracture load will be more than $8F_s$.[12.4]

The radial static-load capacity is computed from Stribeck's equation,

$$F_s = C_s N_b D_b^2, \tag{c}$$

which says that the static load capacity F_s, called the ***basic static load rating***, is proportional to the number of balls (or rollers) N_b and to the square of the ball diameter D_b. The proportionality constant C_s depends on the type of bearing and the materials. For example, the static load capacity of a single-row, deep-groove, radial ball bearing is about $F_s = 5000 N_b D_b^2$.[11.4] Catalogs generally give values of F_s so that the engineer does not need to compute it. If a bearing is to be chosen to withstand a combination of radial F_x and thrust F_z loads, an equivalent static load F_{es}, which causes the same deformation as the combined loads, is computed from

(d) $$F_{es} = C_1 F_x + C_2 F_z,$$

where C_1 and C_2, given in catalogs, depend upon the type of bearing; for the single-row deep-groove, $C_1 = 0.6$, $C_2 = 0.5$, but F_{es} is never taken less than F_x.

12.5 DYNAMIC LOAD CAPACITY. Since a rotating bearing fails by fatigue, the dynamic capacity is different from the static. Palmgren[12.1] first modified Stribeck's equation, considering the variables introduced by rotation; then the ASA,[12.8] on the recommendation of the AFBMA, sanctioned the rating equation now in use,

(e) $$F_d = C N_b^{2/3} (N_r \cos \alpha)^{0.7} D^{1.8},$$

where F_d = the dynamic load capacity of the bearing (= F_r for 1 mr in Table 12.3), N_b = number of balls, N_r = number of rows of balls, D = diameter of a ball ($D \gtreqless 1$ in.), C = a constant that varies somewhat with the type of ball bearing, and α locates the plane of the resultant force, Fig. 12.4, when there is a thrust load.

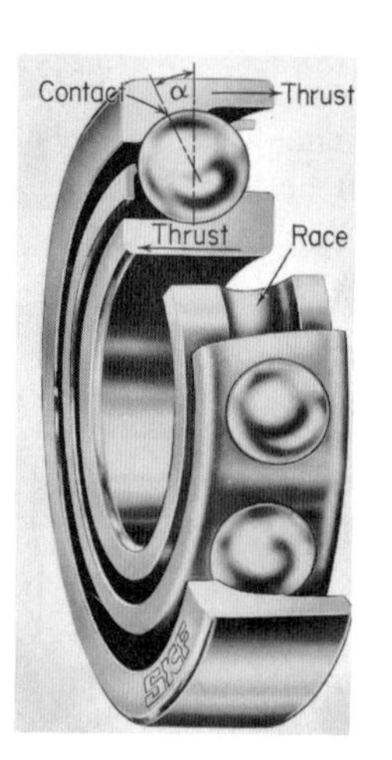

FIGURE 12.4 **Angular Contact Bearing. Observe that the action of the thrust is such as to move the surface of contact away from the center-line plane of the balls. Compare the shape of the grooves in this figure with those of Figs. 12.2 and 12.6. (Courtesy SKF Industries, Inc., Philadelphia).**

12.6 EQUIVALENT DYNAMIC LOAD. Ball bearings and some roller bearings are subjected to both radial and thrust loads. Since the possible combinations of these loads are infinite, manufacturers rate their radial bearings in terms of a radial load only, their thrust bearings in terms of thrust load only. Therefore, it becomes necessary to use an ***equivalent***

load. Not all manufacturers' catalogs give the ASA procedure;[12.8] we shall cover some of its highlights for deep-groove, single- and double-row ball bearings only. The equivalent load F_e is obtained from

(f) $$F_e = C_r F_x \quad [F_z/C_r F_x \lesseqgtr Q] \qquad \text{or} \qquad F_e = 0.56 C_r F_x + C_t F_z, \quad [F_z/C_r F_x > Q]$$

where F_x is the radial load (computed from a force analysis), F_z is the thrust load, C_r is a rotation factor (the evidence is incomplete but $C_r = 1$ for inner race rotating, $C_r = 1.2$ for outer race rotating with respect to a stationary inner race; for a particular number of revolutions, a ball rolls further on the larger outer race than on the smaller inner race), C_t is a thrust factor, obtained from Table 12.2, that applies only when $F_z/(C_r F_x) > Q$; Q is found in Table 12.2. We see that if the thrust load is a low enough fraction of the radial load, it can be ignored. It is a common practice to preload ball bearings with a thrust force by means of an adjusting nut or a spring; this may be done to hold a shaft in position or to reduce vibrations. All bearing manufacturers are eager to help with special problems.

12.7 SELECTING BEARINGS FROM TABLES.

The rated loads in Table 12.3 are on the basis that 90% will survive 10^6 rev. (1 mr) when the equivalent load is used for selection. If there is shock or vibration, the equivalent load should be further modified in accordance with the designer's judgment. For example, for *shafts connected by gears*, multiply the computed values of F_e by a service factor as follows:[12.4]

Rotating machines, no impact; electric motors, rotary compressors, etc.,	1.1 to 1.5;
Reciprocating machines,	1.3 to 1.9;
Machines with pronounced impact, hammer mills, etc.,	1.6 to 4.

Having decided upon the equivalent load F_e, compute the rated load F_r from equation (12.1). It is convenient to work with millions of revolutions (mr); let $B_r = 1$ mr, the rated number for Table 12.3; $k = 3$ for deep-groove ball bearings; then by (12.1),

(g) $$F_r = (B_{10})^{1/3} F_e,$$

where B_{10} mr is the desired number of revolutions before 10% failures have occurred, and the bearing is chosen to have a rated load equal to or better than F_r. For $B_r = 1$ mr, as above, this corresponding F_r is called the ***basic dynamic load rating***. Compute B_{10} from equation **(b)** above. Ordinarily, there will be several types of bearings that meet the requirements; so a decision on the kind of bearing, § 12.15, must be made. Moreover, the load ratings of the different series of the same kind overlap, as seen in Table 12.3; the decision in this regard depends in some measure on the space requirements. The following example will be helpful concerning the details.

TABLE 12.1
DESIGN LIFE FOR ROLLING BEARINGS, HOURS[12.4]

TYPE OF SERVICE	HOURS (90% *life*)
Infrequent use—instruments, demonstration apparatus, sliding doors	500
Aircraft engines	500 to 2000
Intermittent use, service interruptions of minor importance—hand tools, hand-driven machines, farm machinery, cranes, household machines	4000 to 8000
Intermittent use, dependable operation important—work moving devices in assembly lines, elevators, cranes, and less-frequently used machine tools	8000 to 12,000
8-hour service, not fully utilized—gear drives, electric motors	12,000 to 20,000
8-hour service, fully utilized—machines in general, cranes, blowers, shop shafts	20,000 to 30,000
24-hour service, continuous operation—separators, compressors, pumps, conveyor rollers, mine hoists, electric motors	40,000 to 60,000
24-hour service, dependable operation important—machines in continuous-process plants, such as paper, cellulose; power stations, pumping stations, continuous service machines aboard ships	100,000 to 200,000

12.8 EXAMPLE. Select a deep-groove ball bearing to carry a radial load $F_r = 800$ lb. and a thrust load $F_z = 700$ lb. at 1750 rpm. This service is 8 hr. per day, but it is not continuous; design for 18,000 hr., Table 12.1. The operation is smooth with little vibration; the outer ring rotates.

Solution. First, convert 18,000 hr. at 1750 rpm to millions of revolutions B;

$$B_{10} = (18{,}000)(60)(1750)(10^{-6}) = 1890 \text{ mr},$$

the desired life with no more than 10% failures. Not knowing what bearing will be used, we do not at this point know the static rating F_s (Table 12.3), which is needed to decide upon the form of equation **(f)** to use; assume a value of C_t, Table 12.2, and check it later; assume $C_t = 1.8$; $C_r = 1.2$ for the outer ring rotating. The corresponding equivalent load is

$$F_e = 0.56C_rF_r + C_tF_z = (0.56)(1.2)(800) + (1.8)(700) \approx 1800 \text{ lb}.$$

From equation (12.1) with $k = 3$, we get

$$F_r = \left(\frac{B_{10}}{B_r}\right)^{1/3} F_e = (1890)^{1/3}(1800) = 22{,}200 \text{ lb}.$$

TABLE 12.2
THRUST FACTOR FOR DEEP-GROOVE, SINGLE- AND DOUBLE-ROW BALL BEARINGS
(F_z = *thrust load*; F_s = *basic static load rating*)

F_z/F_s	0.014	0.028	0.056	0.084	0.11	0.17	0.28	0.42	0.56
C_t	2.3	1.99	1.71	1.55	1.45	1.31	1.15	1.04	1.00
Q	0.19	0.22	0.26	0.28	0.30	0.34	0.38	0.42	0.44

at $B_r = 1$ mr. Examine Table 12.3 and note that either ball bearing No. 221 or 317 has a larger rated load.

Check 221 first; $F_s = 20{,}100$ lb., $F_r = 23{,}000$ lb. Then

$$\frac{F_z}{F_s} = \frac{700}{20{,}100} = 0.0348, \qquad \frac{F_z}{C_r F_x} = \frac{700}{(1.2)(800)} = 0.729,$$

which is seen in Table 12.2 to be greater than Q ($0.22 < Q < 0.26$) for the foregoing value of F_z/F_s. Interpolate in Table 12.2 for C_t corresponding to $F_z/F_s = 0.0348$ and find $C_t = 1.92$. Recompute the rated load for this value of C_t;

$$F_e = (0.56)(1.2)(800) + (1.92)(700) \approx 1880 \text{ lb.}$$
$$F_r = B_{10}{}^{1/3} F_e = (1890)^{1/3}(1880) = 23{,}200 \text{ lb.}$$

TABLE 12.3

SINGLE-ROW, DEEP-GROOVE BALL BEARINGS

From Ref. *(12.14)*. Basic dynamic load rating F_r is for 1 million revolutions (mr); 90% of a group of bearings should last longer than 1 mr. with the rated loads given. The speed limit given is approximate for oil bath lubrication; for grease, use $\frac{2}{3}$ of these values. More detail in catalogs. Factors limiting speed include lubrication, fit, dynamic balance, vibration. It is possible to exceed the specified limits.[12.16] Higher speeds can be obtained by oil mist lubrication, by circulating and cooling the oil, or by cooling the bearing. Consult manufacturers.

BRG. NO.	200 SERIES (200–222) Balls No.	Balls Dia.	Static F_s, lb.	Rated F_r, lb. 1 mr.	Speed Limit rpm	300 SERIES (300–322) Balls No.	Balls Dia.	Static F_s, lb.	Rated F_r, lb. 1 mr.	200 Roller F_r 1 mr.
00	7	$\frac{3}{16}$	440	805	25,000	6	$\frac{9}{32}$	845	1,400	
01	7	$\frac{15}{64}$	685	1,180	23,000	6	$\frac{5}{16}$	1,040	1,680	
02	8	$\frac{15}{64}$	790	1,320	20,000	7	$\frac{5}{16}$	1,220	1,960	
03	8	$\frac{17}{64}$	1,000	1,650	18,000	7	$\frac{11}{32}$	1,470	2,340	
04	8	$\frac{5}{16}$	1,390	2,210	15,000	7	$\frac{3}{8}$	1,750	2,750	
05	9	$\frac{5}{16}$	1,560	2,420	13,000	7	$\frac{7}{16}$	2,390	3,660	2,980
06	9	$\frac{3}{8}$	2,250	3,360	11,000	8	$\frac{31}{64}$	3,340	4,850	3,970
07	9	$\frac{7}{16}$	3,070	4,440	9,400	8	$\frac{17}{32}$	4,020	5,750	5,900
08	9	$\frac{15}{32}$	3,520	5,040	8,400	8	$\frac{19}{32}$	5,020	7,040	7,670
09	9	$\frac{1}{2}$	4,010	5,660	7,700	8	$\frac{11}{16}$	6,730	9,120	8,070
10	10	$\frac{1}{2}$	4,450	6,070	7,100	8	$\frac{3}{4}$	8,010	10,700	8,440
11	10	$\frac{9}{16}$	5,630	7,500	6,500	8	$\frac{13}{16}$	9,400	12,400	10,300
12	10	$\frac{5}{8}$	6,950	9,070	5,900	8	$\frac{7}{8}$	10,900	14,100	12,600
13	10	$\frac{21}{32}$	7,670	9,900	5,400	8	$\frac{15}{16}$	12,500	16,000	14,900
14	10	$\frac{11}{16}$	8,410	10,800	5,100	8	1	14,200	18,000	14,800
15	11	$\frac{11}{16}$	9,250	11,400	4,800	8	$1\frac{1}{16}$	16,100	19,600	18,200
16	10	$\frac{3}{4}$	10,000	12,600	4,500	8	$1\frac{1}{8}$	18,000	21,300	19,600
17	11	$\frac{25}{32}$	12,000	14,400	4,200	8	$1\frac{3}{16}$	20,100	22,900	22,400
18	10	$\frac{7}{8}$	13,600	16,600	3,900	8	$1\frac{1}{4}$	22,300	24,700	28,600
19	10	$\frac{15}{16}$	15,600	18,800	3,700	8	$1\frac{5}{16}$	24,500	26,400	31,400
20	10	1	17,800	21,100	3,500	8	$1\frac{7}{16}$	29,400	29,900	34,800
21	10	$1\frac{1}{16}$	20,100	23,000	3,300	8	$1\frac{1}{2}$	32,000	31,800	
22	10	$1\frac{1}{8}$	22,500	24,900	3,100	8	$1\frac{5}{8}$	37,600	35,400	46,200

vs. 23,000 lb. in Table 12.3. This is of course borderline and the engineer would have to decide if a slightly larger percentage of failures (than 10%) is permissible in the 18,000 hrs. Actually, of course, the truth is not known with this accuracy, so that if 90% survival is considered satisfactory, bearing 221 is all right.

In checking bearing 317 in the same manner, we find the calculations the same *because the two bearings coincidentally have the same static capacity* and nearly the same dynamic capacity. They are so close to the same capacity that the final choice would be made on bases other than capacity. In general, before a final decision is made, designs should be made or considered for all appropriate types of bearings. See the next two articles for further considerations with respect to a choice.

TABLE 12.4 DIMENSIONS OF ROLLING BEARINGS[12.4]

This table does not give all standard dimensions. The maximum fillet radius *r* is the maximum radius at the shoulder *on the shaft* which is cleared by the corner radius on the bearing. Conversion factors: 0.03937 in./mm.; 25.4 mm/in.

	BORE		OUTSIDE DIAMETER mm			WIDTH OF RACES, mm			MAX. FILLET *r*		
BRG. NO.	*mm.*	*in.*	200 *Series*	300 *Series*	400 *Series*	200 *Series*	300 *Series*	400 *Series*	200 *Series*	300 *Series*	400 *Series*
00	10	0.3937	30	35		9	11		0.024	0.024	
01	12	0.4724	32	37		10	12		0.024	0.039	
02	15	0.5906	35	42		11	13		0.024	0.039	
03	17	0.6693	40	47		12	14		0.024	0.039	
04	20	0.7874	47	52		14	15		0.039	0.039	
05	25	0.9843	52	62	80	15	17	21	0.039	0.039	0.059
06	30	1.1811	62	72	90	16	19	23	0.039	0.039	0.059
07	35	1.3780	72	80	100	17	21	25	0.039	0.059	0.059
08	40	1.5748	80	90	110	18	23	27	0.039	0.059	0.079
09	45	1.7717	85	100	120	19	25	29	0.039	0.059	0.079
10	50	1.9685	90	110	130	20	27	31	0.039	0.079	0.079
11	55	2.1654	100	120	140	21	29	33	0.059	0.079	0.079
12	60	2.3622	110	130	150	22	31	35	0.059	0.079	0.079
13	65	2.5591	120	140	160	23	33	37	0.059	0.079	0.079
14	70	2.7559	125	150	180	24	35	42	0.059	0.079	0.098
15	75	2.9528	130	160	190	25	37	45	0.059	0.079	0.098
16	80	3.1496	140	170		26	39		0.079	0.079	
17	85	3.3465	150	180		28	41		0.079	0.098	
18	90	3.5433	160	190		30	43		0.079	0.098	
19	95	3.7402	170	200		32	45		0.079	0.098	
20	100	3.9370	180	215		34	47		0.079	0.098	
21	105	4.1339	190	225		36	49		0.079	0.098	
22	110	4.3307	200	240		38	50		0.079	0.098	

12.9 CHOOSING BEARINGS WHEN THE PROBABILITY OF SURVIVAL IS DIFFERENT FROM 90%. In equation (12.2), B may represent the life in revolutions, mr, or hours (or any unit of time) at a specified speed, as 3800 hours at 1000 rpm. Changing signs in equation (12.2) and writing them for $P_{10} = 0.90$, probability of 90% survival, and for P, any other probability, we have

$$\ln\left(\frac{1}{P}\right) = \left(\frac{B}{a}\right)^b \qquad \text{and} \qquad \ln\frac{1}{P_{10}} = \left(\frac{B_{10}}{a}\right)^b$$

Divide one by the other and take the b root;

$$\text{(12.3)} \qquad \frac{B}{B_{10}} = \left[\frac{\ln(1/P)}{\ln(1/P_{10})}\right]^{1/b}, \quad [\ln(1/P_{10}) = \ln(1/0.9) = 0.1053]$$

from which the life B for a probability P may be found when the life B_{10} and probability P_{10} at 90% life (or at other state) are known; say $B_{10} = 1$ mr for $P_{10} = 0.9$; or the probability P for a life B may be computed. Shube[12.9] finds a and b by using two points, the catalog 90% life and an assumed median life; for the median life, he used both the generally recognized relation of 5 times 90% life and an experimental value of 4.08 times 90% life (§ 12.3):

(h) For median life = (5)(90% life): $a = 6.84$, $b = 1.17$.

(i) For median life = (4.08)(90% life): $a = 5.35$, $b = 1.34$.

Additionally, Harris[12.11] recommends $b = 1.125$.

More frequently than in former years, it is desired to select bearings with a much greater probability than 0.9 that the bearing will survive a certain life. For example, where human life is at stake and, sometimes, just because the purchaser can afford it, reliabilities better than 99% may be specified. The theoretical equation (12.3) predicts that no life is so short but that some failures will occur if the population is large enough, and no life is so long but that some bearings would run indefinitely. Experience does not bear out these extreme conclusions,[12.11] but tentative calculations can be made with equation (12.3). One could also introduce limits in terms of standard deviation, as the 3σ limits of §§ 3.9–3.12; $\pm 4\sigma$ almost insures certainty. Experimental data suggest that failure proceeds in two stages; first the inception of a crack in the subsurface and then the propagation of the crack until it reaches the bearing surface.[12.11] When the probability is about 0.9 or lower, the first stage lasts much longer than the stage of crack propagation and it is to this phenomenon that the Weibull equation (12.2) applies. At high reliabilities (lower loading), the time for crack propagation lengthens and this stage may last longer than the first. A plot of data suggests that the actual minimum life at virtually 100% reliability is about 5% of 90% life.[12.11] The corresponding reduction in the load for a given life is then predicted by equation (12.1).

A significant aspect of probabilities should not be overlooked. If there are x bearings in a particular system, the probability of a failure of any one is the product of the individual probabilities. If x bearings each have the same probability P, the chances of survival for the specified life becomes P^x. For example, suppose a gear box contains 10 bearings, all having exactly $P = 0.9$ at the design life; the probability of any one of them failing during this life is $0.9^{10} \approx 0.35$. In short, *on average for a large population*, one of the bearings in the system fails at about one-third design life. This characteristic suggests that for "space systems," which are to have an excellent reliability, the individual reliabilities must be quite high. Some examples involving probabilities will be helpful.

12.10 EXAMPLES—PROBABILITIES AND LIVES OF ROLLING BEARINGS. (a) In a certain catalog, we find that the rated load capacity of a No. 212 bearing is $F_1 = 2550$ lb. at a speed of 1000 rpm for 3800 hr., average life basis. What is the corresponding basic dynamic load rating (for 1 mr)?

Solution. Convert the life to mr;

$$B_{av} = (3800)(1000)(60)(10^{-6}) = 228 \text{ mr.}$$

To estimate the 90% survival life, use $228/5 = 45.6$ mr $= B_1$ for $F_1 = 2550$ lb., because 5 is the ratio given in most catalogs. Since the basic dynamic load rating is F_r for $B_r = 1$ mr, use equation (12.1) with $k = 3$ and get

$$F_r = F_1\left(\frac{B_1}{B_r}\right)^{1/3} = 2550(45.6)^{1/3} = 9100 \text{ lb.}$$

Compare with 9070 lb. from Table 12.3, same bearing.

(b) Preliminary calculations have been made just as those in the example of § 12.8; the equivalent load has been computed to be $F_e = 23{,}800$ lb. when the rated capacity is $F_r = 23{,}000$ lb. for No. 221 bearing. The question is: do we dare risk using the 221 bearing?

Solution. Many things need to be known for a final decision, one of them being the probable number of failures in the appointed lifetime. Since F_r is the rated load for 1 mr., we can find the revolutions B_2 at the point of 10% failures for $B_1 = 1$ mr. from ($F_2 = F_e = 23{,}800$ lb.)

$$B_2 = B_1\left(\frac{F_1}{F_2}\right)^3 = \left(\frac{23{,}000}{23{,}800}\right)^3 = 0.901 \text{ mr.,}$$

that is, 10% are expected to fail in 0.901 mr. Since statistical logic tends to become elusive, the problem may be restated now. We have for $B_2 = 0.901$ the probability of $P_2 = 0.9$; for $B_3 = 1$, the design life, what is P_3? Using $b = 1.34$ in equation (12.3), we get

$$\frac{\ln(1/P_3)}{0.1053} = \left(\frac{B_3}{B_2}\right)^b = \left(\frac{1}{0.901}\right)^{1.34},$$

from which $P_3 \approx 0.887$, the probability of a particular bearing surviving the design life.

(c) A reliability of 0.98 (98% survival) in 10 mr. is desired for a roller bearing supporting an equivalent radial load of 5000 lb. What 200-series bearing should be selected?

Solution. The first question is: given $P_1 = 0.98$ for $B_1 = 10$ mr, what is B_2 for $P_2 = 0.9$ (catalog basis)? From equation (12.3), with $b = 1.125$,

$$\frac{B_2}{B_1} = \frac{B_2}{10} = \left[\frac{\ln 1/P_2}{\ln 1/P_1}\right]^{1/b} = \left[\frac{\ln 1/0.90}{\ln 1/0.98}\right]^{1/1.125}$$

from which, $B_2 = 44.3$ mr; this life is now equivalent to the B_{10} life for 0.9 probability and a load of $F = 5$ kips. The next step is to compute the rated load by equation (12.1), using $k = 10/3$ for roller bearings;

$$\frac{F_r}{F} = \frac{F_r}{5} = \left(\frac{B_2}{B_r}\right)^{1/k} = 44.3^{0.3},$$

from which $F_r = 15.6$ kips (for $B_r = 1$ mr); choose roller bearing No. 215, Table 12.3, whose rated capacity is 18.2 kips. This bearing is probably overly conservative because of the statistical tendencies discussed in § 12.9; the chances are good that a 214 bearing would meet the required reliability.[12.11]

12.11 VARIABLE LOADING. If the load on the bearing varies, the practice is to use the cubic mean load (on the assumption that life is inversely proportional to the cube of the load). Thus, if force F_1 acts for n_1 revolutions (or fraction of a revolution), F_2 for n_2 rev., F_3 for n_3 rev., etc., the cubic mean load F_m is given by

$$\text{(j)} \qquad F_m = \left[\frac{F_1{}^3 n_1 + F_2{}^3 n_2 + F_3{}^3 n_3 + \cdots}{\Sigma n}\right]^{1/3} = \left[\frac{\Sigma F^3 n}{\Sigma n}\right]^{1/3},$$

where $\Sigma n = n_1 + n_2 + n_3 + \cdots$ is the total number of revolutions (for a constant rpm, time may be used in place of $n_1, n_2, \ldots$). When the variation of the load against number of revolutions (or time) is a smooth curve, the curve may be approximated by a series of constant forces F_1, F_2, etc., laid off in steps to approximate the curve. Or the curve can be integrated;

$$\textbf{(k)} \qquad F_m = \left[\frac{\int F^3\,dn}{\Sigma n}\right]^{1/3} \qquad \text{or} \qquad F_m = \left[\frac{\int F^3\,d\tau}{\Sigma \tau}\right]^{1/3},$$

where $\Sigma\tau$ stands for the total time. For example, if the variation of F is sinusoidal from 0 to F_{max}, $F_m \approx 0.65F_{max}$ from (**k**).

12.12 MATERIALS AND FINISH. The most commonly used material is SAE 52100, an alloy steel of nominally 1% C and 1.5% chromium, hardened to Rockwell C 58–65. Nickel and molybdenum are also used with chromium as alloys. Hardness is important for wear (surface fatigue); for example, the average life when the hardness is $R_C = 50$ is only about half of that when $R_C = 60$.[11.4] The operating temperature of SAE 52100 and similar steels is generally held to about 300°F (but 200°F is considered as the maximum value for the usual installation). Tool steel is occasionally used for rolling bearings because its temperature can be permitted to rise to about 1000°F without it losing too much hardness. Nonferrous metals are used for rolling bearings for a reason; also phenolic plastic ball bearings (and other plastics, as nylon, Teflon) are available. Glass has some application for balls; and in exceptionally high temperature situations, Pyroceram is exceptionally promising. If parts are made of different materials the coefficients of thermal expansion become important with respect to clearances.

The diameters of the rolling elements in a particular bearing need to be closely the same, say with a tolerance of 50 to 100 microinches, and even finer for the exacting applications, as in instruments and high-speed situations. When a difference in size exists, the load is not well distributed among the elements, the larger ones taking excessive stresses. Manufacturer's catalogs show the various tolerances needed by the designer. The surface finish is the smoothest possible for commercial processes. See Fig. 3.9.

12.13 SIZES OF BEARINGS. There are several series of bearings for which certain key dimensions have been standardized, some of which are shown in Table 12.4. The 200 series is called *light*, the 300 *medium*, and the 400 *heavy*. Since this classification was adopted, bearings heavier than the

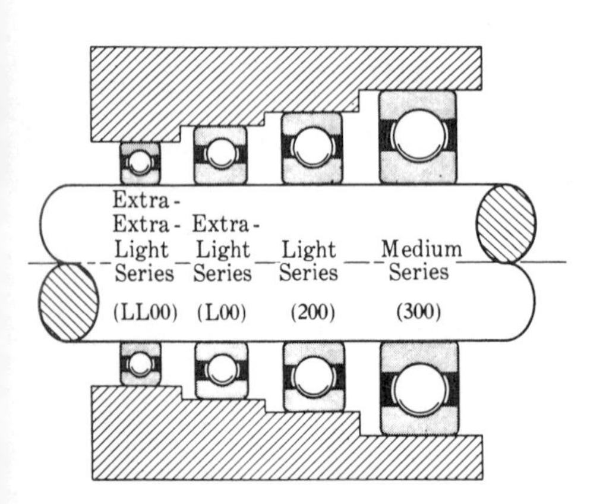

FIGURE 12.5 Various Series of Ball Bearings. All these bearings have the same basic number, as 09. (After New Departure[12.6]).

heavy and lighter than the light have been developed for unusual service—rolling mills and fine instruments, for examples. We observe in Table 12.4 that an 06 bearing has a 30-mm bore in each series. The bearing is made heavier by increasing the outside diameter, Fig. 12.5. The number and size of balls or rollers in these bearings is unstandardized and left to the manufacturer's judgment. A bearing No. 206 must be defined as to type, because it may be a roller bearing, a self-aligning bearing, etc. The various manufacturers use prefix and suffix notations to distinguish their products, but most of them include the standardized numbers. With the exception of the four smallest sizes, the bore in millimeters is equal to five times the basic bearing number, for example, $5 \times 06 = 30$ mm, which is the bore of the 06 bearing.

12.14 FRICTION IN ROLLING BEARINGS. The resistance to motion of these bearings is a combination of rolling resistance and sliding.[12.1,12.17] Tests are generally designed to measure the total frictional *torque* T_f, and a "coefficient of friction" is computed from the equivalent frictional force F_f *at the bore diameter* ($F_f = T_f/r$) divided by the total bearing load (radially loaded). This plan makes these coefficients of friction comparable with those for sleeve bearings of the same bore. For the rolling bearings, the coefficient of friction is greater at start up than at operating speed, it decreases sharply with increasing load to a point and soon tends to become constant—except the deep-groove bearing, which seems to have a minimum value at some load.[11.4] Stribeck found that fewer and larger rolling elements tended to reduce the frictional loss at light loads. With the exceptions noted, the following values of f are due to Palmgren,[12.1] determined for the load that gives a life of 1000 mr, and are suitable only for order of magnitude approximations.

Angular contact[11.4]	$f = 0.0032$
Cylindrical roller, short rollers, flange-guided	$f = 0.0011$
Deep-groove, single-row ball bearing	$f = 0.0015$
Needle bearings (0.0014–0.0022)[12.14]	$f = 0.0045$
Self-aligning ball bearings	$f = 0.0010$
Tapered roller, spherical roller, flange-guided	$f = 0.0018$
Thrust ball bearings	$f = 0.0013$

The values are higher in new bearings, and when too much lubricant is used. Since a lubricant seal, Fig. 12.15, may increase the friction by several hundred per cent,[11.4] these effects must not be overlooked in estimating the energy loss. Considering all effects, we observe that rolling bearings do not necessarily have a lower coefficient of friction than full-film sleeve bearings.

12.15 TYPES OF ROLLING BEARINGS. We can mention only the principal types. The ***deep-groove ball bearing***, Fig. 12.2, with which we have already become familiar, is one in which the balls are assembled by the eccentric displacement of the inner ring. With the inner ring in contact with the outer ring, as many balls as possible are placed in the grooves. The rings are then centered, and the balls are kept in position by a *separator* or *retainer* or *cage*. Bearings loaded in this manner are called the Conrad

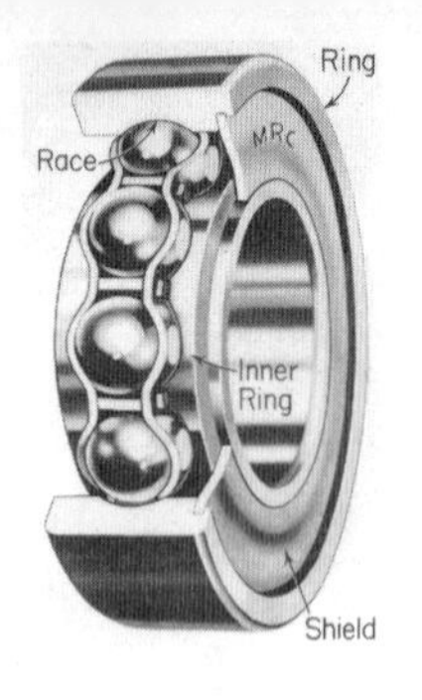

FIGURE 12.6 Single-row Ball Bearing with Shield. The balls in this bearing were loaded with the aid of a filling slot. Notice their spacing. The shield aids in keeping out foreign matter, important in rolling bearings. (Courtesy Marlin-Rockwell Corp., Jamestown, N.Y.).

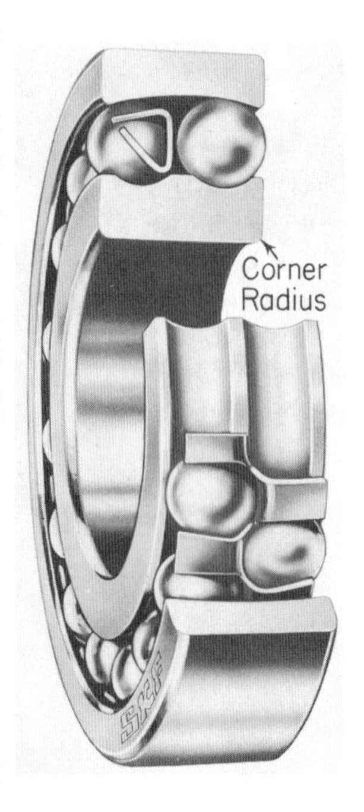

FIGURE 12.7 Self-aligning Ball Bearing. Self-aligning roller bearings are similar. (Courtesy SKF Industries, Inc., Philadelphia).

FIGURE 12.8 Cylindrical Roller Bearing. The rollers run in a groove in the inner ring. (Courtesy Norma-Hoffman Bearings Corp., Stamford, Conn.).

FIGURE 12.9 Needle Bearing. (Courtesy The Torrington Co., Torrington, Conn.).

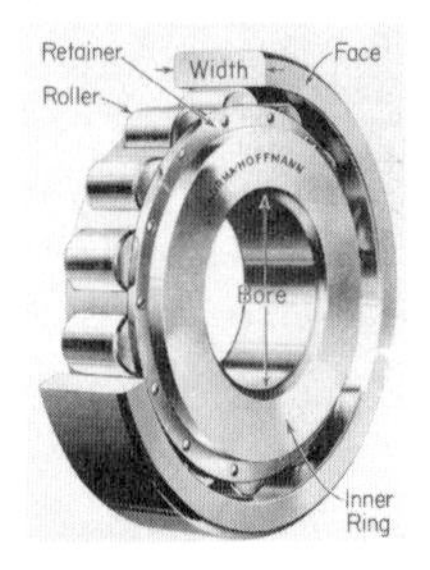

type. Careful alignment is essential, say a maximum misalignment of 0.5°. This bearing will support a relatively high thrust load.

The ***filling-slot type*** of ball bearing, Fig. 12.6, has slots or notches that permit the assembly of more balls, giving a bearing of larger radial load capacity [equation (**e**)]. Since their thrust capacity is small, these bearings are used where the load is principally radial.

Self-aligning ball bearing, Fig. 12.7, compensate for angular misalignments that arise from shaft or foundation deflection or errors in mounting; they are recommended for radial loads and moderate thrust in either direction. Since the outer race has a spherical shape, the shaft may pass through the bearings at a small angle without causing binding.

Angular-contact bearings, so named because the line through the areas carrying the load makes an angle with the plane of the face of the bearing, Fig. 12.4, are intended to take heavy thrust loads. They are often used in opposed pairs and are suited to ***preloading***. Preloading a bearing consists of placing it under an initial axial load that is independent of the working load, for the purpose of maintaining a nearly constant alignment of parts by reducing the axial movement and, to some extent, the radial deflection under working loads.

Double-row ball bearings (not self-aligning) are similar to single-row ball bearings, except that each ring has two grooves. The two rows of balls give the bearing a capacity somewhat less than twice that of a single row.

Cylindrical roller bearings, Fig. 12.8, are made in different styles and weights, but the bore and outside diameters are the same as those given in Table 12.4 for like bearing numbers. Geometrically, the contact is a line instead of a point as in ball bearings, which results in a greater area carrying the load and hence, for a particular size, in a larger radial capacity.

The principal function of the retainer in roller bearings is to keep the roller axes parallel. If the rollers should skew, the frictional loss is greatly increased. Skewing may also occur because the roller diameter is not constant or because one end of the roller carries a greater share of the load due to misalignment.

Self-aligning roller bearings, with spherical rollers running in a double-grooved inner ring, have curved outer rings that look much like the outer ring of a self-aligning ball bearing, Fig. 12.7. They can carry relatively heavy radial and thrust loads and have the usual advantages of the self-aligning feature. Figure 12.15 is a drawing of this type of bearing.

Cylindrical roller bearings are also made with relatively long rollers, a popular type being called a ***needle bearing***, Fig. 12.9; this type has no retainer to hold rollers in alignment. Figure 12.10 shows an application to a universal joint. If the needles run on the surface of the shaft, rather than on an inner race, this surface should be hardened and polished for a reasonable life expectancy. For example, if the surface is cold-rolled shafting with

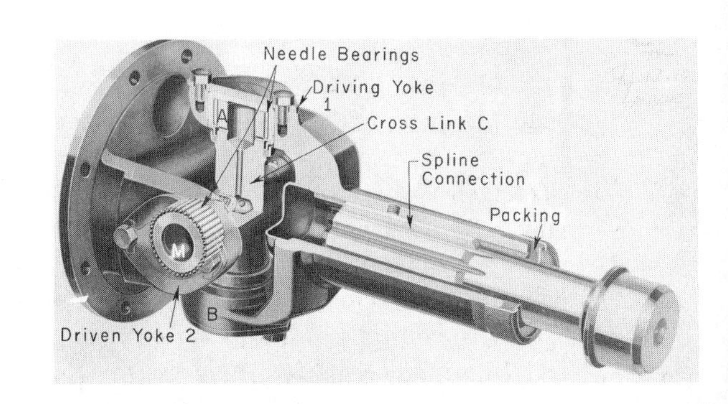

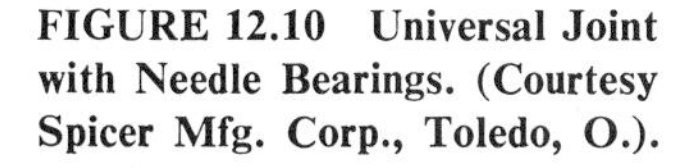
FIGURE 12.10 Universal Joint with Needle Bearings. (Courtesy Spicer Mfg. Corp., Toledo, O.).

a Rockwell C 15, the capacity for the same life is only 3% of the capacity for a Rockwell C 60 surface. Needle bearings are appropriate where their smaller diametral dimensions are advantageous, where the speed is not too high, and where there is oscillating motion.

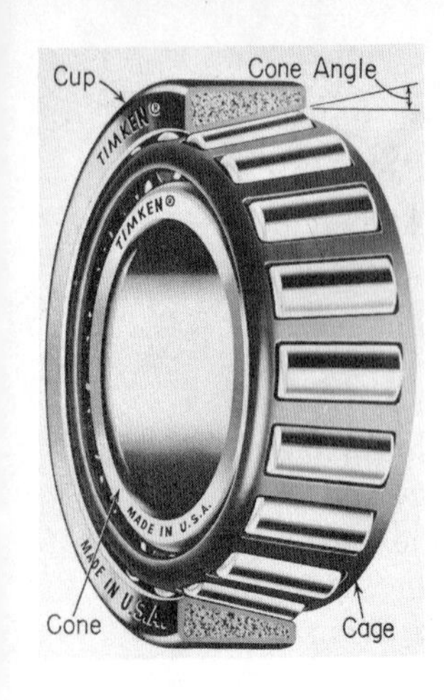

FIGURE 12.11 Tapered Roller Bearing. (Courtesy Timken Roller Bearing Co., Canton, O.).

In ***tapered roller bearings***, Fig. 12.11, the rolling elements are frustums of a cone, so mounted in the races that all of their surface elements and their axes intersect at a point on the axis of the shaft. They are capable of carrying substantial axial and thrust loads. Tapered roller bearings are manufactured in standard sizes in English units (inches rather than millimeters) and in a variety of styles.

The Hyatt ***"spherangular" roller bearing***, Fig. 12.12, is an angular contact bearing with rollers instead of balls. Since the outer race has a spherical surface, the bearing also is self-aligning. With a large contact area, it has a high load capacity.

12.16 THRUST BEARINGS. In the rolling type of thrust bearing, the rolling elements are balls, Fig. 12.13, short cylindrical rollers, tapered rollers, Fig. 12.14, or spherical rollers that run in spherical races and are therefore self-aligning. These bearings may be rigidly supported, or one of the races may be supported in a spherical seat to make it self-aligning.

12.17 HOUSING AND LUBRICATION. Cleanliness is a prime consideration in the life of a rolling bearing. For this reason, these bearings must be protected from airborne particles of matter, as well as from the expected sources of dirt—during handling, for instance. Housings for this purpose are usually designed to fit a particular application, Fig. 12.15, or, if appropriate, rolling bearings may be obtained in pillow blocks, Fig. 12.2, which have shields to keep the lubricant in and dirt out. The designer should study the designs found in various manufacturer's catalogs.

Interference fits for the rings must be tight enough to prevent relative motion but not enough to harm the fit of the rolling elements by excessive deformation of the rings. The stationary ring is also generally made a light press fit in its housing. The bearing is usually mounted against a shoulder whose height makes it possible to force off the bearing without the force being transmitted through the rolling elements.

At low and medium speeds, grease is the most satisfactory lubricant;

it tends to provide a better seal against dirt; it easily can be made suitable for temperatures between $-70°F$ and $210°F$. The bearing and its housing should preferably not be packed tight; two-thirds full is enough.

At higher speeds, oil becomes the desirable lubricant. (Catalogs give

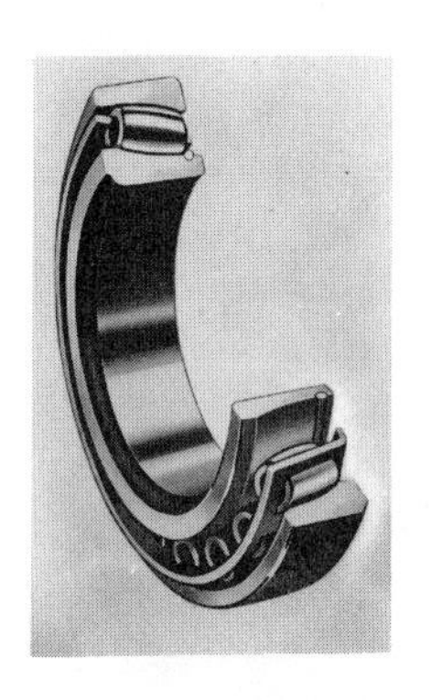

FIGURE 12.12 Spherangular Roller Bearing. (Courtesy Hyatt Roller Bearing Co., Harrison, N.J.).

FIGURE 12.13 Thrust Ball Bearing. (Courtesy Aetna Ball & Roller Bearing Co., Chicago).

FIGURE 12.14 Tapered Roller Thrust Bearing. (Courtesy Timken Roller Bearing Co., Canton, O.).

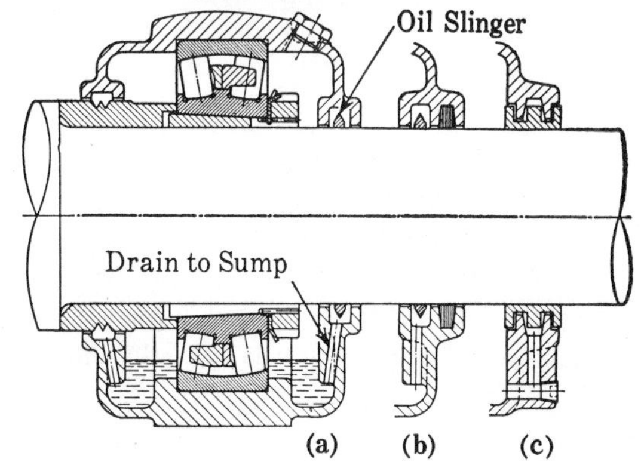

FIGURE 12.15 Typical Housing, with Various Seals. Sometimes, an oil slinger only is sufficient, as at (a). At (b), a slinger and a seal; (c) labyrinth seal. (Courtesy SKF Industries, Inc. Philadelphia).

information concerning the permissible limits of grease.) The oil level in the housing should be kept low, below the center of the lowest rolling element, Fig. 12.15, or excessive churning of the oil at high speed will cause overheating. At unusually high speeds, oil sprayed as a mist to the point of loading results in lower coefficients of friction (absence of churning), lower running temperature. Since the coefficient of friction is nearly constant for a particular method of lubricating, the temperature rise under constant load is nearly proportional to the speed. Bearings that would otherwise run hot may be kept cool by circulating and cooling the oil. Sometimes it is convenient and simple to lubricate rolling bearings by splash from nearby moving parts. Prelubricated bearings with integral seals may be expected to run for the life of the bearing (or machine) without attention, Fig. 12.16, although means of relubrication are provided in some. The seals may be on one or both sides.

The function of the lubricant in protecting the highly finished surfaces from rust and corrosion should also be kept in mind.

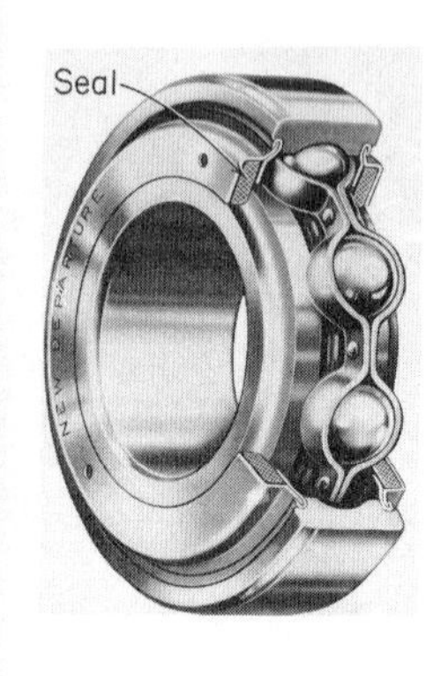

FIGURE 12.16 Ball Bearing with Seals. This type is relubricated by means of a hypodermic-like needle through the small holes on the side. (Courtesy New Departure, Bristol, Conn.).

12.18 OTHER BALL-BEARING DEVICES.

By providing means for the circulation of balls, engineers have adapted the principle of rolling contact to special situations. Application to a screw is shown in Fig. 8.25, p. 250. Figure 12.17 shows a ball bushing, to be used for reciprocating motion.[12.20] A circulating ball adaptation for spline connections is also available.[12.18]

12.19 COMPARISONS OF SLEEVE AND ROLLING BEARINGS.

For sleeve bearings with full-film lubrication, the speed is limited by temperature rise (which in turn is a function of the lubricant), the starting friction is high, the damping is relatively good, large amounts of lubricant are needed, noise is no problem, their life is unlimited. For rolling bearings, the speed is limited largely by dynamic considerations (vibration), the starting friction is low, the damping effect is poor, very small amount of

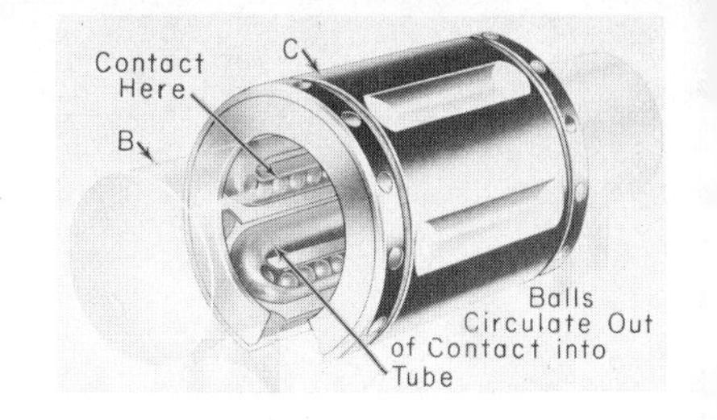

FIGURE 12.17 Ball Bushing. For axial motion. The balls roll on the reciprocating surface *B*, circulating through tubes. (Courtesy Thomson Industries, Inc., Manhasset, N.Y.).

lubricant needed except where lubricant carries away heat, noise may be objectionable, their life is limited. Sleeve bearings tolerate more misalignment than rolling, they occupy less radial but more axial space than rolling bearings, and are likely to be cheaper in quantity.[11.5]

12.20 CLOSURE. One cannot learn what an engineer should know about rolling bearings simply by selecting a few bearings from a catalog—even though this is what he does in the end. This chapter high-lights other engineering information of an important nature and gives leads to more of the same. Reference *(12.22)* contains much important basic information. While it is true that the capacities of a given size and type of bearing are not all the same when taken from different manufacturers' catalogs, a large difference in answers by different catalogs suggests further checking and more knowledge. And of course, there are the out-of-ordinary cases: 30,000 rpm in textile machinery, 40,000 rpm of woodworking spindles, 100,000 rpm for internal grinders and more for small turbines, environments in space where lubrication may present several unusual problems.

13. SPUR GEARS

13.1 INTRODUCTION. Spur gears are toothed wheels whose tooth elements are straight and parallel to the shaft axis; they are used to transmit motion and power between parallel shafts. Kinematic considerations, as well as the terminology, are essential for design decisions, but we must assume that the reader is acquainted with these matters; a few pages of review are included.

13.2 DEFINITIONS. See Fig. 13.1 on page 356. The ***pitch circle*** is the basis of measurement of gears. The size of a gear is its pitch-circle diameter in inches, called the ***pitch diameter***. For interchangeable gear teeth, it is intended that the pitch circles of mating gears be tangent, in which case, these imaginary circles roll on one another without slipping. Gears may not be so mounted; then there may be said to be an *operating pitch circle* and a *standard pitch circle*. The pitch circle is also the trace of the ***pitch cylinder*** (pitch surface) as it intersects a plane normal to the axis.

The ***pitch point*** of meshing gears is the point of tangency of the pitch circles; for an individual gear, the pitch point will be located where the tooth profile cuts the standard pitch circle.

The ***addendum circle*** (also *outside circle*) is the circle that bounds the outer ends of the teeth. The addendum *cylinder* encloses a gear. The

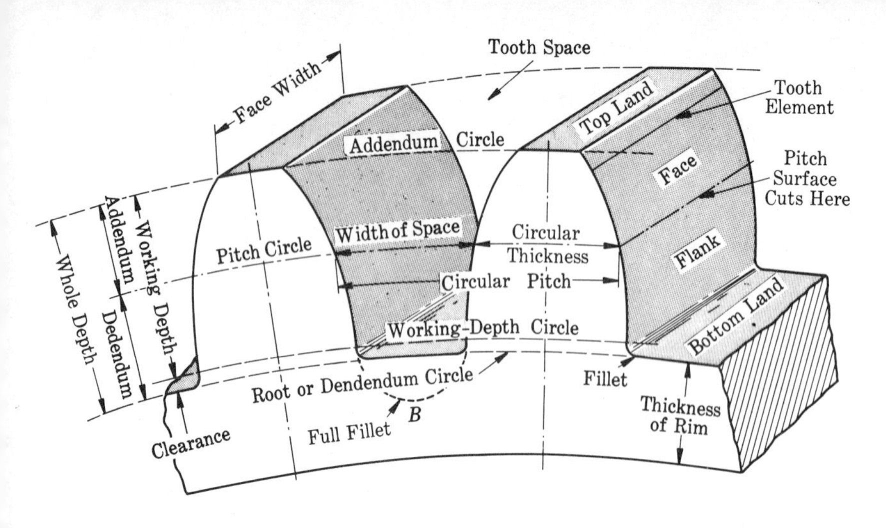

FIGURE 13.1 Dimensions and Nomenclature, Gear Teeth. Full-fillet roots are sometimes used for severe duty because of the reduced stress concentration—as shown dotted at *B*.

diameter of this circle is called the ***outside diameter***. The ***addendum*** a is the radial distance between the pitch circle and the addendum circle.

The ***dedendum circle*** or ***root circle*** is the circle that bounds the bottoms of the teeth. The ***dedendum*** d is the radial distance from the pitch circle to the root circle, that is, to the bottom of the tooth space.

The ***whole depth*** is equal to the addendum plus the dedendum. The ***working depth*** is the radial distance from the addendum circle to the ***working-depth circle*** which marks the distance that the mating tooth projects into the tooth space; it is the sum of the addendums of mating gears.

The ***clearance*** is the radial distance between the working-depth circle and the root circle; it is the dedendum minus the mating addendum.

The ***circular thickness***, also called *tooth thickness*, is the width of tooth measured along the pitch circle. The ***chordal thickness*** is the tooth width measured along the chord at the pitch circle. The ***width of space*** or *tooth space* is the space between teeth measured along the pitch circle.

Backlash is the tooth space minus the circular thickness. When backlash exists between two gears, one gear can be turned through a small angle while the mating gear is held stationary. Backlash is necessary to care for errors and inaccuracies in the spacing and in the form of the tooth, to provide a space between the teeth for lubricant (§ 13.38), and to allow for the expansion of the teeth with a temperature rise. Accurately cut gears should be mounted with a backlash from $0.03/P_d$ to $0.04/P_d$, where P_d is the diametral pitch. Use the larger values for higher speeds. To provide backlash, the cutter generally is fed in somewhat deeper than standard depth on the larger of the two gears.

The ***face width*** is the length of teeth in an axial direction.

The ***face*** of the *tooth* is the surface of the tooth between the pitch cylinder and the addendum cylinder.

The ***flank*** is the surface of the tooth between the pitch and root cylinders.

The ***top land*** is the surface of the top of the tooth.

The ***bottom land*** is the surface of the bottom of the tooth space.

When two gears are in mesh, the smaller is called the ***pinion*** and the larger the ***gear***.

The ***angle of action*** is the angle through which the gear turns from the time a particular pair of teeth come into contact until they go out of contact. The *arc of action* subtends the angle of action.

The ***angle of approach*** is the angle through which the gear turns from the time a particular pair of teeth come into contact until they are in contact *at the pitch point*.

The ***angle of recess*** is the angle through which the gear turns from the time a given pair of teeth are in contact *at the pitch point* until they pass out of mesh. The angle of approach plus the angle of recess is equal to the angle of action.

The ***velocity ratio*** m_ω is the angular velocity of the driver divided by the angular velocity of the driven gear. For spur gears, this ratio varies inversely as the pitch (or base-circle) diameters and as the tooth numbers.

(a)
$$m_\omega = \frac{\omega_1}{\omega_2} = \frac{n_1}{n_2} = \frac{D_2}{D_1} = \frac{N_2}{N_1},$$

where n is angular velocity in, say revolutions per minute, the subscript 1 refers to the driver, and the subscript 2 refers to the driven gear.

The ***gear ratio*** m_g is the number of teeth in the gear divided by the number of teeth in the pinion. When the pinion is the driver, $m_g = m_\omega$.

13.3 BASE CIRCLE AND PRESSURE ANGLE.

The ***base circle*** is the circle from which the involute is generated. In gearing, the expression ***degree of involute*** is used to define the base circle for a particular pitch circle. The line AA, Fig. 13.2 (p. 359), is tangent to the standard pitch circle. If, through the point of tangency, a line BB is drawn, the corresponding base circle is tangent to BB with its center at the center of the pitch circle. The involute obtained from this base circle and used for the interchangeable tooth profile is a ϕ_1-degree involute. An involute generated from the base circle tangent to the line at ϕ_2 degrees with AA would produce a tooth profile of ϕ_2-degree involute. The angle ϕ is called the ***pressure angle***, because the direction of the force acting normal to the profiles is along the line BB (for ϕ_1) when two involute teeth are in contact with the pitch circles of the mating gears tangent to each other. The angle ϕ is also called the ***angle of obliquity***.

The line BB is the ***line of action***; that is, all points of contact between mating teeth properly lie somewhere on this line, which is also the *generating line*. It is the nature of involute gearing that the line of action is always perpendicular to the tooth profiles at the point of contact. Two gear teeth in mesh must have profiles with a common generating line.

The operating pressure angle is determined by the center distance, because once the profiles are established, involute gears may be moved away from each other, thereby increasing the backlash, and still operate correctly with no change of velocity ratio. Since the generating line remains tangent to the base circles, the effects of increasing the center distance are to increase the pressure angle and to move the pitch circles of the gears apart. For spur gears, 20° is the preferred pressure angle; but a 14½° angle was most widely used for years. When we say a gear is a 20° involute gear, we mean that the pressure angle will be 20°, provided that the mating gears are mounted with their pitch circles tangent to each other and that $\phi = 20°$ defines the base circle from which the involute profile is obtained in accordance with Fig. 13.2.

13.4 PITCH.

The pitch of a gear is a measure of the *spacing*, and usually also of the *size*, of the tooth. In this country, there are two pitches in common use, the ***circular*** and the ***diametral*** (di-am′-e-tral).

The ***circular pitch*** P_c is the distance in inches measured along the pitch circle from a point on one tooth to the corresponding point on an adjacent tooth, Fig. 13.1. Let D be the diameter of the pitch circle and N_g be the number of teeth in the gear. Then the circumference of the pitch circle πD divided by the number of teeth N_g is the circular pitch P_c;

$$P_c = \frac{\pi D}{N_g}. \tag{13.1}$$

The ***diametral pitch*** P_d is the ratio of the number of teeth per inch of pitch diameter, or

$$P_d = \frac{N_g}{D}. \tag{13.2}$$

In Fig. 13.3, it is seen that the size *increases* as the diametral pitch decreases. Conversion from circular to diametral pitch is frequently necessary. From (13.1) and (13.2) we have

$$P_c P_d = \left(\frac{\pi D}{N_g}\right)\left(\frac{N_g}{D}\right), \quad \text{or} \quad P_c P_d = \pi. \tag{13.3}$$

Circular pitch is regularly used for cast-tooth gears as a convenience to the pattern maker, though cutters of many circular pitches are available from stock. For cut teeth, the diametral pitch is nearly always used. Therefore, in specifying the diametral pitch, we should try to use one for which cutters are available from stock. One manufacturer's list of standard pitches will not necessarily be the same as another's; for problem purposes select one of the following standard pitches: 2, 2.25, 2.5, 3, 4, 5, 6, 8, 10, 12, 16.

FIGURE 13.2 **Degree of Involute. Just as two involute gears can operate at different pressure angles, a so-called 20° cutter can cut gears of pressure angle other than 20°.**

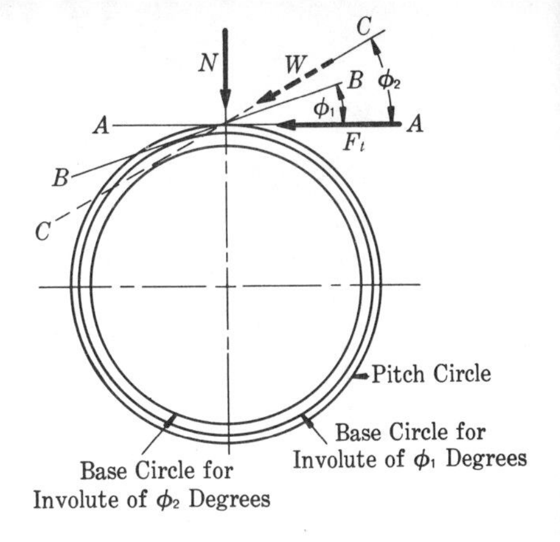

The ***base pitch*** P_b (involute gearing only) is the distance in inches measured along the base circle from a point on one tooth to the corresponding point on an adjacent tooth. It is also the distance between parallel profiles on adjacent teeth measured along the generating line;

$$P_b = \frac{\pi D_b}{N_g} = \frac{\pi D \cos \phi}{N_g} = P_c \cos \phi, \tag{13.4}$$

where D_b is the diameter of the base circle and D is the pitch diameter.

In spur gearing the ***pitch angle*** is the angle subtended by an arc on the pitch circle equal in length to the circular pitch.

FIGURE 13.3 Sizes of Gear Teeth of Different Diametral Pitches—Full Size ($14\frac{1}{2}°$). For $P_d \geqq 20$, the teeth are classified as *fine pitch*, with some proportions different from those of larger teeth. (Courtesy The Barber-Colman Co., Rockford, Ill.)

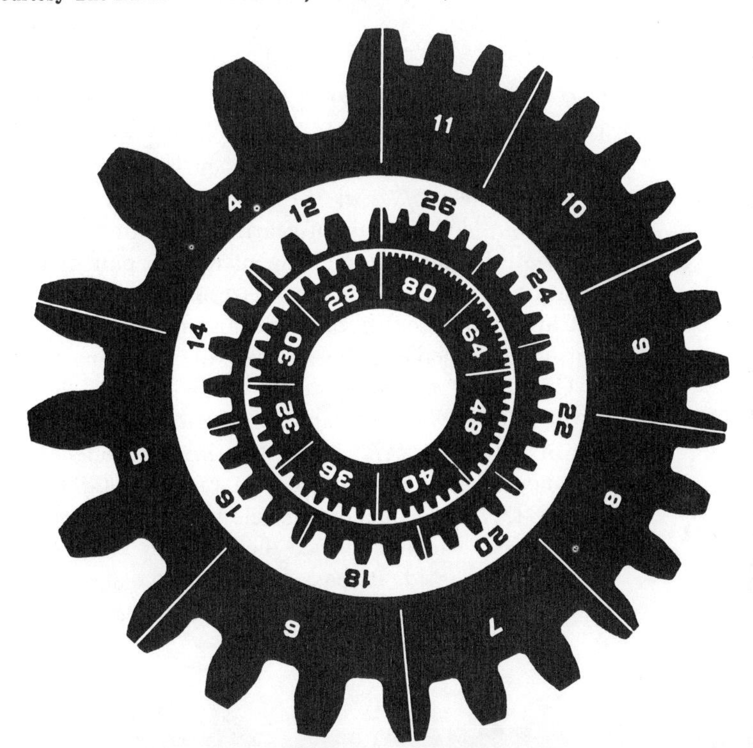

13.5 LENGTH OF ACTION AND CONTACT RATIO.

The ***length of action*** Z is the length of that part of the line of action between the initial and final points of contact and is given by

(b) $$Z = (r_{a2}^2 - r_{b2}^2)^{1/2} + (r_{a1}^2 - r_{b1}^2)^{1/2} - C \sin \phi,$$

[NO INTERFERENCE]

where r_{a1} and r_{a2} are the radii of the addendum circles of the mating gears, r_{b1} and r_{b2} are the base-circle radii, and C is the center distance; this equation is applicable only when the intersections of the addendum circles and the line of action lie between the interference points.[13.2] The *interference points* are the points of tangency of the line of action and the base circles, Fig. 13.4.

The ***contact ratio*** m_c can be defined in several ways: it is the ratio of the angle of action to the pitch angle; the ratio of the *length of action* to the *base pitch*; the ratio of the arc of action to the circular pitch. It may be thought of as the average number of teeth in contact for mating gears. If possible, the contact ratio should be about 1.25 to 1.4 for best running conditions. A contact ratio less than unity means that one pair of teeth go out of contact before the next pair have reached their initial point of contact, and the drive is not continuous. For example, two 12-tooth, full-depth, $14\frac{1}{2}^\circ$ involute gears have a contact ratio slightly less than one, but for low speeds, the operation may be satisfactory.

The contact ratio can now be computed from

(c) $$m_c = \frac{Z}{P_b} = \frac{Z}{P_c \cos \phi}.$$

13.6 LAW OF GEARING AND THE ACTION OF GEAR TEETH.

The law of gearing states that for a pair of engaging teeth to transmit a constant velocity ratio the tooth curves must be such that *the common normal to the profiles at the point of contact will always pass through the pitch point.*[13.2] The law of gearing does not refer to the over-all ratio, but to variations during the engagement of a pair of teeth. These momentary variations in the velocity ratio result in accelerations and decelerations that are manifested in vibration and noise. Curves that satisfy the law of gearing are called ***conjugate curves***, of which any number may be obtained.

Contact of a pair of teeth *starts* between the driver's flank and the driven's face, where the addendum circle of the driven gear intersects the line of action—provided that this intersection lies between the points of tangency (I_1 and I_2, Fig. 13.4) of the line of action and the base circles. In Fig. 13.4, the addendum circle of the driven gear intersects the line of action at M, which is *outside* the tangency point I_1. That part of the tooth on the pinion inside its base circle is drawn as a radial line, and it happens that this line is not conjugate with the involute face of the gear. To prevent

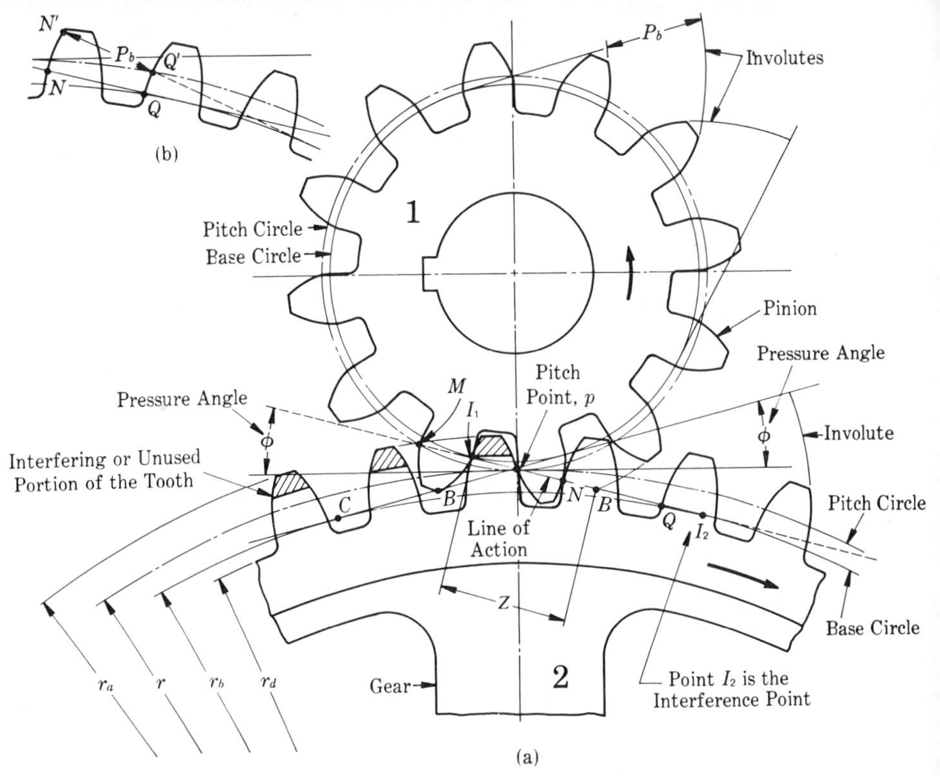

FIGURE 13.4 Action of Gear Teeth. The base pitch $P_b = NQ = N'Q'$, as seen upper left in (b). The radii of curvature of the involute profiles at the pitch circles are: for the gear, Cp; for the pinion, I_1p.

this nonconjugate engagement, the base of the pinion tooth must be undercut, or the face of the gear tooth must be relieved. Both methods are used, so that the teeth should not touch one another until point I_1 is reached.

Contact will *cease* where the addendum circle of the driving gear intersects the line of action, point B, Fig. 13.4, when the face of the driver is in contact with the flank of the driven—provided that this intersection lies between the points of tangency I_1 and I_2 of the line of action and the base circles. Since point B satisfies this provision, it marks the point where a pair of teeth leave contact.

13.7 INTERFERENCE OF INVOLUTE TEETH. The points of tangency of the line of action and the base circles, I_1 and I_2, Fig. 13.4, are called the interference points. The nonconjugate action referred to in the previous article is called interference; when the gear tooth is long enough to project inside of the base circle of the pinion, the tip of the involute profile overlaps the radial flank of the pinion tooth. Interference is a serious disadvantage of involute gears; it is a maximum when the smallest gear mates the largest. The amount of interference decreases as the gear decreases in size (for a particular pinion) or as the pinion increases in size (for a particular gear). In the $14\frac{1}{2}°$, full-depth system, § 13.8, a gear with 32 or more teeth will have no interference with a rack (or any other gear). In the 20°, full-depth system, a gear with 18 or more teeth will not have interference with a rack; for a 20° stub tooth, the corresponding number of teeth is 14. Thus, increasing the pressure angle and decreasing the height of tooth decreases the interference problem.

Gear teeth generated by involute rack cutters are automatically undercut in the flank, enough of the interfering portion being removed to eliminate actual interference. While this solves the interference problem, the tooth is thereby weakened, and the contact ratio may become undesirably low. The best course is to avoid the condition of theoretical interference unless there is some outweighing advantage. The 20° full-depth teeth have the advantages of greater capacity and less interference trouble. Where conditions warrant the higher cost, noninterchangeable teeth, unequal addendum and dedendum, § 13.34, may be used to avoid interference.

13.8 INTERCHANGEABLE INVOLUTE GEAR SYSTEMS.

An interchangeable gear-tooth system is one wherein any gear of a particular pitch will run properly with every other gear of the same pitch. The conditions for interchangeability are: (1) all gears have the same pitch; (2) all gears have the same addendum, which is equal to the dedendum minus the clearance; and (3) all gears are cut with the same angle of obliquity. The following tooth proportions are in use.

(a) **Full-depth Involute System.** The recommended dimensions for full-depth interchangeable teeth are:

Addendum $a = 1/P_d = P_c/\pi$ — Working depth $= 2/P_d$
Clearance (min.) $= 0.25/P_d$ — Whole depth $= 2.25/P_d$
Dedendum $d = 1.25/P_d$ — Outside diameter $= D + 2a$

The AGMA recommended standard pressure angle is 20°; 25° is sometimes used; most full-depth gears now in service are $14\frac{1}{2}°$ involute. Somewhat greater clearance, say $0.35/P_d$, may be needed for tool clearance on shaved or ground tooth profiles. Compare teeth in Fig. 13.5.

(b) **Stub-tooth System.** The usual tooth proportions are (ASA):

Working depth $= 1.6/P_d$, — Clearance (min.) $= 0.2/P_d$,
Addendum $a = 0.8/P_d$, — Whole depth $= 1.8/P_d$,
Dedendum $d = 1/P_d = P_c/\pi$ — Pressure angle $= 20°$.

Full-depth teeth are favored unless the interference problem is especially severe—few teeth on the pinion, many on the gear—and an interchangeable system, rather than unequal addendum and dedendum gears, is desired.

13.9 STRENGTH OF GEAR TEETH.

Without friction, the resultant force W acting on the gear tooth lies on the generating line in involute

FIGURE 13.5 Comparison of Tooth Profiles. Tooth profiles as they appear on 20-tooth gears; (a) is a $14\frac{1}{2}°$ full depth, (b) is a 20° full depth, and (c) is a 20° stub.

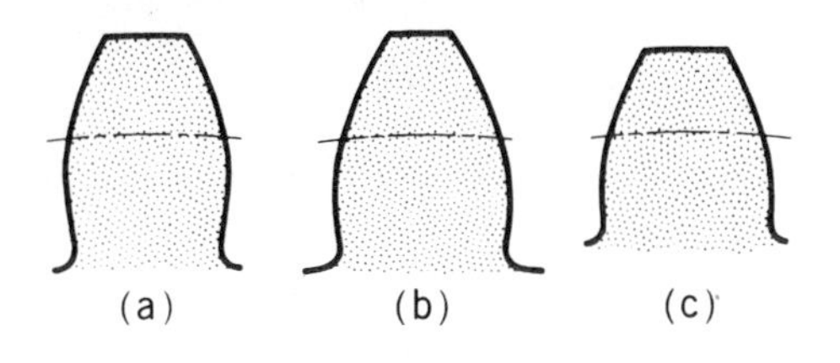

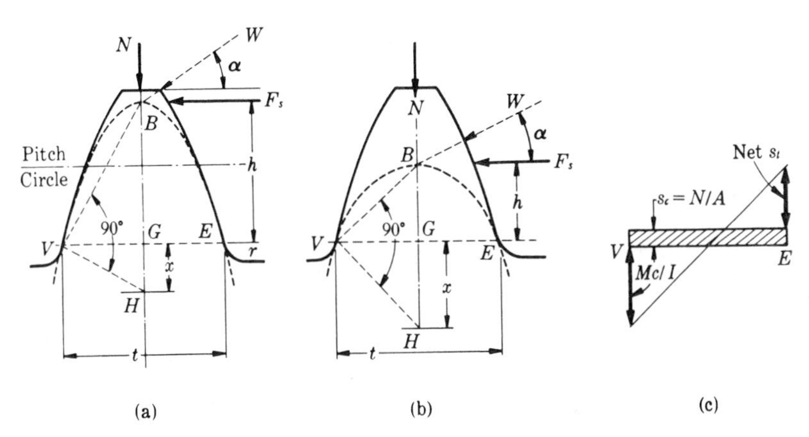

FIGURE 13.6 Forces on a Tooth.

gearing, and its line of application moves from the upper (or lower) part of the tooth to the lower (or upper) part; W, F_s, N, Fig. 13.6, are assumed to be uniformly distributed along the length of tooth. Considering the tooth as a cantilever beam, we find the stress at its maximum when one tooth carries the entire load at its tip, as in Fig. 13.6(a). However, if the contact ratio is greater than one and if the teeth are geometrically accurate, another tooth is sharing the transmission of power when the tip of this tooth is in contact. As the tooth in Fig. 13.6(a) moves through its angle of action, the point of application of W moves down the profile. At some stage of this motion, with a contact ratio less than 2, the tooth will carry the entire load as in Fig. 13.6(b).

At the point where the line of action of W cuts the center line of the tooth, Fig. 13.6, force W is replaced by its radial and tangential components, N and F_s. The force N produces a uniform compressive stress over any section of the tooth, say at VE. The component F_s produces a bending stress: tension at E and compression at V. The uniform compression at E subtracts from the bending tension at E, producing a lower, and therefore safer, net stress at E. The uniform compression at V adds to the bending compression at V to give a higher total compressive stress. If the material is stronger in compression than in tension, as is cast iron, the effect of the compressive stress due to N is to strengthen the tooth. Since the uniform compressive stress is small compared to the bending stress, its effect on the strength of tooth is usually ignored.

With F_s acting through B with a moment arm h, the bending moment at the section VE is $M = F_sh$. With the face width b, Fig. 13.1, the section modulus of the rectangular section at VE is $Z = bt^2/6$. From $M = sZ$,

(d)
$$F_sh = s\frac{bt^2}{6}.$$

The section VE is to be the one where the stress is a maximum from the load F_s and is located by the following reasoning. Let s and b in equation **(d)** be constant and get $h = (sb/6F_s)t^2 = Ct^2$, which is the equation of a parabola. Inscribe this parabola through point B; it is tangent to the tooth

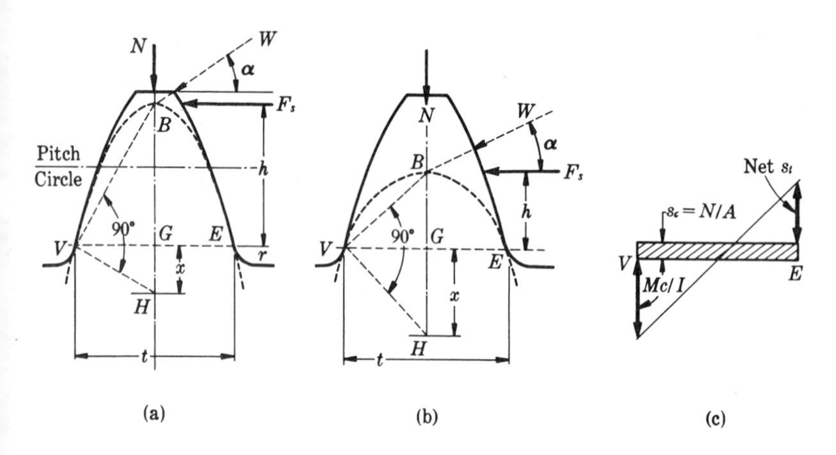

FIGURE 13.6 (Repeated).

profile at V and E (where the tooth width is t) and it will be found to be inside the tooth elsewhere. Since the parabola outlines a beam of uniform strength (s = constant), the tooth is weakest at the section VE. (This argument is not strictly true because the stress-concentration effect results in the maximum stress being in the fillet.) The inconvenient dimensions h and t can be replaced by using the proportion, from the similar triangles BVG and GVH (corresponding sides perpendicular), $x/(t/2) = (t/2)/h$, or $h = t^2/(4x)$. Substitute this value of h into equation **(d)** and obtain

(e)
$$\frac{F_st^2}{4x} = s\frac{bt^2}{6} \quad \text{or} \quad F_s = sb\frac{4x}{6}.$$

Multiply and divide the right hand side of this equation by the diametral pitch P_d and find

(f)
$$F_s = \frac{sb}{P_d}\left(\frac{2xP_d}{3}\right).$$

Consider the parameter $2xP_d/3$ with respect to Fig. 13.6 and note that its magnitude depends upon the shape of the tooth (a function of the number of teeth for a particular ϕ) and the point of application of the load. It is known as the Lewis ***form factor*** and designated by Y; $Y = 2xP_d/3$. Use Y in equation (**f**) and get the equation known as the Lewis equation, after Wilfred Lewis who first derived it in 1893;

$$F_s = \frac{sbY}{P_d} \quad \text{or} \quad F_s = sbP_c y, \tag{13.5}$$

[LEWIS EQUATION]

where we have used $P_d = \pi/P_c$ and $Y = \pi y$; y is the form factor for use with circular pitch P_c, but our Table AT 24 gives values of Y because cut teeth usually have a standard diametral pitch that makes Y more convenient.

Values of Y with the load at the tip of the tooth and with the load near the middle of the tooth are given in Table AT 24. For standard interchangeable teeth, a 20-tooth gear of one pitch has the same value of Y as a 20-tooth gear of any other pitch. If other tooth forms are used, construct the outline of the tooth to large scale, insert the line of action of the load at the desired point, measure x, Fig. 13.6, and then find $Y = 2xP_d/3$. Also see Ref. *(13.1)*.

13.10 STRESS CONCENTRATION. Since there is a reentrant corner at the bottoms of the teeth, where the profile joins the bottom land, a stress concentration exists, Fig. 13.7, a phenomenon ignored by the original Lewis equation. Inasmuch as the load on the tooth is repeatedly applied, we should expect that if breakage occurs, the failure would be due to fatigue. The value of the theoretical stress concentration factor is not simply defined because of the complex geometry. If the teeth are cut by a generating method, the fillets are not arcs of circles, but trochoids, whose form in turn depends on the number of teeth being cut and the radius of

FIGURE 13.7 Stress Concentration in Gear Teeth. (Courtesy T. J. Dolan, Univ. of Ill.).

curvature of the corner of the cutter tips for which there is no single standard; K_t also depends on the position of the point (line) of contact, Fig. 13.6(a) and (b).[13.16] Because of these complications, the stress concentration factors used are ordinarily reasonable estimates of the true values. Black,[13.4] using photoelastic technics, found $1.345 < K_t < 1.47$ on the tensile side and up to $K_t = 1.61$ on the compression side. Most authorities seem to agree that the fatigue crack propagates from the tensile side. Dolan and Broghamer[13.5] devised the following equations that accord with their photoelastic studies:

(g) $$K_t = 0.22 + \left(\frac{t}{r}\right)^{0.2}\left(\frac{t}{h}\right)^{0.4} \quad \text{for } 14\tfrac{1}{2}^\circ,$$

(h) $$K_t = 0.18 + \left(\frac{t}{r}\right)^{0.2}\left(\frac{t}{h}\right)^{0.4} \quad \text{for } 20^\circ,$$

where r is the minimum radius of fillet, h is the height of load above the section VE, Fig. 13.6, and t is the thickness of tooth at this section. In general, the theoretical values should be modified by the notch-sensitivity factor q, § 4.10, but when the gear teeth are hardened, the value of q is nearly unity, except for very small radii (Fig. AF 7). With a strength reduction factor K_f, Lewis' equation becomes

(13.6) $$F_s = \frac{sbY}{K_f P_d}.$$

We shall use both forms of the Lewis equation, (13.5) and (13.6). Buckingham states that if the load is assumed to be acting at the tip of the tooth [and comparison is made with his dynamic load (§ 13.17)], this is so conservative that no strength reduction factor K_f is needed; when the value of Y is chosen for the load near the middle of the tooth, Table AT 24, K_f should be included. When two gears are of the same material, the pinion tooth is weaker, $Y_p < Y_g$. When the materials are different, assume the tooth with the smaller sY to be the weaker (§ 13.11). The value of K_f should fall within the range of 1.2 to 1.7 when the load is applied at the tip; but K_f is higher, say 1.4 to 2, when the load is applied near the middle. The AGMA gives a chart of values of Y/K_f for 20° full depth, cut with a certain rack cutter.

13.11 DESIGN STRESS. In general, a design stress is chosen to agree with experience and the choice is affected by how closely the stress analysis accords with the facts of life and by the assumed loading condition. A common attitude is to use the endurance limit as the design stress, which then corresponds to the expected maximum load, to be called the dynamic load (§ 13.14). Buckingham[13.3] says to use the endurance strength for a *repeated* load ($R = 0$) and says that this is 1.5 times the endurance limit s_n'; however, he uses the endurance limit as $s_u/3$, whereas we are taking it as $s_u/2$. Hence, for steels, Buckingham's endurance strength in one direction is the same as that which we have defined for reversed bending; but note that reversed bending as produced by a reversed load does not necessarily result in the same strength as a rotating beam with a constant load. At any rate, use s'_n as given in our tables for the design stress in order to agree with those recommended by Buckingham and with the dynamic load to be defined later. If test values are not at hand, recall that *for steel only*

(i) $s'_n \approx (250)(\text{BHN})$ psi or $(\text{BHN})/4$ ksi, but not over 100 ksi.

The beam fatigue strength of *carburized teeth*, favorably affected by the high surface strength and the residual compressive stresses as a consequence of the hardening operation, may be taken as $60 + 0.2s_{u\ \text{core}}$ ksi with a tolerance of $\pm 10\%$;[2.1] $s_{u\ \text{core}}$ ksi is the ultimate strength of the core, which can be estimated as BHN/2 ksi if necessary. See Table AT 11. Use the lower range of values for maximum reliability; as usual, a maximum value of 100 ksi is preferred. In general, the use of the core BHN in equation (i) is conservative practice when the fillets are carburized; and this value may be used even if it is larger than the $60 + 0.2s_{u\ \text{core}}$ and it is advised for flame-hardened teeth, but for flame hardening, one must be sure that the fillet areas do not become annealed and therefore weaker. Since gears are so often heat treated, read now §§ 2.6–2.9, 4.23, 4.33, 4.28–4.31, inclusive; also check § 13.24.

For bronze and alloy cast irons, $s_n \approx 0.4s_u$ may be used in the absence of better information. See also § 13.24. If a tooth is subjected to loading in both directions, as the teeth of an idler, some engineers reduce the design stress by 25–30%.

13.12 FACE WIDTH. The derivation of the Lewis equation is based on the assumption that the load is uniformly distributed across the face width. Sometimes this is far from true, due to misalignment or warping of the teeth. One cause of tooth breakage is the concentration of load on one end of a tooth, which results in higher stresses than when the load is distributed. To minimize this kind of trouble, the face width b should not be too great as compared to the thickness (or pitch) of the tooth. In the absence of special considerations, the following proportions are considered good:

(j) $2.5P_c < b < 4P_c$, or $\dfrac{8}{P_d} < b < \dfrac{12.5}{P_d}$, cut teeth.

There are many exceptions to these proportions. For example, automotive transmission gears have shorter faces because of the need for a compact arrangement. In general, the longer the face and the more rigid the material of the teeth, the more accurate should be both the tooth profiles and the alignment of the shaft for long life and trouble-free operation. In order to avoid a concentration of the load on one end of a tooth, spur-gear teeth are sometimes "crowned," Fig. 13.8; that is, the teeth are shaved with an elliptical reduction from the center of the tooth to the end of about 0.0003 in. per in. on each side.[13.1]

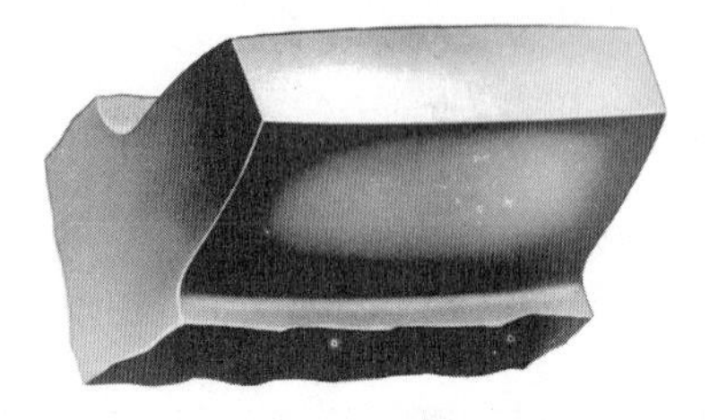

FIGURE 13.8 Crowned Tooth. The amount of crowning is exaggerated in order to highlight the idea. (Courtesy National Broach and Machine Co., Detroit.)

13.13 THE TRANSMITTED LOAD. In design we usually know the transmitted power and the angular speeds of the gears. The *transmitted load* F_t, the average tangential force on the teeth, is then obtained from the horsepower; or from the applied torque. Although the applied force varies a little as the point of application moves from the top to the bottom of the tooth, or *vice versa*, the nominal force acting at the pitch circle is used in design; thus[1.15]

$$F_t = \frac{33{,}000 \text{ hp}}{v_m},$$

where $v_m = \pi D n$ fpm, the pitch-line speed; n rpm; D ft.; and hp is the *input* horsepower. In the case of a variable transmitted load, as in punch presses, shears, etc., the *maximum* transmitted load should be the basis of the design for *strength*.

13.14 DYNAMIC LOADS ON GEAR TEETH. Since gear tooth profiles are not perfect involutes (not precisely conjugate), since the tooth spacing is not exactly right, since the shaft and mountings deflect under load, and since a load will deflect teeth out of shape even if they are initially perfect, the law of gearing is not ideally satisfied and local accelerations inevitably occur. At what are considered normal speeds, two peak loads occur with each tooth engagement, the first accompanying accelerations due to the foregoing imperfections. The driving gear slows down, the driven speeds up, and likely the teeth momentarily cease contact. Then the acting forces bring the teeth together again with an impact that results in a *dynamic load* considerably greater than the transmitted load. Thus, the maximum load acts shortly after the initial engagement and occurs near the middle of the profile.

At high speed, the phenomenon changes. The time of engagement becomes so short that there is only a single peak load, regardless of the initial accuracy of the teeth. (See § 13.38.) Buckingham points out that the dynamic load in this case approaches asymptotically from above a value of twice the transmitted force; that is, as the speed increases, the dynamic load peaks and then decreases with increasing speed, because the point is reached where there is not time enough for the teeth to bounce apart as they do at slower speeds.

The maximum value of the dynamic load depends on the masses of the gears and connected masses ($F = ma$), the speed of operation as pointed out above, and the materials. Materials with a low modulus of elasticity deflect more than high-modulus materials, other things being equal, and therefore absorb the energy of impact with a lower peak load.

If we can compute a dynamic load that is a conservative estimate of the maximum load, then we are justified in using the endurance strength as a design stress, as explained in a previous article, when stress concentration effects are cared for. As stated above, the methods of determining the load, design stress, and the unknown dimensions must be known to result in satisfactory designs. A number of equations have been derived for computing the dynamic load,[13.3,13.12,13.13] but first we shall review a few traditional and easier methods of estimating it.

13.15 DYNAMIC LOAD AS A FUNCTION OF VELOCITY ONLY—METAL TEETH.

After Lewis introduced his equation for the strength of gear teeth, experiment showed that a "velocity factor" needed to be applied in obtaining a design stress in order to get a better agreement between design computations and test results, which is the same thing as saying that the dynamic load is a function of velocity only. It is not, but it is so nearly so over limited ranges of speed and for a particular class of gears that velocity factors are still widely used. *When the service is intermittent and wear is not a factor*, the use of the following values of the dynamic loads F_d for metal teeth, compared with the beam fatigue strength computed from equation (13.6) with K_f, should give satisfactory designs; v_m fpm = pitch-line speed, F_t = transmitted load from equation (1.15):

(k) $$F_d = \frac{600 + v_m}{600} F_t \text{ lb.} \quad \begin{cases} \text{Commercially cut} \\ v_m \lesseqqgtr 2000 \text{ fpm} \end{cases}$$

(l) $$F_d = \frac{1200 + v_m}{1200} F_t \text{ lb.} \quad \begin{cases} \text{Carefully cut} \\ 1000 < v_m < 4000 \text{ fpm} \end{cases}$$

(m) $$F_d = \frac{78 + v_m^{1/2}}{78} F_t \text{ lb.} \quad \begin{cases} \text{Precision cut} \\ v_m > 4000 \text{ fpm} \end{cases}$$

If the speed is over 2000 fpm, carefully cut teeth *should* be used; if $v_m >$ 4000 fpm, the teeth *should* generally be precision cut; the corresponding production methods are described later.* Gear teeth are assumed to be strong enough against fatigue failure when the strength F_s from (13.6) is equal to or greater than the above dynamic load, $F_s \geqq F_d$ with Y for the load at the tip; how much greater depends on the service (§ 13.18).

Velocity factors are still widely used, with AGMA sanction, probably because it is so much easier to design with them than to consider the elasticity and inertia of the whole system, and there are many years of experience with their use. For types of service and ranges of capacity for which they are known to apply, they give satisfactory results, but as is seen from Fig. 13.9, it is not prudent to extrapolate into a region where it is not known that they apply without confirming tests. The AGMA recommends the use of equation (**m**) for high-precision shaved or ground spur gears, and

(**n**)
$$F_d = \frac{50 + v_m^{1/2}}{50} F_t \text{ lb.} \qquad \left\{\begin{matrix}\text{Commercial} \\ \text{Hobbed or} \\ \text{Shaved}\end{matrix}\right.$$

for commercially hobbed or shaved teeth. But in addition to a service factor, § 13.18, there is recommended a load-distribution factor[13.1] whose value ranges from 1.3 to more than 2.2; also other factors. Thus, while Fig. 13.9 indicates that these equations (**m**) and (**n**) tend to give low estimates of F_d, the use of additional "factors" tends to even things up. It should be observed that the velocity factors take no account of the differences of materials, whereas Buckingham's work (§ 13.17) indicates that for a particular quality of manufacture, the dynamic load for steel gears is much larger than for cast iron ($E_{\text{steel}} > E_{\text{C.I.}}$).

13.16 EXAMPLE—SPUR GEARS, INTERMITTENT SERVICE. Design a pair of 20° full-depth involute gears to transmit 20 hp at 1150 rpm. The diameter of the pinion is to be 5 in. and the velocity ratio m_ω should be about 2.5; smooth intermittent service (wear not considered). Let the dynamic load be taken as a function of velocity only.

Solution. The pitch-line speed and transmitted load are

$$v_m = \pi D n = \pi\left(\frac{5}{12}\right)1150 = 1505 \text{ fpm},$$

$$F_t = \frac{33{,}000 \text{ hp}}{v_m} = \frac{(33{,}000)(20)}{1505} = 438 \text{ lb}.$$

The speed, 1505 fpm, is in the upper acceptable range for commercially cut teeth

* To those who used the previous edition: the velocity factors were originally adjusted in this text to match the design stresses recommended by Buckingham. Since the velocity factors are generally quoted in their original form, we have conformed. The introduction of the fatigue strength reduction factor results in virtually the same results.

(§ 13.15), so let us assume that the teeth will be carefully cut. Then the dynamic load is estimated from equation (**l**) as

$$F_d = \frac{1200 + v_m}{1200} F_t = \frac{(1200 + 1505)(438)}{1200} = 988 \text{ lb.}$$

With interchangeable type teeth and the same material, the pinion tooth is the weaker. Part of the designer's responsibility is to decide upon a material. We might try a cast iron for both gears since it is an inexpensive material; say ASTM 25, whose endurance limit may be taken as (0.4)(25) = 10 ksi. Table AT 6 gives the typical $s_n' = 11.5$; hence, 10 ksi is conservative. In Lewis' equation (13.6), there are still four unknowns: b, P_d, Y, K_f; hence, it must be solved by trial. Let $K_f = 1.48$ (this should be fairly close for the load at the tip); let $b = 10/P_d$, which is nicely within the generally desired range (§ 13.12). When equation (**k**), (**l**), or (**m**) is used for the dynamic load, the traditional assumption is that one tooth may take the full load on the tip; therefore, assume an appropriate value of Y for this configuration. Since Y does not vary markedly, some reasonable value, as $Y = 0.32$, Table AT 24, is suitable for the first approximate solution. With these various assumptions,

$$F_s = F_d = 988 = \frac{sbY}{K_f P_d} = \frac{(10{,}000)(10/P_d)(0.32)}{1.48 P_d},$$

from which $P_d = 4.68$. If we assume that readily available cutters have a pitch of 4 or 5, § 13.4, we must now make a choice of one of these pitches and check the strength; choose 5, the nearest. Perhaps as good procedure as any is to solve for the face width b that makes the strength equal to the dynamic load. For $N_p = P_d D_p = (5)(5) = 25$, we find $Y = 0.34$, Table AT 24, use $K_f = 1.48$ as before;

$$F_s = 988 = \frac{sbY}{K_f P_d} = \frac{(10{,}000)(b)(0.34)}{(1.48)(5)},$$

from which $b = 2.15$; use $b = 2\frac{1}{4}$ in. Check the proportion; $bP_d = (2.25)(5) = 11.25$, which is within the range of 8 to 12.5 and is therefore satisfactory. For $m_\omega = 2.5$, we get $N_g = (2.5)(25) = 62.5$; use 62 teeth in the gear. (If an exact velocity ratio of 2.5 were essential, the pinion would have to have an even number of teeth.) *A* solution therefore is

$$P_d = 5, \qquad b = 2\tfrac{1}{4} \text{ in.}, \qquad N_p = 25, \qquad N_g = 62.$$

A single series of computations as given above is not likely to give the best design. In this case, it had already been determined that cast iron *could* be used without having too few teeth on the pinion. The designer should make a number of designs considering different materials and pitches and *sizes*, all satisfying the mathematical requirements, and then choose the one that is most economic for the purpose.

13.17 BUCKINGHAM'S AVERAGE DYNAMIC LOAD FOR METAL TEETH.

The dynamic load is considered to be made up of the

transmitted load F_t plus a dynamic ***increment load*** I, the increment being the consequence of the various inaccuracies and the accompanying accelerations. Since its magnitude depends upon the masses of the gears, a simple equation purporting to give the dynamic load can apply only to a certain class of gear. Thus, for gears of average mass with average connected masses, $F_d = F_t + I$,

$$F_d = F_t + \frac{0.05v_m(bC + F_t)}{0.05v_m + (bC + F_t)^{1/2}} \text{ lb.,} \tag{13.7}$$

where v_m fpm is the pitch-line speed, F_t lb. is the transmitted load, b in. is the face width, and C is a function of the amount of the effective error and of elasticity of the gear materials;

$$C = \frac{kE_gE_p}{E_g + E_p}, \tag{o}$$ [TABLE AT 25]

where $k = 0.107e$ for $14\frac{1}{2}°$ full-depth teeth; $k = 0.111e$ for 20° full-depth teeth; $k = 0.115e$ for 20° stub teeth; e = effective or composite tooth error, which should be smaller at the higher speeds, Fig. AF 19 (choose the design value from Fig. AF 20; see § 13.20); and E_g and E_p are the modulii of elasticity of the materials of the gear and pinion, respectively.

Gears that would not be average are, for example, aeronautical gears, where the moving masses, considering the loads, are less than average and gears connected by short shafts to heavy flywheels where the masses and stiffness are greater than average. Also, small gears transmitting low power are not average; nor are high-speed, lightly-loaded gears. The predominant influence on larger gears at moderate speed is the connected masses. But for small gears, especially on small shafts that easily twist through an angle equivalent to the effective tooth error, the mass of the gears themselves is the predominant factor, and the dynamic load is much less than given by equation (13.7). There is no easy way to define a dividing line, but if the shaft is as small as $\frac{1}{2}$ in. (let up to 2 in. be suspect), the gear on it will probably fall into this (not average) category. Whenever the horsepower is less than about 10 to 20, use of equation (13.7) will be overly conservative. For these small powers, equation **(k)** or **(l)**, § 13.15, is recommended for the dynamic load.

Since the origin of equation (13.7) is different from that of equations **(k)**, **(l)**, and **(m)**, there is no reason to expect that dynamic loads computed from these various equations should check with one another except at certain operation conditions, Fig. 13.9.

An alignment chart in *Problems* will be helpful in solving equation (13.7).

13.18 SERVICE FACTORS. Gears in actual service, like other machine elements, are subject to such a variety of operating conditions, each of which is impossible of detailed analysis, that the only practical

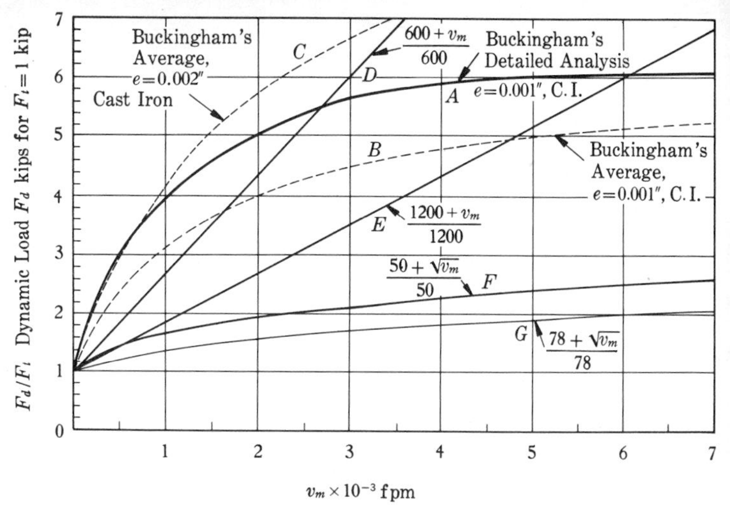

FIGURE 13.9 Dynamic Load by Different Equations. Buckingham's detailed analysis,[13.3] curve A, is for a constant transmitted load of $F_t = 1000$ lb. and the following data: $b = 5$ in., 20° full depth, cast-iron gears, $e = 0.001$ in., $N_p = 30$, $m_g = 4$, connected mass on pinion shaft and the pinion mass are 16 and 2.5 slugs, connected mass on gear shaft and the gear mass are 40 and 13 slugs, respectively. The dotted curves B and C are for Buckingham's average equation (13.7), other data the same. The velocity factors are of course independent of the elasticity of the materials and the face width. Note that curve C for cast iron and $e = 0.002$ in. is the same as would be obtained with steel and $e = 0.001$ (see Table AT 25). The values of F_d from curves A, B, C have the relations shown with those from C, D, E, F, only for $F_t = 1000$; the relative positions of these curves will in general be different for other values of F_t. Similar studies could be made with another independent variable, as F_d plotted against power with constant speed.

alternative is to introduce an experience factor, in essence an increase of the "factor of safety" to cover ignorance. Let N_{sf} represent this service factor; then $F_s \geqq N_{sf}F_d$ in order for the teeth to be safe against breakage. Both the driving machine (electric motor, gasoline engine, etc.) and the driven machine affect the choice of service factor. The service factor may be taken as 1 if an electric motor drives a centrifugal blower (smooth load), whereas if the drive is by gasoline engine, 1.25 may be more appropriate. The following summary from various sources may be helpful when the dynamic load is known with some confidence [see Ref. *(13.1)*]; the machines are the driven machines, driven by electric motors:

$1 < N_{sf} < 1.25$: uniform load without shock; centrifugal machines, hoisting machinery, belt-driven machine tools, textile machinery, smoothly running conveyors.

$1.25 < N_{sf} < 1.5$: medium shock; frequent starts; reciprocating compressors and pumps, pneumatic tools, well-drilling machinery, wire-drawing machinery, portable electric tools, lobe blowers, heavy-duty conveyors, machine tools, kilns.

$1.5 < N_{sf} < 1.75$: moderately heavy shock; dredging machinery, road machinery, railway motor cars, single-cylinder compressor, ore or stone crushers, punch press, tumbling barrels.

$1.75 < N_{sf} < 2$: heavy shock; rolling mill, rock crushers.

For increased reliability, use higher service factors; if limited life is satisfactory, use lower values.

13.19 PERMISSIBLE AND EXPECTED ERRORS. The higher the pitch-line speed of gears, the greater the dynamic reactions arising from the composite tooth error. Moreover, the relative magnitude of the noise of a running pair of metallic gears is an indicator of the magnitude of the errors—in general, the louder the noise for a particular speed, pitch, and environment, the larger the error. So, higher speeds require the greater accuracy for satisfactory operation. The permissible error from Fig. AF 19 is taken as the maximum error for satisfactory operation. We note in Fig. AF 19 that for speeds greater than 5000 fpm, the tooth error should be of the order of 0.0005 in.; Fig. AF 20 suggests that small teeth ($P_d > 5$) are therefore advantageous.

The way to use Figs. AF 19 and AF 20 in design is first to determine the permissible error from Fig. AF 19 after having computed the speed. This is an indicator of the needed accuracy of manufacture. Then with a known or assumed pitch, enter Fig. AF 20 and decide on the quality of manufacture and use the value of e from the next lower curve in computing C in Buckingham's equation. In general, the teeth should not be more accurate than necessary because greater accuracy frequently means greater cost, which suggests using the permissible error in design—if the permissible error is smaller than that corresponding to commercially cut teeth. However, the most economic value can only be determined by knowing how much effort (and cost) is necessary in a particular shop to obtain a particular error; and this is quite variable.

In Fig. AF 20, the curve labeled *First-class Commercial Gears* represents the results that might be expected of gears cut by form cutters, Fig. 13.10(a), or hobs, Fig. 13.11, if reasonable care is taken in performing the work and if the machines and tools are in good order; hardness ranges up to about 350 Brinell.

The curve labeled *Carefully Cut Gears* represents the results that should be obtained with accurately ground hobs or shaping cutters, Fig. 13.12, when the finishing cut is taken on one side at a time. In quantity production, carefully cut gears may be made virtually as cheaply as commercially cut gears.

The curve labeled *Precision Gears* represents the degree of accuracy that may be expected in carefully ground or shaved teeth, or teeth cut with

FIGURE 13.10 (a) Form Cutter. (Courtesy Brown & Sharpe Mfg. Co., Providence, R.I.). (b) Shaving Cutter. (Courtesy National Broach and Machine Co., Detroit). In (a), the profile *abcd* is the shape of the tooth space on the gear being cut. (b) The shaving cutter has many cutting edges; produces accurate profiles.

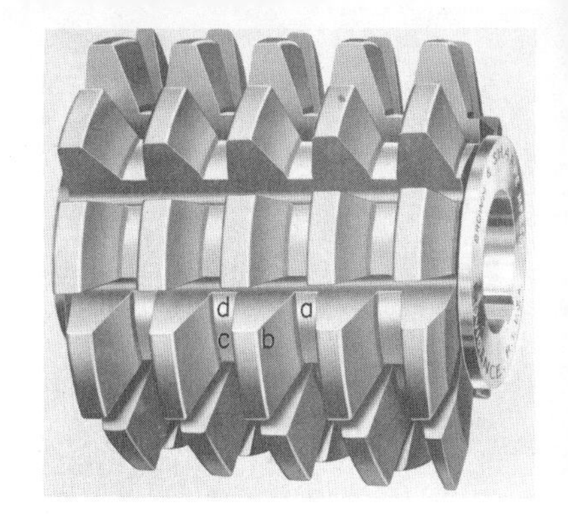

FIGURE 13.11 Hob Cutter. The profile *abcd* is the shape of a rack tooth, except as adjusted for clearance. This cutter *generates* the tooth profile on tne gear being cut. (Courtesy Brown & Sharpe Mfg. Co., Providence, R.I.).

extreme care with accurately ground tools on shapers kept in the best condition. On hardnesses up to 450 Brinell, the teeth may be finish ***shaved***, a precision finishing process. Shaving can produce the involute profile within a tolerance of 0.0002 in. and a spacing error within 0.0003 in. Shaving is preferably done *before* heat treating on material with a machinability rating of 25% or better. Under favorable conditions, the error may be reduced to 0.0003 to 0.0004 in. by lapping or grinding. Shaving is the fastest process and lapping the slowest; see Fig. 13.10(b).

When gears are heat treated *after* the teeth are cut, the profile error is sure to increase. Under the most carefully controlled conditions, the increase in error due to heat treatment has been held to as little as 0.0005 in. For less exacting treatments, the warpage may be considerably more—in some cases, so much more that unground heat-treated teeth have shorter life than teeth not heat-treated.

For exacting, high-capacity service, hardened teeth should be finish ground: (1) by a form grinder whose profile fits the tooth profile after the manner of a form cutter, Fig. 13.10; (2) by a straight-sided grinding wheel that generates the profile as does a rack cutter; (3) by a grinding worm that has a thread section much like that of the hob of Fig. 13.11. Observe

FIGURE 13.12 Gear Shaper. The pinion cutter has teeth the same as on a like size gear, except as adjusted for clearance. The gear being cut and the cutter rotate together as the cutter reciprocates vertically, *generating* the tooth profile on the gear being cut. (Courtesy Fellows Gear Shaper Co., Springfield, Vt.).

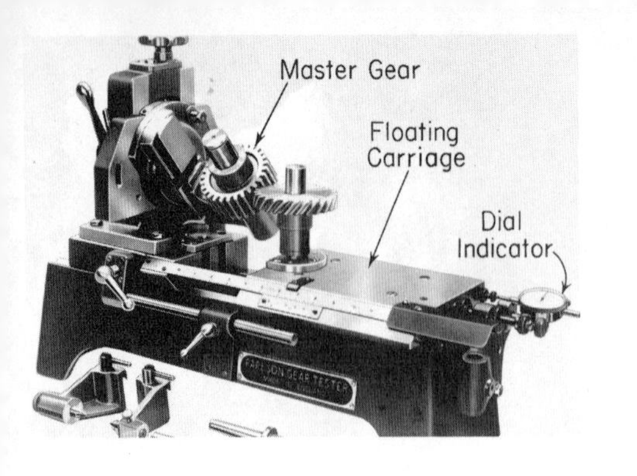

FIGURE 13.13 Gear Tester. Helical gear mounted for test. (Courtesy George Scherr Co., Inc., NYC).

that the surface finish of teeth is either machined (hobbed or shaped), shaved, or ground (or lapped).

There are several machines available for testing the accuracy of the tooth profiles, one of which is shown in Fig. 13.13. If two perfect gears were in mesh, the dial pointer would not change position; the center distance would be the same for every point of engagement; and the floating carriage would not move. This type of machine will indicate when the teeth are not concentric with the bore, when the tooth profiles depart from an involute, when there are variations in thickness of the teeth, and when there are variations in the spacing. See Ref. *(13.1)* for considerable detail of manufacturing processes.

13.20 EXAMPLE—BUCKINGHAM'S EQUATION FOR DYNAMIC LOAD. The data are the same as in § 13.16; design the teeth using Buckingham's equation for average gears.

Solution. From § 13.16, we have $v_m = 1505$ fpm, $F_t = 438$ lb. For $v_m = 1505$ fpm, we find the maximum permissible error from Fig. AF 19 to be $e = 0.0019$ in. Preliminary calculations suggest that a pitch of 5 will not give a large enough tooth according to this method; therefore, assume $P_d = 4$. Entering Fig. AF 20 at $P_d = 4$ and $e = 0.0019$, we decide that the teeth must be carefully cut, which corresponds to an expected error of 0.00125; use $e = 0.00125$ in. With carefully cut teeth and the Buckingham dynamic load, we may assume that the load is shared by two teeth until the point of application has moved to about mid-profile. From Table AT 25, $C = (1.25)(830) = 1037$. For a face $b = 3$ in. ($bP_d = 3 \times 4 = 12$), we get from equation (13.7)

$$F_d = F_t + \frac{0.05v_m(bC + F_t)}{0.05v_m + (bC + F_t)^{1/2}}$$

$$= 438 + \frac{(0.05)(1505)[(3)(1037) + 438]}{(0.05)(1505) + [(3)(1037) + 438]^{1/2}} = 2420 \text{ lb.}$$

The fatigue strength reduction factor for the load applied near the middle of the tooth is higher than with the load applied at the tip; use $K_f = 1.7$. From Table AT 24, $Y = 0.544$ for $P_dD_p = (4)(5) = 20$ teeth and load near the middle of the profile. The strength of the assumed tooth is then

$$F_s = \frac{sbY}{K_fP_d} = \frac{(10{,}000)(3)(0.544)}{(1.7)(4)} = 2400 \text{ lb.}$$

We observe immediately that for these computations there is no margin of safety;

the service factor is virtually unity, $F_s/F_d = 2400/2420$. If this is not satisfactory, and usually some margin of safety is desired, an even larger tooth may have to be used, or a more accurate (and expensive) method of manufacture will reduce the error and the dynamic load. On the other hand, the Buckingham equation gives results on the conservative side. See Fig. 13.9. We do have the evidence from the previous traditional design to substantiate this observation. Therefore, the solution by this method of design is

$$P_d = 4, \qquad N_p = 20, \qquad N_g = 50, \qquad b = 3 \text{ in., carefully cut.}$$

13.21 LIMITING LOAD FOR WEAR.

The kind of wear referred to here is that which occurs because of a fatigue failure of the surface material as a result of high contact stresses, a phenomenon called ***pitting***. Except when they are in contact at the pitch point where pure rolling occurs for an instant, tooth profiles roll and slide on one another while in contact. The frictional force produces a tangential tensile stress, thus acting to increase and complicate the stresses; its adverse effects are of course much reduced when a lubricant is present to reduce the force tangential to the surface. Running-in under light load and thus smoothing the tooth profiles increases the capacity because a thinner oil-film will keep the metal surfaces out of contact (as in sleeve bearings). If a lubricant film exists, the area carrying the load is increased and the film absorbs some of the shock of tooth engagement, which reduces the magnitude of the peak stress. Whatever the offsetting factors may be, Buckingham found a good correlation between surface fatigue failure and the Hertz contact stress when the contact is taken at the pitch point. According to the Hertz equation, the calculated stress is greater where contact starts or ends (r_1 or r_2 may become quite small), but the pitting phenomenon almost invariably occurs first in the vicinity of the pitch surface, when one pair of teeth is carrying the load, thus justifying Buckingham's assumption. If contact occurs at a base circle, one r is zero and the Hertz equation does not apply. To derive Buckingham's wear equation, start with the Hertz equation (1.19) of § 1.27; let the materials of both gears have the same value of Poisson's ratio, $\mu_1 = \mu_2 = 0.3$; the radii of profile curvature at the pitch surface are $r_1 = r_p \sin\phi$ and $r_2 = r_g \sin\phi$, Fig. 13.14; the face width b is the length of contact line. With these modifications, equation (1.19) becomes

$$s^2 = \frac{0.35F(1/r_p + 1/r_g)}{\sin\phi\, b(1/E_p + 1/E_g)},$$

$$\text{(p)} \qquad F = \frac{bs^2 \sin\phi(1/E_p + 1/E_g)}{(0.35)(2)(1/D_p + 1/D_g)},$$

where the subscripts p and g refer to pinion and gear, respectively. Note that the larger the pressure angle ϕ, the larger the load F for a particular stress s, which means that the 20° full-depth tooth has a greater capacity F_w than the $14\frac{1}{2}°$ tooth. The force F in equation (**p**) is strictly the normal force W in Fig. 13.6; $F = W = F_{tan}/\cos\phi$ when the teeth are in contact at

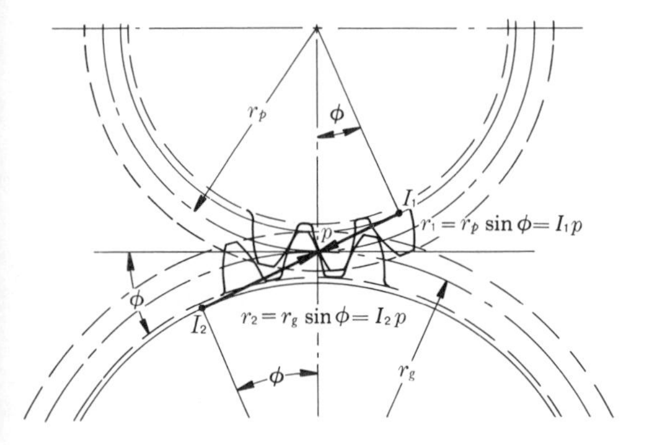

FIGURE 13.14 Curvature of Tooth Profiles. $r_1 = I_1p$ = **radius of curvature of pinion profile at pitch surface;** $r_2 = I_2p$ = **radius of curvature of gear profile at pitch surface.**

the pitch point, where F_{tan} is a tangential load. However, Buckingham compares the F in (**p**) with the tangential dynamic load. Thus, we let

$$\frac{1}{1/D_p + 1/D_g} = \frac{D_pD_g}{D_g + D_p} = \frac{D_p}{2}Q;$$

(**q**)
$$K_g = \left(\frac{s^2 \sin\phi}{1.4}\right)\left(\frac{1}{E_p} + \frac{1}{E_g}\right).$$

Use these in (**p**), which then reduces to

(13.8)
$$F_w = D_p b Q K_g,$$

where $F = F_w$ is the limiting wear load and

(**r**)
$$Q = \frac{2D_g}{D_g + D_p} = \frac{2N_g}{N_g + N_p} = \frac{2m_g}{m_g + 1};$$

the gear ratio $m_g = N_g/N_p = D_g/D_p$; the wear factor for gearing K_g is a material factor* dependent also on the pressure angle ϕ; for indefinite service, s is taken as the limiting value for surface fatigue ($s_{n\,surf}$ in Table AT 26). Most of the values of K_g in Table AT 26 are from "actuarial" tables by Buckingham (in a personal communication), some are the original Buckingham values. Where you find the asterisk *, the values are deduced from Cram[12.22] by reducing by 15% his laboratory results for 0% sliding (except class 20 C.I.). There is some evidence[12.22] that different alloy

* Some give the wear factor K_g other definitions, even to the extent of varying it with the service, thus making it in part a service factor.

steels heat treated to the same hardness may have significantly different values of K_g, but not enough is known to generalize. Also the results reported by Cram suggest that some of the Buckingham values are too high; see those with a double asterisk **, Table AT 26. See Ref. *(13.1)* for different values of $s_{n\,\text{surf}}$. For steel, the surface endurance strength may be estimated from

$$\text{(s)}\quad s_{n\,\text{surf}} = (400)(\text{BHN}) - 10{,}000 \text{ psi} \qquad \text{or} \qquad s = (0.4)(\text{BHN}) - 10 \text{ ksi} \quad [\text{STEEL}]$$

when test values are not available. When F_w is compared with the Buckingham dynamic load F_d, no margin of safety or service factor is needed except for extreme situations, and perhaps for the surface-hardened teeth where there is some conflict in the literature as to whether there *is* a surface endurance *limit*.

As previously discussed, with the contact compressive stress, there is a high shear stress just below the surface, § 1.27, which is about $0.3s$, where s is given by the Hertz equation (**p**). It is believed that surface fatigue failure is a consequence of this shear stress with cracks starting in the vicinity of maximum shear stress,[12.22] and it is evidenced by small particles of metal being sheared away, leaving a pitted surface. For this reason, surface hardening, as carburizing, flame or induction hardening, should extend further than double the depth to the point of maximum shear, given in § 1.27.[13.3] See also Grover.[12.22] It is preferable for there not to be a steep hardness gradient because the boundary area may then be the weak part.

The pinion teeth should be somewhat harder than the gear teeth, since its teeth undergo more contacts than the gear teeth and the harder metal cold works the softer, improving the surface smoothness and mechanical properties. See Fig. AF 4 for conversion of hardness numbers. A reasonable commercial tolerance on BHN would be some 40 or more points. Some designers prefer to use the computed contact stress from (**p**) as a criterion of fatigue, compared with limit values from tests or service.

In general, when indefinite, more-or-less continuous service is desired, it is necessary for both $F_w \gtreqqless F_d$ and $F_s \gtreqqless F_d$. For intermittent or other limited service, it is common to make $F_s \gtreqqless F_d$ only; and for very short service, fatigue limit F_s may be smaller than F_d. If sufficient information were available, both the beam strength and the surface fatigue life could be estimated on an actuarial (probability) basis (as discussed for rolling bearings). See values of K_g for different numbers of cycles of life in Table AT 26. If obtained at some cost, indefinite life is too long when a limited life will do.

Steel teeth are more likely to fail by "wear" (surface fatigue) than by breakage; design steel teeth for wear, check strength. Design cast-iron teeth for fatigue breaking, and check for surface fatigue—with exceptions.

13.22 EXAMPLE—WEAR OF CAST IRON TEETH. Check the limiting wear load for the gear teeth of the example of § 13.20. The data are: $P_d = 4, b = 3$ in., $N_p = 20$, $m_g = 2.5$, $D_p = 5$ in., hp $= 20$ at $n_p = 1150$ rpm, $\phi = 20°$, and $F_d = 2420$ lb.

Solution. Using the conservative value from Table AT 26 for ordinary cast iron, say class 20, and 20° involute, we get $K_g = 112$ from Table AT 26. From equations (**r**) and (13.8), we get

$$Q = \frac{2m_g}{1 + m_g} = \frac{(2)(2.5)}{1 + 2.5} = 1.43,$$

$$F_w = D_p b Q K_g = (5)(3)(112)(1.43) = 2400 \text{ lb}.$$

Since $F_w = 2400$ is virtually the same as $F_d = 2420$, these gears should last indefinitely if properly made, assembled, and cared for.

13.23 GEAR-TOOTH WEAR. In addition to the "wear," called pitting, that is a surface fatigue failure, several other phenomena occur that are described in Ref. *(13.18)*. We shall mention only four more, giving the words the meanings in accordance with ASA and AGMA; see also Ref. *(14.1)*.

Abrasion is caused by foreign matter, such as grit or metal particles, or by a failure of the oil film at low speed. Abrasion from foreign matter is prevented by protecting enclosures and clean lubricants. Heavier oil stops abrasion due to film failure.

Scoring occurs when the oil film fails, but in this case, the load and speed are so high that the surface metal is melted and the metal is smeared down the profile.[13.28] Sometimes there is local spot welding of the teeth together and subsequent separation, leaving a pockmark. Scoring occurs at a lower loading with thin oil than with heavy oil. So-called EP (extreme pressure) lubricants, which contain an additive that forms a film on the surfaces by a chemical reaction between the metal and the additive (sulfur and chlorine compounds, for example), can withstand much heavier loading without permitting scoring. This film is continuously being formed and worn, and under steady operating conditions, arrives at steady-state thicknesses. It is thinner near the pitch surface where rolling occurs and thicker near the tips where sliding is a maximum, leading to the deduction that the heat generated by friction (minimum at pitch surface, maximum at tip), is the factor that brings about the reaction to form the film.[13.28] Scoring may be a problem in the high-surface-hardened, heavily loaded teeth, but cast irons and bronzes with normal lubrication will probably have a tooth breakage before a film failure.

Spalling is a surface fatigue of greater extent than pitting; that is, the flakes are much larger. This type of failure occurs in surface-hardened teeth.

Corrosion causes surface deterioration that in turn results in weakening of fatigue strengths, but presents no problem with clean petroleum oils and a clean environment.

13.24 MATERIALS USED FOR GEARS. All kinds of materials are used for gears. The cheapest is ordinary gray iron, say ASTM 20, which is relatively good from the standpoint of wear and which should be used if it fits the scheme of things. ASTM classes 30 and 40 are frequently used; and the high-strength irons, say class 60 or higher, are also appropriate for gears with the proper heat treatment.

Untreated wrought-steel gears are inexpensive and generally have a carbon content of 0.30-0.50%; low wear capacity. Small gears are frequently cut from solid pieces of cold-finished stock. Cast steel should be annealed to avoid excessive distortion. Both the wrought and cast steels may be heat treated for improved mechanical properties. Oil quenching is not so drastic as water quenching and is generally used because of the consequent lesser distortion.

Also because of less distortion, alloy steels are favored when teeth are to be heat treated. Of course, the blank can be heat treated before the teeth are cut. Ordinarily, the maximum hardness for machining is about 250 BHN, but successful machining is often done on alloy steels of 350 BHN, even to 450 BHN on AISI 4340 (not with ease).

The heat treatments used are described briefly in Chapter 2. When gears are to be through-hardened, the carbon content should be 0.35% to about 0.6%; frequently used steels, roughly in order of cost, are: 1335, 5140, 4037, 4140, 8640, 8740, 3135. The hardness depends on the carbon content, the hardenability, and the heat treatment. See Tables AT 8, AT 9, and Figs. AF 1, AF 2, and AF 3. There is sometimes a problem of excessive brittleness; to avoid this, we may for pedagogical purposes arbitrarily set a limit on the lowest permissible tempering temperature, as 700°F. (This temperature may be too low in some cases; remember the loss of fatigue strength of highly hardened pieces with stress raisers.)

When through-hardening does not produce hard enough surfaces for surface fatigue strength, the surface is especially hardened. Cyaniding, a shallow hardening process, is done on the medium carbon steels. Much the same steels are used for flame-hardened and induction-hardened teeth as are used for through hardening. The higher the carbon content (to a point), the harder the surface may be. In the absence of more specific information, assume that the surface hardness of teeth by flame and induction hardening to be 500 BHN, a value that should be readily attainable if the carbon is 0.4% or more. If the teeth are carburized, a minimum BHN = 600 is a reasonable expectation; use this for design. Automotive gears, which are not subjected to continuous full power operation, were formerly through hardened. The addition of cyaniding reduced troubles from surface fatigue; 5145 through-hardened took 10,000 hr. of operation to pit, whereas the same material cyanided ran 25,000 hr. before pitting.[2.1] Now the automotive industry by carefully controlled processes to keep warping at a minimum[13.33] uses case-hardened alloys without a finish operation after hardening (small gears); in the same experiment just quoted,

8620 case hardened to a depth of 0.035–0.05 in. ran 75,000 hr. without pitting, and if the teeth were shaved first and then hardened, the running time was 150,000 hr. The life of case-hardened teeth improves with depth of case perhaps up to some 0.08 in. thickness; a rule of thumb for case thickness is about 0.1 times full depth of teeth.

Alloy steels for carburizing, roughly in order of cost, are[2.1]

5120, 4118, 8620, 8720, 4720, 4620, 4320, 4820.

The flame-hardening process for gear teeth needs development and testing for particular cases. The heating and quenching is of one tooth at a time; this tends to anneal the fillets of the preceding tooth just hardened, thus reducing the beam fatigue strength. Also, the roots of induction-hardened teeth may not be fully hardened and therefore the teeth may be unexpectedly weak in bending fatigue. If unusual precision of surface hardened teeth is necessary, care must be taken not to lap or grind away too much of the case and lose the residual compressive stress on or near the surface, and not to leave a discontinuity. Grinding tends to leave residual tensile stresses.

In extreme service requirements, nitriding may be used at increased cost. Because the process is carried out at a relatively low temperature, there is not so much trouble with warping. Other surface hardening processes are used (§ 2.8), even peening; but peening the fillets to improve the beam strength or peening to remove scale after heat treatment are the more likely applications of this process.

To summarize the relationships between machining and heat treating we may use gears after the following operations; (1) the blank can be annealed or normalized, the teeth machined complete (low hardness); (2) the blank can be annealed and/or normalized, then heat treated to the final hardness, the teeth machined complete (hardness limited by machine-able hardness; excellent hardenability probably required to obtain the desired hardness all the way through), or for small gears, one might buy suitably heat-treated bars; (3) as in (1) except that teeth are rough cut, then heat treated, then finish cut [can get somewhat higher hardness than in (2)]; (4) as in (2) except that the surface is later hardened by one of the processes previously described, after which a finishing operation of grinding or lapping may be done (necessary in large gears) to improve the profile accuracy. There are of course numerous variations of heat-treatments, but the principal ideas are as outlined. Sintered gears find frequent applications, especially small ones as in appliances. For example, a crank gear (automotive) with teeth is produced from a mixture of 6% Ni, 2% Cu, 92% iron powders, double pressed and sintered. Designing the gears with a pressure angle larger than 20° improves their capacity.[13.26]

13.25 EXAMPLE—DESIGNING STEEL GEARS FOR CONTINUOUS SERVICE. Design the teeth for a pair of steel gears to transmit 40 hp at 1150 rpm of the 5-in. pinion; $m_\omega = m_g = 2.5$; 20° full-depth, interchangeable teeth.

Solution. The pitch-line speed is $\pi(5/12)(1150) = 1505$ fpm (as in § 13.20); the transmitted force is*

$$F_t = \frac{33{,}000 \text{ hp}}{v_m} = \frac{(33{,}000)(40)}{1505} = 876 \text{ lb.}$$

Checking the speed of 1505 with respect to Fig. AF 19, we decide as before that the teeth are preferably carefully cut. As previously pointed out, steel gears without special surface hardening are more likely to wear out than break. Therefore, design for wear first. As usual, there are so many unknowns that iteration is necessary. The following procedure is reasonable.

Assume $P_d = 4$; then $e = 0.00125$ from Fig. AF 20, and from Table AT 25 for steel and steel, we get $C = (1.25)(1660) = 2075$. Let $b = 2.75$ in. ($bP_d = 11$, which is satisfactory). We shall compare the limiting wear load with the Buckingham dynamic load; as computed from (13.7),

$$F_d = F_t + \frac{0.05v_m(bC + F_t)}{0.05v_m + (bC + F_t)^{1/2}}$$

$$= 876 + \frac{(0.05)(1505)(2.75 \times 2075 + 867)}{(0.05)(1505) + (2.75 \times 2075 + 867)^{1/2}} = 4036 \text{ lb.}$$

We make the limiting wear load $F_w = F_d = 4036$ lb. In equation (13.8), $D_p =$ 5 in., $b = 2.75$ in., $Q = 2m_g/(1 + m_g) = 1.43$; solve for K_g, which will indicate something about the material needed. From $F_w = D_pbQK_g$,

$$K_g = \frac{4036}{(5)(2.75)(1.43)} = 206.$$

Look for this value of K_g in Table AT 26 in the 20° column (indefinite service). Interpolating between the total ΣBHN = 600, $K_g = 196$ and ΣBHN = 650, $K_g = 233$, we find the needed total ΣBHN = 614 for $K_g = 206$. Let the gear be $\text{BHN}_g = 250$ because carbon steel can be machined at this hardness and there should be a significant saving on material cost (as compared with an alloy); then the pinion hardness is $\text{BHN}_p = 364$. These values are starting points for consideration of the kind of steel to use. The final decisions are made by studying charts and tables giving mechanical properties, with appropriate metallurgical and heat-treating knowledge in mind. Consider through hardening first. If we stay within the pedagogical limit of 700°F tempering temperature and use only the information in the text, Fig. AF 1 shows that oil-quenched C 1040 can barely qualify; BHN = 248 at 700°F. Since the hardenability of plain carbon steels is low, the teeth would need to be cut before the heat treatment. Tentatively then, the gear is to be made of C 1040, OQT to a minimum hardness of 250 Brinell—provided the teeth prove to be strong enough in bending. (An alloy with a good hardenability, which could be hardened first and the teeth cut afterwards, could prove more economical on an over-all basis.)

The pinion with Brinell of 364 should be an alloy, any one of several being satisfactory, Table AT 9, Figs. AF 2 and 3. For this hardness, the teeth are more

* The data are the same as in § 13.20 except that the horsepower is 40.

likely to be cut first and hardened later. Choose 4140, Table AT 9, and interpolate for the approximate tempering temperature from 800°F, 429 BHN and 1200°F, 277 BHN; this indicates a tempering temperature of about 975°F (for 364 BHN), which should not result in trouble from brittleness. Now check the strength.

For BHN = 250, we estimate $s'_{ng} = 250/4 = 62.5$ ksi. For BHN = 364, we get $s'_{np} = 364/4 = 91$ ksi. For $N_p = 20$ and the load near the middle of the profile, $Y_p = 0.544$. For $N_g = 50$, $Y_g = 0.694$; $Y_p s'_{np} = (0.544)(91) = 49.5$; $Y_g s'_{ng} = (0.694)(62.5) = 43.4$. Therefore, the gear is weaker. Let $K_f = 2$, which should be on the high side.

$$F_s = \frac{(sY)_g b}{K_f P_d} = \frac{(43.4)(2.75)}{(2)(4)} = 14.9 \text{ kips},$$

or a strength of 14,900 lb. compared with a dynamic load of 4036; therefore, breakage is quite unlikely.

Discussion. We observe that there is an infinite number of solutions to this problem by any approach, considering all trivial differences in the kind of steel and heat treatment. The optimum solution, which we may define as the one that satisfies the objectives at the minimum cost of manufacture, can only be decided upon with a complete knowledge of the whole situation. There is no substitute for an actual manufacturing plant in a particular environment (kinds of machine tools, heat-treating facilities, local suppliers, etc.). If the gears are to be bought from a gear manufacturer, bids could be obtained on two or more alternative specifications (perhaps leaving the choice of the particular material up to the supplier who may be prepared to process one material more cheaply than another). One alternative to consider may be to use a pinion of about 330 Brinell and a gear of about 290 Brinell, with the idea of using a material of such good hardenability and machineability that the teeth may be cut after heat treatment. The possibility that this procedure (at least, in the general case) might result in more accurate profiles, therefore lower dynamic load, therefore more economical gears, should not be overlooked. Perhaps the pinion is not necessarily 5 in. in diameter; a cheaper solution may be found in increasing or by decreasing its size. If low weight is one of the objectives, surface hardening would significantly reduce the size of tooth and of pinion needed.

The basic idea of the design procedure outlined by the foregoing solution is to assume a pitch and face width, and then determine the corresponding required hardness; if this hardness is not obviously the best, assume another tooth size and repeat the calculations; etc. Always check the beam strength; sometimes it is significant even for indefinitely continuous service.

The design should be studied further with respect to AGMA recommendations. As a result, it may be found that the Buckingham equation is overly conservative. Then there is the traditional design with velocity factors. Equation (l) gives a dynamic load of $F_d = 1980$ lb. Introducing a factor for safety of 1.5, we get $F_d = 2970$ lb., considerably less than that by the Buckingham equation. This lower value of F_d could mean important savings in heat treatment costs.

13.26 MORE ON THE DESIGN OF GEAR TEETH. Those who design gear teeth may be classified into three groups: (1) the casual designer who, more or less blindly, follows as printed an accepted procedure, (2) the designer of gear reducers who must meet AGMA ratings, but being something of an expert, allows currently available knowledge to affect his decisions, (3) the designer who is concerned with situations that require special judgment and experience, or the extrapolation of experience, as in automotive, aircraft, and space-systems gears. (Of course, these basic ideas of classification are extendable to other machine elements.)

So many variables are involved, including the manufacturing and operating variables, that formulas for special service conditions have proliferated. However, it is to be hoped that eventually reasonably accurate computations of the dynamic load can be easily made, and related to the beam strength via suitable design stresses modified by service factors, and to the limiting wear load (or Hertz stress). In the meantime, the full Buckingham procedure[13.3] (Fig. 13.9), which considers the inertia of the connected masses and the rigidity of the connecting members and which we do not have room to explain in sufficient detail to make useful, is probably the wisest for the designer working in an area where experience is limited. Where applicable, the AGMA rating formulas should produce conservative results; this approach is given in Chapter 15 for bevel gears.

As seen from Fig. 13.9, the estimation of the dynamic load is not nearly so easy as the pat use of a formula suggests. In heavily loaded gears, the deflection of the tooth is greater than the manufacturing profile error (the effective error, § 13.17, is the manufacturing error plus the deflection). In lightly loaded gears, the tooth deflection may be very much less than the profile error.[13.13] Lightly loaded gears operated at high speed may fail because the dynamic load grows with speed until it is destructive. But see § 13.38. On the other hand, Reswick[13.13] says that the dynamic load can be virtually ignored in heavily (transmitted) loaded gears. A heavily loaded tooth is one for which the deflection by the transmitted load is greater than the manufacturing error. Also, Kohler observed (pp. 14–32, Ref. *13.1*) that for his data, the dynamic load did not exceed three times the transmitted load, but the Buckingham analysis predicts much higher ratios when the transmitted load is small compared to the effective inertia of the system. Buckingham points out that heavy *dead* loads (flywheels, rotors) on shafts result in shaft vibrations doing more damage to the teeth than low inertia loading that permits the vibrations to be largely absorbed in the transmission rather than reflected to the teeth.

In extreme situations, as in heavily loaded, carburized, aircraft gears, the profiles are modified so that they have more nearly perfect involutes in a deflected position and also for the purpose of reducing the load on that part of the profile subjected to maximum sliding velocity v_s. Thus it should not be assumed that the applied-dynamic-load-wear-and-strength-resistance concept is necessarily the basic criterion for all situations. The parameter $s_c v_s$ has been used as a criterion for high-speed gear capacity (against scoring), where s_c psi is the Hertz contact stress, and v_s fps is the sliding velocity,[13.9] applied particularly to helical gearing. The limit value of $s_c v_s$ may be from 3×10^6 to 4.5×10^6; but see § 13.38, also Kelley.[12.22] For more detail on the sliding velocity, see Ref. *(13.2.)*

13.27 DESIGN OF NONMETALLIC GEAR TEETH. Various nonmetallic materials, Table AT 5, have long been used for gears, perhaps principally because of their quiet operation. A disadvantage that may be significant is the low heat conductivity of these materials. If the power involved is large enough, the increase in temperature may seriously weaken the nonmetallic gear teeth; the friction and work lost in deforming the profile may not be dissipated to the surroundings at the rate at which it occurs until the temperature is excessive. The teeth for rawhide and laminated phenolics are designed as previously explained, with $F_s \geqq F_d$, and in case of continuous service, $F_w \geqq F_d$. Because of the low modulus of elasticity of these materials, the dynamic load is smaller than for metal teeth and is taken as

$$F_d = \frac{(200 + v_m)F_t}{200 + v_m/4}, \tag{13.9}$$

which is compared with the strength $F_s = sbY/P_d$ when the load is applied at the tip of the tooth; and design stress $s = 6000$ psi (no stress concentration factor); no change in the equation for F_w.

Since the modulus of elasticity E is much less for these materials than for metals, the teeth have a greater deformation under a given load, and therefore the load is more likely to be distributed over two teeth, except when contact is in the vicinity of the pitch point. This fact is recognized in the recommended design procedure for nylon (Zytel) gears. When made of Zytel, the teeth may be either molded (inexpensive in quantity) or cut; they are most commonly of fine pitch ($P_d > 16$); and strength is taken as the determinant. Use the value of Y for the load near the middle of profile in $F_s = sbY/P_d$ and design stresses as follows:[11.55,13.19] for 5×10^8 cycles, $s = 2.3$ ksi for $P_d = 16$, $s = 2.6$ ksi for $P_d = 20$, $s = 3.5$ ksi for $P_d = 32$, $s = 3.7$ ksi for $P_d = 48$; for 10^7 cycles, and P_d of 16, 20, 32, 48, respectively, use s as 4.2, 4.7, 5.8, 6.1 ksi; for other values of s, interpolate directly as the log N, N = number of cycles. The dynamic load is recommended to be a velocity factor (VF) times the transmitted load; $F_d = (\text{VF})F_t$. If the gear teeth are lubricated, $16 < P_d < 48$, use VF as follows:

	Molded Teeth	Cut Teeth
$v_m < 4000$ fpm, VF =	1.0	1.2
$v_m > 4000$ fpm, VF =	1.2	1.4

Let $F_s = F_d$ for smooth loads; include a service factor for other kinds of loads. Nylon gears can be run without a lubricant (preferably initially wetted with one), but the design stress should be much reduced.[11.55] Use 20 or more full-depth, 20°-involute teeth, but as usual, the smallest teeth that will do the job.

Buckingham[13.15] says that mating phenolic gears with steel of BHN < 400 leads to excessive abrasive wear. Also, the large deflection under load may result in the tip of a metal gear tooth digging into the flank of the plastic pinion at the initial point of contact, thereby damaging it. For this reason, the use of unequal addendum and dedendum teeth, § 13.34, even to the extent of eliminating the angle of approach, is recommended.

13.28 EXAMPLE—LAMINATED PHENOLIC GEAR TEETH. What should be the pitch, face, and number of teeth for a pair of 20° full-depth spur gears if they are to transmit 12 hp at 1150 rpm of the pinion; $m_\omega = m_g = 2$; the load is smooth and the service is continuous. Let a Bakelite pinion drive a cast-iron gear, ASTM 20 (but there may be trouble—see the previous paragraph).

Solution. First decide on a suitable number of teeth; let $N_p = 20$ for which there is no interference trouble; then $N_g = 40$. Comparing sY for gear and pinion, we note that the pinion tooth is the weaker. Since neither the diameter nor pitch are known, some iteration is indicated. (You can express velocity, transmitted load, dynamic load, and strength load in terms of P_d; then solve for P_d; try it.) Assume $P_d = 5$; this gives $D_p = N_p/P_d = 20/5 = 4$ in.,

$$v_m = \pi Dn = \pi\left(\frac{4}{12}\right)(1150) = 1205 \text{ fpm},$$

$$F_t = \frac{33{,}000 \text{ hp}}{v_m} = \frac{(33{,}000)(12)}{1205} = 328 \text{ lb.},$$

$$F_d = \frac{(200 + v_m)F_t}{200 + v_m/4} = \frac{(200 + 1205)(328)}{200 + 1205/4} = 920 \text{ lb.}$$

From Table AT 24, find $Y = 0.32$ (at tip) for $N_p = 20$, and then solve for b from Lewis' equation (13.5):

$$F_d = F_s = 920 = \frac{sbY}{P_d} = \frac{(6000)(b)(0.32)}{5}$$

from which $b = 2.39$ in., length of face needed for strength. For the wear equation, we have ($K_g = 64$ from Table AT 26—only value at hand)

$$Q = \frac{2m_g}{1 + m_g} = \frac{(2)(2)}{1 + 2} = 1.333,$$

$$F_w = F_d = 920 = D_p bQK_g = (4)(b)(1.33)(64),$$

from which $b = 2.7$ in., needed for wear; $bP_d = (2.7)(5) = 13.5$, on the high side. If not satisfactory (see E of phenolics), recheck for $P_d = 4$.

Discussion. To emphasize that design must follow a validated combination of loading, design stress, and the method of computing the stress, calculate the stress in the tooth found above if the gear is molded Zytel; that is, follow the design instructions for the nylon gears. Since $v_m < 4000$, the velocity factor is 1; take $F_d = F_t = F_s$; use $Y = 0.544$ for the load near the middle; then

$$s = \frac{F_s P_d}{bY} = \frac{(328)(5)}{(2.375)(0.544)} = 1270 \text{ psi}.$$

This tooth is much larger than the fine pitches for which we have quoted justification in experience with nylon, but the stress is on the low side (see stresses given in § 13.27), leading to the conclusion that if a nonmetallic gear is needed here, nylon should not be ruled out.

13.29 DESIGN OF CAST TEETH. Cast teeth are comparatively rough and inaccurate, and they are therefore more likely to be used in large, slow-speed (below 600 fpm) service, preferably outdoors where noise will not be objectionable. Use equation (**k**), § 13.15, for the dynamic load ($v_m < 600$) and use

$$F_s = 0.054sbP_c \tag{t}$$

for computing the strength. Use the endurance strength for s, but include a service factor for any but the smoothest loads. Because cast teeth have relatively inaccurate outlines, their face width is generally less than for cut teeth, say $2P_c < b < 3P_c$, cast teeth; circular pitch is used for the convenience of the pattern maker.

13.30 HUNTING TOOTH. In the example of § 13.28, we have $N_p = 20$ teeth and $N_g = 40$ teeth. After every two revolutions of the pinion, the same pair of teeth engage. If however, we use 41 teeth in the gear, the pinion will rotate 41 times and the gear 20 before the same pair of teeth engage again. This extra tooth is called a ***hunting tooth***, and its use distributes the wear more evenly. The resulting velocity ratio must be permissible.

13.31 HUBS, METAL GEARS. Hubs of gears are proportioned empirically. The bore of the hub depends on the shaft diameter. The length of the hub depends fundamentally on the length of key, if a key is used. Two keys may sometimes be used to avoid excessive hub length. *The length of the hub should not in general be made less than the face width of the gear.* Hub lengths *usually* vary from about $1.25D_s$ to $2D_s$, where D_s is the *bore*. The following values are reasonable for the *diameter* of the hub:

For cast iron, diameter of hub $= 2D_s$;
For steel, diameter of hub $= 1.8D_s$.

For light, low-speed service, these proportions may be decreased somewhat. The bead at the *hub*, Fig. 13.15, may be allowed to depend on the size of the arms at the hub.

13.32 ARMS AND WEBS. The arms, Fig. 13.15, are designed as cantilever beams. If the speed is not so high that centrifugal force induces significant stresses, it is safe to assume that the dynamic load is the maximum load and design the arms using the endurance strength as the design stress, $M = sI/c = sZ$, and to assume that the load is equally shared by the arms; inclusion of a factor to care for stress concentration and a rough surface would be prudent. For N_a = number of arms, the maximum moment on each arm is $M = FL/N_a$, where L is the distance from the pitch

FIGURE 13.15 Spur Gears. (Courtesy Globe Stock Gear Division, Philadelphia).

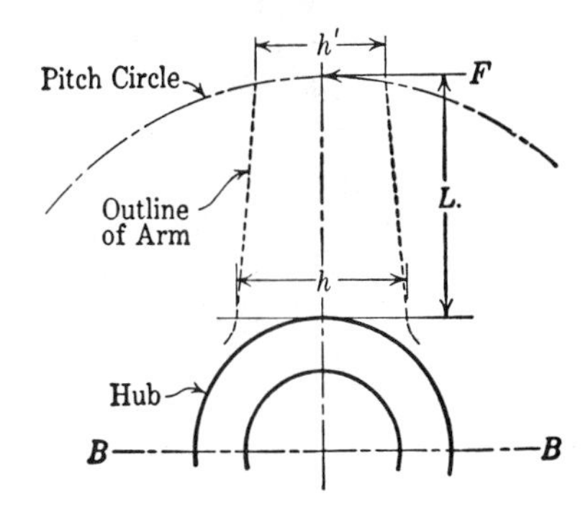

FIGURE 13.16 Moment on Arm. Dimensions at pitch circle may be made (0.7) × (dimensions at hub); or a taper of about $\frac{3}{4}$ in./ft. is reasonable.

circle to the hub, Fig. 13.16, and $F = F_d$ if the design is based on fatigue. The section modulus Z is in accordance with the shape of the section. Shapes of arms include ellipses, rectangles (with corners generously rounded), H section, I sections, and cross sections, Fig. 13.17. For an elliptical section with the major diameter h twice the minor diameter, $h_1 = h' = h/2$, we get $Z = \pi h^3/64$, using Table AT 1. Solve for h from the bending moment equation as explained above.

As a convenience for the shapes of arms shown in Fig. 13.17, solve for h and Z as described above, then compute G and G_1 as given in this illustration.

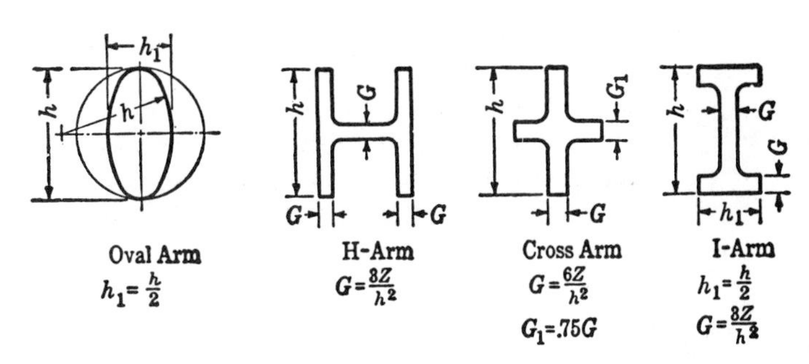

FIGURE 13.17 Proportions for Gear Arms.

FIGURE 13.18 Welded Gear Blank. An example of large gear construction; outside diameter about 9 ft. 7 in.; weight, 25,000 lb; inside hub diameter, $24\frac{3}{4}$ in. (Courtesy Luckenweld Inc., Coatesville, Pa.).

Use six arms whenever possible for gears up to 120 in. in diameter, although four arms do well for small gears. For gears over 120 in. in size, use eight or more arms. Very large gears may be of welded construction, Fig. 13.18.

Whatever the computed dimensions, care must be taken to avoid abrupt changes in the thickness of adjoining parts. If a very thin section (comparatively) adjoins a very heavy section, unequal cooling of the casting in the case of cast gear blanks may result in severe residual stresses that materially weaken the gear.

Small gears are frequently made with webs of a thickness equal to about $0.5P_c$ to $0.6P_c$, or the teeth may be cut from solid stock, sometimes from the shaft forging itself. For minimum stress concentration in a solid gear, locate the keyway on the center line of a tooth *space*.[13.4] If the teeth are integral with the shaft forging, the diameter of the dedendum circle should be somewhat larger than the shaft diameter.

13.33 RIM AND BEAD.

For the assumption recommended for arm design, namely that the load is equally shared by the arms, the rim must be rigid enough to accomplish this. Experienced engineers can decide quite well by appearance what is good proportion. Empirical values that have been used are: make the rim thickness and depth of bead each about $0.56P_c$ (but as small as $0.4P_c$ for small, high-strength, aircraft gears); a bead is recommended for its stiffening value when arms are used.

In case of speeds that produce sizeable centrifugal forces, some estimation of the corresponding stresses is in order.

13.34 UNEQUAL ADDENDUM AND UNEQUAL DEDENDUM TEETH.

Because of interference troubles when pinions with small tooth numbers are used and because of certain operating advantages, noninterchangeable gear teeth are frequently appropriate.[13.24] Basically in these systems, the dedendum on the pinion is decreased in order to reduce or eliminate interfering flank and the addendum is increased, more or less a like amount. To match, the gear's addendum is decreased and its dedendum is increased Fig. 13.19. Ways of accomplishing these changes are explained in engineering texts on kinematics or mechanism.[13.2] The amount of the change of addendum may be adjusted so that the pinion and gear teeth have about the same form factor Y. If interference is otherwise present, this change increases the contact ratio m_c, with the result that the gears run smoother and quieter. For quantity production for a particular system, the disadvantage of noninterchangeability virtually disappears. The principles of design are the same as already explained; the values of the Y's will have to be determined from large layouts of the teeth; but the interested reader will have to review the kinematics of gear teeth for complete understanding.

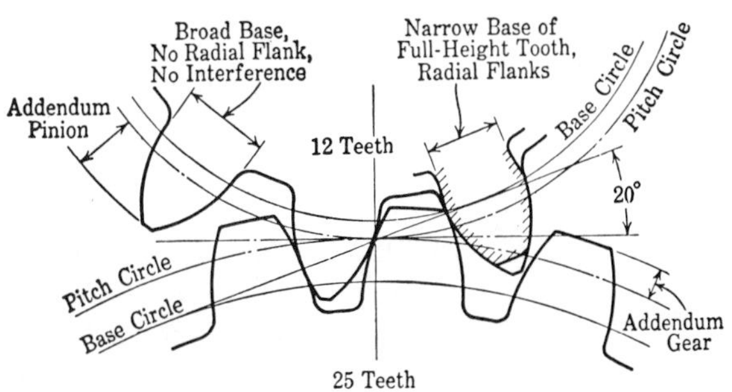

FIGURE 13.19 Unequal Addendum and Dedendum Teeth.

13.35 INTERNAL GEARS.

An ***internal*** or ***annular*** gear, Fig. 13.20, has teeth cut on the inside of the rim instead of on the outside. The shape of the teeth is thus the same as the shape of the space on an external gear of the same pitch diameter, except as modified by clearance requirements. As for external gears, whenever a working part of a tooth extends inside its base circle, either on the gear or pinion or both, interference may occur. If the pinion is too large, as compared to the internal gear, the teeth will "foul." To eliminate fouling, let the *minimum differences* in tooth numbers between the internal gear and pinion, $N_g - N_p$, be:

for $14\frac{1}{2}°$ involute, full-depth, 12 teeth;
for 20° involute, full-depth, 10 teeth; and
for 20° involute, stub, 7 teeth.

If the profiles are not to be modified at all, the difference should be greater than given above.

FIGURE 13.20 Internal Gear. A broached internal gear, 16 pitch, about 3.7 in. in diameter, made of 4140 steel; the ring gear in a planetary transmission. (Courtesy National Broach & Machine Co., Detroit).

Internal gear-and-pinion sets are quieter than like external gears, probably because more teeth are carrying the load which is therefore not shifted so abruptly. The combination is compact, with a short center distance. If the gear and pinion are of the same material, the gear tooth will be stronger; hence the design for strength is based on the pinion tooth.

With a concave surface on the gear tooth in contact with the convex surface of the pinion tooth, the contact area is larger than for two external teeth (two convex surfaces), and the limiting wear load is larger (the radii r_1 and r_2 in the Hertz equation are of opposite signs). An extra factor is that the distribution of the load among more teeth decreases the intensity of the stress. The limiting wear load is computed from $F_w = D_p b Q K_g$, which is the same as before[13.3] except that

$$Q = \frac{2N_g}{N_g - N_p} = \frac{2m_g}{m_g - 1}. \tag{u}$$

13.36 GEAR TRAINS. When the velocity ratio is high, it is desirable to reduce the speed in two or more steps. A high limit for one pair of spur gears is about $m_\omega = 10$; a better limit is 6. Helical and herringbone gears, Chapter 14, are most often used in heavy-duty gear boxes.

When two gears are integral in a double reduction, Buckingham suggests that the profile error in the second pair be assumed to be about 50% more than the method of manufacture is expected to produce. This assumption is made because the rigidity of the connection results in a transmission to the second pair of gears of the effects of the dynamic loads on the first pair. If, however, the two gears on the second shaft in a double reduction are separated by a long length of shafting, the elasticity of the shaft absorbs to a large extent the dynamic loads originating at the first pair of gears.

Internal gears are commonly used in planetary trains. From the horsepower equation, say hp $= Tn/63{,}000$, we conclude, ignoring frictional losses (hp = constant), that the torque on a particular shaft in a train varies inversely as the angular velocity n (or ω). Thus, the torque on any gear can be found from

$$\frac{T_1}{T_2} = \frac{n_2}{n_1}, \tag{v}$$

when the torque on another gear in the train is known, as the input torque; after which the transmitted load F_t is calculated by dividing the torque by the pitch radius. The easiest method of computing the angular velocity of gears in a planetary train is by the equation[13.2]

$$n_L = en_F + n_a(1 - e), \qquad \text{[PLANETARY TRAINS]} \tag{w}$$

where n_a = absolute turns of the arm (planet carrier), positive for CL rotation, negative sign for CC rotation,

n_F = absolute turns of the first gear in the planetary train, same sign convention as for n_a,

n_L = absolute turns of the last gear in the planetary train, same sign convention, CL+, CC−,

e = train value when the first gear in the planetary train is taken as a driver; evaluated with the arm held stationary (ordinary gear train) as the product of the tooth numbers of the driving gears *divided* by the product of the tooth numbers of the driven gears; given a positive sign if, with the arm held stationary, the first and last gears turn in the same sense, a negative sign if they turn in opposite directions.

Correct algebraic signs are essential in the use of (**w**). If properly handled, it does not make any difference which end of the *planetary* train is taken as the first gear; let it be the input gear. Then the gear for n_L can be *any other* gear in the train with e figured to include the one considered to be the last one.

As you recall from the study of kinematics, three or more planet gears are common. This permits designing the teeth for much less than the total power or torque, although it is probably overly optimistic to assume that the power is equally divided among the planets; something more like total power divided by the number of planets plus 15% would be more prudent. This same idea is used on occasion with ordinary gear trains, several pinions spaced round the gear delivering power to it.

13.37 EFFICIENCY OF GEARS AND THERMAL CAPACITY. The loss per pair of spur, helical, or bevel gears in an ordinary train should not exceed 2%, giving an efficiency of about 98%, or more. The loss *in the mesh* of spur and helical gears should be less than 1%. Ordinarily, the amount of the loss in spur and helical gear boxes presents no overheating problem, but as the speed increases with more or less the same force on the teeth (power increasing), the dissipation of the loss as heat to the surroundings may become a problem. It should be investigated when the pitch line speed is greater than about 2000 fpm,[13.1] earlier when there are multiple pinions each with an input. Follow the plan outlined in § 16.6, using the thermal coefficient from Fig. AF 21. Also, the AGMA gives rating formulas for thermal capacity.

13.38 LUBRICATION OF GEAR TEETH. Fortunately, there is a tendency for the hydrodynamic wedge to form between gear teeth, making it possible to have thick-film lubrication. However, at a particular speed and oil viscosity, the tooth load may be so high as to cause the oil film to be squeezed out enough to allow metal-to-metal contact; if so and if the speed is below some value ($v_m \approx 800$ fpm pitch line, in one research), *abrasive* wear occurs (§ 13.23); if the speed is higher, *scoring* occurs. If the load and lubricant are such that only the high spots on the surface are touching, enough friction is likely to be generated locally to result in a lowering of viscosity and a consequent early scoring or abrasion; in short, the safe load for very smooth surfaces (20 μin.) is much higher than for rough surfaces (100 μin.), at all speeds. Once scoring or abrasion has occurred, the capacity of the gears is much reduced.

Other findings include: the smaller the teeth, for the same face width, the higher the load capacity before scoring, which is explained by the fact that the maximum sliding velocity v_s for the same pitch-line speed is higher on larger teeth; the capacity of the oil film to stand up decreased as the profile error increased, but the effect of the error was not so great as predicted by Buckingham's F_d; with SAE 30 oil, the load capacity decreased with temperature increase to about 350°F, but increased somewhat as the temperature went above 400°F because of the formation of carbonaceous deposits on the working surfaces; a synthetic oil of the same viscosity as a mineral oil does not have the same load carrying characteristics as the mineral oil, but in both cases, load capacity increases with viscosity; relieving the tip of the tooth had a beneficial effect on load capacity up to a relief of 0.0008 in.; directing the oil to the initial point of the mesh gave a higher capacity than for other points of lubrication.

One relationship that might not be expected—the load capacity *decreased* as the speed increased (up to about 4000 fpm pitch line in the tests cited—mineral oil), and then increased moderately (to some 15,000 fpm), after which there was a rapid increase in load capacity (maximum test speed, about 22,000 fpm). At speeds over 15,000 fpm (speeds of 30,000 fpm, maybe more, are now in use, introducing some unusual considerations[13.32]), the scoring load for the heavier oils was never reached. The conclusions are, first, that the conventional design procedures are not applicable at very high speeds, and that small gears can transmit great power; in the tests, a pair of gears rated at 10 hp (AGMA) actually transmitted 500 hp without surface damage. At these high speeds, the thick lubricant film remains intact, first, because the contact time is shorter than a deformation pulse and the fluid is acting as a plastic solid, and second, because of the squeeze effect (it takes a definite time for the lubricant to be squeezed out). The facts of this article and some of the opinions are taken from Borsoff.[13.28,13.29] The actual speeds cited are not intended as general measures; they are known to apply only to the particular tests. See Electrofilm, § 11.27.

13.39 CLOSURE. As we have seen, there are so many variables involved in a gear drive that no simple procedure has been evolved that gives generally correct solutions. For the student's sake, we regret that it cannot be said to do it in a certain way, but it *is* inherent that there are question marks in *every* engineering answer; there are just more of them with respect to gearing than for most any other machine element. When quantity production is involved, laboratory and service tests should be used to prove the design. There is much highly specialized literature that can help. In important cases without adequate reassuring past experience, the American designer may care to follow a British procedure as a check.[13.25]

The design of the gear case is no small matter in the success of the gears; it must have enough rigidity to maintain the gears in the expected alignment.

14. HELICAL GEARS

14.1 INTRODUCTION. In accordance with our general plan, we shall not give as detailed explanations of the engineering of other types of gears as we have of spur gears, but most of the generalizations of the previous chapter on kinematics, materials, wear, fatigue strength, dynamic load, etc. apply as well to any gear system. Differences and changes will of course be highlighted.

As seen in Fig. 14.1, the elements of helical teeth are cylindrical helices; one end of the tooth is advanced circumferentially over the other end.

FIGURE 14.1 Helical Gears. See also Fig. 14.7. (Courtesy The Falk Corp., Milwaukee).

Without a modifying adjective or other inference by context, "helical gears" refer to ones on parallel shafts (see § 14.9). Since the contact on a particular tooth starts at one end, always with other teeth in contact, and then succeeding tooth sections make contact, the tooth picks up the load gradually. The line of contact is never across the whole tip, as it is in spur gears, but is always diagonal. When the midsections of perfect teeth are in contact at the pitch point, the line of contact runs from the addendum circle near one end of the tooth to the working depth near the other end. The consequence is that the bending moment on the tooth is only a little over half as much as it would be for the entire load at the tip. Helical teeth (parallel shafts) thus have a greater breaking strength. Moreover, the gradual transfer of load results in quieter running, lower dynamic load, and higher permissible speeds. A commercial grade of helical gear runs about as quietly as precise spur gears, at less cost. Because of these advantages, helical and herringbone gears are nearly always used in heavy-duty gear boxes, as well as in many others.

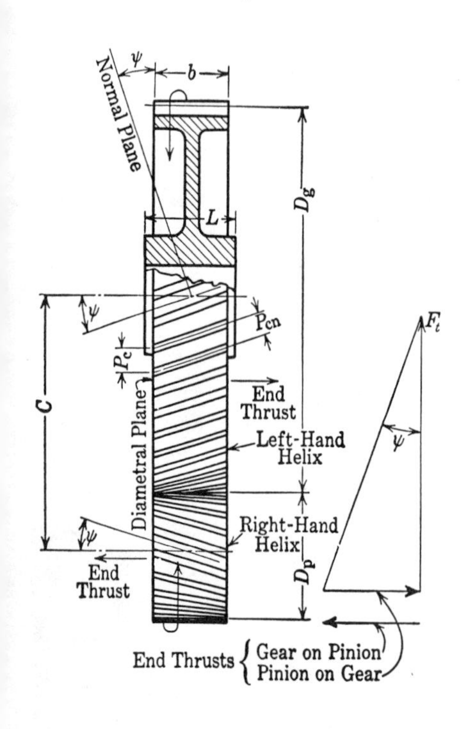

FIGURE 14.2 Helical-gear Notation. A right-hand helix is one that slopes around the pitch cylinder in the manner of a right-hand screw thread; that is, the tooth on the front side of the gear slopes up to the right with the *axis vertical*; apply the right-hand rule.

14.2 HELIX ANGLE AND FACE WIDTH. The helix angle ψ, Fig. 14.2, is the angle between a tangent to the pitch helix and an intersecting axial element of the pitch cylinder. There are other helix angles, for example the base helix angle ψ_b, which is the angle between a tangent to the base helix and an element of the base cylinder (which contains the base circles). Since helical gears are inherently not interchangeable—a right-hand helix meshes with a left-hand helix when the shafts are parallel—there are no standard helix angles. Angles of $\psi = 15°$ to $25°$ are typical, but for wide-face gears, the helix angle may be less than 15°. Since the end thrust increases as the helix angle increases, Fig. 14.2, the helix angle is sometimes limited for this reason. On the other hand, the larger the helix angle for a particular face width, the more the overlap of the teeth and the more gradual the transfer of load. For significant advantage from the overlapping of tooth engagement, the face width should be[13.1]

(a) $$b \geqq 2P_a$$

($b_{min} = 1.15P_a$ has been used in the past), where $P_a = P_c/\tan\psi$ in. is the axial pitch [equation (**d**) below]. The ***face contact ratio*** m_f is the advance of the tooth in the face width $b \tan\psi$ divided by the circular pitch; $m_f = b\tan\psi/P_c = b/P_a$. Without the adjective, *contact ratio* has the meaning previously defined, § 13.5, and in particular is called the *profile contact ratio*. Basically, the face width is determined by the loading to be carried, but the helical gears often have large active face widths compared to spur gears of the same pitch.

14.3 PITCHES. *The* pitch of a helical gear is the pitch in the diametral (transverse) plane, P_c for circular pitch, P_d for diametral pitch. As seen in Fig. 14.2, the distance between the teeth measured on the pitch surface along a normal to the helix, which is called the ***normal circular pitch*** P_{cn}, is less than P_c measured around the pitch circle:

(b) $$P_{cn} = P_c \cos\psi \qquad \text{and} \qquad P_{dn} = \frac{P_d}{\cos\psi},$$

where the latter form is obtained by using $P_c = \pi/P_d$ and $P_{cn} = \pi/P_{dn}$. If N_t is the number of teeth in a particular gear and D is its pitch diameter,

(c) $$P_{cn} = P_c \cos\psi = \frac{\pi D \cos\psi}{N_t} \qquad \text{and} \qquad P_{dn} = \frac{P_d}{\cos\psi} = \frac{N_t}{D\cos\psi}.$$

If the helical gears are manufactured with standard *hobs*, a common practice, the *normal pitch* P_{dn} is standard. It follows that the pitch in the diametral plane P_d and also the pitch diameter will contain a decimal fraction. For example, suppose $N_t = 20$, $P_{dn} = 6$ and $\psi = 23°$; then

$$P_d = P_{dn} \cos\psi = (6)(\cos 23°) = (6)(0.9205) = 5.523,$$

and $D = N_t/P_d = 20/5.523 = 3.62$ in. *Shaper cutters* for helical gears are based on the pitch P_d in the diametral plane. Thus helical gears may have a normal pitch that is standard or a pitch in the diametral plane that is standard.

The axial pitch P_a is the distance between corresponding points on adjacent teeth measured in an axial direction; it is seen in Fig. 14.3 that $P_c/P_a = \tan\psi$, or

(d)
$$P_a = \frac{P_c}{\tan\psi} = \frac{\pi D}{N_t \tan\psi} = \frac{\pi}{P_d \tan\psi}.$$

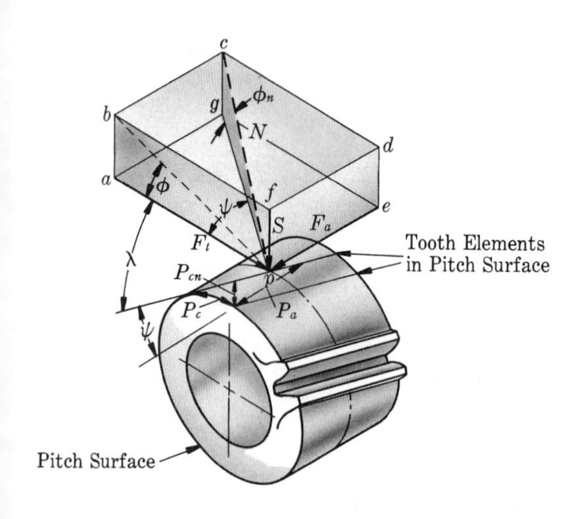

FIGURE 14.3 Force System of Helical Gear. Friction is neglected; $N = cp$ is the total force normal to the tooth at the point of contact, with components F_t = tangential load, S = separating force, F_a = axial thrust.

14.4 PRESSURE ANGLES. The pressure angle in the normal plane ϕ_n is different from that in the diametral plane. In Fig. 14.3, note that $\tan\phi_n = gc/pg$ and that $gc = ab$; then multiply and divide by ap;

(e)
$$\tan\phi_n = \frac{gc}{pg} = \frac{ab}{pg}\frac{ap}{ap} = \frac{ab}{ap}\frac{ap}{pg} = \tan\phi\cos\psi.$$

(See also § 16.7). Thus, if a standard 20° hob is used, the normal pressure angle is $\phi_n = 20°$; and the diametral pressure angle with a 23° helix angle is

$$\phi = \tan^{-1}\left(\frac{\tan\phi_n}{\cos\psi}\right) = \tan^{-1}\left(\frac{0.364}{0.9205}\right) = 21.57°.$$

The interchangeable tooth proportions with 20° pressure angle are often used for helical gears, even though there is no reason other than manufacturing convenience, except when interference would occur. Where appropriate, unequal addendums can be used. Note that for a constant pressure angle in the diametral plane, the top lands get narrower as the helix angle increases. (Sketch this to get the idea.) This leads to heat-treating difficulties; when the teeth are nearly pointed, the tip comes out excessively hard and brittle and causes trouble by breaking off. One action to take in this event is to stub the teeth somewhat, thereby making the pinion tip blunter (which is the usual practice in spiral bevel gears, § 15.4). At least one manufacturer uses a 25° pressure angle and full-depth proportions, which tends to improve capacity and to give smoother kinematic action (as compared to a stubbed tooth).

14.5 DYNAMIC LOAD, HELICAL GEARS.

Equations (**m**), (**n**), § 13.15, and others with velocity factors, plus operating and service factors, are used for estimating the dynamic load on these teeth, which will in general be smaller than that on similarly loaded straight teeth. For "average" gears, Buckingham's equation is

$$F_d = F_t + \frac{0.05v_m(F_t + Cb\cos^2\psi)\cos\psi}{0.05v_m + (F_t + Cb\cos^2\psi)^{1/2}}\ \text{lb.,} \tag{14.1}$$

where the symbols have the usual meanings. For precision helical gears running at above 5000 fpm, the dynamic load can be taken as the transmitted load.

14.6 STRENGTH OF HELICAL TEETH.

The Lewis equation is used to compute the strength of helical teeth. The pitch is that in the normal plane. Thus if s is the endurance strength, the limiting tooth strength is taken as (Y is for load near middle if K_f is used; §13.10)

$$F_s = \frac{sbY}{K_f P_{dn}}\ \text{lb.,} \tag{14.2}$$

where the symbols have the usual meanings, except that Y is chosen in accordance with the ***equivalent*** (also called *formative* or *virtual*) number of teeth N_e. To understand the significance of this, imagine a plane cutting a right cylinder obliquely, as the normal plane cuts the pitch cylinder,

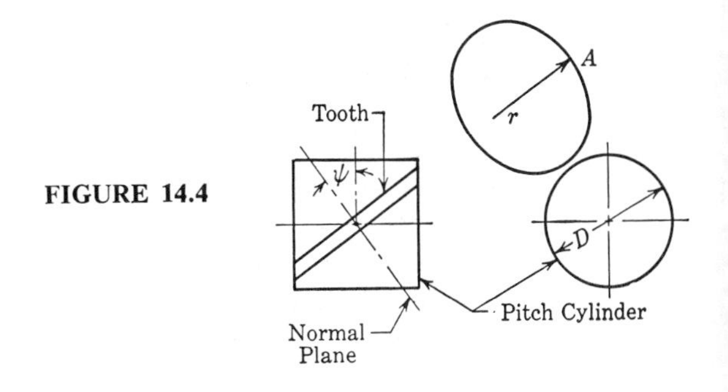

FIGURE 14.4

Fig. 14.4. The shape of the cut section is an ellipse whose minor diameter is the pitch diameter D, and the profile of the tooth in the plane of the ellipse is the normal profile. Let the radius of curvature of the ellipse at A be r. The equivalent number of teeth is the number on a gear of radius r with a

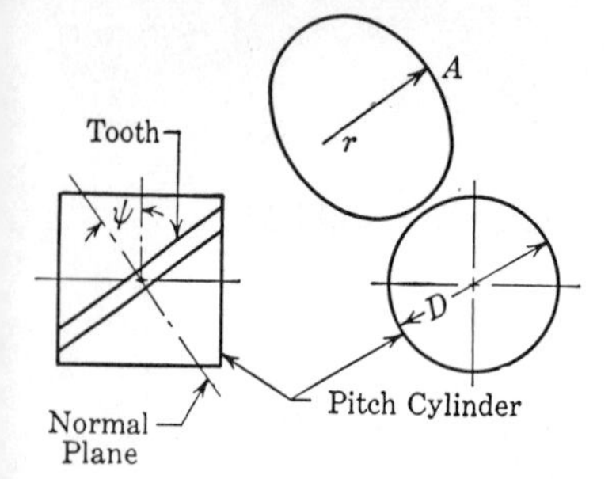

FIGURE 14.4 (Repeated).

pitch equal to the normal pitch. From analytical geometry, we find that $r = D/(2 \cos^2 \psi)$, Fig. 14.4. The equivalent number of teeth $N_e = P_{dn}(2r)$, or

$$\text{(f)} \qquad N_e = P_{dn} \frac{D}{\cos^2 \psi} = \left(\frac{P_d}{\cos \psi}\right)\left(\frac{D}{\cos^2 \psi}\right) = \frac{N}{\cos^3 \psi},$$

where N is the actual number of teeth on the gear, $P_d D$. The equivalent number of teeth, rather than the actual number, defines the form of the tooth in the normal plane.

Since wear is usually the determinant for steel gears in continuous service, a reasonable estimate for Y is ordinarily satisfactory; for example, Y for N_e interchangeable type teeth in accordance with the closest approximate standard from Table AT 24, load near the middle. Otherwise, the true profile must be developed and Y found as indicated in Fig. 13.6.

14.7 LIMITING WEAR LOAD. Buckingham[13.3] gives the limiting wear load for helical gears on parallel shafts as

$$\text{(14.3)} \qquad F_w = \frac{b D_p Q K_g}{\cos^2 \psi} \text{ lb.},$$

where the symbols have the same meanings as for spur gears, ψ is the helix angle, K_g is based on the pressure angle ϕ_n in the normal plane, and the actual numbers of teeth are used to find Q. Follow the same plan of design as detailed for spur gears.

14.8 HERRINGBONE GEARS. All of the foregoing discussion applies to herringbone gears, Figs. 14.5 and 14.6, which are double helical gears. The object of the right-hand and left-hand helices is to absorb the axial thrust within the gear, thus eliminating the necessity of providing against this thrust in the bearings. In order to divide the load equally between the two parts of the herringbone gear, one shaft is mounted so that it floats in the axial direction. Large helix angles, 30° to 45°, are used; hence, the thrust forces are surely large enough to keep both sides in driving contact. The face width b in equations (14.1), (14.2) and (14.3) should be the *active* width; that is, the width of groove, Fig. 14.6, should not be included in b; minimum $b = 4P_a$ [see equation (**a**) above].

FIGURE 14.5 ***(left)*** **Sikes Herringbone Gears. The large gear is manufactured from a casting. The teeth are the continuous variety; that is, they are cut up to the center of the gear. The helix angle is 30°; pressure angle is 20°. (Courtesy of Link-Belt Co., Chicago).**

FIGURE 14.6 ***(right)*** **Herringbone Gear Made from a Welded Built-up Blank. A particular machine is necessary to cut teeth to the center of the blank as shown in Fig. 14.5. The space in the center of this gear is to allow clearance for the cutting tool. (Courtesy Lukenweld Inc., Coatesville, Pa.).**

The hardness of helical and herringbone teeth cut after heat treatment will generally fall between the limits of 210 and 300 Brinell for the gear, with the pinion being some 40 to 50 Brinell points harder than the gear. A double-reduction helical gear box is shown in Fig. 14.7.

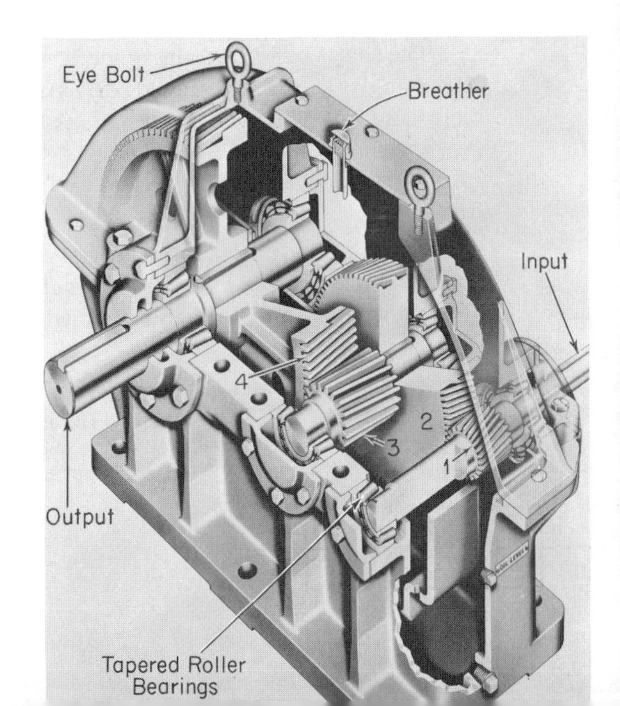

FIGURE 14.7 Helical Gear Box. These teeth are hobbed after heat treatment, unequal addendum and dedendum, full depth. Small gears made from about 1045 forged blanks; largest gear, cast steel, annealed and heat treated. Gear bore and shaft surfaces at the gear are ground finished for close limits. (Courtesy Westinghouse Electric Corp., Pittsburgh).

FIGURE 14.8 Crossed Helical Gears.

14.9 CROSSED HELICAL GEARS. Helical gears mounted on non-parallel shafts are called ***crossed helical gears*** (formerly *spiral gears*). If we imagine such gears working together, Fig. 14.8, we see that the teeth rub *across* one another (as a worm thread rubs across a meshing gear tooth), an action that is fundamentally different from that of gears on parallel shafts where the only sliding is up or down the profile. Contact occurs only at a point, theoretically, and the cross rubbing is much greater than up-and-down sliding with the result that the capacity of such gears is relatively small—light power at reasonable speed. Running-in under light load until a bright line is visible along the tooth length materially increases the capacity by increasing the area taking the load and improving the surface for the lubricant.

The shafts can be at any angle Σ, but 90° is the most common, and the helix angles ψ_1 (driver) and ψ_2 (driven) can have innumerable values, either left hand or right hand, the relations being[13.2]

$$\Sigma = \psi_1 + \psi_2 \quad \text{[SAME HANDS]} \qquad \text{or} \qquad \Sigma = \psi_1 - \psi_2 \quad \text{[OPPOSITE HANDS]}$$

The conditions for operating together are that the gears have the same normal pitch P_{cn}, P_{dn} and pressure angle ϕ_n. When the shafts are not parallel, the velocity ratio m_ω, which is defined as the angular velocity (n, ω) of the

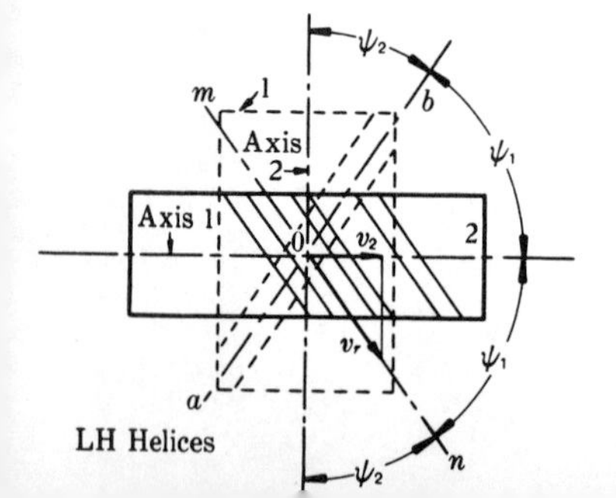

FIGURE 14.9 Crossed Helicals. Diagrammatic, showing ψ_1 and ψ_2. The dotted outline is driver 1, imagined on top of 2. The tangent line *ab* through *0* of gear 1 takes the position *mn* after 180° turn of 1.

driver divided by the angular velocity of the driven, is not a ratio of pitch diameters; but it *is* the ratio of the tooth numbers, N_1, N_2. Equation (**c**) applied to the driver 1 and driven 2 gives

$$\text{(g)} \qquad N_1 = \frac{\pi D_1 \cos \psi_1}{P_{cn}} \quad \text{and} \quad N_2 = \frac{\pi D_2 \cos \psi_2}{P_{cn}},$$

$$\text{(h)} \qquad m_\omega = \frac{n_1}{n_2} = \frac{\omega_1}{\omega_2} = \frac{N_2}{N_1} = \frac{D_2 \cos \psi_2}{D_1 \cos \psi_1},$$

and the center distance $C = (D_1 + D_2)/2$. For the usual case, $\psi_1 + \psi_2 = \Sigma = 90°$, Fig. 14.9; thus, $\cos \psi_2 = \cos (90 - \psi_1) = \sin \psi_1$ and $\cos \psi_1 = \cos (90 - \psi_2) = \sin \psi_2$. Therefore, for the shafts at right angles, equation (**h**) can be written

$$\text{(i)} \qquad m_\omega = \frac{\omega_1}{\omega_2} = \frac{D_2 \sin \psi_1}{D_1 \cos \psi_1} = \frac{D_2 \tan \psi_1}{D_1} = \frac{D_2}{D_1 \tan \psi_2}, \qquad [\Sigma = 90°]$$

Multiply through the middle terms by D_1/D_2 and note that $D_1\omega_1/(D_2\omega_2) = v_1/v_2 = \tan \psi_1$; similarly, using the last term, we find $v_2/v_1 = \tan \psi_2$; where v_1 and v_2 are pitch-line velocities of driver and driven, respectively. The rubbing speed v_r is in the tangential direction and is $v_r = v_2/\cos \psi_1 = v_1/\cos \psi_2$, Fig. 14.9.

The force arrangement on these gears is as described in detail for worm gears, Chapter 16; therefore, the equation for their efficiency may be derived in just the same manner as in § 16.8. Adapted to crossed helicals at right angles and with the same hand, it becomes (f = coefficient of friction)

$$\text{(j)} \quad e = \cot \psi_1 \left[\frac{\cos \phi_n \sin \psi_1 - f \cos \psi_1}{\cos \phi_n \cos \psi_1 + f \sin \psi_1}\right] = \frac{\cos \phi_n - f \cot \psi_1}{\cos \phi_n + f \tan \psi_1}. \quad [\Sigma = 90°]$$

For coefficients of friction, see § 16.9. The literature often quotes Buckingham's[13.3] approach to design for particular power transmission, to which the interested reader should turn; for considerable detail on designing crossed helicals, see Ref. *(14.1)*.

14.10 CLOSURE. Helical gear transmissions are the ones most likely to be running at very high speeds; thus the discussions of the phenomena associated with speed in Chapter 13 are appropriately considered in this context. Engineers concerned with the design of commercial gear boxes will check their designs against the AGMA ratings; see Ref. *(14.4)*.

15. BEVEL GEARS

15.1 INTRODUCTION. Bevel gears are used to connect intersecting shafts, usually but not necessarily at 90°, Fig. 15.1. Bevel-gear teeth are subjected to much the same action as spur and helical teeth; the total maximum load on a tooth is compounded of the transmitted load and a dynamic increment arising from profile and tooth-spacing inaccuracies, and the maximum surface compressive stress is the principal criterion of wear resistance. Since bevel gears are inherently not interchangeable, they are designed in pairs. This chapter also mentions briefly a few of the skew gears.

15.2 BEVEL-GEAR NOMENCLATURE. Bevel-gear teeth are built with respect to a ***pitch cone,*** rather than to a pitch cylinder as in spur gears. The elements of the pitch cone intersect at the ***cone center,*** Fig. 15.2. When two bevel gears are properly mounted, their cone centers are coincident.

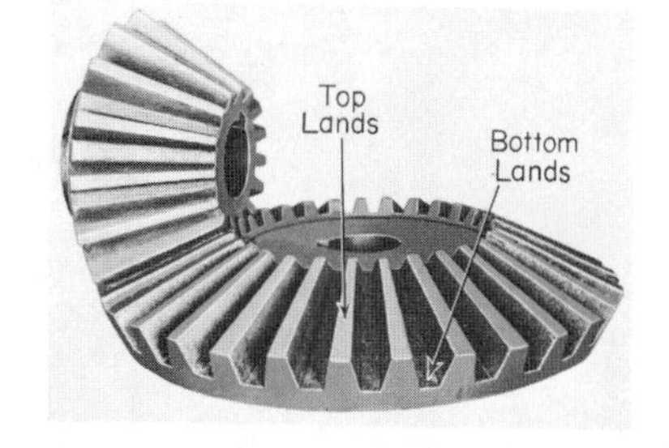

FIGURE 15.1 Straight Bevel Gears. (Courtesy Link-Belt Co., Chicago).

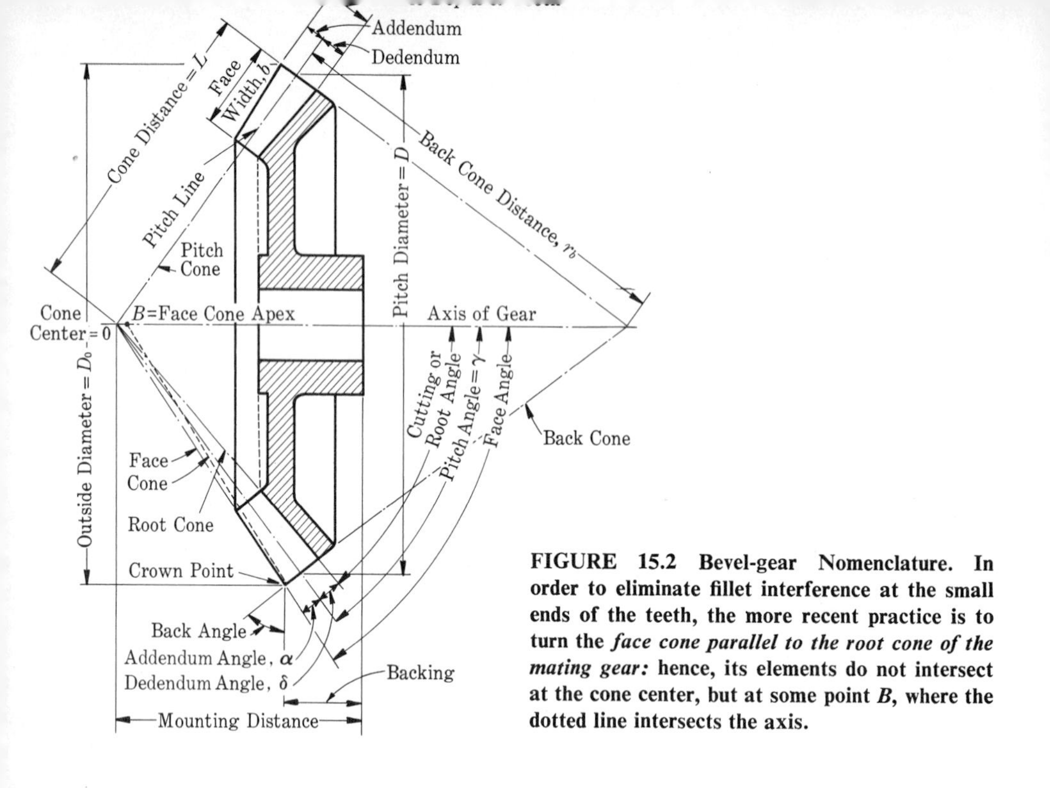

FIGURE 15.2 Bevel-gear Nomenclature. In order to eliminate fillet interference at the small ends of the teeth, the more recent practice is to turn the *face cone parallel to the root cone of the mating gear:* hence, its elements do not intersect at the cone center, but at some point *B*, where the dotted line intersects the axis.

The length of a pitch-cone element is called the ***cone distance,*** dimension L Fig. 15.2. ***Top lands*** and ***bottom lands,*** Fig. 15.1, mean the same as before. The cone formed by the elements of the top lands is called the ***face cone;*** the cone formed by the elements of the bottom lands is called the ***root cone.*** The angle between a pitch element and the axis is called the ***pitch angle*** (do not confuse with the pitch angle in spur gears, § 13.4). See Fig. 15.2 for the meanings of ***root, face, addendum,*** and ***dedendum angles.***

Since a section of the tooth decreases in size as it approaches the cone center, there is not a single value of the pitch, pitch diameter, addendum, dedendum, whole depth, etc. Unless otherwise indicated, however, these names used alone refer to the largest tooth section, Fig. 15.2. *The* pitch of a bevel gear and the *pitch diameter* are those at the large end. The ***back cone*** is an imaginary cone whose elements are perpendicular to the pitch-cone elements at the large end of the tooth. The length of a back-cone element is the ***back-cone distance,*** r_b in Fig. 15.2.

The pitch angle γ is computed as follows (Σ = angle between shafts):

$$\Sigma = 90^\circ \qquad \tan \gamma_g = \frac{N_g}{N_p} = m_g \qquad \tan \gamma_p = \frac{1}{m_g};$$

$$\Sigma < 90^\circ \qquad \tan \gamma_g = \frac{\sin \Sigma}{N_p/N_g + \cos \Sigma} = \frac{m_g \sin \Sigma}{1 + m_g \cos \Sigma}$$

The pitch angle on the pinion γ_p is equal to $\Sigma - \gamma_g$; N_p denotes the number of teeth on the pinion, N_g the number of teeth on the gear. The addendum angle α and dedendum angle δ are given by

$$\alpha = \tan^{-1}\frac{\text{addendum}}{\text{cone distance}} \quad \text{and} \quad \delta = \tan^{-1}\frac{\text{dedendum}}{\text{cone distance}}.$$

15.3 STRENGTH OF STRAIGHT BEVEL-GEAR TEETH.

A ***straight bevel gear,*** Fig. 15.1, is one whose tooth profiles consist of straight elements that converge to a point at the cone center. Since the force on a bevel-gear tooth varies from point to point along the face, we may resort to the calculus to determine the tooth strength in terms of a force that can be computed from the horsepower and the speed. See Fig. 15.3. The differential force dF_x acts on an increment of the face dx at a distance x from the cone center O. Since dx is very small, the variation of force along this length may be neglected and the Lewis equation applied; $F = sbY/P_d$,

$$\text{(a)} \qquad dF_x = \frac{sY\,dx}{P_{dx}},$$

where P_{dx} is the diametral pitch at the increment dx. Multiplying both sides of (**a**) by r_x, the radius of the gear at the distance x from O;

$$\int r_x\,dF_x = \int \frac{sYr_x dx}{P_{dx}}.$$

The sum of the product of all elemental forces by their radii $\int r_x dF_x$ is equal to the torque T transmitted by the gear. Therefore,

$$\text{(b)} \qquad T = \int \frac{sYr_x\,dx}{P_{dx}}.$$

Since all elements of the teeth on a straight bevel gear converge to the cone center O, the circular thickness and, consequently, the circular pitch vary as the distance from O. Thus $P_{cx}/P_c = x/L$, where P_{cx} is the circular

FIGURE 15.3 Strength of Bevel-gear Teeth. The value of the form factor Y depends on the form of the tooth in the plane normal to a pitch element. The theory of equation (f) is off slightly when constant clearance is used.

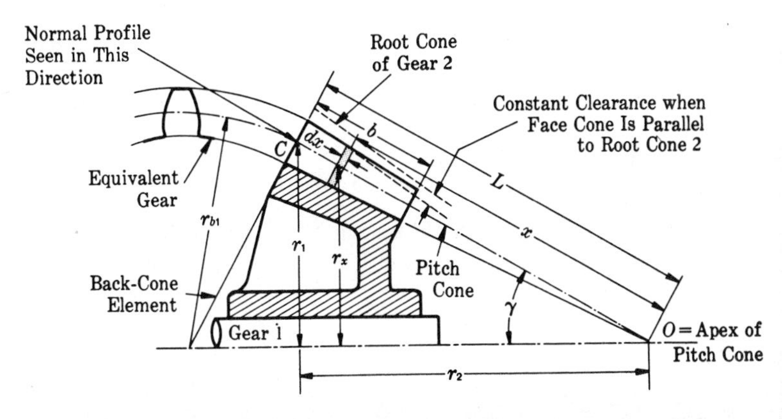

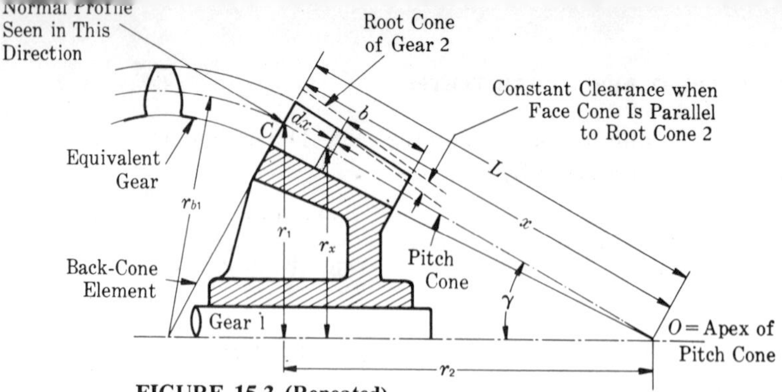

FIGURE 15.3 (Repeated).

pitch at a distance x from the cone center. Since the diametral pitch varies inversely as the circular pitch, we get

$$\textbf{(c)} \qquad \frac{P_{cx}}{P_c} = \frac{P_d}{P_{dx}} = \frac{x}{L}, \qquad \text{or} \qquad P_{dx} = P_d\left(\frac{L}{x}\right).$$

Moreover, from similar triangles in Fig. 15.3, we find the proportion

$$\textbf{(d)} \qquad \frac{r_x}{r} = \frac{x}{L}, \qquad \text{or} \qquad r_x = r\left(\frac{x}{L}\right).$$

Substituting the values P_{dx} and r_x from **(c)** and **(d)** into **(b)**, we obtain

$$\textbf{(e)} \qquad T = \frac{sYr}{P_d L^2}\int x^2\,dx,$$

in which, s, Y, r, P_d, and L for a particular gear are each constant; r is *the* pitch radius of the gear, P_d *the* diametral pitch, and L the cone distance. Since the origin of the coordinates is O, the limits of the integral in **(e)** should be from $(L - b)$ to L to include the whole face b. Integration then gives

$$\textbf{(f)} \qquad T = \frac{sYr}{P_d L^2}\left[\frac{x^3}{3}\right]_{L-b}^{L} = \frac{sYrb}{P_d}\left[1 - \frac{b}{L} + \frac{b^2}{3L^2}\right].$$

Although the teeth could just as well be designed from this equation, it is customary to use the force F obtained from the horsepower equation for the pitch-line velocity *at the large end.* Dividing both sides of **(f)** by the radius r at the large end, we find F in the form

$$\textbf{(g)} \qquad \frac{T}{r} = F = \frac{sYb}{P_d}\left[1 - \frac{b}{L} + \frac{b^2}{3L^2}\right].$$

The last term within the bracket is negligible when $b \lesseqgtr L/3$;

$$(15.1) \qquad F_s = \frac{sYb(L-b)}{P_d L},$$

where F_s represents the strength of the tooth, stress concentration neglected. Inasmuch as it is practically impossible to get the load distributed as assumed above, modern practice in bevel gearing is to plan for localized contact, as suggested by Fig. 15.9. Moreover, there are well recognized relations between b and L (§ 15.4), so that the factor $(L - b)/L$ can be cared for in a blanket fashion.

15.4 TOOTH PROPORTIONS, BEVEL GEARS. In bevel gears that should have good conjugate action, the teeth are generated. Since bevel gears are not interchangeable and since the generating process is adaptable, unequal addendum tooth forms are used.[15.1] The desired qualities in gears are *quietness*, *strength*, and *durability*. Good kinematic design can contribute materially to quietness. The addendum on the pinion is increased, its dedendum is decreased, § 13.34, eliminating (or perhaps nearly so) interference conditions and undercut flanks, but keeping the working depth at $2/P_d$ for straight and Zerol bevels, Table 15.1. Although bevel gears with other pressure angles can be obtained ($14\frac{1}{2}°$, $17\frac{1}{2}°$, etc.), $\phi = 20°$ has been adopted as standard in the interests of simplification. The combination of shortened dedendums on the pinion and the 20° pressure angle seems to give a reasonably good kinematic compromise. Recess action increases with the velocity ratio, so that a larger proportion of the angle of action is recess, which contributes to quietness (smoothness).

Spiral bevel gears, § 15.11, have a working depth of $1.7/P_d$, because the spiral angle reduces the width of top lands and shorter teeth counteract this effect (§ 14.4).

The face widths that should be observed for the design procedure to be described are:

Straight and spiral bevels: max. of $L/3$ or $10/P_d$, whichever is smaller; recommended $b = 0.3L$.

Zerol bevels (Fig. 15.10): max. of $0.275L$ or $10/P_d$, whichever is smaller; recommended $b = 0.25L$.

15.5 FORM FACTOR. Looking in a direction parallel to the *axis* of a bevel gear, Fig. 15.3, one sees the tooth foreshortened. The value of Y in equation (15.1) depends on the shape of the profile normal to a pitch-cone element; if this profile is an interchangeable type, the value of Y would be chosen from Table AT 24 in accordance with the ***equivalent*** (also called *formative* or *virtual*) number of teeth N_e. The equivalent gear is one that has a radius equal to the back-cone radius r_b, Fig. 15.4, with a pitch equal to the pitch at the large end (heel). Thus, for the pinion,

$$\textbf{(h)} \qquad N_{ep} = 2r_bP_d = \frac{2r_pLP_d}{r_g},$$

where we have used r_b from $r_b/L = r_p/r_g$, and $L = (r_p^2 + r_g^2)^{1/2}$, as seen from Fig. 15.4. The equivalent number of teeth for the gear $N_{eg} = 2r_g L P_d / r_p$ is found similarly. If the bevel gear teeth are cut with a milling cutter, they may be an interchangeable type but inaccurate; in this event, let the dynamic load be $F_d = F_t(600 + v_m)/600$ and make $F_s \gtreqless F_d$, as outlined above with Y chosen for the load at the tip. Also v_m fpm at the large end should be less than 1000.

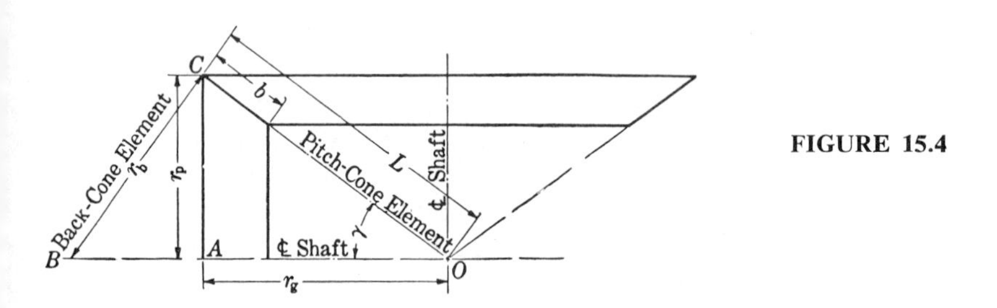

FIGURE 15.4

If, however, accurately generated, unequal-addendum teeth are used, it is appropriate to assume that the load is shared by two teeth until contact is carried by one tooth in the vicinity of the middle of the profile, as explained for spur gears, at which point the value of Y is considerably larger than that when the load is at the tip, but there are other factors; see the geometry factor J, § 15.7.

15.6 DYNAMIC LOAD FOR GENERATED BEVEL GEARS.

We shall let the subsequent calculations for bevel gear teeth serve to illustrate the AGMA approach, although minor rearrangements will be made in order to conform with our dynamic-load point of view. The dynamic load is computed from

$$F_d = (VF) N_{sf} K_m F_t, \tag{15.2}$$

where the transmitted load $F_t = 33{,}000\ \text{hp}/v_m$ lb. and $v_m = \pi D n$ fpm is the pitch-line speed at the large end, and K_m = load distribution factor, attached to the computation of F_d for convenience. Easy means of determining K_m are given[13.1,15.1] when the deflections of the gear and pinion are known; bevel gears are sensitive to misalignment, including that due to deflection. Experience has shown that the life of bevel gears increases as the rigidity of the mounting is increased. In the absence of detailed knowledge, choose values as follows:[13.1,13.34]

Both gears straddle mounted (Fig. 15.18) . .	$K_m = 1$ to 1.1
One gear straddle, one not	$K_m = 1.1$ to 1.25
Both gears overhung	$K_m = 1.25$ to 1.4

N_{sf} = service factor, chosen from Table 15.2, and VF = velocity factor. Use $VF = 1$ if the dynamic increment is negligible. For commercial straight bevels, use the factor in equation (**n**), § 13.15, which is

$$VF = \frac{50 + v_m^{1/2}}{50}. \tag{i} \quad \text{[STRAIGHT]}$$

For commercial-quality spiral bevels,

$$VF = \left[\frac{78 + v_m^{1/2}}{78}\right]^{1/2}. \tag{j} \quad \text{[SPIRAL]}$$

Straight and Zerol bevel gears should not be used when the pitch-line speed is greater than about 1000 fpm.[15.1] Spiral bevel, Fig. 15.11 and hypoid gears Fig. 15.13, are recommended when the pitch-line speed exceeds 1000 fpm or where angular speeds exceed 1000 rpm. Of course, they may also be used for lower speeds when quietness is important. When the pitch-line speed is above 8000 fpm, the teeth should be ground after hardening. Because of the differences between spiral angles, hypoid pinions are larger than spiral-bevel pinions of the same capacity, which makes it possible to use a larger shaft on the hypoid pinion than on the spiral pinion.

15.7 RATED STRENGTH OF BEVEL GEARS.

The strength equation to match F_d in equation (15.2) is[13.1]

$$F_s = \frac{s_d bJ}{P_d} \frac{K_l}{K_s K_t K_r}, \tag{15.3}$$

where K_s = size factor; for bevel gears, choose K_s according to pitch:

P_d:	1	2	3	4	6	8	10	over
K_s:	1.00	0.84	0.76	0.71	0.64	0.59	0.56	0.50

K_l = life factor for strength; for indefinite life, 6×10^7 cycles or more, use $K_l = 1$; for 10^6 cycles, $K_l = 1.4$; for 10^5 cycles, $K_l = 2.1$; for 10^4 cycles, $K_l = 3.1$; for 10^3 cycles, $K_l = 4.6$;

K_t = temperature factor; use unity if $t°\text{F} < 160°\text{F}$; acceptable value, $K_t = (460 + t°\text{F})/620$.

K_r = reliability factor; for 1 failure in 3, $K_r \approx 0.75$; for 1 in 100, $K_r \approx 1.12$; for high reliability, $K_r = 1.5$ or more; for maximum safety, $K_r = 3$.

J = geometry factor for strength, Fig. 15.5 for straight bevels, Fig. 15.6 for spiral bevels with $\psi = 35°$; 20° pressure angle, 90° shaft angle. J cares

TABLE 15.1 TOOTH DIMENSIONS FOR STRAIGHT AND ZEROL BEVEL GEARS

The addendum of the gear is $a_g = A/P_d$; of the pinion, $a = 2/P_d - a_g$ where A is taken from the tabulation below. Whole depth $= 2.188P_d + 0.002$ in.; working depth $= 2/P_d$. Dedendum = whole depth − addendum.

***For drawing purposes only*, gear and pinion teeth may be considered of equal thickness as measured halfway between the top and bottom of the tooth. Approximate values for *spiral bevel gears* can be obtained by multiplying these values by 0.85.**

GEAR RATIOS			GEAR RATIOS		
From	*To*	*A*	*From*	*To*	*A*
1.00	1.00	1.000	1.42	1.45	0.760
1.00	1.02	0.990	1.45	1.48	0.750
1.02	1.03	0.980	1.48	1.52	0.740
1.03	1.04	0.970	1.52	1.56	0.730
1.04	1.05	0.960	1.56	1.60	0.720
1.05	1.06	0.950	1.60	1.65	0.710
1.06	1.08	0.940	1.65	1.70	0.700
1.08	1.09	0.930	1.70	1.76	0.690
1.09	1.11	0.920	1.76	1.82	0.680
1.11	1.12	0.910	1.82	1.89	0.670
1.12	1.14	0.900	1.89	1.97	0.660
1.14	1.15	0.890	1.97	2.06	0.650
1.15	1.17	0.880	2.06	2.16	0.640
1.17	1.19	0.870	2.16	2.27	0.630
1.19	1.21	0.860	2.27	2.41	0.620
1.21	1.23	0.850	2.41	2.58	0.610
1.23	1.25	0.840	2.58	2.78	0.600
1.25	1.27	0.830	2.78	3.05	0.590
1.27	1.29	0.820	3.05	3.41	0.580
1.29	1.31	0.810	3.41	3.94	0.570
1.31	1.33	0.800	3.94	4.82	0.560
1.33	1.36	0.790	4.82	6.81	0.550
1.36	1.39	0.780	6.81	∞	0.540
1.39	1.42	0.770			

for: the form factor Y, the strength reduction factor K_f, effect of radial load, load sharing (m_c), effective face width, tooth taper.[15.1]

b = face width, and P_d = pitch at the large end.

s_d = design flexural stress, as follows:

Steel, Min. BHN:	140	150	180	200	300	400	450	55–63 R_c
s_d ksi:	11	12	14	15	19	23	25	30

R_c = 55–63 is for carburized teeth. Flame or induction hardened with unhardened root fillets, surface 450–500 BHN, s_d = 13.5 ksi. For cast iron: Class 20, 2.7 ksi; Class 30, 4.6 ksi; Class 40, 7 ksi. For nodular iron, use the value for steel at the same BHN less 10%.

Compare (15.3) with the strength equation (15.1) and note the factors not involved in the strength theory. For no breakage $F_s \geqq F_d$. Observe that each gear has its geometry factor J; check and design for the tooth with the lower value.

TABLE 15.2 SERVICE FACTORS N_{sf}, BEVEL GEARING

For speed-increasing drives, add $0.01m_g^2$ to the factors given.[15.3] Character-of-load examples are suggestive only. See also § 13.18.

POWER SOURCE	UNIFORM (*centrifugal mach., belt conveyor*)	MODERATE SHOCK (*lobe blowers, multicyl. pumps, machine tools*)	HEAVY SHOCK (*ore crusher, single-cyl. compressor, punch press*)
Uniform (elec. motor, turbine) .	1.00	1.25	1.75 +
Light Shock (multicyl. ICE) .	1.25	1.50	2.00 +
Medium Shock (single-cyl. ICE) .	1.5	1.75	2.25 +

15.8 RATED WEAR LOAD FOR BEVEL GEARS. The equation for the rated wear load originates with the Hertz contact-stress equation as does the Buckingham equation. To be compared with F_d from (15.2), it is

$$F_w = D_p b I \frac{s_{cd}^2}{C_e^2}\left(\frac{C_l}{K_t C_r}\right)^2, \tag{15.4}$$

where D_p = pitch diameter of the pinion, b = face width, K_t is as defined in § 15.7.

s_{cd} = design contact stress, whose values (and s_{cd}^2) are found in Table 15.3.

C_l = life factor for wear; for indefinite life, 10^7 cycles or more, $C_l = 1$; for 10^6 cycles, $C_l = 1.15$; for 10^5 cycles, $C_l = 1.3$; for 10^4 cycles, $C_l = 1.5$.

TABLE 15.3 DESIGN CONTACT STRESSES, s_{cd}

For bevel gears; BHN's are minimum. Reduce s_{cd} by 10% unless the manufacturing and materials are highest quality[13.1,13.34,15.1]

MATERIAL (BHN)	s_{cd} ksi	s_{cd}^2 (ksi)^2
Steel (180) . .	95	9,025
Steel (240) . .	115	13,200
Steel (300) . .	135	18,200
Steel (360) . .	160	25,600
Steel (440) . .	190	36,100
Steel (500) . .	190	36,100
Steel (625) carb. .	225	50,625
Cast iron, 20 . .	30	900
Cast iron, 30 (175) .	50	2,500
Cast iron, 40 (200) .	65	4,230
Bronze (10–12% tin) .	30	900
Al. bronze, HT .	65	4,230
Nodular iron (use 0.9 × steel value at same BHN) .		

TABLE 15.4 ELASTIC COEFFICIENTS

For bevel gears. Values of E ksi used are: steel (3×10^4), cast iron (1.9×10^4), aluminum bronze (1.75×10^4), tin bronze (1.6×10^4).

MATERIAL COMBINATIONS	C_e (psi)$^{1/2}$	C_e^2 ksi
Steel and		
Steel . . .	2800	7840
Cast iron . .	2450	6000
Al. bronze . .	2400	5760
Tin bronze . .	2350	5530
Cast iron and		
Cast iron . .	2250	5060
Al. bronze . .	2200	4840
Tin bronze . .	2150	4620
Al. bronze and		
Al. bronze . .	2150	4620
Tin bronze . .	2100	4410

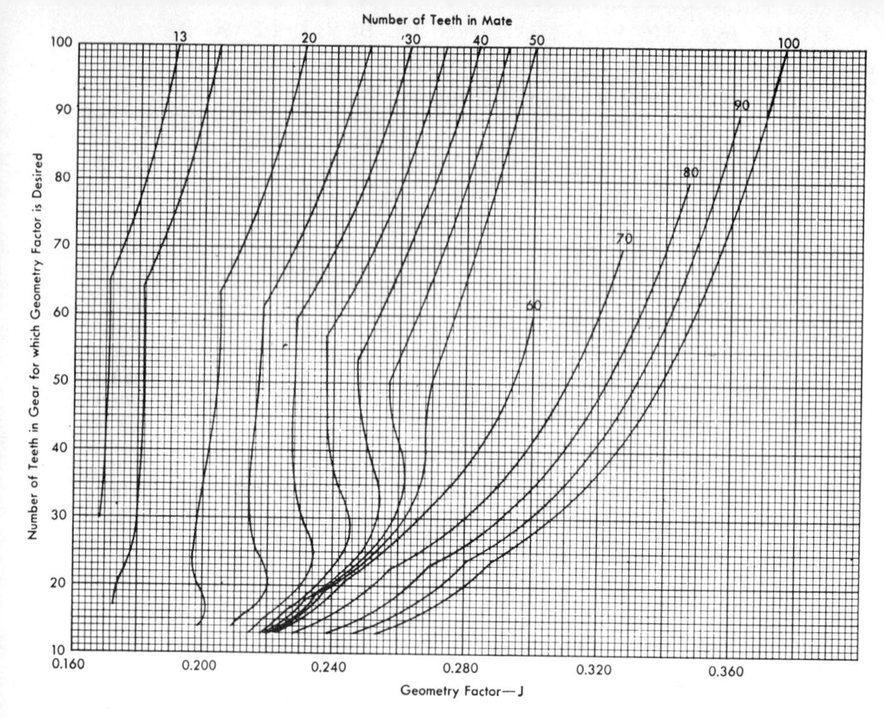

FIGURE 15.5 Geometry Factor J for Coniflex Straight Bevel Gears ($\phi = 20°$, $\Sigma = 90°$). (Courtesy Gleason Works, Rochester, N.Y.).

FIGURE 15.6 Geometry Factor J for Spiral Bevel Gears ($\phi = 20°$, $\psi = 35°$, $\Sigma = 90°$). (Courtesy Gleason Works, Rochester, N.Y.).

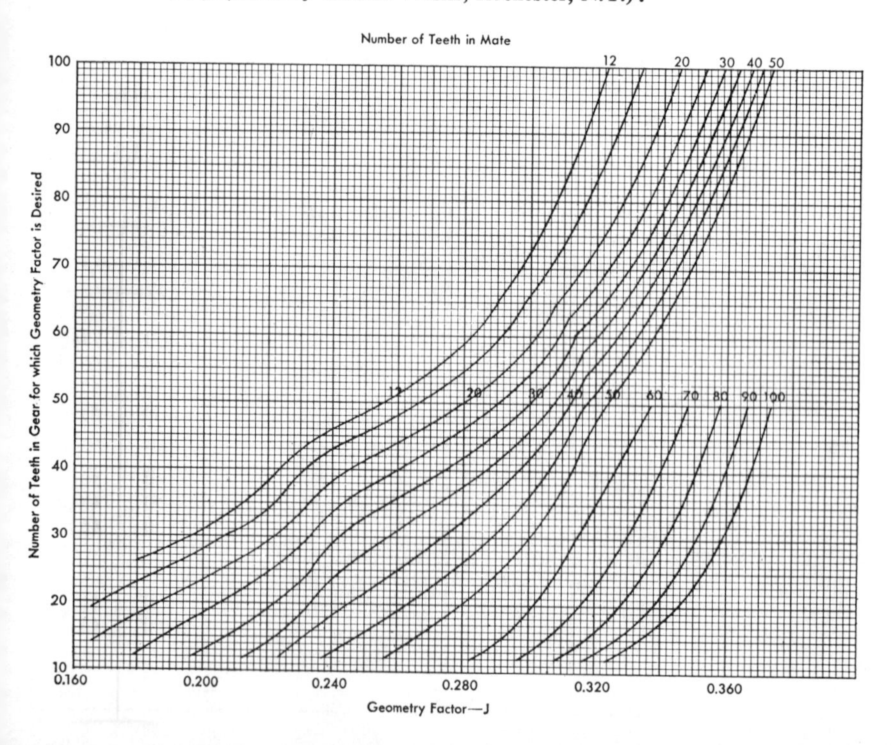

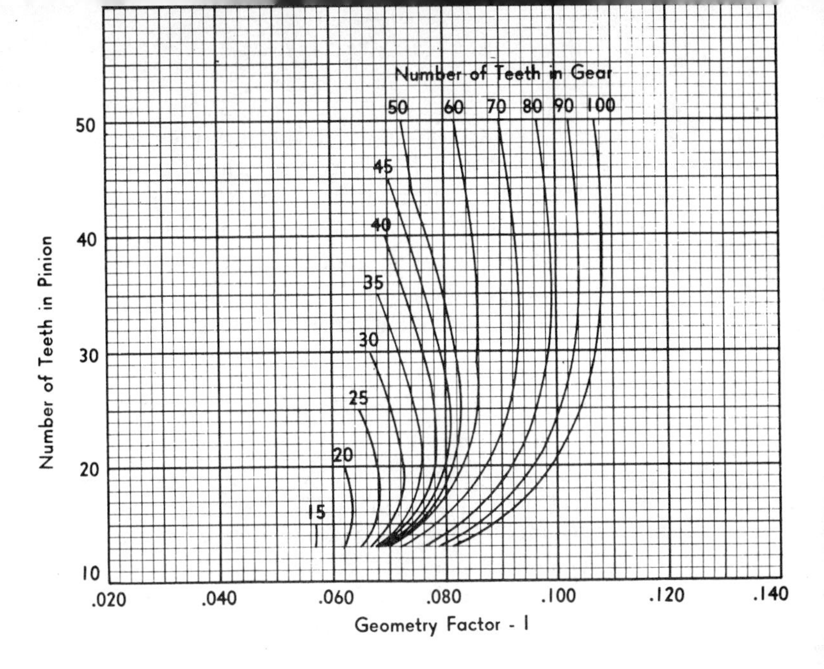

FIGURE 15.7 Geometry Factor I for Coniflex Straight Bevel Gears ($\phi = 20°$, $\Sigma = 90°$). (Courtesy Gleason Works, Rochester, N.Y.).

FIGURE 15.8 Geometry Factor I for Spiral Bevel Gears ($\phi = 20°$, $\psi = 35°$, $\Sigma = 90°$). (Courtesy Gleason Works, Rochester, N.Y.).

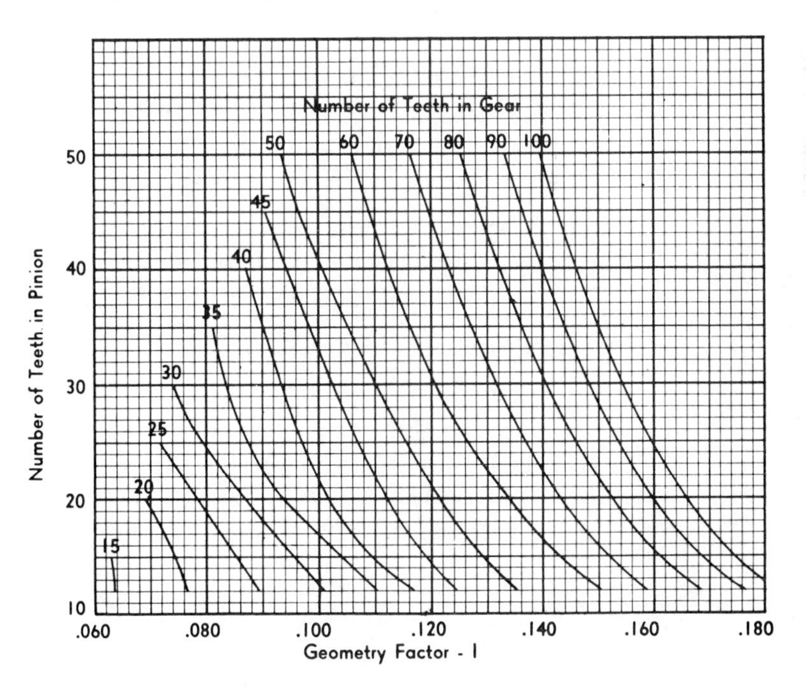

C_r = reliability factor for wear; for high reliability, let $C_r = 1.25$ or more; for fewer than 1 failure in 100, use $C_r = 1$.

I = geometry factor for wear; Fig. 15.7 for straight bevels; Fig. 15.8 for spiral bevels, $\psi = 35°$; 20° pressure angle, 90° shaft angle. I cares for profile curvature at the contact point, the location and sharing of the load, and the effective face width.

C_e = elastic coefficient, given for bevel gears by

$$\text{(k)} \qquad C_e^2 = \frac{1.5/\pi}{(1 - \mu_p^2)/E_p + (1 - \mu_g^2)/E_g}$$

psi or ksi, depending on the units of E. The factor C_e is seen to come from the Hertz equation (its constant makes it appropriate for localized contact, Fig. 15.9), and values of C_e^2 are available in Table 15.4.

Other factors that may be inserted at a later date include size factor, hardness ratio factor, surface roughness factor; but at this time, not enough is known to say what they should be.

15.9 EXAMPLE—HORSEPOWER FOR BEVEL GEARS. A motor turning at 1160 rpm drives a conveyor via straight bevel gears with 19 and 58 teeth, $P_d = 6.75$, $D_p = 2.8148$ in., $D_g = 8.5926$ in., $b = 1\frac{3}{8}$ in. The material of the gears is AISI 3140, QT to 300 BHN. The pinion is overhung, the gear is straddle mounted. The service is such that a service factor of $N_{sf} = 1.2$ applies; operating temperature is normal. What maximum horsepower may be transmitted for indefinite life?

Solution. The pitch line speed is $v_m = \pi Dn = \pi(2.8148/12)(1160) = 855$ fpm. For straight bevels, the velocity factor from equation (i) is

$$VF = \frac{50 + v_m^{1/2}}{50} = \frac{50 + \sqrt{855}}{50} = 1.585.$$

The dynamic load can be found in terms of the transmitted load. From § 15.6 choose $K_m = 1.15$ on the basis of good construction; then from equation (15.2)

$$F_d = (VF)N_{sf}K_mF_t = (1.585)(1.2)(1.15)F_t = 2.19F_t.$$

The fatigue strength F_s from equation (15.3) is

$$F_s = \frac{s_d bJ}{P_d}\frac{K_l}{K_sK_tK_r} = \frac{(19)(1.375)(0.2)}{6.75}\frac{1}{(0.62)(1.1)} = 1.13 \text{ kips},$$

where s_d is taken from § 15.7 for 300 BHN, J from Fig. 15.5 (note that it is for the *gear*), $K_l = 1$ for indefinite life, $K_s = 0.62$ from § 15.7, $K_t = 1$ for $t < 160°$F, $K_r = 1.1$ for reasonable reliability. The surface fatigue strength from equation (15.4) is

$$F_w = D_pbI\frac{s_{cd}^2}{C_e^2}\left(\frac{C_l}{K_tC_r}\right)^2 = (2.8148)(1.375)(0.081)\left(\frac{18{,}200}{7840}\right)\left(\frac{1}{1.1}\right)^2 = 0.602 \text{ kips},$$

where I is taken from Fig. 15.7, s_{cd}^2 from Table 15.3, C_e^2 from Table 15.4, $C_l = 1$ for indefinite life, $K_t = 1$ as before, and the reliability factor C_r is taken as 1.1. Since $F_w < F_s$ (0.602 < 1.13), wear governs for indefinite life. Equating $F_w = F_d$, we get $F_t = 0.602/2.19 = 0.275$ kips, with which the power is

$$\text{hp} = \frac{F_t v_m}{33{,}000} = \frac{(275)(855)}{33{,}000} = 7.12 \text{ hp}.$$

15.10 CONIFLEX AND ZEROL BEVEL GEARS.

The generated teeth on ***Coniflex bevel gears,*** developed by the Gleason Works, are slightly relieved lengthwise, as in Fig. 13.8, p. 368, in order to localize the tooth pressure in the middle portion. There is less chance of the load being concentrated on an end of a Coniflex tooth, Fig. 15.9, because of deflection or small misalignment.

Zerol bevel gears, Fig. 15.10, also developed by Gleason, have curved teeth, as in spiral bevels, Fig. 15.11, but with a zero spiral angle. These teeth are generated as are spiral bevels, Fig. 15.12; they are used in place

FIGURE 15.9 Coniflex Gears. The dark spots indicate the areas of contact. (Courtesy Gleason Works, Rochester, N.Y.).

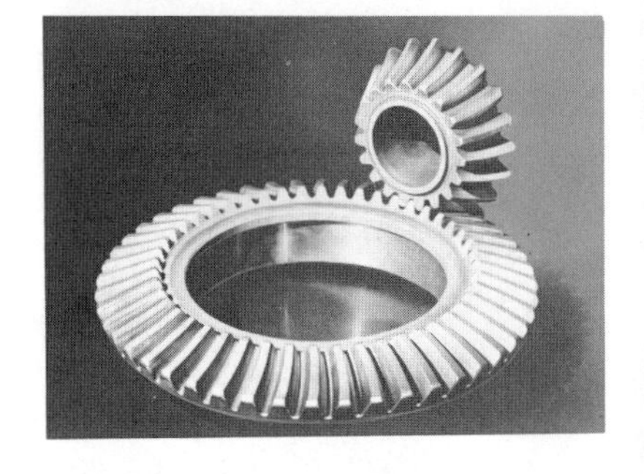

FIGURE 15.10 Zerol Bevel Gears. (Courtesy Gleason Works, Rochester, N.Y.).

of straight bevels, and they have localized tooth pressure as in Coniflex gears. Also, since Zerol teeth can be accurately finish-ground, they are suitable where hardened gears of extreme accuracy are required. Their thrusts and tooth action are the same as in straight bevels, and they may therefore be used in the same mountings.

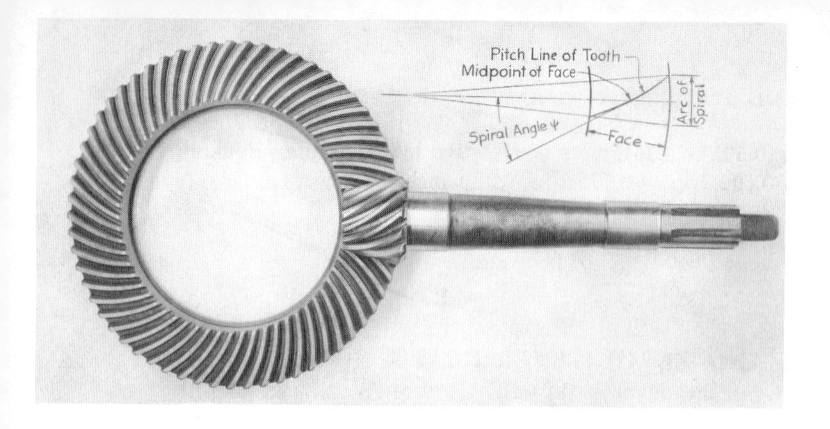

FIGURE 15.11 **Spiral Bevel Gears. The inset in the upper right-hand corner shows the meaning of the spiral angle ψ. The pinion has a left-hand spiral, the gear a right-hand spiral; apply the right-hand rule. (Courtesy Gleason Works, Rochester, N.Y.).**

FIGURE 15.12 **Machine for Cutting Spiral-bevel or Hypoid Teeth. An automatic machine. In generating a tooth profile, "a relative rolling motion takes place between the gear . . . and the rotating cutter; the action is as though the gear . . . were rolling with a mating crown gear of which the cutter represents a tooth. . . . At the completion of one tooth space, the cradle and work reverse direction and the work is withdrawn and indexing takes place." The sequence is repeated for each space. (Courtesy Gleason Works, Rochester, N.Y.).**

15.11 SPIRAL BEVEL GEARS. These gears, Fig. 15.11, which are to straight bevel gears as helical gears are to straight-tooth spur gears, have much the same advantages as helical gears; smoother tooth engagement, quietness of operation, greater strength, and higher permissible speeds. (Read § 14.1 again). The profiles are modified slightly to obtain the advantages of localized contact, which also contributes to the good running qualities of these gears. Since the teeth are cut with a rotary cutter, Fig. 15.12, a good rate of production of accurately cut gears is attainable. The standard pressure angle is 20°; also used are $14\frac{1}{2}$°, 16°, $17\frac{1}{2}$°. The spiral angle

ψ, Fig. 15.11, should be large enough that the arc of spiral is some $1.25P_c$ to $1.4P_c$ (face contact ratio 1.25 to 1.4). Typical spiral angles are around 30° to 35°.

15.12 HYPOID GEARS.

Looked at individually, hypoid gears, Fig. 15.13, appear much like spiral bevels, but their pitch surfaces are not the same. Instead of a pitch cone, the pitch surface is a hyperboloid. When two hyperboloidal pitch surfaces are in contact along a common element, their axes do not intersect, which is the way in which they most obviously differ from spiral bevels. These gears are generally used in automotive rear-axle drives because their use makes it possible to lower the center of gravity of the car. Since a pair of mating pitch circles do not roll on one another, there is sliding lengthwise along the tooth (analogous to crossed helicals and worm gears) which results in considerably more heat generation. On this account, an EP lubricant (§ 13.38) is needed for a good power capacity. The offset of the shafts may be so great that the shafts may continue past each other; this characteristic is used for providing multiple take-offs from a single shaft on which several drive pinions are mounted.

Skew hypoid gears, Fig. 15.14, are similar to curved-tooth hypoids except that the teeth are straight. Since generated curved-tooth hypoids are readily manufactured, this variety is not so widely used.

Spiral bevel and hypoid gears are recommended when the pitch-line speed exceeds 1000 fpm or where angular speeds exceed 1000 rpm. Of course, they may also be used for lower speeds when quietness is important. When the pitch-line speed is about 8000 fpm, the teeth should be ground

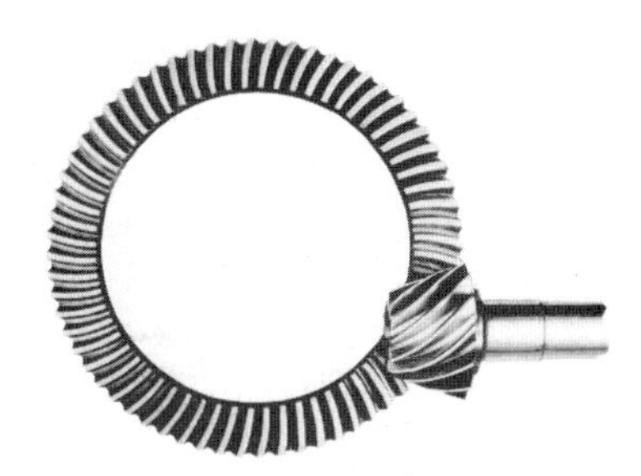

FIGURE 15.13 Hypoid Gears. (Courtesy Gleason Works, Rochester, N.Y.).

FIGURE 15.14 Skew Hypoid Gears, Straight Teeth. (Courtesy Gleason Works, Rochester, N.Y.).

after hardening. Because of the differences between spiral angles, hypoid pinions are larger than spiral-bevel pinions of the same capacity, which makes it possible to use a larger shaft on the hypoid pinion than on the spiral pinion. This characteristic may sometimes suggest the use of hypoid gears, even though it is not necessary to offset the shafts for any other reason.

Skewed gears have appeared in a number of forms, another of which is the Spiroid gears,[16.2] Fig. 15.15. Virtually, the pinion has become a conical worm, so that this drive is appropriate for high velocity ratios. See also Ref. *(15.4)*.

FIGURE 15.15 Spiroid Gears. (Courtesy Illinois Tool Works, Chicago).

FIGURE 15.16 Angular Gears. (Courtesy Link-Belt Co., Chicago).

15.13 OTHER TYPES OF BEVEL GEARS.

When a pair of bevel gears of the *same size* are on shafts intersecting at right angles, they are called ***miter gears.*** The velocity ratio is of course one, and $\gamma_p = \gamma_g = 45°$.

Angular gears are bevel gears mounted on intersecting shafts at angles of other than 90°, Fig. 15.16.

A ***crown gear*** is one in which the pitch angle, Fig. 15.2, is 90°; that is, the pitch cone has become a plane.

15.14 FORCES ON A BEVEL GEAR.

(a) **Straight Bevel Gear.** The resultant of all the forces acting along the face of the tooth may be assumed without serious error to act at the midpoint of the face. The direction of the total force is along the pressure line, which is in a *plane* perpendicular to a pitch-cone element at that point. The pressure line makes an angle equal to the angle of obliquity ϕ with the tangent to the back-cone circle. The average driving force F_A, Fig. 15.17, is found from the horsepower equation, where the velocity *at the mean point* of the face A is used.

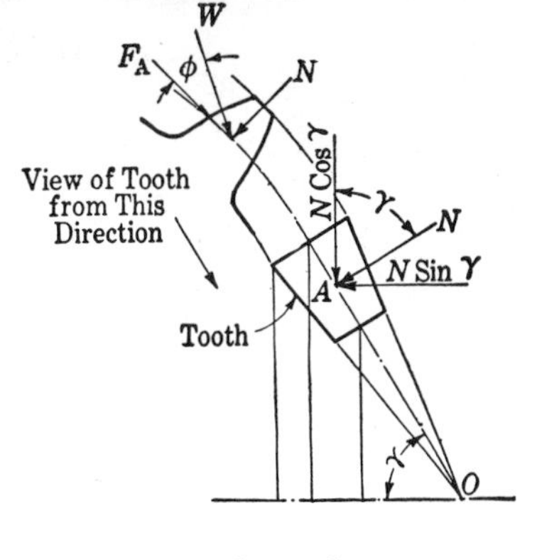

FIGURE 15.17 Forces on a Straight Bevel-gear Tooth. The forces N and F_A are components of the force W; $N \cos \gamma$ and $N \sin \gamma$ are components of N. The force F_A acts at A normal to the plane of the paper.

The force $N = F_A \tan \phi$ acts normal to the pitch-cone element. The end thrust on the shaft that must be cared for is $N \sin \gamma$. The bending force on the shaft *in the diametral plane* is the resultant of F_A and $N \cos \gamma$, which are forces at right angles. These forces are useful in shaft and bearing design.

(b) Axial thrust on Spiral Bevel Gear. If the driving member has a *right-hand* spiral (see Fig. 15.11) and rotates *clockwise*, or if it has a *left-hand* spiral and rotates *counterclockwise*, the end thrusts are as follows:

$$\text{Driving member,} \qquad F_T = F_A\left[\overleftarrow{\tan \psi \cos \gamma} - \frac{\overrightarrow{\tan \phi \sin \gamma}}{\cos \psi}\right],$$

$$\text{Driven member,} \qquad F_T = F_A\left[\overrightarrow{\tan \psi \cos \gamma} + \frac{\overrightarrow{\tan \phi \sin \gamma}}{\cos \psi}\right].$$

If the driving member has a *right-hand* spiral and rotates *counter-clockwise*, or, if it has a *left-hand* spiral and rotates *clockwise*, the end thrusts are as follows:

$$\text{Driving member,} \qquad F_T = F_A\left[\overrightarrow{\tan \psi \cos \gamma} + \frac{\overrightarrow{\tan \phi \sin \gamma}}{\cos \psi}\right],$$

$$\text{Driven member,} \qquad F_T = F_A\left[\overleftarrow{\tan \psi \cos \gamma} - \frac{\overrightarrow{\tan \phi \sin \gamma}}{\cos \psi}\right],$$

where F_T = axial thrust,
F_A = average transmitted load tangent to the pitch circle at midpoint of the face,
ψ = spiral angle, Fig. 15.11,
γ = pitch angle of the gear whose thrust is being computed, Fig. 15.2,
ϕ = pressure angle in a plane normal to a cone element,
$\leftarrow$ indicates that the thrust is *toward* the cone center, and
$\rightarrow$ indicates that the thrust is *away* from the cone center.

The resultant direction of F_T depends upon which term is greater, $\rightarrow$ or $\leftarrow$. The direction of rotation is the direction to an observer looking from the back of the member toward the cone center. Since the end thrusts may be

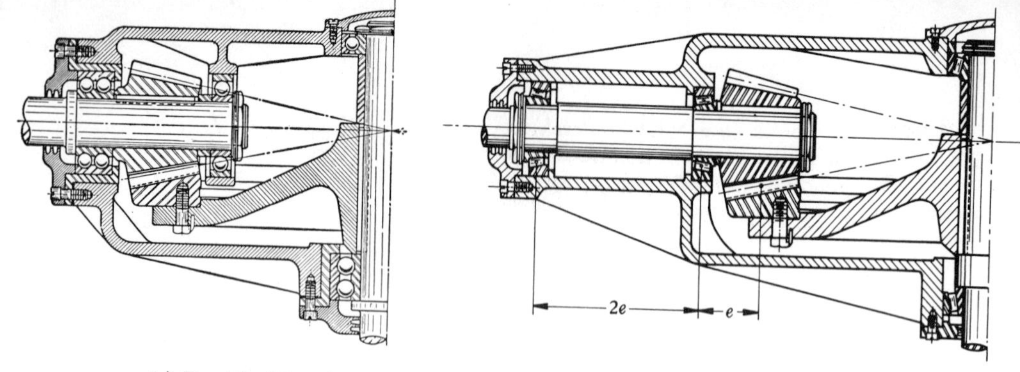

(a) Straddle Mounting

(b) Overhung Mounting

FIGURE 15.18 Bevel-gear Assemblies. This figure shows the steel ring gear, with teeth cut on it, bolted to a cast flange. Since the straddle mounting in (a) is relatively expensive, the pinion is frequently overhung as in (b). The distance between the bearings on an overhung pinion, as in (b), should not be less than 2*e*, where *e* is the distance from the point of application of the load on the tooth to the nearest bearing. (Courtesy Gleason Works, Rochester, N.Y.).

very large, often as much as half of the transmitted load, they must always be cared for in the bearing design.

15.15 DESIGN DETAILS. The design shown in Fig. 15.18, in which the gear is a steel ring attached to a separate hub or center, is especially adapted to bevel gears. A steel ring, preferably with a flat back, has the most suitable shape for a hardening operation, and the center can be re-enforced with ribs if necessary. However, a center having a conical shape of web is generally best, especially for high speeds, because the casting strains are less, it is easier to machine and balance, and because the oil in the gear case will not be so severely churned.

The rim, bead, and hub proportions may be fixed according to the instructions for spur gears, except that the hub should be long enough to give a positive ***backing,*** Fig. 15.2. Bevel-gear mountings should include some arrangement, such as a spacing washer, for adjusting or locating the gears correctly and should provide for lubricating and guarding both the bearings and the gear teeth.

Accurate alignment in mounting and good support against deflections are major requirements for bevel gears of all types. For this reason, ball and roller bearings are recommended; they can be arranged to resist a change of position, either axially or radially. The distribution of bearing pressure between the teeth and the amount of the backlash should be checked.

15.16 MATERIALS FOR BEVEL GEARS. Although cast-iron and ordinary steel bevel gears are used extensively, the tendency is to employ heat-treated, alloy-steel gears whenever strength and wearing qualities are important and the sizes of the gears are within the limits for heat treatment. An excellent construction is a carburized alloy steel, such as 2512, 4120, 8620, etc., with the surface hardened to a minimum of Rockwell C 60 (see Table 15.3) and with a finish grinding in order to insure accurate profiles. However, excellent gears are obtained from oil hardening or flame hardening such materials as 4640 or 8640. Nitrided gears are also used. Nonferrous metals, plastics, die castings, and sintered materials are used when there is reason. The discussion of materials in § 13.24 is also applicable here.

15.17 CLOSURE. Since there are many different gear forms, no one book covers all of them, much less a general text on machine design. If it is desired to check a design against the Buckingham approach, see Ref. *(13.3)*. Observe that with the life factor, one can design for limited life. The design procedure outlined in this chapter has been adapted to other types of gearing.[13.1]

16. WORM GEARING

16.1 INTRODUCTION. Worm gearing, Fig. 16.1, is used to transmit power between nonintersecting shafts, nearly always at right angles to each other. Comparatively high velocity ratios may be obtained satisfactorily in a small space, though perhaps at a sacrifice in efficiency as compared to other types of gears. The impacting contact at engagement of spur and other gears is absent in worm gearing. Instead, the threads slide into contact with the gear teeth, an action that results in quiet running if the design, manufacture, and operation are proper. The greater sliding sometimes introduces a problem in heat from friction. Under extreme loading conditions, any gear case may overheat. The principles discussed later in this chapter regarding the heat dissipation of worm-gear cases may be applied to others. We see then that for worm gears, we must check our designs not only for strength and wear but also for overheating.

The section of a worm thread in a diametral plane is generally straight-sided, as is the section of an involute rack tooth. If the worm were moved without rotation in a straight line perpendicular to the axis of the worm wheel, the action of the teeth in a plane through the worm axis and normal

FIGURE 16.1 Worm and Worm Gear. (Courtesy The Grant Gear Works, Inc., Boston).

to the gear axis would be similar to the tooth action of a rack and gear. The straight sides of the worm thread facilitate production from the standpoints both of quantity and accuracy. The worm threads may be cut on a lathe or with dies, or they may be milled, shaped, rolled, or hobbed. The worm wheel should be hobbed to match, Fig. 13.11.

16.2 PITCH AND LEAD. The meaning of the terms *pitch* and *lead* for a worm are the same as explained for power screws. See § 8.18 and Fig. 8.22, p. 247, now. There tends to be a slight confusion of terms since, on the worm, the pitch is the ***axial pitch*** P_a, which when the shafts are at right angles, is the same as the circular pitch P_c on the worm gear; in this chapter $P_a = P_c$, since we shall have no occasion to use the axial pitch of the gear.

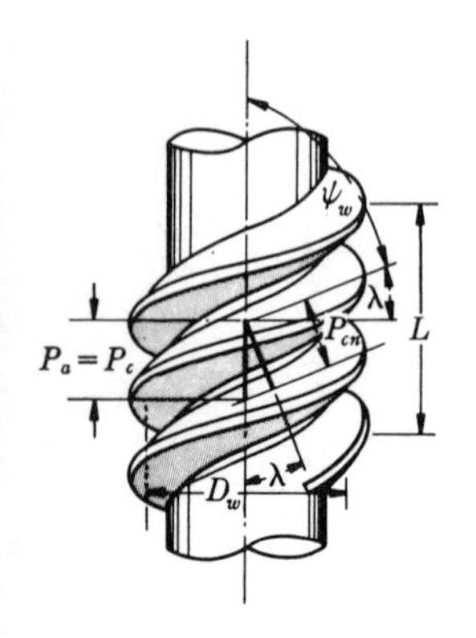

FIGURE 16.2 Triple Thread.

As for helical gears, worms have a normal pitch P_{cn}, Fig. 16.2, but the lead angle of the worm λ, which is the angle between a tangent to the pitch helix and the plane of rotation, is more convenient than the helix angle ψ_w of the worm (§ 14.2). From Fig. 16.2,

(a) $$P_{cn} = P_a \cos \lambda = P_c \cos \lambda.$$

The axial distance that a helix advances in one turn is the lead L. Whether on the pitch cylinder, whose diameter is D_w, or on the addendum or base cylinder, etc., all helices making up the thread have the same advance L, but the *lead angle* changes. If we imagine one turn of a *pitch* helix unwrapped from Fig. 16.2 into a plane, we have the diagram of Fig. 16.3, where we see that

(b) $$\tan \lambda = \frac{\text{Lead}}{\pi D_w} = \frac{L}{\pi D_w},$$

and that $\lambda + \psi_w = 90°$; L = lead of the worm thread. The pitch helix on the worm and the pitch helix on the gear each have a helix angle and a lead angle; λ_g = lead angle of the gear, ψ = helix angle of gear, and $\lambda_g + \psi = 90°$. If the shaft angle is 90°, the lead angle on the worm $\lambda = \psi$, the helix angle on the gear; also $\psi_w = \lambda_g$. Equation (**b**) applies to the gear with these modifications.

Whereas power screws are generally single-threaded, worms usually have multiple threads, unless mechanical advantage is more important than efficiency. When a worm makes one turn, a point on the pitch circle of the

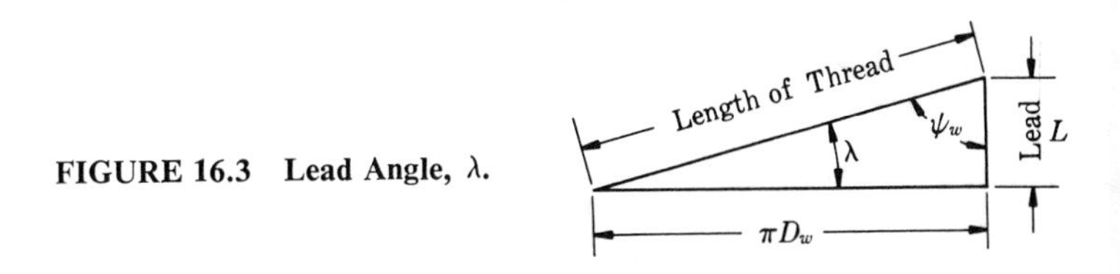

FIGURE 16.3 Lead Angle, λ.

worm wheel moves through an arc equal to the lead, $L = N_t P_a$, but for computing the velocity ratio, we find it easier to use the number of threads N_t (or starts) on the worm and the number of teeth on the gear N_g; $m_\omega = N_g/N_t$. Also equation (**h**) of § 14.9, which is $m_\omega = D_2 \cos \psi_2/(D_1 \cos \psi_1)$, applies with $\psi_2 = \lambda$, the worm's lead angle, and $\cos \psi_1 = \cos (90 - \lambda) = \sin \lambda$; hence, with the worm driving,

$$\text{(c)} \qquad m_w = \frac{\omega_w}{\omega_g} = \frac{N_g}{N_t} = \frac{D_g \cos \lambda}{D_w \sin \lambda} = \frac{D_g}{D_w \tan \lambda},$$

where $N_t = 1$ for a single-threaded worm, $N_t = 2$ for a double thread, etc. If high-efficiency power transmission is the objective, four, six, or more threads may be used on the worm. See § 16.8.

Standard pitches. Consider the following axial pitches as standard: $\frac{1}{4}$, $\frac{5}{16}$, $\frac{3}{8}$, $\frac{1}{2}$, $\frac{5}{8}$, $\frac{3}{4}$, 1, $1\frac{1}{4}$, $1\frac{1}{2}$, $1\frac{3}{4}$, and 2 inches. Diametral pitches are also used in worm gearing.

16.3 STRENGTH OF WORM-GEAR TEETH.

The teeth on the worm gear are weaker than the threads on the worm. The Lewis equation (13.5) without K_f is usually applied:

$$\text{(d)} \qquad F_s = \frac{sYb}{P_{dn}} = \frac{sYbP_{cn}}{\pi}.$$

The normal circular pitch P_{cn} is introduced into the equation because circular pitch is most often used for worm gearing. When $s = s_n$, the force F_s as obtained from this formula conservatively represents the bending endurance strength of the tooth on the gear; b is the effective face width of the gear. If the sum of the teeth in the worm and worm gear is greater than forty, the following blanket values of Y may be used with safety.[13.3]

For: $\phi_n = 14\frac{1}{2}°$, $Y = 0.314$; $\phi_n = 25°$, $Y = 0.470$;
$\phi_n = 20°$, $Y = 0.392$; $\phi_n = 30°$, $Y = 0.550$.

16.4 DYNAMIC LOAD FOR WORM GEARS. The dynamic load on worm gears is often estimated from

$$F_d = \left(\frac{1200 + v_{mg}}{1200}\right) F_t \text{ lb.,} \tag{16.1}$$

where F_t is the transmitted load computed from the horsepower equation applied to the *gear*; $F_t = 33{,}000\ \text{hp}/v_{mg}$, and v_{mg} fpm the pitch-line speed of the gear, where strictly hp is the *output* power, but since the efficiency is unknown at the outset of design, it is conventional to use the input power; make $F_s \gtreqless F_d$. As given by (16.1), F_d is likely to be on the high side.

16.5 WEAR LOAD FOR WORM GEARS. Buckingham gives an approximate relation for the limiting wear load F_w as[14.1]

$$F_w = D_g b K_w, \tag{16.2}$$

where D_g in. is the pitch diameter of the gear, b in. is the effective face width, and K_w is a factor obtainable from Table AT 27, which depends upon the materials used, the pressure angle, and the lead angle. The approximate values of Table AT 27 ignore change due to change of lead angle. If the material of the worm is soft or rough, wear will be rapid. A hard, smooth worm smooths and cold works the gear's softer tooth surfaces, and for this reason, the surface of the worm should be made as smooth as possible. The best combination has been generally considered to be a case-hardened (carburized) and ground worm (though polishing helps) with a chilled-bronze gear (but see high-test cast iron and same in Table AT 27). Recall that the theoretical contact stress is calculated from the *normal* force and that therefore there is always a certain amount of approximation in using the tangential force, as we have been doing, especially in the case of the large lead angles common in worm drives. See § 16.11 for suggested b values.

If the service is continuous, F_w should be equal to or greater than F_d.

16.6 THERMAL CAPACITY. If the gear case gets too hot, the oil may become too thin; it is then easily squeezed from the contact surfaces. If this happens, friction increases; more heat is generated; and finally serious abrasion and scoring occur. Extreme-pressure (EP) lubricants reduce such troubles when driver and driven are made of steel (see § 16.12).

In many situations, the horsepower capacity is dictated by the radiating capacity of the *housing*, which in turn depends upon environmental conditions.

The heat Q to be dissipated from any gear case is equal to the loss due to friction, which is taken as the horsepower input hp_i times $(1 - e)$ where e is the efficiency of the drive as a fraction (§ 16.8); thus

(e) $Q = (1 - e)(hp_i)$ hp or $= (1 - e)(hp_i)(33{,}000)$ ft-lb/min.
$= (1 - e)(hp_i)(2544)$ Btu/hr., etc.

The amount of heat that the housing dissipates by convection and radiation depends upon the following: the area of the housing, the temperature difference of the housing and the surroundings, and the transmittance or coefficient of heat transfer $\boldsymbol{h}_{cr}$, which in turn is a function of the temperature, the velocity of air across the housing, and other variables. See § 11.18. Since the environmental temperatures encountered ordinarily vary little, $\boldsymbol{h}_{cr}$ varies most significantly with the size of the housing and the air velocity. Values of $\boldsymbol{h}_{cr}$ ft-lb/min-sq. in.-°F are shown in Fig. AF 21 for average conditions of natural ventilation. The heat-dissipating capacity of the housing (gear case) is then expressed as

(f) $$Q_c = \boldsymbol{h}_{cr} A \, \Delta t \text{ ft-lb/min.},$$

where A sq. in. is the radiating area of a smooth housing (the area of the base is not counted) and Δt is the temperature rise of the lubricant above *the ambient temperature*. The maximum temperature of the lubricant should preferably not exceed 190°F (150°F in other types of gear cases). For heavy-duty worm-gear speed-reducers with integral worms, the AGMA recommends a minimum area of housing exclusive of the base, flanges, and fins of

(g) $$A_{\min} = 43.2C^{1.7} \text{ sq. in.},$$

where C in. is the center distance. In case the natural heat-dissipating capacity of the housing is not enough to maintain the temperature at a reasonable level, extra cooling is obtained by (1) circulating the lubricant and cooling it outside of the housing, (2) circulating water in cooling coils inside of the housing, or (3) blowing air across the housing, as illustrated in Fig. 16.4 (p. 432); in this case fins for extra heat radiating area are helpful. The amount of coolant (water) circulated in coils and the surface area of coils needed may be computed on the assumption that the coolant carries away all the heat of friction.

16.7 RELATION OF NORMAL AND DIAMETRAL PRESSURE ANGLES. A study of the contact forces will lead to an expression for efficiency that is theoretically correct for any type of thread-fastening screws,

FIGURE 16.4 Worm-gear Speed Reducer. This is an air-cooled model, the air circulating around the outside of the oil sump, which is ribbed to increase the heat-transfer surface. Observe the large lead angle on the worm. Shaft extensions are standardized to take standard couplings with a press fit. (Courtesy Cleveland Worm and Gear Co., Cleveland).

power screws, or worm threads. The total plane reaction between the worm and the gear is the resultant of the force N normal to the tooth and the frictional force $F_f = fN$. Since, up to a point at least, the frictional force is independent of the area, we may imagine the force system acting at a point O, Fig. 16.5; N, being normal to the surface at that point, leans away from the z axis in the direction of the y axis (because of the pressure angle) and in the direction of the x axis (because of the angle λ of the helix). The plane rectangle $abcd$ is perpendicular to the z axis; angle dOc is ϕ; angle aOb is the normal pressure angle ϕ_n (this plane is normal to the thread at O). Write $\tan \phi_n = ab/bO$, let $ab = cd$, multiply and divide by cO, and get

(h)
$$\tan \phi_n = \frac{ab}{bO}\frac{cO}{cO} = \frac{cd}{cO}\frac{cO}{bO} = \tan \phi \cos \lambda.$$

16.8 EFFICIENCY OF WORM GEARING. As explained above, the forces N and F_f are components of the total plane reaction shown acting *on* the worm. The frictional force F_f is tangent to the helix at O and lies in the xz plane. However, if anything, the rectangular components in the x, y, and z directions are more useful; they are W_t, which is the tangential (to pitch cylinder) transmitted force on the worm, F_t, the transmitted force on the worm gear, and S, the separating force tending to push the members apart. Since these forces are respectively the x, z, and y components of the resultant of N and F_f (notice that F_f has no y component), we shall equate F_t to the sum of the z components of N and F_f, and W_t to the sum of the x components. The component of N along line Ob is $N \cos \phi_n$; the component of $N \cos \phi_n$ in the z direction is $N \cos \phi_n \cos \lambda$, acting down. The z component of F_f is $F_f \sin \lambda = fN \sin \lambda$; thus

(i)
$$F_t = N \cos \phi_n \cos \lambda - fN \sin \lambda,$$

acting down in Fig. 16.5, where F_t is the driving force on the worm gear, obtainable from the horsepower equation (*output*) applied to the gear.

The horizontal component of the total plane reaction (N and fN) is

(j) $$W_t = N \cos \phi_n \sin \lambda + fN \cos \lambda,$$

where W_t is the driving force on the worm, obtainable from the horsepower equation (*input*) applied to the worm. (W_t is an *axial* force on the gear.) Eliminating N from equations **(i)** and **(j)**, we find the relation between W_t and F_t,

(k) $$W_t = F_t \left[\frac{\cos \phi_n \sin \lambda + f \cos \lambda}{\cos \phi_n \cos \lambda - f \sin \lambda}\right].$$

Considering efficiency from the basic notion of output/input, we have the input work as $W_t v_w$ ft-lb/min., where v_w fpm is the pitch-line speed of the worm. The frictional force F_f acts through the distance of the length of thread, Fig. 16.3; now with $v_w = \pi D_w n_w$ as the pitch-line speed, the rubbing speed $v_r = v_w/\cos \lambda$ fpm, the distance through which F_f acts in 1 min. Hence, the work of F_f is $F_f v_r = fNv_w/\cos \lambda$, and the output work becomes $W_t v_w - fNv_w/\cos \lambda$. Using the value of W_t from **(j)**, we get

$$e = \frac{(N \cos \phi_n \sin \lambda + fN \cos \lambda)\, v_w - fNv_w/\cos \lambda}{(N \cos \phi_n \sin \lambda + fN \cos \lambda)\, v_w}.$$

Cancelling N and v_w and manipulating the remainder, we find

(16.3) $$e = \tan \lambda \left[\frac{\cos \phi_n \cos \lambda - f \sin \lambda}{\cos \phi_n \sin \lambda + f \cos \lambda}\right] = \tan \lambda \left[\frac{\cos \phi_n - f \tan \lambda}{\cos \phi_n \tan \lambda + f}\right].$$

This equation also gives the theoretical efficiency of an Acme power screw (§ 8.15) or a fastening thread; moreover, let $\phi_n = 0$ to get a square thread

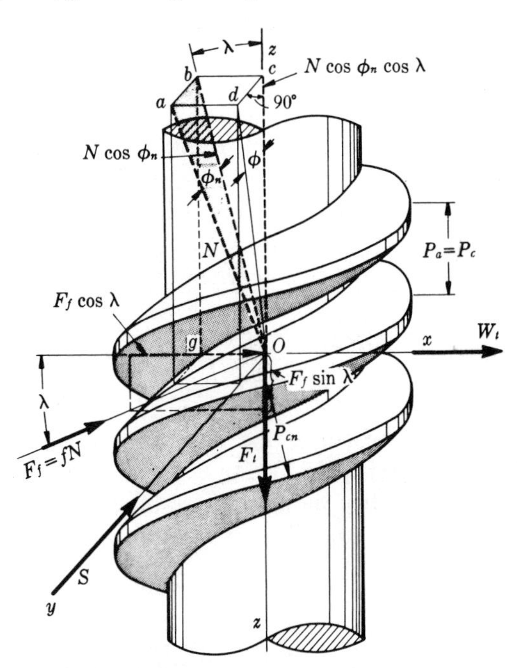

FIGURE 16.5 Forces for Worm Gearing. Not a free body; all forces shown as acting on the worm.

and see that the equation reduces to the form of equation (j), § 8.22. Observe that equation (16.3) *does not include the bearing losses*, which may be relatively small. Finally note that the foregoing equations apply to crossed helicals on shafts at right angles; worm driver, $\psi_1 = \psi_w$, $\psi_2 = \psi_g = \lambda$.

For the best efficiencies, the lead angle λ should be greater than 30°. For a given pitch, the smaller the worm diameter, the larger λ, but to get the highest efficiencies, it is necessary to use multiple-threaded worms with up to 24 threads. However, for a particular velocity ratio and pitch, the gear must be three times as large for a triple, for example, as for a single thread; but note that the approximate limiting wear load (16.2) depends only on D_g, the face width, and K_w. See typical curves for the variation of efficiency in Fig. 16.6.

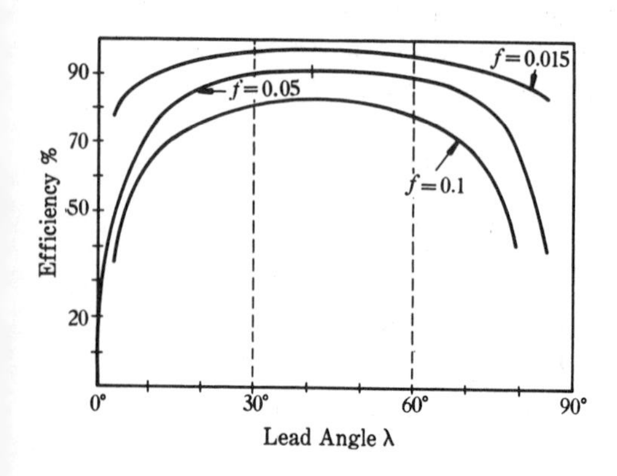

FIGURE 16.6 Efficiency of Worm Drives. Observe that the efficiency falls off rapidly at low lead angles; that, especially for low friction coefficients, the curve is relatively flat for $30° < \lambda < 60°$. The maximum efficiency occurs in the general vicinity of $\lambda = 45°$. For comparison, the efficiency of the ball bearing screw, Fig. 8.25, is said to increase to better than 90% within a lead angle of 2°.

16.9 COEFFICIENT OF FRICTION, WORM GEARING. The efficiency is seen in (16.3) to vary with the coefficient of friction, as well as with ϕ_n and λ. As usual, the coefficient of friction varies, sometimes widely and seemingly capriciously; it is certainly affected by the condition of the surfaces and the materials, by the adequacy of the lubricant and its temperature, by the rubbing speed v_r, and by alignment and workmanship in general. Not all sources of experimental data for f agree. The following equations represent compromises of data in Ref. *(13.1)* and should give conservative values for a carburized and ground worm driving a phosphor bronze gear when the design and manufacture is good:

$$(16.4) \qquad f = \frac{0.155}{v_r^{0.2}} \quad [3 < v_r < 70 \text{ fpm}] \qquad \text{or} \qquad f = \frac{0.32}{v_r^{0.36}}, \quad [70 < v_r < 3000 \text{ fpm}] \qquad [\text{BRONZE GEAR}]$$

where the rubbing speed $v_r = \pi D_w n_w/(12 \cos \lambda)$ fpm when D_w is in inches (Fig. 16.3). For other metals, increase these values by 25%, in which case, they should be suitable also for crossed helicals well run in.

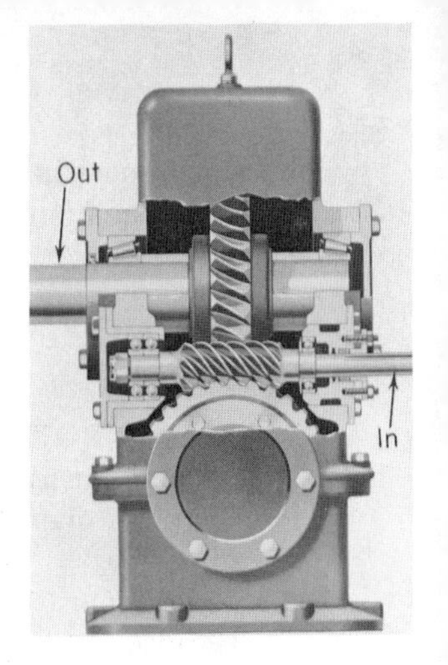

FIGURE 16.7 End View of Double-reduction Worm Drive. The second worm is behind the small gear, invisible. Observe the large increase in shaft size from input to output shaft. (Courtesy Cleveland Worm and Gear Co., Cleveland).

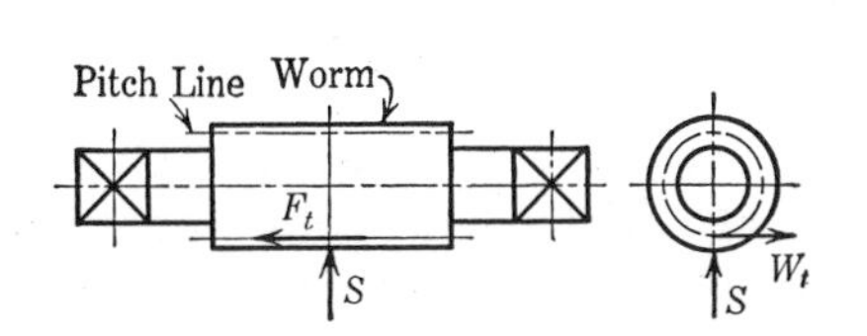

(a) Forces on Worm. F_t is the end thrust; the transmitted force W_t produces torsion and bending on the shaft, S produces bending; F_t, being eccentric, also produces bending that should not be neglected.

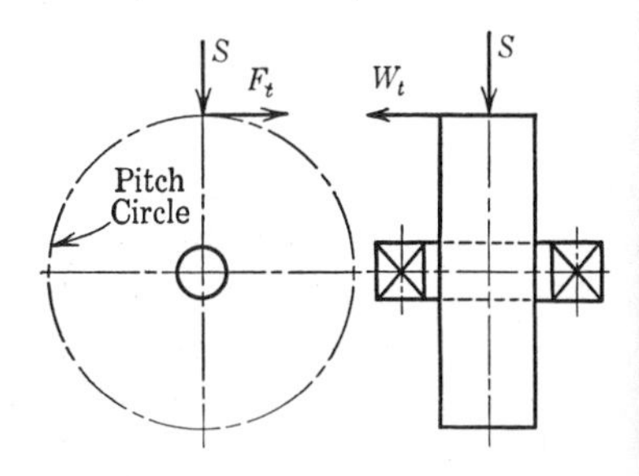

(b) Forces on Worm Gear. W_t produces end thrust and bending. The transmitted forced F_t produces torsion and bending; the separating force S produces bending.

FIGURE 16.8 Forces on Worm and Worm Gear. This is the simplest arrangement for finding bearing reactions and bending moments.

The efficiency tends to fall off as the velocity ratio increases, so that with high ratios, it may pay to use a double reduction, Fig. 16.7. When it is impossible for the gear to drive the worm, the drive is said to be ***self-locking***. This condition exists when the coefficient of friction is larger than *approximately* $\tan \lambda$ (actually also dependent on ϕ). This may mean a lead angle of 6° or less; if locking is *required*, let $\lambda < 5°$.

16.10 SEPARATING FORCE BETWEEN WORM AND GEAR. The driving forces W_t on the worm and F_t on the gear have been discussed in detail above and are shown in Fig. 16.8. The separating force S, Fig. 16.8, is the component of N in the y direction, radial to the *worm* (F_f has no component in the y direction). Thus $S = N \sin \phi_n$. Using the values of N from equations (i) and (j) in this expression, we get

(l) $$S = \frac{F_t \sin \phi_n}{\cos \phi_n \cos \lambda - f \sin \lambda} = \frac{W_t \sin \phi_n}{\cos \phi_n \sin \lambda + f \cos \lambda}.$$

16.11 PROPORTIONS FOR WORM GEARING. The meanings of certain dimensions are shown in Figs. 16.9 and 16.10; note what is meant by the pitch diameter D_g of the gear, Fig. 16.10; D_{wo} = outside worm diameter, D_{go} = outside gear diameter. The choice of the pressure angle to be used must be made with some knowledge of the size of the lead angle, because if the lead angle is very much larger than the pressure angle, there will be excessive undercutting of the gear tooth on the leaving side. The following limits are recommended.[13.3]

for $\phi_n = 14\frac{1}{2}°$, $\lambda_{max} = 16°$; for $\phi_n = 20°$, $\lambda_{max} = 25°$;
for $\phi_n = 25°$, $\lambda_{max} = 35°$; for $\phi_n = 30°$, $\lambda_{max} = 45°$;

but this is not to prohibit the use of a 20° pressure angle with a lead angle of 15°, for example. There are no standard tooth proportions used for all worm drives; an interchangeable type of tooth with an addendum of $1/P_d = 0.3183P_c$ has been widely used for single- and double-thread worms, and an addendum of $0.3183P_{cn}$ on double, triple, etc. threads. Dudley[13.20] suggests using the normal circular pitch for all worm drives that use the interchangeable type tooth: $a = 0.3183P_{cn}$, whole depth $= 0.7P_{cn}$. However, as the lead angle increases for a given whole depth and pitch in

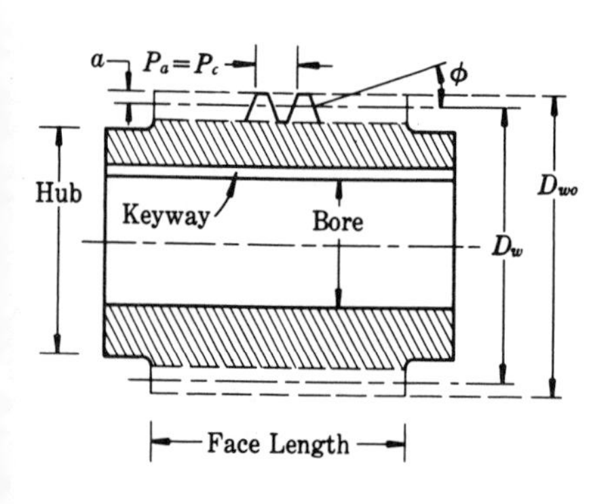

FIGURE 16.9 Shell Worm.

the diametral plane, the top land becomes narrower and the problem is not to have the teeth become too pointed. Since worm gears are inherently non-interchangeable, there is no imperative reason to use an interchangeable type.

There are two types of construction for the worm, the *shell type*, Fig. 16.9, designed to be keyed or splined to its shaft, and the *integral type*, Fig. 16.7, which is a part of its shaft. A proper worm diameter is a decision to be made; for the shell type,[0.3] $D_w \approx 2.4P_c + 1.1$ may be taken as a guide for up to quintuple (perhaps more) threads; for integral worms, try $D_w = 2.35P_c + 0.4$, or, if a center distance C is known,

$$D_w = \frac{C^{0.875}}{2.2} \text{ in.,} \tag{m}$$

say, $\pm$ 25%. Equation (**m**) is intended to give a worm size which is approximately optimum for maximum power for particular materials and centers. It is easier to get high efficiency (higher λ) with the integral worm than with the shell type. The *face length* of worms, Fig. 16.9, may be of the order of[0.3]

$$2[2a(D_g - 2a)]^{1/2}, \text{ where } a = \text{addendum} \quad \text{or } P_c(4.5 + N_g/50).$$

Dudley[13.20] recommends allowing about 6° of lead angle per start; for example, a lead angle of 30° would require a 5-thread worm according to this rule; but for another reason, 8 or more may be desired.

The face width of the gear is partly a matter of good proportion, but since the load is never uniformly distributed across the face, the value of the peak load enters the decision. The wider the face, the larger the difference between the peak load and the average load; on this account, Buckingham[14.1] recommends $b_{max} = 0.5D_{wo}$. Dudley[13.20] suggests a value *mn*, Fig. 16.10, which is defined by the intercepts of the outside circle with a tangent to the worm's pitch circle. Small worm gears may be made from a solid round with a hole for the shaft (and key). The proportions in Fig. 16.10 are intermediate, with a solid web. In larger gears, the web may be

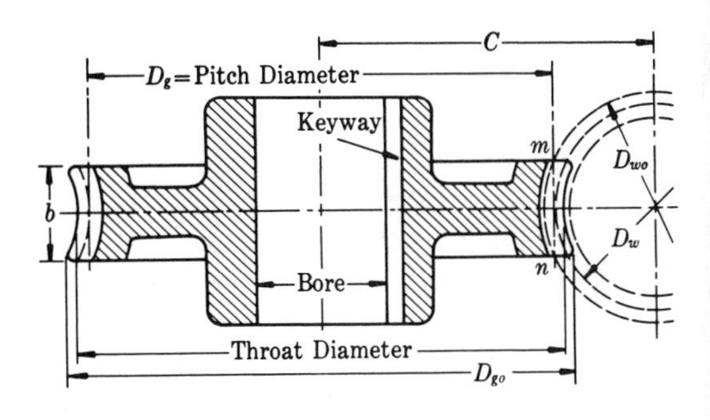

FIGURE 16.10 Throated Worm Gear

lightened by having holes in it. Also, in the larger sizes, costs may be significantly lowered by using a cast-iron or cast-steel spider to which is attached a bronze rim in which the teeth are cut; or weight and inertia might be reduced by making the spider of aluminum or titanium. An important attribute of the gear should be a good lateral and rim stiffness; the rim may be equal to about $0.6P_c$. Buckingham[14.1] suggests "dishing" the web, and he proposed the interesting hub design of Fig. 16.11. The bores

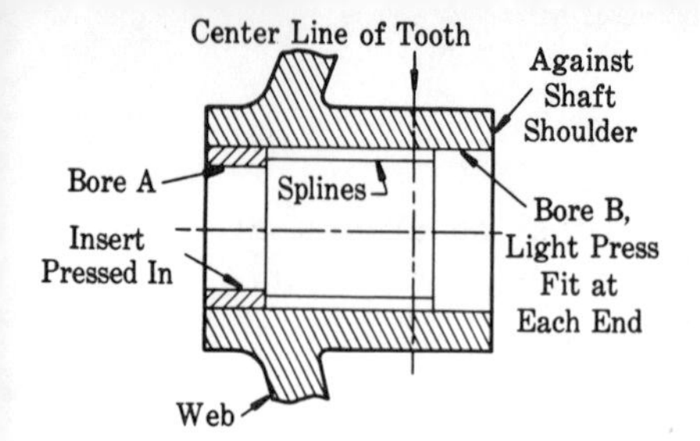

FIGURE 16.11 Hub Design for Large-diameter Gear.

A and *B* should be quite concentric, with press fits on the shaft that contribute to the lateral stability of the fit; these bores are needed because the splines are not adapted to good centering. Ordinary keys are often unsatisfactory.

There is nothing compelling about any of the proportions proposed above; some of them may be helpful as starting points. The basic factor for the worm size is that the shaft be of the right size to carry the load, either for strength or for deflection; then for an integral worm, the root diameter should be somewhat larger than the shaft. For a shell worm, a hub diameter of about (1.8) (bore) should be adequate with a root diameter somewhat larger, Fig. 16.9.

16.12 GENERAL REMARKS ON DESIGN OF WORM GEARING.

Because of the curvature of the enveloping surfaces and the consequent shape of the contacting surfaces, tooth action is complex. The details of the profile shape depend on the method of cutting the threads,[14.1] and the gear teeth should be hobbed by a hob that matches the worm to be used. Partly for this reason, a starting point in design is often a list of hobs (sizes) available, followed by an endeavor to use one of these. Nonstandard teeth can be cut with standard hobs. For highly refined designs with all-recess angle of action, see Ref. (*14.1*).

Worm drives have typically been used where a relatively large velocity ratio is desired, but they are being made with low velocity ratios where the right-angle relationship is advantageous and more power must be transmitted than can be handled by crossed helicals (worm gears have theoretical line contact versus point contact for helicals).

There are situations where a step up in speed is required, the gear driving the worm; for this direction of power flux, the frictional force in Fig. 16.5 acts opposite to the direction shown (opposite sense of rotation), which requires another force analysis.

If the worm is steel and the gear bronze, EP lubricants are not as beneficial as for steel and steel, The recommended lubricant[13.1] is a steam cylinder oil or an oil with 3% to 10% acidless tallow or other appropriate animal fat. The customary system of lubrication is for the gear to dip into an oil, Fig. 16.7, but if the oil must be circulated for cooling, it may just as well be directed at the mesh on the return.

We have considered only the case where the gear envelops the worm, Fig. 16.10, but the worm may also envelop the gear, as it does in Cone gears,

Fig. 16.12. Both of the gears in this drive have straight-sided teeth, giving surface contact.

16.13 DESIGN PROCEDURE. Since the boundary conditions vary so widely from job to job, it is impossible to set up a single design procedure. If a certain limited space is available, this in effect limits the size of the gear and center distance. One could then estimate a worm diameter,

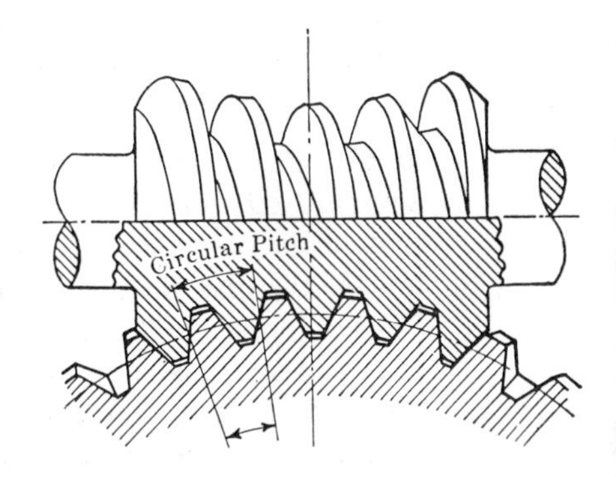

FIGURE 16.12 Cone Worm. (Courtesy Michigan Tool Co., Detroit).

§ 16.11, and proceed from there. In any event, iteration is inevitable. If significant power is being transmitted, the objectives are a limiting wear load equal to or greater than the dynamic load, adequate gear-tooth strength, and a good efficiency ($\lambda \gtreqless 30°$). Since our estimate of the dynamic load is inherently in error, the tooth design can reasonably be made ignoring the efficiency in computing F_t, which is on the safe side. At this point, at least the approximate size of shaft needed for the worm should be known. Given a certain power and speed for the worm, one may proceed somewhat as follows.

(a) Assume the number of starts (threads) for the worm and, for a given velocity ratio, compute N_g.

(b) Set up expressions for v_{mg}, F_t, and F_d in terms of P_c.

(c) Since wear is the probable determinant, if not heating, express F_w in terms of P_c. Although there is not always an apparent reason, the face width on worm gears is generally narrower in terms of circular pitch than on spur gears (§ 13.12); for the first approximation, try $b = 2P_c$, or less. In the end, b must have a satisfactory relation with D_w.

(d) Choose materials and select K_w. Equate $F_d = F_w$ with P_c as the only unknown. Solve for P_c (which is now a point of departure), judge its appropriateness, choose a standard value, check to see that the various proportions are satisfactory. This necessitates knowing the worm diameter. Note that the number of starts on the worm can be changed with little effect on the wear capacity if D_g and D_w are kept about the same (no change of materials), but watch that the teeth do not become too pointed.

(e) Other details of the iteration process are best not standardized. Compute the efficiency (or λ) early, because this is a factor, and the temperature rise with natural ventilation. A complete set of computations that validate the final choices is necessary.

16.14 MATERIALS FOR WORM GEARING. Preferred gear materials are the bronzes, especially the tin and nickel-tin bronzes (which may or may not be chilled cast, a process that produces the hardest bronze surface); but others, as leaded bronze (for high speed) and aluminum and silicon bronzes (for heavily-loaded, low-speed gears), are often used. To reduce costs, epecially for large gears, use a bronze rim attached to a cast-iron or cast-steel spider. See also § 16.5.

16.15 CLOSURE. Since the tooth dimensions cannot be said to be standardized for worm gearing and since the hob and the worm should match, the designer is limited only by his knowledge of involute geometry and of the manufacture of unequal addendum teeth; but further detail is beyond the scope of this book. Backlash is commonly made larger on worm gearing than on spur gears, and customarily it is obtained by reducing the thread thickness on the worm only; if there is a significant temperature rise, investigate the effect of the differential expansion of the different materials on backlash. There are of course many applications of worms in which the power is not significant, as in instruments and servomechanisms, for which the design criteria are not at all the same.

17. FLEXIBLE POWER-TRANSMITTING ELEMENTS

17.1 INTRODUCTION. As is true of other machine elements, flexible connectors for transmitting power appear in many different forms: flat belts, V-belts, flat-V, "toothed" belts, ropes (manila, cotton, wire). Also to be covered in this chapter are chain drives, which are much less flexible except in the sense of having turning joints. The flexible drives have distinct advantages that are on occasion overwhelming: they absorb vibration and shock, tending to transmit only a minimum to the connected shaft; they are suitable for relatively large center distances; they are quiet; when properly maintained, they can be designed to have a long trouble-free life.

17.2 NET BELT PULL AND THE VARIATION OF STRESS IN BELTS. If the smaller pulley is driving clockwise, Fig. 17.1, the force F_1 on the approaching belt is greater than that F_2 on the receding (slack) side. These forces produce the resisting torque $(F_1 - F_2)\,r_s = T_s$. The difference of the forces $F_1 - F_2$ is called the *net belt pull.* The driving torque on the larger (driven) pulley is $(F_1 - F_2)r_l = T_l$. The net belt pull (or torque) is computed from the horsepower equation (1.15). With $F_1 - F_2 = F$, $\text{hp} = Fv_m/33{,}000 = Fv_s/550$; also $\text{hp} = Tn/63{,}000$ where T in-lb. and n rpm apply to the same pulley.

We speak of belts wearing out, but the actual failure is quite analogous

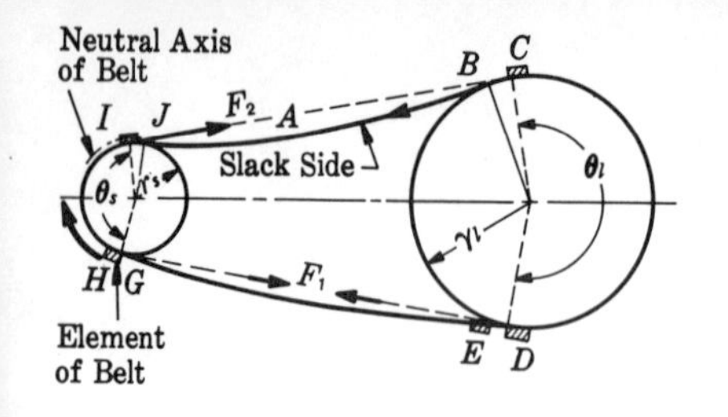

FIGURE 17.1 Open Belt. The tight side (larger tension F_1) is preferably on the bottom side of the pulleys because, as you see, the natural sag on the slack side on top results in the actual angle of contact being greater than the theoretical θ.

to fatigue. To trace out the variation of stress in a belt, accept first that with a particular point on the belt following a curved path, there is a centrifugal force that induces a stress s_{cf}, more or less uniform and so depicted in Fig. 17.2. Think of an element of belt as it makes its rounds, as at I, C, D, E, H. At position J, Figs. 17.1 and 17.2, just as a section of belt leaves the small pulley, the curvature of the belt is relatively negligible and the stress s_2 is that due to the slack tension F_2. Between B and C, the element moves onto the larger pulley, is bent to a radius r_l, which thereby induces a bending tensile stress at points outside of the neutral plane (see at I, Fig. 17.1); total stress is $s_2 + s_{b2}$. (If the action were elastic, the stress due to curving to radius r would be $s_b = Ec/r$ (§ 17.25), but ordinarily the strains are not elastic.) As the element moves about the larger pulley to D, the force on it due to the transmitted power increases more or less gradually to the value F_1; total stress is $s_1 + s_{b2}$. Between D and E, the bending stress drops to a negligible amount, but as the element moves onto the smaller pulley G to H, it is bent to its sharpest curvature making this the point of maximum stress $s_1 + s_{b1}$. As the element moves from H to I, the force due to power decreases to F_2, and the cycle repeats with every complete revolution of the *belt*. A belt can stand so many of these applications of the peak stress, depending on the magnitude of the peak value and probably on the magnitude of the mean value. We may conclude that if the bending on a pulley (or sheave) is sharp enough, the belt may eventually fail from the flexing with no power transmitted.

17.3 CAPACITY OF A FLAT BELT.

Make a free body of an element of belt, Fig. 17.3, whose length is $dL = r\,d\theta$. The analysis is for the case of *impending slipping* (in order to eliminate N), so that the frictional force is $f\,dN$, where dN is the normal force of the pulley on the element (or one can think of f as being simply a ratio of the actual frictional force to the normal force, and not a true coefficient). The belt pull on the slack side is F, on the tight side of the element, $F + dF$. A centrifugal (reversed effective) force dS acts radially outward. The sum of the forces is any direction is zero.

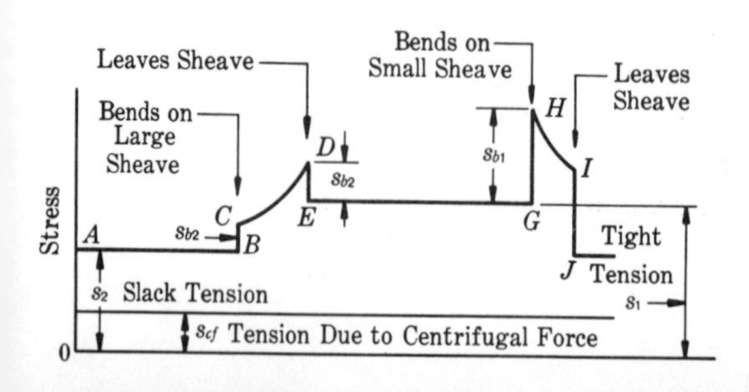

FIGURE 17.2 Stress Variation in Belt. A simplified model. After W. S. Worley.[17.3]

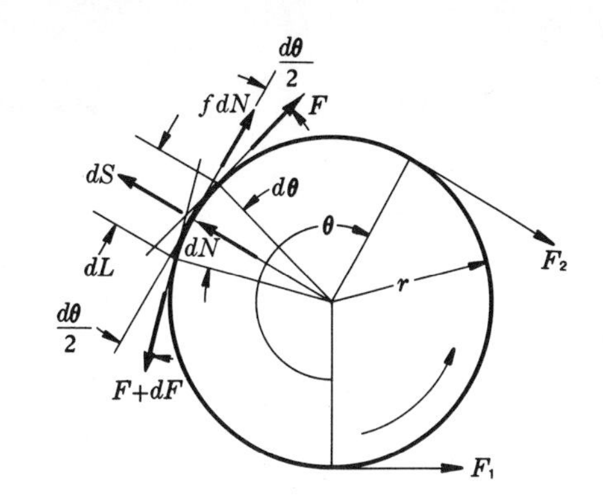

FIGURE 17.3 Forces on an Element of Belt on Driven Pulley.

Summing forces normal to the element, letting $\sin d\theta/2 = d\theta/2$, and dropping the product of two infinitesimals, we have

$$\text{(a)} \quad \Sigma F_n = dS + dN - F \sin\frac{d\theta}{2} - (F + dF)\sin\frac{d\theta}{2} = dS + dN - F d\theta = 0,$$

from which $dN = F\,d\theta - dS$. Summing forces tangential to the element, letting $\cos d\theta/2 = 1$, and substituting the value of dN just found, we get

$$\Sigma F_t = f\,dN + F\cos\frac{d\theta}{2} - (F + dF)\cos\frac{d\theta}{2}$$

$$\text{(b)} \qquad = fF\,d\theta - f\,dS - dF = 0.$$

The centrifugal force ($F = ma$) on a particle (the element) is its mass times the normal acceleration $a_n = v_s^2/(r/12)$, where the belt speed v_s is in fps and r in. divided by 12 is $r/12$ ft. to match $g_o = 32.2$ fps^2 to be introduced next. The volume of the element for a belt thickness t in. and belt width b in. is $bt\,dL = btr\,d\theta$ cu. in.; let the density ρ be lb/cu. in., so that the mass is $\rho\,btr\,d\theta$ lb. or $\rho\,btr\,d\theta/g_o$ slugs (since the pound is used for force in mechanics by engineers, the mass must be in slugs to be in a consistent system of units with $g_o = 32.2$ fps^2). Thus,

$$\text{(c)} \qquad dS = dm\,a_n = \frac{\rho btr\,d\theta\,v_s^2}{g_o(r/12)} = \frac{12\rho btv_s^2}{g_o}\,d\theta = K\,d\theta,$$

where $K = 12\,\rho btv_s^2/g_o$, a constant, is used for convenience. Substitute the value of dS from **(c)** into **(b)**, separate the variables, and integrate from the smaller to the larger, F_2 to F_1;

$$\int_{F_2}^{F_1} \frac{dF}{F - K} = f\int_0^{\theta} d\theta \qquad \text{or} \qquad \ln\frac{F_1 - K}{F_2 - K} = f\theta.$$

In exponential form,

$$\text{(d)} \qquad \frac{F_1 - K}{F_2 - K} = e^{f\theta}.$$

Observe that for $v_s = 0$ ($K = 0$), this equation reduces to $F_1/F_2 = e^{f\theta}$, a familiar form from analytic mechanics. Since the speed and mass are often low enough that the effect of the centrifugal force is negligible, K may be dropped from any of these equations when this is so. Invert (**d**), subtract one from each side, change signs, and solve for the net belt pull $F_1 - F_2$;

$$F_1 - F_2 = (F_1 - K)\left(\frac{e^{f\theta} - 1}{e^{f\theta}}\right), \tag{17.1}$$

where K is the load on the belt due to the centrifugal force, and is sometimes conveniently expressed in terms of a mass per foot of length, w lb/ft.; $w = 12\,bt\rho$ lb., where 12 in. is 1 ft., so that

$$K = \frac{12\rho btv_s^2}{g_o} = \frac{wv_s^2}{32.2}. \tag{e}$$

Since F_1 is the maximum tension, it has been the custom in flat belt design to let $F_1 = sA = sbt$, where it is recognized from the above discussions that s is not the real maximum stress; design stresses that have been found to give good designs are used. Making this substitution for F_1 in (17.1) and using K from (**e**), we get

$$F_1 - F_2 = bt\left(s - \frac{12\rho v_s^2}{32.2}\right)\left(\frac{e^{f\theta} - 1}{e^{f\theta}}\right), \tag{17.2}$$

where $e = 2.718 \cdots$. For design, use

$$\rho = 0.035 \text{ lb/cu. in. for leather,}$$
$$\rho = 0.045 \text{ lb/cu. in. for flat rubber belting.}$$

The theory of this derivation is impeccable, but as usual, there are facts of life that cannot be incorporated into a simple theory. (If this were not so, there would be little use for engineers.) We shall now discuss those factors, some in equation (17.2), some not, that affect belt design.

17.4 BELT THICKNESS AND WIDTH. Since the repeated flexing of the belt about the pulleys is a significant determinant of the life of the belt and since the thicker the belt, the greater the maximum stress induced by the flexing, there should be some consideration of belt thickness as related to the smaller pulley diameter. The thickness of a leather belt depends on the number of plies and on the thicknesses of the hides used in its manufacture. Average values are given in Table 17.1, which also shows the recommended minimum pulley diameter to be used with each belt thickness. If larger pulleys are permissible and used, longer life with the same b and t for the belt may be expected. With the qualification on pulley sizes, thick, narrow belts are more economical than thin wide ones. When equation (17.2) is used for design, the normal action is to solve for the belt width b

needed after decisions have been made on the various other factors involved. Let the standard widths be as follows:

The width varies: by $\frac{1}{8}$ in. increments from $\frac{1}{2}$ to 1 in.,
by $\frac{1}{4}$ in. increments from 1 to 4 in.,
by $\frac{1}{2}$ in. increments from 4 to 7 in.,
by 1 in. increments from 7 to 12 in.

Larger sizes are obtained by special order. Single leather belts should be specified in widths not over 8 in., because this represents about the maximum width of suitable leather than can be cut from a hide.

17.5 COEFFICIENT OF FRICTION. Not only does the coefficient of friction vary widely with operating conditions and not only is it difficult to know what it is under various operating conditions, but if any of the friction belts of this chapter were run continuously with the operating net belt tension at the point of limiting friction, their life would usually be uneconomically short. This situation can be allowed for arbitrarily by using a low value of f [or by using a value of the net tension ratio, equation (**d**) above, that is known to be satisfactory]. Another approach is to use a low design stress s in equation (17.2)—§ 17.6; and of course, one can use safe values of both f and s.

The coefficient of friction varies with the amount of slip. A part of the total slip is ***creep***, which exists because the driving pulley receives a longer (stretched) belt than it delivers and the driven pulley receives a shorter belt than it delivers (Fig. 17.1), giving rise to relative motion between the belt and the pulley. As the load increases, there is, in addition, slippage in the customary sense of the word. When used without qualification, *slip* or *total slip* means the total amount including creep. If the slip becomes excessive, leather belting "squeals," giving a warning. As the total slip increases, the coefficient increases, Fig. 17.4; 2% slip is satisfactory for leather on steel or iron. For normal design conditions for flat belts and the design stress of the next article, use the following:

Leather on iron or steel, $f = 0.3$;
Leather on paper pulleys, $f = 0.5$.

FIGURE 17.4 Typical Friction-slip Curves for Three Kinds of Pulleys.[17.5]

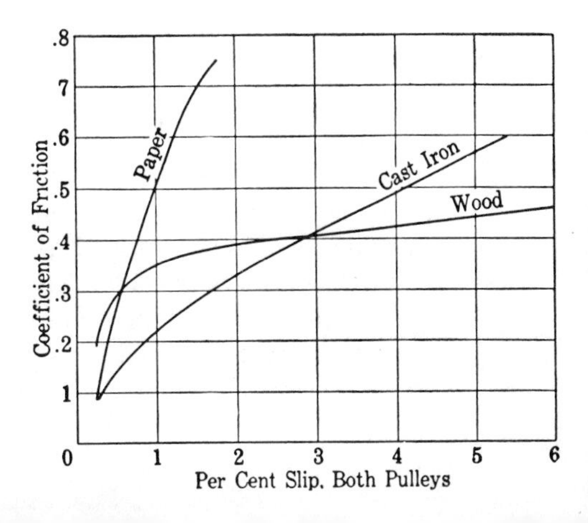

17.6 STRENGTH OF LEATHER. Since the hides of no two steers are likely to be of the same quality and since the strength of leather also depends upon the method of tanning, the variability is expected to be large. The breaking strength of *oak-tanned* belting varies from 3 to more than 6 ksi. *Mineral tanned* leather is stronger, say 7 to 12 ksi. To make the belt endless, the ends must be joined: by cementing ($\eta = 1.00$); by wire lacing with machine ($\eta = 0.88$); by metal hooks ($\eta = 0.35$); and by other methods; where η is the relative strength of the joint. For normal good operating conditions and pulley sizes larger than the minimums of Table 17.1, a design stress of

$$s_d = 400\eta \qquad \text{(f)} \quad \text{[LEATHER]}$$

in equation (17.2) should result in a belt capable of a satisfactory life. Since the strength *per se* is not the criterion for belt capacity, the relative strength η probably has little to do with the belt's life unless it is rather low (belt damaged by the connection). Reduce the design stress for unusually long life or for unfavorable conditions (or introduce service factors—see Table 17.7).

17.7 LENGTH OF BELTS. Exact lengths are as detailed in Figs. 17.5 and 17.6; closely, the belt lengths may be computed from

$$L \approx 2C + 1.57\,(D_2 + D_1) + \frac{(D_2 - D_1)^2}{4C}. \qquad (17.3)$$

[OPEN BELT, FIG. 17.5]

$$L \approx 2C + 1.57\,(D_2 + D_1) + \frac{(D_2 + D_1)^2}{4C}, \qquad \text{(g)}$$

[CROSSED BELT, FIG. 17.6]

where C is the center distance, D_2 is the diameter of the larger pulley, D_1 is the diameter of the smaller pulley, and all dimensions are in inches. If the center distance C for a given endless belt is desired, solve for it from (17.3); or see equation (**o**), § 17.17. A crossed belt has the advantage of a large contact angle and it does very well at the lower speeds. The wear due to the rubbing at the point of crossing is a disadvantage.

Although it does not appear in equation (17.1), the length of belt is an important engineering factor. The longer the belt for a given belt speed, the less frequently it is flexed about a pulley or sheave. Since the flexing is often the significant damaging action, doubling the center distance, for instance, would be expected to increase the belt's life materially. Especially for flat belts on horizontal drives, a long belt (force of gravity) contributes to maintaining the initial tension (§ 17.10). However, not only is a longer belt more expensive, but the space it occupies is costly. Other more economical means of maintaining belt tension are available (§ 17.13).

17.8 ANGLE OF CONTACT. For an ***open belt,*** the angles of contact are

$$\theta = \pi \pm 2 \sin^{-1} \frac{R - r}{C} \approx \pi \pm \frac{D_2 - D_1}{C} \text{ radians,} \tag{17.4}$$

from Fig. 17.5, where it is seen that $\alpha = \sin^{-1}(R - r)/C$. Use the *plus* sign for the *larger* pulley and the *minus* sign for the *smaller* pulley. For ***crossed belts,*** the angles of contact are the same on both pulleys, Fig. 17.6,

$$\theta = \pi + 2 \sin^{-1} \frac{R + r}{C} \text{ radians,} \tag{h}$$

where R = the radius (D_2 = diameter) of the larger pulley, r = the radius (D_1 = diameter) of the smaller pulley, C = the distance between pulley centers, and θ = the angle of contact in radians; all dimensions in inches.

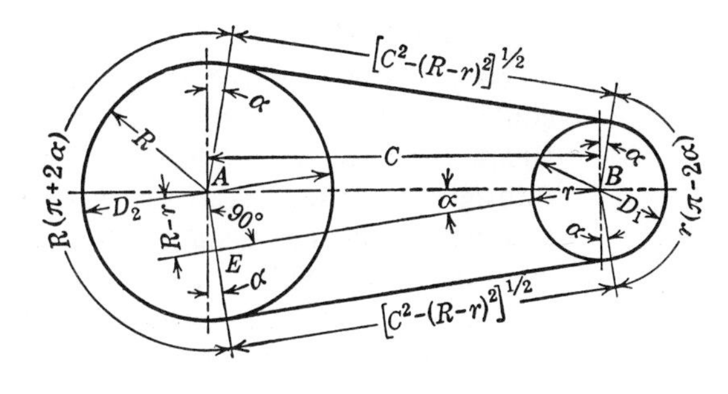

FIGURE 17.5 Open Belt.

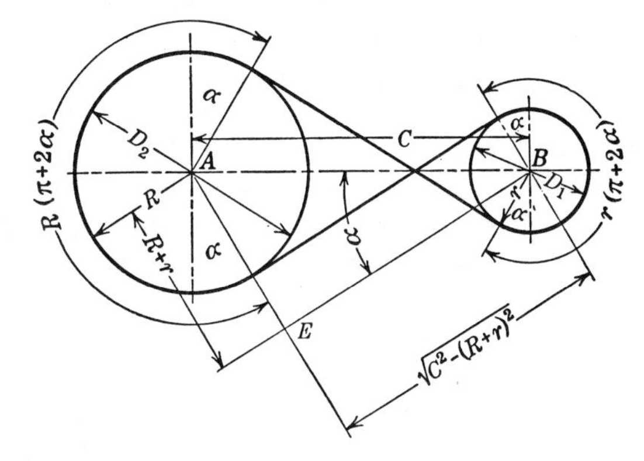

FIGURE 17.6 Crossed Belt.

Considering the term $(e^{f\theta} - 1)/e^{f\theta}$ in equation (17.1), we see that significant changes in the minimum angle of contact, as would occur for a large velocity ratio and relatively short center distance or for the case of a tension pulley, Fig. 17.7, will bring about significant changes in capacity. However, there is evidence that the actual effect is not so great as predicted by the theory, so that one could use some average applicable value for $(e^{f\theta} - 1)/e^{f\theta}$ for the more usual configurations. In general, the design by

FIGURE 17.7 Tension Pulley for Use on Short-center Drives. (Courtesy Link-Belt Co., Chicago).

equation (17.1) or (17.2) is done for *the minimum value of* $f\theta$, which is for the smaller pulley when the pulleys are of the same material (no tension pulley).

17.9 BELT SPEED. The tension in the belt due to centrifugal force increases rapidly above about 2500 fpm. If every thing holds together, equation (17.3) shows that a speed is reached where the belt is stressed to whatever design stress s has been chosen, at which point no power can be transmitted without exceeding this stress. Thus, given a certain belt size and permissible stress, we find that there is a certain speed at which maximum power would be transmitted (see *Problems* for problems illustrating this idea). Experience suggests that the most economical designs are obtained for a belt speed of 4000 to 4500 fpm,[17.4] but of course, any particular application may *require* some other speed. For leather, speeds of 7000 to 8000 fpm are in use (but consult the vendor). Flat fabric belts operate up to 20,000 fpm and more. On high-speed centrifugal blowers, it has been observed that the arc of contact θ is reduced from 180° at rest to 110° in motion, one of the side effects of speed.

We might well keep in mind that doubling the belt speed, for example, doubles the number of flexes of the belt on the pulley in a given time, which suggests that the higher the speed the more important the minimum pulley size.

To find the stress due to centrifugal force alone, let $F_1 - F_2 = 0$; that is, assume no power is transmitted. It follows then that the bracket, $s - 12\rho v_s^2/32.2$, must be equal to zero. Equating this expression to zero, assuming $s = 400$ psi, $\rho = 0.035$ lb/cu. in., and solving for the velocity, we get $v_m = 10{,}500$ fpm (nearly 2 mi/min.), which is the speed at which the centrifugal force alone theoretically stresses the belt to 400 psi. This computation does not mean that a belt cannot, or even should not, transmit power at this, or a higher, speed, as some have thought. In the first place, experiments[17.7] show that the effect of centrifugal force is not as great as it is indicated by equation (17.1); but even so, it would still be possible to transmit power if values of s greater than 400 were used.

17.10 INITIAL TENSION. In order to transmit power, the belt must have some *initial tension* (tension at rest); Taylor's[17.4] recommendation is 71 lb/in. of width. The tighter the tension, the more power can be transmitted without excessive slipping; the initial tension may be so low that the belt will not transmit its power. As the initial tension is increased, the value of the tight tension F_1 also increases. It not only becomes damaging to the belt, but since $F_1 + F_2$ (a bending force on the shaft) also increases, the operating conditions may become dangerous for the shaft and bearings. A relation F_1/F_2 is often assumed for design purposes (see Fig. 17.10), say $F_1/F_2 = 3$ with $\theta = 180°$. The effect of initial tension is not recognized by equation (17.1).

17.11 RATED CAPACITY OF LEATHER BELTS. The foregoing discussion of the details is included because an engineer is eager to consider all factors that might effect his designs and because most of the ideas apply to all kinds of belt drives (some are even easily interpreted for chain drives). But there is no denying the many inherently involved intangible variants that only experience of a certain kind might successfully evaluate. Thus, for the common case, a simple design procedure using rating tables and experience factors will give good answers to many problems; say, those of Tables 17.1 and 17.2 for flat leather belts; the values in Table 17.1 are based on the general belt equation (17.2). Interpolations may be made between the speeds given, with the experience factors of Table 17.2 applied as follows:

$$\text{(i)} \qquad \text{hp} = (\text{hp/in., Table 17.1})(bC_mC_p)(C_{f1}C_{f2}\ldots).$$

which is the nominal horsepower that the belt is to transmit (or the name-plate horsepower for an electric motor), and

- b in. is the belt width,
- C_m is the correction factor for the type of drive; except for electric motors, use $C_m = 1$;
- C_p is the correction factor for the size of the smallest pulley; the smaller the pulleys, the greater is the flexure,
- C_f is the correction factor for the environmental conditions, and more than one of these factors may apply; thus a vertical drive in a dusty atmosphere and subjected to shock loads would correspond to a total factor of $C_f = (0.83)(0.74)(0.71)$.

Other deviations from desirable practice that should be considered include abnormally short center distance and tight side on top. The life expectancy of belts selected in accordance with these ratings should be some 5 to 7 years. Observe, at the bottom of Table 17.1, that smaller pulleys are permissible at the lower speeds, because the belt does not bend as often. The strength of the belt joint is not considered in this design procedure. For leather belting, mount the hair or grain side next to the pulley, because

TABLE 17.1 RATED HORSEPOWER PER INCH OF WIDTH, LEATHER BELTS[17.9]

For belt speeds over 6000 fpm, consult a leather belting manufacturer; t is the average thickness of leather belting.

Belt Speed fpm	Single Ply $t = \frac{11}{64}$ Med.	Single Ply $\frac{13}{64}''$ Heavy	Double Ply $\frac{18}{64}''$ Light	Double Ply $\frac{20}{64}''$ Medium	Double Ply $\frac{23}{64}''$ Heavy	Triple Ply $\frac{30}{64}''$ Medium	Triple Ply $\frac{34}{64}''$ Heavy
600	1.1	1.2	1.5	1.8	2.2	2.5	2.8
800	1.4	1.7	2.0	2.4	2.9	3.3	3.6
1000	1.8	2.1	2.6	3.1	3.6	4.1	4.5
1200	2.1	2.5	3.1	3.7	4.3	4.9	5.4
1400	2.5	2.9	3.5	4.3	4.9	5.7	6.3
1600	2.8	3.3	4.0	4.9	5.6	6.5	7.1
1800	3.2	3.7	4.5	5.4	6.2	7.3	8.0
2000	3.5	4.1	4.9	6.0	6.9	8.1	8.9
2200	3.9	4.5	5.4	6.6	7.6	8.8	9.7
2400	4.2	4.9	5.9	7.1	8.2	9.5	10.5
2600	4.5	5.3	6.3	7.7	8.9	10.3	11.4
2800	4.9	5.6	6.8	8.2	9.5	11.0	12.1
3000	5.2	5.9	7.2	8.7	10.0	11.6	12.8
3200	5.4	6.3	7.6	9.2	10.6	12.3	13.5
3400	5.7	6.6	7.9	9.7	11.2	12.9	14.2
3600	5.9	6.9	8.3	10.1	11.7	13.4	14.8
3800	6.2	7.1	8.7	10.5	12.2	14.0	15.4
4000	6.4	7.4	9.0	10.9	12.6	14.5	16.0
4200	6.7	7.7	9.3	11.3	13.0	15.0	16.5
4400	6.9	7.9	9.6	11.7	13.4	15.4	16.9
4600	7.1	8.1	9.8	12.0	13.8	15.8	17.4
4800	7.2	8.3	10.1	12.3	14.1	16.2	17.8
5000	7.4	8.4	10.3	12.5	14.3	16.5	18.2
5200	7.5	8.6	10.5	12.8	14.6	16.8	18.5
5400	7.6	8.7	10.6	12.9	14.8	17.1	18.8
5600	7.7	8.8	10.8	13.1	15.0	17.3	19.0
5800	7.7	8.9	10.9	13.2	15.1	17.5	19.2
6000	7.8	8.9	10.9	13.2	15.2	17.6	19.3

Minimum Pulley Diameters, Inches

Belt Speed fpm	Single Ply Med.	Single Ply Heavy	Double Ply Light	Double Ply Medium	Double Ply Heavy	Triple Ply Medium	Triple Ply Heavy
up to 2500	$2\frac{1}{2}$	3	4	5*	8*	16**	20**
2500–4000	3	$3\frac{1}{2}$	$4\frac{1}{2}$	6*	9*	18**	22**
4000–6000	$3\frac{1}{2}$	4	5	7*	10*	20**	24**

* For belts over 8 in. wide, add 2 in. to the minimum diameters shown.

** For belts over 8 in. wide, add 4 in. to the minimum diameters shown.

TABLE 17.2 EXPERIENCE FACTORS

Type of Drive	C_m
Any except electric motor	1
Squirrel cage, compensator starting	0.67
Squirrel cage, line starting	0.5
Slip ring, and high starting torque	0.4
Pulley Size, in.	C_p
4 or less	0.5
$4\frac{1}{8}$ to 8	0.6
9 to 12	0.7
13 to 16	0.8
17 to 30	0.9
Over 30	1.0
Operating Conditions	C_f
Oily, wet or dusty atmosphere	0.74
Vertical drives	0.83
Jerky loads	0.83
Shock and reversing loads	0.71

the flesh side is tougher and can stand better the stretching it gets on the outside.

17.12 EXAMPLE—FLAT LEATHER BELT. A squirrel-cage, line-starting motor is rated at 30 hp at 1750 rpm and delivers its power to a horizontal, flat leather belt located in a dusty atmosphere. The motor pulley is 10 in. in diameter. What width of medium double-ply belt is needed?

Solution. The belt speed is

$$v_m = \pi Dn = \pi\left(\frac{10}{12}\right)(1750) = 4580 \text{ fpm}.$$

Corresponding to this speed and to a medium double-ply belt, we get 11.97, say 12 hp/in., from Table 17.1. From Table 17.2, we read $C_m = 0.5$, $C_p = 0.7$, and $C_f = 0.74$. Substituting these various values into (**i**), we get

$$hp = (\text{hp/in.})\, bC_mC_pC_f$$
$$30 = (12)(b)(0.5)(0.7)(0.74),$$

from which $b = 9.65$; use a 10-in. belt.

17.13 MAINTAINING INITIAL TENSION. All belts (and chains) elongate with service. Thus, although some types of belts may be shortened, it is convenient to have the motor mounted on an adjustable base. If it is bolted down in slots on the base, it can be moved away from the driven member from time to time to reestablish a good initial tension. Similarly, the motor may be mounted on runners with an attached spring designed to maintain the tension, but since a large stretch of the belt may result in a spring force too low, an adjustment to take care of this effect is also necessary.

An early idea was the ***tension pulley,*** Fig. 17.7, which not only will maintain initial tension but also greatly increases the arc of contact on the nearest pulley, preferably the smaller one. In Fig. 17.7, an arm with adjustable weights (not visible) provides the tension; a spring may be used instead of weights. The tension pulley *should not* be smaller than that permitted by

FIGURE 17.7 (Repeated).

FIGURE 17.8 Flat Belt Drive with Econ-o-matic Base. If the smaller pulley exerts clockwise torque, the motor frame exerts an equal and opposite counterclockwise torque, with the supporting frame pivoting about the point indicated. Also used on V-belt drives. (Courtesy American Pulley Co., Philadelphia).

the belt thickness, because the belt does bend about it, and moreover in the usual location, it results in a *reversed bending.*

When a motor is delivering power, the torque on the stator (frame) is equal and opposite to that on the rotor—action and reaction are equal. Hence, the frame tends to turn in the opposite direction to the motor shaft, and this tendency is utilized to provide the belt tension by mounting the motor on a cradle which is free to move through a small arc, Fig. 17.8. The higher the delivered torque, the greater is the effort of the motor frame to turn and the tighter it draws the belt, because the driving pulley moves away slightly from the driven pulley. Thus, this system can be designed to provide automatically an adequate, but not constant, amount of tension for any load. As a result, if the loading varies, the average net and total tensions will be lower than for a constant pull device.

If the motor mounting is pivoted, Fig. 17.9, the center of gravity of the motor can be so located that its weight provides belt tension. In this type of drive, the design may be based on an appropriate value of the ratio of the tensions F_1/F_2, say from Fig. 17.10, which are recommended by Tatnall[17.10] for pivoted drives with a paper driving pulley, Fig. 17.36.

17.14 ANALYSIS OF PIVOTED-MOTOR DRIVE. Having decided upon a suitable value of F_1/F_2, one may solve directly for the corresponding location of the pivot axis.[17.10] But due to space limitations, we shall proceed on a trial-and-error basis in accordance with the following outline:

(a) Decide upon a suitable value of F_1/F_2, Fig. 17.10.

(b) Compute the net tension $F_1 - F_2$ from the horsepower equation.

(c) From the simultaneous solution of these two equations, find the values of F_1 and F_2.

(d) The location and size of the pulleys are known or are determined from space and velocity-ratio considerations affecting the drive.

(e) Lay out the pulleys in size and location to scale. Since the dimension h,

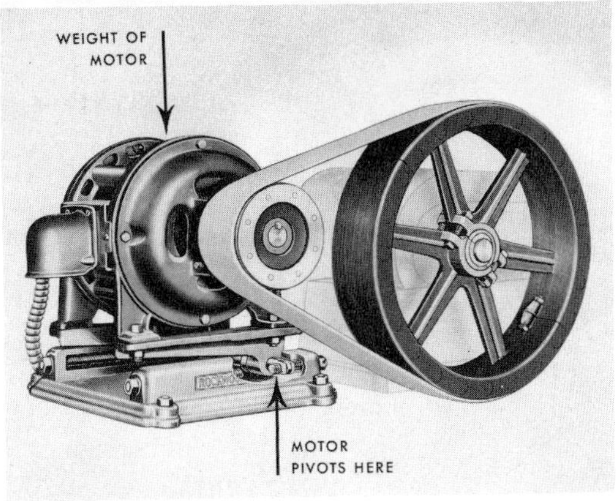

FIGURE 17.9 Pivoted-motor Drive. (Courtesy The Rockwood Mfg. Co., Indianapolis, Ind.)

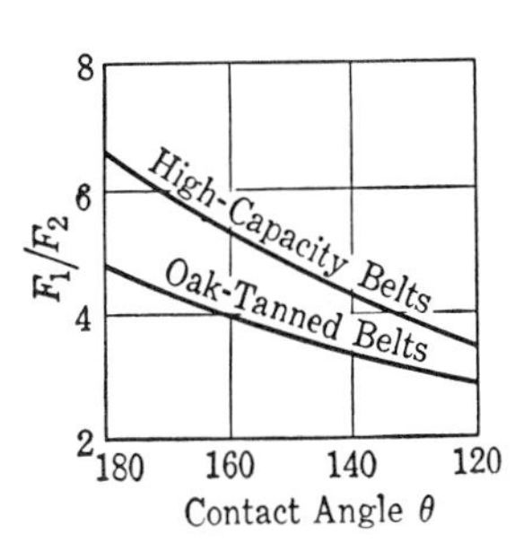

FIGURE 17.10 Ratio of Tensions. With paper driving pulley.

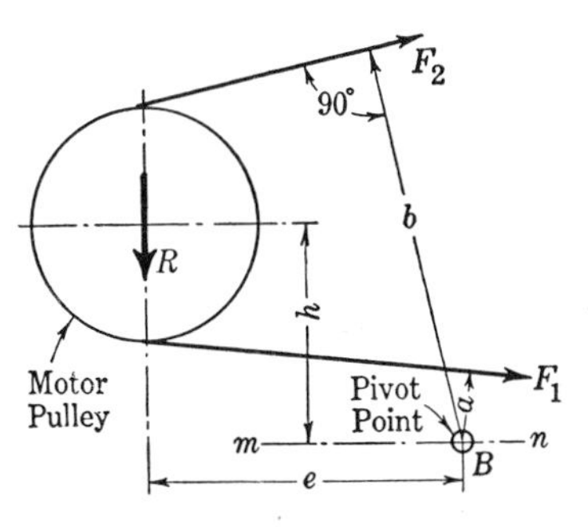

FIGURE 17.11 Location of Pivot Axis.

Fig. 17.11, is known from the dimensions of the motor and pivoting base, the line *mn* may be drawn. The lines of action of F_1 and F_2 are taken tangent to the pulleys. The line of action of the resultant weight of the motor assembly R is usually taken through the center line of the shaft of the motor, although it actually passes through the center of gravity of the assembly.

(f) Now assume a location for the pivot point B on the line *mn*, and measure the moment arms, a, b, and e. Then, a sum of the moments of F_1, F_2, and R about the pivot B should be equal to zero. Thus

$$Re - F_1a - F_2b = 0. \tag{j}$$

If this equation is not satisfied, assume another location of B and try (j) again.

In the pivoted drive, the tight tension F_1 should be between the pivot B and the slack tension F_2, but such a location is not essential in all cases. In general, keep the moment arm a as small as possible. Base the design on the maximum output of the motor, which may be taken as (1.4)(rated hp of motor) if the electric motor is compensator-started and (2.5)(rated hp) if it is started across the line.

17.15 RUBBER BELTING. In addition to leather, a number of other materials are used for belting, principally rubber, natural or synthetic. Rubber belts are made in layers (say 3 to 12), called plies, of canvas duck impregnated with rubber which is later vulcanized. A 6-ply rubber belt has six layers of duck of about 32-ounce weight. Also rubber belts are made of cord construction which is more stretch resistant. Since excessive heat causes a slow deterioration of rubber and since oil is particularly harmful, a natural rubber belt should not be used in very hot places or where oil is spattered, unless rapid deterioration is acceptable. Neoprene belts, or neoprene covered, less susceptible to damage by oil, are available. Since the belts are preferably endless, stock sizes should of course be chosen; for pedagogical purposes, let b vary by $\frac{1}{4}$ in. to 2 in. width, by $\frac{1}{2}$ in. to 5 in., by 1 in. to 10 in., by 2 in. to 24 in. Minimum recommended pully diameters[17.14] are: 3 ply, 3 in.; 4 ply, 5 in.; 5 ply, 7 in.; 6 ply, 10 in.; 7 ply, 14 in.; 8 ply, 18 in.; 9 ply, 22 in. According to the Goodrich Company, the permissible net belt pull is 13.75 lb/ply per inch of width, which gives a simple means for quick estimation;

$$\text{(k)} \qquad F_1 - F_2 = 13.75\, bN_p, \qquad \text{or} \qquad hp = \frac{bv_m N_p}{2400}. \qquad \text{[FABRIC PLIES]}$$

From either of these equations, compute the width of belt b in. for N_p plies. Correction should be made for other than smooth service and for arcs of contact other than 180°. Choose the arc of contact factor K_θ as follows:[17.14]

θ	220	200	180	170	160	150	140	130	120
K_θ	1.13	1.07	1.00	0.96	0.92	0.86	0.82	0.78	0.74

Examples of service factors, not necessarily the same as the rubber belt manufacturer's recommendations, are found in Tables 17.2 and 17.7. Rubber belts with cord construction have a higher capacity than those with fabric ply construction. Rubber belts should be mounted with an initial tension of 15 to 20 lb. per inch-ply, which is approximated if the belt is $\frac{1}{8}$ inch per foot of belt length shorter than the steel tape measurement around the pulleys.

17.16 FLAT BELT DRIVES FOR NONPARALLEL SHAFTS. Belt-connected shafts do not have to be parallel, but the ***law of belting*** must be satisfied, to wit: *the approaching side must approach the pulley in a direction perpendicular to the pulley's axis.*

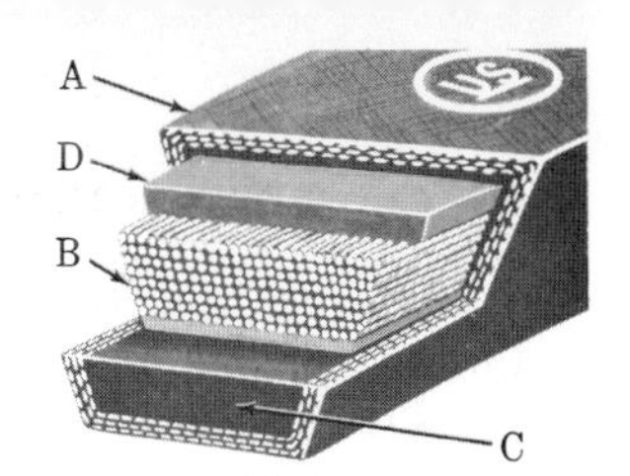

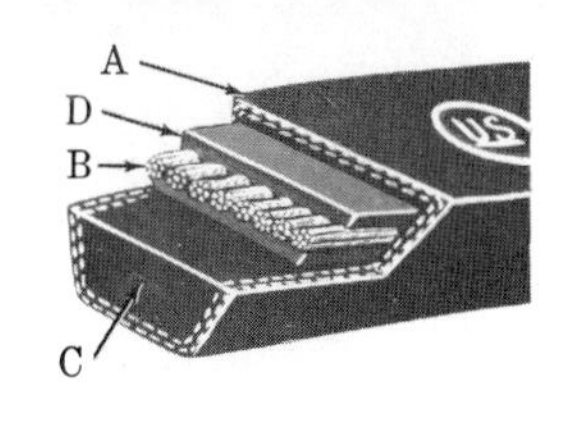

FIGURE 17.12 V-Belt Construction. The minimum components in V-belt construction are: a rubberized, woven jacket material *A*; tensile members *B* that carry the principal load, which is sometimes a fabric, sometimes cords; a resilient cushion or compression material *C* that provides the side-wall pressure; the tensile layer *D*, capable of withstanding the repeated flexing. (Courtesy United States Rubber Co., N.Y.C.).

(a) **Intersecting Shafts.** If two intersecting shafts are to be belt connected, two ***guide*** pulleys, one on the approaching side and one on the receding side of the belt, must be used. Each guide pulley must be so located that it receives the belt from a direction perpendicular to its axis and delivers the belt in a direction perpendicular to the axis of the next-approached pulley in accordance with the law of belting. See the guide pulley at the top of Fig. 17.15.

(b) **Shafts at Right Angles.** Two shafts at right angles to each other may be connected by the quarter-turn arrangement. The planes of the pulleys are of course perpendicular to each other. The relation of the pulleys should be such that a center-line plane through each pulley is tangent to the face of the other pulley. This drive satisfies the law of belting for only one direction of rotation. If rotation is to be reversed, a guide pulley may be located to direct the belt properly.

17.17 V-BELTS. Since they are always used for various auxiliary drives under the hood of an automobile, the appearance of V-belts is familiar to nearly every one. Two typical internal constructions, which vary in detail among different manufacturers, are shown in Fig. 17.12. Inasmuch as rubber has a low strength and stretches easily, invariably there are tension members *B*, Fig. 17.12, usually either fabric or cord construction. The cushion material *C* is natural or synthetic rubber. The grooved wheels that V-belts run on are called ***sheaves*** (Fig. 17.38).

The factors that effect the capacity of a V-belt are much the same as those previously described, except that the wedging action, Fig. 17.13, results in a much larger normal force N for the same belt tension and therefore a larger tangential (transmitted) force is obtainable for the same coefficient of friction. By the same token the reasonably attainable tension ratio F_1/F_2 is greater than for flat belts [the ratings and the arc correction K_θ below are for a net tension ratio of 5, equation (**d**), § 17.3]. Also because of the wedging, V-belts do well on short center distances without

FIGURE 17.13 The groove angle 2ϕ is 34°, 36°, or 38°.

2ϕ
b
Belt
B
ϕ
t
ϕ
$\frac{N}{2}$
$\frac{N}{2}$
A
Sheave
$N \sin \phi$

frequent adjustment for initial tension. In case one belt in a multiple-belt drive breaks, the remaining belts will carry the load without the necessity of an immediate shut down (no lost production time). In this case, install a complete new set of belts with lengths especially matched (close tolerance).* One belt much shorter than the others will take an excessive share of the load and soon fail.

The theoretical equation (17.1) for the assumptions made in the flat-belt derivation apply to V-belts if, instead of f, you insert $f/\sin\phi$ (2ϕ = groove angle). However, the practice is to use a standardized rated horsepower, equation (17.5) below. The second term in the brackets provides an allowance for bending, the effect of which is greater for the thicker V-belt.[17.17] The third term in the brackets is a correction for centrifugal action. For one belt,

$$\text{(17.5)} \qquad \text{Rated hp} = \left[a\left(\frac{10^3}{v_m}\right)^{0.09} - \frac{c}{K_d D_1} - e\frac{v_m^{\,2}}{10^6}\right]\frac{v_m}{10^3},$$

where a, c, and e are constants for a particular belt section, D_1 is the pitch diameter of the smaller sheave, K_d is a small-diameter factor, obtained from Table 17.4 for the given velocity ratio, and v_m fpm is the belt speed.[17.11]

The most commonly used V-belt sizes in general industrial applications are designated by letters A, B, C, D, E, each with standard, nominal, cross-sectional dimensions b and t, Fig. 17.13, as given in Fig. 17.14. Premium belts with a greater capacity are available. A narrower design, designated 3V ($b = \frac{3}{8}$), 5V ($b = \frac{5}{8}$), and 8V ($b = 1$) is gaining favor, and there are other special-purpose belts. Equation (17.5) for one belt with a B section is

$$\text{(l)} \qquad \text{Rated hp} = \left[4.737\left(\frac{10^3}{v_m}\right)^{0.09} - \frac{13.962}{K_d D_1} - 0.0234\frac{v_m^{\,2}}{10^6}\right]\frac{v_m}{10^3},$$

and values of a, c, and e for A, C, and D sections are given in Table 17.3 below the columns of standard lengths.

Manufacturer's catalogs have voluminous tables with the rated horsepowers already calculated, but we cannot spare the space. In any case, the first step is to compute the *design horsepower* by mutliplying the horsepower to be transmitted (or the nameplate horsepower on the motor) by a proper service factor N_{sf}. Service factors are a consequence of engineering experience and judgments; those in Table 17.7 may be used here.

$$\text{(m)} \qquad \text{Design hp} = N_{sf}\ (\text{transmitted hp}).$$

With the design horsepower and the rpm of the smaller sheave, enter Fig. 17.14 with these values and decide upon the cross section to be used, as suggested by the dotted lines, which indicate a C-section. If the point is close to a dividing line, either section is reasonable; design for both and decide upon which to use on the basis of cost or other consideration.

* At least one manufacturer claims that the manufacturing tolerance on the length of his belts is so close that special selection of matched lengths is not necessary.

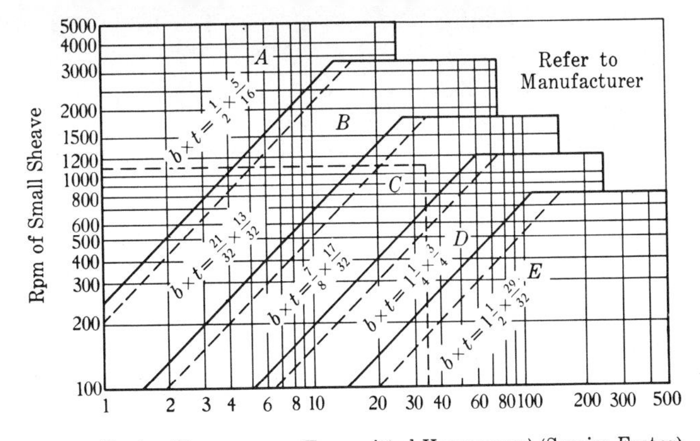

FIGURE 17.14 Belt Sections from Horsepower and Speed. The design horsepower is equal to the transmitted horsepower times the service factor. The solid diagonal lines are from ASA;[17.11] the dotted diagonal lines closely agree with more than one manufacturer's catalog.[17.13]

The horsepower from the rating equations must be corrected for length of belt (K_L) and arc of contact (K_θ); that is

(n) $$\text{Adjusted rated hp} = K_\theta K_L \text{ (rated hp)},$$

where K_θ is a correction factor for an arc of contact other than 180°. Since the arc of contact is a function of $(D_2 - D_1)/C$, as equation (17.4) shows, this value is convenient to use in choosing K_θ from Table 17.5, the easy way to make the correction; D_2 = larger diameter, D_1 = smaller diameter, C = center distance. This factor K_θ is basically a correction of the net tension ratio equation (**d**), § 17.3, because the permissible value of this ratio decreases as θ decreases.[1.15] If the sheave sizes are not known, they must be assumed, as well as an approximate center distance. If possible, stay above the minimum sheave diameters D_s given at the top of the columns of Table 17.3.

Short belts are flexed more often about the sheaves than long belts. The factor K_L in equation (**n**) corrects for the effect of the length and is obtained from Table 17.6. The pitch length of a belt is computed from equation (17.3), § 17.7; choose a standard length from Table 17.3. Or a length of belt may be known for which the center distance is desired; in this case, solve for C from equation (17.3). With $B = 4L - 6.28(D_2 + D_1)$, we get

(o) $$C = \frac{B + \sqrt{B^2 - 32(D_2 - D_1)^2}}{16}.$$

If the size of the smaller sheave is assumed and the velocity ratio m_ω is known, choose a minimum center distance as[17.13]

(p) $$C = \frac{D_1 + D_2}{2} + D_1 \qquad \text{or} \qquad C = D_2, \qquad \text{[MINIMUMS]}$$

whichever is larger. The center distance is made adjustable so that the belts

TABLE 17.3 STANDARD V-BELT LENGTHS; HORSEPOWER CONSTANTS[17.13]

See equation (l) for the constants a, c, e, for a B belt.
Minimum D_s is the smallest sheave pitch diameter that should be used with that section. If a smaller sheave is used, short belt life should be expected; L in inches.

A SECTION		B SECTION		C SECTION		D SECTION	
Min. D_s = 3 *in.*		*Min.* D_s = 5.4 *in.*		*Min.* D_s = 9 *in.*		*Min.* D_s = 13 *in.*	
Belt No.	*Pitch Length*	*Belt No.*	*Pitch Length*	*Belt No.*	*Pitch Length*	*Belt No.*	*Pitch Length*
A26	27.3	B35	36.8	C51	53.9	D120	123.3
A31	32.3	B38	39.8	C60	62.9	D128	131.3
A35	36.3	B42	43.8	C68	70.9	D144	147.3
A38	39.3	B46	47.8	C75	77.9	D158	161.3
A42	43.3	B51	52.8	C81	83.9	D173	176.3
A46	47.3	B55	56.8	C85	87.9	D180	183.3
A51	52.3	B60	61.8	C90	92.9	D195	198.3
A55	56.3	B68	69.8	C96	98.9	D210	213.3
A60	61.3	B75	76.8	C105	107.9	D240	240.8
A68	69.3	B81	82.8	C112	114.9	D270	270.8
A75	76.3	B85	86.8	C120	122.9	D300	300.8
A80	81.3	B90	91.8	C128	130.9	D330	330.8
A85	86.3	B97	98.8	C144	146.9	D360	360.8
A90	91.3	B105	106.8	C158	160.9	D390	390.8
A96	97.3	B112	113.8	C173	175.9	D420	420.8
A105	106.3	B120	121.8	C180	182.9	D480	480.8
A112	113.3	B128	129.8	C195	197.9	D540	540.8
A120	121.3	B144	145.8	C210	212.9	D600	600.8
A128	129.3	B158	159.8	C240	240.9	D660	660.8
		B173	174.8	C270	270.9		
		B180	181.8	C300	300.9		
		B195	196.8	C330	330.9		
		B210	211.8	C360	360.9		
		B240	240.3	C390	390.9		
		B270	270.3	C420	420.9		
		B300	300.3				
Rated hp. Constants: $a = 2.684$ $b = 5.326$ $e = 0.0136$				Rated hp. Constants: $a = 8.792$ $c = 38.819$ $e = 0.0416$		Rated hp. Constants: $a = 18.788$ $c = 137.7$ $e = 0.0848$	

SOME STOCK SHEAVE DIAMETERS

A	B	C	D
Varies by 0.2 in. from 2.6 through 5.2 in; then by 0.4 to 6.4; then 7, 8.2, 9, 10.6, 12, 15, 18 in.	Varies by 0.2 in. to 4.6; then 5, 5.2, 5.4, 5.6, 6, 6.4, 6.8, 7.4, 8.6, 9.4, 11, 12.4, 15.4, 18.4, 20, 25, 30, 38 in.	Varies by 0.5 in. from 7 to 11 in.; then by 1 to 14; by 2 to 20; then 24, 30, 36, 44, 50 in.	Varies by 0.5 in. from 13 to 16 in.; then 18, 22, 27, 33, 40, 48, 58 in.

TABLE 17.4 SMALL-DIAMETER FACTORS K_d[17.11]

D_2/D_1	K_d
1.000–1.019	1.00
1.020–1.032	1.01
1.033–1.055	1.02
1.056–1.081	1.03
1.082–1.109	1.04
1.110–1.142	1.05
1.143–1.178	1.06
1.179–1.222	1.07
1.223–1.274	1.08
1.275–1.340	1.09
1.341–1.429	1.10
1.430–1.562	1.11
1.563–1.814	1.12
1.815–2.948	1.13
2.949 and over	1.14

TABLE 17.5 ARC-OF-CONTACT FACTORS, K_θ[17.13]

$\frac{D_2 - D_1}{C}$	K_θ	
	VV	*V-Flat*
0.00	1.00	0.75
0.10	0.99	0.76
0.20	0.97	0.78
0.30	0.96	0.79
0.40	0.94	0.80
0.50	0.93	0.81
0.60	0.91	0.83
0.70	0.89	0.84
0.80	0.87	0.85
0.90	0.85	0.85
1.00	0.82	0.82
1.10	0.80	0.80
1.20	0.77	0.77
1.30	0.73	0.73
1.40	0.70	0.70
1.50	0.65	0.65

TABLE 17.6 LENGTH CORRECTION FACTORS K_L[17.11]

STD. LENGTH DESIGNATION	BELT CROSS SECTION				
	A	*B*	*C*	*D*	*E*
26	0.81	..	..	..	..
31	0.84	..	..	..	..
35	0.87	0.81	..	..	..
38	0.88	0.83	..	..	..
42	0.90	0.85	..	..	..
46	0.92	0.87	..	..	..
51	0.94	0.89	0.80	..	..
55	0.96	0.90	..	..	..
60	0.98	0.92	0.82	..	..
68	1.00	0.95	0.85	..	..
75	1.02	0.97	0.87	..	..
80	1.04	..	..	..	..
81	..	0.98	0.89	..	..
85	1.05	0.99	0.90	..	..
90	1.06	1.00	0.91	..	..
96	1.08	..	0.92	..	..
97	..	1.02	..	..	..
105	1.10	1.04	0.94	..	..
112	1.11	1.05	0.95	..	..
120	1.13	1.07	0.97	0.86	..
128	1.14	1.08	0.98	0.87	..
144	..	1.11	1.00	0.90	..
158	..	1.13	1.02	0.92	..
173	..	1.15	1.04	0.93	..
180	..	1.16	1.05	0.94	0.91
195	..	1.18	1.07	0.96	0.92
210	..	1.19	1.08	0.96	0.94
240	..	1.22	1.11	1.00	0.96
270	..	1.25	1.14	1.03	0.99
300	..	1.27	1.16	1.05	1.01
330	..	..	1.19	1.07	1.03
360	..	..	1.21	1.09	1.05
390	..	..	1.23	1.11	1.07
420	..	..	1.24	1.12	1.09
480	..	..	..	1.16	1.12
540	..	..	..	1.18	1.14
600	..	..	..	1.20	1.17
660	..	..	..	1.23	1.19

TABLE 17.7 SERVICE FACTORS, N_{sf}

Add 0.2 to the values given for each of the following conditions: continuous (over 16 hr/day) service; wet environment; idler in drive; speed-up drives. Subtract 0.2 if the operation is quite intermittent or seasonal. These factors represent compromises of those found in the literature, with the most attention to Refs. *(17.1, 17.11)*, and are primarily for V-belts; but they may serve as a guide for other transmission elements.

DRIVEN MACHINES	DRIVING MACHINES	
	Electric Motors *AC Split Phase* *AC Normal Torque Squirrel Cage, and Synchronous* *DC Shunt Wound* *Water Wheels* *Turbines, Steam and Water* *Internal Combustion Engines (Hydraulic Drive)*	*Electric Motors* *AC Single Phase Series Wound* *AC High Torque or High Slip* *AC Slip Ring* *AC Repulsion Induction* *AC Capacitor* *DC Compound Wound* *Steam Engines and Line Shafts* *Clutch on Driver or Driven Shaft*
Agitators, liquid Cam Cutters Conveyors, package Drill presses, Lathes Screw machines Small fans to 10 hp	1.1	1.2
Compressors and blowers, (rotating) Conveyors, ore, sand Generators Line Shafts Machine tools (other) Printing machinery Pumps (rotating) Shears	1.2	1.4
Ball mills Beaters (paper) Circular saws Compressors (recip.) Conveyors, bucket, apron, screw, drag Crushers, jaw, etc. Hammer mills Pulverizers Pumps (recip.) Punches, presses Propellers Revolving screens Tube mills	1.4	1.6
Hoists Mine fans Positive blowers Spinning frames Tumbling barrels Twisters (textile)	1.6	1.8

can be mounted into the grooves without harmful stretching and so that the initial tension can be maintained. Sometimes idler pulleys are used on open V-belt drives. Observe that if possible; not only standard belt lengths are used, but also standard sheave sizes, approaching as closely as possible the desired velocity ratio; some standard sizes are given at the bottom of Table 17.3. Of course, the specified sheave sizes and center distance must match a standard belt length. Now, the number of belts needed is

$$\textbf{(q)} \qquad \text{No. of belts} = \frac{\text{Design hp, equation } \textbf{(m)}}{\text{Adjusted rated hp, equation } \textbf{(n)}}.$$

Use the next larger whole number. More detail on a stress analysis of a V-belt is found in Ref. *(17.17)*.

17.18 V-FLAT DRIVES AND OTHERS. V-belts can be used on nonparallel shafts, if the law of belting (§ 17.16) is satisfied, as seen in Fig. 17.15 where a guide pulley is required (in the rear at top of illustration). More commonly advantageous is the V-flat drive, in which the large wheel is a flat pulley (or flywheel), Fig. 17.16. Not only is a plain pulley less expensive than a grooved sheave, it often happens that an existing flywheel or pulley may be utilized in a conversion job. The pulley face must be wide

FIGURE 17.15 V-Belts on Angled Shafts. (Courtesy Gates Rubber Co., Denver).

FIGURE 17.16 V-Flat Drive. Air compressor drive: 50 hp at 855 rpm of 13-in. motor sheave; 54-in. flywheel; 8 D-section belts. (Courtesy Gates Rubber Co., Denver).

enough to accommodate the required number of belts. The power transmitting capacity of V-flat drives may be computed as explained for VV drives except that there is a change of arc correction factor K_θ, Table 17.5; this difference in K_θ is accounted for by the fact that the net tension ratio, equation (**d**), § 17.3, should be less when the belts run over a flat pulley.[1.15]

Double V-belts, designed to flex in both directions, are available; they can be used for driving a sheave by the "back" side for instance. Also wide belts with multiple longitudinal V-grooves that run with matching grooves on the sheave (sometimes called *ribbed belts*) have been developed to serve the same purposes as multiple V-belts. V-belts with steel-cable tension members can be obtained for high-capacity needs.

The operating temperature of V-belts should be less than 200°F, preferably less than 160°F. V-belts that eliminate static electricity for use in explosive atmospheres (flour mills, for example) are available. Finally, the service factor for speed-up drives should be increased over the usual values.

17.19 VARIABLE-SPEED TRANSMISSIONS. Variable-speed mechanisms play an important and indispensible role in modern industry. They are widely employed to regulate feeds and speeds, for example: annealing furnaces, assembly conveyors, automatic welders, cement kilns, inspection tables, metering systems, printing presses, pumps, stokers, traveling bread ovens, and wire-coiling machines. Sheaves with movable sides are used with V-belts to give a variable-speed adjustment. In Fig. 17.17, let the upper shaft A drive the lower shaft B. The setting is for high speed on the driven shaft B (small radius at B, large one at A). When the speed control operates via the member C, it can move the sides of the lower sheaves toward each other, causing the V-belt to climb to a larger radius, and at the same time separate the sides of the upper sheaves, causing the belt to move

FIGURE 17.17 Variable-speed V-Drive. (Courtesy Allis-Chalmers Mfg. Co., Milwaukee).

to a smaller radius. These movements are simultaneous, the belt fitting the sheaves at any setting.

The Reeves variable-speed transmission, Fig. 17.18, operates on the same principle with movable conical disks which can be controlled manually or automatically. On each wide side of a rubber belt are bolted hardwood blocks, as seen in Fig. 17.18, whose ends are beveled and tipped with leather in order to improve the coefficient of friction. A variety of variable-speed mechanisms have been developed, many of which operate on the principle of adjustable V-sheaves as described above. Some drives of this kind are integrally combined with motors to give in one unit a standard motor, a variable-speed control, and a speed reducer.

17.20 TOOTHED BELTS.

These belts, Fig. 17.19, are called *timing belts*, and have other trade names. Since the usual tension members are steel cables, they stretch very little under load and in service; therefore, the initial tension may be low (low bearing loads) and tensioning devices are not necessary. The backing and teeth are made of neoprene and covered with a nylon duck facing. Since the drive is through the teeth, motions of different shafts can be synchronized, as is so often necessary. Other notable characteristics include: large capacity in small space, can be designed for exceptionally light weight by using nonferrous pulleys, quiet operation at low speeds, no lubrication needed, can tolerate low arc of contact (a minimum of 6 teeth in contact is required for rated power), there is no chordal action (as described for roller chains below); but they do transmit shock loads; if bent over too small pulleys, fatigue will occur.[10.5,17.14] They have been used up to 600 hp. Centrifugal action at high speed reduces

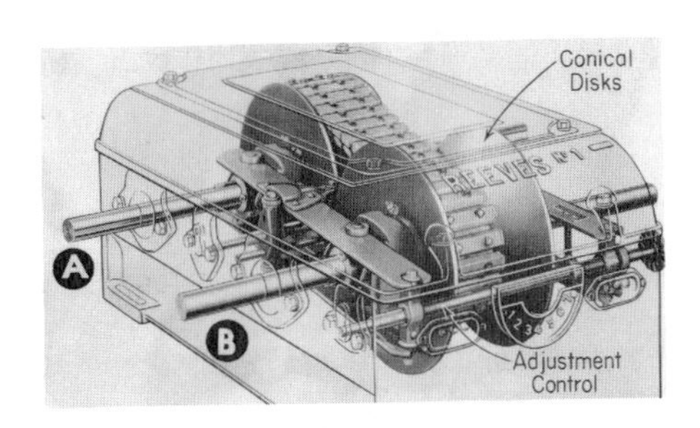

FIGURE 17.18 Reeves Variable-speed Transmission. Constant-speed shaft is A. Adjustment in illustration is for maximum speed of driven shaft B. (Courtesy Reeves Pulley Co., Columbus, Ind.).

FIGURE 17.19 Timing Belt. (Courtesy United States Rubber Co., N.Y.C.).

the force available for transmitting power [equation (**d**), § 17.3], so that for the $\frac{7}{8}$- and $1\frac{1}{4}$-in.-pitch belts, speeds above 8000 fpm are generally impracticable; higher speeds may be utilized with the small pitches. Since these belts are made endless and run on toothed wheels designed for them, one has the problem of adjusting pulley sizes, center distance, and belt length to accommodate stock parts if possible.

Although toothed belts should undoubtedly be considered in any drive design, we cannot afford the space to reproduce the catalog instructions.[17.14]

17.21 ROLLER CHAIN DRIVES.

The most widely used of the various types of chain power-drives is a roller chain, whose principal parts are named in Fig. 17.20. The forces acting on the chain are much the same as those for the other drives of this chapter, except that significant impact (dynamic) load may occur when a roller makes contact with a sprocket tooth and except that inertia forces arise because of ***chordal action,*** described next.

The sprocket in Fig. 17.21 has much fewer teeth than actual sprockets usually have in order to accentuate the chordal phenomenon. In Fig. 17.21(a), the roller A has just seated and the center line of the chain is at a radius r_s. This radius r_s is smaller than the radius r after a rotation through an angle θ, as shown in Fig. 17.21(b). If we assume that this sprocket drives at a constant angular speed of n rpm, the speed of the center line of the chain changes from $v = 2\pi r_s n$ to $v = 2\pi rn$ and back to the lower speed during every cycle of tooth engagement. We recall that change of velocity means an acceleration, and acceleration means force, in accordance

FIGURE 17.20 Roller Chain. (Courtesy Link-Belt Co., Chicago).

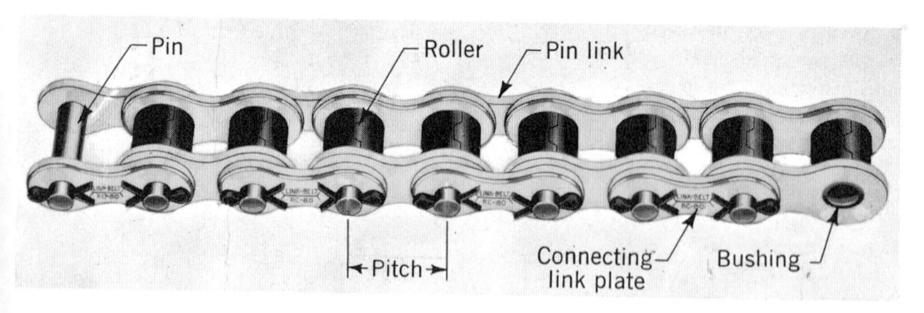

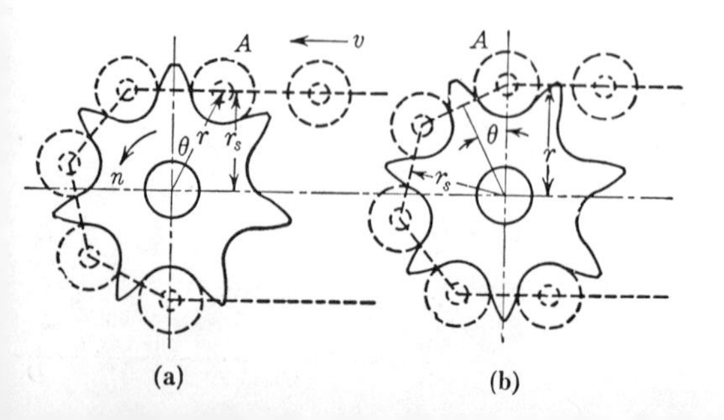

FIGURE 17.21 Chordal Action. The angle $\theta = 180/N_t$ deg.; $\sin\theta = P/(2r)$, or pitch diameter is $D = P/\sin\theta$.

with Newton's law ($F = ma$). Moreover, note the vertical accelerations, because the chain naturally does some bouncing with this changing radius. Since $r_s = r \cos \theta$, the change of radius is $r(1 - \cos \theta)$ and is seen to decrease as the number of teeth increases, thus reducing chordal action, which becomes almost negligible when there are 25 or more teeth.

For very slow speeds, the recommended minimum number of teeth on the smaller sprocket is $N_{min} = 12$; for slow speeds, $N_{min} = 17$; for moderate speeds, $N_{min} = 21$; for high speeds, $N_{min} = 25$; for speed increasing drives, $N_{min} = 23$.[17.15] With odd tooth numbers on the smaller sprocket and an even number of pitches in the chain, the frequency of contact between a particular tooth and a particular roller is a minimum, presumably better distributing the wear.

As for a belt, the centrifugal forces induce a tension in a chain of $K = wv_s^2/g_o$, equation **(e)**, § 17.3, and therefore the power that can be transmitted for a particular maximum tension F_1 ($F_2 \approx 0$ for chains) increases with speed to a peak and then decreases. Additionally for chains, the phenomenon that results in failure changes as the speed increases. At lower speeds, Fig. 17.22, failure is more likely to occur by fatigue of the link plate AB; at higher speeds, roller impact and joint \ ear will limit capacity BC; finally, a point is reached where the capacity drops rapidly to zero when the load is great enough to break down the joint lubrication. The ratings, as in Table 17.9, are for a service factor of 1 and an expected life of 15,000 hr. (where the chain length has presumably increased 3 % maximum), and are predicated on adequate lubrication, the requirements being designated as Type I, Type II, etc. Type I ($v_{max} = 300$ fpm) is manual, oil being applied periodically with a brush, or can; in Type II ($v_{max} = 1300$ fpm), oil is supplied from a drip lubricator to link plate edges; Type III ($v_{max} = 2300$ fpm) is an oil bath or oil slinger disk, but a long length of chain should not be immersed; in Type IV, oil is pumped and directed to the inside of the lower strand (Fig. 17.26). Each size chain has its own speed limitations, the limits for Type III lubrication being given in Table 17.8. The linear (chain) speed may be computed from

$$v_m = \pi\left(\frac{D}{12}\right)n = \frac{PN_t n}{12} \text{ fpm},$$

where D in. is the pitch diameter of the sprocket whose angular speed is n rpm, P in. is the pitch, N_t is the number of teeth in the sprocket that runs at n rpm; PN_t in. is the circumference of the sprocket. The rating tables are

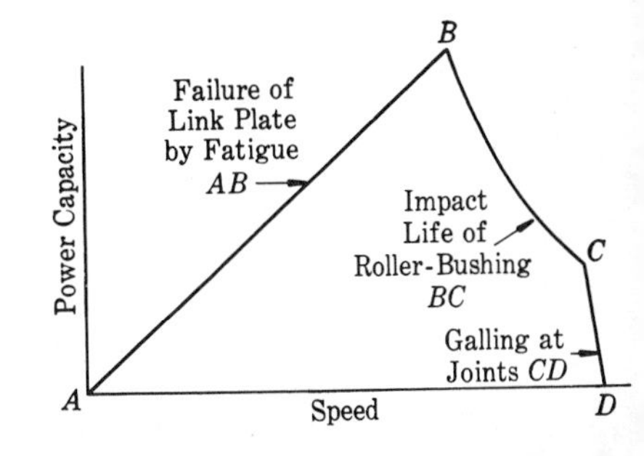

FIGURE 17.22 Capacity vs. Speed [after Ref. (17.21)].

FIGURE 17.23 Multiple-chain Drive. Total of 9 strands of $1\frac{1}{2}$-in. pitch, transmitting 300 hp. (Courtesy Whitney Chain Co., Hartford, Conn.).

entered with a design horsepower = (N_{sf})(nominal transmitted hp), or the design horsepower divided by the proper factor for multiple chain widths (Fig. 17.23), given in the heading of Table 17.9. Since the service factors of Table 17.7 are in general conservative as compared with the recommendations of the chain catalogs, their use is permissible here. Extra capacity results in longer life, which may be desired.

The equations used to compute the rated power of the tables are:[17.33]

(r) $$\text{hp} = 0.004N_{ts}^{1.08}n^{0.9}P^{3-0.07P} \quad \text{and} \quad \text{hp} = K_r\left(\frac{100N_{ts}}{n}\right)^{1.5}P^{0.8},$$

[LINK PLATE FATIGUE] [ROLLER-BUSHING IMPACT]

whichever is smaller; N_{ts} = number of teeth on the smaller sprocket, n = rpm of the smaller sprocket, P in. = chain pitch, hp is treated as explained for the table values, and it is assumed that the limiting speed for galling is not exceeded; $K_r = 17$ for all chain numbers 40 to 240, *except* for No. 41, $K_r = 3.4$, for Nos. 25 and 35 chains, $K_r = 29$.

The center distance for chain drives may of course be relatively short, but a minimum wrap of 120° is desirable; this condition is inevitably met when $m_\omega < 3$. An average good center distance would be $D_2 + D_1/2$, where D_2 is the pitch diameter of the larger sprocket, D_1 of the smaller. The approximate length of chain is

(s) $$L \approx 2C + \frac{N_1 + N_2}{2} + \frac{(N_2 - N_1)^2}{40C} \text{ pitches},$$

where C is in pitches (may contain a fraction). The length should be an even number of pitches to avoid using an *offset link*. There is the usual matter of adjusting chain length, center distance, and sprocket sizes so that everything fits. Contrary to belting practice, the slack side of the chain is preferred on the bottom of horizontal drives, especially for long centers, because, if the slack strand is on top, the strands may touch after the chain has lengthened in service. Lengthening occurs because of wear in the joints.

While chains of special materials, for example nonmagnetic or corrosion-resistant metals, are available, the usual materials will be heat-treated alloy steels. Good practices include: pins and bushings carburized

FIGURE 17.24 Improving Fatigue Strength.

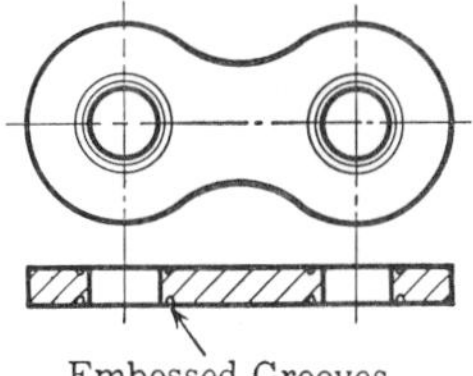

and case-hardened for wear resistance; shot peened pins, rollers, link plates, and bushings; pressing balls through the holes of the link plates for cold working to improve fatigue strength; and pressing around holes as suggested by Fig. 17.24. When necessary, the sprocket teeth are hardened (induction, flame, carburizing).

For a single reduction, the velocity ratio should be kept below about 5, but as high as 10 has been used. Reference *(2.1)* has pertinent information on wear in chains.

17.22 EXAMPLE—ROLLER CHAIN DRIVE. A 15-hp motor, running at 1180 rpm, is to drive a 2-stage (reciprocating) hydrogen compressor. The velocity ratio is to be about $m_\omega = 3.2$; service is nominally 8 hr/day. It has been decided to use a $\frac{1}{2}$-in. pitch (No. 40) chain (because this is the table at hand). Determine the number of strands and sprocket sizes.

Solution. Use the recommended $N_1 = 21$ teeth for the smaller sprocket, and $N_{sf} = 1.4$ from Table 17.7. The design horsepower is then $(1.4)(15) = 21$ hp. Assume 4 strands. Divide by the factor 3.3, from Table 17.9, for 4 strands and get $21/3.3 = 6.36$ hp/strand, rated value needed. From Table 17.9 for $n_1 = 1180$ rpm, find the table value for 21 teeth to be 8 hp; therefore 4 strands is the answer. (A check for 3 strands shows that the chain would be about 4% overloaded. If a life a little shorter than 15,000 hr. should be satisfactory, 3 strands would do.) For

$$v_m = \frac{PN_s n}{12} = \frac{(0.5)(21)(1180)}{12} = 1030 \text{ fpm},$$

compared with $v_{max} = 1300$ fpm for Type II lubrication, we see that Type II is satisfactory. For $m_\omega = 3.2$, $N_2 = 3.2 \times 21 = 67.2$, say 67 if this size is available from stock; $D_1 = (0.5)(21)/\pi = 3.34$ in., $D_2 = (0.5)(67)/\pi = 10.35$ in., approximately. Without conflicting factors, use the recommended center distance; this computation and that for the length of chain is left to the student.

17.23 INVERTED-TOOTH CHAINS. Inverted-tooth chains, commonly called silent chains, are widely used for power transmission under much the same conditions as are roller chains, so that many of the remarks

TABLE 17.8 PROPERTIES OF STANDARD ROLLER CHAINS (REGULAR)

*** Rollerless. † Limiting chainspeed with oil bath (Type III) lubrication.[17.16] The fatigue strength of the chain may be taken as $F_u/4$.**

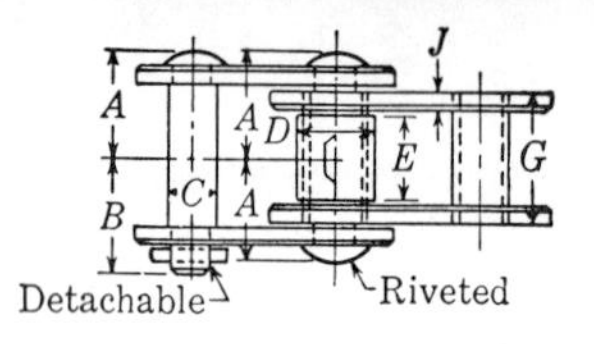

CHAIN NO.	PITCH	AVERAGE ULTIMATE STRENGTH lb.	APPROX. w lb. per ft.	LIMITING SPEED fpm	A	B	C	D	E	J
25*	$\frac{1}{4}$	875	0.09	3500	0.155	0.19	0.0905	0.130	$\frac{1}{8}$	0.030
35*	$\frac{3}{8}$	2100	0.21	2800	0.231	0.283	0.141	0.200	$\frac{3}{16}$	0.050
41	$\frac{1}{2}$	2000	0.26	2300	0.26	0.37	0.141	0.306	$\frac{1}{4}$	0.050
40	$\frac{1}{2}$	3700	0.42	2300	0.314	0.357	0.156	$\frac{5}{16}$	$\frac{5}{16}$	0.060
50	$\frac{5}{8}$	6100	0.68	2000	0.398	0.434	0.200	0.400	$\frac{3}{8}$	0.080
60	$\frac{3}{4}$	8500	1.00	1800	0.489	0.574	0.234	$\frac{15}{32}$	$\frac{1}{2}$	0.094
80	1	14,500	1.73	1500	0.615	0.741	0.312	$\frac{5}{8}$	$\frac{5}{8}$	0.125
100	$1\frac{1}{4}$	24,000	2.5	1300	0.754	0.882	0.375	$\frac{3}{4}$	$\frac{3}{4}$	0.156
120	$1\frac{1}{2}$	34,000	3.69	1200	0.940	1.116	$\frac{7}{16}$	$\frac{7}{8}$	1	0.187
140	$1\frac{3}{4}$	46,000	5.0	1100	1.022	1.210	$\frac{1}{2}$	1	1	0.219
160	2	58,000	6.5	1000	1.228	1.383	0.562	$1\frac{1}{8}$	$1\frac{1}{4}$	0.250
180	$2\frac{1}{4}$	76,000	9.06	950	1.362	1.718	0.687	1.406	$1\frac{13}{32}$	0.281
200	$2\frac{1}{2}$	95,000	10.65	900	1.546	1.827	0.781	$1\frac{9}{16}$	$1\frac{1}{2}$	0.312

on roller chains also apply to inverted-tooth chains. Horsepower ratings are given in manufacturers' catalogs. The regular type of inverted-tooth chains has links whose contact faces are straight, Fig. 17.25, and these faces contact a straight-tooth profile on the sprocket.

The Link-Belt construction of the joint consists of case-hardened bushings made in two parts that extend for the width of chain, with a case-hardened pin joining the links. The Morse chain has a joint made up of the two rockers seen in Fig. 17.25; as the joint works, these rockers roll on one another. Morse also makes a premium silent-chain drive, called Hy-Vo, for which the sprocket tooth has an involute profile; this plan reduces the chordal effect materially, permitting quite high speeds when there are 25 or more teeth on the smaller sprocket (over 13,000 fpm for $1\frac{1}{2}$-in. pitch). Some means must be provided to keep the chain on the sprockets; in Fig. 17.26, this is done with central guide links running in grooves on

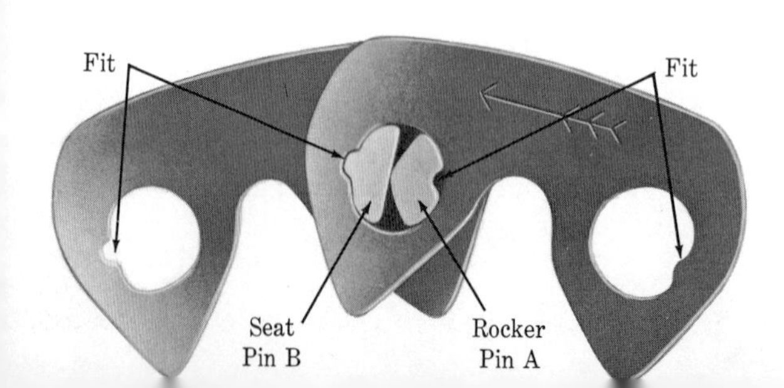

FIGURE 17.25 Links for Inverted-tooth Chain. As the chain bends around the sprocket, the rocker pin rolls on the seat pin. (Courtesy Morse Chain Co., Ithaca).

TABLE 17.9 RATED HORSEPOWER OF SINGLE-STRAND, No. 40 ROLLER CHAIN

Courtesy Hewitt-Robins, Inc. For a service factor of unity. Multiply by the following factors for multiple strands: 1.7 for 2 strands; 2.5 for 3 strands; 3.3 for 4 strands. Especially for the higher speeds in the Type IV lubrication region, consider inverted-tooth chains.

No. of Teeth Small Sprocket	Revolutions Per Minute—Small Sprocket																		
	50	200	400	600	900	1200	1800	2400	3000	3500	4000	4500	5000	5500	6000	6500	7000	7500	8000
	TYPE I			TYPE II				TYPE III					TYPE IV						
11	0.23	0.80	1.50	2.16	3.11	4.03	4.66	3.03	2.17	1.72	1.41	1.18	1.01	0.87	0.77	0.68	0.61	0.55	0.50
12	0.25	0.88	1.65	2.37	3.42	4.43	5.31	3.45	2.47	1.96	1.60	1.34	1.15	1.00	0.87	0.77	0.69	0.62	0.57
13	0.28	0.96	1.80	2.59	3.73	4.83	5.99	3.89	2.79	2.21	1.81	1.52	1.29	1.12	0.98	0.87	0.78	0.70	0.64
14	0.30	1.04	1.95	2.80	4.04	5.23	6.70	4.35	3.11	2.47	2.02	1.69	1.45	1.25	1.10	0.98	0.87	0.79	0.71
15	0.32	1.12	2.10	3.02	4.35	5.64	7.43	4.82	3.45	2.74	2.24	1.88	1.60	1.39	1.22	1.08	0.97	0.87	
16	0.35	1.20	2.25	3.24	4.66	6.04	8.18	5.31	3.80	3.02	2.47	2.07	1.77	1.53	1.34	1.19	1.07	0.96	
17	0.37	1.29	2.40	3.45	4.98	6.45	8.96	5.82	4.17	3.31	2.71	2.27	1.94	1.68	1.47	1.31	1.17	1.05	
18	0.39	1.37	2.55	3.68	5.30	6.86	9.76	6.34	4.54	3.60	2.95	2.47	2.11	1.83	1.60	1.42	1.27	1.15	
19	0.42	1.45	2.71	3.90	5.62	7.27	10.5	6.88	4.92	3.91	3.20	2.68	2.29	1.98	1.74	1.54	1.38	1.25	
20	0.44	1.53	2.86	4.12	5.94	7.69	11.1	7.43	5.31	4.22	3.45	2.89	2.47	2.14	1.88	1.67	1.49	1.34	
21	0.46	1.62	3.02	4.34	6.26	8.11	11.7	7.99	5.72	4.54	3.71	3.11	2.66	2.30	2.02	1.79	1.60	1.45	
22	0.49	1.70	3.17	4.57	6.58	8.52	12.3	8.57	6.13	4.87	3.98	3.34	2.85	2.47	2.17	1.92	1.72		
23	0.51	1.78	3.33	4.79	6.90	8.94	12.9	9.16	6.55	5.20	4.26	3.57	3.05	2.64	2.32	2.06	1.84		
24	0.54	1.87	3.48	5.02	7.23	9.36	13.5	9.76	6.99	5.54	4.54	3.80	3.25	2.81	2.47	2.19	1.96		
25	0.56	1.95	3.64	5.24	7.55	9.78	14.1	10.4	7.43	5.89	4.82	4.04	3.45	2.99	2.63	2.33			
28	0.63	2.20	4.11	5.93	8.54	11.1	15.9	12.3	8.80	6.99	5.72	4.79	4.09	3.55	3.11	2.76			
30	0.68	2.38	4.43	6.38	9.20	11.9	17.2	13.6	9.76	7.75	6.34	5.31	4.54	3.93	3.45				
32	0.73	2.55	4.75	6.85	9.86	12.8	18.4	15.0	10.8	8.54	6.99	5.86	5.00	4.33	3.80				
35	0.81	2.80	5.24	7.54	10.9	14.1	20.3	17.2	12.3	9.76	7.99	6.70	5.72	4.96					
40	0.93	3.24	6.05	8.71	12.5	16.3	23.4	21.0	15.0	11.9	9.76	8.18	6.99						
45	1.06	3.68	6.87	9.89	14.2	18.5	26.6	25.1	17.9	14.2	11.7	9.76							
50	1.18	4.12	7.70	11.1	16.0	20.7	29.8	29.4	21.0	16.7	13.6								
55	1.31	4.57	8.53	12.3	17.7	22.9	33.0	33.9	24.2	19.2									
60	1.44	5.02	9.37	13.5	19.4	25.2	36.3	38.6	27.6										

Operation at speeds beyond those shown, results in joint galling, regardless of the volume of lubricant applied.

the sprockets. For maximum quietness, use sprockets with 27 or more teeth. Both roller and silent chains may require an idler (or other means) to take up the slack.

17.24 WIRE ROPE. Wire ropes are made from cold-drawn wires that are first wrapped into ***strands;*** the strands are then wrapped into helices about a core or central element, which is usually hemp or pulp,

FIGURE 17.26 Silent-chain Drive. Preferably, there is one stream of oil per inch of chain width. (Courtesy Socony-Vacuum Oil Co., N.Y.C.).

Fig. 17.27. Often, the central element is an independent wire rope core (IWRC), Fig. 17.28(e), which makes the rope much more resistant to crushing. Other factors that may make the IWRC preferred include: high temperature that may ruin a hemp core, its 7.5% greater strength (Table AT 28), and the smaller elongation under load. The rope is made with either: a ***regular lay,*** in which the wires and strands are twisted in opposite directions, Fig. 17.27(a); or ***lang lay,*** in which the wires and strands are twisted in the same direction, Fig. 17.27(b); the wires and strands may form

(a) Regular Lay. These strands are right hand, the wires are wound left hand.

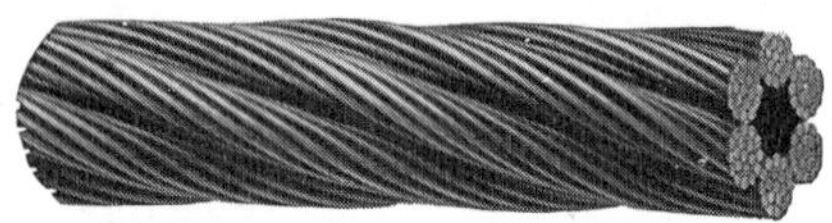

(b) Lang Lay. Both strands and wire are right hand.

FIGURE 17.27 Lay of Wire Rope. (Courtesy John A. Roebling's Sons Co., Trenton, N.J.).

either right-hand or left-hand helices. All wire rope was originally *non-preformed*, the wires and strands being twisted and bent into place, resulting in high stresses in a straight, unloaded rope. At this time, most wire ropes are ***preformed,*** the individual strands having been mechanically shaped ahead of time into the helical configuration they have in the rope. Preformed ropes are more flexible and spool easier.

The size of a wire rope D_r is the diameter of the circle that just contains the rope. In general, the greater the number of wires in a strand, the more flexible the rope; conversely, the fewer wires, the stiffer the rope. Thus, ropes made of small wires are suitable for sharp bends. However, the outside wires are subjected to wear as they rub surfaces (pass over a sheave), and small wires will wear through quicker than large ones. The construction is indicated by two numbers, the first giving the number of strands, the second the number of wires in each strand. For example, a 6×19 wire rope has 6 strands each with 19 wires, Fig. 17.28(b). There are many styles

FIGURE 17.28 Cross Sections of Wire Ropes. (Courtesy Jones & Laughlin Steel Corp., Pittsburgh).

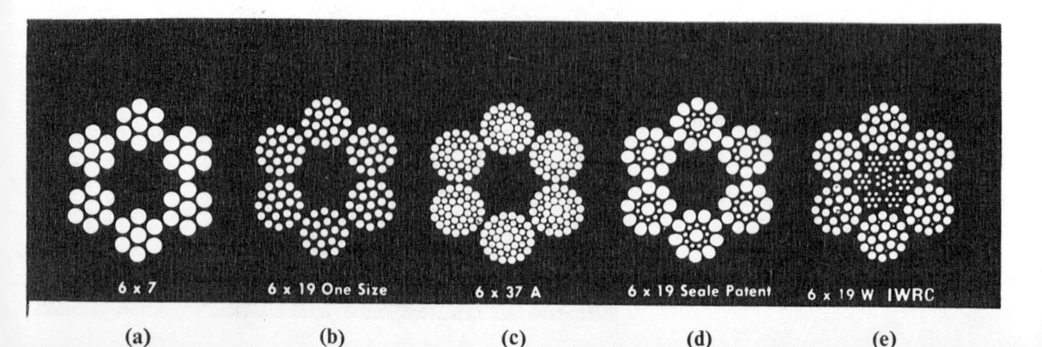

of cross sections, a few of which are seen in Fig. 17.28. The following remarks indicate something of the uses of the different types, any one of which may have a wire rope core (IWRC).

6 × 7, Fig. 17.28(a), being made of heavy wire, provides maximum resistance to abrasion and wear; would be 7 × 7 with IWRC; used for haulages, rigging, guard rails.

6 × 19, Fig. 17.28(b), being a compromise of flexibility and wear resistance, is one of the most popular styles; a good general purpose rope. Special constructions, such as the Seale, Fig. 17.28(d), are designed for good wear by having large wires on the outside, and for good flexibility by having small wires in the inner layer. Uses, including all the various cross sections: scraper and shovel cables, draglines, logging ropes, haulage, hoists.

6 × 37, Fig. 17.28(c), is an extra-flexible rope and therefore useful where abrasion is not severe and where relatively sharp bends must be tolerated. Without other designation, a 6 × 37 rope has wires all the same diameter (as in Table AT 28); not as the 6 × 37 A of Fig. 17.28(c). Used for winch lines, hawsers, overhead cranes, and hoists.

The regular materials for wire rope are high-carbon, with most rope made of improved plow steel (IPS), which has an ultimate tensile strength of s_u between 240 and 280 ksi, the higher values applying to the smaller wires (Chapter 6). Several other grades of steel are used:[17.28] plow steel (PS) ($210 < s_u < 240$ ksi); mild plow steel (MPS) ($180 < s_u < 210$); traction ($180 < s_u < 190$ ksi); iron, with a lower carbon (about 0.1%) content ($s_u < 100$), useful in undemanding situations; very-high-strength (VHS) ($280 < s_u < 340$ ksi), for the premium jobs, about 15% stronger in rope form than IPS; and also used for a reason are galvanized wire ropes in various grades of steel, phosphor bronze, stainless steel.

The strength of the rope is always less than the sum of the strengths of the wires, and is commonly stated in tons of force to break, Table AT 28. Other information in Table AT 28: the weight per foot of length of rope w; the minimum permissible sheave diameter D_s; the desirable minimum sheave diameter; the approximate wire diameter D_w for the style of rope in which *all wires have the same diameter*; the cross-sectional area of metal in each rope A_m; and the modulus of elasticity E_r of the rope (§ 17.25).

17.25 DESIGN CONSIDERATIONS FOR WIRE ROPE.

There are a number of applications where the design will be covered by code, often a legal requirement; for example, elevators. We shall assume that if this is true, the designer will meet the requirements of the code, and give our brief attention to basic considerations. There is of course some tensile force F_t, on the rope due to the loading on it (including inertia loading). In addition, since the rope is often bent about a sheave, more stress is incurred on this account. Use the elastic curve equation $M = EI/r$, in which r is the radius to which a uniform, straight, homogeneous member is bent by the moment M, equate this value of M to that in $M = sI/c$, and get $s = Ec/r$. If this equation is applied to a wire, for which the distance from the neutral axis to the farthest fiber is $c = D_w/2$, bent about a sheave of pitch radius $r = D_s/2$, we get the approximate stress in the wire;

(t)
$$s_b = \frac{Ec}{r} = \frac{ED_w}{D_s} \text{ psi or ksi}$$

where E = modulus of elasticity (3×10^4 ksi for steel)* and approximate values of D_w are found in Table AT 28. This value of s multiplied by the area of metal A_m gives the force on the rope that presumably would produce this stress, called the *equivalent bending load* $F_b = s_b A_m$. If the sheave size is the "desirable diameter" in Table AT 28, this equivalent force may be well cared for by an overall design factor, but of course it increases as the sheave size decreases.

The modulus E_r of the rope is much less than for the material,[17.26] because as the rope is loaded there is considerable sliding of the wires on one another. Thus, to get the extension of the wire rope under load, use E_r in equation (1.3) to estimate

(u)
$$\delta = \frac{sL}{E_r} = \frac{FL}{A_m E_r} \text{ in.,}$$

where L in. is the length of the rope with a uniform stress F/A_m; see Table AT 28.

The Roebling *Handbook*[17.23] suggests minimum design factors of: guys, 3.5; miscellaneous hoisting equipment, 5; haulage ropes, cranes, and derricks, 6; small hoists, 7; hot ladle cranes, 8; when the factor is defined as $N = F_u/F_t$, where F_u is the breaking strength, Table AT 28, and F_t is the maximum working load on the rope (§ 17.26). However, it seems prudent to include the effect of bending the rope in the definition of N, so unless it is stated otherwise in the problems, use

(v)
$$N = \frac{(\text{Breaking strength, } F_u) - (\text{Equivalent bending load, } F_b)}{\text{Tensile force in the rope, } F_t} = \frac{F_u - F_b}{F_t},$$

where the numerator is approximately the available strength left to carry an external load F_t; $F_u = NF_t + F_b$.

The trouble with the conventional design procedure as described above is that it is not in accord with the manner in which most wire ropes fail. It is a static approach, whereas, if the rope is continually flexed, failure would be expected to be a fatigue failure (except as abrasion may wear out

* The reader may, if he wishes, conform to past conventional practice by using the modulus of elasticity E_r of the rope (Table AT 28) in equation (**t**), but since in the ideal model, all wires bend to a radius $D_s/2$, we may as well be theoretically consistent here and use E for the wire material; especially inasmuch as this static approach tends to underestimate the fatigue anyway.

a rope). Drucker and Tachau[17.25] found that the sum of the stresses due to bending, $s_b = ED_w/D_s$, and due to the load, $s = F_t/A_m$, had only a fair correlation with fatigue failure (6 × 7 ropes not included), and that the limit of this total stress was about 60 to 70 ksi. However, they found a better correlation with the bearing pressure per square inch of projected area of the rope on the sheave, $p = 2F_t/(D_r D_s)$, Fig. 17.29, where the contact angle is taken as 180°. Fatigue failure seemed to be well predicted by the ratio p/s_u (plotted against number of bending cycles to failure in Fig. 17.30), where s_u is the ultimate strength of the wires. Introducing a design factor N, dividing both sides of the pressure equation by s_u and solving for $D_r D_s$, we get

$$D_r D_s = \frac{2NF_t}{(p/s_u)s_u}, \tag{w}$$

where an appropriate value of p/s_u is taken from Fig. 17.30. We note that indefinite life may be obtained when $p/s_u \lesseqgtr 0.0015$ and that p/s_u may be much greater than this if only a limited life is needed, but as usual a design factor should be used to be on the safe side. The actual experimental points fall surprisingly close to these curves, so that a large design factor does not appear necessary if operating conditions are good. With minimum and desirable values of D_s in terms of D_r given in Table AT 28, we may quickly investigate the effect of varying D_s with equation (**w**) in the form shown (unless the wire size is involved in F_t, § 17.26).

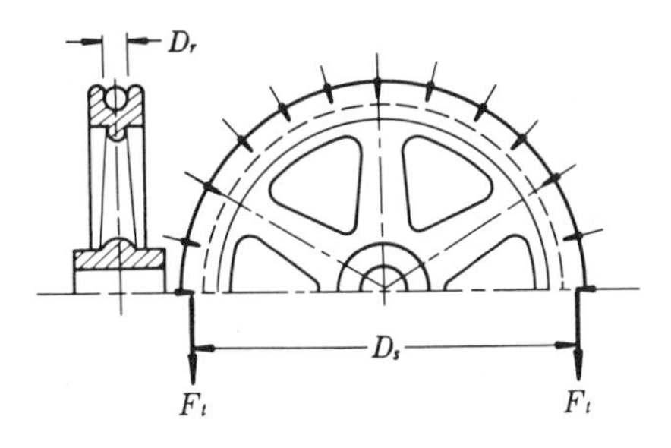

FIGURE 17.29

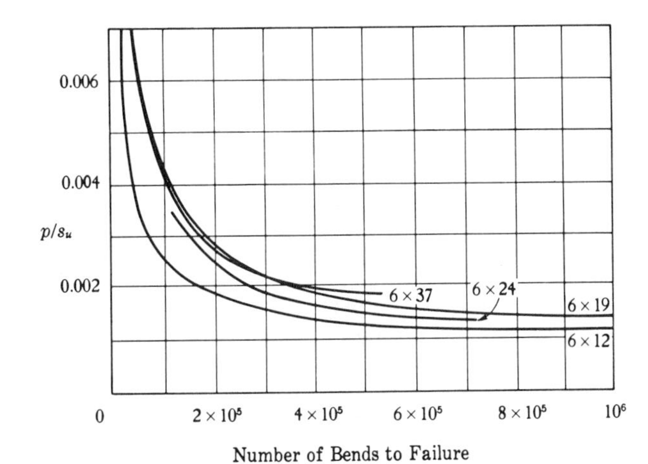

FIGURE 17.30 Pressure Ratio vs. Cycles to Failure. (After Drucker and Tachau.[17.25])

A theoretical and experimental study of the contact stresses, wire on wire, by Starkey and Cress[17.27] confirms the foregoing design criterion in that it would be normal to expect a direct relation between the pressure p and the Hertz contact stress; the Hertz stress appears to be the significant one, rather than tension plus bending. Starkey and Cress postulate a failure theory: the relative motion of the wires, particularly during bending, together with the high contact stresses, causes fretting at points where fatigue cracks were observed to start and propagate.

If corrosion or other unfavorable operating conditions are involved, fatigue life will be shortened. The groove of the sheave should be wide enough not to pinch the rope, which rests on the bottom. In general, the softer the groove, the longer is the rope's life; but the pressure of the rope on the sheave may be too high. To avoid an excessive rate of wear (abrasion), Roebling recommends limiting the pressure p for 6×19 rope as follows: on cast-iron sheave, 500 psi; on cast steel, 900 psi; on manganese steel, 2500 psi.

Since a typical fatigue failure starts with the breaking of individual wires, proper maintenance procedures can obviate a disaster, because the rope is replaced when broken wires are discovered. However, this is expensive and design for fatigue may be the better solution. Of course, there are situations where the cycles of loading are few enough that static strength is sufficient. Loads suddenly applied to a slack rope can snap it in two. The shorter the rope, the less energy it can absorb within its breaking strength ($F\delta/2$ for elastic energy). A good operator is sure that the slack is taken up gradually before applying full power. During manufacture, the rope's core is saturated with a lubricant for the purposes of reducing wear from the wires rubbing on each other and the sheaves and of protecting the rope from rust and corrosion; for best service, it should be kept clean and lubricated from time to time. Recent tests suggest consideration of molybdenum disulfide as a lubricant.

17.26 EXAMPLE—WIRE ROPE FOR MINE HOIST. A skip for a mine shaft weighs 2000 lb. and is to lift a maximum load of 3000 lb. from a depth of 1000 ft. The maximum speed of 20 fps is attained in 5 sec. (a) What sizes of 6×19, IPS rope and sheave should be used for indefinite life and for $N = 1.3$ on the basis of fatigue? (b) What is the design factor on the static basis? (c) What sizes of rope and sheave are required if the number of cycles (bending and unbending) in the desired life is 200,000? What is the corresponding static design factor? (d) What is the elongation of the rope found in (c) if the 3000 lb. load is added while the hoist hangs free at the bottom?

Solution. (a) The acceleration, which introduces inertia effects, depends on the torque characteristics of the motor (or on the brake properties if the maximum loading occurs during stopping as the hoist is moving down); but the conventional approach is to use the average; that is, $a = \Delta v/\Delta t = 20/5 = 4$ fps^2.

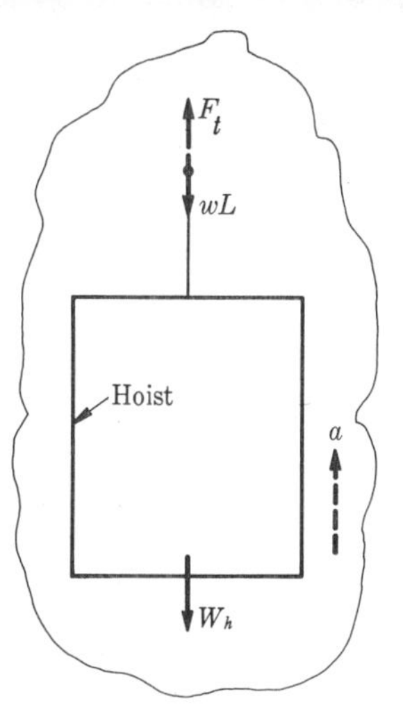

FIGURE 17.31 Free Body of Hoist.

Use kips for convenience. The free-body diagram, Fig. 17.31, shows the load $W_h = 2 + 3 = 5$ kips, the weight of the rope, which is long enough to be significant, as $wL = (1.6)(1000)(10^{-3})D_r^2 = 1.6D_r^2$ kips (Table AT 28), and F_t is the force on the straight rope at the top. Summing forces with the direction of the acceleration (up) as positive, we have $\Sigma F = ma$,

$$F_t - 5 - 1.6D_r^2 = \left(\frac{5 + 1.6D_r^2}{32.2}\right)(4),$$

$$F_t = 5.62 + 1.8D_r^2.$$

Use this value of F_t, the minimum $s_u = 240$ ksi from § 17.24, $N = 1.3$, $p/s_u = 0.0015$ for indefinite life from Fig. 17.30, and equation (**w**), and get

$$D_r D_s = \frac{2NF_t}{(p/s_u)s_u} = \frac{(2)(1.3)(5.62 + 1.8D_r^2)}{(0.0015)(240)}.$$

At this point, one can assume a value of D_s and solve directly for D_r. We see that the larger D_s, the smaller D_r, and since the rope is long, there may be an economic advantage in using the desirable $D_s = 45D_r$, Table AT 28. With this value as the first trial, we find $D_r = 1.125$ in. from the foregoing equation, which is exactly a $1\frac{1}{8}$-in. rope; $D_s = (45)(1.125) = 50.6$, say 50 in. A check is always desirable (usually, the standard sizes will depart further from those first computed), so solve for N;

$$N = \frac{(0.0015)s_u D_r D_s}{2F_t} = \frac{(0.0015)(240)(1.125)(50)}{2(5.62 + 1.8 \times 1.125^2)} = 1.28,$$

which, as expected, is close enough to the desired 1.3. Use $1\frac{1}{8}$-in. rope on 50 in. sheave. At a mine shaft there should be room enough for this size sheave. Let the reader try other ratios of D_s/D_w.

(b) To get the design factor defined by (**v**), we have, from Table AT 28, $F_u = 52.6$ tons or 105.2 kips, $D_w \approx 0.067D_r$, $A_m \approx 0.4D_r^2$. Then with equation (**t**),

$$F_b = s_b A_m = \frac{ED_w A_m}{D_s} = \frac{(3 \times 10^4)(0.067D_r)(0.4D_r^2)}{50} = 22.9 \text{ kips}$$

for $D_r = 1.125$. Since $F_t = 5.62 + 1.8D_r^2 = 7.9$ kips,

$$N = (F_u - F_b)/F_t = (105.2 - 22.9)/7.9 = 10.4.$$

This result indicates that if the design had been on the static basis with a more normal design factor (say 5), it could not have been expected to withstand an indefinite number of flexings.

(c) If we could study the operation, or if a similar operation were at hand, we could estimate the number of flexings that might be expected in some desirable lifetime. Certainly we should note that it would take a while to load, hoist, dump, return, and therefore indefinite life may be unnecessary. Suppose we guess 6 round trips per hour; for 300 working days of 8 hr. per year, and a life of 7 yr. (which may well be too long for a mine shaft at a given level), this gives about 10^5 round trips, or flexings. Now double this, in case means are found to speed up the operation or in case of extra shifts at times, and get 200,000 cycles as specified in the problem statement. Enter Fig. 17.30 with 200,000, find $p/s_u \approx 0.028$, substitute this value in equation (**w**) with $D_s = 45D_r$;

$$45D_r^2 = \frac{2NF_t}{(p/s_u)s_u} = \frac{(2)(1.3)(5.62 + 1.8D_r^2)}{(0.0028)(240)}$$

from which $D_r = 0.756$; use $\frac{3}{4}$ in.; $D_s = (45)(0.75) = 33.8$, say 34 in. Check as before. From Table AT 28: $F_u = 23.8$ tons or 47.6 kips,

$$F_b = \frac{ED_wA_m}{D_s} = \frac{(3 \times 10^4)(0.067D_r)(0.4D_r^2)}{34} = 9.97 \text{ kips},$$

$$F_t = 5.62 + 1.8D_r^2 = 5.62 + (1.8)(0.75^2) = 6.63 \text{ kips},$$

$$N = \frac{F_u - F_b}{F_t} = \frac{47.6 - 9.97}{6.63} = 5.68,$$

which is observed to be more in line with the design factors given in § 17.25.

(d) The elongation due to a static load of 3000 lb. on the $\frac{3}{4}$-in. rope, equation (**u**) and Table AT 28, is approximately

$$\delta = \frac{FL}{A_mE_r} = \frac{(3000)(1000)(12)}{(0.4 \times 0.75^2)(12 \times 10^6)} = 13.35 \text{ in.}$$

17.27 TRACTION DRIVES. The friction between the sheave and rope is often used as a driving force; for example, if shaft B, Fig. 17.32, is driven by a motor (likely via a speed reduction), if W represents the load and CW represents the counterweight, and if load W is moving up (not

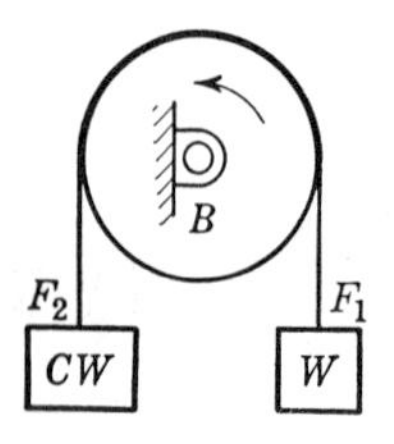

FIGURE 17.32 Traction Drive.

slowing down), then F_1 is the "tight" tension and $F_2 < F_1$. The driving frictional force is $F_1 - F_2$, which can be computed from the power and speed at B. The maximum value of the ratio of the forces (at the point of limiting friction), with negligible centrifugal effects, as would be likely on hoists, elevators, etc., is

$$\text{(x)} \qquad \frac{F_1}{F_2} = e^{f\theta}.$$

Compare with equation (**d**), § 17.3. Find F_1 and F_2 only by free bodies of W and CW, including inertia. Some values of f in equation (**x**) for this application are:[17.23]

Iron or steel sheave: greasy rope, 0.07; wet rope, 0.085; dry rope, 0.12.

Wood-lined sheave: greasy rope, 0.14; wet rope, 0.17; dry rope, 0.235.

Rubber- or leather-lined sheave: greasy rope, 0.205; wet rope, 0.4; dry rope, 0.495.

17.28 FITTINGS FOR WIRE ROPE. There are a number of fittings that have been developed for wire rope,[17.28] of which the hook is shown in Fig. 17.33, and a thimble with clip in Fig. 17.34. The connection with the rope is a socket fitting in Fig. 17.33; the wires are separated and spread into a conical shape in the socket after which zinc is poured in. Properly done, this connection should be as strong as the rope. The thimble is frequently used for connections; the rope may be turned over the thimble with its end spliced to the main part of the rope (not illustrated), or clips, of which one is shown in Fig. 17.34, may be used.

17.29 PULLEYS AND SHEAVES. Various materials are used for pulleys and sheaves, the most common being cast iron and steel. There are notable differences in the construction of the pulleys and sheaves of Figs. 17.35–17.40, which a careful examination will reveal. Wood pulleys have a use but they need to be shellacked and varnished at intervals to keep out moisture. Paper pulleys, Fig. 17.36, are very popular in small sizes because of their good coefficient of friction. Steel pulleys, Fig. 17.37, are lighter than cast iron pulleys and can be safely run at higher speeds (without

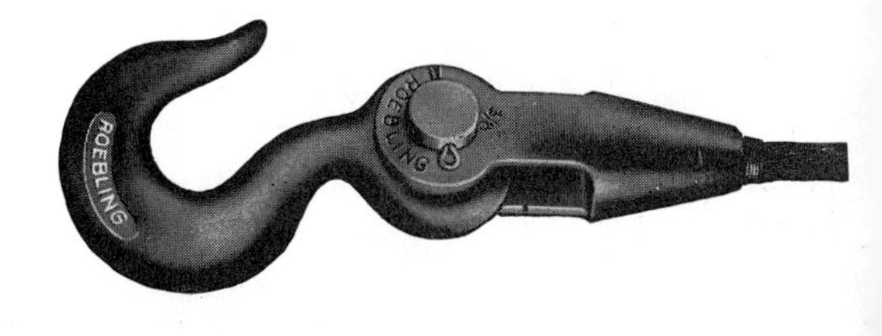

FIGURE 17.33 Hook and Open Socket, Drop-forged Steel. (Courtesy John A. Roebling's Sons Co., Trenton, N.J.).

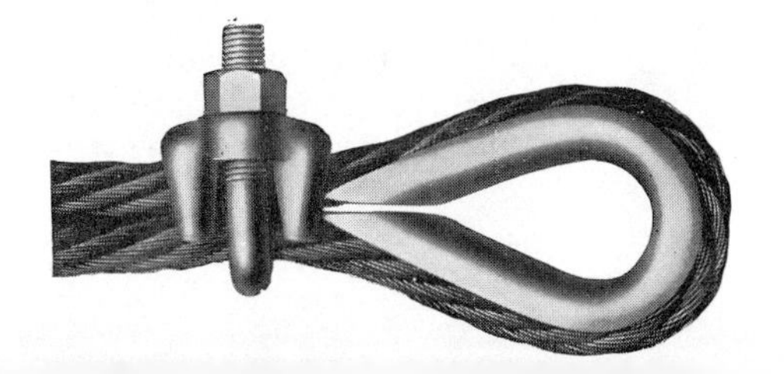

FIGURE 17.34 Thimble with a Clip Connection. (Courtesy John A. Roebling's Sons Co., Trenton, N.J.).

FIGURE 17.35 Wooden Step Cone. The diameters of matching steps on step cones must be such that each step requires the same length of belt, equation (17.3). (Courtesy Dodge Mfg. Corp., Mishawaka, Ind.).

FIGURE 17.36 Fiber (Paper) Pulley. Notice the metal sleeve and metal rim plate. Pulleys are crowned (this one larger at the middle) to keep the belt from running off.[17.31] (Courtesy The Rockwood Mfg. Co., Indianapolis).

FIGURE 17.37 Split Steel Pulley. Some steel pulleys are welded. Desired size of bushing inserted. (Courtesy Dodge Mfg. Co., Mishawaka, Ind.).

FIGURE 17.38 V-Belt Sheave. Cast iron, with split, tapered bushings available for different sizes of shafts. (Courtesy Allis-Chalmers Mfg. Co., Milwaukee).

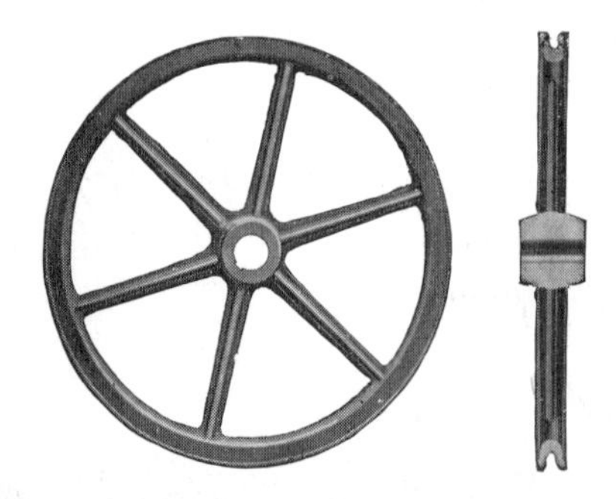

FIGURE 17.39 Cast-iron Wire-rope Sheave. (Courtesy John A. Roebling's Sons Co., Trenton, N.J.).

FIGURE 17.40 Roller-chain Sprocket Wheel. (Courtesy Link-Belt Co., Chicago).

danger of flying apart from centrifugal force). The separate bushing of Fig. 17.38 is an economic idea applied to all kinds of pulleys; the main body of the member can be manufactured in larger quantities with a common bore, and a bushing that fits a large number of different sizes of pulleys can be bored to different shaft diameters. Sheave grooves for wire rope have rounded bottoms and should, in general, not pinch the rope, contrary to V-belt practice. Pulleys, etc. that run at high speeds should be well balanced dynamically as well as statically.

17.30 HARMONIC DRIVE.

The Harmonic Drive,[17.35] a new and novel power-transmitting device, does not fit into any of the traditional classifications. It is made in many configurations, but the basic elements are a rigid circular spline, which, in Fig. 17.41, is the external member A, a flexible spline B, called a "flexspline," and a rotating elliptical member C (or other element) whose function is to thrust the teeth on the flexspline into mesh with the teeth on the rigid spline at two or more sectors (two in Fig. 17.41). The member C is called a wave generator. The teeth are standard 30° involute splines, § 10.7, and are the same pitch on both members. The flexspline in its undeformed shape is circular and it has fewer teeth than the external member A (which has internal teeth); the difference in the number of teeth is always a multiple of the number of lobes in the wave generator C—say, 2 or 4 in this case. To understand its operation, note that the teeth are making contact in the vicinity of M, M, at the major diameter of the wave generator rotating clockwise. The design may be such that either

FIGURE 17.41 Harmonic Drive. As C rotates, teeth are thrust into engagement with a minimum of sliding, forcing B to rotate if A is held stationary. (Courtesy United Shoe Machinery Corp., Beverly, Mass.)

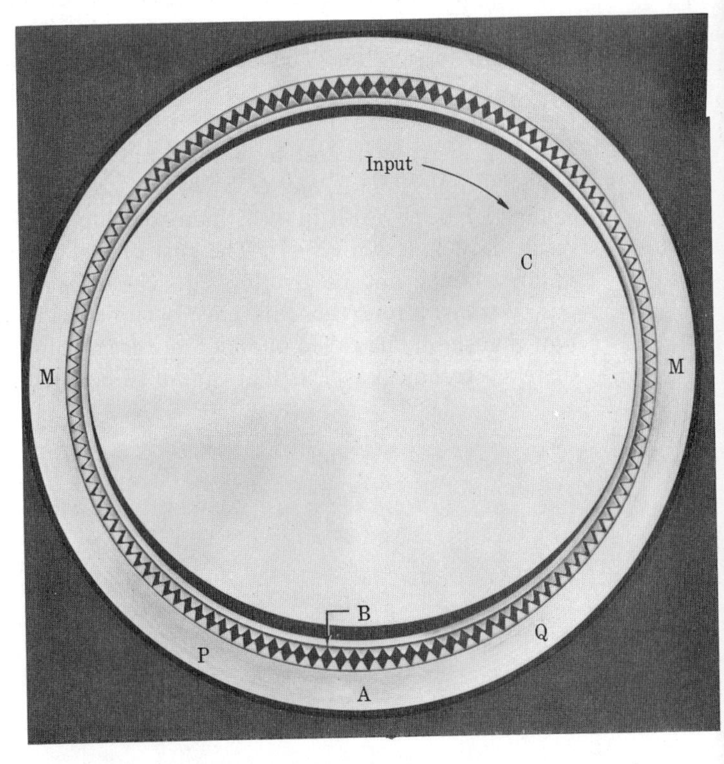

the rigid spline or the flexspline may be stationary. Let the rigid spline be stationary. As C rotates, other teeth come into mesh and the flexspline moves. Examining the relations of the teeth between P and Q, Fig. 17.41, note that the flexspline moves counterclockwise, or in the sense opposite to C. The velocity ratio is

$$\text{(y)} \qquad m_\omega = \frac{\text{No. of teeth on the output member}}{\text{Difference in No. of teeth on the two splines}}.$$

Suppose the rigid spline has 132 teeth, the flexspline 130; since the output member is the flexspline, the velocity ratio is $130/2 = 65$. The consequence of this development is that a high velocity ratio can be obtained in a small "package." Other characteristics claimed by the manufacturer include, high torque capacity, accurate angular positioning, high efficiency, low noise level, and collinear input and output shafts.

17.31 CLOSURE. There are a number of operating variables, initial tension, for example, that affect the efficiency of these drives, and their efficiency will also vary because of large differences in the amount of power transmitted. Properly installed and maintained, efficiencies better than 95% should be attained. Reflect on the discussions of this chapter and note that design generally is against fatigue failure because of the continually varying stress, especially due to flexing (except the flexing of the chain causes wear at the joints). Also, in each case, an undersized element results in shorter life (not necessarily an immediate breakage), an oversized element results in longer life. The percentage increase in life is generally much greater than the increase in cost, so it just depends on whether or not the longer life is desired. For example, a 100% increase in wear life is obtained for 20% increase in cost by going from a $\frac{1}{2}$-in., No. 40 roller chain to a $\frac{5}{8}$-in., No. 50.[17.21] If, say 6 V-belts are required by normal design procedures, 7 belts result in 40% increased life for 15% greater cost, 5 belts result in 35% less life.[17.14] The cost of lost production owing to a drive failure should not be overlooked. The manufacturers of these various power transmitting elements have accumulated much engineering know-how that should be called upon if there are any unusual factors. And finally, there are technical aspects that cannot be covered in this brief review.[17.32]

18. BRAKES AND CLUTCHES

18.1 INTRODUCTION. This chapter deals largely with ***brakes***, which are friction devices used to regulate the motion of bodies (slowing them down, holding their speed constant, holding them at rest, etc.) and with ***clutches***, which are friction devices used to connect shafts (speeding up the driven bodies to the same angular velocity as the driving shaft). We may classify such friction devices into types as follows: external shoe on drum, internal shoe on drum, disk on disk, cone in conical surface, wrapping bands on drums, and expanding bands on drums. Any one of these fundamental types might be either a brake or a clutch, although the details of the brake design would be different from the details of the clutch design. However, each type has a characteristic use; an external shoe on a drum is most likely to be a brake, a disk on disk is most likely to be a clutch, but the cones and disks and others are widely used for both. We shall discuss a few of these friction devices, not exhaustively, but briefly in their most likely forms, secure in the belief that the reader can make a transition from a brake to a clutch, or vice versa, as needed.

18.2 FRICTIONAL WORK AND POWER. Frictional work is simply work done by a frictional force F. If a brake is used on a moving body, it brings about a decrease of kinetic energy KE or it opposes a loss of potential energy PE, or both, the first consequence being an increase in the internal molecular energy of the contacting bodies (mostly near the contacting surfaces). This is to say that the temperatures of the bodies increase, which results in the frictional work being eventually dissipated to the surroundings as heat. Getting rid of the energy as heat takes time, and for this reason, the capacity of a brake is often stated in terms of the amount of frictional work U_f it can absorb in a particular time or in terms of frictional horsepower (fhp). Sometimes a brake has to be applied steadily over long periods of time, in which case it must be able to radiate and convect the heat to the atmosphere at such a rate that the steady-state temperature is below a damaging value, § 18.4. Sometimes the brake is applied for short intervals of time intermittently, with enough time between applications for it to cool to a value close to the environmental temperature, in which case it may safely absorb energy at a much higher time rate, the energy being mostly used to heat the metal parts. The rate at which heat is conducted through the metal is much greater than the rate at which it is radiated and convected to the atmosphere. If the human element is a primary factor, there is no telling what the pattern of brake application may be.

The theory of this chapter is from elementary mechanics. An important parameter for brakes and clutches is the frictional torque T_f (a rotating body is nearly always involved), and brake capacity is often expressed in terms of T_f. The reader will be reminded later of force analyses of several configurations, from which T_f is found. Recall now that work is equal to "force times distance," and for a rotating body, work is $T\,d\psi$, where $d\psi$ radians is the angle the body turns while subjected to any torque T. If T_f is constant or assumed to be some average value, the work of a frictional force on the surface of a rotating drum of diameter D in., Fig. 18.1, is

$$\textbf{(a)} \qquad U_f = F\pi D N_t = T_f(2\pi N_t) = T_f\psi \qquad \text{or} \qquad T_f = \frac{U_f}{2\pi N_t},$$

where N_t is the number of turns of the drum while F and T_f are acting ($\psi = 2\pi N_t$ radians), and U_f and T_f have the same units, usually either in-lb. or ft-lb. Also, for a constant angular acceleration α rad./sec.^{2}, $T_f = I\alpha$ ft-lb., where I slug-ft.^{2} (for feet, seconds) is the polar moment of inertia of rotating members about the axis of rotation.

The power associated with T_f is $T_f\omega$, where ω rad./unit of time is the angular velocity of the drum; $\omega = 2\pi n$, where n is in rpm or rps. Sometimes a brake is used to keep a body moving downward with more or less constant speed, but more often the speed changes, frequently to a state of rest. We write

$$\textbf{(b)} \qquad \text{fhp} = \frac{(T_f \text{ ft-lb.})\omega_m}{33{,}000} = \frac{(T_f \text{ in-lb.})n}{63{,}000}$$

for ω_m rad./min., n rpm, where the frictional horsepower fhp is for some instantaneous speed ω_m, n; often, an average fhp is computed for an average speed. The limiting capacity of a brake is commonly expressed in terms of the maximum instantaneous rate of energy absorption, Fv_m or fhp, but the total amount of energy involved is also vital. For example, certain disk brakes on railway cars could be subjected to 1000 fhp maximum for short-time application without damage, but to only 75 fhp for steady-state application.[18.8] Stated another way, this brake absorbs 7×10^6 ft-lb. of energy without damage in 20 sec., but it takes some 8 min. for most of this energy to be radiated and convected to the atmosphere.

18.3 COMPUTING THE ENERGY TO BE ABSORBED.

The body being braked may undergo a change of potential energy $-\Delta PE$,

$$-\Delta PE = W(h_1 - h_2), \tag{c}$$

where a body of weight W moves from an elevation h_1 to h_2, each measured from the same datum level. The minus sign appears before ΔPE because the conventional interpretation of Δx is $x_2 - x_1$, which convention makes the energy absorbed by the brake a negative number as computed from the change of motion of the braked body; that is, the only purpose of the minus sign is to make the energy quantities positive—but $-\Delta PE$ may be positive or negative, depending on the direction of motion. The change of kinetic energy ($mv^2/2$) of a translating body of weight W lb. is

$$-\Delta KE = \frac{W}{2g}(v_{s1}{}^2 - v_{s2}{}^2) \text{ ft-lb.,} \qquad \text{[TRANSLATION]} \tag{d}$$

where v_{s1} fps is the initial speed, v_{s2} fps is the final speed, g fps^2 is the local acceleration [$W/g = m$ slugs (for g fps^2) is the mass of the body]; on or near the surface of the earth, use $g = 32.2$ fps^2. For a rotating body, the change of kinetic energy is

$$-\Delta KE = \frac{I}{2}(\omega_1{}^2 - \omega_2{}^2) \text{ ft-lb.,} \qquad \text{[ROTATION]} \tag{e}$$

where I slug-ft.^{2} is the moment of inertia of the body about its axis of rotation, ω_1 and ω_2 rad./sec. are, respectively, the initial and final angular velocities. For a rolling body,

$$-\Delta KE = \frac{W}{2g}(\bar{v}_{s1}{}^2 - \bar{v}_{s2}{}^2) + \frac{\bar{I}}{2}(\omega_1{}^2 - \omega_2{}^2) \text{ ft-lb.,} \tag{f}$$

[ROLLING BODY]

where the bar over the symbols means that the velocities are for the center of gravity and the moment of inertia is with respect to that gravity axis perpendicular to the plane of rotation. Energy is a scalar quantity; the total energy U_f that a brake absorbs is the decrease of the stored mechanical energy brought about by the braking. Use equation (**a**) to find the corresponding frictional force or torque, and (**b**) to convert to horsepower. If more than one body has a significant energy change, find the change for each and add algebraically to get the total energy to be absorbed by the brake.

18.4 PERMISSIBLE ENERGY ABSORPTION AND OTHER DESIGN DATA.

For design values of the coefficient of friction, see Table AT 29; also § 18.14 for some remarks on it.

The permissible normal pressures between the braking surfaces depend in varying degree on the brake-lining material, the coefficient of friction, and on the maximum rate at which energy is to be absorbed. *The higher the pressure, the greater is the rate of wear* and the energy absorbed at a particular speed. Recommended design values of p are given in Table AT 29, but the stated values are exceeded on occasion. Moreover, they may be lower than shown when application periods may be frequent and long. In short, experience with given materials and class of service tells what is best for design.

The pressure p psi is for the projected area A of the shoe; $p = N/A$, where N is the total normal force (detail later). The actual rate of energy dissipation is $fNv_m = Fv_m = fpAv_m$ ft-lb/min. at v_m fpm, from which we see that it is proportional to pv_m ft-lb/sq. in.-min. (or $pv_m/33{,}000$ hp/sq. in.) for a particular brake. On this account, design values of pv_m are found in the literature. Typical values for industrial shoe brakes are:[18.14] frequent application, $pv_m = 5500$; average use, $pv_m = 16{,}500$; infrequent short-time use, $pv_m = 49{,}500$. Raybestos-Manhattan[13.13] states that most manufacturers' design values of pv_m will fall within the range of 1500 and 10,000.

The following values, in agreement with Hütte, are widely quoted and reproduced here for their informational value:

$pv_m \gtreqless 55{,}000$ for intermittent applications of the load, comparatively long periods of rest, poor dissipation of heat (wood blocks); for $f = 0.25$, this value is equivalent to about 0.42 fhp/sq. in. absorbed;

$pv_m \gtreqless 28{,}000$ for continuous application of the load as in lowering operations, poor heat dissipation (wood blocks); equivalent to 0.21 fhp/sq. in. for $f = 0.25$;

$pv_m \gtreqless 83{,}000$ for continuous application of the load, good heat dissipation (oil bath); equivalent to about 0.63 fhp/sq. in. for $f = 0.25$.

Compare these various values with some peak values found in the literature: automobiles, (100 psi)(2500 fpm) = 250,000 at 600°F; earth moving tractors (50 psi)(5000 fpm) = 250,000 at 800°F; railway cars (100 psi)(10,000 fpm) = 1,000,000.

For open and exposed *band brakes*, § 18.9, Rasmussen[18.11] recommends an energy adsorption capacity of 0.2 to 0.3 fhp per square inch of brake contact area.

The temperatures are limited by the properties of the materials (Table AT 29). Calculations that purport to give the surface temperatures can only approximate some kind of average, but estimations must be made.

The heating effects are naturally quite complex and local temperatures at points of contact may be very high. That part of the total frictional energy that is stored in the brake parts, principally in the drum or disk, has been variously estimated at 75% up, but it actually varies with the duration of the application as well as with the quantity of energy. For very short applications of the brake, say a few seconds, the percentage may be nearly 100% at the instant that the brake is released. An assumption made in brake-drum design for airplanes was that the drum would absorb 92% of the kinetic energy and be at a temperature of 1075°F at the stop.[18.20] After a while, the energy stored in the parts by the braking will be rejected to the sink. If long continuous applications are necessary, the drum and other adjacent parts keep rising in temperature until the amount of energy lost to the surroundings is equal to the rate at which frictional energy is generated. Thus, as the magnitude of pv_m increases, the duration of the braking must decrease in order to avoid overheating.

For peak short-time requirements, it is generally assumed that all the frictional energy is absorbed by the adjacent metal in the drum wheel, which gives a temperature increase Δt of the metal as

$$\text{(g)} \qquad \Delta t°\text{F} = \frac{U_f \text{ ft-lb.}}{w_m c},$$

where w_m lb. is the mass of the metal absorbing the energy U_f, and c ft-lb/lb-°F is the average specific heat of the metal for the temperature range (for a large temperature range, c varies markedly). Suitable for ordinary calculations: for cast iron, $c \approx 101$; for steel, $c \approx 93$; for Al, $c \approx 195$.

In automobile brakes (Fig. 18.5), a desirable maximum instantaneous loading is about 2.2 fhp/sq. in. of brake rubbing surface on the drum. One way to compute such an "instantaneous" value is from a known or assumed deceleration. For example, 15 fps^2 may be taken as a reasonable value for automobiles, which means that the speed decreases 15 fps in one second; the fhp loading can be computed from any top speed during, say, one second.

The effectiveness of the brake may greatly decrease shortly after it begins to act continuously, a phenomenon called ***fade***. This is basically due to a significant decrease in the coefficient of friction at the high surface temperatures induced, and it can be combated to some extent by the design of the braking system; for example, by searching for a configuration such that the frictional moment T_f divided by the applied moment Wa (T_f/Wa) shows a minimum variation when plotted against the coefficient of friction (see *mechanical advantage* below), or by more effective dissipation of heat.

18.5 EXAMPLE—DRUM TEMPERATURE AND fhp. A 3000-lb. automobile, moving on a level road at 60 mph, is braked so that the deceleration is constant at 20 fps^2 (close to a normal maximum) until it stops. (If it were going downhill, the loss of *PE* would be added to the loss of *KE*; if uphill, gravity would help slow down the car.) Each cast-iron brake drum is $D = 9$ in. in diameter, $b = 2\frac{1}{4}$ in. face width, and $t = \frac{3}{16}$ in. thick. (a) Assuming that the radiated and convected heat is negligible, compute the temperature rise of the drum. (b) What is the average rate of power absorption (fhp) during the first second? During the last second? (c) What is the over-all average fhp?

Solution. (a) Since the process of conduction in iron is relatively fast, we get a reasonable estimate of conditions by assuming that all the energy of friction is absorbed by the drum, keeping in mind that temperature differences are sure to exist. From Table AT 6, we find the density of cast iron is 0.253 lb/cu. in. Thus the approximate mass of 4 drums is $[V = \pi Dbt + (\pi D^2/4)t]$

$$w_m = \rho V = (4)(0.253)\left[\pi(9)(2.25)(\tfrac{3}{16}) + \frac{\pi(81)(3)}{(4)(16)}\right] = 24.1 \text{ lb.}$$

(Considering attached metal, the effective mass may be greater.) The speed of 60 mph is the same as (60)(5280)/3600 = 88 fps. The 3000-lb. car loses all of its kinetic energy (rotational KE of wheels neglected on the assumption that it is a small part of the total)

$$-\Delta KE = \frac{Wv_s^2}{2g} = \frac{(3000)(88)^2}{64.4} = 361{,}000 \text{ ft-lb.}$$

With $U_f = -\Delta KE = 361{,}000$ in equation (g), the average temperature rise is

$$\Delta t = \frac{U_f}{w_m c} = \frac{361{,}000}{(24.1)(101)} = 149°\text{F.}$$

If the ambient temperature is 100°F, the drum is at 249°F average.

(b) During the first second, the velocity drops from 88 fps to 68 fps ($a = -20$ fps^2), and therefore the loss of kinetic energy per second is

$$-\Delta KE = \frac{W}{2g}(v_{s1}^2 - v_{s2}^2) = \frac{3000}{64.4}(88^2 - 68^2) = 145{,}000 \text{ ft-lb/sec.}$$

$$\text{fhp} = \frac{145{,}000}{550} = 264 \text{ hp average for 1 sec.}$$

The approximate area of the 4 rubbing surfaces is

$$A = (4)(\pi Db) = (4)(\pi)(9)(2.25) = 254 \text{ sq. in.;}$$

264/254 = 1.04 fhp/sq. in. (But note that during the first fraction of a second, the rate is higher than 1.04.) During the final second, the car stops from a speed of 20 fps, and fhp is $-\Delta KE/550$ or

$$\text{fhp} = \frac{-\Delta KE}{550} = \frac{Wv_s^2}{(2g)(550)} = \frac{(3000)(20)^2}{(64.4)(550)} = 33.9 \text{ hp.,}$$

or 33.9/254 = 0.133 fhp/sq. in.

(c) For constant acceleration, $a = \Delta v/\Delta\tau$ or $\Delta\tau = \Delta v/a = 88/20 = 4.4$ sec. to stop the car. With an initial $KE = 361{,}000$ ft-lb., the over-all time average loss is at the rate of

$$\text{fhp} = \frac{361{,}000}{(4.4)(550)} = 149 \text{ hp.,}$$

or 149/254 = 0.587 fhp/sq. in. of rubbing surface on the drum.

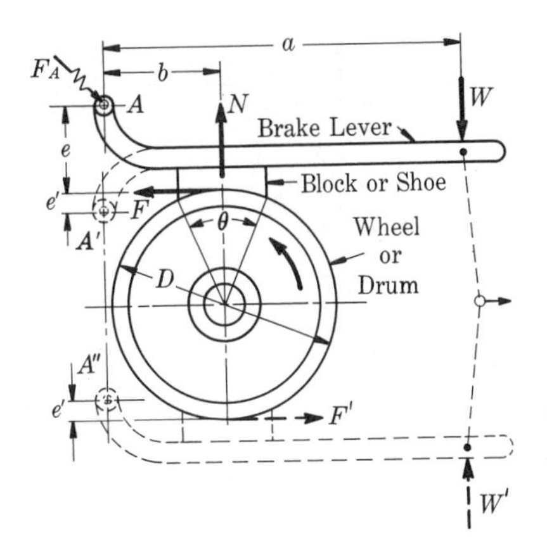

FIGURE 18.1 Block Brake.

18.6 BLOCK BRAKE, SMALL BLOCKS. If the block is short enough (small θ, Fig. 18.1), it is reasonably accurate to assume that the resultant frictional force F is tangent to the drum at the center of the block, an assumption often made for quick calculations. The free body of the brake lever and block, solid outline, Fig. 18.1, serves to illustrate the procedure for all such cases and shows all the forces: W, the force applied to the lever shown parallel to the normal force N (its components parallel and perpendicular to N would be the most convenient if W were at another angle); $F = fN$, where f is the "kinetic" coefficient of friction; and F_A, unknown in magnitude and direction, the pin reaction, which would be used to design the pivot connection A. Assuming that the reader is qualified to find F_A and use it, we shall concentrate on the braking aspects, and eliminate F_A by choosing A as the center of moments;

(h) $$\Sigma M_A = Wa - Nb + Fe = Wa - \frac{F}{f}b + Fe = 0.$$

The location of the pivot A with respect to the line of action of the frictional force F is significant in brakes. For the position as given, we see that the sense of the moment of F is the same as that for W; that is, the frictional force helps to apply the brake. When this condition exists, the brake is said to have *self-actuating properties* or to be ***self-energizing***. Considering Fig. 18.1, note that if the pivot is located at the more likely point A', the moment of the frictional force *opposes* the moment of the applied force W and there is no self-energizing effect; one may think of this as a *negative* self-energizing. Thus, it takes a larger applied force W to produce a particular braking force F when the pivot is at A' than when it is at A, other conditions remaining the same. However, if this is a double-block brake with another lever and block pivoted at A'', the more likely construction, we see that F' provides self-actuation of the lower lever while F opposes the applied moment on the upper lever; the effects are reversed if the rotation is reversed. In actual brake analysis, one must make a free body of each lever and its block; F' is not necessarily equal to F and the braking effects are generally different, as explained. For practice, the student should write

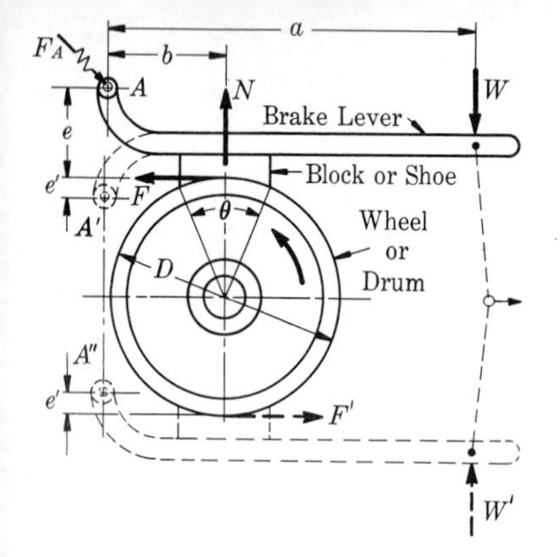

FIGURE 18.1 **(Repeated).**

the equivalent of equation (**h**) for different locations of the pivot A and for different directions of rotation.

Notice in equation (**h**) that if $b/f = e$, then $W = 0$, a limiting (and quite unstable) condition meaning that no force is necessary to apply the brake. If $e > b/f$, W is negative, which means that a force is necessary to disengage the brake, once it is engaged, or that less than no force is required to cause braking; consequently, the brake is ***self-locking***. A few numbers in equation (**h**) would soon demonstrate that it would be necessary to have an unusual configuration (large e) to make a simple block brake self-locking; but if self-locking were desired, as it sometimes is, § 18.9, other kinds of brakes are more adaptable for this purpose.

In addition to a brake's capacity to absorb energy, its frictional or ***braking torque*** T_f is used for rating. For the single lever of Fig. 18.1, solve for F from equation (**h**), multiply both sides by the drum radius $D/2$, and get the braking torque T_f for an applied force W;

(**i**)
$$T_f = \frac{FD}{2} = \frac{WDfa}{2(b - fe)}.$$

If there are two or more blocks on the same drum, the safe procedure is to find T_f for each block and add the results.

In the actual design of a brake, one would probably draw curves, showing how the action of the brake varies with changes in certain dimensions and in the coefficient of friction, which is inherently variable. A dimensionless parameter used for such studies, sometimes called the ***mechanical advantage*** MA, is defined as the ratio of the braking torque T_f divided by the applied moment; for example, in Fig. 18.1, for the single lever, $MA = T_f/Wa$, where T_f is given by (**i**).

The advantage of using opposed double blocks or two-shoe brakes is that the normal forces more or less balance one another, significantly reducing the bearing loads and the bending moment on the drum shaft. The details of the linkage designs vary widely, Fig. 18.2 being an illustration; since this is so, there is no recourse except for the engineer to apply the principles of mechanics in finding the magnitudes of the various forces and moments for the particular configuration involved.

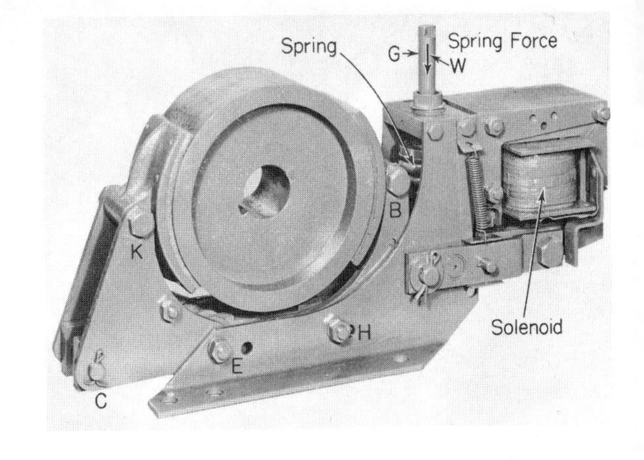

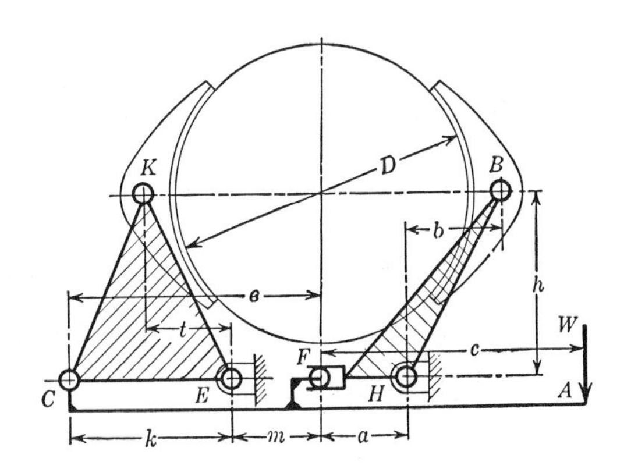

FIGURE 18.2 Two-shoe Brake With Solenoid. A spring applies the brake with force *W* and can be adjusted somewhat to increase or decrease the braking moment. When the current is on, flowing through the motor, it flows through the solenoid too, and the solenoid keeps the force off the brake; braking occurs when current ceases to flow, as when an elevator or hoist is stopped. In a force analysis, consider the free bodies of links *AFC*, *BHF*, and *CKE*, as shown in the kinematic representation. See §18.7. The bolts *K* and *B* may be considered to be tight enough to make a rigid connection. (Courtesy Westinghouse Electric Corp., East Pittsburgh).

18.7 FORCES FOR LONG SHOES. The error resulting from the assumptions of uniform pressure and the frictional force tangent to rim at the center of the shoes, as for the short shoe above, becomes greater as the arc subtended by the brake-shoe surface increases. Typical values of θ, Fig. 18.3, are in the vicinity of 90°. To improve the analysis, one may assume something that is expected to approximate actual conditions—the distribution of the normal pressure p on the shoe, for example. For the configurations shown in Fig. 18.3, the assumption that wear is proportional to the pressure results in a cosine distribution of pressure with respect to the line OB; so we might as well assume this distribution, $p = P \cos \alpha$, where P is the proportionality constant and the value of p when $\cos \alpha = 1$. Such a distribution is suggested by experimental evidence. Should the shoe and drum be relatively flexible, the pressure distribution would be more nearly uniform. Actually, the distribution changes with lining wear and with the magnitude of the applied forces. The following analysis applies to pivoted shoes, Fig. 18.3(a), where the axis OB from which α is measured passes through the pin whose center is B (not as drawn), and it applies to Fig. 18.3(b) when the axis OB is perpendicular to OH; with the additional limitation that the contact surface is symmetric about OB, an unnecessary limitation that the reader can easily change by changing the limits of integration.

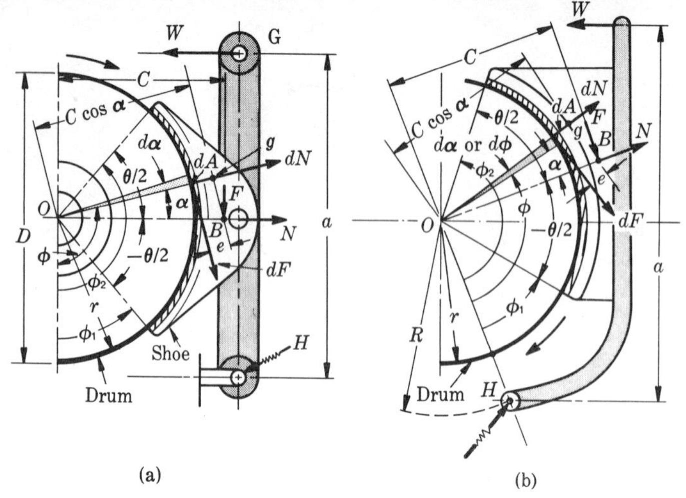

FIGURE 18.3 Long Shoes. In (a), there is a pivoted shoe; for $p = P \cos \alpha$, force N acts through the pin and B; force F on the *shoe* acts through B. Notice that for CL rotation and as constructed in (a), the resultant F has a moment about H that makes it self-energizing, and that the force F on a matching shoe on the opposite side of the drum makes it negative self-energizing. In (b) with CL rotation, F as shown results in negative self-energizing, but the opposite matching shoe would be self-energizing. When braking is largely for one direction of motion, the shoe that is not self-energizing can be made with a smaller contact surface, or the linkage system could be designed to produce a larger normal force on the negative self-energizing shoe than on the other. The brake shoe is ordinarily made symmetric about the horizontal center line; in (b), it is intentionally as shown in order to accentuate the conditions for the formulas as derived.

Consider the area $dA = b(r\,d\alpha)$, Fig. 18.3, where b is the face width of the braking surface and $r\,d\alpha = ds$. For a pressure p, the normal and frictional forces are

(j) $dN = pbr\,d\alpha = Pbr \cos \alpha\,d\alpha$ and $dF = fpbr\,d\alpha = fPbr \cos \alpha\,d\alpha,$

where we have substituted $p = P \cos \alpha$. The moment of dF about some point B, Fig. 18.3, with f assumed constant, is $(Og = C \cos \alpha)$

$$dM_{F/B} = (dF)e = dF(C \cos \alpha - r) = fPbr(C \cos \alpha - r)\cos \alpha\,d\alpha,$$

(k) $$dM_{F/B} = fPbr(C \cos^2\alpha - r \cos \alpha)d\alpha.$$

Since P, b, r, and C are constants for a particular brake, the integral of this expression **(k)** between the limits of $-\theta/2$ to $+\theta/2$ gives the moment of F about some point B as

$$M_{F/B} = fPbr\left[C\left(\frac{\alpha}{2} + \frac{\sin 2\alpha}{4}\right) - r \sin \alpha\right]_{-\theta/2}^{\theta/2}.$$

The moment of F about a point on its own line of action is zero; therefore if $M_{F/B} = 0$, the resulting value of C locates the line of action of F. Since only the parts within the brackets can equal zero, we have

(l) $$C = \frac{4r \sin \theta/2}{\theta + \sin \theta} = \frac{2D \sin \theta/2}{\theta + \sin \theta},$$

where C is the distance from O to the line of action of F, either construction, and it has the same units as r or D. In brake parlance, point B is called the ***center of pressure***. Thus, *if* the pivot pin *is* located at point B, there is no tendency for the shoe to tip and concentrate the pressure on the toe or heel of the lining, and the lines of action of N and F *pass through point* B. However, point B is ordinarily so close to the drum surface that there is not room enough for a pin (though other designs might be used to get the pivot virtually at B). Hence, for the construction shown in Fig. 18.3(a), the resultant F tends to turn the shoe CC, producing a greater pressure at the top end of the shoe than at the bottom; and the sinusoidal distribution with the center at OB is therefore no longer true. (If the pin, Fig. 18.3(a), is fairly close to B, the equations for T_f, etc., give reasonable answers for the pivoted shoe.)

The normal force N is the sum of the components in the direction OB, $N = \int dN \cos \alpha$, with dN from equation (**j**);

$$\text{(m)} \qquad N = \int dN \cos \alpha = Pbr \int_{-\theta/2}^{+\theta/2} \cos^2 \alpha \, d\alpha = Pbr\left(\frac{\theta + \sin \theta}{2}\right).$$

The frictional torque T_f is the moment of F about O, $M_{F/O} = T_f = \int r \, dF$, with the value of dF from (**j**);

$$\text{(n)} \qquad T_f = \int r \, dF = fPbr^2 \int_{-\theta/2}^{+\theta/2} \cos \alpha \, d\alpha = 2fPbr^2 \sin \frac{\theta}{2},$$

where the units are in.-lb. or ft-lb., depending on the linear units used. Equations (**m**) and (**n**) apply to both configurations in Fig. 18.3 if the pressure distribution is sinusoidal as explained; as usual $F = fN$. It may be convenient to note that $T_f = FC$, where C is defined by (**l**). The maximum pressure P may be computed from any of the foregoing equations containing it, say equation (**m**)—boundary conditions as defined for Fig. 18.3.

For the lever as a free body in Fig. 18.3(a), the forces F and N act through the center of the pin (assuming that the pin connection can offer no turning moment). However, for the rigidly attached shoe of Fig. 18.3(b), one must integrate for the moment of dF about H, or use the moment arm to H of the F vector through B; forces W, F, N, H as shown in Fig. 18.3(b) constitute the free body for the link and shoe. If the lining does not extend as far as the line of action of N ($\phi_2 < 90°$), the actual maximum pressure is that at the end of the lining ($P \sin \phi_2$).

For best understanding, the reader should solve some problems (different from the foregoing) by integration, running checks as possible by the stated principles. One variation consists of deriving the equations in terms of ϕ_1, ϕ_2, Figs. 18.3 and 18.4, for which the sinusoidal pressure is $p = P \sin \phi$. See Fig. 18.4 and show that, for $p = P \sin \phi$,

$$\text{(o)} \qquad T_f = M_{F/O} = fbr^2P(\cos \phi_1 - \cos \phi_2),$$

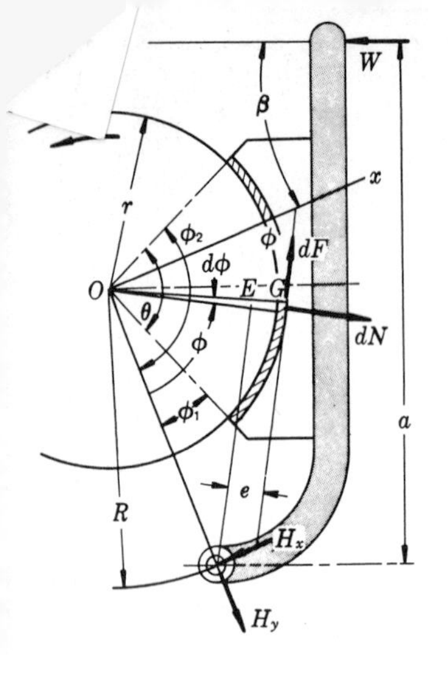

FIGURE 18.4 Unsymmetric External Shoe.

(p) $$M_{F/H} = fbrP\left[r(\cos\phi_1 - \cos\phi_2) - \frac{R}{2}(\sin^2\phi_2 - \sin^2\phi_1)\right],$$

(q) $$M_{N/H} = \frac{brRP}{4}[2(\phi_2 - \phi_1) - (\sin 2\phi_2 - \sin 2\phi_1)],$$

where angles that stand alone, as $\phi_2 - \phi_1 = \theta$, are of course in radians and where the equations are set up to give positive answers for each moment. With $M_{F/H}$ and $M_{N/H}$ as positive numbers, the sum of the moments with respect to H in Fig. 18.4 is

(r) $$\Sigma M_H = Wa + M_{F/H} - M_{N/H} = 0.$$

For opposite rotation, the sign of $M_{F/H}$ would reverse. The reaction H at the pin is obtained from the sums of components in the x and y directions; for the force system in Fig. 18.4,

(s) $$\Sigma F_x = -H_x - W\cos\beta + \int dN\sin\phi + \int dF\cos\phi = 0,$$

(t) $$\Sigma F_y = -H_y + W\sin\beta - \int dN\cos\phi + \int dF\sin\phi = 0,$$

where the integrations are made as previously explained; $H = (H_x^2 + H_y^2)^{1/2}$. It is worth noting that the equilibriant of W and H is equal to the resultant of F and N (the lines of action of three forces in equilibrium intersect at a point).

18.8 INTERNAL SHOE. The problem of braking an automobile has led to the development of several types of brakes with pivoted internal shoes. In some brakes, the shoes are on a fixed pivot, as in Fig. 18.5; in

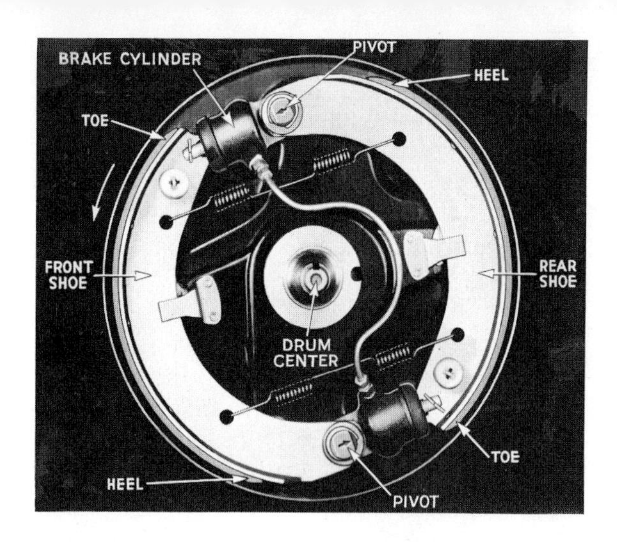

FIGURE 18.5 Internal Shoe Brake. (Courtesy Chrysler Corp., New York).

others, the shoe pivot is attached to another link and is free to move within a short arc. With this considerable variation in detail, a number of analyses of internal-shoe brakes have been made.[18.2,18.3,18.4] For a fixed pivot, the braking torque, etc., may be determined on the basis of a sinusoidal pressure distribution ($p = P \cos \alpha$ when α is measured from OB, Fig. 18.3, the perpendicular to the radial line from the pivot to the drum center; or $p = P \sin \phi$, where ϕ is measured from the radial line from the pivot to the drum center). The moment of the frictional force and of the normal force about the fixed pivot of the shoe can be found by integration, as suggested by the procedure in § 18.7. Horsepower capacities are mentioned in § 18.4.

In the two-shoe automotive brake of Fig. 18.5, observe that if the car is moving forward, the frictional forces on the shoes have counterclockwise moments about their respective pivots and that these moments help to apply the brakes—that is, both shoes are self-energizing, a good arrangement for front-wheel brakes, which do most of the braking. If the car is backing, the moments of the frictional forces about the pivots tend to reduce the pressure on the brake drum. The result is that the brakes are less effective (for a particular foot force) when the car is moving backwards than when it is moving forward. Maybe you have noticed.

18.9 BAND BRAKES.

In a band brake, the band wraps partly around the brake wheel or drum, and braking action is obtained by pulling the band tight onto the wheel, Fig. 18.6. The braking force F is the difference

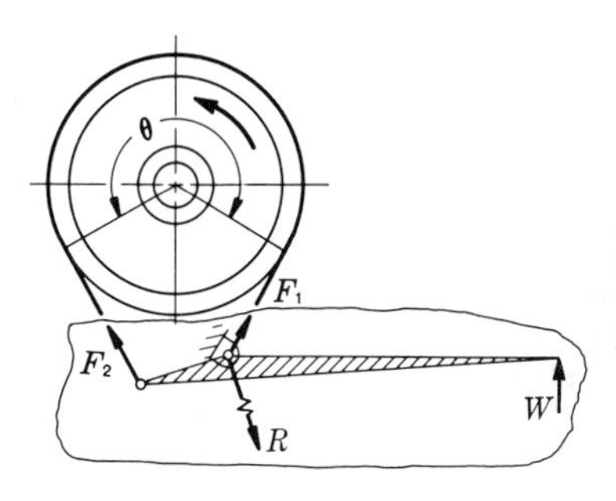

FIGURE 18.6 Band Brake.

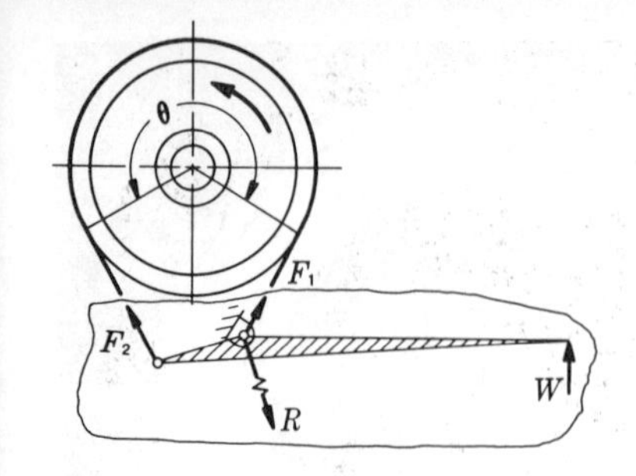

FIGURE 18.6 (Repeated).

between the tensions at the two ends of the band, $F = F_1 - F_2$. The force analysis is similar to that given for belts, except that the centrifugal force is absent. Directly from equation (**d**), § 17.3, we get

$$\frac{F_1}{F_2} = e^{f\theta}, \tag{18.1}$$

where F_1 is the tension of the *tight* side of the band, F_2 the *smaller* tension, and θ the angle of contact between the band and the wheel in radians.

The conceptual tools for computations on band brakes include equation (18.1), $F = F_1 - F_2$, and an equilibrium equation, as equation (**u**) below for the differential band brake; the detail is made clear by the analysis below. In Fig. 18.6, if the rotation is reversed, the larger tension F_1 will oppose W, resulting in a smaller braking torque for a particular W.

In the differential band brake, Fig. 18.7, the tension on one end of the brake band aids in applying the brake. Such brakes are self-energizing and may indeed be self-acting, Fig. 18.8. A force analysis, with the brake lever as the free body, Fig. 18.7, will suggest some characteristics of this type of brake. Using the fulcrum B as the center of moments in order to eliminate the pin reaction at B from the moment equation, Fig. 18.7, we get

$$\Sigma M_B = Wa + F_1 b - F_2 c = 0. \tag{u}$$

Hence $F_1 b$ is in the same sense as Wa, and the externally applied force W is assisted by the tight tension F_1 in putting on the brake. To get the relation of the frictional force F and the applied force W, substitute the value of $F_1 = F_2 e^{f\theta}$ into $F_1 - F_2 = F$, and find

$$F_2 = \frac{F}{e^{f\theta} - 1}, \qquad \text{and then,} \qquad F_1 = \frac{Fe^{f\theta}}{e^{f\theta} - 1}. \tag{v}$$

Use these values of F_1 and F_2 in equation (**u**) and solve for W;

$$W = \frac{F(c - be^{f\theta})}{a(e^{f\theta} - 1)}, \tag{w}$$

which shows that W will be negative if $be^{f\theta} > c$; that is, the brake will be self-acting and grab once the frictional force begins to act.

Since the braking torque is $T_f = FD/2$, where $D/2$ is the radius of the

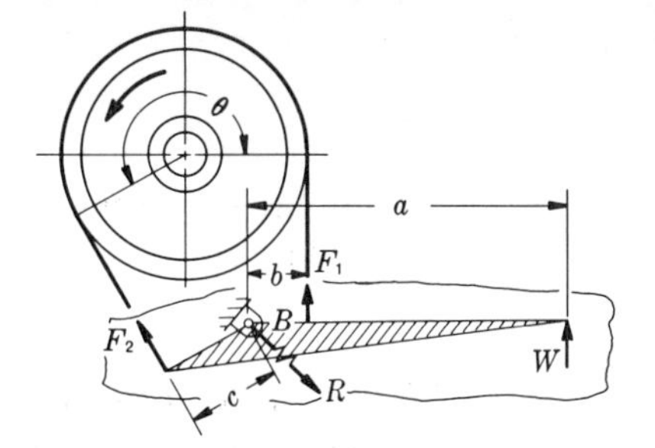

FIGURE 18.7 Differential Band Brake.

FIGURE 18.8 Differential Back Stop. This is a band brake used as a stop to prevent reverse motion. Compare with Fig. 18.7. It is used on the motor shaft or countershaft of an elevator, hoist, or conveyor.

drum, we may multiply both sides of equation (**w**) by $D/2$, solve for $FD/2 = T_f$, and find

(**x**) $$T_f = \frac{WDa(e^{f\theta} - 1)}{2(c - be^{f\theta})}.$$

The pressure (see Table AT 29) on a band brake is computed as

(**y**) $$p = \frac{F}{fA}$$

in which A is the contact area between the band and the drum. According to Hagenbook,[18.6] suitable drum diameters may fall between (T_f in-lb.),

(**z**) $$\left(\frac{T_f}{5}\right)^{1/3} < D < \left(\frac{T_f}{4}\right)^{1/3}.$$

Another relationship that may help size the drum is

(**a**) $$(60\text{ fhp})^{1/3} < D < (80\text{ fhp})^{1/3},$$

where fhp is taken as the maximum horsepower to be dissipated in any fifteen-minute period. See also fhp capacity for band brakes mentioned in § 18.4. Although the empirical values given by the independent equations (**z**) and (**a**) are presumed to be reasonable, new designs should, if possible, follow known successful practice.

18.10 DISK FRICTIONAL TORQUE. Disk surfaces used for clutches or brakes generally have an annular shape, Fig. 18.9. The theoretical equation for T_f may be based on either of two assumptions: (1) the pressure is uniformly distributed over the surface, or (2) the wear on the surface is uniform. Since (2) is a little more conservative than (1), and gives a simpler result besides, we shall use uniform wear.

The wear may be expected to be proportional to the intensity of pressure p and the rubbing speed. In turn, the rubbing speed is proportional to the distance from the center of rotation ρ, Fig. 18.9. Thus, wear is proportional to $p\rho$; and if wear is uniform, $p\rho = C$, a constant. Consider the differential area $dA = \rho\, d\theta\, d\rho$, Fig. 18.9; the normal force on it is $dN = p\, dA = p\rho\, d\theta\, d\rho$. Letting $p\rho = C$ and integrating, we get

(**b**) $$N = \int_{r_i}^{r_o}\int_0^{2\pi} C\, d\theta\, d\rho = 2\pi C(r_o - r_i), \qquad \text{or} \qquad C = \frac{N}{2\pi(r_o - r_i)},$$

where the limits have been taken for an annular area of outside radius r_o and an inside radius r_i. To get the frictional torque, multiply dN by f to get $dF = f\, dN$, the moment of which about the central axis of the area is

$$dT_f = \rho\, dF = \rho f\, dN = fp\rho^2\, d\theta\, d\rho = fC\rho\, d\theta\, d\rho.$$

Integrate with f constant [see Ref. *(18.7)* for f variable];

(**c**) $$T_f = Cf\int_{r_i}^{r_o}\int_0^{2\pi} \rho\, d\theta\, d\rho = \pi Cf(r_o^2 - r_i^2).$$

Substituting the value of C from (**b**) into (**c**), we get

(18.2) $$T_f = \frac{fN(r_o + r_i)}{2} = fNr_m, \qquad \text{[1 PAIR FACES]}$$

where r_m is the mean radius of the annular surface, $r_m = (r_o + r_i)/2$, and N the axial force between a pair of faces in contact. The total torque that can be transmitted when more than one pair of faces is in contact with a normal force N on each pair is the value from (18.2) multiplied by the number of pairs in contact.

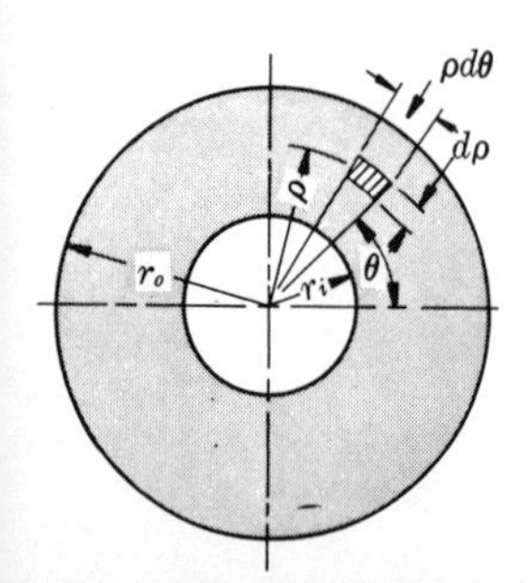

FIGURE 18.9 Disk Friction.

Disk brakes have found favor in some areas of heavy duty because they can be designed to radiate heat faster with less trouble from fading. Chrysler claims that their self-energizing disk brakes operate at temperatures 35% lower than usual (no detail given). Self-energizing is attained by virtue of the wedging action of steel balls that move up ramps between the disks and press outward on the friction surfaces when the brake is applied.

FIGURE 18.10 **Plate Clutch. The release bearing *A* is mounted on the shaft *K* to the transmission, and when the clutch pedal is depressed, the bearing *A* moves toward the flywheel *B*, contacting the inner ends of the release levers *C*. The release levers are pivoted at *D*, on pins mounted in the eyebolts *E*. The outer ends of the release levers engage the pressure plate lugs *F*, pulling the pressure plate *G* away from the driven plate *H*, compressing the several coil springs *I*, and disconnecting the drive from the flywheel (which is attached to the engine shaft) to the shaft *K*. When the foot pressure is removed from the clutch pedal, the springs *I* force the pressure plate *G* forward against the driven plate *H*, and power is transmitted through two pairs of friction faces, one pair on each side of the driven plate *H*, which is mounted on a splined connection *M*. (Courtesy Borg and Beck, Chicago).**

18.11 GENERAL REMARKS ON DISK CLUTCHES.

Think of equation (18.2) as applied to a clutch; it shows that for a given N and f, maximum torque is transmitted when r_i is a maximum, but if $r_i \approx r_o$, the pressure (psi) would be quite high and the clutch would rapidly wear out. Although an approximate optimum balance of wear and torque is obtained for $r_o = 2r_i$, such an optimum may not be the best solution for a particular application; more typical, $r_o \approx 1.3r_i$.[18.7] For example, a certain heavy tractor clutch has $r_o = 1.33r_i$, in which the friction disks are lubricated sintered metal for which wear was not a serious factor; on the other hand, a large tractor had a dry friction disk clutch for which $r_o = 2.2r_i$.

An automotive type of disk clutch, called a *single-plate clutch*, is shown in Fig. 18.10, in which it is seen that a single-plate clutch has *two* pairs of faces in contact, each with the normal force N produced by the springs I. In the multiple-disk clutch of Fig. 18.11, all surfaces are subjected to virtually the same normal force, and there are 4 pairs of clutch surfaces. Multiple disks, while frequently needed to handle the load, have the disadvantage of heavier rotating masses (with correspondingly greater inertia).

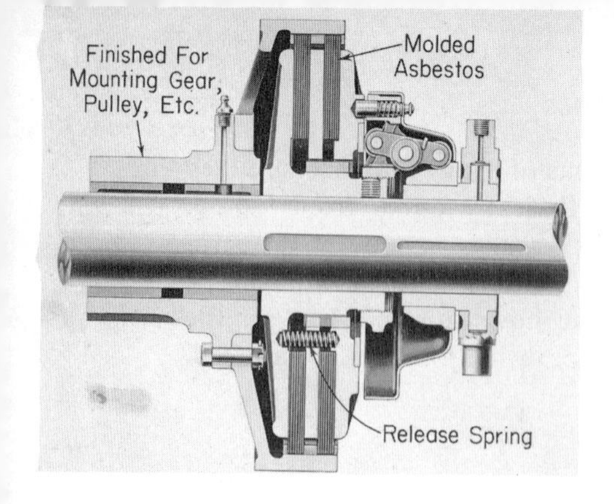

FIGURE 18.11 Multiple-disk Clutch. (Courtesy Link-Belt Co., Chicago).

To be on the safe side, design values of the coefficient of friction for clutches should probably be somewhat lower than those for brakes, say about 0.2 to 0.3 for dry asbestos facing and about 0.1 or less in oil. The normal pressure between the disks may be of the order of 25 to 30 psi for commercial vehicles, but some 30 to 37 psi for passenger-car service. Pressures are as low as 15–18 psi for bus service and other heavy-duty service where starts are frequent. Another manufacturer of industrial clutches (pedal-operated) shows pressures varying from 42 psi for the 6-in. size to 26 psi for the 14-in. size, and on a type with a relatively large internal radius r_i, Fig. 18.9, pressures up to 60 psi are used. In addition to varying the pressure to fit the service, apply a service factor to the computed torque if operating conditions are not favorable.

For clutches that cycle frequently, there may be a heating problem, inasmuch as it takes time to dissipate the frictional energy. The work done by the frictional torque will be equal to the energy (kinetic, plus potential if a body moves upward) that the driven body gains while the clutch faces have relative motion (driven parts brought up to speed); this torque may or may not be the maximum that can be delivered. When engagement is complete, the frictional work is theoretically zero and no longer a problem. In addition to frictional energy, electromagnetic clutches must dissipate the electrical losses.[18.19] If the rate of frictional work during engagement is high, excessive surface temperatures may result in welding of metallic plates, in disintegration of friction materials, warping, large variation of the coefficient of friction, and early failure.[18.7]

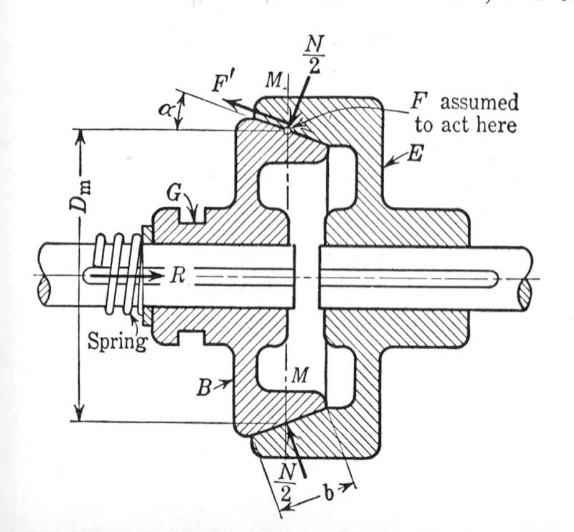

FIGURE 18.12 Cone Clutch. Forces shown are acting on the male member. The normal force N is assumed to be symmetrically distributed; so that the resultant of that part of the normal force on the upper half is in the plane of the paper and equal to $N/2$. The resultant *frictional* force F on either half will then be normal to the plane of the paper.

18.12 CONE CLUTCH. In the diagrammatic representation of Fig. 18.12, a forked lever engaging the groove G moves the part B in and out of contact with E, thus connecting and disconnecting the shafts. Think of a free body of the male member B. The resultant normal force on each half is shown, $N/2$, Fig. 18.12; the frictional force F is also distributed, $F = fN$. Assuming that the resultant frictional force acts at the mean diameter, we find the frictional torque to be

$$\text{(d)} \qquad T_f = \frac{FD_m}{2} = \frac{fND_m}{2} = \frac{fD_mR}{2(\sin\alpha + f\cos\alpha)},$$

where the value of N has been substituted from **(e)** below. The normal force N arises from the spring force R, which needs to be large enough to overcome the frictional force F' during engagement, as well as to produce the normal force necessary for the power transmission; F' is also taken as fN. Sum forces horizontally and get

$$\text{(e)} \qquad R = N\sin\alpha + F'\cos\alpha = N(\sin\alpha + f\cos\alpha).$$

While the capacity T_f increases as α decreases, α should be greater than 8° if the clutch is not to have a strong tendency to grab; typical $\alpha = 12°$.

18.13 BRAKING MATERIALS. Stress relieved cast iron, with alloys to qualify it for about class 30, makes an excellent brake drum material. Some manufacturers use nodular iron; steel, stainless steel, monel, and aluminum are sometimes appropriate. Since the conductivity of aluminum alloys is some twice that for iron or steel, the frictional energy generated at the braking surface is conducted away faster, tending toward a lower surface temperature. It is thought that this phenomenon reduces the loss of braking from fading by avoiding some reduction in f. Presuming sufficient strength, we may say that the life of a brake drum depends on its resistance to thermal fatigue and to wear. The braking members are subjected to temperature gradients, often to the limit of their capacity. The hottest layers expand the most; if they are on the inside of a drum, the outside layers of the drum are in tension and the inside ones in compression. If the highest stress in the stress gradient exceeds the yield strength at its operating temperature, permanent deformations and residual stresses are the result; if yielding occurs at the hot inside, the stresses are reversed after ambient temperature is again reached, tension inside, compression outside. Brake drums that have been subjected to severe cycling of this sort develop evident "temperature checks," which may lead to a fatigue failure. A smooth drum surface is desirable to avoid excessive wear on the lining, but it is soon unavoidably roughened somewhat from galling.

Brake linings are made of many materials, but most of them are of some mixture of asbestos and a binder, molded or woven, with or without wire or metal chips or metal powder. Since the metal is for the purpose of improving the heat conduction, it is usually brass or copper. For higher temperatures (above 400°F) and heavier loads, sintered metal linings, say with either a copper or iron base with lead, tin, graphite, and/or silica in the mixture, will absorb energy faster with less checking of the drum. For the temperature range of 750° to 1000°F, sintered mixtures containing ceramics are used.[18.13] Any of the general classes of linings may be used either dry or in oil. Oil is sometimes used only as a coolant; at other times, it is necessary as a lubricant for other parts. When used in oil, the pressure must be greatly increased (and the lining must be able to withstand the pressure), Table AT 29, for perhaps two reasons: first, a heavy pressure is needed to break through the oil film between band and drum or disk, and second, a large pressure offsets the much lower coefficient of friction.

18.14 COEFFICIENT OF FRICTION. Data on the coefficient of friction in general seem to be so disparate that the only safe generalization is that the coefficient of friction varies. For clutch and brake material, the conventional rule for friction, to wit, that the coefficient of friction is independent of the total normal force, does not apply. The curves of Fig. 18.13 depict the results for a typical lining for automatic-transmission use.

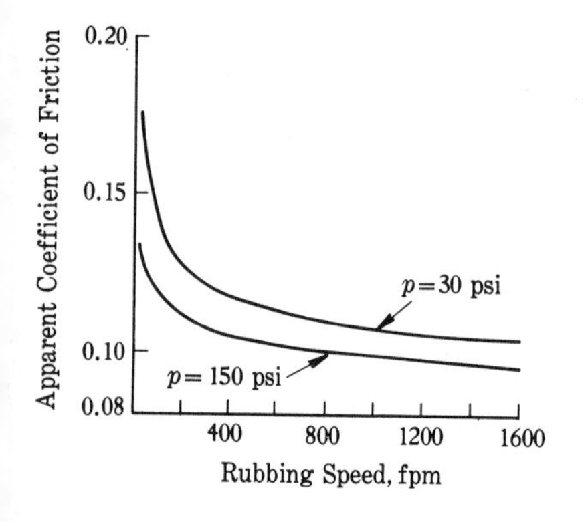

FIGURE 18.13 Coefficient of Friction. After Ref. (18.21). The variation of f for a cork-facing material for a clutch in oil (automatic transmission).

Rabins and Harker, who may be referred to for a brief review of theories regarding friction [also Ref. *(12.22)*], found a much larger spread for dry molded material, from $f \approx 2$ for a load of 2 lb. to $f \approx 0.6$ for a load of 80 lb., though the change is asymtotic with a relatively small difference in some cases for loads of 20 lb. and 80 lb. (area $= 0.5 \times 0.625$). Some other findings reported by these experimenters are: the coefficient of friction is affected by the presence of a film deposited on the metal surface by the friction material; for the molded materials tested and the speed range (40–240 fpm), f changed little with sliding velocity; the variation of f with temperature when $t < 500$°F was small (decrease of f at higher temperatures may be associated with changes in the deposited film); after run-in, f is different for different metals (for examples, at a load of 40 lb., $f \approx 0.4$ for Al, $f \approx 0.5$ for C.I., $f = 0.7$ for steel).

Miscellaneous factors that bring about changes in f include: materials, surface finish, surface temperature, pressure, rubbing speed, foreign matter including the deposited film on the rubbing surface, moisture in the atmosphere, number of cycles of operation, and changes of properties of the fluid, if any, on the braking surfaces. As in any other design, a margin of safety is needed; that is, the design is such that the designer is convinced that the desired braking or clutching effect can be obtained, or that his design has properties as close to the desired ones as possible. As a safety measure, some designers use $f = 0.25$ in design when the typical experimental coefficient for the usual friction materials is about 0.4.

For metal on metal, it may be more safely assumed that the coefficient of friction is independent of the pressure; however, with significant changes of speed, f may vary widely. One series of tests for steel on steel (with Fe_3O_4 film) showed f decreasing from 0.4 at 500 fpm to 0.1 at 5000 fpm;[12.22] with a film of molybdenum disulfide MoS_2 for the same speeds, f varied from about 0.13 to 0.09. Even though a good frictional force is desired, the latter combination with low f's may be preferred in some applications because of the lesser variation of f. Starting with quite clean steel-on-steel surfaces, we find the coefficient varies through a large range, often from values well above 1, to much lower values as the surface oxydizes, as it does in the usual operating surroundings. With some materials (e.g., polyethylene, polytetrafluoroethylene), the coefficient decreases markedly at low speeds as the speed decreases[18.23]—contrast with Fig. 18.13. And so it goes.

18.15 OTHER TYPES OF BRAKES AND CLUTCHES. In addition to the fact that the various basic elements previously described can be adapted to either brake or clutch operation, there are a number of special designs, often patented. Grooved wheels, much like a V-belt sheave, are used when the wedging effect of the groove is helpful; water brakes are indispensable in some circumstances; electrical and magnetic clutches and brakes (dynamometer) are quite common, as are fluid couplings. In one class, centrifugal force causes a friction member to move outward and press against a drum on the driven member, the higher the speed, the greater the frictional force. Representative of this category is the friction-band clutch of Fig. 18.14, where a driving arm attached to the driving shaft tows the weighted friction band which is free to exert radial pressure in accordance with the law of centrifugal force ($mv^2/r = mr\omega^2$). Thus, the torque that may be transmitted depends upon the weight of the friction band (with its attached material), the speed of the driving shaft, and the inside diameter of the drum. Interesting characteristics of centrifugal clutches include:

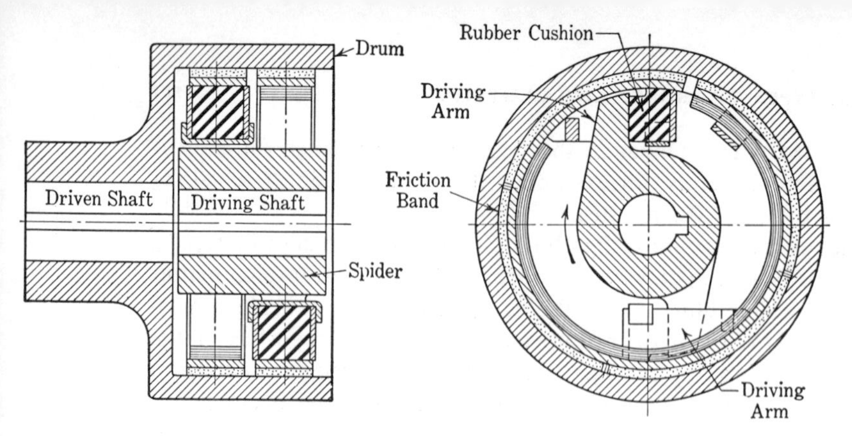

FIGURE 18.14 Friction-band Clutch. This starter is applicable for one direction of drive. Starters designed for reversible drives are available. They may also be obtained in conjunction with flat-belt, V-belt, chain, and gear drives. Rigid shoes subjected to centrifugal force are also used in a similar manner. (Courtesy J. P. Madden, Bethlehem Steel Co.)

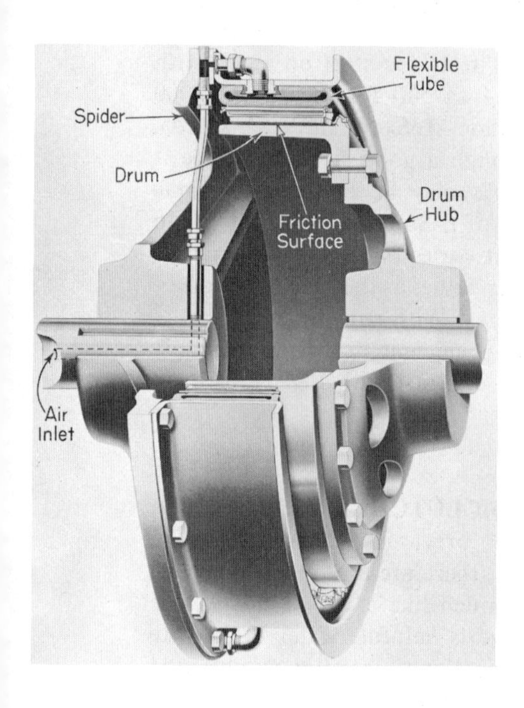

FIGURE 18.15 "Airflex" Clutch. (Courtesy Federal Fawick Corp., Cleveland).

(1) the driving shaft almost reaches its operating speed before the driver is subjected to a significant torque; (2) consequently, the size of a prime mover may be smaller than would be necessary if high starting torque were required at low speed; and (3) the clutch may be designed to slip before a dangerous overload occurs.

The "Airflex" idea, Fig. 18.15, is used for both brakes and clutches. The normal force on the friction surface is applied by air pressure in the flexible tube. This type of clutch is made in ratings from 5 hp to 2880 hp at 100 rpm. The idea of air-pressure operation is also used on disks.

18.16 CLOSURE. Brakes and clutches, like some of the other machine elements, can often be selected from the product lines of specialized manufacturers. Both of these elements deserve the use of a service factor to care for extreme operating conditions. We have said nothing of the design of the links, pins, and shafts for strength (and the design of bearings) on the assumption that, with a knowledge of all the forces, the reader is equipped to design these things by the principles already covered.

19. DESIGN OF WELDED JOINTS

19.1 INTRODUCTION. Welding, rather than being a machine element, is a manufacturing process, which reminds us that there are many facets of design in addition to stress analysis. In fact, stress analysis and sizing of parts likely consume only a minor part of the total design time. In most instances, designs are affected in some significant way by manufacturing processes, which must be learned about elsewhere, perhaps with help from the instructor. However, since the conventional stress analyses of welds often take on an esoteric flavor, they are worth attention. The effect of this manufacturing process on design is great enough to give welded machines and machine elements a distinctive appearance (see Figs. 2.16, § 2.24, 13.18, § 13.32, and Fig. 19.1). The designer must exercise his ingenuity in applying welding advantageously in his own designs; but there is considerable help in the literature.

Whether to weld, cast, or forge, etc., is an economic problem that may be answered correctly in different ways, depending upon local circumstances. Welding may be the least expensive process where the pattern cost for castings would be a large percentage of the total cost or where there are unusual machining or casting difficulties. Large weldments are built up from easily fabricated parts, as in the gear blank of Fig. 13.18. To meet the demand, there are available a number of special rolled shapes which are produced particularly for weldments, special screws and studs designed to be welded in place, etc.

FIGURE 19.1 A Frame of Welded Construction. (Courtesy Lukenweld, Inc., Coatesville, Pa.)

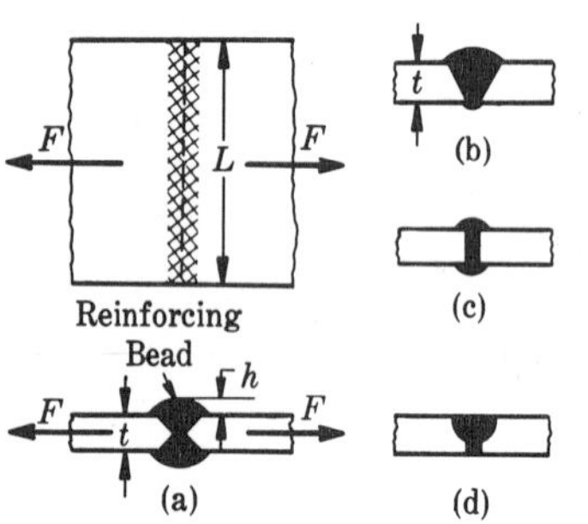

FIGURE 19.2 Butt Joint. In (a), a double-V groove, with reinforcement on both sides; in (b), a single-V groove, reinforced; in (c), a square groove, reinforced; in (d), single-U, reinforcing removed.

19.2 BUTT JOINT. The plates for butt joints, Fig. 19.2, may be unbeveled (square groove), beveled on one side only, or beveled on both sides; the groove is given various shapes such as V, U, J, for reasons associated with the economics of the job. A butt weld may be reinforced on both sides, on one side, or not at all. When there is no reinforcement, the usual practice is to build up the bead and then machine or grind it flush. With or without reinforcement, the nominal average stress is computed from

(a) $$F = s_t tL,$$

where L is the length of weld and t is the plate thickness, Fig. 19.2. Tests suggest that reinforced butt welds *on mild steel* made with filler rod of about the same mechanical properties as the parent metal have about the same *static* strength as the plates, but it would be safer to assume a relative strength (§ 17.6), or efficiency, of $\eta = 90\%$ or less; note the stress concentration due to the discontinuity of the bead.

19.3 FILLET WELDS. Fillet welds may be loaded transversely, Fig. 19.3(a) and (b), or longitudinally (parallel welds, that is parallel to the vector F), Fig. 19.3(d), or of course at some other angle with vector F.

The standard full fillet weld has a section of an isosceles right triangle, as shown, often with the legs b of the triangle equal to the plate thickness, although they are also often less than this; moreover, one leg may be longer than the other. A reinforced fillet weld, Fig. 19.3(a), is one that has a throat dimension $t' > b \cos 45°$. For a particular amount of metal, a fillet weld with a concave surface (not shown) is relatively weak. However, the sharp corner where the weld joins the surface of the plate at B, Fig. 19.3(a), and, to a somewhat lesser extent, the junction of a standard fillet weld at C, are points of stress concentration. If the joint is subjected to repeated loading, it may be worth the cost to make this junction less abrupt; however, do not overlook consideration of discontinuities such as P, P', Fig. 19.3.

The nominal stress as computed for the configurations of Fig. 19.3 is taken to be a shear stress, and the area assumed to be resisting the load is always the *throat area*, because weld failures are more often across the throat—but the *size of the weld is its leg dimension b*. Throat areas for Fig. 19.3(a) are tL at C and $t'L$ at B; but assume that both welds are standard, each with area tL; then for Figs. 19.3(a), (b), and (d), the nominal stress is computed from

(b) $$F = s_s(2tL) = 2s_sLb \cos 45°,$$

where $t = b \cos 45°$. Although, as you readily recognize by now, this simple equation does not care for the complications of the actual stress

FIGURE 19.3 Fillet Welds. The various dotted boundaries about the welds suggest the penetration of weld metal. The points P designate theoretical points of stress concentration at the heel. When the base metal has melted and solidified with filler metal, the stress concentration points become such as P', suggested in (a), (b) and (d), which, it might be observed, may become very sharp discontinuities. In (a), transverse loading, the upper weld is standard, the lower one is reinforced; usually the welds are made alike. Fatigue loaded, the reinforced fillet weld would not be expected to be any stronger, if as strong, as the standard weld. Lap welds are sometimes single-welded. In (b) and (c) are indicated closed T-joints, one (b) with no bevel, one (c) with a single bevel. Double V and other groove shapes are used. In (b), note the sharp stress concentration point P', which difficulty, for varying loading, can be alleviated by double bevel, welded from both sides with the opposite welds joining each other. In (b), the full strength of the plate M can be attained (static loading) by welds with $b \approx 0.75\ h$. For the load F as shown in (c), it is evident that a symmetric disposition of weld metal with respect to F would be preferred, but it is not always possible to weld from both sides. In (d) are shown standard fillet welds, loaded longitudinally. Drawing symbols for designating welds are found in various books.[19.3]

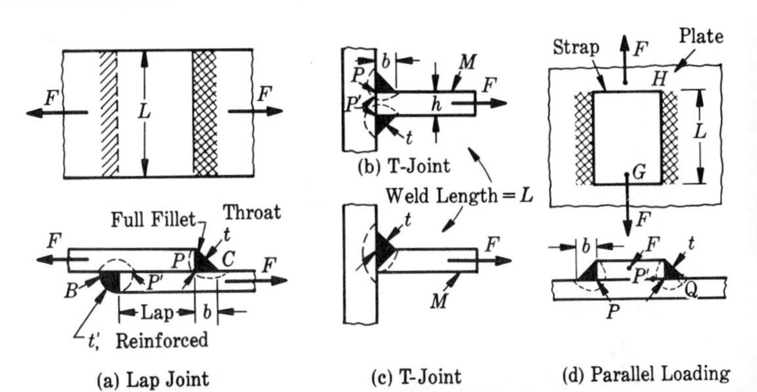

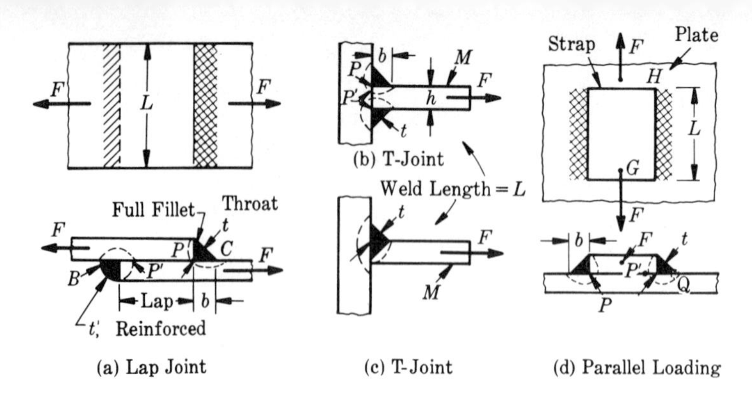

FIGURE 19.3 (Repeated).

patterns, it is suitable enough for static loads when the recommended design stresses are used.

In the lap joint, we observe that the forces F are not collinear and that therefore the weld is subjected to bending. This fact may cause the designer to be somewhat more conservative in his choice of design stress, or better, try to design a joint with the weld metal symmetric with respect to F—perhaps a butt joint.

The situation is equally bad for the parallel weld, Fig. 19.3(d). It is known that the stress at the end of the weld G where the applied load acts is much greater than the average, one experiment[19.7] giving a ratio of about 1.9 to 1 (two 4-in. lengths of 0.35-in. weld). This ratio would be expected to vary materially, being about unity for very short longitudinal welds; also it is a function of the ratio of the sectional area of the plate to the sectional area of the strap,[19.8] Fig. 19.3(d). With some proportions, the stress tends toward a minimum near the mid-length of weld;[19.8] with others, the stress at the end H, Fig. 19.3(d), is the minimum. However, the true peak stress, if the elastic limit has not any where been exceeded, is localized at a reentry discontinuity, point P or Q. If a small part of the metal here is stressed beyond the yield strength, no harm would be done for static or nearly static loads. Nevertheless, if the weld is other than "short," a higher design stress may be used for transverse loading than for parallel, Table AT 30.

19.4 FILLET WELDS, ECCENTRIC LOADING.

Since there are many ways to impose eccentric loading, the cases that follow are suggestive only of ways to make stress analyses.[19.20]

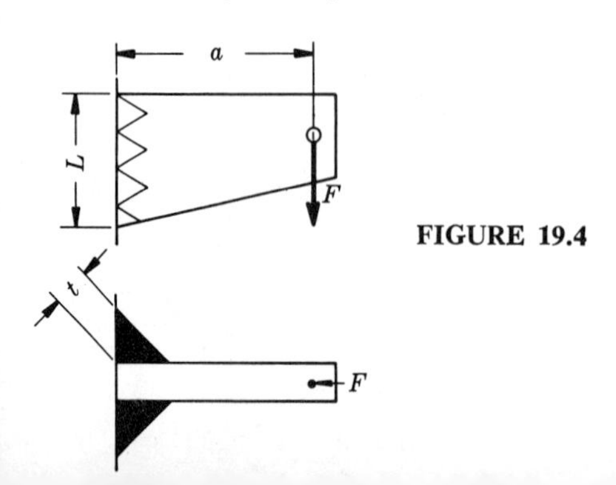

FIGURE 19.4

(a) Case 1, Fig. 19.4. Imagine the force F shown replaced by a force F at the weld, that produces a shear stress, and a couple Fa, which is the bending moment; $M = Fa$. The section modulus through the throat, the conventional approach, is $tL^2/6$ for each weld or $Z = tL^2/3$ for both sides. These values substituted into $s = Mc/I$ give the normal stress,

$$s = \frac{M}{Z} = \frac{3Fa}{tL^2} = \frac{3Fa}{bL^2 \cos 45^\circ} = \frac{4.24Fa}{bL^2}. \tag{c}$$

Assuming a uniformly distributed shearing stress, we get

$$s_s = \frac{F}{A} = \frac{F}{2tL} = \frac{F}{2Lb \cos 45^\circ} = \frac{0.707F}{Lb}. \tag{d}$$

The maximum shearing stress theory then gives

$$\tau = \left[s_s^2 + \left(\frac{s}{2}\right)^2\right]^{1/2} = \left[\left(\frac{F}{2tL}\right)^2 + \left(\frac{3Fa}{2tL^2}\right)^2\right]^{1/2}, \tag{e}$$

from which the required length of weld L for a particular design stress τ may be found, or vice versa.

(b) Case 2, Fig. 19.5. There are a number of configurations that would be analyzed in accordance with the principles used in this case. Conventionally, we consider the welds as lines (adjacent to the plate being welded), find the centroid C of the "weld lines," replace the eccentric force F by a force $F' = F$ through the centroid and a couple Fe, Fig. 19.5,

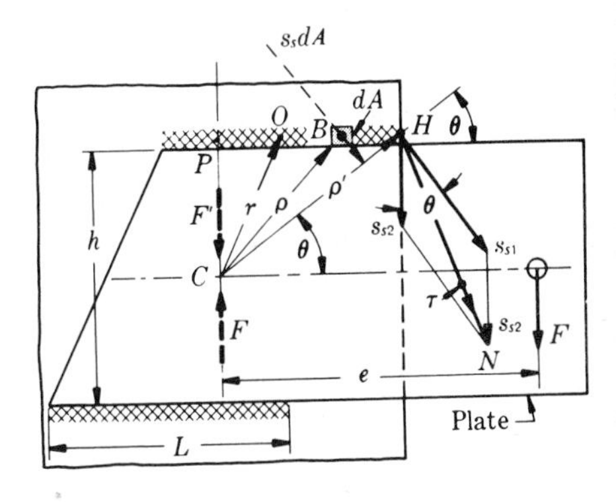

FIGURE 19.5 Eccentric Load.

and then compute a nominal stress that is intended to be the maximum. The stress s_s due to the moment Fe at any point B of the weld, Fig. 19.5, is assumed to be proportional to its distance from C; that is, $s_s/\rho = s_{s1}/\rho'$, where s_{s1} is the maximum stress, which occurs at the maximum radius ρ', point H. Thus, at B, the shearing force perpendicular to ρ is taken as $s_s\, dA$, and the resisting moment of this force about C is $\rho s_s\, dA$, where dA is arbitrarily taken as an infinitesimal part of the throat area. Using $s_s =$

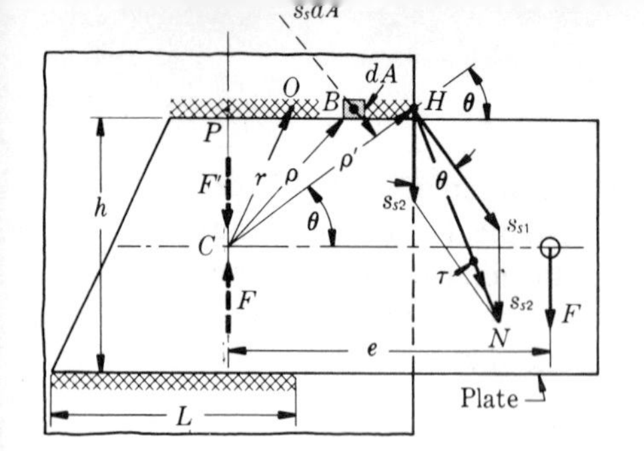

FIGURE 19.5 (Repeated).

$\rho s_{s1}/\rho'$ and equating the applied moment Fe to the resisting moment, we get

(f)
$$Fe = \int \rho s_s \, dA = \frac{s_{s1}}{\rho'} \int \rho^2 \, dA,$$

where we note that $\int \rho^2 \, dA$ is a polar moment of inertia J_C of an area with respect to C. Equation **(f)** may therefore be written

(g)
$$Fe = \frac{s_{s1} J_C}{\rho'} \qquad \text{or} \qquad s_{s1} = \frac{Fe\rho'}{J_C},$$

where J_C is computed as follows. The moment of inertia of a long slender area with respect to a centroidal axis O, perpendicular to the area is closely $\bar{J} \approx AL^2/12$, where L is the length of the area A and the other (throat) dimension is small compared to L.[1.6] Let A be the throat area tL and apply the parallel axis theorem ($J = \bar{J} + Ad^2$), Fig. 19.5, to get the polar moment of inertia about C. For each strip of weld,

(h)
$$J_{C1} = \bar{J} + Ad^2 = \frac{AL^2}{12} + Ar^2,$$

where r is the distance between the axis O of a weld line and the axis C, which marks the centroid of all the weld lines. In equation **(g)**, J_C should be the moment of inertia of all the throat areas about C, which is the sum of the moments of the individual strips of welding; all moments of inertia are positive.

In addition to the stress from equation **(g)**, which is in a direction perpendicular to the radius ρ', Fig. 19.5, a uniform shear stress is assumed to be induced by the force $F' = F$;

(i)
$$s_{s2} = \frac{F}{A},$$

where A is taken as the total throat area and the direction of the stress is downward. The stresses s_{s1} and s_{s2} add vectorially, or from the cosine law, Fig. 19.5,

(j)
$$\tau = (s_{s1}^2 + s_{s2}^2 + 2 s_{s1} s_{s2} \cos \theta)^{1/2},$$

which is taken as the maximum shearing stress. The foregoing analysis is not only approximate, but it presumes that there is no tendency for the plate, Fig. 19.5, to twist.

The odd distribution of weld metal in Fig. 19.5 was chosen deliberately to be sure that the principles involved were not obscured by the simplicity of a symmetric arrangement. If a connection were to be made similar to Fig. 19.5, it would be more like Fig. 19.6, perhaps with additional weld on the left-end short dimension—and probably with some modification to reduce or evade a twisting of the plate. Reports of experimental justifications of the various equations used in designing eccentrically loaded welded joints are scarce. In one experiment,[19.7] which was limited and therefore inconclusive, with welds as in Fig. 19.6 and two plates to avoid twisting, the indication was that the actual maximum stress is higher than that predicted by equation (j). Both theory and experiment give mostly an increasing stress gradient from left to right, Fig. 19.6, but the experiment indicated a relatively higher stress at B than theory, for reasons that are unclear—possibly some twisting augmenting the stress. Otherwise, the experiment could be said to agree with the theory. If care were taken in actual design and fabrication to apply the load F so as not to twist the plate, then the small twist that may actually exist could be allowed for during design by arbitrarily moving the center of moments away from the load by say 5 to 8% of the theoretical e; that is, use a new center C' (not shown) at $e' = 1.05e$ to $1.08e$ for all calculations, which is a shift in the conservative direction.

The configuration of Fig. 19.7 presents the same problem as far as the theory is concerned. The maximum $Fe\rho'/J_C$ stress is at the point of maximum ρ', but one must be certain that the resultant shear τ used is the maximum. Note in Fig. 19.7 that τ_B and τ_A are each less than τ_G by virtue of the directions of the vectors.

19.5 EXAMPLE—ECCENTRICALLY LOADED FILLET WELD. In Fig. 19.6, $h = 6$ in., $e = 24$ in., $L = 10$ in. and the welds are $\frac{1}{4}$ in. (leg dimension). What would be a safe variable load F?

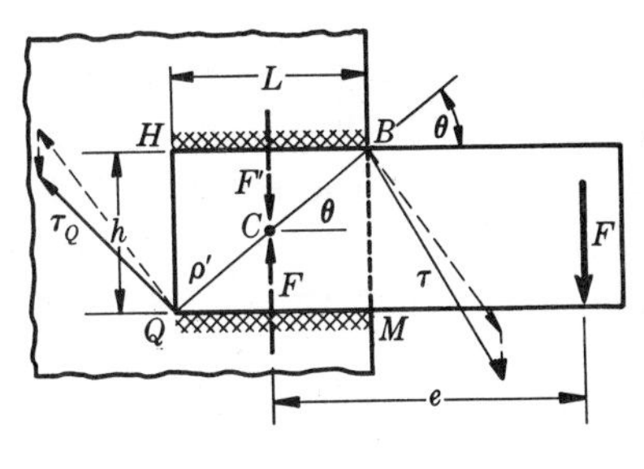

FIGURE 19.6 Observe that the larger h, the greater the resistance of a particular length of weld.

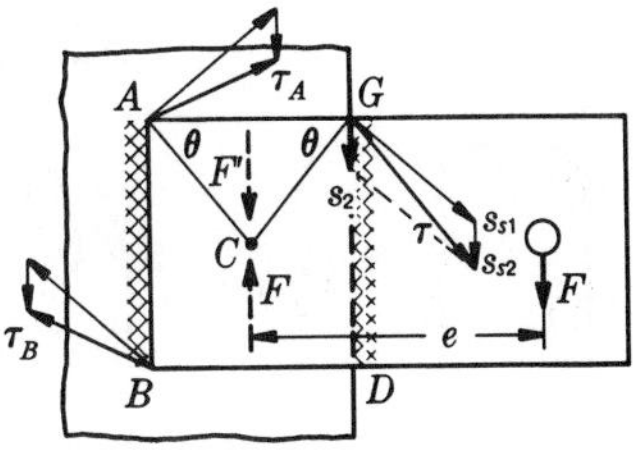

FIGURE 19.7

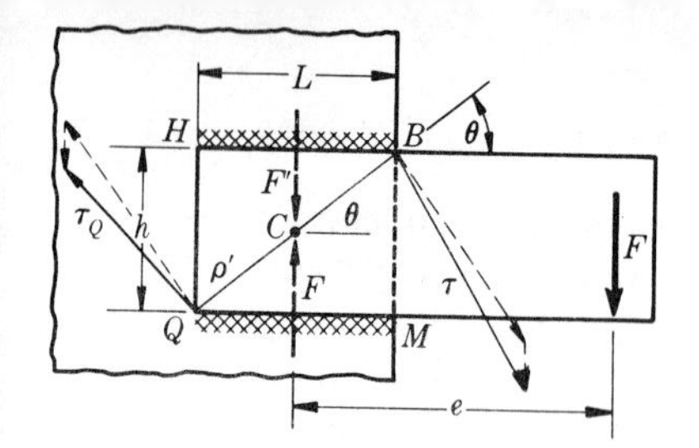

FIGURE 19.6 (Repeated).

Solution. This connection is neither parallel nor transversely loaded. We decide to use the Jennings recommendation in Table AT 30 for fillet welds ($s_a = 14$) and a strength reduction factor for stress concentration of $K_f = 1.4$ because the load is varying; a safe calculated stress is then $14/1.4 = 10$ ksi. (See § 19.8.) With $t = 0.707b = (0.707)(0.25) = 0.177$ and $A = tL$ for one weld, we find from **(h)**

$$J_C = 2\left[\frac{AL^2}{12} + A\left(\frac{h}{2}\right)^2\right] = 2\left[\frac{(0.177)(10)^3}{12} + (0.177)(10)(3)^2\right] = 61.4 \text{ in.}^4$$

The centroid C of the weld lines is located by symmetry; the maximum ρ' is

$$\rho' = \left[\left(\frac{L}{2}\right)^2 + \left(\frac{h}{2}\right)^2\right]^{1/2} = (25 + 9)^{1/2} = 5.84 \text{ in.}$$

The stress s_{s1} at the weld ends due to the couple whose moment is Fe and the uniform stress s_{s2} are

$$s_{s1} = \frac{Fe\rho'}{J_C} = \frac{F(24)(5.84)}{61.4} = 2.28F.$$

$$s_{s2} = \frac{F}{A} = \frac{F}{2tL} = \frac{F}{(2)(0.177)(10)} = 0.282F.$$

Considering the directions of these stresses at the corners H, B, M, Q, Fig. 19.6, we conclude that they combine to give the maximum resultant at B (or M) and that equation **(j)** applies, with $\cos\theta = L/(2\rho') = 5/5.84 = 0.856$;

$$\tau = (s_{s1}^2 + s_{s2}^2 + 2s_{s1}s_{s2}\cos\theta)^{1/2}$$
$$= [(2.28F)^2 + (0.282F)^2 + (2)(2.28)(0.282)F^2(0.856)]^{1/2}$$

or $\tau = 2.53F = 10$ ksi (10 ksi = the allowable stress); then $F = 10/2.53 = 3.96$ kips or 3960 lb.

19.6 ANNULAR FILLET WELD IN BENDING.

One more case will throw additional light on the method of using differential areas. Let the annular weld of Fig. 19.8 be subjected to a moment M only; $r\,d\theta$ is the differential length; then $dA = tr\,d\theta$, when the throat area is the basis of the calculations. The force dF resisting the bending at this point is normal to the page; $dF = s\,dA = str\,d\theta$. If this stress s is assumed to be proportional to its distance from the neutral plane (horizontal center line on end view), which is $r\sin\theta$, we can write $s_1/r = s/(r\sin\theta)$ or $s = s_1\sin\theta$, where

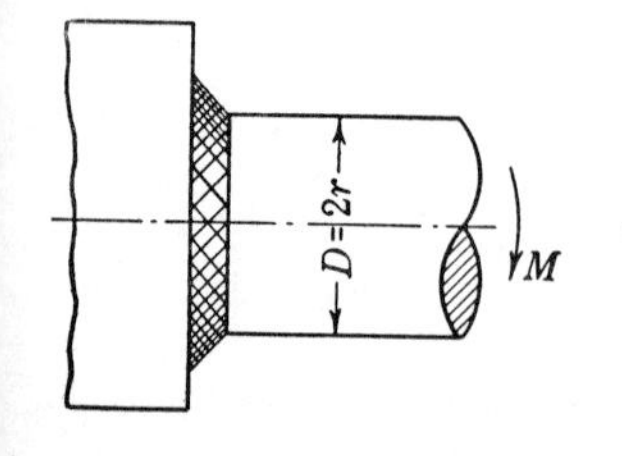

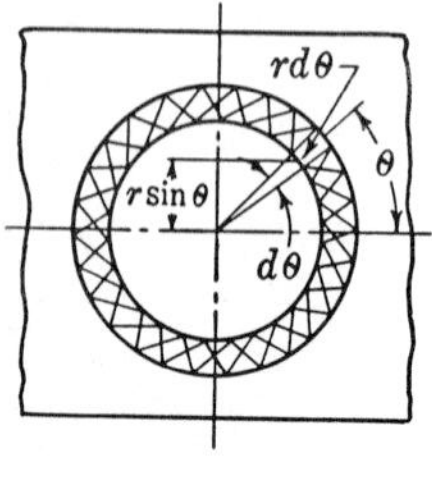

FIGURE 19.8

s_1 is the maximum stress (at top). Substitute this value of s into the expression for dF above and get

$$dF = s_1(\sin\theta)tr\, d\theta.$$

Multiply both sides by $r \sin\theta$ to get the moment dM; then

$$\text{(k)} \qquad M = \int dM = \int (dF)(r\sin\theta) = s_1 tr^2 \int \sin^2\theta\, d\theta,$$

$$M = s_1 tr^2 \int_0^{2\pi} \sin^2\theta\, d\theta = s_1 tr^2\pi,$$

$$\text{(l)} \qquad s_1 = \frac{4M}{\pi t D^2} = \frac{4M}{\pi(b\cos 45°)D^2} = \frac{5.66M}{\pi b D^2},$$

where s_1 may be taken as a normal stress. If the moment M is produced by an eccentric load, the stresses may be combined after the fashion of § 19.4(a). (The weld may be considered as a line; that is, the moment of inertia or section modulus of the line that represents the weld may be found and used in the basic equations, with the throat dimension introduced later; helpful for repetitious calculations.[19.3])

19.7 DESIGN STRESSES. The design of many structures such as bridges and building frameworks are covered by codes, which are commonly a legal requirement. Although the machine designer is not bound by these specifications, their permissible stresses, and other specified practices obtainable from the codes themselves, constitute evidence for design that need not be ignored. Table AT 30 gives excerpts that may be helpful when "regular" welding rod (as 60XX) and base metal of less than 0.3% C are used; the welding rods have some 0.15% C.[19.4] The quenching action of the cold base metal tends to harden the steel and will actually do so if the carbon content is too high. For alloy steels, if hardening and the resulting brittle zone are to be avoided, the carbon should be less than 0.2%, unless preheating and/or tempering after welding are employed. When a high-strength rod, as E 10016, ($s_u = 100$ ksi), is used on alloy steel (it would not otherwise be used), a tempering at about 1150°F, usually called *stress relieving* in this context, is necessary to obtain the benefits from the alloy. Higher carbon steels and many alloys need to be preheated (200–800°F, depending on carbon content and thickness), in order to avoid a rapid cooling that hardens the material in a zone that therefore becomes brittle. Moreover, the rapid cooling and extreme temperature range of the molten metal next to the cold base metal may result in a cracking of the weld, which is eliminated by preheat. Manganese and silicon adversely affect the weld and an *equivalent* carbon content of $C + Mn/4 + Si/4$ is a substitute criterion for the carbon content alone;[0.2] C, Mn, Si, being respectively the percentages by mass of each.

Most electric arc welds are made with coated electrodes (constituents used for coatings: cellulose, feldspar, mica, potassium silicate, calcium carbonate, and others, in various combinations), whose function is to form a protective environment about the molten metal during welding, thus keeping oxygen and nitrogen away (see § 19.11). The steel electrodes have code numbers (by ASTM and AWS) such as E 6010, where the first two (or three) digits (60) indicate a *minimum tensile strength* (ksi) and the last two digits in effect specify other variables such as the coating, current supply, position of the weld, etc. Unshielded arc welding, done with bare rod with nothing to protect the molten metal from air, results in a penetration of O_2 and N_2 that seriously impairs the quality of the weld.

There are over 12 rods in the E 60XX series; 6 or more in the E 70XX series;[6.11] numerous others for special purposes as for gas welding, brazing, welding nonferrous metals and cast iron, corrosion resistance, unusually high-strength, low-alloy-steel rod (as E 10016, a Ni, Mo, V alloy [19.18]). If the higher-strength electrodes are used, the designer must decide on a design stress with less help. In general, it would seem that a design factor of

(m) $N = 3.75$ on ultimate strength of electrode, steady load

should be adequate. A lower value may be all right if the highest quality weld is certain. The transverse static strengths of inert-gas fillet welds of certain aluminum alloys are:[19.25] for 3003, 16 ksi; for 5086, 37 ksi; for 5154, 33 ksi.

19.8 DESIGN FOR FATIGUE. If a weld is subjected to varying loading, the natural discontinuities, surface roughness, and residual tensile stresses will encourage the cautious designer to look for points of maximum stress and make checks as possible for fatigue strength—using the principles previously explained (mostly in Chapter 4). Factors that unfavorably affect the fatigue strength of a weld include: lack of penetration of the weld metal, shrinkage cracks, slag and other inclusions, porosity and gas pockets, undercutting (unfilled groove melted in base metal); all of which can be corrected by proper welding techniques and materials.

As the molten metal solidifies and cools, residual stresses are rampant, and their magnitude cannot be predicted with any degree of certainty. The rule that the last part to cool has tensile residuals is significant. This leads us to conclude that the weld metal in a butt joint as-welded has a large residual tensile stress, in a complex pattern (balanced elsewhere by residual compressive stresses, since the internal forces are in equilibrium), and this is found to be true. If a welded part is to be machined with close tolerances, it should be stress relieved by normalizing or annealing before machining, otherwise the deformations as stressed material is removed will spoil the dimensions. In structural work, removal of the reinforcement on a butt weld would be rare indeed, but it is at least less rare in welds for machines.

With the butt weld flush with the plate, the strength of the joint in fatigue is increased. Jennings[19.1] suggests the following strength reduction factors:

Reinforced butt weld, $K_f = 1.2$ Toe of transverse fillet weld, $K_f = 1.5$
T-weld, sharp corners, $K_f = 2.0$ End of longitudinal fillet weld, $K_f = 2.7$

There is a certain uncertainty attached to these numbers. Consider values

TABLE 19.1 SELECTED FATIGUE DATA

Strengths in ksi, average values.
(a) W. M. Wilson.[0.2] (b) Standard cylindrical test specimens, cut transverse of butt weld; E 10015 (Ni, Mo, V) electrode; rotating beam fatigue. (c) Weld metal as welded. (d) Tempered at 1150°F, s_n at 2×10^7. (e) Annealed at 1650°F. (f) Double-V butt, $\frac{3}{8}$-in. plate, axial load. The fatigue strength of welded aluminum is in about the same range for all the alloys used. (g) From Ref. *(19.6)*; failure of weld metal.

MATERIAL AND JOINT	BASE METAL		ENDURANCE STRENGTH s_n, ksi							
			$R = -1$		$R = 0$			$R = 0.5$		
			No. of cycles		*No. of cycles n_c*			*No. of cycles n_c*		
	s_u	s_y	2×10^6	10^5	2×10^7	2×10^6	10^5	2×10^7	2×10^6	10^5
BUTT JOINTS										
$\frac{7}{8}$-in. plate, carbon steel (a)	61.5	34.2								
With bead, as welded			14.5	22.1		22.7	33		36.8	53.3
Ditto, tempered 1200°F			14.3	21.4		23.3	31.9		38.4	55
Bead mch. off, no H.T.			17.6	28.7		28.7	48.8		44	
Bead off, 1200°F temper			16.6	28		27.7	49.3		42.5	
Alloy steel (b)	106(c)	97(c)								
As welded			58	78						
Stress relieved (d)			66	86						
Annealed (e)			48	62						
WQT 1050°F			61	79						
Al 5083-H113 (f)										
As welded	47.3	26.2			8	9.2	18	13.2	15.7	
Bead removed					16.2	16.6	24.5			
FILLET WELDS										
Carbon steel (a)										
Single pass transverse			11.7	16.7		18.1	30.2		40.1	46.1
Single pass, parallel			11.8	15.2		20.6	27.5		37*	
ASTM-A7 steel (g)	60–72	33								
Plug weld			6.5	11.3		12.6	23.1			

* Very wide spread, 28–45 ksi.

of K_f that can be determined from Table 19.1. Computing the values of K_f for the first material in Table 19.1, the $\frac{7}{8}$-in. plate, we find a range of K_f from 1.16 to 1.31; for the aluminum alloy 5083-H 113, Table 19.1, we find K_f values of 1.36 to 2.02. Not only are some of these values significantly different from 1.2, but they vary in different ways. For the steel, K_f seems to increase as the number of cycles decreases, while for the aluminum, it definitely decreases as the number of cycles decreases (the normal expectation); moreover, these tests indicate that the aluminum is much more affected by the discontinuity of the bead (more notch sensitive—higher K_f). On the other hand, there are some tests that indicate little improvement in fatigue with the removal of the discontinuity. If the bead is ground off, the grinding leaves a residual tensile stress on the surface that would be expected to result in a fatigue strength lower than if the residual were not present. Also the surface left exposed may have a significant tensile residual stress left from the progressive solidification of the weld metal. We read that peening, which leaves a surface compressive stress, has no effect on the fatigue strength of the joint,[19.6] usually, no doubt, with the weld as welded in mind. Nevertheless, it is reasonable to expect that if shot peening is applied to the ground surface, it would be possible to make the joint virtually as strong as the parent metal—if the welding rod and parent metal are of about the same material. Tests would show. The peening as commonly practiced on welds is hammer peening that often results in cracks and therefore stress-concentration points in the weld. Hammer peening of the first or last layers of weld has been demonstrated to be harmful for pressure vessel welds and is prohibited by code. If the expense of grinding is warranted, the additional cost of shot peening would be minor. As a matter of principle, welds should be located as remote as possible from points of large bending moment. Heat treated alloy steels would naturally need special treatments.

Since some weld metals become brittle after an annealing operation, this method must be used with discretion as a means of reducing tensile residuals. Also in this connection, observe the results in Table 19.1 for the alloy steel; tempering at 1150°F improves the fatigue strength, presumably by releasing *most* of the residual tensile stresses without a significant reduction of mechanical properties, whereas annealing at 1650°F reduces the fatigue strength to less than the as-welded value. This experiment also indicated that water quenching and tempering was not as effective as the ordinary stress-relieving heating. Available experiments suggest that stress-relieving plain carbon steel ($< 0.25\%$ C) does not improve fatigue strength. Yet temperature stress-relieving of welded pressure vessels is recommended practice (required for some materials), and is important in restoring ductility. Welding a strap across a good butt weld invariably lowers its fatigue strength.

From the experiment on the aluminum alloy of Table 19.1, we have the following observations.[19.19] Analysis of the origination and propagation of fatigue cracks showed that the prime factor causing these failures was the notch effect. The notch effect of the double-V weld was greater than that for the single-V weld. (Since the metal in the big end of the V is the last to solidify and cool, one would expect a very unbalanced residual stress pattern when the base plate is restrained from distorting.) Manual welding resulted in fatigue strengths lower than those for automatic welding. Porosity, oxide inclusions, and poor penetration significantly reduced fatigue strength; 79% of the broken specimens had defects of which about half were revealed by X ray.

One must be careful about attachments welded to main bodies. For example,[19.6] a plain carbon plate, $\frac{5}{8} \times 14$ in., had a fatigue strength (2×10^6 cycles) of 22.8 ksi; with a piece attached on one side with transverse fillet welds (to form T), the fatigue strength of the plate was 18.9 ksi; with pieces on each side, opposite each other, $\frac{5}{16}$-in. transverse fillet welds, the fatigue strength dropped to 13.1 ksi. The same experiment run with low alloy steel ($s_u = 78.9$), no change in configurations, resulted in fatigue strengths of 26.4, 23.9, 10.1 ksi, respectively. While the principal cause of fatigue-strength reduction is probably the discontinuity, the residual compressive stresses that balance the tensile residuals in the weld may, by superposition, seriously weaken a column. (NOTE. The evidence is not all in, and there is not a unanimity of opinion concerning the effect of residual stresses for structural welds.) Besukladov *et al.*[4.28] on research related to ship fatigue failures, concluded that the usual steels used for hulls all had a fatigue of plate in the vicinity of a weld of about 10.7 ksi ($n_c = 5 \times 10^6$), and that the sn_c curve that bounded the lower limit of all the failure points is defined by

$$\text{(n)} \qquad n_c = 11.5 \times 10^6\left(\frac{1}{s - 9.95} - 0.054\right),$$

where n_c is the cycles of life to be expected for a stress of s ksi.

Since the residual tensile stress in the direction of the weld length is of the order of magnitude of the yield strength, another maneuver of stress relieving is sometimes practiced. A band of the base metal on each side of the weld is heated to some 350–400°F, by oxyacetylene torches or electrical resistance, causing the bands to tend to expand (and be in compression by the action of the adjacent unheated parts). The force of this deformation elongates the length of the weld, which, already being at yield, undergoes little or no additional stress, but is plastically deformed. Therefore, when the heated bands have cooled back to ambient temperature, the weld, having been stretched, returns more or less elastically to a lower residual tensile stress in the longitudinal direction; and there is evidence of a reduction in the transverse residual tension also.[19.6] This process is used on seams in large plates, as on ships, and on girth seams of pressure vessels.

19.9 OTHER TYPES OF WELDS. The configurations in which welding may appear are quite varied. The following are worth noting.

A ***corner joint***, Fig 19.9(a), may have weld metal placed on either the inside, outside, or both. It is much cheaper to bend a plate to form a corner, but if welding is necessary, see Ref. (*19.22*).

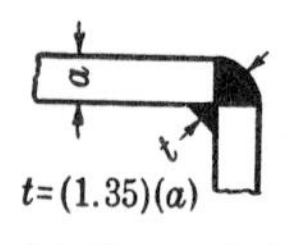

(a) Corner Joint

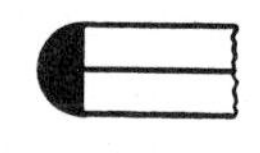

(b) Edge Joint

FIGURE 19.9

An ***edge joint***, Fig. 19.9(b), is made along the edges of two or more parallel, or nearly parallel, plates.

If one plate with holes in it lies on a second plate, a ***plug weld*** is obtained when the holes are filled or partially filled by weld metal fusing with the second plate.

Spot welds are resistance welds (§ 19.11), usually round, in the same form as the electrodes that press the sheets together; used on thin material only; see Ref. (*19.23*).

An ***intermittent weld*** consists of short lengths (2–3 in. long) of welds with space between, as 6 in. on centers. The minimum length should be at least $4b$ and not less than 1 in. This plan saves expensive weld metal when a continuous weld is not needed.

A ***tack weld*** is an intermittent weld, lightly done to hold the members in position for assembly purposes or for the principal welding.

Electroslag welding,[19.12] a method of welding rather than a type of weld, is applicable to quite heavy sections; to be investigated by those concerned with such a welding problem.

19.10 MINIMUM FILLET-WELD SIZE. If the leg dimension of a fillet weld is doubled, the amount of weld metal is increased by four times; which suggests that a light continuous weld is preferred to a heavier intermittent. However, even though strength requirements might permit a very small fillet weld, experience suggests the following minimum sizes according to plate thickness:

Plate thick., in. .	$\frac{1}{8}$–$\frac{3}{16}$	$\frac{1}{4}$–$\frac{5}{16}$	$\frac{3}{8}$–$\frac{5}{8}$	$\frac{3}{4}$–1	$1\frac{1}{8}$–$1\frac{3}{8}$	$1\frac{1}{2}$ up
Min. b in. . .	$\frac{1}{8}$	$\frac{3}{16}$	$\frac{1}{4}$	$\frac{3}{8}$	$\frac{1}{2}$	$\frac{3}{4}$ in.

19.11 TYPES OF WELDING PROCESSES. Only a few of the most common processes will be mentioned. Electric welding may be *arc welding*,

FIGURE 19.10 Automatic Welding Head. (Courtesy General Electric Corp., Schenectady).

resistance welding, or *induction welding*. Arc welding is done either with a carbon electrode on steel (tungsten electrode on non-ferrous) or with a metal electrode. The metal electrode supplies the filler metal as it melts and may be fed in manually or automatically, Fig. 19.10. When a carbon or tungsten electrode is used, a separate rod supplies the filler metal, either manually or automatically.

In addition to coated rods, inert gases, usually argon or helium or a mixture of both, are used to shield molten metal from atmospheric oxygen and nitrogen (by displacement). Inert-gas methods are very common for stainless steel and nonferrous metals; for example, aluminum and titanium, with a nonconsumable tungsten electrode and separate filler rods as appropriate.

In ***submerged arc welding***, Fig. 19.11, the arc is covered with a welding composition, and bare electrode wire is fed automatically. This process is excellent for automatic welding of flat welds (also girth seams) and is used extensively on pressure vessels. Its high speed recommends it for production jobs.

FIGURE 19.11 Submerged Welding (Unionmelt). (Courtesy Linde Air Products Co., New York).

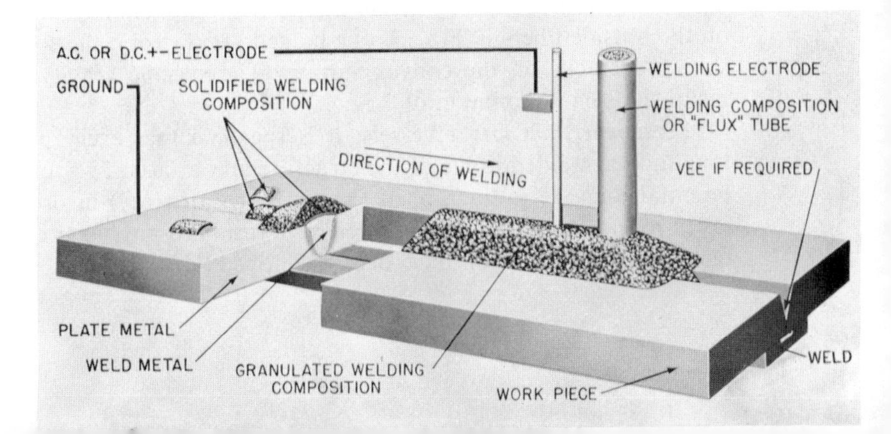

In the ***atomic-hydrogen process***, the energy from the arc is used to break the hydrogen molecules into atoms rather than to melt the metal. Then the hydrogen recombines to form molecules, releasing energy at high temperature (7200°F), and it is this released energy that fuses the metal. The hydrogen also provides the shield to avert porosity and brittleness from nitrogen and oxygen. This process is used on a wide variety of metals, manually and automatically.

The welds mentioned above are called ***fusion welds*** because the metals are joined by fusion. ***Resistance welding*** depends upon the resistance to the flow of electricity at the points to be joined. In ***spot welding***, two electrodes press the sheets of metal together, and at the spot where the pressure is exerted, the resistance to the flow of current causes a heating which, together with the pressure, results in a weld. If two copper rollers are used as electrodes and if two sheets of metal are passed between the rollers, a seam is welded where the rollers press the sheets together, in a resistance-welding process called ***seam welding***.

In ***gas welding***, a hot flame and a metal rod are used. The oxyacetylene process uses acetylene burned in oxygen. The flame heats the parts to be joined, maintaining a puddle of molten metal at the junction of the parts, and filler metal is added to form the joint. Gas welding is particularly suited to welding thin metal and is applied to many different metals.

19.12 TESTING WELDED JOINTS. When practicable, tests of welded joints should be made to ascertain the quality of the weld. In important applications, such as pressure vessels, careful tests are made. Test plates are welded at the same time that the joint is welded. From these test plates, specimens for tensile tests, for a bending test, and for determining the density are made. Some of the requirements of a weld on *Class-1* pressure vessels will be suggestive. The tensile strength of the *joint* and of the deposited weld metal must be at least equal to the minimum tensile strength expected of the plate material. The elongation of the deposited weld metal is to be a minimum of 20% in 2 in. In the bend test, the surfaces are machined flush with the plate, and the specimen is bent cold until the least elongation measured within or across approximately the entire weld on the outside fibers is 30%. To be judged satisfactory, the specimen must show no cracks on the convex surface. The specific gravity of the weld metal shall be a minimum of 7.8.

Moreover, in *Class 1* vessels, it is specified that every portion of all longitudinal welded joints of the structure and at least 25% of the circumferential joins be radiographed (X ray or gamma ray) in order to reveal excessive porosity and points of defective fusion, if any. A number of other techniques of inspection have been developed.[19.6]

19.13 OTHER METHODS OF JOINING METALS. Ordinary soldering, using a mixture of lead and tin, which is the cheapest, is applied only where strength is unimportant. Whereas soldering is done on cold parts, brazing is done with parts heated to a temperature above the melting point of the nonferrous filler material. Silver soldering is a process of this type, used where the surfaces to be joined are brought into intimate contact with the silver alloy on the fluxed surfaces. Heating the base metal until the silver alloy melts and then allowing it to solidify results in a strong connection, especially in joining copper surfaces. Heating may be done in a furnace, with an acetylene flame, by induction, etc. Alloys other than silver alloys are used in this manner. In another method of silver soldering, the surfaces to be joined are placed close together and heated in the presence of flux and silver alloy, which enter the space between the surfaces by capillary action.

In braze welding, nonferrous filler metal is melted into holes, grooves, or as a fillet on the base metal; the base metal is not melted. This process is often used for joining cast-iron parts, and is widely used in repair work.

Plastic adhesives are also available for joining metal parts.

19.14 CLOSURE. For weldment designs, there are in print many practical suggestions that represent accumulated experience, which should be consulted. For example, a liberal use of bent members and rolled sections, such as angles and X shapes, is advised. *Welding metal in place costs considerably more per pound* than structural steel.

Little has been said or implied about the human element in welding. It has been noted that manually applied welds do not average as strong as machine-laid welds, and the variability is greater too. Internal flaws, if any, are as likely to be sources of fatigue failures as the known points of stress concentration; hence, if there is any question concerning the quality of the weld, allowance should be made. Eliminating "stress raisers" by persistent design effort and manufacturing care is probably more rewarding in increasing fatigue strength than using higher strength materials.

20. MISCELLANEOUS PROBLEMS

20.1 INTRODUCTION. In this chapter, a few additional problems that appear with some frequency are presented briefly. Additional details on some of these topics, on some of the topics previously discussed, and on some not mentioned at all are found in books on advanced strength of materials and theory of elasticity.

20.2 THIN CYLINDRICAL SHELLS UNDER EXTERNAL PRESSURE. If the external pressure is large enough, the needed thickness of shell will be so great that the thick-cylinder formulas apply as explained in § 8.26. When the needed thickness is small, the thin-shell equation, which assumes uniform distribution of stress, § 1.25, accurately predicts the stress produced by the external pressure; the value of this compressive stress in a tangential direction is $s_c = pD/(2t)$; see equation (1.18). However, this stress for thin shells generally indicates nothing concerning the safety against failure, because the failure is one of elastic instability, quite analogous to long columns. Thus, there is a critical pressure that causes collapse, given by Saunders and Windenburg[20.2] as

$$p_c = \frac{2.60E(t/D)^{5/2}}{L/D - 0.45(t/D)^{1/2}} \text{ psi} \tag{20.1}$$

for E psi; where the critical pressure p_c is seen to depend on the unsupported length L in. of a vessel whose diameter is D in. and thickness of shell t in. When collapse occurs, characteristic lobes or bulges are formed, the number depending on the ratios L/D and t/D. See Figs. 20.1 and 20.2. The lower the L/D and t/D ratios, the greater the number of lobes.

If the L/D ratio is above a certain value that is within the range of $8 < L/D < 15$, the collapse is a two-lobe failure, Fig. 20.1, and is independent of the length. Economic employment of material is obtained by

FIGURE 20.1 Two-lobe Collapse. Typical of long tubes and cylinders; the lobes are not necessarily as regular as shown.

using thin plate and short unsupported lengths, rather than a large t/D. The short length is obtained by using circumferential stiffeners, the unsupported length L being the center distance between stiffening rings, Figs. 20.2 and 20.3. The ASME Code[1.10] specifies that the operating pressure $p \gtreqless p_c/5$, and in no case should the vessel be designed for a working pressure of less than 15 psi, corresponding to a minimum critical pressure of 75 psi. Moreover, the computed maximum tangential stress should not exceed $s_y/5$. The following procedure may be used: (1) Calculate the plate thickness for stress from $t = p_c D/(2s_y)$. For this thickness t, it takes approximately 5 times the working pressure to stress the plate to the yield strength and this t is taken as the minimum permissible value. (2) Compute the

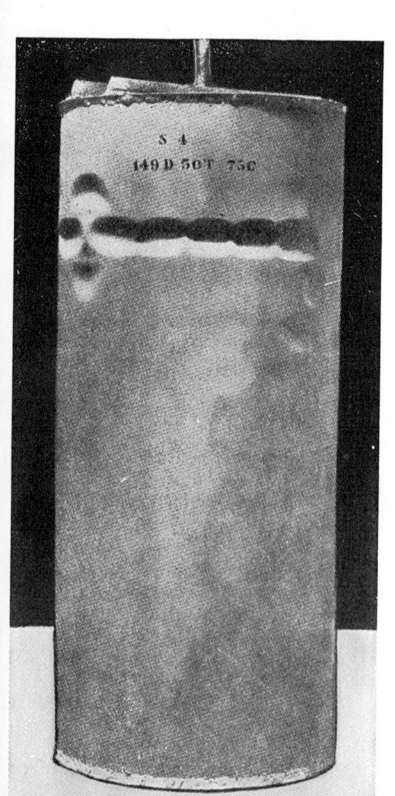

FIGURE 20.2 Thin Cylindrical Vessel with Closely Spaced Internal Stiffening Rings. This shows the formation of lobes between stiffening rings, the result of a collapsing external pressure. If the lobes had formed entirely around the circumference, there would have been 16, the theoretical number for the particular L/D ratio. (Saunders and Windenburg[20.2]).

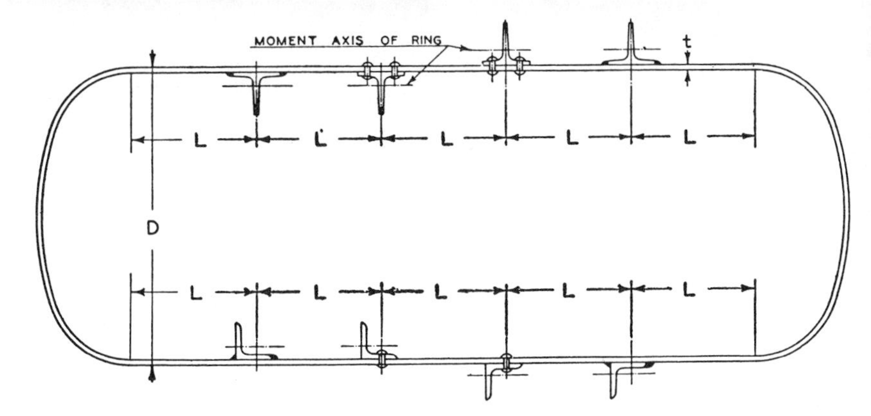

FIGURE 20.3 Stiffeners for Tanks under External Pressure. This figure shows L the length of unsupported shell. The heads serve as stiffeners at the ends. The rivets used to rivet stiffening rings on the *outside* should have a diameter not less than the thickness of plate. If a single row of rivets is used, as on an angle, the spacing or pitch should not be greater than $7t$. If a double row of rivets is used, as on a tee or I section, the maximum pitch should be $5t$ (t = plate thickness) and the rivets should be staggered. If the stiffening rings are welded on the *outside*, the welding should be intermittent with the maximum spacing *between* welds of $8t$. If the stiffening rings are *inside*, they should be secure enough to maintain their proper position under normal operating conditions. (From Rules for Construction of Unfired Vessels Subjected to External Pressure).

unsupported length L from equation (20.1), using the value of t from calculation (1). This value of L is the maximum spacing of stiffening rings if there is to be no collapse at a pressure $p_c = 5\,p$. For practical reasons, a closer spacing may be used; but if for any reason the spacing is increased above the value found in (2), then the thickness of plate must be increased to a value t that satisfies equation (20.1). The optimum economic design is the one using the minimum permissible plate thickness t.

Since the modulus of elasticity E is constant only to the proportional limit, the Code contains a chart taking into account its subsequent variation.

The stiffening rings, which may be either inside or outside the vessel, Fig. 20.3, should probably be strong enough to support the load even if a collapse occurs. The load that one ring supports depends on the length L, the smaller L, the lighter the section needed. The rings may be rolled shapes, such as angles, tees, I-section, or rectangular sections whose properties are found in various handbooks; they may be riveted or welded on. The moment of inertia I of the sectional area of a stiffener ring with respect to its centroidal axis parallel to the longitudinal axis of the vessel should be

(a)
$$I = \frac{0.035 D^3 L p_c}{E} \text{ in.}^4$$

For a refinement of this computation, see Ref. *(20.2)*. The amount of out-of-roundness permissible is defined in the Code. The supports for the vessel should impose no concentrated loads on the shell, and, in horizontal vessels, they should be placed at the heads or at stiffening rings.

Equations for a number of other members subject to elastic instability are found in Roark.[0.7]

20.3 STEEL TUBES SUBJECTED TO EXTERNAL PRESSURE. Continuing the analogy with the elastic instability of a column, we recall that the strength of a very short column, L/k less than about 30, is dependent on the yield strength of the material, a situation corresponding to a "thick cylinder," § 8.26. The thin-cylinder discussion of § 20.2 is analogous to the Euler column. In between these extremes, as in the case of columns, there is an area where the failure depends on both the buckling strength and yield strength. As for columns, this area for cylindrical members is handled by largely empirical equations, and it is into this category that most commercial tubes fall. The collapsing pressure for these tubes depends on the L/D and t/D ratios, the variations of wall thickness, the yield strength of the material, the elastic constants E and μ, and the variations in roundness.[20.3] Based on many tests, Stewart[20.5] proposed the following formula for long commercial lap-welded steel tubes, when $D < 40t$,

(b) $$p_c = 50{,}200{,}000\left(\frac{t}{D}\right)^3 = 50.2\left(\frac{100t}{D}\right)^3 \text{ psi},$$

where D is the outside tube diameter and t the wall thickness. According to charts by Jasper and Sullivan,[20.3] one may conclude that if $D \lesseqgtr 10t$, the thick-cylinder formula applies. It is worth noting that equation **(b)** is of the same form as the theoretical equation for long thin-shell tubes,[1.2,1.7] which is

(c) $$p_c = \frac{2Et^3}{(1 - \mu^2)D^3} \text{ psi},$$

where μ is Poisson's ratio and the other symbols have the usual meanings.

20.4 FLAT PLATES. Loaded flat plates appear in innumerable configurations; Roark[0.7] gives formulas for stresses and deflections for 70 cases without exhausting the possibilities, plus some tabular help in solving the equations; see also Refs. *(0.2, 20.6)*. Hence, the purpose of including this topic is to make the student aware of the problem and to give leads to further help.

(a) Circular Plate Supported at Edge with Uniformly Distributed Load. A diametral section may be considered as the critical section. Taking

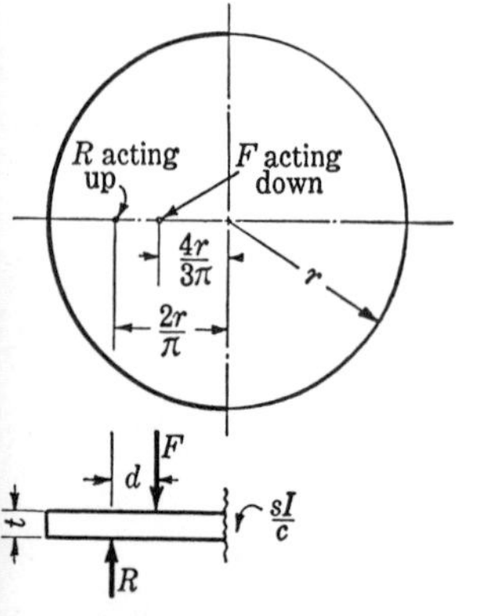

FIGURE 20.4 Circular Plate with Uniformly Distributed Pressure. The total load on half the plate is uniformly distributed. Therefore, the line of action of the resultant F is down through the centroid of the area. The reaction of the edge is assumed to be uniformly distributed around the edge, and hence the resultant reaction R acts up through the centroid of the semicircle.

half of the circular plate for analysis,[1.7] Fig. 20.4, we represent the forces acting on it as F, acting down through the centroid of the semicircular area, and R, acting up through the centroid of the semicircle. For equilibrium, $R = F$, and the moment of the couple formed by R and F is Fd, opposed by the resisting moment at the diametral section. This resisting moment is only approximately sI/c because the stress is not uniformly distributed along the section, but is some 25% higher than average at the center. Equating the applied moment to the resisting moment, we get

(d) $$F\left(\frac{2r}{\pi} - \frac{4r}{3\pi}\right) = \frac{sI}{c}, \qquad \text{or} \qquad \frac{2rF}{3\pi} = \frac{sI}{c}.$$

If the unit pressure on the surface is p psi, the load F on the semicircular area is $p\pi r^2/2$; at the diametral section, $I = (2r)t^3/12$; $c = t/2$. Substituting these values into the above equation, we get

(e) $$\frac{pr^3}{3} = \frac{srt^2}{3}, \qquad \text{or} \qquad s = p\left(\frac{r}{t}\right)^2 \text{ psi},$$

a reasonably good estimate of the average stress in a diametral section. We have assumed bending in only one plane, an untrue assumption although warranted by experience.

(b) Rectangular Plate Uniformly Loaded and Supported Along the Edges. Experiments on square and rectangular plates supported along all four edges, uniformly loaded and made of ductile materials, indicate that

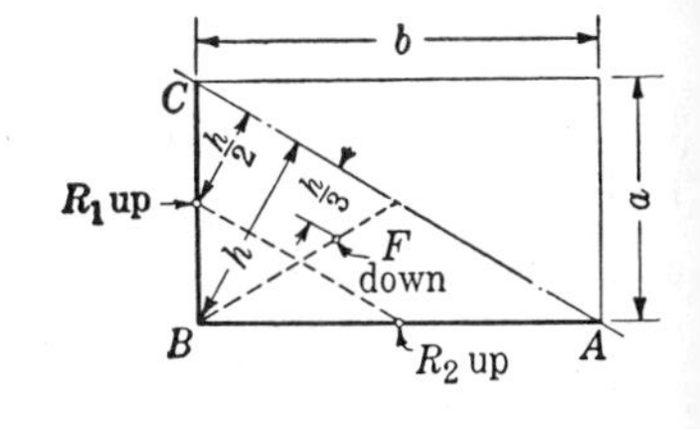

FIGURE 20.5 Rectangular Plate. The resultant reaction along each edge (R_1 and R_2) is assumed to act up through the mid-point of the edge. Since the total load on the half of the plate under consideration is uniformly distributed, the resultant F acts through the centroid of the triangle ABC.

$$h = \frac{ab}{(a^2+b^2)^{1/2}},$$

obtained by establishing a proportionality between h and the intercepts made by h on the diagonal AC.

the significant stress is closely that obtained by an analysis similar to the one above for the round plate (a is not very small compared to b). Proceeding as before with help from Fig. 20.5, sum moments about the diagonal AC, substitute the values of F, R_1, and R_2 in terms of the uniform pressure p psi, and get[1.7]

(f) $$s = \frac{a^2b^2p}{2t^2(a^2 + b^2)} \text{ psi},$$

where s is the *average stress* along the diagonal (the stress at the corners A and C is some 10% more than the average). If $b = a$, the above equation applies to a square plate similarly loaded and supported.

20.5 CAMS. Cams are among the most useful and important machine elements, especially in automatic machinery for the control of movements of parts. The details of the design of the profile or shape of a cam are covered in texts on kinematics or mechanism;[13.2] there is at least one comprehensive book on the subject.[20.10] Since it is assumed that the reader is familiar with this phase of the design, this article will concern principally the contact stresses.

The contact (Hertz) stress depends upon the force between the cam and follower, the shapes and sizes of the mating surfaces, and the modulii of elasticity of the materials (also μ). In turn, the force depends, sometimes predominantly, on the acceleration imparted to the follower system, because the so-called inertia force (ma) varies with acceleration. The accelerations are determined by the design and manufacturing accuracy of the cam profile and by the elasticities and clearances of the components of the system. Among the more common motions designed into cams are the parabolic motion (a = constant), harmonic motion, cycloidal motion, and various polynomial motions. In any event, if an equation for the displacement can be written, then the usual differentiations can be made to obtain velocity, acceleration, and jerk; $v = dx/d\tau$, $a = dv/d\tau = d^2x/d\tau^2$, $j = da/d\tau = d^3x/d\tau^3$. Jerk is seen to be the time rate of change of acceleration and it has a pronounced effect on the actual maximum acceleration.

The theoretical curves for the variation of the acceleration for certain popular motions are shown in Fig. 20.6. The curves OAB and CDE are purely suggestive of the high acceleration that would be characteristic of a cam that was designed to start a follower at some constant speed; but since the change in speed is $\Delta v = \int a \, d\tau$, which is proportional to the area

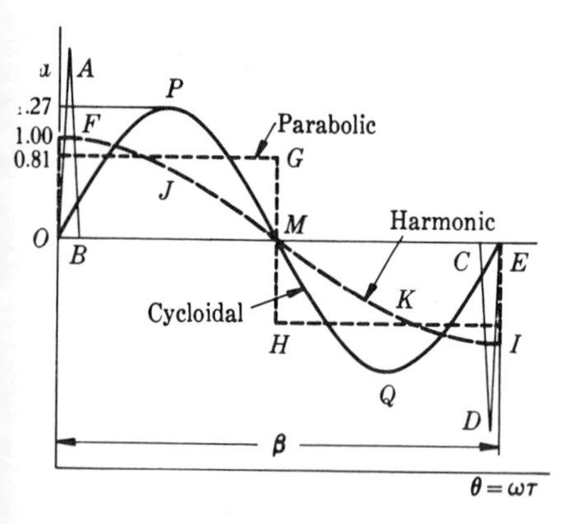

FIGURE 20.6 Variation of Acceleration. Acceleration for the parabolic motion is constant to *G*, where it suddenly becomes negative at *H* (see Fig. 20.7 where this change occurs). Acceleration for the harmonic motion, *FJKI*, starts at its maximum positive value and ends at its maximum negative value, *I*.

under the curves of Fig. 20.6, the follower speed for these curves is not nearly so large as for the others shown. Assuming that the follower is at rest at the beginning of the accelerations in Fig. 20.6, we observe that for the parabolic and harmonic motions, the motion starts with a finite value

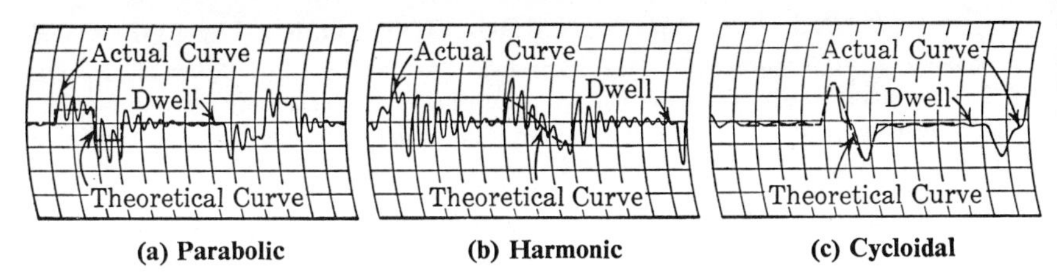

(a) Parabolic **(b) Harmonic** **(c) Cycloidal**

FIGURE 20.7 Comparison of Actual vs. Theoretical Accelerations. (Courtesy D. B. Mitchell (20.7) and E. I. du Pont de Nemours and Company).

of acceleration, which means that for this instant the rate of change of acceleration, or jerk, is quite high. Under these circumstances, it has been found[20.7,20.8] that the actual acceleration rises much higher than the simple motion equations (Table 20.1) predict. Table 20.1 gives equations for three popular motions conveniently in terms of cam angle θ, which for constant angular velocity of the cam, is proportional to time; $\theta = \omega\tau$. Let the accelerations computed from these equations be the "theoretical accelerations." The comparative actual accelerations are shown in Fig. 20.7; the theoretical curves are visible with a hard look. NOTE: that the actual maximum values for parabolic and harmonic motions are about twice the theoretical values because they are suddenly applied; that vibrations carry over into the dwell periods; that the vibrations (and a and ma) are damped, so that the actual acceleration would eventually equal the theoretical if the duration were long enough. When the acceleration starts from zero at the beginning of the motion, as for cycloidal, we see that the amplitude of the vibrations is low enough that the actual acceleration is reasonably comparable to the theoretical. Let the index number 1 represent the *theoretical* maximum acceleration for parabolic acceleration ($a_a = 1$) then the theoretical maximum values for harmonic and cycloidal motions are $a_h = 1.23$ and $a_c = 1.57$. However, for an undamped system, Mitchell[20.7] found the *actual* accelerations to be in the ratio $a_a : a_h : a_c = 1:0.834:0.584$; which is to say that the *actual* maximum acceleration for the cycloidal motion in this case is only some 60% of that for constant acceleration, even though *theoretically* it is the reverse relation that holds. With damping, the differences are not so great. In any event, the designer is justified, when the inertia force is a significant part of the total load, in making adjustments of the theoretical acceleration on the basis of any available experimental data. In making the free body of a force system, there is some advantage in using the reversed effective force F_{re} if we say $F = ma$, then $F_{re} = -ma$, where F_{re} is in a sense opposite to that of a. Lacking better evidence for computing the contact stresses, multiply the reversed effective forces by a factor q ($F_{re} = -qma$) to account for actual accelerations as follows: parabolic motion, $F_{re} = -3ma_a$; harmonic motion, $F_{re} = -2ma_h$ following a dwell; cycloidal and other zero acceleration starts following a dwell, $F_{re} = -1.1ma_c$; where a_a, a_h, and a_c are the computed theoretical values. See also Ref. *(20.10)*. See Figs. 20.6 and 20.7 for locations of maximum a. For a particular cam, the numerical values of the accelerations increase with increases in cam angular velocity.

TABLE 20.1 MOTION EQUATIONS FOR CAM FOLLOWERS

Symbols: L = total movement of the follower in one stroke; x = follower displacement at any point of the movement; θ rad. = angular displacement of the cam during any follower movement x; β rad. = angular displacement of the cam during the total follower movement L; v = speed of the follower; a = acceleration of follower; j = jerk of follower; ω = angular velocity of the cam; suggested consistent units, inches and seconds.

For the sinusoidal curves, the maximum values occur when the trigonometric function is unity. When motion starts from dwell with a finite acceleration, the jerk is theoretically infinite for an instant—a fact not revealed by the jerk equations.

EQUATION FOR	MOTION		
	*Parabolic** *Limit:* $\theta = \beta/2$	*Harmonic*	*Cycloidal*
Displacement .	$x = \frac{2L}{\beta^2}\theta^2$	$x = \frac{L}{2}\left(1 - \cos\frac{\pi}{\beta}\theta\right)$	$x = \frac{\theta}{\beta}L - \frac{L}{2\pi}\sin\frac{2\pi}{\beta}\theta$
Speed . .	$\dot{x} = \frac{4L\omega}{\beta^2}\theta$	$\dot{x} = \frac{L\pi\omega}{2\beta}\sin\frac{\pi}{\beta}\theta$	$\dot{x} = \frac{L\omega}{\beta}\left(1 - \cos\frac{2\pi\theta}{\beta}\right)$
Acceleration .	$\ddot{x} = L\left(\frac{2\omega}{\beta}\right)^2$	$\ddot{x} = \frac{L}{2}\left(\frac{\pi\omega}{\beta}\right)^2\cos\frac{\pi}{\beta}\theta$	$\ddot{x} = \frac{2\pi L\omega^2}{\beta^2}\sin\frac{2\pi}{\beta}\theta$
Jerk . . .	$\dddot{x} = 0$	$\dddot{x} = -\frac{L}{2}\left(\frac{\pi\omega}{\beta}\right)^3\sin\frac{\pi}{\beta}\theta$	$\dddot{x} = \frac{4\pi^2L\omega^3}{\beta^3}\cos\frac{2\pi}{\beta}\theta$

* The limit of $\theta = \beta/2$ applies when it is assumed that the deceleration has the same numerical value as the positive acceleration. The simplest way to get the values for the last half of this motion is to use the same equations in a reversed image.

Do not conclude that the designer should design all cams to provide an initial acceleration of zero. In the first place, instantaneous changes of acceleration introduce difficulties only as the speeds become relatively high (which is of course often desired in order to increase the output of a machine) and in the second place, such cams are much more expensive to manufacture because the cam curve changes so gradually at the beginning and end that only precision manufacturing methods would approximate such curves satisfactorily. Thus, one uses zero acceleration starts only when there is a worthwhile advantage.

The load between the cam and follower arises from: the force of gravity (unless the relevant motion is horizontal), the spring force (if a spring is used), frictional forces, inertia forces (which may or may not involve the acceleration of a connected body), vibratory forces (arising from operation near resonant conditions, the periodic changes of force $F = ma$, impacts, and instantaneously infinite jerk as described above). If a roller operates in a groove (and for other positive-motion drives), impact occurs when the resultant force reverses direction; the magnitude of the impact can be controlled to some extent by control of clearances, but clearances are necessary. If a follower operates on an open-face (plate) cam, it may be difficult or impossible at high speeds to keep the follower in continuous contact with some cam curves, in which case, the immediately adjacent point on the follower does not perform the motion designed into the cam. Furthermore, if the follower system is elastic (long links involved), the

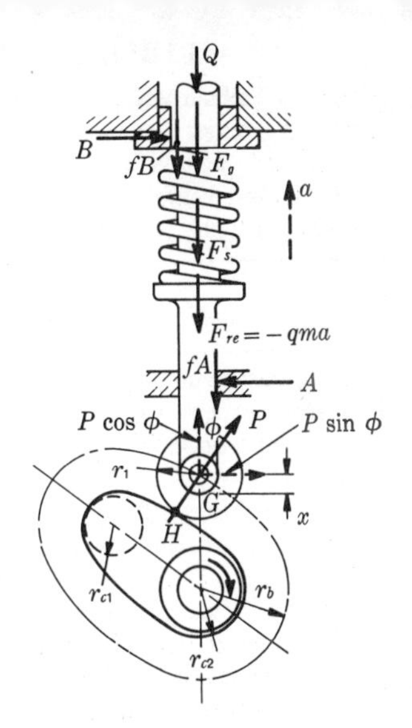

FIGURE 20.8 Forces on Follower. A and B are normal forces at guides, fA and fB are the corresponding frictional forces; F_g = force of gravity (weight); Q = external load, which may not be constant; F_s = spring force; F_{re} = reversed effective (inertia) force with a factor q shown to care for actual acceleration when a is the theoretical acceleration; P = normal force on contact surfaces; friction at G and H assumed to have neglible effect on N, as does fA and fB for low friction coefficient. Curvature r_{c2} = $r_b - r_1$.

final point on the follower mechanism may have a motion quite different from that of the contact surface, especially at high accelerations, because of the lag in the transmission of the force due to deformations. This phenomenon can be allowed for,[20.10] at least on a theoretical basis.

A free-body analysis of the follower system is necessary for determining the contact force, and since there are innumerable systems, the only possible general advice is to follow the principles of mechanics; Fig. 20.8 shows one arrangement. From an acceleration diagram,[13.2] one may compute the inertia forces at enough points to plot an inertia force F_{re} diagram; then the spring force $F_s = k\delta$ (k = scale of the spring and δ = deflection of the spring from its free length) can be plotted; the force of gravity F_g is constant, as is perhaps load Q; finally, the contact force P, Fig. 20.8, perpendicular to the profiles, is determined at key points where the contact stress may be a maximum. The experimental results are related to the Hertz contact-stress equation, and by using a wear factor K_c (as in gearing), the corresponding total force P is given by

$$(20.2) \qquad P = \frac{K_c b}{N(1/r_1 \pm 1/r_2)} \text{ lb.,}$$

where r_1 in. = radius of the follower roller ($r_1 = \infty$ for flat face follower); r_2 in. is the radius of curvature of the cam surface at the point of contact (use the positive sign if the surface is convex, negative sign if its is concave): b in. is the effective face width (length of contact); N = design factor; K_c is the wear-load factor (called a *load-stress factor* elsewhere—surface fatigue as in gearing) for cylindrical rollers, but used for the design of cam systems. The relation between the roller K_c and gear K_g wear factors is

$$\text{(g)} \qquad K_g = \frac{K_c \sin \phi_g}{4},$$

where ϕ_g is the pressure angle for the gear teeth; equation **(g)** may be verified

TABLE 20.2 WEAR FACTORS FOR CYLINDRICAL ROLLS

Sources: Talbourdet,[13.31] Buckingham,[13.3] Cram.[12.22] Hardness numbers are minimum limits. K_c values are average for 10^8 cycles, laboratory results. Copious lubrication with mineral oil, approximately SAE 20. (a) Through hardened to $R_C = 60$. (b) Class 35. (c) Phosphate coated. (d) Case thickness = 0.045 in. Values of K_{c3} with 300% sliding are: (e) 390, (f) 2000, (g) 750.

MATERIALS	K_{c1} *Rolling*	K_{c2} *9% sliding*
Gray cast iron and same, class 20, BHN = 130 (e)	1,300	1,050
Ditto except class 30, austempered, BHN = 270 (f)	4,200	3,400
Nodular iron, 80-60-03 and same, BHN = 207 (g)	3,400	1,850
Tool steel (a), and gray cast iron 20, BHN = 140	1,000	900
Tool steel (a), and gray cast iron, BHN = 225 (b)	2,300	2,100
Tool steel (a), and austempered class 30, BHN = 255	3,100	2,500
Tool steel (a), and SAE 4150, OQT to 270 BHN (c)	9,000	6,700
Ditto, Parco-Lubrite coated	12,000	7,900
Tool steel (a) and carburized 1020, case R_C = 50 (d)	13,000	8,500
Tool steel (a) and SAE 4340 induction h'dn'd to R_C = 50	13,000	9,000
Tool steel (a) and SAE 65 phosphor bronze, BHN = 67	1,000	
Tool steel (a) and laminated phenolic, grade L	880	830
SAE 39 cast Al, BHN = 60, and cast iron, OQT to 340 BHN	300	

by comparing the derivations of equations (13.8) and (20.2). Some values of K_c obtained under laboratory conditions are given in Table 20.2, where it is observed that the load capacity decreases materially as sliding increases, because the friction significantly increases the contact stress. The percentage sliding in this table is defined by:[12.22]

(h) $$\frac{v_1 - v_2}{v_2},$$

where v_1 is the peripheral speed (ips, say) of the driving roller and v_2 is the for driven roller. We have found no values of K_c for flat-face followers for which $v_2 = 0$, but it is seen that design values may be of the order of 20% to 30% of those for rolling. For rollers expected to roll, it would be conservative to use K_{c2} values for 9% sliding with a minimum design factor of perhaps $N = 1.15$; those who prefer to live more dangerously may find satisfaction from the use of K_{c1} values, especially when rolling-type bearings are used for the roller, Fig. 20.9. If the alignment and deflections are not the most favorable, use higher values of N with K_c from Table 20.2.

There are two frictional moments on the roller, the one from the frictional force at the bearing and the one at the point of contact. If the

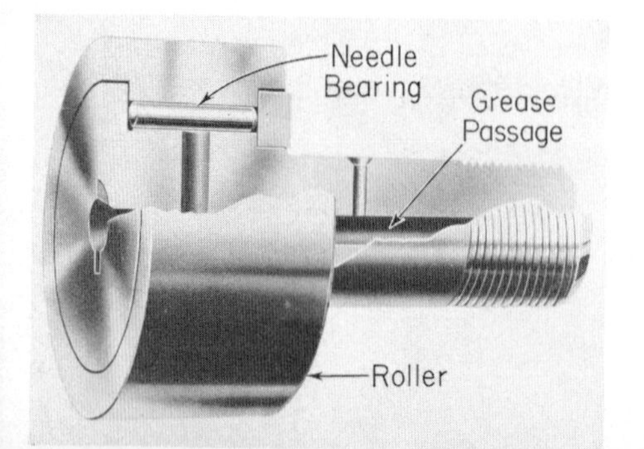

FIGURE 20.9 Cam Follower Roller. An example of stock rollers—with a needle bearing. Diameters from 0.5 in. to 4 in.; widths from $\frac{3}{8}$ in. to $2\frac{1}{4}$ in. (Courtesy McGill Mfg. Co., Valparaiso, Ind.)

frictional moment at the bearing is greater than that at the surface, there will be a large percentage of sliding and relatively rapid wear. If a journal bearing is used, Talbourdet[13.31] recommends a roller diameter of 3 times the bore diameter in order to obtain a moment of the contact frictional force adequate to insure mostly rolling.

For a *radial follower* with roller, Fig. 20.8, the curvature r_c of the cam surface at any displacement x in. of the follower from its "lowest" position is given by:[20.10]

$$\textbf{(i)} \qquad r_c = \frac{[(r_b + x)^2 + (v_f/\omega)^2]^{3/2}}{(r_b + x)^2 + 2(v_f/\omega)^2 - (r_b + x)a_f/\omega^2} \text{ in.,}$$

where r_b in. = radius of the base circle, Fig. 20.8, v_f ips = $\dot{x}$ = speed of the follower at displacement x, ω rad./sec. = angular speed of the cam, and a_f ips^2 = $\ddot{x}$ = acceleration of the follower at displacement x, and a_f is to be considered algebraic, negative when in the opposite sense to v_f on the outward motion. Equation **(i)** may be used for estimation purposes on other than radial roller followers.[20.10]

Cam shafts, linkages, bearings, etc., may be designed in accordance with methods previously discussed.

20.6 FLYWHEELS. A flywheel is a rotating member that acts as a storage reservoir for energy when work is not "consumed" at as fast a rate as the power is supplied. When the work being done is greater than the work input, the flywheel gives up some of its stored energy to supply the deficiency. The kind of energy stored in the flywheel is kinetic energy, which for a rotating member is $KE = I\omega^2/2$, from mechanics; kinetic energy is all available, 100% convertible into work without friction.[11.27] Thus, to serve its purpose, the angular velocity of the flywheel necessarily varies, the amount of the fluctuation for a given variability of the power input and consumption being a function of its moment of inertia I.

Reciprocating prime movers are characterized by power pulses; in Diesel engines, for example, most of the power to be delivered is generated shortly after the fuel ignites, but a properly designed flywheel keeps the speed fluctuation within suitable limits, and makes it possible for the engine to deliver work at almost a constant rate. Also, many driven machines require power in fluctuating amounts. The work to drive a reciprocating compressor varies for much the same reasons as explained for reciprocating prime movers. The variable consumption of power is obvious in many other machines, as punches, shears, machine tools. In the case of machines, such as punches, that have a high peak power requirement for a fraction of a revolution, a smaller driving motor can be used, so small in fact that the flywheel may be the major source of immediate energy needed during the working part of the cycle (§ 20.7). The prime mover then again stores energy during the remainder of the cycle by bringing the flywheel back to its original speed.

The amount of speed fluctuation that is permissible is an engineering decision that depends upon the application. Companies making certain kinds of machinery have their experience as a guide, and the conventional way of specifying the fluctuation is by the ***coefficient of fluctuation*** C_f, which is defined by

$$\text{(j)} \qquad C_f = \frac{\omega_1 - \omega_2}{\omega} = \frac{n_1 - n_2}{n} = \frac{v_{s1} - v_{s2}}{v_s},$$

where the subscript 1 designates the maximum and subscript 2 designates the minimum speed; the units of each term in a particular expression must

TABLE 20.3
COEFFICIENTS OF FLUCTUATION, FLYWHEELS

Typical values, taken from various sources.

DRIVEN MACHINE	C_f
D-C generators, direct drive	0.002
A-C generators, direct drive	0.0035
Punching, shearing, pressing machines .	0.05–0.1
Stamp mills, crushers	0.2
Reciprocating pumps, compressors . .	0.03–0.05
Machine tools, looms, paper mills . .	0.025
Spinning mills, fine to coarse thread . .	0.01–0.02
Geared drives	0.02

be the same and we shall use ω rad./sec., n rpm, and v_s fps, the linear speed of some chosen point on the flywheel; the speeds ω, n, v_s in the denominators are the average values, as $\omega = (\omega_1 + \omega_2)/2$. Values of C_f in Table 20.3 may be used as a guide; also see mechanical engineering handbooks.

The change of kinetic energy of the flywheel as a positive number is $\Delta KE = I(\omega_1^2 - \omega_2^2)/2$ ft-lb. for I slug-ft.^{2} [(lb-sec.^{2}/ft.)(ft.^{2})]. Since in mechanics we use the pound for force, we must use the mass in slugs for consistent units (with feet and seconds). The moment of inertia for any mass may be written $I = mk^2$, in this case with respect to the axis of rotation, where k ft. is the radius of gyration, and $m = w/g_o$ slugs for a mass w lb.; the standard acceleration $g_o \approx 32.2$ fps^2 becomes also a conversion constant 32.2 lb./slug. Since a large proportion of the mass of a flywheel may be in its rim (but see § 20.9) and since the rim is farthest from the axis of rotation, the contribution of the hub and arms to the moment of inertia may be only some 5% to 10% of the whole I.[1.6] On this account, engineers frequently neglect the effect of the hub and arms, an approximation that means that the fluctuations of speed will be slightly less than if the total I were used. With this simplification, $\Delta KE = I(\omega_1{}^2 - \omega_2{}^2)/2$ can be reduced as explained next.

If the rim thickness t is small compared with the diameter D of the flywheel, the mean radius of the rim is nearly equal to the rim's radius of gyration k. Let v_s be the average speed of a point at the radius k and v_{s1} and v_{s2} the limiting speeds at this radius; then $v_s = r\omega = k\omega$ fps (for $g_o = 32.2$ fps^2) and

$$\text{(k)}\quad \Delta KE = \frac{mk^2}{2}(\omega_1{}^2 - \omega_2{}^2) = \frac{w}{2g_o}(v_{s1}{}^2 - v_{s2}{}^2) = \frac{w(v_{s1} - v_{s2})(v_{s1} + v_{s2})}{2g_o}$$

ft-lb. for units as defined. Using $v_{s1} - v_{s2} = C_f v_s$ from equation (**j**) and $v_s = (v_{s1} + v_{s2})/2$ in (**k**), we get

$$\text{(l)}\qquad w = \frac{32.2\ \Delta KE}{C_f v_s{}^2}\ \text{lb.},$$

which is the mass of flywheel rim required to care for a change of kinetic energy ΔKE ft-lb. with a coefficient of fluctuation C_f at an average speed v_s fps.

Computation of the variation of energy ΔKE is sometimes a lengthy problem. For a reciprocating engine, a force diagram plotted for a revolution showing the resulting tangential force on the crankpin, in which both fluid pressures and inertia forces from the accelerations are accounted for, is needed. This work can be shortened significantly if average values of the work-output fluctuation of a particular type of engine is known; some such values are found in the literature. Some other situations may be satisfactorily estimated as illustrated in the next example. When a machine has a rotor whose moment of inertia is significant, as perhaps an electric motor, allowance for this in deciding on the size of the flywheel is appropriate.

20.7 EXAMPLE—FLYWHEEL RIM FOR PUNCH PRESS. Determine the size and mass of the rim of a flywheel for a punching machine which is to punch a maximum of one $\frac{3}{4}$-in. hole in C1020 plate, normalized, $\frac{1}{2}$-in. thick. The mean speed of the flywheel is to be 150 rpm during the punching.

Solution. Experiments show that the variation of the force exerted during the punching operation will be similar to the solid surve of Fig. 20.10. The area included in this curve ABC, which represents the work done ($\int F_x\,dx$), is given approximately by the triangular area ADE. Thus, the energy to punch this hole for an ultimate shearing stress of 54 ksi, Table AT 7, is

$$\frac{1}{2}Ft = \frac{1}{2}sAt = \frac{(54{,}000)[\pi(3/4)(1/2)(1/2)]}{(2)(12)} = 1325\ \text{ft-lb.}$$

In a machine of this kind, the motor is large enough to overcome frictional losses and to bring the flywheel back to speed soon after each punching operation. Since the time to punch is usually short compared with the time required for a complete

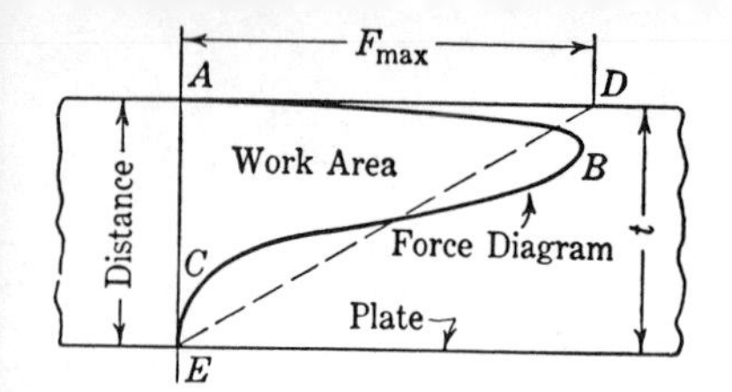

FIGURE 20.10

cycle, the work supplied by the motor during punching will be small, and it might be assumed that all of the energy for punching comes from the loss of kinetic energy of the flywheel; that is, ΔKE = 1325 ft-lb. In terms of the dimensions of the flywheel, any number of solutions is possible. Suppose space limitations are such that a flywheel with a mean rim diameter of 42 in. ($k \approx 21$ in.) can be used. Then

$$v_s = \pi D n_s = \pi\left(\frac{42}{12}\right)\left(\frac{150}{60}\right) = 27.5 \text{ fps.}$$

Using $C_f = 0.1$ from Table 20.3 and equation (I), we get

$$w = \frac{g_o \Delta KE}{C_f v_s^2} = \frac{(32.2)(1325)}{(0.1)(27.5)^2} = 565 \text{ lb.,}$$

which is the mass required for a 42-in. mean diameter. The dimensions should be in good proportion. Assume the width of rim b as 5 in. (If the flywheel is used as a pulley, this width may be governed by the width of belt needed.) The approximate volume of the material in the rim is $V = \pi Dbt = \pi(42)(5)t$ cu. in.; the corresponding mass is $\rho V = 0.284 V$, where the density $\rho = 0.284$ lb/cu. in. is taken from Table AT 7. Equating the mass of the rim obtained from its dimensions to the mass required for energy purposes, we have

$$w = (0.284)(\pi)(42)(5)t = 565,$$

from which the thickness of the rim t = 3.02 in.; use 3 in. The outside diameter of the flywheel is then 42 + 3 = 45 in.

20.8 STRESSES IN FLYWHEEL RIMS.

The centrifugal force that is produced by the rim of a flywheel may be large at high speeds, and therefore its effect should be checked unless it is known to be negligible. The deformation of the rim under the action of this load, Fig. 20.11, results in a tensile load on the arms and bending moments at the sections where the arms join the rim; but the stress calculated on the assumption of a free

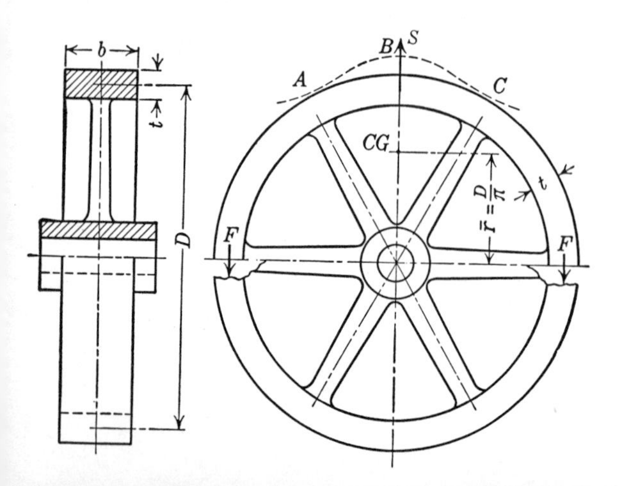

FIGURE 20.11 Flywheel. The size of a flywheel is its outside diameter in inches. The dimension *D* shown is used in computations. Centrifugal force tends to deform the rim somewhat as exaggerated *ABC*.

rotating ring is often a satisfactory criterion.[1.2, vol. II] On this basis, take half of the rim as a free body, ignoring the arms, and note that the force system consists of F acting down on each side, and the centrifugal force $S = m\bar{v}_s^2/\bar{r}$, where $\bar{v}_s$ fps (for $g_o = 32.2$ fps^2) is the velocity of the center of gravity of the body, $\bar{r}$ ft. is the radius to the c.g., and $m = w/32.2$ slugs (and w lb.) is the mass of the half rim. If the rim thickness t, Fig. 20.11, is small enough, the center of gravity of a half rim may be taken to be in the same position as the centroid of a semicircular arc;[1.6] $\bar{r} = D/\pi$ from the center. Thus, $\bar{v}_s = \bar{r}\omega = D\omega/\pi$ and $S = wD^2\omega^2/(\pi g_o D) = wD\omega^2/(\pi g_o)$. From the sum of the vertical forces, the force F is $F = \frac{1}{2}wD\omega^2/(\pi g_o)$. The mass of *half* the rim $w = V\rho = A\pi(D/2)\rho$, where ρ lb/ft.^{3} is the density of the material. Then from $s = F/A$, we get the average stress as

$$\textbf{(m)} \qquad s = \frac{A\pi(D/2)\rho D\omega^2}{2\pi g_o A} = \frac{D^2\omega^2\rho}{4g_o} = \frac{v_s^2\rho}{g_o} \text{ psf} \qquad \text{or} \qquad \frac{v_s^2\rho}{144g_o} \text{ psi},$$

where $v_s = D\omega/2$ fps is taken as the velocity of a point on the mean circumference of the rim.

Rules of thumb from experience specify the conventional limits of operation; for examples, 6000 fpm for cast iron and 10,000 fpm for cast steel are often given as limiting speeds for these materials. For the cast-iron limit speed, the hoop stress from (l) is $s = (100^2)(0.26)(12)/g_o = 970$ psi, which appears to be unreasonably low. Yet if the casting is not stress relieved, there are undoubtedly severe tensile residual stresses, and some unbalance is sure to exist; therefore, while much higher limiting speeds are often used, one should observe the foregoing restrictions unless it is known that higher speeds are safe in a particular situation. If the flywheel is made in halves, the strength of the connections for the rim sections is or may be the determinant of the safe operating speed.

If the flywheel acts as a sheave or pulley, the bending moment on all the arms [(total torque)/(no. of arms) on each] is approximately equal to the torque transmitted to or from the shaft; hence, the torque on the shaft is sometimes used as the bending moment on the arms for design purposes whether or not the flywheel is serving as a pulley. However, if there are momentary high accelerations (as from a sudden stop), the inertia of the rim results in a tangential reversed effective force of approximately ma, where m slugs is the mass of the rim and a fps^2 is the tangential acceleration, which may induce a greater bending moment than the torque. Also, see handbooks for empirical approaches.

20.9 ROTATING DISKS. Rotating disks or cylinders are common elements of machines, and are used as flywheels. The moment of inertia of a hollow cylinder about its axis is $I = m(r_o^2 + r_i^2)/2$, where r_o is the outside radius, r_i the inside radius, Fig. 20.12, and m is the mass of the hollow cylinder (slugs for r ft. and g_o fps^2); then the kinetic energy of the

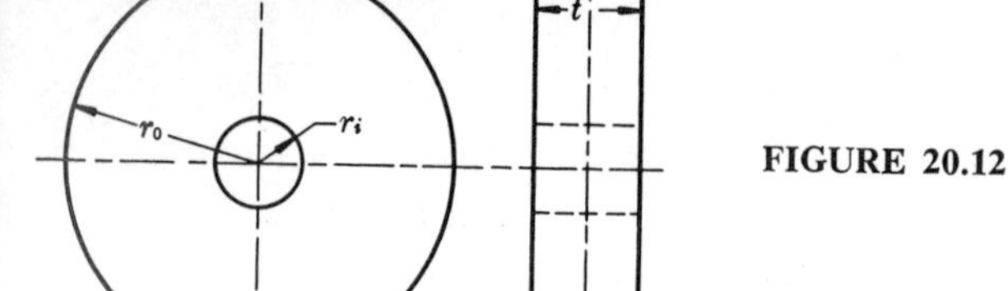

FIGURE 20.12

disk is $I\omega^2/2$. If the shaft on which the disk or cylinder is mounted rotates, it would certainly be satisfactory to let $r_i = 0$, and perhaps include the moment of inertia of other attached bodies.

The maximum stress caused by centrifugal action is the tangential stress at the perimeter of the hole, given by[0.7,1.2]

(n)
$$s_{t\,\max} = \frac{\rho\omega^2}{4g_o}[(3 + \mu)r_o^2 + (1 - \mu)r_i^2] \text{ psf},$$

where the density ρ lb/ft.^{3}, g_o fps^2, r_o, r_i, ft. result in s_t psf; μ is Poisson's ratio; if all linear units are inches ($g_o = 386$ ips^2), the stress is then s_t psi. If the hole becomes so small that $r_i \approx 0$, the maximum stress from **(n)**, becomes

(o)
$$s_t = \frac{\rho\omega^2 r_o^2(3 + \mu)}{4g_o} = \frac{\rho v_s^2(3 + \mu)}{4g_o},$$

where $v_s = r_o\omega$ fps (for s_t psf) is the peripheral speed of a point at radius r_o, Fig. 20.12. The maximum stress from centrifugal action in a rotating *solid* cylinder or disk is at the center and is[1.2]

(p)
$$s_t = \frac{\rho v_s^2(3 + \mu)}{8g_o},$$

from which, by comparison with **(o)**, we conclude that the stress is doubled by putting a pinhole in the center of a solid disk. As Timoshenko[1.2] points out, the central region of large forgings, where the stress is a maximum, is likely to contain defects (stress raisers); hence, it is common practice to drill or bore a central hole, which serves the additional purpose of permitting inspection for defects. If the disk or cylinder is made of ductile material, it can be run at overspeeds to stress the metal adjacent to the hole beyond the yield strength, thus resulting in a residual compressive stress, the phenomenon being as described for autofrettage, § 4.23, except that centrifugal force is producing the overstressing.

We should note that the maximum stress, from **(o)** or **(p)**, increases as the square of the size r_o of the disk, and as the square of its angular velocity. If the disk is of uniform thickness t, Fig. 20.12, the tangential stress decreases toward the outer periphery. For this reason, rotors are often tapered in the outward radial direction to approach more closely a uniform stress condition. See Refs. *(1.2)* and *(0.7)*.

If the rotating member is subjected to forces other than centrifugal, the stresses induced by these other forces are superposed on those described above. For example, the stress at the inner perimeter arising from a force fit is computed as explained in §§ 8.26 and 8.27.

20.10 CLOSURE. Since there are innumerable special problems in machine design, this book could be extended indefinitely. We hope that answers of one kind or another can be found herein for the most common engineering questions related to the application of the science of mechanics to machine design. A number of related topics, such as vibration and its control, corrosion, and others, are so involved that a brief discussion may be more misleading than informative; also true of some items that have been discussed. Besides, all things must eventually come to an end.

LIST OF REFERENCES

This is not a bibliography. It includes books and papers for which acknowledgment to *specific* material could be made, and it also includes works that should be helpful to those who wish to pursue further study. The references whose first digit is zero are general, many of which should be at hand in a design office.

0.1 ASME (S. L. Hoyt, ed.), *Metals Properties*, McGraw-Hill.

0.2 ASME (O. J. Horger, ed.), *Metals Engineering—Design*, McGraw-Hill.

0.3 ASME (J. Huckert, ed.), *Engineering Tables*, McGraw-Hill.

0.4 ASME (R. W. Bolz, ed.), *Metals Engineering—Processes*, McGraw-Hill.

0.5 Baumeister, T., ed., *Marks' Mechanical Engineers Handbook*. 6th ed., McGraw-Hill.

0.6 Carmichel, C., ed., *Kent's Mechanical Engineers Handbook*, 12th ed., Wiley.

0.7 Roark, R. J., *Formulas for Stress and Strain*, 3rd ed., McGraw-Hill.
Proceedings of Society for Experimental Stress Analysis.
Transactions of the American Society of Mechanical Engineers.

1.1 Timoshenko and MacCullough, *Elements of Strength of Materials*, Van Nostrand.

1.2 Timoshenko, S., *Strength of Materials*, 2 vols., Van Nostrand.

1.3 Higdon, Ohlsen, Stiles, *Mechanics of Materials*, Wiley.

1.4 Beedle, L. S., *Plastic Design of Steel Frames*, Wiley.

1.5 Grinter, L. E., *Theory of Modern Steel Structures*, Macmillan.

1.6 Faires and Chambers, *Analytic Mechanics*, 3rd ed., Macmillan.

1.7 Seeley and Smith, *Advanced Mechanics of Materials*, 2nd ed., Wiley.
1.8 Slupek, Main, and Fugenschuh, *Diameter and Length of a Pin for a Pin-Bushing Type*, Reprints from General Motors Engineering Journal, 1959.
1.9 ASME, *Boiler Construction Code.*
1.10 ASME, *Rules for Construction of Unfired Pressure Vessels.*
1.11 Maulbetsch and Hetényi, *Stresses in Pressure Vessels*, ASME Design Data, Book 1, p. A 29.
1.12 Boley and Weiner, *Theory of Thermal Stresses*, Wiley.
1.13 Manson, S. S., *Thermal stresses*, Machine Design, Vol. 33, Nos. 23–26, incl.
1.14 Marin, Joseph, *Mechanical Behavior of Engineering Materials*, Prentice-Hall.
1.15 Slaymaker, R. R., *Mechanical Design and Analysis*, Wiley.
1.16 van den Broek, J. A., *Theory of Limit Design*, Wiley.
1.17 Hall, Holowenko, Laughlin, *Theory and Problems of Machine Design*, Schaum Pub. Co. A good study aid.
1.18 Crandall and Dahl, *The Mechanics of Solids*, McGraw-Hill.
1.19 Hinkle, R. T., *Design of Machines*, Prentice-Hall.
1.20 Miller and Wright, *Contact stresses*, Machine Design, Vol. 35 No. 17, p. 185.
1.21 Asimow, M., *Introduction to Design*, Prentice-Hall.
1.22 Vidosic, Bogardus and Durden, *Curved beams with eccentric boundaries*, ASME Trans., Vol. 79, p. 1317.
1.23 Vidosic, J. P., *Machine Design Projects*, Ronald.
1.24 Myatt, D. J., *Machine Design Problems*, McGraw-Hill.

2.1 *Metals Handbook*, 8th ed., by Am. Soc. for Metals, Metals Park, Novelty, O. (The most complete single reference on metals for mechanical engineers.)
2.2 Soc. Auto. Eng. *Handbook.*
2.3 Publications of the International Nickel Co., Inc.
2.4 *Molybdenum Steels*, pub. by Climac-Molybdenum Co.
2.5 *Vanadium Steels and Irons*, pub. by Vanadium Corp. of Am.
2.6 *Modern Steels and Their Properties*, pub. by Bethlehem Steel Co.
2.7 *Carilloy Steels*, pub. by U.S. Steel.
2.8 Publications of Am. Soc. for Testing Materials.
2.9 *Mechanical Properties of Metals and Alloys*, Circular No. C 447, Bur. of Standards.
2.10 *Steel Handbook*, Republic Steel Corp.
2.11 *Stainless Steel Handbook*, pub. by Allegheny Ludlum Steel Corp.
2.12 *Reynolds Aluminum Data Book*, Reynolds Metals Co.
2.13 Publications of Aluminum Company of America.
2.14 Publications of Am. Foundrymen's Assoc., 222 W. Adams St., Chicago.
2.15 Publications of Gray Iron Founders' Society, 33 Public Square, Cleveland 13, O.
2.16 *Steel Casings Handbook*, by Steel Founders' Soc. of Am., 920 Midland Bldg., Cleveland.
2.17 *Am. Malleable Iron*, by Malleable Founders' Soc., Cleveland.

2.18 *Heat Treatment of Stainless Steels*, pub. by Rustless Iron & Steel Corp., Baltimore.
2.19 Simonds, Weith, Bigelow, *Handbook of Plastics*, Van Nostrand.
2.20 *Plastics Engineering Handbook*, Soc. of the Plastics Industry, Reinhold.
2.21 Smith, P. C., *Plastics for Production*, Chapman & Hall.
2.22 Lennie, A. M., *Designing for Magnesium, Metals Handbook*, 1948, p. 999.
2.23 *Die Casting for Engineers*, pub. by New Jersey Zinc Co.
2.24 *Metal Quality*, pub. by the Drop Forging Assoc., 605 Hanna Bldg.,Cleveland.
2.25 Evans, Ebert, Briggs, *Fatigue properties of cast and comparable wrought steels*, Proc. ASTM, vol. 56.
2.26 Hausner, H. H., ed., *Modern Materials*, 2 vol., Academic Press.
2.27 Kinney, G. F., *Engineering Properties . . . of Plastics*, Wiley.
2.28 Symposium on *Basic Mechanisms of Fatigue*, ASTM Sp. Tech. Pub. No. 237.
2.29 Finnie and Heller, *Creep of Engineering Materials*, McGraw-Hill.
2.30 Mantell, C. L., *Engineering Materials Handbook*, McGraw-Hill.
2.31 ASTM Special Technical Publications Nos. 151, 170, 199 (high temperature).
2.32 ASTM Special Technical Publication No. 158 (low temperature).
2.33 Gillespie, Saxton, and Chapman, *New design data for FEP and TFE*, Machine Design, Vol. 32, No. 2, p. 126.
2.34 Vanden Berg, R. V., *Anodic coatings*, Machine Design, Vol. 34, No. 6, p. 155.
2.35 *Plastics Book*, Machine Design, Vol. 36, No. 22.
2.36 Gaston, S. P., *Cold-drawn parts*, Machine Design, Vol. 35, No. 10, p. 142.
2.37 Briggs and Parker, *Alloy steels*, Machine Design, Vol. 35, No. 13, p. 153.
2.38 *The Ferrous Metals Book*, Machine Design, (updated periodically).
2.39 *Nonferrous Metals Book*, Machine Design, Vol. 35, No. 22.

3.1 ASA standard B 4.1—1955, *Preferred Limits and Fits for Cylindrical Parts*.
3.2 Burgess, A. R., *Selective interchangeability*, Texas Industrial Engineer, Vol. 2.
3.3 Shainin, D., *Quality control methods, their use in design*, a series of nine articles in Machine Design, beginning Vol. 24, No. 7. Good for an introduction.
3.4 Mallett & Lundberg, *Tolerances and allowances for interchangeable assembly*, Product Engineering, Vol. 15, p. 477.
3.5 Mikelson, W., *Determining surface roughness*, Mech. Eng., Vol. 69, p. 391.
3.6 Sawyer, J. W., *Surface finish literature*, Machine Design, Vol. 24, No. 9, p. 147.
3.7 Bolz, R. W., *Production Processes*, Penton Pub. Co. This book contains considerable information on manufacturing processes of value to a designer.
3.8 Broadston, J. A., *Standards for surface quality and machine finish designations*, Product Engineering, Vol. 15, p. 622.
3.9 ASA Standard, B46.1–1955, *Surface Roughness, Waviness, and Lay*.

3.10 Trowbridge, R. P., *Surface finish and the designer*, Product Engineering, Vol. 21, No. 9, p. 122.

3.11 Nielsen, L. M., *Shop run tolerances*, Product Engineering, Vol. 19, Nos. 5 and 6.

3.12 *Standard Practices and Tolerances for Impression Die Forgings*, Drop Forging Assoc.

3.13 ASA Standard, B 4.1–1947, *Limits and Fits for Engineering and Manufacturing.*

3.14 ASTM, *Manual of Quality Control of Materials.*

3.15 Grant, E. L., *Statistical Quality Control*, McGraw-Hill. Good for a beginner on this subject.

3.16 Brooks, K. A., *Statistical dimensioning program*, Machine Design, Vol. 33, No. 19, p. 140.

3.17 Hagen and Linberg, *Instrumentation for determining surface roughness*, G. M. Engineering Journal, Vol. 1, No. 7, p. 18.

3.18 Geuder and Ebeling, *Press fits with plastic parts*, Machine Design, Vol. 34, No. 9, p. 121.

3.19 Ring, W. A., *Determining practical tolerances*, Machine Design, Vol. 23, No. 3, p. 121.

3.20 Bolz, R. W., *Standard fits and tolerances*, reprinted from Machine Design, January 1955.

3.21 Bowker and Lieberman, *Engineering Statistics*, Prentice-Hall.

4.1 Sines and Waisman, ed., *Metal Fatigue*, McGraw-Hill.
Contains a wealth of detailed references; a good recent summary of the state of the art. Also Ref. *(4.62)*.

4.2 Lipson, Noll, Clock, *Stress and Strength of Manufactured Parts*, McGraw-Hill. Many charts in this one.

4.3 Battelle Memorial Institute, *Prevention of Fatigue of Metals*, John Wiley.

4.4 Karpov, A. V., *Fatigue problems in structural designs*, Metals and Alloys, December 1939.

4.5 Moore and Kommers, *The Fatigue of Metals*, McGraw-Hill.

4.6 McDowell, *The fatigue endurance of killed, capped, and rimmed steels*, Metals and Alloys, January 1940.

4.7 Buckwalter, Horger, Sanders, *Locomotive axial testing*, ASME Trans., Vol. 60, p. 335.

4.8 Morkovin and Moore, *Effect of size of specimen on fatigue strength*, Proc. ASTM, Vols. 43 and 44.

4.9 Peterson, R. E., *Fatigue tests of small specimens*, Trans. ASST, Vol. 42.

4.10 Hetényi, M., *et al.*, *Handbook of Experimental Stress Analysis*, John Wiley.

4.11 Murray, W. M., *et al.*, *Fatigue and Fracture of Metals*, John Wiley.

4.12 Moore, H. F., *et al.*, *Surface stressing of metals*, ASM.

4.13 Soderberg, C. R., *Factor of safety and working stresses*, ASME Trans., Vol. 52., p. 13.

4.14 Hinkle, R. T., *A simple method of presenting the combined variable-load equations*, Jour. ASEE, Vol. 41, p. 409.

4.15 Wahl and Beeuwkes, *Stress concentration produced by holes and notches*, ASME Trans., Vol. 56, p. 617.

4.16 Frocht and Hill, *Stress-concentration factors around a central circular hole in a plate loaded through a pin in the hole*, ASME Trans., Vol. 62, p. A-5.

4.17 Peterson, R. E. *Stress concentration phenomena in fatigue of metals*, ASME Trans., Vol. 55.

4.18 Peterson and Wahl, *Two- and three-dimensional cases of stress concentration*, ASME Trans., Vol. 58, p. A-15.

4.19 Frocht, M. M., *Factors of stress concentration photoelastically determined*, ASME Trans., Vol. 57, p. A-67.

4.20 Neugebauer, G. H., *Stress concentration factors and their effect on design*, Product Engineering, Vol. 14, p. 82.

4.21 Peterson, R. E., *Stress Concentration Design Factors*, John Wiley. A fine collection of factors; highly recommended for all design offices.

4.22 Horger and Neifert, *Shot peening to improve fatigue resistance*, Proc. SESA, Vol. II, Nos. 1 and 2.

4.23 Moore, H. F., *Size effect and notch sensitivity*, Proc. ASTM, Vol. 45, p. 507.

4.24 Cazaud, R., *Fatigue of Metals*, translated by A. J. Fenner, Chapman & Hall, London.

4.25 Lessells, J. M., *Strength and Resistance of Metals*, Wiley.

4.26 Stulen, Cummings, Schulte, a series of articles in Machine Design, Vol. 33, Nos. 9–14, incl.

4.27 ASTM, *Manual for Fatigue Testing*, Pub. No. 91.

4.28 Proceedings, *International Conference on Fatigue of Metals*, 1956, The Inst. of Mech. Eng., London.

4.29 ASTM, *Symposium on Fretting Corrosion*, Pub. No. 144.

4.30 Templin, R. L., *Fatigue of aluminum*, H. W. Gillet Memorial Lecture (ASTM), distributed by Alcoa.

4.31 Grover, Gordon, and Jackson, *Fatigue of Metals and Structures*, (NAVWEPS 00–25–534) Supt. Doc., US Govt. Print. Office., Washington 25. Has an extensive summary of experimental fatigue strengths.

4.32 Buckwalter, Horger, Sanders, *Locomotive axial testing*, ASME Trans., Vol. 60, p. 335.

4.33 ASTM, *Metals*, Sp. Pub. No. 196.

4.34 Guhse, D. E., *Effect of Two Stress Raisers Acting at a Point*, thesis, USN Postgraduate School.

4.35 Bridge, J. A., Jr., *Fatigue Strength Reduction Caused by Two Discontinuities at a Point*, thesis, USN Postgraduate School.

4.36 Parker, E. R., *Brittle Behavior of Engineering Structures*, Wiley.

4.37 Rassweiller and Grube, ed., *Internal Stresses and Fatigue in Metals*, Elsevier Pub. Co.

4.38 Pope, J. A., ed., *Metal Fatigue*, Chapman and Hall.

4.39 Frocht, M. M., *Photoelasticity*, Wiley.

4.40 Frankland, J. M., *Effects of impact on simple elastic structures*, Proc. SESA, Vol. VI, No. 2, p. 7.

4.41 Hudson, G. E., *A method of estimating equivalent static loads in simple elastic structures*, Proc. SESA, Vol. VI, No. 2, p. 28.

4.42 Dohrenwend and Mehaffy, *Dynamic loading in design*, Machine Design, Vol. 15, June, p. 99.

4.43 Hagenbook, L. D., *Impact loads by Brinell techniques*, Product Engineering, Vol. 14, p. 300.

4.44 Marco and Starkey, *A concept of fatigue damage*, ASME Trans., Vol. 76, p. 627.

4.45 Tavernelli and Coffin, *Experimental support for generalized equation predicting low cycle fatigue*, ASME paper 61–WA–199.

4.46 Gatts, R. R., *Cumulative fatigue damage with random loading*, ASME paper 61–WA–31.

4.47 Dieter, Horne, Mehl, *Statistical study of overstressing steel*, NACA Tech. Note No. 3211.

4.48 *Grinding Stresses*, collected papers, pub. by Grinding Wheel Institute, Keith Bldg., Cleveland.

4.49 ASTM, *Effect of Temperature on the Brittle Behavior of Metals*, Sp. Tech. Pub., No. 158.

4.50 Vitovec, F. H., *Superalloys under combined stress at elevated temperatures*, Machine Design, Vol. 34, No. 11, p. 157.

4.51 Langer, B. F., *Design of pressure vessels for low-cycle fatigue*, ASME paper 61–WA–18.

4.52 Gwinn, J. T., Jr., *Stop over-designing for impact loads*, Machine Design, Vol. 33, No. 16, p. 105.

4.53 Ellis and Karbowniczek, *Absorbing shock loads*, Machine Design, Vol. 34, No. 6, p. 176.

4.54 ASTM, *Symposium on Statistical Aspects of Fatigue*, Sp. Tech. Pub. No. 121.

4.55 ASTM, *Symposium on Basic Mechanisms of Fatigue*, Sp. Tech. Pub. No. 237.

4.56 ASTM, *Tentative Guide for Fatigue Testing and the Statistical Analysis of Fatigue Data*, Sp. Tech. Pub. No. 91-A.

4.57 Peterson, R. E., *Analytical Approach to Stress Concentration Effect in Fatigue of Aircraft Materials*, U.S Air Force Sc. paper No. 10–0509–6–P2.

4.58 Fuchs and Hutchinson, *Shot peening*, Machine Design, Vol. 30, No. 3, p. 116.

4.59 Henry, D. L., *A theory of fatigue-damage accumulation in steel*, ASME Trans., Vol. 77, p. 913.

4.60 Miscellaneous articles, in Materials Research and Standards, Vol. 3, No. 2, ASTM.

4.61 Coffey, R. L., *Effect of superposition of stress raisers on axial-load fatigue*, thesis, USN Postgraduate School.

4.62 Heywood R. B., *Designing against Fatigue of Metals*, Reinhold.

4.63 Danforth and Starkey, *The effects of high-stress-amplitude fatigue damage on the endurance properties of high-performance stainless-steel and titanium alloys*, ASME paper, 62–WA–230.

4.64 Almen and Black, *Residual Stresses and Fatigue in Metals*, McGraw-Hill.

4.65 Lipson and Juvinall, *Handbook of Stress and Strength*, Macmillan.

5.1 *Unified Screw Threads*, ASA B1.1–1960, ASME.

5.2 *Nomenclature, Definitions, and Letter Symbols for Screw Threads*, ASA B1.7–1949, ASME.

5.3 *Class 5 Interference-Fit Thread*, ASA B1.12, Tentative, ASME.
5.4 *Unified Miniature Screw Threads*, ASA B1.10–1958, ASME.
5.5 *Acme Screw Threads*, ASA B1.5–1952, ASME.
5.6 *Stub Acme Screw Threads*, ASA B1.8–1952, ASME.
5.7 *Buttress Screw Threads*, ASA B1.9–1953, ASME.
5.8 Seaton and Routhewaite, *Marine Engineer's Pocket Book*.
5.9 *Fasteners Data Book*, Industrial Fasteners Institute, Cleveland 15, Ohio.
5.10 Maney, G. A., *Predicting bolt tension*, Fasteners Data Book.
5.11 Crane Catalog, No. 53, Engineering Data Section.
5.12 Laughner and Hargan, *Handbook of Fastening and Jointing of Metal Parts*, McGraw-Hill.
5.13 Wilhelm, J., *Torque-tension standards*, Machine Design, Vol. 34, No. 4, p. 159.
5.14 *The Fasteners Book*, pub. by Machine Design; also contains numerous advertisements showing a variety of products and suppliers; Vol. 35, No. 7.
5.15 Almen, J. O., *Tightening is a vital factor in bolt endurance*, Machine Design, Vol. 16, No. 2, p. 158.
5.16 Dolan and McClow, *The influence of bolt tension and eccentric tensile loads on the behavior of a bolted joint*, Proc. SESA, Vol. VIII, No. 1, p. 29.
5.17 Pickel, W. F., *Tightening characteristics of nut and stud assemblies*, Product Engineering, Vol. 20, p. 98.
5.18 Dolan, T. J., *Load relations in bolted joints*, Mechanical Engineering, Vol. 64, p. 607.
5.19 Radzimovsky, E. I., *Bolt design for repeated loading*, Machine Design, Vol. 24, No. 11, p. 135.
5.20 Moslander, K. D., *Machine fastenings in design*, Machine Design, Vol. 13, No. 1, p. 39.
5.21 Hetényi, M., *A photoelastic study of bolt and nut fastenings*, ASME Trans. Vol. 65, p. A-93.
5.22 Wesstom and Bergh, *Effect of internal pressure on stresses and strains in bolted flanged connections*, ASME Trans., Vol. 73, p. 553.
5.23 Roberts, I., *Gaskets and bolted joints*, ASME Jour. App. Mech., Vol. 72, p. 169.
5.24 Rossheim and Markl, *Gasket loading constants*, Mech. Eng., Vol. 65, p. 647.
5.25 Armstrong Cork Co., *Gasket Design Manual*.
5.26 *The Seals Book*, pub. by Machine Design.
5.27 Oest, L., *Selecting and applying tapping screws*, Machine Design, Vol. 32, No. 5, p. 115.
5.28 Ollis, Ray Jr., *Self-aligning nuts*, Machine Design, Vol. 34, No. 14, p. 176.
5.29 Wilson and Oliver, *Tension Tests of Rivets*, Bull 210., Eng. Exp. Sta., Univ. of Illinois.
5.30 Crum, R. G., *Fatigue in metal joints*, Machine Design, Vol. 33, No. 7, p. 108.
5.31 Doughtie and Carter, *Bolted assemblies*, Machine Design, Vol. 22, No. 2, p. 127.
5.32 Creech, M. D., *Screw-thread torque*, Machine Design, Vol. 35, No. 12, p. 173.

5.33 Pfeiffer, W., *Bolted-flange assemblies*, Machine Design, Vol. 35, No. 15, p. 193.

5.34 AISC, *Steel Construction*, 6th ed.

6.1A Wahl, A. M., *Mechanical Springs*, Penton Pub. Co.

6.1B Wahl, A. M., *Mechanical Springs*, 2nd ed., Wiley (appeared after this chapter was written).

6.2 Associated Spring Corp., *Mechanical Springs*.

6.3 *The Mainspring*, a publication by the Associated Spring Corp., Bristol, Conn.

6.4 Bittner, E. T., *Alloy spring steels*, Trans. ASM, Vol. 40.

6.5 Tatnall, R. R., *Fatigue of helical springs*, Mech. Eng., April 1940.

6.6 Votta, F. A. Jr., *The theory and design of long-deflection constant-force spring elements* (Neg'ator), Trans. ASME, Vol. 74, p. 439.

6.7 Nolde, G. V., *Designing leaf springs for precision timing and indexing devices*, Machine Design, Vol. 24, No. 6, p. 125.

6.8 Johnson and Crooks, *Helical springs for high temperature*, Machine Design, Vol. 35, No. 19, p. 127.

6.9 Meyers, O. G., *Working stresses for helical springs*, Machine Design, Vol. 23, No. 11, p. 135.

6.10 Chandler, R. D., *Compression spring design*, Machine Design, Vol. 33, No. 22, p. 138.

6.11 ASTM Standards, Part 1, *Ferrous Metals*.

6.12 Personal communication, W. R. Johnson, Associated Spring Corp.

6.13 Technical Bulletin, T-35, International Nickel Co.

6.14 Atterbury and Diboll, *Effect of presetting helical compression springs*, ASME J. Eng. for Ind., Vol. 82, p. 41.

6.15 Hunter Spring Company, *Spring Design Data*.

6.16 Hinkle and Morse, *Design of helical springs for minimum weight, volume, and length*, ASME J. of Eng. for Ind., Vol. 81, p. 37.

6.17 Chandler, R. D., *Double-nested compression springs*, Machine Design, Vol. 34, No. 19, p. 177.

6.18 Almen and Laszlo, *The uniform section disk spring*, ASME Trans., Vol. 58, p. 305.

6.19 Weibel, E. E., *The correlation of spring-wire bending and torsion fatigue tests*, ASME Trans., Vol. 57, p. 501.

6.20 Wang and Worley, *Load-deflection behavior of conical spiral compression springs*, ASME Trans. Paper No. 61–WA–128.

6.21 Groesberg, S. W., *Zero-gradient spring system*, Machine Design, Vol. 32, No. 2, p. 143.

6.22 Heldt, P. M., *Motor Vehicles and Tractors*.

6.23 Smith, J. F. D., *Rubber mountings*, ASME Trans., Vols. 60 and 61.

6.24 Brown, R. W., *Engineering properties of rubber in compression*, SAE Trans., Vol. 47, p. 432.

6.25 Votta, F. A., *Constant-force springs*, Machine Design, Vol. 35, No. 3, p. 102.

6.26 Krotz, A. S., *Elastomeric torsion springs*, Machine Design, Vol. 33, No. 4, p. 130.

6.27 Weiss, K., *Helical compression springs*, Machine Design, Vol. 34, No. 28, p. 173 (graphical solution).

7.1 Stephenson and Cloninger, *Stress Analysis and Design of Steel Columns*, Tex. Eng. Exp. Sta. Bull. No. 129, College Station. Contains help in the form of tables and charts for solving the secant column formula.
7.2 Young, D. H., *Rational design of steel columns*, ASCE Trans., Vol. 101, p. 442.
7.3 Merriam, K. G., *Dimensionless coefficients applied to the solution of column problems*, Jour. Aero. Sciences, Vol. 7, p. 478.
7.4 Hoeschel, H. G., *Simplified column design*, Machine Design, Vol. 31, No. 7, p. 135.
7.5 Salmon, E. H., *Columns*, Hodder & Stoughton.
7.6 *Guide to Design Criteria for Metal Compression Members*, by Column Research Council, Crosby Lockwood and Son, Ltd., London.
7.7 Newman and Forray, *Beam columns*, Machine Design, Vol. 35, No. 5, p. 145.
7.8 Dow, James, *Columns loaded between supports*, Machine Design, Vol. 33, No. 4, p. 167.

8.1 Nadai, A., *Theories of strength*, ASME Trans., Vol. 55, p. 111.
8.2 Nadai and Wahl, *Plasticity*, McGraw-Hill.
8.3 Van den Broek, *Elastic Energy Theory*, John Wiley.
8.4 Den Hartog, J. P., *Advanced Strength of Materials*, McGraw-Hill.
8.5 Soderberg, C. R., *Working Stresses*, ASME Trans., Vol. 55, p. 131.
8.6 *Hot Rolled Carbon Steel Structural Shapes*, U.S. Steel Corp.
8.7 Ham and Ryan, *An Experimental Investigation of the Friction of Screw Threads*, Bull. 247, Eng. Exp. Sta., Univ. of Ill.
8.8 ASA Standard, B 1.5–1952, *Acme Screw Threads.*
8.9 ASA Standard, B 1.8–1952, *Stub Acme Screw Threads.*
8.10 Findley, W. N., *Fatigue of metals under combinations of stresses*, ASME Trans., Vol. 79, p. 1337.
8.11 Findley, Mitchell, Strohbeck, *Effect of range of stress in combined bending and torsion fatigue tests of 25 S-T6 aluminum alloy*, ASME Trans., Vol. 78, p. 1481.
8.12 Findley, W. N., *A theory for the effect of mean stress on fatigue of metals under combined torsion and axial load or bending*, ASME J. Eng. for Ind., Vol. 81, p. 301.
8.13 Cox and Field, *The initiation and propagation of fatigue cracks in mild steel pieces of square section*, Aeronaut. Quarterly, Vol. 4, part I, p. 1.
8.14 Murphey, Glenn, *Advanced Mechanics of Materials*, McGraw-Hill.
8.15 Crossland, Jorgensen, and Bones, *The strength of thick-walled cylinders*, ASME Paper 58–PET–20.
8.16 Davidson, Eisenstadt, and Reiner, *Fatigue characteristics of open-end thick-walled cylinders under cyclic internal pressure*, ASME Paper 62–WA–164.
8.17 Sessler and Weiss, *Low-cycle fatigue damage in pressure-vessel materials*, ASME Paper 62–WA–233.

8.18 Baugher, J. W., Jr., *Transmission of torque by means of press and shrink fits*, ASME Trans., Vol. 53, p. 85 MSP.
8.19 Horger and Nelson, *Design of press- and shrink-fitted assemblies*, Design Data and Methods, pub. by ASME.
8.20 Faupel, J. H., *Designing for shrink fits*. Machine Design, Vol. 26, No. 1, p. 114.
8.21 ASA Standard, B1.9–1953. *Buttress Screw Threads*, ASME.
8.22 Vidosic, J. P., *Design stress factors*, J. Eng. Ed., Vol. 38, No. 9, p. 653.

9.1 *Code for Design of Transmission Shafting*, ASME.
9.2 Hopkins, R. B., *Stepped shafts and nonuniform beams*, Machine Design, Vol. 33, No. 4, p. 159.
9.3 Griffel, Wm., *Beams*, Machine Design, Vol. 34, No. 17, p. 185.
9.4 Griffel, Wm., *Critical speed of multiple loaded shafts*, Machine Design, Vol. 34, No. 11, p. 153.
9.5 ASA Standard, B 17.1–1943, *Shafting and stock keys.*
9.6 Machinery's Handbook, 14th ed.
9.7 Timoshenko, S., *Vibration Problems in Engineering*, Van Nostrand.
9.8 Kimball, A. L., *Vibration Prevention in Engineering*, McGraw-Hill.
9.9 Thorn and Church, *Simplified Vibration Analysis by Mobility and Impedance Methods*, Penton Pub. Co.
9.10 Freberg and Kemler, *Elements of Mechanical Vibration*, Wiley.
9.11 Den Hartog, J. P., *Mechanical Vibration*, McGraw-Hill.
9.12 Hansen and Chenea, *Mechanics of Vibration*, Wiley.
9.13 Church, A. H., *Mechanical Vibrations*, Wiley.
9.14 Tsien, V. C., *Tubular shaft design*, Machine Design, Vol. 32, No. 20, p. 128.
9.15 Kano, C., *Influence of geometrical design factors on the bending fatigue strength of crankshafts*, ASME paper 62–WA–134.

10.1 *INCO*, Vol. 23, No. 2, p. 10, International Nickel Co.
10.2 ASA Standard B 5.15–1950, *Involute Splines.*
10.3 ASA Standard B 5.26–1950, *Involute Serrations.*
10.4 Armitage, J. B., *Straight-sided splines*, Mech. Eng., Vol. 70, p. 738.
10.5 *Mechanical Drives Book*, pub. by Machine Design, Vol. 34, No. 29.
10.6 *Flexible Couplings*, in Lubrication, Vol. 47, Nos. 10, 11, pub. by Texaco, Inc.
10.7 Gensheimer, J. R., *How to design flexible couplings*, Machine Design, September 14, Vol. 33, p. 154.
10.8 Condon, W. T., *Universal joints*, Machine Design, Vol. 33, No. 24, p. 172.

11.1 Fuller, D. D., *Theory and Practice of Lubrication for Engineers*, Wiley.
11.2 Slaymaker, R. R., *Bearing Lubrication Analysis*, Wiley.
11.3 Clower, J. I., *Lubricants and Lubrication*, McGraw-Hill.
11.4 Shaw and Macks, *Analysis and Lubrication of Bearings*, McGraw-Hill.
11.5 Wilcock and Booser, *Bearing Design and Application*, McGraw-Hill.
11.6 Pinkus and Sternlicht, *Theory of Hydrodynamic Lubrication*, McGraw-Hill.
11.7 Raimondi and Boyd, *A solution for the finite journal bearing and its application to analysis and design*, in three parts, ASLE Trans., Vol. 1, No. 1, p. 159.

11.8 *The Bearings Book*, Machine Design, Vol. 35, No. 14.

11.9 Howarth, H. A. S., *The loading and friction of thrust and journal bearings*, ASME Trans., Vol. 57, p. 169.

11.10 Howarth, H. A. S., *Graphical analysis of journal-bearing lubrication*, ASME Trans., Vols. 45, 46, 47, pp. 421, 809, 1073.

11.11 Boyd and Raimondi, *Applying bearing theory to the analysis and design of journal bearings*, ASME Jour. of Applied Mech., Vol. 73, pp. 298 and 310.

11.12 Norton, A. E., *Lubrication*, McGraw-Hill.

11.13 Radzimovsky, E. I., *Lubrication of Bearings*, Ronald.

11.14 Hersey, M. D., *Theory of Lubrication*, Wiley.

11.15 Needs, S. J., *Effects of side leakage in 120° centrally supported journal bearings*, ASME Trans., Vol. 56, p. 721.

11.16 Kingsbury, A., *Optimum conditions in journal bearings*, ASME Trans., Vol. 54, p. 123.

11.17 McKee and McKee, *Pressure distribution in oil films of journal bearings*, ASME Trans., Vol. 54, p. 149.

11.18 Bradford, L. J., *Oil film pressures in an end-lubricated sleeve bearing*, Tech. Bull. No. 14, Penna. State Col.

11.19 Karelitz, G. B., *Performance of oil-ring bearings*, ASME Trans., Vol. 52, p. APM–52–5–57.

11.20 Karalitz and Kenyon, *Oil-film thickness of transition from semifluid to viscous lubrication*, ASME Trans., Vol. 59, p. 239.

11.21 Dennison, E. S., *Film lubrication theory and engine-bearing design*, ASME Trans., Vol. 58, p. 25.

11.22 Howarth, H. A. S., *Current* [1934] *practices in pressures, speeds, clearances, and lubrication of oil-film bearings*, ASME Trans., Vol. 56, p. 891.

11.23 Burwell, Kaye, van Nymegen, and Morgan, *Effects of surface finish*, ASME Trans., Vol. 63, p. A-49.

11.24 Rippel, H. C., A series of informative articles in Machine Design, September 17, October 1, October 29, November 12, 1959.

11.25 Publications of Johnson Bronze Co., New Castle, Pa.

11.26 McKee and McKee, *Friction of journal bearings as influenced by clearance and length*, ASME Trans., Vol. 51, p. 161.

11.27 Faires, V. M., *Thermodynamics*, Macmillan.

11.28 McAdams, W. H., *Heat Transmission*, McGraw-Hill.

11.29 McKee and McKee, *Journal bearing friction in the region of thin-film lubrication*, SAE Jour., Vol. 31, p. 371.

11.30 McKee, S. A. *Journal bearing design as related to maximum loads, speeds and operating temperatures*, National Bur. St'd. Research paper, No. 1295, p. 491.

11.31 McKee, S. A., *The effect of running in on journal-bearing performance*, Mech. Eng., December, 1927.

11.32 McKee and McKee, *Friction of journal bearings as influenced by clearance and length*. ASME Trans., Vol. 51.

11.33 Baudry, R. A., *Some thermal effects in oil-ring journal bearings*, ASME Trans., Vol. 67, No. 2, p. 117.

11.34 Karelitz, G. B., *Grooving bearings in machines*, ASME Trans., 1929.

11.35 McKee and White, *Oil holes and grooves in plain journal bearings*, ASME Trans., Vol. 72, p. 1025.

11.36 Busse and Denton, *Water-lubricated soft-rubber bearings*, ASME Trans., Vol. 54.

11.37 Gillespie, Saxton, Chapman, *New design data for FEP and TFE*, Machine Design, Vol. 32, No. 4, p. 156.

11.38 Carlyon, G., *Plastic sleeve bearings*, Machine Design, Vol. 32, No. 21, p. 168.

11.39 Rentschler, L. M., *Plastic bearings*, Machine Design, Vol. 34, No. 23, p. 156.

11.40 Walker, G., *Bonded solid-lubricant coatings*, Machine Design, Vol. 35, No. 2, p. 182.

11.41 O'Rourke, Lewis, and Lewis, *Performance of Teflon fluorocarbon resins as bearing materials*, ASME Paper 61–WA–334.

11.42 Kingsbury, A., *On problems in the theory of fluid-film lubrication, with an experimental solution*, ASME Trans., Vol. 53, p. 59.

11.43 Tichvinsky and Courtel, *On the development of theories of dry and boundary friction*, ASME Paper 63–LUB–7.

11.44 Raimondi and Boyd, *Applying bearing theory to the analysis and design of pad-type bearings*, ASME Trans., Vol. 77, p. 287.

11.45 Raimondi, A. A., *The influence of longitudinal and transverse profile on the load capacity of pivoted pad bearings*, Trans. ASLE, Vol. 3, No. 2, p. 265.

11.46 Raimondi and Boyd, *An analysis of orifice- and capillary-compensated hydrostatic journal bearings*, Lubrication Engineering, January, 1957.

11.47 Raimondi, A. A., *A numerical solution for the gas lubricated full journal bearing of finite length*, Trans. ASLE, Vol. 4, p. 131.

11.48 Gross, W. A., *Gas Film Lubrication*, Wiley.

11.49 Katto and Soda, *Theoretical contributions to the study of gas-lubricated journal bearings*, ASME Paper 61–LUB–8.

11.50 Lemon, J. R., *Analytical and experimental study of externally pressurized air lubricated journal bearings*, ASME Paper 61–LUB–15.

11.51 Adams, C. R., *High-capacity gas step bearings*, Machine Design, Vol. 34, No. 5, p. 118.

11.52 Burwell, J. T., *The calculated performance of dynamically loaded sleeve bearings*, ASME Trans., Vol. 73, p. 393.

11.53 McKee, White, and Swindells, *Measurements of the combined frictional and thermal behavior in journal bearing lubrication*, ASME Trans., Vol. 70, p. 409.

11.54 Needs, S. J., *Influence of pressure on film viscosity in heavily loaded bearings*, ASME Trans., Vol. 60, p. 347.

11.55 *Designing with Zytel nylon resins*, a bulletin of the Du Pont Co.

11.56 Hays, D. F., *A variational approach to lubrication problems and the solution of the finite-journal bearing*, ASME J. Basic Eng., Vol. 81, No. 1, p. 13. Contains charts worthy of note.

11.57 Ocvirk, F. W., *Short bearing approximation for full journal bearings*, NACA Tech. Note 2808.

11.58 Du Bois and Ocvirk, *Experimental investigation of eccentricity ratio, friction, . . .* NACA Tech. Note 2809.

11.59 Du Bois and Ocvirk, *Analytical derivation and experimental evaluation of short-bearing approximation for full journal bearings*, NACA rept. 1157.

11.60 Du Bois, Ocvirk, and Wehe, *Experimental investigation of eccentricity ratio, friction, and oil flow of long and short journal bearings . . .*, NACA Tech. Note 3491.

11.61 Halberstadt, L., *Polyurethane bearings*, Machine Design, Vol. 35, No. 13, p. 151.

11.62 Ruffini, A. J., *Bearing noise*, Machine Design, Vol. 35, No. 12, p. 158.

11.63 Bowen and Hickam, *Outgassing characteristics of dry lubricant materials in a vacuum*, Machine Design, Vol. 35, No. 16, p. 119.

11.64 Klaus, E. E., *New developments in liquid lubricants*, ASME Paper 63–M–27.

11.65 Rippel, H. C., *Hydrostatic bearings*, Machine Design, a series of nine articles, starting in Vol. 35, No. 18.

11.66 Bowen, Boes and McDowell, *Self-lubricating composite materials*, Machine Design, Vol. 35, No. 20, p. 139.

11.67 Popovich and Hering, *Fuels and Lubricants*, Wiley.

12.1 Palmgren, A., *Ball and Roller Bearing Engineering*, pub. by SKF Industries, Philadelphia.

12.2 Styri, H., *Friction torque in ball and roller bearings*, Mechanical Engineering, Vol. 62, p. 886.

12.3 Styri, H., *Fatigue strength of ball bearing races and heat-treated* 52100 *steel specimens*, ASTM Proc., 1951.

12.4 SKF, *General Catalog and Engineering Data.*

12.5 The Timken *Engineering Journal.*

12.6 New Departure *General Catalog.*

12.7 AFBMA Standard, *Method of evaluating load ratings of annular ball bearings.*

12.8 ASA Standard B 3.11–1959, *Load Ratings for Ball and Roller Bearings.*

12.9 Shube, Eugene, *Ball-bearing survival*, Machine Design, Vol. 34, No. 17, p. 158.

12.10 Jones, A. B., *Ball motion and sliding friction in ball bearings*, ASME J. of Basic Eng., Vol. 81, No. 1, p. 1.

12.11 Harris, T. A., *Predicting bearing reliability*, Machine Design, Vol. 35, No. 1, p. 129.

12.12 ASA Standard B 3.7–1960, *Terminology and Definitions for Ball and Roller Bearings.*

12.13 ASA Standard B 3.5–1960, *Tolerances for Ball and Roller Bearings.*

12.14 Daniels, C. M., *High-load oscillating bearings*, Machine Design, Vol. 32, No. 15, p. 136.

12.15 Baniak and Kohl, *Fatigue life of ball bearings*, Machine Design, Vol. 33, No. 12, p. 190.

12.16 Accinelli, J. B., *Grease lubrication of ultra-high-speed rolling contact bearings*, ASLE Trans., Vol. 1, No. 1, p. 10.

12.17 Dunk and Hall, *Resistance to rolling and sliding*, ASME Paper 57–SA–9.

12.18 Rowland, D. R., *Ball-bearing splines*, Machine Design, Vol. 33, No. 15, p. 142.

12.19 Rumbarger, J. H., *Clearance in rolling bearings*, Machine Design, Vol. 33, No. 12, p. 145.

12.20 Havemeyer, H. R., *Linear ball bushings*, Machine Design, August 22, 1957.

12.21 Bidwell, J. B., ed., *Rolling Contact Phenomena*, Elsevier Pub. Co.

12.22 Lipson and Colwell, ed., *Handbook of Mechanical Wear*, Univ. of Mich. Press.

13.1 Dudley, D. W., ed., *Gear Handbook*, McGraw-Hill.
13.2 Faires, V. M., *Kinematics*, McGraw-Hill.
13.3 Buckingham, Earle, *Analytical Mechanics of Gears*, McGraw-Hill.
13.4 Black, P. H., *An investigation of relative stresses in solid spur gears by the photoelastic method*, Bull. No. 288, Eng. Exp. Sta., Univ. of Ill.
13.5 Dolan and Broghamer, *A photoelastic study of stresses in gear tooth profiles*, Bull. No. 335, Eng. Exp. Sta., Univ. of Ill.
13.6 Straub, J. C., *Shotpeening as a factor in the design of gears*, Product Engineering, 1953 Handbook.
13.7 Connell, C. B., *High-speed industrial gear drives*, presented to AGMA, 1948.
13.8 Tuplin, W. A., *Gear Load Capacity*, Wiley.
13.9 Botstiber and Kingston, *High-capacity gearing*, Machine Design, Vol. 24, No. 12, p. 129.
13.10 ASME Research Report, *Dynamic Loads on Gear Teeth.*
13.11 Candee, A. H., *Introduction to the Kinematic Geometry of Gear Teeth*, Chilton Co.
13.12 Attia, A. Y., *Dynamic loading of spur gear teeth*, ASME Paper 58–SA–32.
13.13 Reswick, J. B., *Dynamic loads on spur and helical gear teeth*, ASME Trans., Vol. 77, p. 635.
13.14 AGMA publications.
13.15 Buckingham, Earle, *Gear Tooth Loads*, pub. by Mass. Gear and Tool Co.
13.16 Jacobson, M. A., *Bending stresses in spur gear teeth*, Proc. I. Mech. E., Vol. 169, p. 587.
13.17 ASA Standard, B 6.10–1954, *Gear Nomenclature.*
13.18 ASA Standard B 6.12–1954, *Nomenclature of Gear Tooth Wear and Failure.*
13.19 Hall and Alvord, *Zytel spur gears*, Mechanical Engineering, Vol. 81, No. 5, p. 50.
13.20 Dudley, D. W., *Practical Gear Design*, McGraw-Hill.
13.21 Fellows Gear Shaper Co., *The Involute Curve and Involute Gearing.*
13.22 Gatcombe and Prowell, *Rocket motor—gear tooth stress analysis*, ASME Paper 59–A–256.
13.23 Gatcombe, E. K., *Lubrication characteristics of involute spur gears*, ASME Trans., Vol. 67, p. 177.
13.24 Kinsman, F. W., *Designing nonstandard spur gears*, Machine Design, Vol. 27, No. 6, p. 195.
13.25 Tuplin, W. A., *Gear Load Capacity*, Wiley.
13.26 Martin, L. D., *Powder-metal gears*, Machine Design, Vol. 33, No. 23, p. 200.
13.27 Borsoff, Accinelli, and Cattaneo, *The effect of oil viscosity on the power-transmitting capacity of spur gear*, ASME Trans., Vol. 73, p. 687.
13.28 Borsoff, V. N., *On the mechanism of gear lubrication*, ASME J. Basic Eng., Vol. 81, No. 1, p. 79.
13.29 Borsoff, V. N., *Gear lubrication*, Machine Design, Vol. 34, No. 16, p. 154.
13.30 Fellows Gear Shaper Co., *The Internal Gear.*

13.31 Talbourdet, G. J., *Progress report on surface endurance limits of engineering materials*, ASME Paper 54–LUB–14.

13.32 Rideout, T. R., *High-speed gearing*, Machine Design, Vol. 35, No. 19, p. 162; also ASME Paper 63–MD–46.

13.33 Hense and Buswell, *Economic factors affecting the steel selection and heat treatment for automotive gears*, General Motors Eng. Jour., Vol. 2, No. 5, p. 2.

13.34 Lipson, Charles, *Wear considerations in design*, a series of articles in Machine Design, beginning Vol. 35, No. 25, p. 156.

13.35 Wellauer, E. J., *Applying new gear strength and durability formulas to design*, Paper given to 1964 ASME Design Conference. (Changes in some details of AGMA calculations after we went to press.)

NOTE: Most of the references in chapter 13 are appropriate also for chapters 14, 15, or 16.

14.1 Buckingham and Ryffel, *Design of Worm and Spiral Gears*, Industrial Press.

14.2 Welch and Boron, *Thermal instability in high speed gearing*, ASME Paper 59–A–118.

14.3 Glover, J. H., *Nonstandard crossed helical gears*, Machine Design, Vol. 33, No. 12, p. 156.

14.4 Wellauer, E. J., *Helical and herringbone gears*, Machine Design, Vol. 33, No. 28, p. 125.

15.1 The Gleason Works, *Bevel and Hypoid Gear Design.*

15.2 Coleman, Wells, *Improved method for estimating fatigue life of bevel and hypoid gears*, SAE Trans., Vol. 6, No. 2.

15.3 Coleman, Wells, *Designing bevel gears*, Machine Design, Vol. 33, No. 24, p. 127.

15.4 Baxter, M. L., Jr., *High-reduction hypoids*, Machine Design, Vol. 33, No. 9, p. 142

15.5 Spear, King, Baxter, *Helixform bevel and hypoid gears*, ASME J. of Eng. for Ind., Paper 59–A–90.

16.1 Larson, G. A., *Helical gear and worm sets*, Machine Design, Vol. 35, No. 6, p. 170.

16.2 Nelson, W. D., *Spiroid gearing*, Machine Design, Vol. 33, Nos. 4 and 5, pp. 136 and 93.

16.3 Popper and Pessen, *The twinworm drive—a self-locking worm-gear transmission of high efficiency*, ASME J. of Eng. for Ind., Paper 59–A–75.

16.4 Vanick, J. S., *The Ni-Vee bronzes*, pub. International Nickel.

17.1 *Power Transmission Handbook*, pub. by Power Publishing Co., Cleveland.

17.2 Greenwood, D. C., *Mechanical Power Transmission*, McGraw-Hill.

17.3 Wormey, W. S., *Design of V-belt drives for mass-produced machines*, Product Engineering, September, 1953, p. 154.

17.4 Taylor, F. W., *Notes on belts*, ASME Trans., Vol. 15, p. 204.

17.5 Sawdon, W. M., *Tests of the Transmitting Capacities of Different Pulleys in Leather Belt Drives.*

17.6 Bolz, R. W., *et al.*, *High speeds in design*, Machine Design, Vol. 22, No. 4, p. 148.

17.7 Norman, C. A., *High-Speed Belt Drives*, Ohio State Eng. Exp. Sta. Bull. No. 83.

17.8 Norman, C. A., *Tests on V-belt drives and flat belt crowning*. ASME Trans., Vol. 71, p. 335.

17.9 Publications of American Leather Belting Association.

17.10 Tatnall, R. R., *The pivoted-motor drive*, Mechanical Engineering, Vol. 57, p. 287.

17.11 ASA Standard B 55.1–1961, *Specifications of Multiple V-Belt Drives*, Pub. by ASME.

17.12 *Specifications for narrow V-belts*, pub. by The Rubber Mfg. Asso., 444 Madison Ave., NYC.

17.13 Catalog of Gates Rubber Co.

17.14 Catalogs of U.S. Rubber Co.

17.15 Catalog of Link-Belt Co.

17.16 Catalog of Diamond Chain Co.

17.17 Marco, Starkey, and Hornung, *A quantitative investigation of the factors which influence the fatigue life of a V-belt*, ASME Paper 59–SA–18.

17.18 ASA Standard B 29.1–1950, *Transmission Roller Chains and Sprocket Teeth.*

17.19 Stamets, W. K. Jr., *Dynamic loading of chain drives*, ASME Trans., Vol. 73, p. 655.

17.20 Gerla, M. K., *Improving fatigue life of roller chain link plates*, Machine Design, Vol. 25, No. 1, p. 171.

17.21 New Horsepower Rating Charts, Bull. 462, Diamond Chain Co., Indianapolis, who may be willing to furnish copies for school use.

17.22 ASA Standard B 29.2–1950, *Inverted Tooth Chains and Sprocket Teeth*,

17.23 John A. Roebling's Sons Co., *Handbook*.

17.24 Jones & Laughlin Steel Corp., *Wire Rope*.

17.25 Dricker and Tachau, *A new design criterion for wire rope*, ASME Trans., Vol. 67, p. A-33.

17.26 Howe, J. F., *Stresses in wire ropes*, ASME Trans., Vol. 40.

17.27 Starkey and Cress, *An analysis of critical stresses and mode of failure of a wire rope*, ASME Paper 58–A–63.

17.28 Hilsher, J. E., *Wire-rope assemblies*, Machine Design, Vol. 33, No. 16, p. 88.

17.29 Rudolph and Imse, *Designing sprocket teeth*, Machine Design, Vol. 34, No. 3, p. 102.

17.30 Lavoie, F. J., *Tensioning V-belts*, Machine Design, Vol. 35, No. 16, p. 94.

17.31 Renner, E. J., *Belt pulley crown*, Machine Design, Vol. 32, No. 20, p. 106.

17.32 Fazekas, G. A. G., *On the lateral creep of flat belts*, ASME Paper 62–WA–54, J. Eng. for Ind.

17.33 Jackson and Moreland, *Roller and Silent Chains*, pub. by Am. Sprocket Chain Mfg. Asso.

17.34 Kerr, R. W., *Plastic roller chains*, Machine Design, Vol. 35, No. 17, p. 164.

17.35 Publications of the United Shoe Machinery Corp.

18.1 Schrader, H. J., *The friction of railway brake shoes*, Univ. of Ill. Bull. No. 72, Vol. 25.

18.2 Frehse, A. W., *Fundamentals of brake design*, SAE Jour., Vol. 27.

18.3 von Mehren, O., *Internal shoe clutches and brakes*, ASME Trans., Vol. 69, p. 913.

18.4 Rasmussen, A. C., *Internal friction, blocks and shoes*, Product Engineering, Vol. 18, No. 3, p. 133. Also other articles in same volume, May, July, October, December.

18.5 Gould, G. G., *Determination of the dynamic coefficient of friction for transient conditions*, ASME Trans., Vol. 73, p. 649.

18.6 Hagenbook, L. D., *Design of brakes and clutches of the wrapping band type*, Product Engineering, Vol. 16, p. 321.

18.7 Jania, Z. J., *Friction clutch transmissions*, Machine Design, Vol. 30, Nos. 23, 24, 25, 26.

18.8 Eksergian, C. L., *High speed braking*, ASME Trans., Vol. 73, p. 935.

18.9 Lowey, F. J., *Powdered-metal friction material*, Mechanical Engineering, Vol. 70, p. 869.

18.10 Loewenberg, F., *Automotive brakes with servo action*, ASME Trans., Vol. 52(1), p. 185 APM.

18.11 Rasmussen, A. C., *Heat radiation capacity of clutches and brakes*, Product Engineering, Vol. 2, p. 529. *Temperature calculations for clutches and brakes*, Product Engineering, Vol. 3, p. 282.

18.12 Siroky, E., Wagner Electric Corp., personal communications.

18.13 Engineering Information, Bull. 501, Raybestos Manhattan, Inc.

18.14 Borchardt, H. A., *Designing external-shoe brakes*, Machine Design, Vol. 32, No. 13, p. 163.

18.15 Rabins and Harker, *The dynamic frictional characteristics of molded friction materials*, ASME Paper 60–WA–35.

18.16 Fazekas, G. A. G., *Graphical shoe-brake analysis*, ASME Trans., Vol. 79, p. 1322.

18.17 Huntress, H. B., four articles in Machine Design: *Friction fundamentals*, Vol. 27, No. 7, p. 151; *Sintered-metal friction materials*, Vol. 27, No. 11, p. 187; *Friction brakes and clutches*, Vol. 28, No. 7, p. 113; *Multiple-disk clutches and brakes*, Vol. 28, No. 5, p. 82.

18.18 Harker and Rabins, *Torque capacity of friction devices with nonlinear frictional elements*, ASME Trans., Paper 62–WA–77.

18.19 Kotnik, R. L., *Electromagnetic disc clutches*, Machine Design, Vol. 32, No. 16, p. 113.

18.20 Walls, F. J., *Brake drum materials*, distributed by International Nickel Co.

18.21 Armstrong Cork Co. pamphlet.

18.22 Bette, A. J. *Friction materials*, Machine Design, Vol. 32, No. 20, p. 141.

18.23 Westover and Vroom, *A variable-speed frictionometer for plastics, rubbers, metals, and other materials*, ASME Paper 62–WA–321.

18.24 Courtel and Tichvinsky, . . . *Friction*, Mechanical Engineering, Vol. 85, Nos. 9, 10, pp. 55, 33, respectively.

19.1 Jennings, C. H., *Welding design*, ASME Trans., Vol. 58, p. 497.

19.2 Wilson and Wilder, *Fatigue tests of butt welds in structural steel plates*, Univ. of Ill. Bull. No. 42.

19.3 *Procedure Handbook of Arc Welding Design and Practice*, 11th ed., The Lincoln Electric Co.

19.4 Jefferson and Mackenzie, *The Welding Encyclopedia*, 14th ed., McGraw-Hill.
19.5 Udin, Funk, and Wulff, *Welding for Engineers*, Wiley.
19.6 Phillips, A. L., ed., *Welding Handbook*, Am. Weld. Soc.
19.7 Young, C. H., *Experimental Study of Stresses in Welded Joints*, Thesis, USN Postgraduate School.
19.8 Smith, J. H., *Stress-strain characteristics of welded joints*, Jour. Am. Weld. Soc., September, 1929.
19.9 *Efficient Machine Design in Welded Steel*, Lincoln Electric Co.
19.10 Churchill and Austin, *Weld Design*, Prentice-Hall.
19.11 Clauser, H. R., *Welding electrodes and rods*, Materials and Methods, December, 1950.
19.12 Paton, B. E., ed., transl. from Russ., *Electroslag Welding*, Reinhold.
19.13 Nagaraja Rao and Tall, *Residual stresses in welded plate*, Weld. J., Vol. 40, No. 10, p. 468s.
19.14 Yao and Munse, *Low cycle fatigue of metals*, Weld. J., Vol. 41, No. 4, p. 182s.
19.15 *Alcoa Structural Handbook*, Alum. Co. of Amer.
19.16 Blodgett, O. W., *Design of Weldments*, James F. Lincoln Arc Weld. Found.
19.17 Cornwall, E. P., *Welded joints for hard-vacuum systems*, Machine Design, Vol. 35, No. 19, p. 135.
19.18 Franks and Wooding, *Some dynamic mechanical properties of heat-treated low-alloy weld deposits*, Weld. J., Vol. 35, No. 6, p. 291s.
19.19 Mindlin, H., *Fatigue Properties of Welded Joints of Aluminum-Magnesium Alloys*, Reynolds Metals.
19.20 Sparagen and Claussen, *Fatigue strength of welded joints*, Weld. J., Vol. 16, p. 1.
19.21 Sparagen and Rosenthal, *Fatigue strength of welded joints*, Weld. J., Vol. 21, p. 297s.
19.22 Toprac, A. A., *An investigation of corner connections loaded in tension*, Weld. J., Vol. 40, No. 11, p. 521s.
19.23 Choquet, Krivobok, and Welter, *Effects of prestressing on fatigue strength of spot-welded stainless steels*, Weld J., Vol. 33, No. 10, p. 509s.
19.24 *Aluminum . . . Product Information*, Kaiser Alum. & Chem. Co.

20.1 Windenberg, D. F., *Vessels under external pressure*, Mech. Eng., August, 1937.
20.2 Saunders and Windenberg, *Strength of thin cylindrical shells under external pressure*, ASME Trans., Vol. 53, p. 207 APM.
20.3 Jasper & Sullivan, *The collapsing strength of steel tubes*, ASME Trans., Vol. 53, p. 219 APM.
20.4 Bryan, G. H., *Application of the energy test to the collapse of a long thin pipe under external pressure*, Cambridge Phil. Soc. Proc., Vol. VI, p. 287.
20.5 Stewart, R. T., *Collapsing pressure of Bessemer steel lap-welded tubes*, ASME Trans., Vol. 27, p. 730.
20.6 Timoshenko, S., *Theory of Plates and Shells*, McGraw-Hill.
20.7 Mitchell, D. B., *Tests on dynamic response of cam-follower systems*, Mechanical Engineering, Vol. 72, p. 467.

20.8 Hrones, J. A., *An analysis of the dynamic forces in a cam-driven system*, ASME Trans., Vol. 70, p. 473.

20.9 Stoddard, D. A., *Polydyne cam design*, Machine Design, Vol. 25, Nos. 1, 2, and 3.

20.10 Rothbart, H. A., *Cams*, Wiley.

20.11 Kloomok and Muffly, *Determination of radius of curvature for radial and swinging follower cam systems*, ASME Paper 55–SA–29.

20.12 Cram, W. D., *Cam design*, Proceedings of Third Mechanisms Conference, Purdue Univ., 1956.

20.13 Moon, C. H., *Cam radius of curvature*, Machine Design, Vol. 34, No. 18, p. 123.

APPENDIX

The order of arrangement
is the same as that of
the text's subject matter.

The ability to understand a question from all sides meant one was totally unfit for action. Fanatical enthusiasm was the mark of the real man.

Thucydides on the Athenian mood during the Peloponnesian wars (circa 455-400 B.C., the eve of the decline of Athens' power).

TABLE AT 1 PROPERTIES OF SECTIONS

I_x = moment of inertia about the axis x–x, J = polar moment of inertia about the centroidal axis, $Z = I/c$ = rectangular section modulus, about x–x, $Z' = J/c$ = polar section modulus, $k = \sqrt{I/\text{area}}$ = radius of gyration.

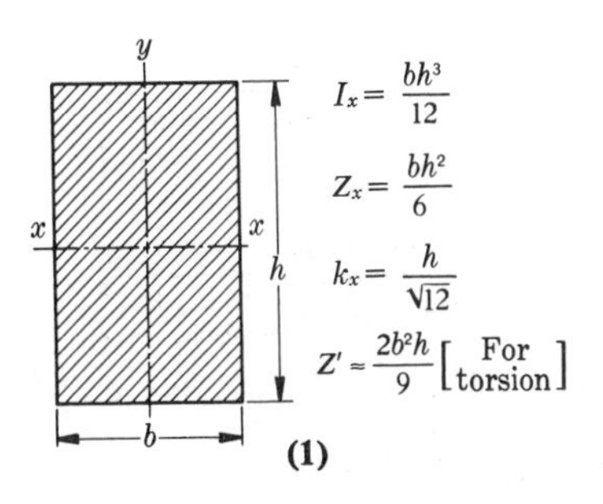

$$I_x = \frac{bh^3}{12}$$

$$Z_x = \frac{bh^2}{6}$$

$$k_x = \frac{h}{\sqrt{12}}$$

$$Z' \approx \frac{2b^2h}{9} \left[\text{For torsion}\right]$$

(1)

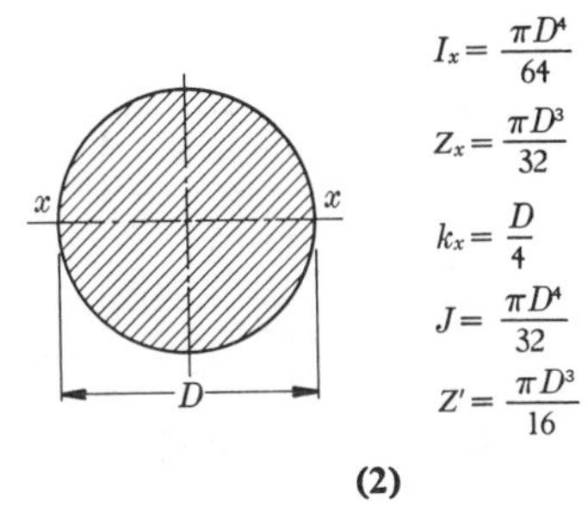

$$I_x = \frac{\pi D^4}{64}$$

$$Z_x = \frac{\pi D^3}{32}$$

$$k_x = \frac{D}{4}$$

$$J = \frac{\pi D^4}{32}$$

$$Z' = \frac{\pi D^3}{16}$$

(2)

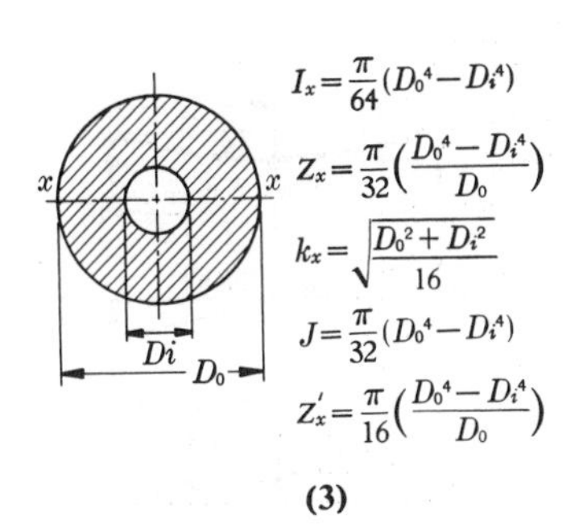

$$I_x = \frac{\pi}{64}(D_0^4 - D_i^4)$$

$$Z_x = \frac{\pi}{32}\left(\frac{D_0^4 - D_i^4}{D_0}\right)$$

$$k_x = \sqrt{\frac{D_0^2 + D_i^2}{16}}$$

$$J = \frac{\pi}{32}(D_0^4 - D_i^4)$$

$$Z_x' = \frac{\pi}{16}\left(\frac{D_0^4 - D_i^4}{D_0}\right)$$

(3)

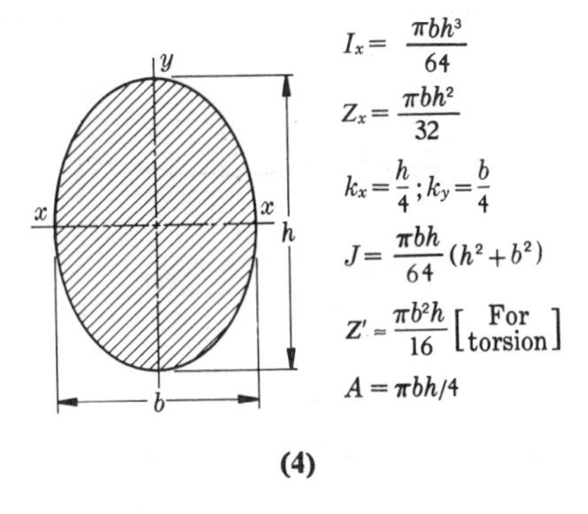

$$I_x = \frac{\pi b h^3}{64}$$

$$Z_x = \frac{\pi b h^2}{32}$$

$$k_x = \frac{h}{4};\ k_y = \frac{b}{4}$$

$$J = \frac{\pi b h}{64}(h^2 + b^2)$$

$$Z' \approx \frac{\pi b^2 h}{16} \left[\text{For torsion}\right]$$

$$A = \pi bh/4$$

(4)

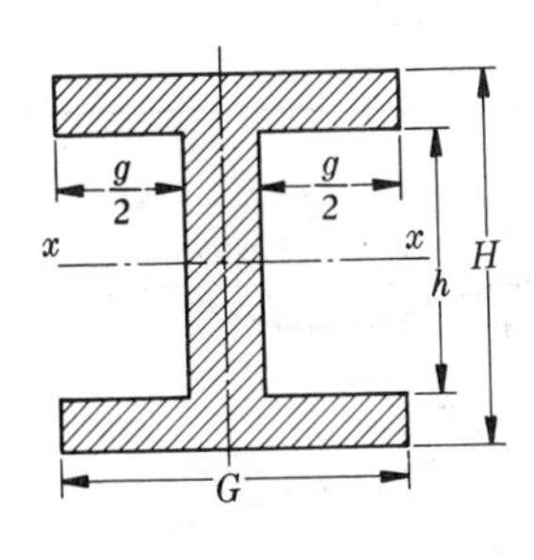

$$I_x = \frac{1}{12}(GH^3 - gh^3)$$

$$Z_x = \frac{GH^3 - gh^3}{6H}$$

$$k_x = \sqrt{\frac{1}{12}\left[\frac{GH^3 - gh^3}{GH - gh}\right]}$$

(5)

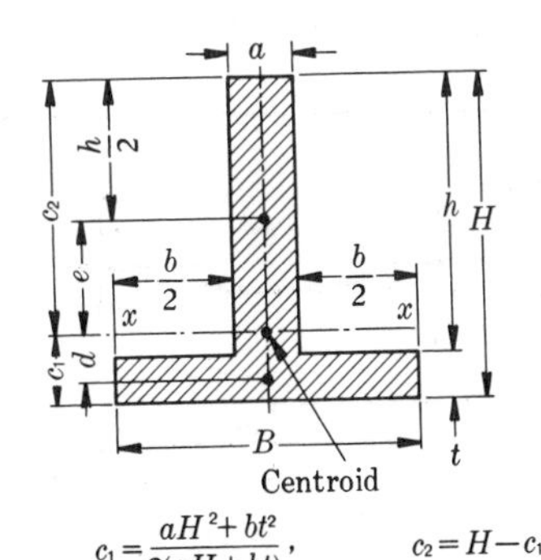

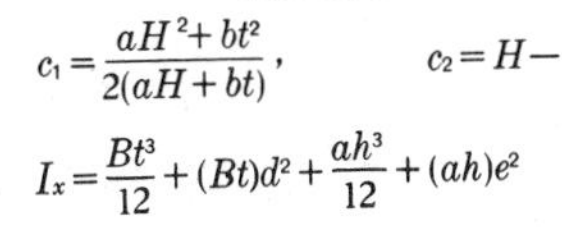

$$c_1 = \frac{aH^2 + bt^2}{2(aH + bt)}, \qquad c_2 = H - c_1$$

$$I_x = \frac{Bt^3}{12} + (Bt)d^2 + \frac{ah^3}{12} + (ah)e^2$$

$$\text{Area} = Bt + a(H - t); \quad k = \sqrt{I/A}$$

(6)

TABLE AT 2 MOMENTS AND DEFLECTIONS IN BEAMS

F lb. = applied force; w = pounds per inch of length; $F = wL$, where L in. = length; E psi = modulus of elasticity, tension; I in.^{4} = moment of inertia; y in. = deflection; θ radians = slope. For other beams of uniform strength, see § 6.24.

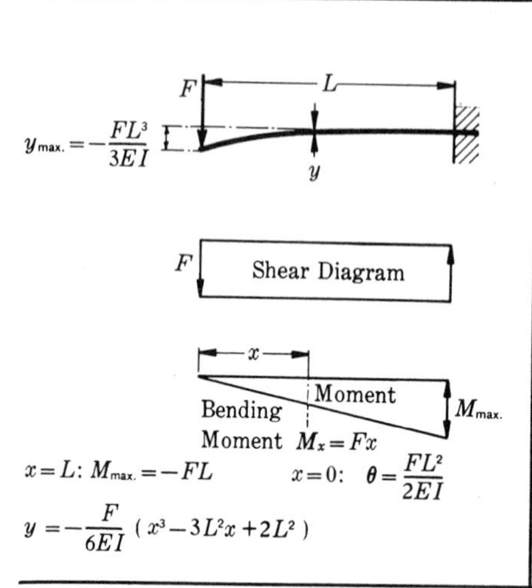

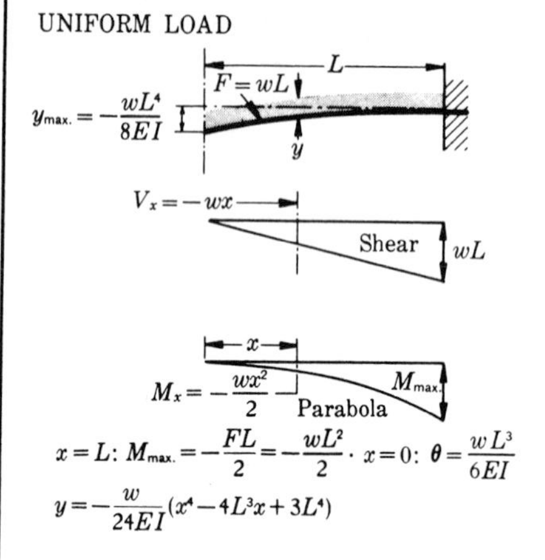

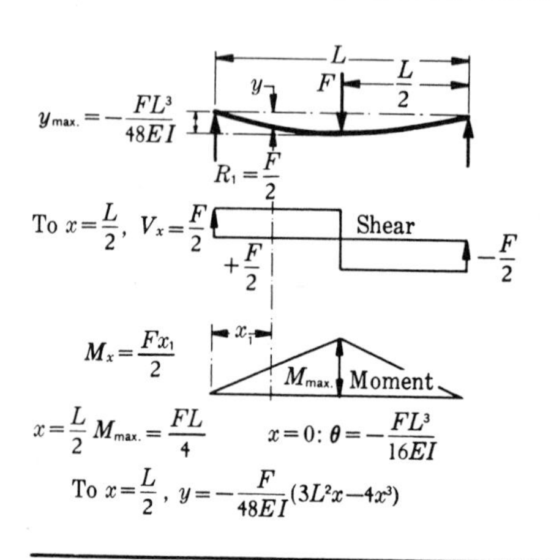

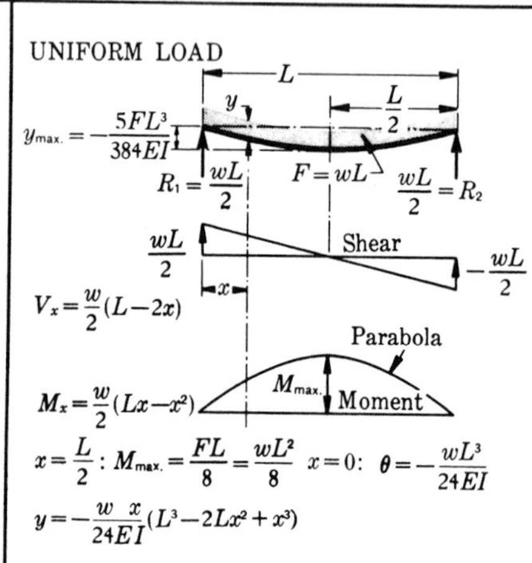

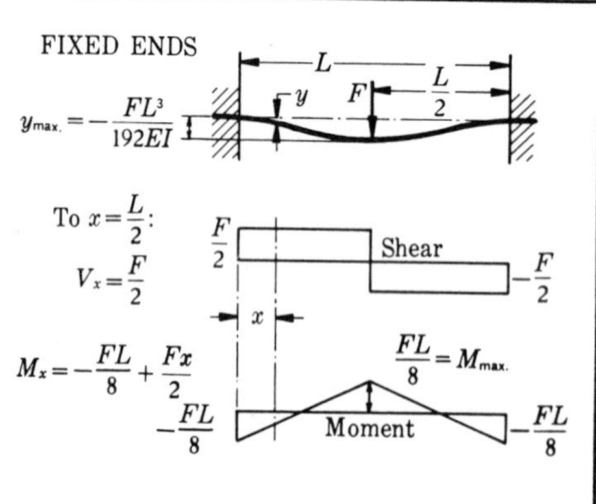

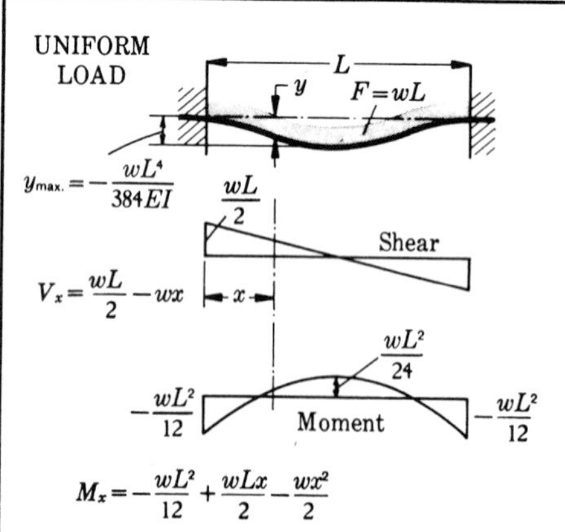

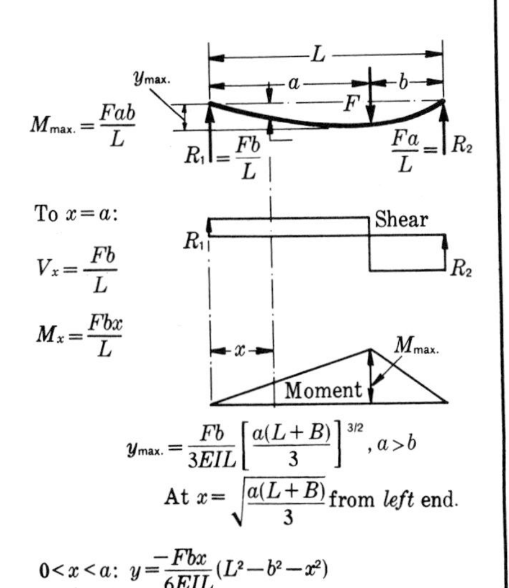

$M_{max.} = \frac{Fab}{L}$

To $x=a$:

$V_x = \frac{Fb}{L}$

$M_x = \frac{Fbx}{L}$

$y_{max.} = \frac{Fb}{3EIL}\left[\frac{a(L+B)}{3}\right]^{3/2}, a>b$

At $x = \sqrt{\frac{a(L+B)}{3}}$ from *left* end.

$0<x<a$: $y = \frac{-Fbx}{6EIL}(L^2-b^2-x^2)$

$a<x<L$: $y = -\frac{Fa(L-x)}{6EIL}\left[L^2-a^2-(L-x)^2\right]$

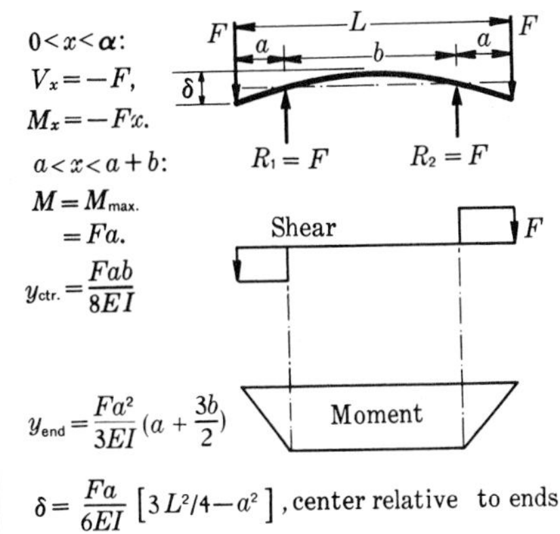

$0<x<a$:

$V_x = -F,$

$M_x = -Fx.$

$a<x<a+b$:

$M = M_{max.}$

$= Fa.$

$y_{ctr.} = \frac{Fab}{8EI}$

$y_{end} = \frac{Fa^2}{3EI}\left(a+\frac{3b}{2}\right)$

$\delta = \frac{Fa}{6EI}\left[3L^2/4 - a^2\right]$, center relative to ends

UNIFORM STRENGTH, CANTILEVER

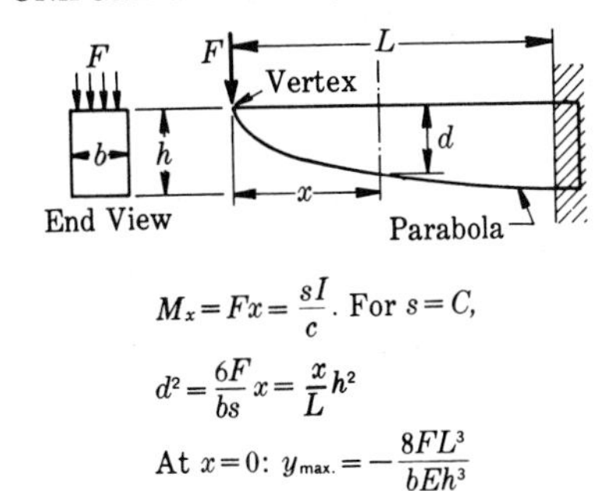

$M_x = Fx = \frac{sI}{c}$. For $s=C$,

$d^2 = \frac{6F}{bs}x = \frac{x}{L}h^2$

At $x=0$: $y_{max.} = -\frac{8FL^3}{bEh^3}$

UNIFORM STRENGTH, SIMPLE BEAM

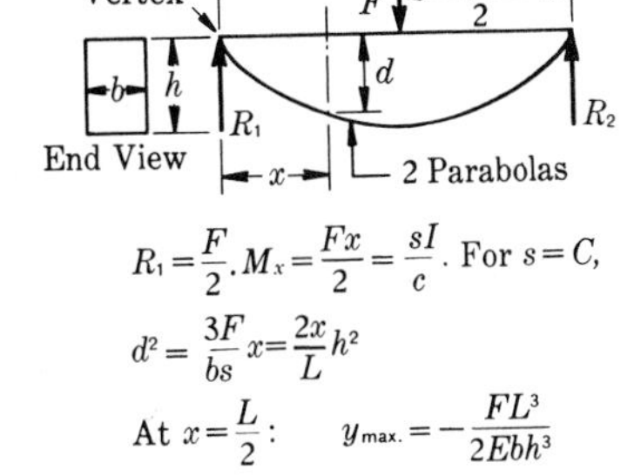

$R_1 = \frac{F}{2}$. $M_x = \frac{Fx}{2} = \frac{sI}{c}$. For $s=C$,

$d^2 = \frac{3F}{bs}x = \frac{2x}{L}h^2$

At $x = \frac{L}{2}$: $y_{max.} = -\frac{FL^3}{2Ebh^3}$

TABLE AT 3 TYPICAL PROPERTIES OF OF SOME NONFERROUS METALS[0.1,2.1,2.8,2.9,2.12]

For *Aluminum alloys*, let Poisson's ratio $\mu = 0.33$; torsional ult. $= 0.65s_u$; torsional yield str. $= 0.55s_y$. Extruded forms over $\frac{3}{4}$ in. will have s_u about 15% greater.

For *Magnesium alloys*, let the flexural strength (symmetric sections) be the average of the tensile and compressive strengths. See note (k). Let $\mu = 0.35$.

***Abbreviations:* H, hard; $\frac{1}{4}$H, $\frac{1}{4}$ hard; H14, temper designation meaning $\frac{1}{2}$ hard; HT, heat treated; T4, temper designation meaning solution heat treated.**

Notes: (a) At 0.5% total elongation under load. (b) Cold reduction of 11%. (c) At 0.2% offset. (d) BHN. (e) BHN with 500 kg. load. (f) Minimum. (g) Flat 0.04 in. thick. (h) End.

MATERIAL (*ASTM No.*)	CONDITION	ULT. STR., *ksi* s_u	ULT. STR., *ksi* s_{us}	TEN YD. s_y *ksi*	END. STR., s_n *ksi at No. cycles*
COPPER ALLOYS					
Admiralty brass (B111)	Annealed	53		22(a)	18 at 10^7
Aluminum bronze (B150-2)	Extruded rod (b)	120		70(a)	32 at 3×10^8
Aluminum bronze (B150-1)	Annealed, 800°F	100			28 at 8×10^7
Beryllium copper (B194)	HT	175		130	35 at 10^8(h)
Cartridge brass (B134-6)	$\frac{1}{2}$H, rod	70	42	52(a)	22 at 5×10^7
Commercial bronze (B134-2)	Spring H(g)	72	42	62(a)	21 at 15×10^6
Free-cutting brass (B16)	2″ Rod, $\frac{1}{2}$H	55	32	44(a)	20 at 10^8
Manganese bronze (B138-A)	$\frac{1}{2}$H, rod	75	48	65(a)	30 at 10^8(g)(q)
Naval brass (B124-3)	$\frac{1}{4}$H, rod	70	43	48(a)	15 at 3×10^8(b)
Nickel silver B	H(g)	100		85(a)	19
Phosphor bronze (B139C)	$\frac{1}{2}$H, rod	80		65(a)	31 at 10^8(r)
Silicon bronze (B98-B)	H, 1″ rod	65	45	35(a)	25 at 10^8
Yellow brass (B36-8)	$\frac{1}{8}$H, 1″ rod	55	36	40(a)	11 at 10^8(z)
Yellow brass (B36-8)	$\frac{1}{2}$H (g)	61	40	50(a)	
ALUMINUM ALLOYS					
3003-H14 (M1A)	Strain hard.	22	14	21(c)	9 at 5×10^8
2014-T6 (CS41A)	HT, aged	70	42	60(c)	18 at 5×10^8
2024-T4 (CG42A)	HT, aged	68	41	47(c)(p)	20 at 5×10^8
6061-T6 (p) (GS11A)	HT, aged	45	30	40(c)	14 at 5×10^8
7075-T6 (ZG62A)	HT, aged	82	48	72(c)	23 at 5×10^8
360	Die casting	40	27	24(c)	17 at 5×10^8
355-T6	Sand casting	35	28	25(c)	9 at 5×10^6
MAGNESIUM ALLOYS					
AZ61A-F	Extruded bar	45	20	33(c,k)	17 at 5×10^8
AZ80A-T5	Forged, aged	50	23	36(c,k)	16 at 5×10^8
AZ91C-T6	Sand casting	40		19(c,k)	14 at 5×10^8 (z)
LEAD AND TIN ALLOYS					
Babbitt (B23-46T-8)	At 68°F (l)	10	$s_{yc} = 3.4$	(m)	3.9 at 2×10^7
Babbitt (B23-46T-8)	At 212°F (l)	5.4	$s_{yc} = 1.7$	(m)	
Tin babbitt (B23-49-1)	At 68°F (l)	9.3	$s_{yc} = 4.4$	(m)	3.8 at 2×10^7
MISCELLANEOUS					
Hastelloy B	Sand cast	90		50(c)	66 at 10^8(n)
K Monel	Cold dr., aged	140	98	100(c)	42 at 10^8
Platinum alloy	Annealed	45			
Titanium (B265, gr. 5) (t)	Annealed (s)	135		130(c)	42 (f)
Titanium (B265, gr. 5) (t)	Hardened (s)(v)	170		158(c)	61 (f))
Zinc (AC41A)	Die cast (w)	47.6	38		8 at 10^8

strength in reversed torsion, 25 ksi. (i) α in./in.–°F, coef. of thermal expansion; room temp. (j) Varies with size of test specimen. (k) Yield point in compression; alloy AZ91C–T6, 19 ksi; AZ61A–F, 19 ksi; AZ80A–T5, 28 ksi. (l) Chill cast. (m) In compression, at 0.125% set. (n) At 1200°F, after water quench and aging. (o) Estimated. (p) Used for rolled structural shapes. (q) For manganese gear bronze, use s_n = 17 ksi. (r) For phosphor gear bronze, SAE 65, use s_n = 24 ksi. (s) Normal temp.; see § 2.21. (t) Sheet. (u) About 17.5 in compression. (v) Water quenched and aged at 975°F. (w) s_{uc} = 87 ksi, Charpy = 48. (x) After 1 yr. (y) Pure platinum. (z) Die castings.

MOD. EL. $E \times 10^{-6}$	MOD. EL. SH. $G \times 10^{-6}$	ELONG. % 2 *in.* (*j*)	ROCK. HARD.	DENSITY *lb/in.*^{3}	$\alpha \times 10^6$ (*i*)	PERCENTAGES OF ELEMENTS
15	5.8	65	F75	0.308	11.2	71 Cu, 28 Zn, 1 Sn
16	6.5	12	B100	0.274	9	81.5 Cu, 9.5 Al, 5 Ni, 2.5 Fe, 1 Mn
15	6.5	25	B90	0.274	9.2	91 Cu, 9 Al
19	7.3	5	C37	0.297	9.3	1.9 Be, 0.2 Ni or Co
16	6	30	B80	0.308	11.1	70 Cu, 30 Zn
17	6.4	3	B78	0.318	10.2	90 Cu, 10 Zn
14	5.3	32	B75	0.307	11.4	61.5 Cu, 35.5 Zn, 3 Pb
16	6	25	B80	0.302	11.8	58 Cu, 39 Zn, plus
15	5.6	25	B80	0.304	11.8	60 Cu, 39.25 Zn, 0.75 Sn
18		3	B91	0.314	9.3	55 Cu, 27 Zn, 18 Ni
16	6	33	B85	0.318	10.1	92 Cu, 8 Sn
17	6.4	10	B80	0.316	9.9	97.7 Cu, 1.5 Si, plus
15	5.6	48	B55	0.306	11.3	65 Cu, 35 Zn
15	5.6	23	B70	0.306	11.3	65 Cu, 35 Zn
10	3.85	16	40(e)	0.099	12.9	1.0 Mn, others
10.6	4.0	13	135(e)	0.101	12.8	3.9 Cu, 0.5 Si, 0.4 Mn, 0.2 Mg
10.6	4.0	20	120(e)	0.098	13.0	3.8 Cu, 1.2 Mg, 0.3 Mn
10	3.75	17	95(e)	0.100	12.7	0.15 Cu, 0.8 Mg, 0.4 Si
10.4	3.9	10	150(e)	0.101	12.9	5.1 Zn, 2.1 Mg, 1.2 Cu
10.3	3.85	1.8	70(e)	0.095	11.7	9 Si, 0.4 Mg
10.3	3.85	3	80(e)	0.098	11.7	1 Cu, 4.5 Si, 0.4 Mg
6.5	2.4	16	E72	0.065	14.4	6 Al, 1 Zn, 0.2 Mn
6.5	2.4	6	E82	0.065	14.4	8.5 Al, 0.5 Zn, 0.15 Mn
6.5	2.4	5	E77	0.066	14.4	9 Al, 0.7 Zn, 0.2 Mn
4.2		5	20(e)	0.36	13.3	80 Pb, 15 Sb, 5 Sn
		27	10(e)	0.36	13.3	80 Pb, 15 Sb, 5 Sn
7.3		2	17(e)	0.265		91 Sn, 4.5 Sb, 4.5 Cu
26.5		10	B93	0.334	5.55	62 Ni, 28 Mo, 5 Fe
26		20	C30	0.306	7.8	66 Ni, 29 Cu, 3 Al
		35	90(e)	0.722	5.0(y)	10 Rhodium
15(u)		12		0.160	5.8	{ 6 Al, 4 V
15(u)		7		0.160	5.8	{ Hi. temp. aero. service
2(x)		7	91(e)	0.24	15.2	4 Al, 1 Cu, 0.04 Mg

TABLE AT 4 TYPICAL PROPERTIES OF SOME STAINLESS STEELS[2.1,2.3,2.11,2.18]

Notes: (a) Coef. of thermal expansion near room temp., α in./in.-°F. (b) Approx. average values of ult. strength of 403, 410, and 416 are given by $s_u = 5 + 0.465$ (BHN) in ksi.[2.3] (c) Varies with details of heat treatment and cold working. (d) Cold worked, full hard. (e) Endurance limits for stainless steels may be estimated at $0.4s_u$, up to tensile strength

MATERIAL *AISI No.*	ULT. STR. s_u, *ksi* (*c*)	TEN. YD. s_y, *ksi* (*c*)	END. LIM. s_n' (*e*)	MOD. EL. $E \times 10^{-6}$ (*f*)	ELONG. 2 *in.*% (*c*)	RED AREA,% (*c*)
301, ¼ hard	125(h)	75(h)	30(g)	28	25(h)	
302, annealed	90	37	34	28	57	65
302, ¼ hard	125(g)(h)	75(h)	70(d)	28	12(h)	
303, annealed	90	35	35	28	50	55
304, annealed	85	35		28	50	70
316, cold worked (i)	90	60	40	28	45	65
321, annealed	87	35	38	28	50	65
347, annealed	90	40	39	28	50	65
403, 410, heat treated (b)	110(h)	85(h)	58	29	20	65
410, cold worked (b)	100(h)	85	53	29	17	60
416, annealed (b)	75	40	40	29	30	60
430, annealed	75	45	40	29	25	65
431, OQT 1000 (b)	150	130		29	18	60
17–7 PH rod (j)	175	155	41	29	6(h)	34

TABLE AT 5 TYPICAL PROPERTIES OF A FEW PLASTICS[2.19,2.21,2.27]

Notes: (a) TS, thermosetting; TP, thermoplastic. (b) National Electrical Mfg. Assoc. grades. (c) Flatwise. (d) For ⅓ to 1-in. dia. Reduce 15% for sizes 1 to 2 in. (e) For ⅛ to 1-in. dia. Reduce 10% for sizes 1 to 2 in. (f) Min. values. (g) Bending strength, symmetric sections. (h) Specific gravity. (i) Aver. water absorption, 24-hr., ¼-in. thickness, per cent.

MATERIAL	TYPE (*a*)	CONDITION (*k*)	ULT. STR., s_u *ksi*	COMP. ULT. STR., s_{uc}	FLEX STR., s_f (*g*) *ksi*
Phenol-formaldehyde					
Grade X (b) (l)	TS	L. sheet	14	35	23
Grade XX (b) (l)	TS	L. rod	8.5(d)	20	15(e)
Grade C (b) (l)	TS	L. rod	7.5(d)	20	17(e)
Grade A (b) (l)	TS	L. rod	6(d)	15	10(e)
Wood flour filler (p)	TS	M	6(f)	24	9(f)
Urea-formaldehyde	TS	M	9	25	10
Polyvinyl chloride	TP	M	8	10	
Polyvinyl chloride (n)	TP	M	8	13	
Polymethyl Methacrylate	TP	M	8	14	9
Polystyrene (f)	TP	M	5	11.5	6
Polyamide (m)	TP	M	11.8(s)	4.9(q)	13.8
Cellulose Acetate	TP	M	4.5	20	
Polyethelene (f)	TP	M	1.7	0.4(m)	1.7
Polytetrafluoroethylene (m)	TP	M	3.8(u)	1.8(t)	2
Polyvinylidene chloride	TP	M	5	2.4	
Polychlorotrifluoroethylene	TP	M	6	5	

of about 160 ksi[2.18] (f) Varies some with condition: annealed, cold worked, stress relieved. In shear, for cold drawn spring wires, $G \approx 10.6 \times 10^6$ psi. (g) 0.058-in. strip. (h) Minimum. (i) 1-in. bars. (j) PH, precipitation hardened; Republic Steel TH 1050; guaranteed $s_{u\min} = 170$ ksi; s_n' at 10^8.

BHN (*aver.*) (*c*)	DENSITY *lb/in.*^{3}	IZOD *ft-lb.* (*c*)	$\alpha \times 10^6$, (*a*)	REMARKS
260	0.286		9.4	(17% Cr, 7% Ni) General use; trim, structural.
150	0.286	90	9.6	Austenitic. Hardenable by cold work.
260	0.286		9.6	302, 303 are 18-8 stainless steels.
160	0.286	80	9.6	Austenitic. Hardenable by cold work.
150	0.286	110	9.6	Austenitic. Hardenable by cold work.
190	0.286		8.9	Austenitic. Hardenable by cold work.
150	0.290	110	9.3	Stabilized by Ti.
160	0.286	100	9.3	Austenitic. Hardenable by cold work.
225	0.279	75	5.7	Martensitic. Hardenable by HT.
205	0.279	80	5.7	Martensitic. Max. hardness.
155	0.278	70	5.7	Martensitic. Hardenable by HT.
160	0.277	35	5.8	Ferritic. Not hardened by HT.
325	0.28	50	6.5	Martensitic. Hardened by HT to high strength.
390	0.276		5.6	(17% Cr, 7% Ni, 1.15% Al) Solution annealed, etc.

(j) 48-hr. immersion. (k) L, laminated; M, molded. (l) When used for gears, let $s_n = 6000$ psi. (m) Yield strength. (n) Unplasticized. (p) General purpose. (q) At 1% def. (r) At failure. (s) At 73°F. (t) At 5% strain. (u) Rupture.

ELONG. % (*r*)	ROCK. HARD.	MOD. EL., $E \times 10^{-5}$	SP. GR. (*h*)	IZOD, *ft-lb.* (*f*)	%H_2O *Ab-sorp.*(*i*)	A FEW TRADE NAMES®
	M100	4–20	1.35	1.3(c)	1.4	Bakelite, Durez, Formica, Textolite, Micarta, Synthane, Durite.
	M100	4–20	1.35	1.0(c)	0.65	
	M100	3.5–15	1.35	3.2(c)	1.2	
	M90	3.5–15	1.65	1.8(c)	0.65	
0.4–0.8	M100	10	1.4	0.4	0.8	
0.6	M118	15	1.45	0.24	0.4	Beetle, Sylplast, Plaskon.
30	R65	3	1.2	0.8	0.05	Geon, Vinylite, Marvinol.
10	M70	8	1.41	0.4	0.1	Exon, Pliovic, Ultron.
8	M100	4	1.16	0.4	0.3(j)	Lucite, Plexiglass, Perspex.
1.2	M85	0.5	1.06	0.2	0.03	Lustrex, Styron, Styrene, Pliolite.
60(s)	R118	3.5	1.14	0.9(s)	1.5	Nylon, Zytel, 101.
20	R100	2	1.27	4	1.5–2.9	Plastacele, Celanese, Kodapak.
30–500	R11	0.15	0.92		0.01	Dylan, Alathon, Orizon.
100–200	R20	0.6	2.2	2–4	none	Teflon (TFE).
200	M55	0.7	1.7	0.7	0.1	Saran.
200	R110	2.5	2.1	4	none	Kel-F, Fluorothene.

TABLE AT 6 TYPICAL PROPERTIES OF CAST FERROUS METALS[2.1,2.8,2.14,2.16,2.17]

Notes: Approximate *coefficients of thermal expansion*, in./in.-°F are: gray iron, 5.6 × 10^{-6}; malleable iron, 6.6 × 10^{-6}; nodular iron, 6.7 × 10^{-6}; cast steel, 6.5 × 10^{-6} (but varies significantly with composition). *Poisson's ratio*: gray iron, 0.211 (min.); malleable iron, 0.265; nodular iron, 0.16; cast steel, 0.27. (a) ASTM and SAE specifications are not identical. (b) Machinability, relative values, AISI B1112 = 100%. (c) 1.2-in. dia., 18-in. supports. (d) Test results suggest that the flexural strength of cast iron in *symmetric* sections, computed from $s_f = M/Z$, is about $1.9s_u$ to $2s_u$. Use $1.9s_u$. (e) Estimated. (f) Minimum values. Typical values may range 10–40% higher. (g) ASTM 35 and higher are considered to be high-strength, and are definitely more expensive. (h) For cast iron, at 25% of ult. stress; varies with section size and chemical analysis. (i) Reversed bending. For

MATERIAL, SPEC. NO.	ULT. STRENGTH, *ksi*				TRANSV. STRENGTH	END. LIM.	TEN. YD.,
	s_u	s_{uc}	s_{us}	Tors.	*lb.* (c)	s_n', *ksi* (*i*)	s_y *ksi*
GRAY IRON (g) (as cast)	(d)	(d)				(e)	
ASTM SAE(a)							
20 110	20(f)	83	32	26	1850	10	
25	25(f)	97	35	32	2175	11.5	
30 111	30(f)	109	41	40	2525	14	
35(g) 120	35(f)	124	49	48.5	2850	16	
40(g) 121	40(f)	140	52	57	3175	18.5	
50(g)	50(f)	164	64	73	3600	21.5	
60(g)	60(f)	187	60	88.5	3700	24.5	
Ni-Resist, Inco K-6	25(f)	100(f)					
Meehanite, (w)	35(f)						28(f)
MALLEABLE IRON							
ASTM Grade							
A47-52 32 510	52	(o)	48	58		25.5	34
A47-52 35 018	55	(o)	43	58		27	36.5
NODULAR CAST IRON (j)							
60-45-10 (annealed) (q)	70	(s)		57		30	55
80-60-03 (as cast) (p)	88	(s)		73		40	65
100-70-03 (heat treat) (r)	110	(s)		88(e)		44(e)	80
CAST STEEL			MAX. CARBON AND HEAT TREAT.				
ASTM SAE(a)							
A27-58(t)	60(f)	60(f)	0.3% C, Annealed			25	30(f)
A27-58(t) 0030(k)	65(f)	65(f)	0.3% C, Normalized			28	35(f)
A27-58	70(f)	70(f)	0.35% C, Normalized			31	36(f)
A27-58	70(f)	70(f)	0.25% C, Normalized				40(f)
A148-58 080	80(f)	80(f)	N&T			35	40(f)
A148-58	80(f)	80(f)	WQT			35	50(f)(v)
A148-58 090	90(f)(v)	90(f)	N&T			41	60(f)
A148-58 0105	105(f)	105(f)	WQT			49	85(f)
A148-58	120(f)	120(f)	WQT			55	95(f)(w)
A148-58 0150	150(f)	150(f)	WQT			65	125(f)
A148-58 0175	175(f)	175(f)	WQT			77	145(f)

gray iron, $0.4s_u < S_n' < 0.6s_u$. (j) The number indicates minimum properties; e.g., 80-60-03 indicates $s_u = 80$ ksi, $s_y = 60$ ksi (0.2% offset), and 3% elongation, minimum, in approximately 1-in. section. (k) 0.3% C, max. (l) N&T, normalized and tempered. Properties of steel castings vary with carbon and alloy contents and with heat treatment, as for wrought steel; min. $s_n' \approx 0.4s_u$. (m) Charpy impact, keyhole notch, 70°F, ft-lb. (n) Charpy impact, V-notched. (o) Take as equal to s_u. (p) ASTM A339-55. (q) ASTM A395-56T. (r) ASTM A396-58. (s) For design, assume compressive ultimate and yield strengths of nodular iron to be the same as s_u and s_y, respectively. (t) Common commercial grades. (u) Tempered at 1200°F. (v) Typical $s_u \approx 96$ ksi, $s_y = 73$ ksi when WQT 1200. (w) General purpose type.

MOD. ELAS., *psi* $E \times 10^{-6}$	SHEAR MOD., *psi* $G \times 10^{-6}$	BHN	IZOD *ft-lb.*	DENSITY *lb/in.³*	MCH. (*b*)	REC. MIN. WALL THICK.	
(h)							
9.6(f)	3.9(f)	156		0.253		$t = \frac{1}{8}$ in.	
11.5(f)	4.6(f)	174		0.253		$t = \frac{1}{4}$ in.	
13(f)	5.2(f)	201	23	0.254	80	$t = \frac{3}{8}$ in.	
14.5(f)	5.8(f)	212	25	0.257	65	$t = \frac{3}{8}$ in.	
16(f)	6.4(f)	235	31	0.262	55	$t = \frac{5}{8}$ in.	
18.8(f)	7.2(f)	262	65	0.269	50	$t = \frac{3}{4}$ in.	
20.4(f)	7.8(f)	302	75	0.269		$t = 1$ in.	
12(f)		145	100			$t = \frac{1}{8}$ in.	
12(f)		190				$t = \frac{1}{8}$–$\frac{7}{8}$ in.	
						ELONG. 2 *in.*, %	RED. *Area*, %
25	10.7	120	12	0.262	120	12.5	
25	10.7	130	16	0.262	120	20	
23	9.5	160	9–20(n)	0.26		18	
23	9.9	230	2–8(n)	0.26		6	
23	9.9	270	2–6(n)			5	
30	11.5	120	18(m)	0.284	55	30	50
30	11.5	130	23(m)	0.284	60	30	53
30	11.5	140	19(m)	0.284	65	26	40
30	11.5		30(m)	0.284			
30	11.5	160	22(m)	0.284	70	27	42
30	11.5	170	30(m)(u)	0.284		28(u)	68(u)
30	11.5	190	20(m)(u)	0.284	70	24	50
30	11.5	235	28(m)	0.284	60	18	42
30	11.5	269	25(m)	0.284		14(f)	30(f)
30	11.5	310		0.284		9(f)	22(f)
30	11.5	390	12(m)	0.284		8	15

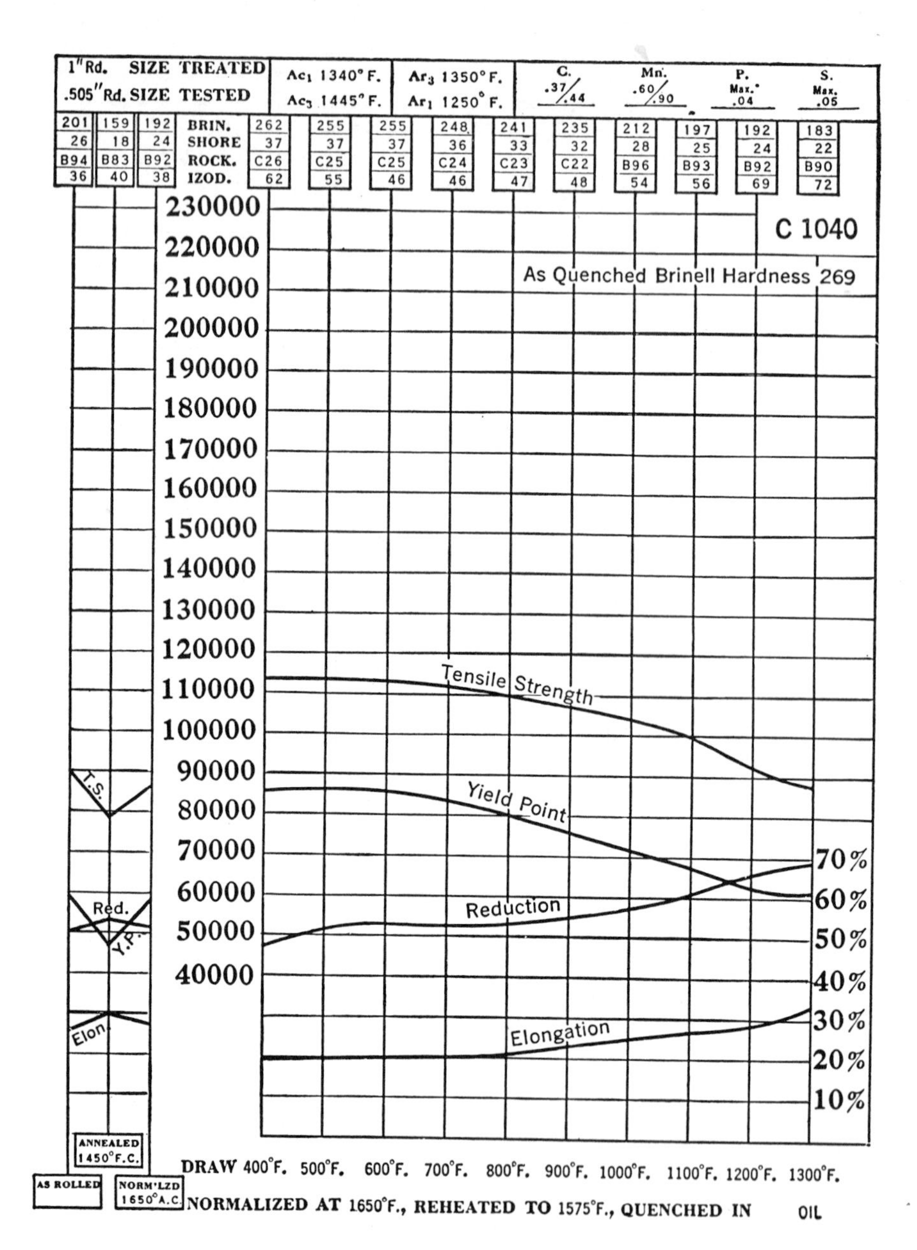

FIGURE AF 1 Properties of Heat-treated AISI C1040. Abscissa is the tempering temperature. Average values. Such charts as these are guides to mechanical properties that may be expected when the diameter is, say, $\frac{1}{4}$ in. to $1\frac{1}{2}$ in. See Tables AT 8 and AT 9. Observe that the BHN increases with tensile strength and that $s_u \approx (500)(\text{BHN})$—not so accurate at very low tempering temperatures. (Courtesy Bethlehem Steel Co., Bethlehem, Pa.).

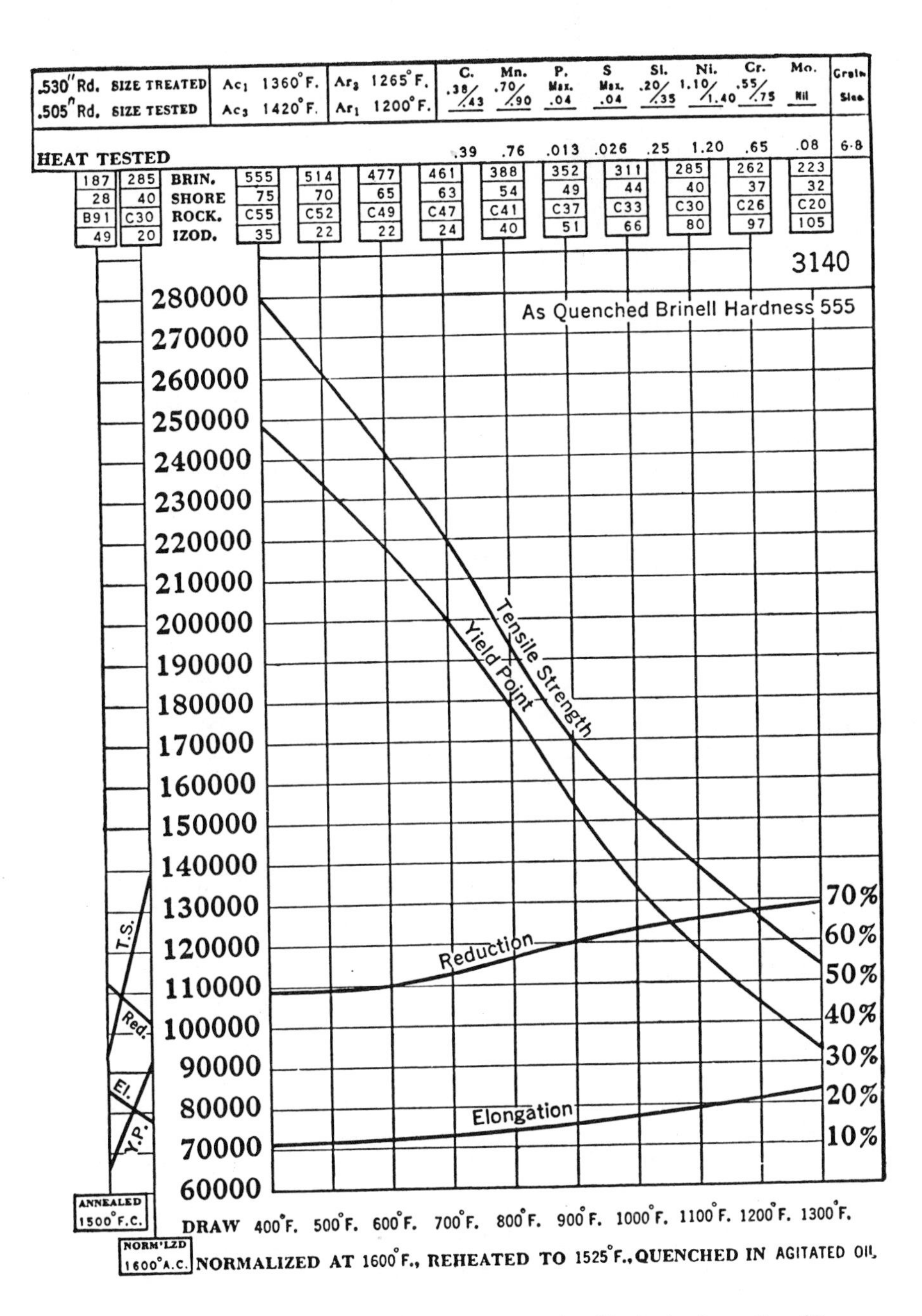

	BRIN.	SHORE	ROCK.	IZOD.
Annealed	187	28	B91	49
Normalized	285	40	C30	20
400°F.	555	75	C55	35
500°F.	514	70	C52	22
600°F.	477	65	C49	22
700°F.	461	63	C47	24
800°F.	388	54	C41	40
900°F.	352	49	C37	51
1000°F.	311	44	C33	66
1100°F.	285	40	C30	80
1200°F.	262	37	C26	97
1300°F.	223	32	C20	105

FIGURE AF 2 Properties of Heat-treated AISI 3140. Single heat results. (Draw = Temper.) Notice the specified heat treatment and the size of the specimen. The ultimate strength $s_u \approx (500)(\text{BHN})$ in psi. This material is widely used for heat-treated parts. For $R_c = 28$, decarburization of surface reduces endurance strength by 50%. For $R_c = 48$, decarburization of surface reduces endurance strength by 75%, from about $s_n' = 82$ ksi but the percentage is unusually high. See Fig. AF 3. (Courtesy Bethlehem Steel Co., Bethlehem, Pa.).

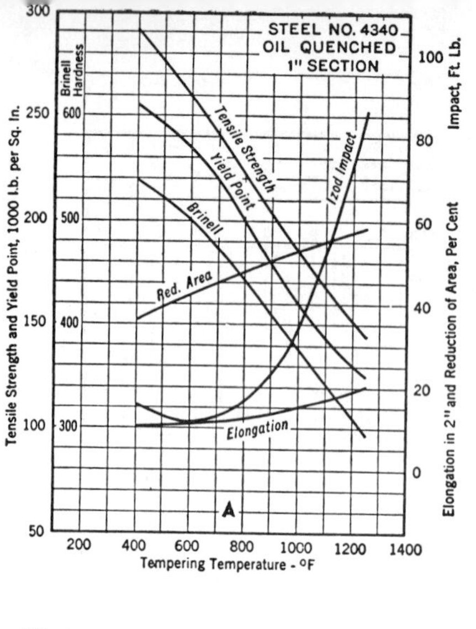

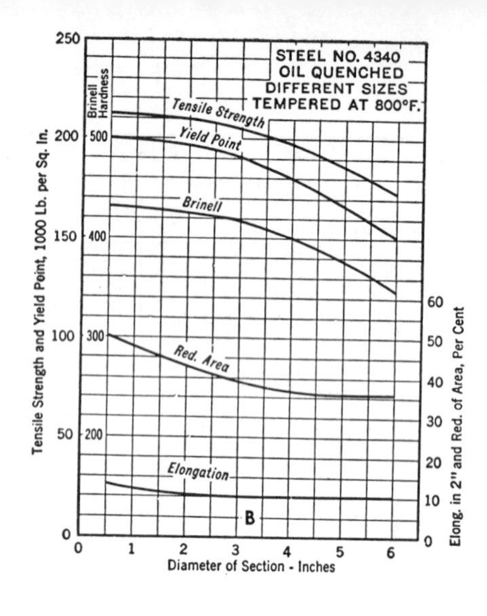

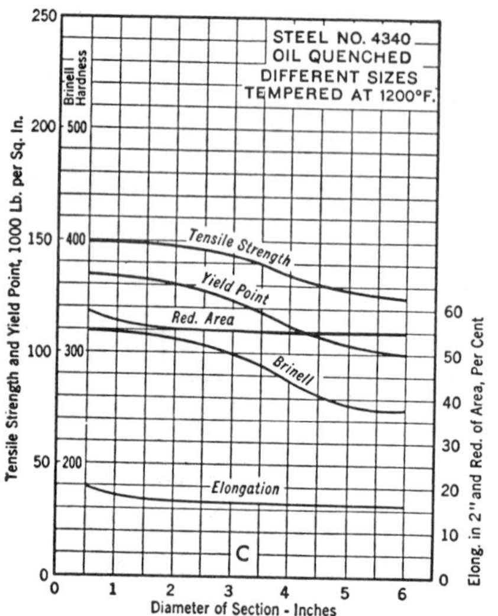

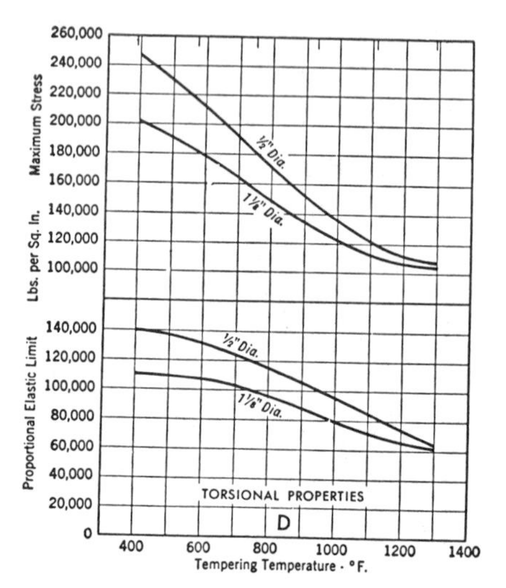

FIGURE AF 3 Properties of Heat-treated AISI 4340. Another manner in which mechanical properties might be depicted; charts A, B, and C are tensile properties; chart D gives torsional properties. An excellent general purpose alloy. Other miscellaneous endurance strengths of this steel follow. For $s_u \approx 270$ ksi:

Surface not decarburized, $s_n = 89$ ksi. Surface decarburized to 0.03 in., $s_n = 40$ ksi. Decarburized surface, shot peened, $s_n = 95$ ksi. OQT 1075, 0.625 dia., nitrided surface, $s_n = 120$ ksi. Test specimen from rolled stock, transversely, $s_n =$ 45–70 ksi.

For $s_u \approx 160$ ksi; variation of endurance strengths with temperature.

Temp.	Reversed s_n', $R = -1$ (ksi)	Repeated, $R = 0$ (ksi)
70°F	70	117
600°F	61	96
800°F	59	82
1000°F	39	65

(Courtesy International Nickel Co., N.Y.)

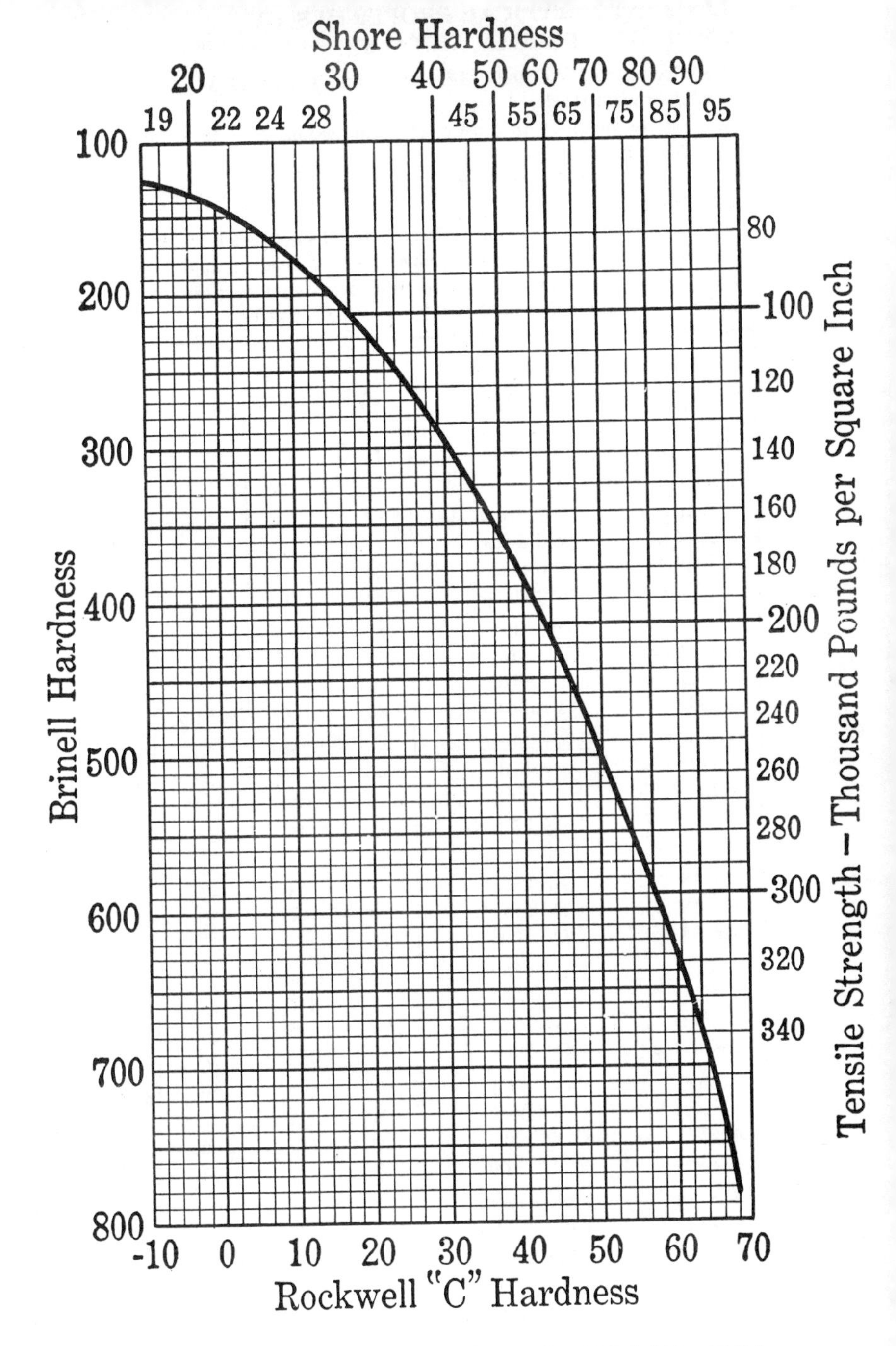

FIGURE AF 4 Relation of Hardness Numbers. (Courtesy International Nickel Co., N.Y.)

TABLE AT 7 TYPICAL PROPERTIES OF WROUGHT FERROUS METALS[1.1,2.1,2.3,2.6,2.10]

(See also charts for C1040, 3140, 4340; and Tables AT 8–AT 10, incl.).
For all wrought steels:
Modulus of elasticity **in tension or compression, $E = 30 \times 10^6$ psi (For Wrought iron, $E = 28 \times 10^6$ psi).**
Modulus of elasticity in shear **or torsion, $G = 11.5 \times 10^6$ psi (For wrought iron, $G = 10 \times 10^6$ psi).**
Yield strength in torsion (shear) **frequently falls between $0.5s_y$ and $0.6s_y$. Use $0.6s_y$.**
Endurance limit **of a polished specimen, reversed bending, approximately $s_u/2$; Table AT 10.**
Endurance limit in reversed torsion, **approximately $0.6s_n'$; see § 4.7.**
Poisson's ratio **is about 0.25 to 0.33. Use 0.3 for steel.**

MATERIAL *AISI No.*	CONDITION (*c*)	ULT. STR., *ksi* s_u	ULT. STR., *ksi* $s_s(d)$	TEN. YD. s_y, *ksi*	ELONG. *2 in.* %	RED. *Area* %	BHN
Wrought Iron	As rolled	48(a)	36	25(a)	35		
Wrought Steel							
C1010 (k)	Cold drawn	67	50	55	25	57	137
C1015 (k)	Cold drawn	77	58	63	25	63	170
C1020	As rolled	65	49	48	36	59	143
C1020	Normalized	64	54*	50	39	69	131
C1020	Annealed	57	43	42	36.5	66	111
C1020 (k)	Cold drawn	78	58	66	20	59	156
C1022	As rolled	72	54	52	35	67	149
C1030	As rolled	80	60	51	32	56	179
C1035	As rolled	85	64	55	29	58	190
C1045	As rolled	96	72	59	22	45	215
C1095	Normalized	141	105	80	8	16	285
B1113 (k)	Cold finished	83	62	72	14	40	170
B1113	As rolled	70		45	25	40	138
C1118	As rolled	75	56	46	32	70	149
C1118 (k)	Cold drawn	80	60	75	16	57	180
C1144	OQT 1000	118	88	83	19	46	235
1340	OQT 1200	113	84	92	21	61	229
13B45	OQT 800	187	140	175	16	56	
2317(e)	OQT 1000	106	79	71	27	72	220
2340(e)	OQT 1000	137	103	120	22	60	285
3150	OQT 1000	151	113	130	16	54	300
3250(e)	QT 1000	166	122	146	16	52	340
4063	OQT 1000	180	135	160	14	43	375
4130	WQT 1100	127	95	114	18	62	260
4130(e)	Cold drawn	122	91	105	16	45	248
4340(e)	Cold drawn	122	91	105	15	45	248
4640(e)	OQT 1000	152	104	130	19	56	310
5140(e)	OQT 1000	150	113	128	19	55	300
5140(e)	Cold drawn	105	79	88	18	52	212
8630	Cold drawn 10%	115	86	100	22	53	222
8640	OQT 1000	160	120	150	16	55	330
8760	OQT 800	220	165	200	12	43	429
9255	OQT 1000	180	135	160	15	32	352
9440	OQT 1000	152	104	135	18	61	311
9850	OQT 1100	180	135	158	15	48	360

Density is about 0.284 lb/in.³ (0.28 lb/in.³ for wrought iron).

Coefficient of thermal expansion (linear) is 0.000 007 in./in.-°F (0.000 0006 5 for wrought iron). Varies significantly with large temperature change. See § 2.22 for cryogenic applications.

Notes: B preceding AISI No. indicates Bessemer, as B1113; C preceding indicates open hearth, as C1020. (a) Minimum values. (b) Annealed. (c) QT 1000 stands for "quenched and tempered at 1000°F," etc. (d) *Ultimate stress in shear* has been arbitrarily taken as (.75) (tensile ultimate); except starred * values which are test values. (e) 1-in. specimen. (f) Torsion. (g) Mill annealed. (h) Cold drawn. (i) See § 2.2 for definition. (j) Charpy V-notch, 70°F. (k) Properties depend on amount of cold work.

ROCK.	IZOD ft-lb.	MACHIN-ABILITY (*i*)	SOME TYPICAL USES; REMARKS
B60		50(h)	ASTM A85-36, A41-36
		50	Bars, strips, sheet, plate. Cold drawn shapes.
	137(j)	50	Bars, sheets. Table AT 8. For carburizing: Table AT 11.
B79	64	64	Structural steel; plate, sheet, strip, wire.
B74	72		Carburizing grade, Table AT 11.
B66	80		General purpose.
B83		62	Misc. machine parts are cold forged; bars.
B81	60	70(h)	General purpose.
B88	55	60	Machinery parts. Table AT 8.
B91	45	57	Machine parts. May be heat treated. Table AT 9.
B96	30	51	Large shafts.
C25	3	39	Tools, springs. Usually heat treated. Table AT 9.
B87		135	Free cutting; high sulfur.
B76			Free cutting; high sulfur.
B81	80	82	Free cutting; not usually welded. Carburized, Table AT 11.
	110(j)	85	Table AT 8 for C1117.
C22	36	65(b)	Free cutting. High sulfur. Tables AT 8 and AT 9 for C1137.
C31	95	45(g)	(1.75% Mn). Manganese steel.
C42			1345 with boron for improved hardenability.
B97	85	55(h)	(3½% Ni)—Gears, pump liners, etc.
C30	50	31	(3½% Ni)—Gears, etc.
C32	46		(1.25% Ni, 0.8% Cr.) Gears, bolts, shafts, etc.
C36.5	30	55(b)	(1.85% Ni, 1.05% Cr) Gears, etc.
C40	59		(0.25% Mo) Shafts, bars, etc.
C25	85	65(b)	(0.95% Cr, 0.20% Mo) Shafts, forgings, pins, aircraft tubes.
		45(g)	(1.85% Ni, 0.8% Cr, 0.25% Mo) General purpose. Fig. AF3.
C33	41	55(b)	(1.85% Ni, 0.25% Mo).
C32		60(b)	(0.80% Cr) Gears, shafts, pins, etc.
		60(g)	
			(0.55% Ni, 0.5% Cr, 0.2% Mo). Table AT 9.
C35	36	60(b)	(0.55% Ni, 0.50% Cr, 0.20% Mo).
C46	19	50(b)	(0.55)% Ni, 0.50% Cr, 0.25% Mo) Tools, gears, bolts.
C36	7	45(b)	(2.00% Si, 0.82% Mn) Springs, chisels, tools.
C33	73	60(b)	(0.45% Ni, 0.4% Cr, 0.11% Mo).
C37	50	50(b)	(1% Ni, 0.8% Mn, 0.8% Cr. 0.25% Mo) Heavy duty; general.

TABLE AT 8 TYPICAL PROPERTIES OF STEEL—VARIOUS SIZES AND CONDITIONS[2.6]

(a) Turned. (b) 10%. (c) Inconsistent—from different mill.

AISI NO.	CONDITION	ROD DIA. *in.*	ULT. STR. s_u *ksi*	TEN YD. s_y *ksi*	ELONG. 2 *in.*, %	RED. Area, %	BHN	IZOD *ft-lb.*
C1015	As rolled	1/2	61	45.5	39	61	126	81
	Annealed	1	56	42	37	69.7	111	83
	Normalized	1/2	63	48	38.6	71	126	85
	Normalized	1	61.5	47	37	69.6	121	85
	Normalized	2	60	44.5	37.5	69.2	116	86
	Normalized	4	59	41.8	36.5	67.8	116	83
C1117	As rolled	1/2	70.6	44.3	33	63	143	60
	Annealed	1	62	40.5	32.8	58	121	69
	Normalized	1/2	69.7	45	34.3	61	143	70
	Normalized	2	67	41.5	33.5	64.7	137	83
	Normalized	4	63.7	35	34.3	64.7	126	84
C1030	As rolled	1/2	80	51	32	54	179	55
	Annealed	1	67	49	31	57.9	126	51
	Normalized	1/2	77.5	50	32	61.1	156	69
	Normalized	4	72.5	47	29.7	56.2	137	61
	WQT 1000	1	88	68	28	68.6	179	92
C1137	As rolled	1/2	93	55	26	63	192	61
	Annealed	1	85	50	27	54	174	37
	Normalized	1/2	98	58	25	58	201	69
	Normalized	2	96	49	22	51	197	21
	Cold drawn	1	103	93	15	56	217	
C1045	Annealed	1	90	55	27	54	174	32(c)
	Normalized	1	99	61	25	49	207	48(c)
	Hot rolled(a)	1	87	54	27	56	187	51(c)
	Cold drawn(b)	2	100	85	19	45	235	
	WQT 1000	1/2	130	110	16	56	260	75(c)
	WQT 1200	1/2	110	84	23	61	220	
	WQT 1000	2	110	70	23	50	205	85(c)
	WQT 1200	2	98	64	26	58	190	
	WQT 1000	4	94	59	25	49	180	62(c)
	WQT 1200	4	93	55	28	55	186	
C1050	As rolled	1/2	102	58	18	37	229	23
	Annealed	1	92	53	23.7	40	187	12
	Normalized	1/2	111	62	21.5	45	223	17
	Normalized	4	100	56	21.7	41.6	201	20
	Cold drawn	1	113	95	12	35	229	
	OQT 1100	1/2	122	81	22.8	58	248	22
	WQT 1100	1/2	119	88	21.7	60	241	51
	OQT 1100	2	112	68	23	55.6	223	20
	WQT 1100	2	117	78.5	23	61	235	24
	OQT 1100	4	101	58.5	25	54.5	207	21
	WQT 1100	4	112	68	23.7	55.5	229	15

TABLE AT 9 TYPICAL PROPERTIES OF HEAT-TREATED STEELS[2.3,2.6]

Values in this table have been taken from charts such as those of Figs. AF 1-AF 3. To get the strength or Brinell number for any other tempering (drawing) temperature, make a straight line interpolation between the values given. Extrapolation to lower temperatures might sometimes give a reasonable estimate, but extrapolation should not be relied upon. (a) Do not interpolate using this value.

AISI NO. (*Quenching medium*)	SIZE	TEMPERED AT, °F	ULT. STR. s_u *ksi*	TEN. YD. s_y *ksi*	BHN	ELONG. 2 *in.* %	IZOD, *ft-lb.*
C1035 (water)	1″	600	118	87	240	11	40
	1″	1000	102	73	200	22	57
	1″	1300	85	57	170	29	93
C1095 (oil)	½″	800	176	112	363	11	6
	½″	1100	145	88	293	17	6
	4″	1100	130	65	262	17	5
C1137 (oil)	½″	700	135	115	277	12	13(a)
	½″	1000	111	88	229	23	61
	2″	1000	105	63	217	23	31
2330 Nickel Steel (water)	½″	600	210	195	429	13	39
	½″	1000	135	126	277	20	77
	½″	1300	107	91	217	26	109
	4″	1000	105	85	207	26	87
4140 Cr-Mo (oil)	½″	500	270	241	534	11	8(a)
	½″	800	210	195	429	15	21
	½″	1200	130	115	277	21	83
	4″	1200	112	83	229	23	87
4150 Cr-Mo (oil)	½″	800	228	215	444	10	12(a)
	½″	1200	159	141	331	16	53(a)
5150 Chromium (oil)	½″	800	210	195	415	11	17(a)
	½″	1000	160	149	321	15	39
	½″	1200	127	117	269	21	59
6152 Cr-V (oil)	½″	700	246	224	495	10	9(a)
	½″	1000	184	173	375	12	30
	½″	1200	142	131	293	18	65
	2″	1200	121	94	241	21	45(a)
8630 Ni-Cr-Mo (water)	½″	800	185	174	375	14	58
	½″	1100	137	125	285	20	95
	4″	1100	96	72	197	25	104
8742 Ni-Cr-Mo (oil)	1″	700	226	203	455	11	14(a)
	1″	1200	130	110	262	21	67(a)
	4″	1200	118	91	235	22	
9261 Si-Mn (oil)	½″	800	259	228	514	10	12
	½″	900	215	192	429	11	13
	½″	1200	147	124	311	17	35(a)
9840 Ni-Cr-Mo (oil)	1″	700	237	214	470	11	10(a)
	1″	1200	140	120	280	19	65(a)
	6″	1000	151	131	302	16	

TABLE AT 10 MISCELLANEOUS ENDURANCE LIMITS AND ENDURANCE STRENGTHS[2.1,2.3,2.5,2.9,2.12,2.16]

Specimens 0.5 in. or smaller. See also Tables AT 3, AT 4, AT 6, AT 7, Fig. AT 3. Endurance ratio s_n/s_u decreases as size of section increases, to as low as 0.35 for 6 in. dimension in cast steel.

Notes: (a) Manganese steel. (b) Number of cycles is indefinitely large unless specified. Cy. = cycles. (c) By analogy (not a test value). (d) Depends on the number of cycles. (e) Permanent mold.

MATERIAL	CONDITION	s_n KSI AT NO. OF CY. (*b*)	$\frac{s_n}{s_u}$ (*d*)	s_y KSI	$\frac{s_y}{s_n}$
Wrought iron	Longitudinal	23	0.49	28	1.22
Wrought iron	Transverse	19	0.55	25(c)	1.31
Cast iron	ASTM 30	12	0.38		
Cast iron	ASTM 30	16 at 10^5			
Cast iron	ASTM 30	21 at 10^4			
Cast steel, 0.18% C	As cast	31.5	0.45	36	1.14
Cast steel, 0.18% C	Cast and annealed	34.5	0.45	37	1.07
Cast steel, 0.25% C	Cast and normalized	35	0.46	45	1.29
Cast steel, 1330(a)	Cast, N&T 1200	48	0.49	61	1.27
Cast steel, 1330(a)	Cast, WQT (269 BHN)	58	0.48	106	1.83
Cast steel, 4340	Cast, WQT 1100	64	0.40	148	2.32
Cast steel, 8630	Cast, N&T 1200	54	0.49	85	1.57
Cast steel, 8630	Cast, WQT (286 BHN)	65	0.47	125	1.92
Wrought Steel 1015					
1015	Cold drawn (10% work)	40	0.57	63	1.58
1020	As rolled	45 at 10^4		48	1.08
1020	As rolled	40 at 10^5		48	1.20
1020	As rolled	33 at 10^6		48	1.45
1035	Cold drawn	46(c)	0.50	78	1.69
1035	In air	40.6	0.46	58	1.43
1035	In brine	24.6		58	2.36
1035	In sulfur	10.6		58	5.48
1040	Cold drawn (10% work)	54	0.54	85	1.57
1040	Cold drawn (20% work)	59	0.5	92	1.56
1117	Cold drawn	40(c)	0.50(c)	68	1.70
1141	Cold drawn	50	0.46	90	1.8
13B45	OQT 1100	68	0.54	112	1.65
1144	Elevated temp. drawn (ETD)	72	0.48	140	1.95
2317	In air	52	0.61	50	0.96
2317	In brine	31.6		50	1.58
2317	In sulfur	23.9		50	2.09
2320	Hot rolled rod	48	0.50	51	1.06
2320	Carburized, case hardened	90	0.53	140	1.56
3120	Carburized, case hardened	90	0.64	100	1.11
4340	At 1000°F (OQT 1150)	40			
6150	Heat treated	96	0.46	190	1.98
8630	Cold drawn (20%)	62	0.51	107	1.73
94B40	OQT 1100	70	0.51	119	1.70
Nitralloy N	Nitrided	124	0.65	180	1.45
Nitralloy 135, modified	Un-nitrided	45			
Nitralloy 135, modified	Nitrided	90	0.66	140	1.56
Nitralloy 135, modified	Notched and un-nitrided	24			
Nitralloy 135, modified	Notched and nitrided	80	0.59	140	1.75
Stainless steel 316	Annealed bar	38	0.37	35	0.92

MATERIAL	CONDITION	s_n KSI AT NO. OF CY. (*b*)	$\frac{s_n}{s_u}$ (*d*)	s_y KSI	$\frac{s_y}{s_n}$
Stainless steel 403 .	Annealed bar	40	0.57	37	0.67
Stainless steel 403 .	Bars, heat treat. to $R_B = 97$	55	0.50	85	1.54
Stainless steel 410 .	Bars, OQT to $R_B = 97$	58	0.52	85	1.47
Stainless steel 410 .	ditto, except at 850°F	43			
Stainless steel 418 .	OQT 1200	75	0.54	108	1.43
Stainless steel 430 .	Annealed and cold drawn; 185 BHN	46	0.61	50	1.09
Aluminum 2011 . .	Wrought, T3	18 at 5×10^8	0.33	43	2.39
Aluminum 2014 . .	Wrought, T4	20 at 5×10^8	0.32	42	2.10
Aluminum 2014 . .	Wrought, T6	18 at 5×10^8	0.26	60	3.33
Aluminum 2014 . .	Ditto	30 at 10^6	0.43	60	2.00
Aluminum 2014 . .	Ditto, 500°F	5 at 5×10^8	0.45	8.5	1.70
Aluminum 2014 . .	T6, reversed axial	15 at 5×10^8	0.21	60	4.00
Aluminum 5052 . .	Cold worked, H32	17 at 5×10^8	0.51	28	1.65
Aluminum 5052 . .	Cold worked, H36	19 at 5×10^8	0.47	35	1.84
Aluminum 6063 . .	Wrought, T5	10 at 5×10^8	0.37	21	2.1
Aluminum 7079 . .	Wrought, T6	23 at 5×10^8	0.30	68	2.96
Aluminum, 142 alloy .	Sand casting, T77	10.5 at 5×10^8	0.35	23	2.19
Aluminum, 142 alloy .	Casting, T61(e)	9.5 at 5×10^8	0.20	42	4.42
Alum. bronze (10%)	Extruded, heat treated	34 at 7×10^7	0.44	50(c)	1.47
Alum. bronze (10%)	Sand cast, annealed	28 at 8×10^7	0.34	40	1.43
Cartridge brass (70-30)	0.08″ spring wire	22 at 10^8	0.17	65(c)	2.96
Cartridge brass (70-30)	Half hard, 1″ rod	22 at 5×10^7	0.31	52	2.36
Free-cutting brass .	Half hard, 2″ rod., SAE 72	14 at 3×10^8	0.25	44	3.14
Commercial bronze	0.08″ hard wire	23 at 10^8	0.31	60(c)	2.61
Leaded tin bronze .	Sand cast, alloy 2A(Navy M)	11 at 10^8	0.29	18	1.64
Low brass (80-20) .	Spring hard, 0.04″ strip	24 at 2×10^7	0.26	65	2.70
Low brass (80-20) .	0.08″ spring wire	26 at 10^8	0.21	88(c)	3.38
Manganese bronze .	Sand cast, alloy 8A	21.2 at 10^8	0.30	28	1.32
Manganese bronze .	Sand cast, alloy 8C	25 at 10^8	0.24	70	2.8
Silicon bronze, type A	Half-hard rod	30 at 3×10^8	0.39	45	1.50
Silicon bronze, type B	Hot rolled	19 at 5×10^7			
Silicon bronze, type B	Extruded	20 at 5×10^7	0.29	55(c)	2.75
Silicon bronze, type B	Cold drawn, 72% reduction	30 at 3×10^8	0.32	69(c)	2.30
Silicon bronze, type B	0.08″ hard wire	25 at 10^8	0.28	67	2.68
Magnesium (AZ63A).	Cast, T5	11 at 5×10^8	0.38	15	1.36
Magensium (AZ31B).	Extruded bar	15 at 5×10^8	0.41	22	1.47
Inconel (Ni-Cr) . .	Cold drawn	40 at 10^8	0.38	80	2.00
Inconel	As forged or hot rolled	38 at 10^8	0.42	35	0.92
Monel (67 Ni, 30 Cu)	Annealed rod	31 at 10^8	0.41	30	0.97
Monel	Cold drawn rod	42 at 10^8	0.42	75	1.78
Monel	Annealed. In brackish water	21 at 10^8	0.28	30	1.43
K-Monel (3 Al) .	Cold drawn, age hardened	45 at 10^8	0.30	110	2.44
Titanium (5 Al, 2.5 Sn)	Formed; ground finish	60	0.5	110	1.83

TABLE AT 11 TYPICAL CORE PROPERTIES OF CARBURIZED STEELS[2.3]

Carburizing is done at about 1700°F. A tempering temperature of 300°F produces maximum case hardness; 450°F results in improved impact strength.

Notes. (a) Nominal size of specimen, 1 in. (b) ½-in. specimen. (c) 2-in. specimen. (d) 4-in. specimen. (e) Abbreviations: "SOQT 450," single oil quench and temper at 450°F; "DWQT 300," double water quench and temper at 300°F; Q, quench; P, pot. (f) Of the order of other hardnesses shown. (g) Case thickness depends on temperature and time of carburizing; for example, at 1700°F for 4 hr., the case should be of the order of 0.05 in.; at 1700°F for 8 hr., about 0.06 in. As seen from the values given, these are not hard and fast rules.

AISI NO.	CONDITION (*e*)	CORE PROPERTIES						CASE	
		Ult. Str. s_u *ksi*	*Ten. Yd.* s_y *ksi*	*Elon.* % 2″	*Red. Area* %	*BHN*	*Izod ft-lb.*	*Rock. Hard.* R_C	*Thick. in.* (8*hr.*)
C1015(b)	SWQT 350	73	46	32	71	149	91	C62	0.048
C1020(a)	DWQT 300	85	55	33	65	170		(f)	(g)
C1020(a)	SWQT 300	80	50	30	60	160		(f)	(g)
C1117(b)	SWQT 350	96	59	23	53	192	33	C65	0.045
2115(a)	DO(or W)QT 300	90	60	30	70	185	70	(f)	(g)
2317(a)	DOQT 300	95	60	35	65	195	85	(f)	(g)
2317(a)	DWQT 300	100	65	30	60	210	70	(f)	(g)
2515(a)	DOQT 300	170	130	14	50	352	40	(f)	(g)
3115(a)	DOQT 300	100	70	25	55	212	55	(f)	(g)
3215(a)	SOQT 300	141	110	17	50		45	(f)	(g)
E3310(b)	SOQT 450	180	149	14.5	58	363	57	C57.5	0.047
E3310(b)	DOQT 300	177	143	15.3	58	352	47	C61	0.047
3415(a)	SOQT 300	130	95	18	52	285	55	(f)	(g)
3415(a)	DOQT 300	135	105	19	55	300	60	(f)	(g)
4320(b)	Direct OQ from P 300	217	159	13	50	429	32	C60.5	0.060
4320(b)	DOQT 450	145	94	21.8	56	293	48	C59	0.075
4620(b)	DOQT 300	122	77	22	56	248	64	C62	0.060
4620(b)	DOQT 450	115	77	22.5	62	235	78	C59	0.060
4820(b)	SOQT 300	207	167	13.8	52	415	44	C61	0.047
4820(b)	SOQT 450	205	184	13	53	415	47	C57.5	0.047
8620(b)	SOQT 300	188	149	11.5	51	388	26	C64	0.075
8620(b)	SOQT 450	167	120	14.3	53	341	29	C61	0.076
8620(b)	DOQT 300	133	83	20	56	269	55	C64	0.070
E9310(b)	Direct OQ from P 300	179	144	15.3	59	375	57	C59.5	0.039
E9310(b)	SOQT 300	173	135	15.5	60	363	61	C62	0.047
E9310(b)	DOQT 300	174	139	15.3	62	363	54	C60.5	0.055
E9310(a)	SOQT 300	159	122	15.5	57	321	68	(f)	(g)
E9310(c)	SOQT 300	145	108	18.5	66	293	93	(f)	(g)
E9310(d)	SOQT 300	136	94	19	62	277	93	(f)	(g)

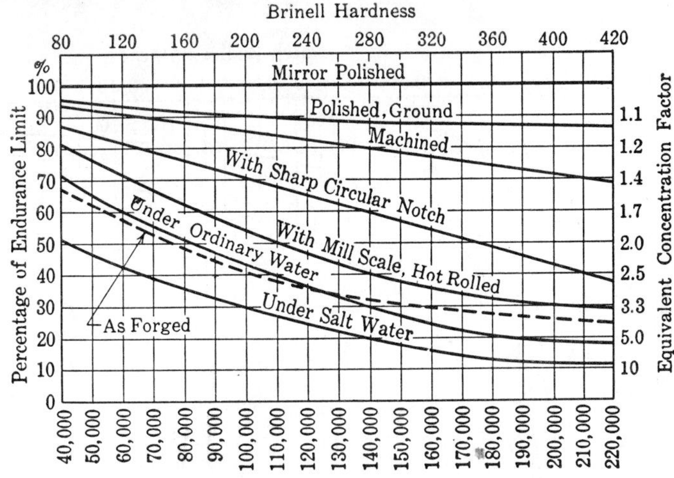

FIGURE AF 5 Reduction of Endurance Strength of Steel.[4.4] Some effects are stress raisers, some are strength reducers. See § 3.14 for the approximate roughnesses of the corresponding surfaces. A ground surface is not expected to have a roughness greater than 100 microinches. The curve for "as forged," adapted from Lipson, *et al.*[4.2], assumes decarburization of the surface. The machined surface is a good one, light cut, fine feed. Polishing leaves a residual compressive stress, helpful against fatigue.

TABLE AT 12 VALUES OF K_f FOR SCREW THREADS [4.2] For tension or bending. Not K_t.

KIND OF THREAD	ANNEALED *Rolled*	ANNEALED *Cut*	HARDENED *Rolled*	HARDENED *Cut*
Sellers, Amer. Nat'l., Sq. Th.	2.2	2.8	3.0	3.8
Whitworth Rounded Roots	1.4	1.8	2.6	3.3
Dardelet	1.8	2.3	2.6	3.3

TABLE AT 13 VALUES OF K_f FOR KEYWAYS.[4.2] *See* § 10.4

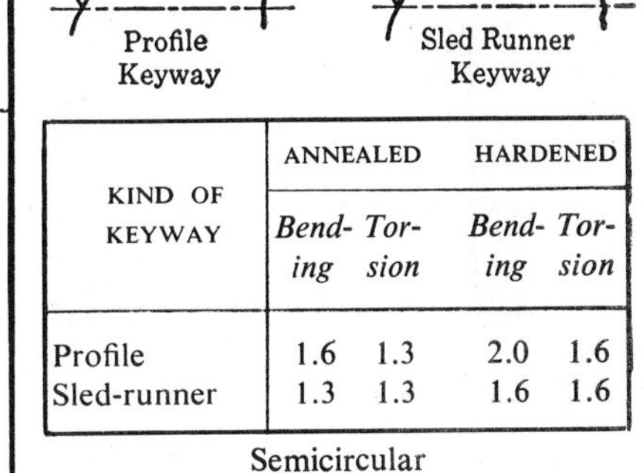

KIND OF KEYWAY	ANNEALED *Bending*	ANNEALED *Torsion*	HARDENED *Bending*	HARDENED *Torsion*
Profile	1.6	1.3	2.0	1.6
Sled-runner	1.3	1.3	1.6	1.6

Semicircular

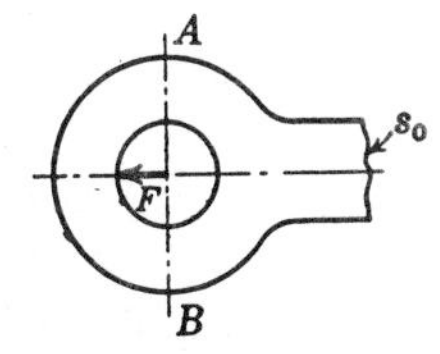

Approx. $K_t = 2.8$. $s_{max} = K_t s_o$ at *AB inside* hole.

INTERFERENCE FITS

$K_f = 1.5$ to 4, Examples: Cold-rolled shaft $K_f = 1.9$. Heat-treated shaft $K_f = 2.6$.7-in as forged $K_f = 3$.

FIGURE AF 6 K_t and K_f for T-head and Miscellaneous.[4.2]

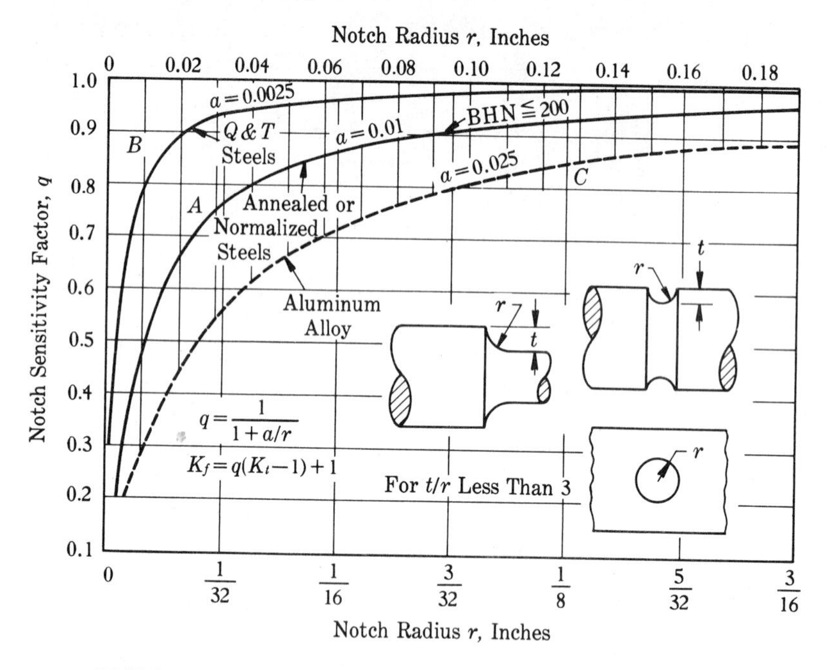

FIGURE AF 7 Average Notch Sensitivity Curves. Applicable particularly to normal stresses; used also for shear stresses. (After R. E. Peterson)[4.1,4.57]

FIGURE AF 8 Flat Plate with Central Hole—Tension and Bending.[4.2,4.16,4.21] Use solid curve for bending in plane of paper; and for rod ($D = h$) with a hole (d) in tension. Symmetric loading. (Some evidence[4.61] that values for a tensile rectangular body are overly conservative.)

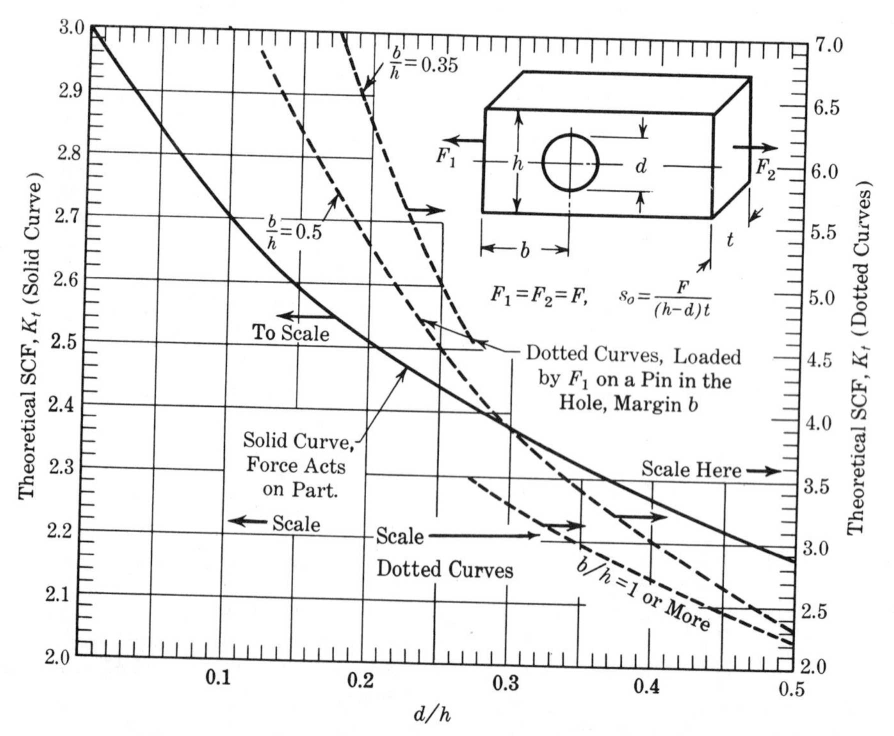

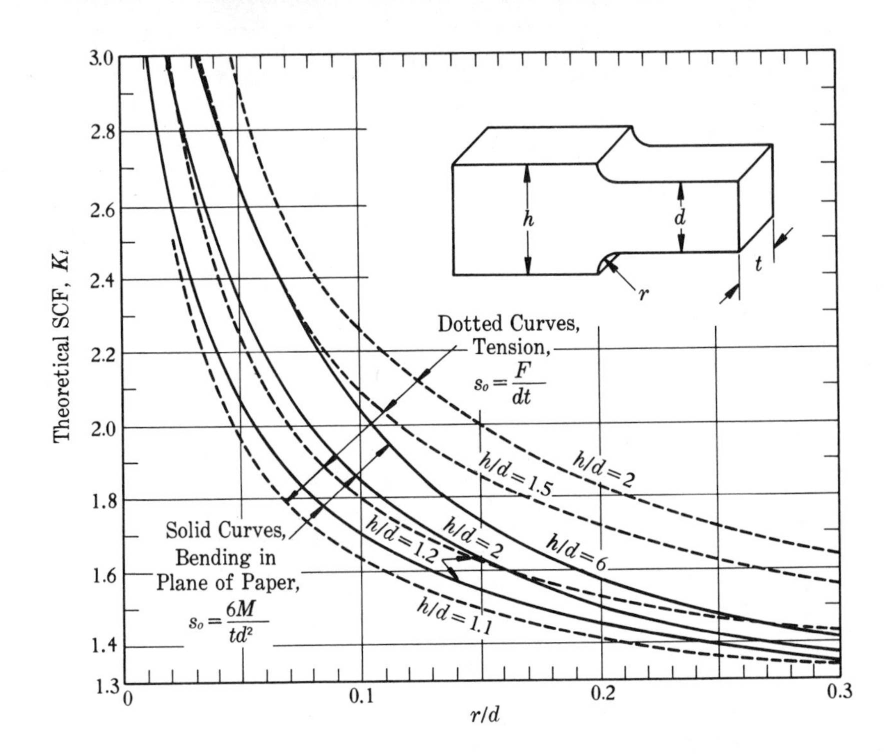

FIGURE AF 9 Flat Plate with Fillets. The tensile load is central. For $h/d = 1.1$, tensile and bending curves are very close together down to $r/d = 0.04$. (After R. E. Peterson)[4.21]

FIGURE AF 10 Flat Plate with Grooves.

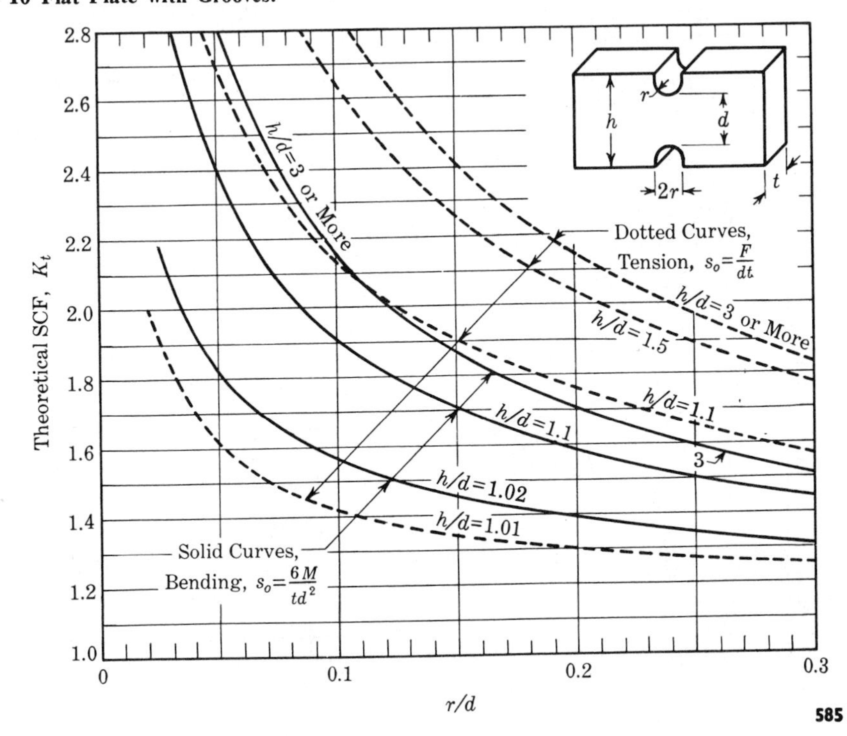

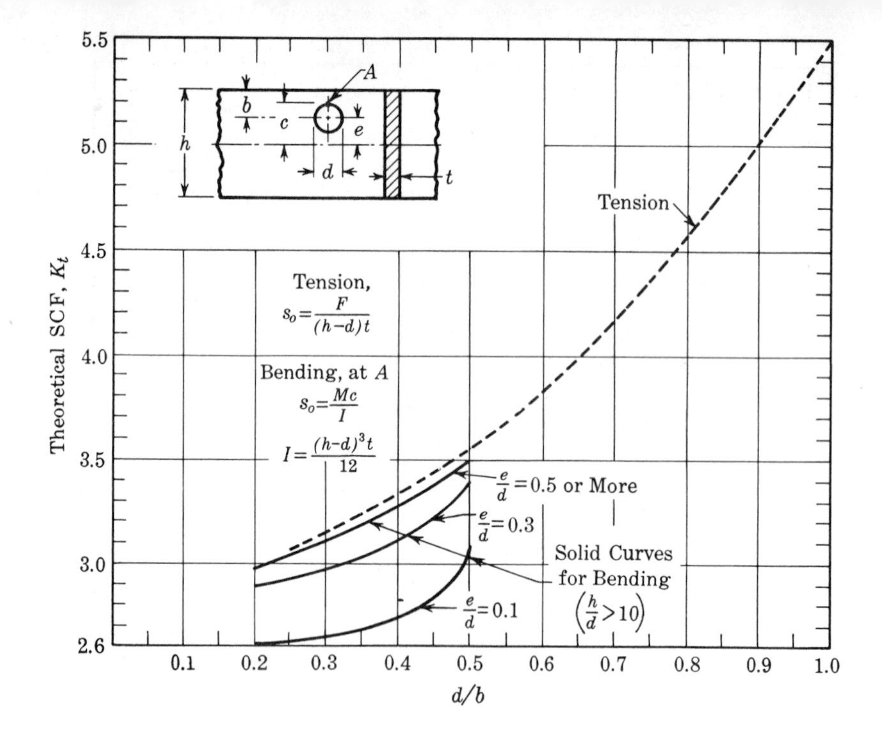

FIGURE AF 11 Flat Plate with Eccentric Hole. [4.2, 4.20] **For tension, if $h < 20d$, K_t is somewhat smaller than shown. (After R. E. Peterson).** [4.21]

FIGURE AF 12 Shaft with Fillet. The tensile load is central. Torsion curve $D/d = 1.2$ approximates the bending curve for $D/d = 1.01$; torsion curve $D/d = 2$, approximates the bending curve for $D/d = 1.02$ (down to $r/d \approx 0.04$). Bending curve $D/d = 1.1$ approximates the tensile curve $D/d = 1.1$. (After R. E. Peterson). [4.21]

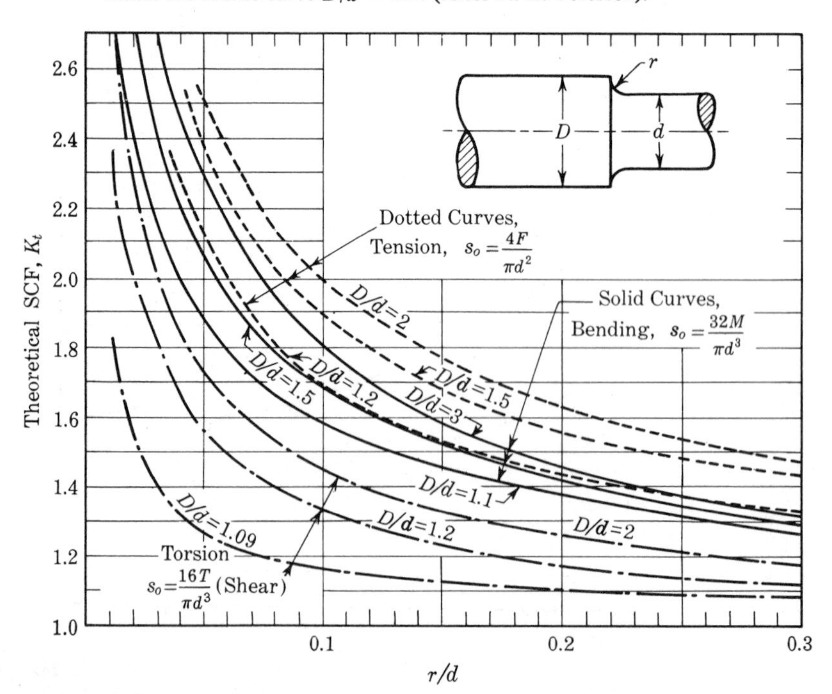

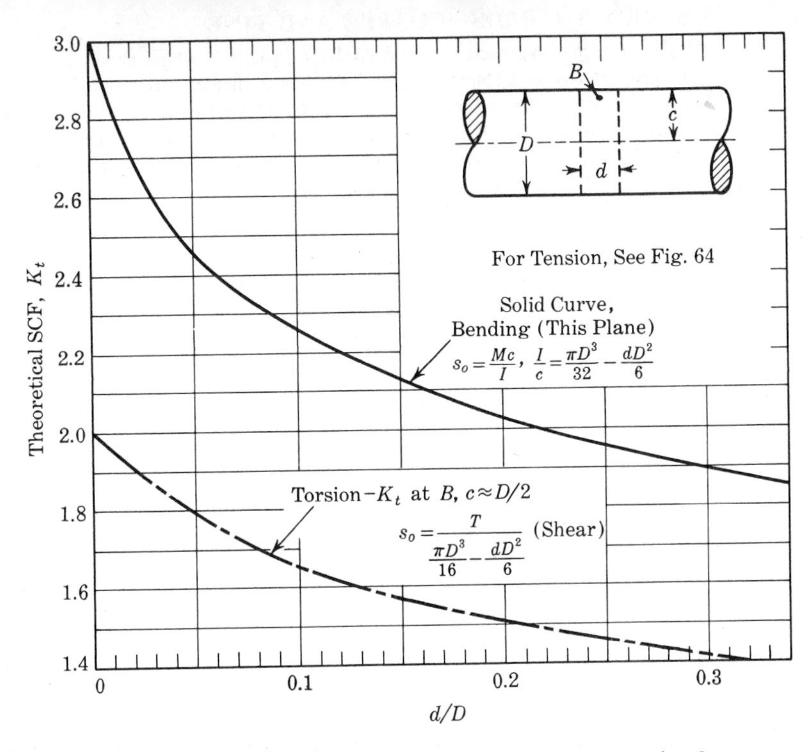

FIGURE AF 13 Shaft with Radial Hole—Bending and Torsion. Maximum torsional stress for this case falls slightly inside the shaft diameter on the inside surface of the hole at some point B. (After R. E. Peterson).[4,21]

FIGURE AF 14 Shaft with Groove. Use the solid curve for $D/d = 1.01$ for tension as well as bending (approximate).

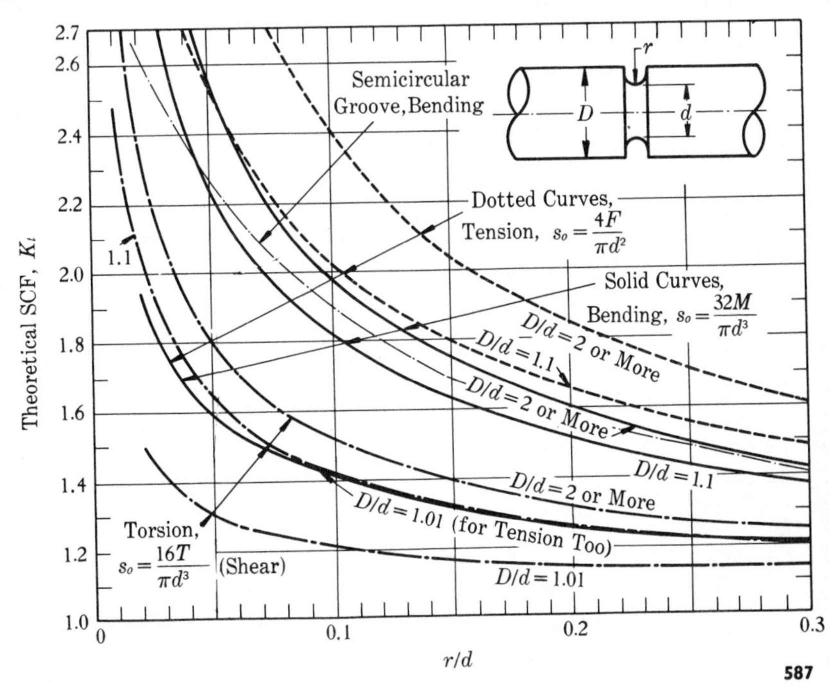

SCREW THREADS—COARSE AND FINE

Selected values abstracted from ASA B1.1-1960,[5.1] which should be referred to for details on proportions and tolerances, and for other thread series; dimensions for a class 3 fit, external thread. The minor diameter for the internal thread is not quite the same as for the external thread. The *tensile stress area* is that area corresponding to a diameter that is approximately the average of the pitch and minor diameters; detail in Ref. *(5.1)*. Selected values of 12-thread series above $1\frac{1}{2}$-in. size.

SIZE	BASIC MAJOR DIA. *in.*	COARSE (UNC)			FINE (UNF), AND 12 UN			WIDTH ACROSS FLATS, REGULAR, UNFINISHED; *A in.* *(Selected from ASA B 18.2–1952.)*	
		Th./In.	*Minor Dia. Ext. Th.*	*Stress Area, A_s sq. in.*	*Th./In.*	*Minor Dia. Ext. Th.*	*Stress Area, A_s sq. in.*		
0	0.0600				80	0.0447	0.00180	For square heads and nuts, *A* is different in the smaller sizes (below 7/8) from these values. See detail in the standard.	
1	0.0730	64	0.0538	0.00263	72	0.0560	0.00278		
2	0.0860	56	0.0641	0.00370	64	0.0668	0.00394		
3	0.0990	48	0.0734	0.00487	56	0.0771	0.00523		
4	0.1120	40	0.0813	0.00604	48	0.0864	0.00661		
5	0.1250	40	0.0943	0.00796	44	0.0971	0.0083		
6	0.1380	32	0.0997	0.00909	40	0.1073	0.01015		
8	0.1640	32	0.1257	0.0140	36	0.1299	0.01474		
10	0.1900	24	0.1389	0.0175	32	0.1517	0.0200		
12	0.2160	24	0.1649	0.0242	28	0.1722	0.0258		
								Head	*Nut*
$\frac{1}{4}$	0.2500	20	0.1887	0.0318	28	0.2062	0.0364	$\frac{7}{16}$	$\frac{7}{16}$
$\frac{5}{16}$	0.3125	18	0.2443	0.0524	24	0.2614	0.0580	$\frac{1}{2}$	$\frac{1}{2}$
$\frac{3}{8}$	0.3750	16	0.2983	0.0775	24	0.3239	0.0878	$\frac{9}{16}$	$\frac{9}{16}$
$\frac{7}{16}$	0.4375	14	0.3499	0.1063	20	0.3762	0.1187	$\frac{5}{8}$	$\frac{11}{16}$
$\frac{1}{2}$	0.5000	13	0.4056	0.1419	20	0.4387	0.1599	$\frac{3}{4}$	$\frac{3}{4}$
$\frac{9}{16}$	0.5625	12	0.4603	0.182	18	0.4943	0.203	$\frac{13}{16}$	$\frac{7}{8}$
$\frac{5}{8}$	0.6250	11	0.5135	0.226	18	0.5568	0.256	$\frac{15}{16}$	$\frac{15}{16}$
$\frac{3}{4}$	0.7500	10	0.6273	0.334	16	0.6733	0.373	$1\frac{1}{8}$	$1\frac{1}{8}$
$\frac{7}{8}$	0.875	9	0.7387	0.462	14	0.7874	0.509	$1\frac{5}{16}$	$1\frac{5}{16}$
1	1.0000	8	0.8466	0.606	12	0.8978	0.663	$1\frac{1}{2}$	$1\frac{1}{2}$
$1\frac{1}{8}$	1.125	7	0.9497	0.763	12	1.0228	0.856	$1\frac{11}{16}$	$1\frac{11}{16}$
$1\frac{1}{4}$	1.2500	7	1.0747	0.969	12	1.1478	1.073	$1\frac{7}{8}$	$1\frac{7}{8}$
$1\frac{3}{8}$	1.375	6	1.1705	1.155	12	1.2728	1.315	$2\frac{1}{16}$	$2\frac{1}{16}$
$1\frac{1}{2}$	1.5000	6	1.2955	1.405	12	1.3978	1.581	$2\frac{1}{4}$	$2\frac{1}{4}$
$1\frac{3}{4}$	1.7500	5	1.5046	1.90	12	1.6478	2.1853	$2\frac{5}{8}$	$2\frac{5}{8}$
2	2.0000	$4\frac{1}{2}$	1.7274	2.50	12	1.8978	2.8892	3	3
$2\frac{1}{4}$	2.2500	$4\frac{1}{2}$	1.9774	3.25	12	2.1478	3.6914	$3\frac{3}{8}$	$3\frac{3}{8}$
$2\frac{1}{2}$	2.5000	4	2.1933	4.00	12	2.3978	4.5916	$3\frac{3}{4}$	$3\frac{3}{4}$
$2\frac{3}{4}$	2.7500	4	2.4433	4.93	12	2.6478	5.5900	$4\frac{1}{8}$	$4\frac{1}{8}$
3	3.0000	4	2.6933	5.97	12	2.8978	6.6865	$4\frac{1}{2}$	$4\frac{1}{2}$
$3\frac{1}{4}$	3.2500	4	2.9433	7.10	12	3.1478	7.8812	$4\frac{7}{8}$	
$3\frac{1}{2}$	3.5000	4	3.1933	8.33	12	3.3978	9.1740	$5\frac{1}{4}$	
$3\frac{3}{4}$	3.7500	4	3.4433	9.66	12	3.6478	10.5649	$5\frac{5}{8}$	
4	4.0000	4	3.6933	11.08	12	3.8978	12.0540	6	

TABLE AT 15 NOMINAL DIMENSIONS OF VARIOUS GAGES

The Washburn and Moen (W & M) gage, called also the steel-wire gage, is used for steel wire. The American Wire or Brown and Sharpe (B & S) gage is used for monel, bronze, copper, aluminum, and brass wires. Standard wire sizes other than those in the table include multiples of $\frac{1}{32}$ up to $\frac{9}{16}$ in. The tendency is to specify the decimal size of the wire. There is also a music-wire gage. And much smaller sizes than those listed are available.

GAGE NO.	WIRE DIAMETER, IN. *W & M* *Ferrous*	*B & S* *Nonferrous*	PLATE THICKNESS, IN. *U.S. Standard*
7–0	0.4900		0.500
6–0	0.4615		0.469
5–0	0.4305		0.438
4–0	0.3938	0.460	0.406
3–0	0.3625	0.401	0.375
2–0	0.3310	0.365	0.344
0	0.3065	0.325	0.313
1	0.2830	0.289	0.281
2	0.2625	0.258	0.266
3	0.2437	0.229	0.250
4	0.2253	0.204	0.234
5	0.2070	0.182	0.219
6	0.1920	0.162	0.203
7	0.1770	0.144	0.188
8	0.1620	0.128	0.172
9	0.1483	0.114	0.156
10	0.1350	0.102	0.141
11	0.1205	0.091	0.125
12	0.1055	0.081	0.109
13	0.0915	0.072	0.094
14	0.0800	0.065	0.078
15	0.0720	0.057	0.070
16	0.0625	0.051	0.063
17	0.0540	0.045	0.056
18	0.0475	0.040	0.050

TABLE AT 16 APPROXIMATE FREE LENGTHS AND SOLID HEIGHTS

(P = pitch of coils, N_c = number of *active* coils, D_w = diameter of wire)

TYPE OF ENDS	FREE LENGTH	TOTAL COILS	SOLID HEIGHT
Plain	$PN_c + D_w$	N_c	$D_wN_c + D_w$
Plain ground	PN_c	N_c	D_wN_c
Squared	$PN_c + 3D_w$	$N_c + 2$	$D_wN_c + 3D_w$
Squared and ground	$PN_c + 2D_w$	$N_c + 2$	$D_wN_c + 2D_w$

MATERIAL	E $\times 10^{-6}$ *psi*	G $\times 10^{-6}$ *psi*	DESIGN STRESS, s_{sd} *ksi. Light Load*	MIN. TENSILE, s_u *ksi. (Uncoiled)*	MAX. "SOLID" s_s *(Approx.* s_{ys}*)*	END. STR'TH. s_{no} *ksi.* $(R = 0)$
Column No. →	(1)	(2)	(3)	(4)	(5)	(6)
Oil Tempered; ASTM A229	29	11.5	(a)	$\frac{146}{D_w^{0.19}}$ (b) $[0.032 < D_w < 0.5]$	$0.6s_u$ (c) $[Q = 87.5, x = 0.19]$	$\frac{47}{D_w^{0.1}}$ (d)(e) $[0.041 < D_w < 0.15]$
Hard Drawn; ASTM A227	29	11.5	Use 0.85 times constants in Note (a)	$\frac{140}{D_w^{0.19}}$ (b) $[0.028 < D_w < 0.625]$	$0.5s_u$ (c) $[Q = 70, x = 0.19]$	$\frac{30}{D_w^{0.34}}$ (d)(e) $[0.15 < D_w < 0.625]$
Music Wire; ASTM A228	30	12	(a)	$\frac{190}{D_w^{0.154}}$ (b) $[0.004 < D_w < 0.192]$	$0.5s_u$ (c) $[Q = 95, x = 0.154]$ $[0.03 < D_w < 0.192$; 190 ksi max.]	$\frac{50}{D_w^{0.154}}$ (d) $[0.018 < D_w < 0.18$; 92 ksi max.]

TABLE AT 17 MECHANICAL PROPERTIES OF WIRE FOR COIL SPRINGS

For extension springs, use 0.8 times the value in column (5) for the maximum occasional stress.

The stress is $s = Q/D_w^x$ wherever this form appears; to be used with equation (6.1); always include the curvature factor except for the mean stress in fatigue design. The values given apply when the spring is not preset and not peened, except as stated. For live loads, stress values for the *steels* may be increased by 25% for shot-peened coils. For preset *steel* springs, the static stress and "solid stress" may be some 40–50% greater than given by columns (3) and (5). See notes (f) and (n) below. The computed design stresses are not to be interpreted as being exact values. Reduce design stresses 50% for shock loads (analogous to hammer blows). Where a maximum stress is given, use it for wire sizes smaller than the specified limits. The limits given for D_w apply only to the equations; for many materials, wire sizes smaller or larger than the limits shown are frequent.

Notes: (a) For *light service*, use design $s_{sd} = 0.405s_u$. For *average service*, use $s_{sd} = 0.324s_u$. For *severe service*, use $s_{sd} = 0.263s_u$. These results agree closely with Westinghouse recommendations, as reported by Wahl. (b) Equations for approximate minimum tensile strength as specified by ASTM. (c) Agrees closely with Alco recommendations; since they are higher than the stresses recommended by some authorities, a small factor of safety may be advisable, unless the spring manufacturer agrees. (d) Derived from Hunter Spring Co. data.[6.15] Value for indefinite life from 0 to max.; for 10^5 cycles for steel wire (except stainless), multiply this value by 1.4, for example. *Use minimum* $N = 1.15$. (e) Use both expressions for oil tempered; for *hard-drawn* wire, multiply by 0.9. (f) By analogy with music wire. Also, Associated Spring recommends the safe design range for valve spring quality as defined by the triangle *ABO*, Fig. 6.9, for $D_w < 0.207$ and unpeened; by triangle *CBO* when peened. (g) Conservative in the larger sizes. (h) In accordance with INCO.[6.13] (i) Light service, use $s_s = 0.32s_u$; average service, $0.26s_u$; severe service, $0.21s_u$. (j) Adapted from Associated Spring data. [6.2] (k) A straight-line interpolation between $s_u = 85$ ksi for $D_w = 0.5$ in. and formula limit is probably satisfactory. (l) Multiply by 0.8 for average service. (m) Stress relieved; decrease 10% if as-drawn. (n) Increase 10% if preset; 25–35% for shot peening wires larger than 0.062 in.[6.13] (o) Age hardened. (p) INCO gives 20 ksi for 10^8 cycles. (q) Multiply by 1.33 for 10^5 cycles, 0-max. (r) INCO data suggest that this K-Monel is somewhat stronger than the Monel, but detail lacking. (s) Valve spring quality.

MATERIAL	$E \times 10^{-6}$ psi	$G \times 10^{-6}$ psi	DESIGN STRESS, s_{sd} ksi. Light Load	MIN. TENSILE, s_u ksi. (Uncoiled)	MAX. "SOLID" s_s (Approx. s_{ys})	END. STR'TH. s_{no} ksi. ($R = 0$)
Column No. →	(1)	(2)	(3)	(4)	(5)	(6)
Carbon Steel, VSQ (s); ASTM A230	30	11.5	(a)	$\frac{182}{D_w^{0.1}}$ (b) $[0.093 < D_w < 0.25]$	$0.5s_u$ (c) $[Q = 91, x = 0.1]$ $[0.093 < D_w < 0.25]$	$\frac{49}{D_w^{0.15}}$ (d)(f) $[0.093 < D_w < 0.25]$
Cr-V Steel, VSQ (s); ASTM A232	30	11.5	(a)	$\frac{168}{D_w^{0.166}}$ (b) $[0.032 < D_w < 0.437]$	$0.6s_u$ (c) $[Q = 100, x = 0.166]$	Same as for A230 (g) $[0.028 < D_w < 0.5]$
Cr-Si Steel ASTM A401	29	11.5	(a)	$\frac{202}{D_w^{0.107}}$ (b) $[0.032 < D_w < 0.375]$	$0.6s_u$ $[Q = 121, x = 0.107]$	Same as for A230 (g) $[0.032 < D_w < 0.375]$
Stainless Steel (Cr-Ni) ASTM A313	26	10	(i)	$\frac{170}{D_w^{0.14}}$ (b) $[0.01 < D_w < 0.13]$ $\frac{97}{D_w^{0.41}}$ (b) $[0.13 < D_w < 0.375]$	$0.47s_u$ (h) $[Q = 80, x = 0.14]$ $[Q = 45.6, x = 0.41]$	$\frac{30}{D_w^{0.17}}$ (d)(q) $[0.01 < D_w < 0.375]$
Beryllium Copper	18.5	7	Use 0.8 times values for A229	160–200	$0.5s_u$ (h)	$\frac{35}{D_w^{0.2}}$ (d) $[0.09 < D_w < 0.5$; 56 ksi max.]
Spring Brass	14.5	4.5	Use 0.35 times values for A229	$\frac{88}{D_w^{0.1}}$ (j)(k) $[0.03 < D_w < 0.20$; 125 ksi max.]	$\frac{42}{D_w^{0.26}}$ (j) $[0.08 < D_w < 0.5$; 68 ksi max.]	$\frac{11.5}{D_w^{0.2}}$ (d) $[0.09 < D_w < 0.5$; 19 ksi max.]
Phosphor Bronze	14.5	6	Use 0.5 times values for A229	$\frac{106}{D_w^{0.08}}$ (j) $[D_w \leq 0.5$; 145 ksi max.]	$0.45s_u$ (h) $[Q = 47.5; x = 0.08]$	$\frac{15.3}{D_w^{0.2}}$ (d) $[0.09 < D_w < 0.5$; 28 ksi max.]
Monel (m)	24.5	9.3	$\frac{52}{D_w^{0.1}}$ (h)(l) $[0.058 < D_w < 0.625$; 70 ksi max.]	$\frac{129}{D_w^{0.1}}$ (h) $[D_w \leq 0.625$; 170 ksi max.]	$0.4s_u$ (h)(n) $[Q = 51.5; x = 0.1]$ $[D_w \leq 0.625$; 68 ksi max.]	$\frac{18}{D_w^{0.2}}$ (d)(n) $[D_w \leq 0.625$; 29 ksi max.(p)]
K-Monel (o)	24.5	9.3	75 ksi (h)(l) 80 ksi, preset $[D_w > 0.058]$	$\frac{158}{D_w^{0.048}}$ (h) $[D_w \leq 0.625$; 180 ksi max.]	$0.4s_u$ (h)(n) $[Q = 63, x = 0.048]$ $[D_w \leq 0.625$; 72 ksi max.]	Use same as Monel (r) [29 ksi max. (p)]

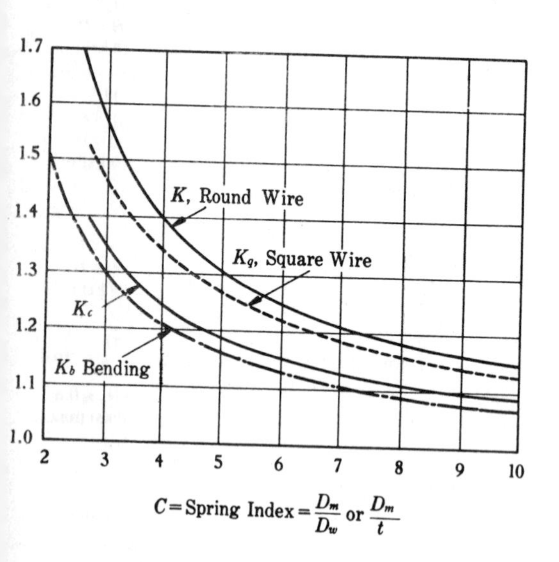

FIGURE AF 15 Stress Factors (Wahl). Use D_m/D_w as the spring index for round-wire springs, D_m/t for rectangular-wire springs, where t is the dimension perpendicular to the axis of the spring. It is advisable that D_m/D_w be not less than 3, preferably greater than 4.

TABLE AT 18 STRESS FACTORS FOR CURVED BEAMS

Extracted with permission of publisher, John Wiley & Sons, Inc., from Seely and Smith.[1.7] For a hollow circular section, the values for a solid section may be used with little error when $r/c \geqq 1.8$; K_{ci} for point on inside of curvature: K_{co} for outside.

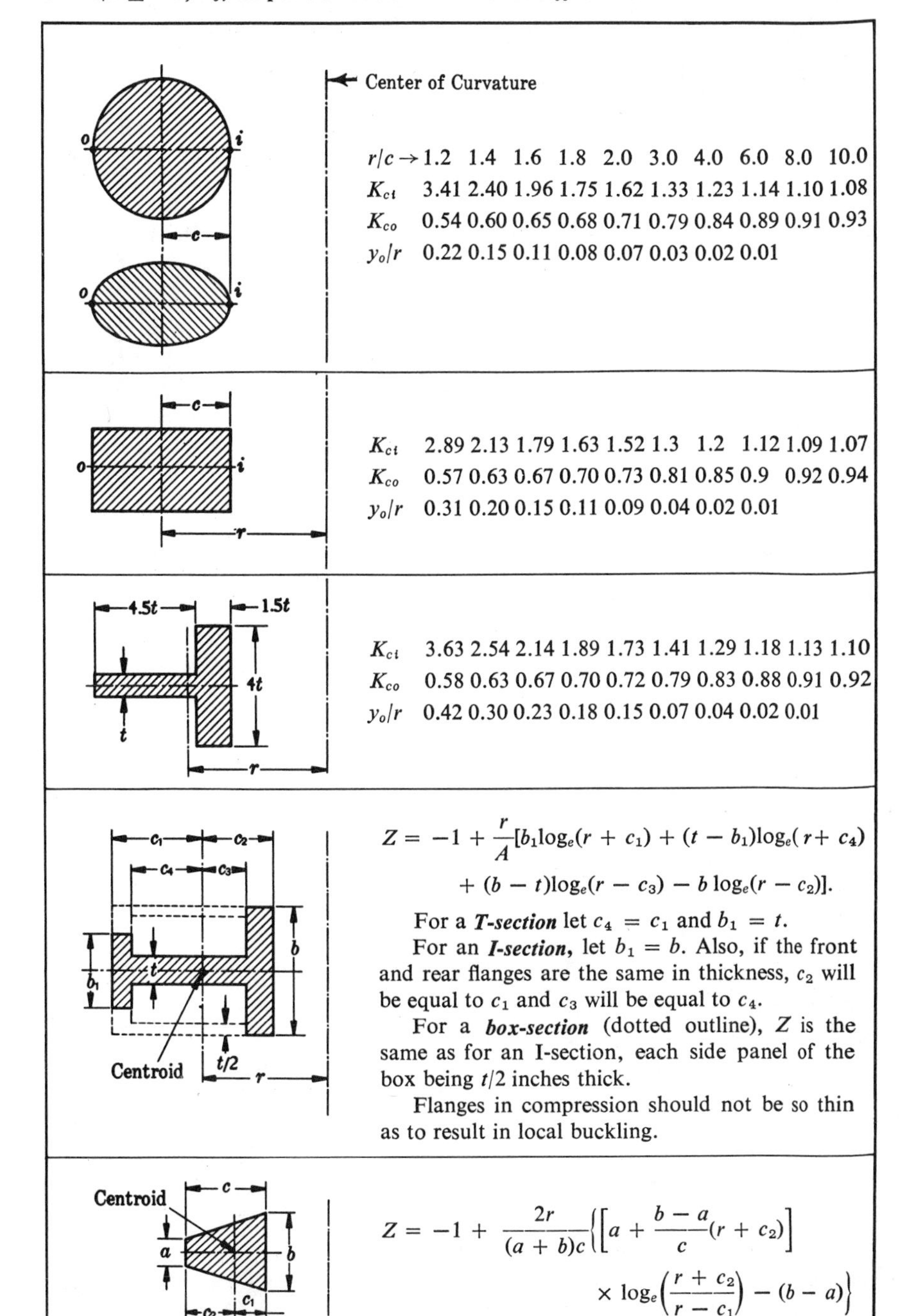

← Center of Curvature

Section											
Circle / ellipse	$r/c \rightarrow$	1.2	1.4	1.6	1.8	2.0	3.0	4.0	6.0	8.0	10.0
	K_{ci}	3.41	2.40	1.96	1.75	1.62	1.33	1.23	1.14	1.10	1.08
	K_{co}	0.54	0.60	0.65	0.68	0.71	0.79	0.84	0.89	0.91	0.93
	y_o/r	0.22	0.15	0.11	0.08	0.07	0.03	0.02	0.01		
Rectangle	K_{ci}	2.89	2.13	1.79	1.63	1.52	1.3	1.2	1.12	1.09	1.07
	K_{co}	0.57	0.63	0.67	0.70	0.73	0.81	0.85	0.9	0.92	0.94
	y_o/r	0.31	0.20	0.15	0.11	0.09	0.04	0.02	0.01		
T-section (4.5t, 1.5t, 4t, t)	K_{ci}	3.63	2.54	2.14	1.89	1.73	1.41	1.29	1.18	1.13	1.10
	K_{co}	0.58	0.63	0.67	0.70	0.72	0.79	0.83	0.88	0.91	0.92
	y_o/r	0.42	0.30	0.23	0.18	0.15	0.07	0.04	0.02	0.01	

I-, T- and box-sections (dimensions c_1, c_2, c_3, c_4, b, b_1, t, $t/2$, r; Centroid):

$$Z = -1 + \frac{r}{A}[b_1 \log_e(r + c_1) + (t - b_1)\log_e(r + c_4) + (b - t)\log_e(r - c_3) - b \log_e(r - c_2)].$$

For a ***T-section*** let $c_4 = c_1$ and $b_1 = t$.

For an ***I-section***, let $b_1 = b$. Also, if the front and rear flanges are the same in thickness, c_2 will be equal to c_1 and c_3 will be equal to c_4.

For a ***box-section*** (dotted outline), Z is the same as for an I-section, each side panel of the box being $t/2$ inches thick.

Flanges in compression should not be so thin as to result in local buckling.

Trapezoidal section (dimensions a, b, c, c_1, c_2, r; Centroid):

$$Z = -1 + \frac{2r}{(a + b)c}\left\{\left[a + \frac{b - a}{c}(r + c_2)\right] \times \log_e\left(\frac{r + c_2}{r - c_1}\right) - (b - a)\right\}$$

TABLE AT 19 KEY DIMENSIONS

See Figs. 10.1 and 10.2 for *b* and *t*. From ASA standard B17.1-1943. Other sizes available. The tolerances on *t* may be numerically the same as given, negative on plain keys, positive on tapered keys.

SHAFT DIAMETER (*Inclusive*)	b	t	TOLERANCE ON b, *in.*
$\frac{1}{2}-\frac{9}{16}$	$\frac{1}{8}$	$\frac{3}{32}$	−0.0020
$\frac{5}{8}-\frac{7}{8}$	$\frac{3}{16}$	$\frac{1}{8}$	−0.0020
$\frac{15}{16}-1\frac{1}{4}$	$\frac{1}{4}$	$\frac{3}{16}$	−0.0020
$1\frac{5}{16}-1\frac{3}{8}$	$\frac{5}{16}$	$\frac{1}{4}$	−0.0020
$1\frac{7}{16}-1\frac{3}{4}$	$\frac{3}{8}$	$\frac{1}{4}$	−0.0020
$1\frac{13}{16}-2\frac{1}{4}$	$\frac{1}{2}$	$\frac{3}{8}$	−0.0025
$2\frac{5}{16}-2\frac{3}{4}$	$\frac{5}{8}$	$\frac{7}{16}$	−0.0025
$2\frac{7}{8}-3\frac{1}{4}$	$\frac{3}{4}$	$\frac{1}{2}$	−0.0025
$3\frac{3}{8}-3\frac{3}{4}$	$\frac{7}{8}$	$\frac{5}{8}$	−0.0030
$3\frac{7}{8}-4\frac{1}{2}$	1	$\frac{3}{4}$	−0.0030
$4\frac{3}{4}-5\frac{1}{2}$	$1\frac{1}{4}$	$\frac{7}{8}$	−0.0030
$5\frac{3}{4}-6$	$1\frac{1}{2}$	1	−0.0030

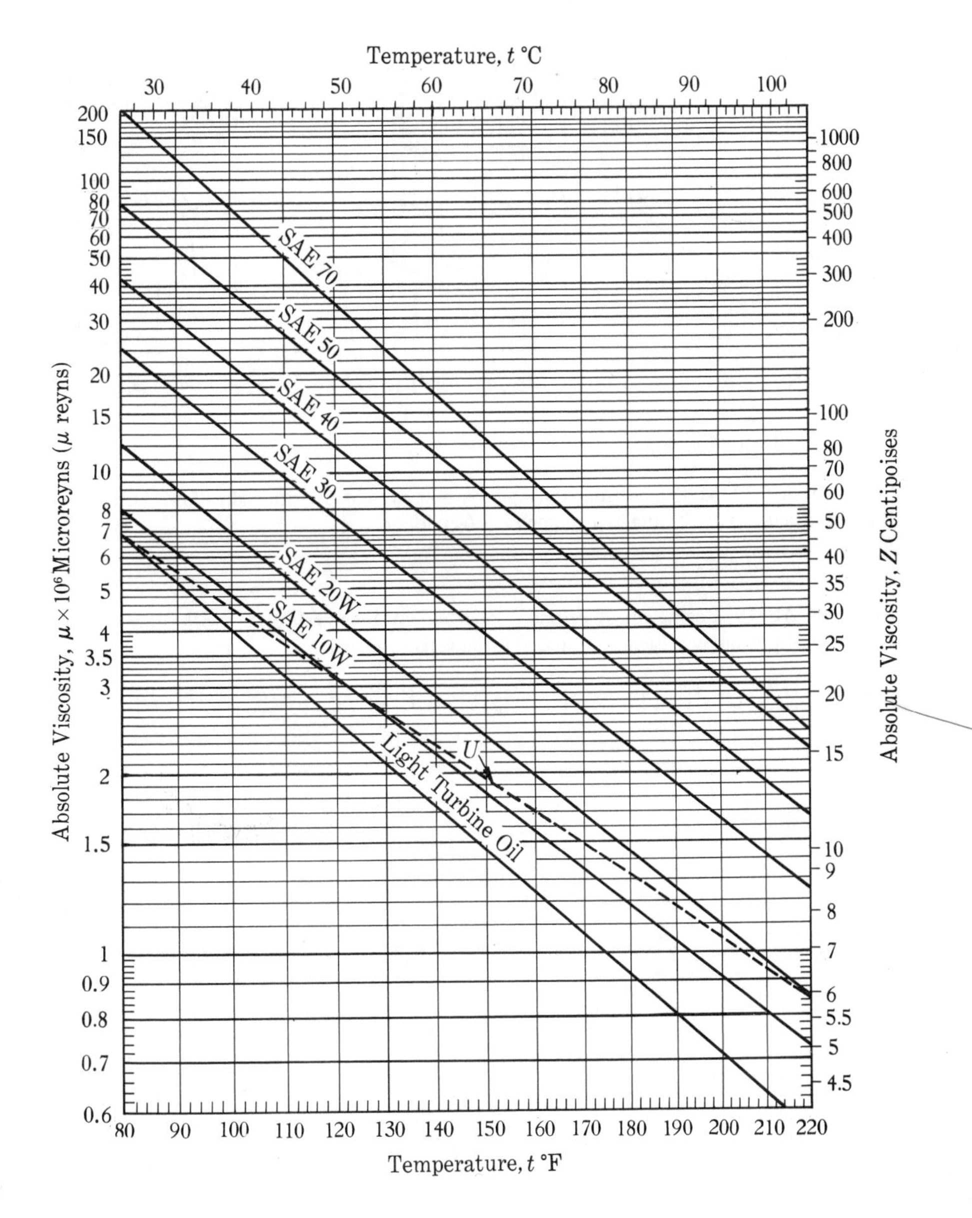

FIGURE AF 16 Typical Viscosities of Oils. For ring-oiled bearings, usually SAE 20 (or the equivalent) or lighter. SAE 70 and chart paper by courtesy of Westinghouse Electric Corp. Dotted curve U is for a high viscosity-index oil, Uniflow—typical test values, Standard Oil of N.J. Other data from The Texas Co. On average, an SAE 10W–30 oil has a viscosity a little lower than SAE 30 at 210°F, a little higher than SAE 10 at 100°F.

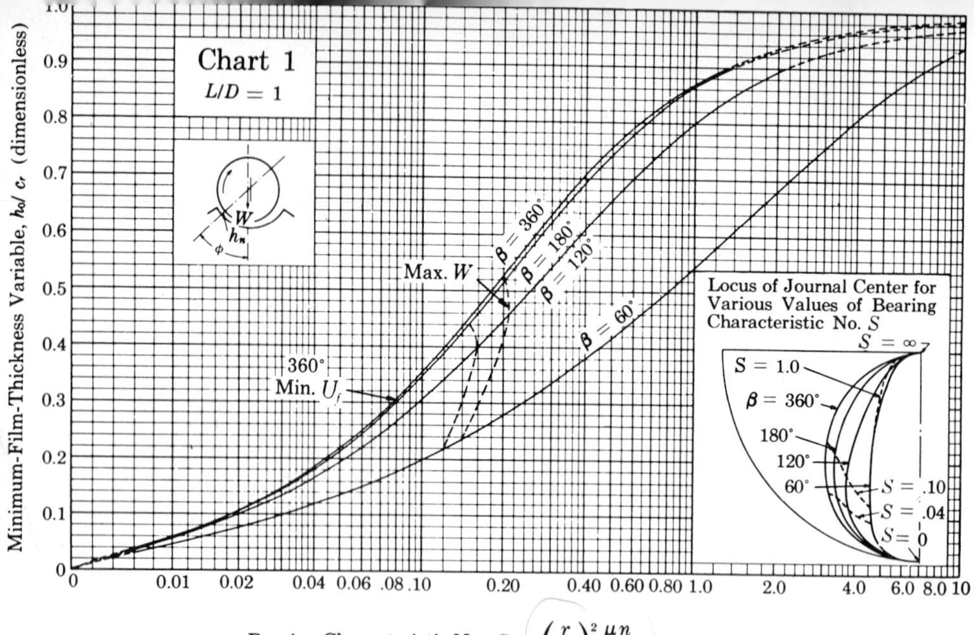

FIGURE AF 17 Minimum-Film Variable vs. Sommerfeld S (with Side Flow). (Courtesy Raimondi and Boyd[11.7] and Westinghouse Electric).

FIGURE AF 18 Coefficient-of-Friction Variable vs. Sommerfeld S (with Side Flow). (Courtesy Raimondi and Boyd[11.7] and Westinghouse Electric).

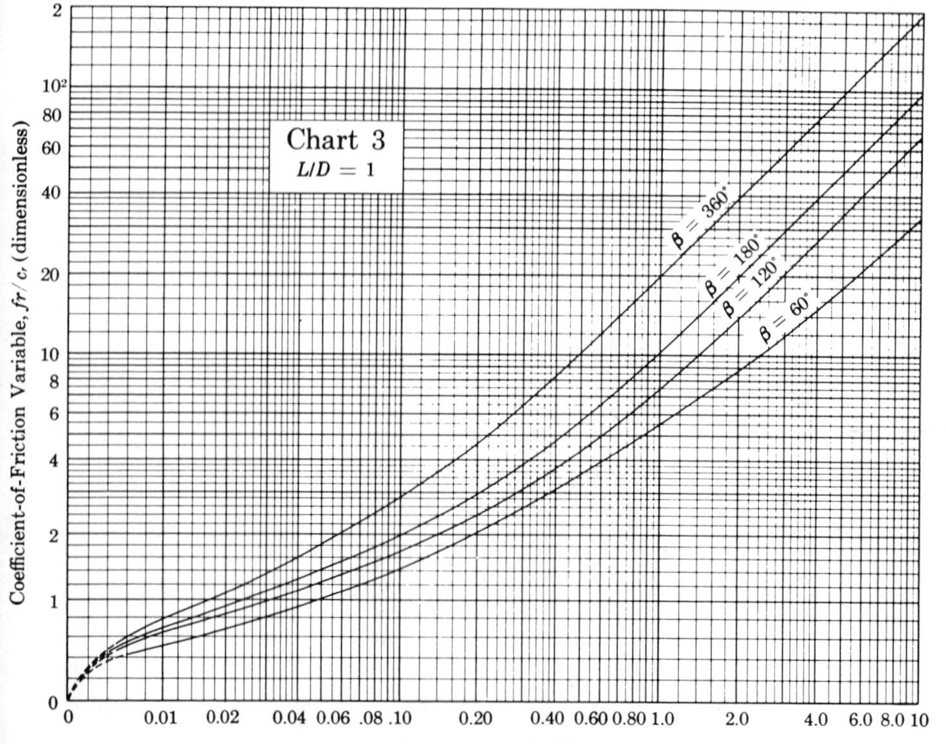

TABLE AT 20 DIMENSIONLESS PERFORMANCE PARAMETERS FOR FULL JOURNAL BEARINGS WITH SIDE FLOW

Courtesy Raimondi and Boyd[11.7] and Westinghouse Electric. Values of h_o/c_r for optimum bearings, maximum load and minimum friction, respectively: for $L/D = \infty$, 0.66, 0.60; for $L/D = 1$, 0.53, 0.30; for $L/D = 0.5$, 0.43, 0.12; for $L/D = 0.25$, 0.27, 0.03.

L/D	ϵ	$\frac{h_o}{c_r}$	S	ϕ	$\frac{r}{c_r}f$	$\frac{q}{rc_r n_s L}$	$\frac{q_s}{q}$	$\frac{\rho c \Delta t_o}{p}$	$\frac{p}{p_{max}}$
∞	0	1.0	∞	(70.92)	∞	π	0	∞	—
	0.1	0.9	0.240	69.10	4.80	3.03	0	19.9	0.826
	0.2	0.8	0.123	67.26	2.57	2.83	0	11.4	0.814
	0.4	0.6	0.0626	61.94	1.52	2.26	0	8.47	0.764
	0.6	0.4	0.0389	54.31	1.20	1.56	0	9.73	0.667
	0.8	0.2	0.021	42.22	0.961	0.760	0	15.9	0.495
	0.9	0.1	0.0115	31.62	0.756	0.411	0	23.1	0.358
	0.97	0.03	—	—	—	—	0	—	—
	1.0	0	0	0	0	0	0	∞	0
1	0	1.0	∞	(85)	∞	π	0	∞	—
	0.1	0.9	1.33	79.5	26.4	3.37	0.150	106	0.540
	0.2	0.8	0.631	74.02	12.8	3.59	0.280	52.1	0.529
	0.4	0.6	0.264	63.10	5.79	3.99	0.497	24.3	0.484
	0.6	0.4	0.121	50.58	3.22	4.33	0.680	14.2	0.415
	0.8	0.2	0.0446	36.24	1.70	4.62	0.842	8.00	0.313
	0.9	0.1	0.0188	26.45	1.05	4.74	0.919	5.16	0.247
	0.97	0.03	0.00474	15.47	0.514	4.82	0.973	2.61	0.152
	1.0	0	0	0	0	—	1.0	0	0
$\frac{1}{2}$	0	1.0	∞	(88.5)	∞	π	0	∞	—
	0.1	0.9	4.31	81.62	85.6	3.43	0.173	343.0	0.523
	0.2	0.8	2.03	74.94	40.9	3.72	0.318	164.0	0.506
	0.4	0.6	0.779	61.45	17.0	4.29	0.552	68.6	0.441
	0.6	0.4	0.319	48.14	8.10	4.85	0.730	33.0	0.365
	0.8	0.2	0.0923	33.31	3.26	5.41	0.874	13.4	0.267
	0.9	0.1	0.0313	23.66	1.60	5.69	0.939	6.66	0.206
	0.97	0.03	0.00609	13.75	0.610	5.88	0.980	2.56	0.126
	1.0	0	0	0	0	—	1.0	0	0
$\frac{1}{4}$	0.0	1.0	∞	(89.5)	∞	π	0	∞	—
	0.1	0.9	16.2	82.31	322.0	3.45	0.180	1287.0	0.515
	0.2	0.8	7.57	75.18	153.0	3.76	0.330	611.0	0.489
	0.4	0.6	2.83	60.86	61.1	4.37	0.567	245.0	0.415
	0.6	0.4	1.07	46.72	26.7	4.99	0.746	107.0	0.334
	0.8	0.2	0.261	31.04	8.80	5.60	0.884	35.4	0.240
	0.9	0.1	0.0736	21.85	3.50	5.91	0.945	14.1	0.180
	0.97	0.03	0.0101	12.22	0.922	6.12	0.984	3.73	0.108
	1.0	0	0	0	0	—	1.0	0	0

q in.^{3}/sec. $\rho \approx 0.03$ lb/in.^{3} $c = 3734$ in–lb/lb–°F. $\rho c = 112$.

TABLE AT 21 DIMENSIONLESS PERFORMANCE PARAMETERS FOR 180° BEARING, CENTRALLY LOADED, WITH SIDE FLOW

Courtesy Raimondi and Boyd[11.7] and Westinghouse Electric. Values of h_o/c_r for optimum bearings, maximum load and minimum friction, respectively: for $L/D = \infty$, 0.64, 0.6; for $L/D = 1$, 0.52, 0.44; for $L/D = 0.5$, 0.42, 0.23; for $L/D = 0.25$, 0.28, 0.03.

L/D	ϵ	$\frac{h_o}{c_r}$	S	ϕ	$\frac{r}{c_r}f$	$\frac{q}{rc_rn_sL}$	$\frac{q_s}{q}$	$\frac{\rho c\Delta t_o}{p}$	$\frac{p}{p_{max}}$
∞	0	1.0	∞	90.0	∞	π	∞	∞	—
	0.1	0.9	0.347	72.90	3.55	3.04	0	14.7	0.778
	0.2	0.8	0.179	61.32	2.01	2.80	0	8.99	0.759
	0.4	0.6	0.0898	49.99	1.29	2.20	0	7.34	0.700
	0.6	0.4	0.0523	43.15	1.06	1.52	0	8.71	0.607
	0.8	0.2	0.0253	33.35	0.859	0.767	0	14.1	0.459
	0.9	0.1	0.0128	25.57	0.681	0.380	0	22.5	0.337
	0.97	0.03	0.00384	15.43	0.416	0.119	0	44.0	0.190
	1.0	0	0	0	0	0	0	∞	0
1	0	1.0	∞	90.0	—	π	0	∞	—
	0.1	0.9	1.40	78.50	14.1	3.34	0.139	57.0	0.525
	0.2	0.8	0.670	68.93	7.15	3.46	0.252	29.7	0.513
	0.4	0.6	0.278	58.86	3.61	3.49	0.425	16.5	0.466
	0.6	0.4	0.128	44.67	2.28	3.25	0.572	12.4	0.403
	0.8	0.2	0.0463	32.33	1.39	2.63	0.721	10.4	0.313
	0.9	0.1	0.0193	24.14	0.921	2.14	0.818	9.13	0.244
	0.97	0.03	0.00483	14.57	0.483	1.60	0.915	6.96	0.157
	1.0	0	0	0	0	—	1.0	0	0
$\frac{1}{2}$	0	1.0	∞	90.0	∞	π	0	∞	—
	0.1	0.9	4.38	79.97	44.0	3.41	0.167	177.0	0.518
	0.2	0.8	2.06	72.14	21.6	3.64	0.302	87.8	0.499
	0.4	0.6	0.794	58.01	9.96	3.93	0.506	42.7	0.438
	0.6	0.4	0.321	45.01	5.41	3.93	0.665	25.9	0.365
	0.8	0.2	0.0921	31.29	2.54	3.56	0.806	15.0	0.273
	0.9	0.1	0.0314	22.80	1.38	3.17	0.886	9.80	0.208
	0.97	0.03	0.00625	13.63	0.581	2.62	0.951	5.30	0.132
	1.0	0	0	0	0	—	1.0	0	0
$\frac{1}{4}$	0	1.0	∞	90.0	∞	π	0	∞	—
	0.1	0.9	16.3	81.40	163.0	3.44	0.176	653.0	0.513
	0.2	0.8	7.60	73.70	79.4	3.71	0.320	320.0	0.489
	0.4	0.6	2.84	58.99	35.1	4.11	0.534	146.0	0.417
	0.6	0.4	1.08	44.96	17.6	4.25	0.698	79.8	0.336
	0.8	0.2	0.263	30.43	6.88	4.07	0.837	36.5	0.241
	0.9	0.1	0.0736	21.43	2.99	3.72	0.905	18.4	0.180
	0.97	0.03	0.0104	12.28	0.877	3.29	0.961	6.46	0.110
	1.0	0	0	0	0	—	1.0	0	0

TABLE AT 22 DIMENSIONLESS PERFORMANCE PARAMETERS FOR 120° BEARING, CENTRALLY LOADED, WITH SIDE FLOW.

Courtesy Raimondi and Boyd[11.7] and Westinghouse Electric. Values of h_o/c_r for optimum bearings, maximum load and minimum friction, respectively: for $L/D = \infty$, 0.53, 0.5; for $L/D = 1$, 0.46, 0.4; for $L/D = 0.5$, 0.38, 0.28; for $L/D = 0.25$, 0.26, 0.06. When $1 - h_o/c_r \neq \epsilon$, the trailing end of the bearing does not reach h_o as defined in Fig. 11.6; that is, h_o in this table is h_{min}.

L/D	ϵ	$\frac{h_o}{c_r}$	S	ϕ	$\frac{r}{c_r}f$	$\frac{q}{rc_rn_sL}$	$\frac{q_s}{q}$	$\frac{\rho c \Delta t_o}{p}$	$\frac{p}{p_{max}}$
∞	0	1.0	∞	90.0	∞	π	0	∞	—
	0.1	0.9007	0.877	66.69	6.02	3.02	0	25.1	0.610
	0.2	0.8	0.431	52.60	3.26	2.75	0	14.9	0.599
	0.4	0.6	0.181	39.02	1.78	2.13	0	10.5	0.566
	0.6	0.4	0.0845	32.67	1.21	1.47	0	10.3	0.509
	0.8	0.2	0.0328	26.80	0.853	0.759	0	14.1	0.405
	0.9	0.1	0.0147	21.51	0.653	0.388	0	21.2	0.311
	0.97	0.03	0.00406	13.86	0.399	0.118	0	42.4	0.199
	1.0	0	0	0	0	0	0	∞	0
1	0	1.0	∞	90.0	∞	π	0	∞	—
	0.1	0.9024	2.14	72.43	14.5	3.20	0.0876	59.5	0.427
	0.2	0.8	1.01	58.25	7.44	3.11	0.157	32.6	0.420
	0.4	0.6	0.385	43.98	3.60	2.75	0.272	19.0	0.396
	0.6	0.4	0.162	35.65	2.16	2.24	0.384	15.0	0.356
	0.8	0.2	0.0531	27.42	1.27	1.57	0.535	13.9	0.290
	0.9	0.1	0.0208	21.29	0.855	1.11	0.657	14.4	0.233
	0.97	0.03	0.00498	13.49	0.461	0.694	0.812	14.0	0.162
	1.0	0	0	0	0	—	1.0	0	0
$\frac{1}{2}$	0	1.0	∞	90.0	∞	π	0	—	—
	0.1	0.9034	5.42	74.99	36.6	3.29	0.124	149.0	0.431
	0.2	0.8003	2.51	63.38	18.1	3.32	0.225	77.2	0.424
	0.4	0.6	0.914	48.07	8.20	3.15	0.386	40.5	0.389
	0.6	0.4	0.354	38.50	4.43	2.80	0.530	27.0	0.336
	0.8	0.2	0.0973	28.02	2.17	2.18	0.684	19.0	0.261
	0.9	0.1	0.0324	21.02	1.24	1.70	0.787	15.1	0.203
	0.97	0.03	0.00631	13.00	0.550	1.19	0.899	10.6	0.136
	1.0	0	0	0	0	—	1.0	0	0
$\frac{1}{4}$	0	1.0	∞	90.0	∞	π	0	∞	—
	0.1	0.9044	18.4	76.97	124.0	3.34	0.143	502.0	0.456
	0.2	0.8011	8.45	65.97	60.4	3.44	0.260	254.0	0.438
	0.4	0.6	3.04	51.23	26.6	3.42	0.442	125.0	0.389
	0.6	0.4	1.12	40.42	13.5	3.20	0.599	75.8	0.321
	0.8	0.2	0.268	28.38	5.65	2.67	0.753	42.7	0.237
	0.9	0.1	0.0743	20.55	2.63	2.21	0.846	25.9	0.178
	0.97	0.03	0.0105	12.11	0.832	1.69	0.931	11.6	0.112
	1.0	0	0	0	0	—	1.0	0	0

TABLE AT 23 DIMENSIONLESS PERFORMANCE PARAMETERS FOR 60° BEARING, CENTRALLY LOADED, WITH SIDE FLOW

Courtesy Raimondi and Boyd[11.7] and Westinghouse Electric. Values of h_o/c_r for optimum bearings, maximum load and minimum friction, respectively: for $L/D = \infty$, 0.25, 0.23; for $L/D = 1$, 0.23, 0.22; for $L/D = 0.5$, 0.2, 0.16; for $L/D = 0.25$, 0.15, 0.1. When $1 - h_o/c_r \neq \epsilon$, the trailing end of the bearing does not reach h_o as defined in Fig. 11.6; that is, h_o in this table is h_{min}.

L/D	ϵ	$\frac{h_o}{c_r}$	S	ϕ	$\frac{r}{c_r}f$	$\frac{q}{rc_rn_sL}$	$\frac{q}{q_s}$	$\frac{\rho c\Delta t_o}{p}$	$\frac{p}{p_{max}}$
∞	0	1.0	∞	90.0	∞	π	0	∞	—
	0.1	0.9191	5.75	65.91	19.7	3.01	0	82.3	0.337
	0.2	0.8109	2.66	48.91	10.1	2.73	0	46.5	0.336
	0.4	0.6002	0.931	31.96	4.67	2.07	0	28.4	0.329
	0.6	0.4	0.322	23.21	2.40	1.40	0	21.5	0.317
	0.8	0.2	0.0755	17.39	1.10	0.722	0	19.2	0.287
	0.9	0.1	0.0241	14.94	0.667	0.372	0	22.5	0.243
	0.97	0.03	0.00495	10.58	0.372	0.115	0	40.7	0.163
	1.0	0	0	0	0	0	0	∞	0
1	0	1.0	∞	90.0	∞	π	0	∞	—
	0.1	0.9212	8.52	67.92	29.1	3.07	0.0267	121.0	0.252
	0.2	0.8133	3.92	50.96	14.8	2.82	0.0481	67.4	0.251
	0.4	0.6010	1.34	33.99	6.61	2.22	0.0849	39.1	0.247
	0.6	0.4	0.450	24.56	3.29	1.56	0.127	28.2	0.239
	0.8	0.2	0.101	18.33	1.42	0.883	0.200	22.5	0.220
	0.9	0.1	0.0309	15.33	0.822	0.519	0.287	23.2	0.192
	0.97	0.03	0.00584	10.88	0.422	0.226	0.465	30.5	0.139
	1.0	0	0	0	0	—	1.0	0	0
$\frac{1}{2}$	0	1.0	∞	90.0	∞	π	0.0	∞	—
	0.1	0.9223	14.2	69.00	48.6	3.11	0.0488	201.0	0.239
	0.2	0.8152	6.47	52.60	24.2	2.91	0.0883	109.0	0.239
	0.4	0.6039	2.14	37.00	10.3	2.38	0.160	59.4	0.233
	0.6	0.4	0.695	26.98	4.93	1.74	0.236	40.3	0.225
	0.8	0.2	0.149	19.57	2.02	1.05	0.350	29.4	0.201
	0.9	0.1	0.0422	15.91	1.08	0.664	0.464	26.5	0.172
	0.97	0.03	0.00704	10.85	0.490	0.329	0.650	27.8	0.122
	1.0	0	0	0	0	—	1.0	0	0
$\frac{1}{4}$	0	1.0	∞	90.0	∞	π	0	∞	—
	0.1	0.9251	35.8	71.55	121.0	3.16	0.0666	499.0	0.251
	0.2	0.8242	16.0	58.51	58.7	3.04	0.131	260.0	0.249
	0.4	0.6074	5.20	41.01	24.5	2.57	0.236	136.0	0.242
	0.6	0.4	1.65	30.14	11.2	1.98	0.346	86.1	0.228
	0.8	0.2	0.333	21.70	4.27	1.30	0.496	54.9	0.195
	0.9	0.1	0.0844	16.87	2.01	0.894	0.620	41.0	0.159
	0.97	0.03	0.0110	10.81	0.713	0.507	0.786	29.1	0.107
	1.0	0	0	0	0	—	1.0	0	0

TABLE AT 24 VALUES OF FORM FACTOR Y IN LEWIS' EQUATION

FD = full depth.

NO. TEETH	LOAD AT TIP			LOAD NEAR MIDDLE		NO. TEETH	LOAD AT TIP			LOAD NEAR MIDDLE	
	$14\frac{1}{2}°$ *FD*	20° *FD*	20° *Stub*	$14\frac{1}{2}°$ *FD*	20° *FD*		$14\frac{1}{2}°$ *FD*	20° *FD*	20° *Stub*	$14\frac{1}{2}°$ *FD*	20° *FD*
10	0.176	0.201	0.261			32	0.322	0.364	0.443	0.547	0.617
11	0.192	0.226	0.289			33	0.324	0.367	0.445	0.550	0.623
12	0.210	0.245	0.311	0.355	0.415	35	0.327	0.373	0.449	0.556	0.633
13	0.223	0.264	0.324	0.377	0.443	37	0.330	0.380	0.454	0.563	0.645
14	0.236	0.276	0.339	0.399	0.468	39	0.335	0.386	0.457	0.568	0.655
15	0.245	0.289	0.349	0.415	0.490	40	0.336	0.389	0.459	0.570	0.659
16	0.255	0.295	0.360	0.430	0.503	45	0.340	0.399	0.468	0.579	0.678
17	0.264	0.302	0.368	0.446	0.512	50	0.346	0.408	0.474	0.588	0.694
18	0.270	0.308	0.377	0.459	0.522	55	0.352	0.415	0.480	0.596	0.704
19	0.277	0.314	0.386	0.471	0.534	60	0.355	0.421	0.484	0.603	0.713
20	0.283	0.320	0.393	0.481	0.544	65	0.358	0.425	0.488	0.607	0.721
21	0.289	0.326	0.399	0.490	0.553	70	0.360	0.429	0.493	0.610	0.728
22	0.292	0.330	0.404	0.496	0.559	75	0.361	0.433	0.496	0.613	0.735
23	0.296	0.333	0.408	0.502	0.565	80	0.363	0.436	0.499	0.615	0.739
24	0.302	0.337	0.411	0.509	0.572	90	0.366	0.442	0.503	0.619	0.747
25	0.305	0.340	0.416	0.515	0.580	100	0.368	0.446	0.506	0.622	0.755
26	0.308	0.344	0.421	0.522	0.588	150	0.375	0.458	0.518	0.635	0.779
27	0.311	0.348	0.426	0.528	0.592	200	0.378	0.463	0.524	0.640	0.787
28	0.314	0.352	0.430	0.534	0.597	300	0.382	0.471	0.534	0.650	0.801
29	0.316	0.355	0.434	0.537	0.602	Rack	0.390	0.484	0.550	0.660	0.823
30	0.318	0.358	0.437	0.540	0.606						

TABLE AT 25 VALUES OF C FOR $e = 0.001$ in.

For other values of e, multiply the value given by the number of thousandths that e is; for example, for cast iron and cast iron, $14\frac{1}{2}°$ full depth, and $e = 0.004$ in., $C = (4)(800) = 3200$. For other materials, use $C = kE_gE_p/(E_g + E_p)$. Values of C for bronze are virtually the same as for the gray iron (modulii of elasticity about the same). FD = full depth.

MATERIAL	$14\frac{1}{2}°$ FD ($k = 0.107e$)	20° FD ($k = 0.111e$)	20° STUB ($k = 0.115e$)
Gray iron and gray iron . .	800	830	860
Gray iron and steel . . .	1100	1140	1180
Steel and steel	1600	1660	1720

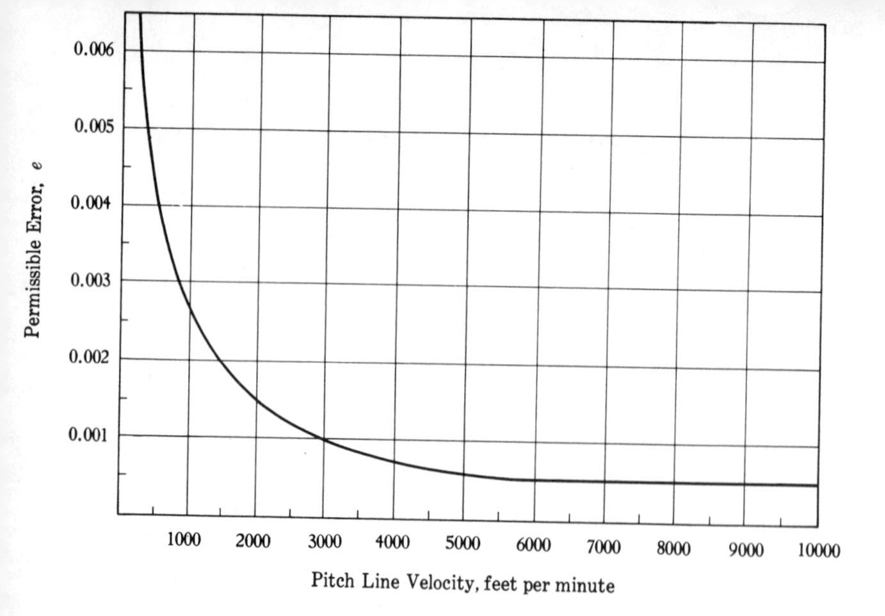

FIGURE AF 19 Maximum Permissible Errors in Gear-tooth Profiles. Extreme quietness will require a smaller error than indicated by this curve.

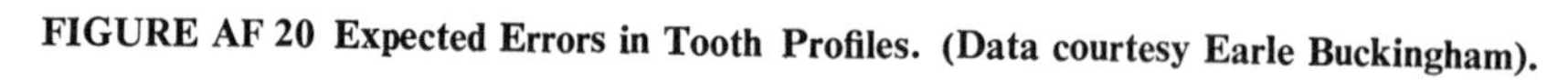

FIGURE AF 20 Expected Errors in Tooth Profiles. (Data courtesy Earle Buckingham).

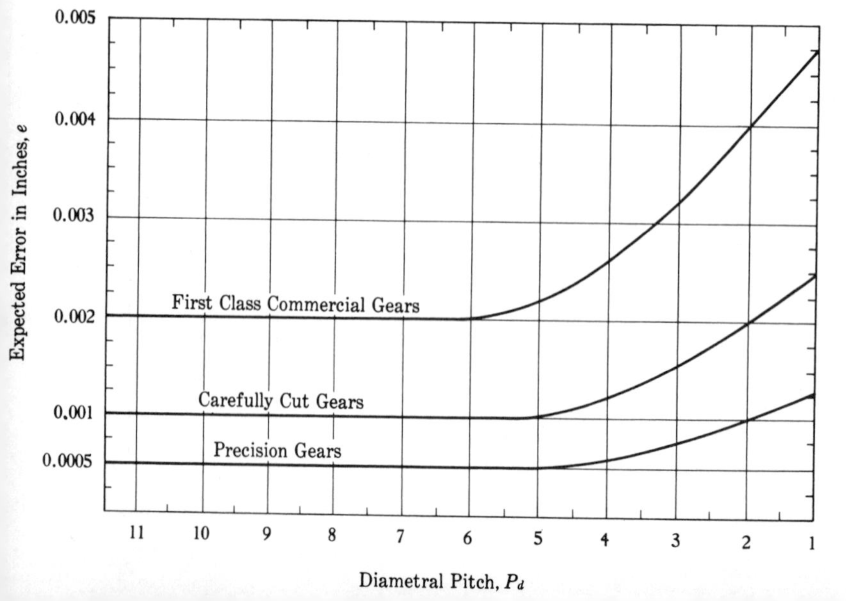

TABLE AT 26 VALUES OF LIMITING WEAR-LOAD FACTOR K_g

Specified BHN's are minimums. Values are for indefinite life unless otherwise indicated. Straight line interpolations on the sum of BHN's are permissible when the difference in BHN's is less than 100 points.

COMBINATIONS OF MATERIALS (BHN) AND LIFE	s_{nsurf} ksi	K_g $14\frac{1}{2}°$	K_g 20°
BOTH GEARS STEEL:			
Sum of BHN = 300, 10^6 cycles		63	86
Ditto 10^7 cycles		40	54
Ditto 4×10^7 cycles or more	50	30	41
Sum of BHN = 350	60	43	58
Sum of BHN = 400, 10^6 cycles		119	162
Ditto 10^7 cycles		75	102
Ditto 4×10^7 cycles or more	70	58	79
Sum of BHN = 450	80	76	103
Sum of BHN = 500	90	96	131
Sum of BHN = 550	100	119	162
Sum of BHN = 600, 10^6 cycles		292	400
Ditto 10^7 cycles		185	252
Ditto 4×10^7 cycles or more	110	144	196
Sum of BHN = 650	120	171	233
Sum of BHN = 700	130	196	270
Sum of BHN = 750	140	233	318
Sum of BHN = 800	150	268	366
Steel (500) and steel (350)	145	250	342
Steel (450) and same	170	344	470
Steel (500), induction hardened, and same, 10^7 cycles		880	1190
Ditto 10^8 cycles		670	920
Ditto 10^{10} cycles		405	555
Steel (600), carburized case hardened, and same, 10^7 cycles		1230	1680
Ditto 10^8 cycles		940	1280
Ditto 10^{10} cycles		550	750
Steel (150) and cast iron	50	44	60
Steel (250) and Ni cast iron, HT	90	150	205
Steel (630) and SAE 65 phosphor bronze (67)*		53	72
Steel (250 and over) and chilled phosphor bronze	83	128	175
Steel (630) and laminated phenolic*		46	64
Cast iron, class 20, and same*		81	112
Cast iron and same, 10^6 cycles		376	515
Ditto 10^7 cycles		212	290
Ditto 4×10^7 cycles†		150	205
Cast iron with steel scrap and same		170	230
Cast iron, class 30, austempered (270) and same*		224	306
G. M. Meehanite (190) and same*		104	142
Nodular iron casting, 80–60–03 (210) and same*		180	248
Cast iron and phosphor bronze	83	170	234
Cast iron, class 30 (340) and cast aluminum, SAE 39 (60)*		16	22

* These values adapted from Cram.[12.22]

† These values are not consistent with those from Cram, and probably should be discounted.

TABLE AT 27 WEAR FACTOR K_w FOR WORM GEARS

Taken from Buckingham and Ryffel,[14.1] with permission of the publisher, The Industrial Press.

MATERIALS		THREAD ANGLE ϕ_n			
Worm	*Gear*	$14\frac{1}{2}°$	20°	25°	30°
†Hardened steel	Chilled bronze	90	125	150	180
†Hardened steel	Bronze	60	80	100	120
Steel, 250 BHN (min.)	Bronze	36	50	60	72
High-test C.I.	Bronze	80	115	140	165
*Gray iron	Aluminum	10	12	15	18
*High-test C.I.	Gray iron	90	125	150	180
High-test C.I.	Cast steel	22	31	37	45
High-test C.I.	High-test C.I.	135	185	225	270
Steel 250 BHN (min.)	Laminated phenolic	47	64	80	95
Gray iron	Laminated phenolic	70	96	120	140

* For steel worms, multiply given values by 0.6. † Over 500 BHN surface.

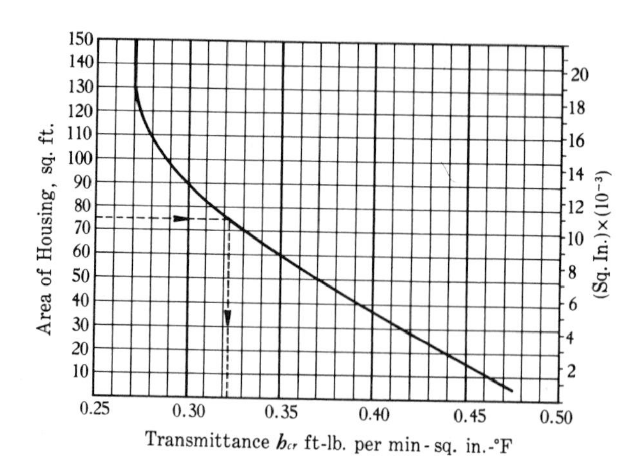

FIGURE AF 21 Transmittance, Gear Cases. To use: Determine the area of the housing; enter chart with this area, move rightward to curve, downward to abscissa, which is the transmittance h_{cr}—as shown by dotted lines.

TABLE AT 28 PROPERTIES OF WIRE ROPE

Approximate weight of rope is w lb. per ft.; D_r = diameter of rope; D_s = diameter of sheave; A_m = cross-section area of metal, sq. in.; E_r = modulus of elasticity of the rope; I.P.S. stands for improved plow steel; P.S. for plow steel; M.P.S. for mild plow steel. All values are for rope with fiber core. It may occasionally be useful in *preliminary* computations to *estimate* the ultimate strengths of terms in D_r; for example, for 6 × 19 rope: VHS, $F_u \approx 48D_r^2$ tons; IPS, $F_u \approx 42D_r^2$; PS $F_u \approx 36D_r^2$; MPS $F_u \approx 32D_r^2$ tons. Multiply the values given by 1.075 to obtain strength of IWRC. For 6 × 19 traction steel, $F_u = (0.87)(F_u$ for MPS) will be safe.

DIA. OF ROPE D_r., *in.*	6 × 7 Wire Rope $w = 1.52D_r^2$ lb/ft. Min. $D_s = 42D_r$ in. Desirable $D_s = 72D_r$ in. $D_w \approx 0.111D_r$ $A_m \approx 0.38D_r^2$ sq. in. $E_r \approx 13 \times 10^6$ psi			6 × 19 Wire Rope $w \approx 1.6D_r^2$ lb/ft. Min. $D_s = 30D_r$ in. Desirable $D_s = 45D_r$ in. $D_w \approx 0.067D_r$ $A_m \approx 0.4D_r^2$ sq. in. $E_r \approx 12 \times 10^6$ psi IWRC: $w \approx 1.76D_r^2$ lb/ft.			6 × 37 Wire Rope $w = 1.55D_r^2$ Min. $D_s = 18D_r$ in. Desirable $D_s = 27D_r$ in. $D_w \approx 0.048D_r$ in. $A_m \approx 0.4D_r^2$ $E_r \approx 12 \times 10^6$ IWRC: $w \approx 1.71D_r^2$	
	NOMINAL BREAKING STRENGTH IN TONS OF 2000 LB., F_u							
	I.P.S.	*P.S.*	*M.P.S.*	*I.P.S.*	*P.S.*	*M.P.S.*	*I.P.S.*	*P.S.*
$\frac{1}{4}$	2.64	2.30	2.00	2.74	2.39	2.07	2.59	2.25
$\frac{5}{16}$	4.10	3.56	3.10	4.26	3.71	3.22	4.03	3.50
$\frac{3}{8}$	5.86	5.10	4.43	6.10	5.31	4.62	5.77	5.02
$\frac{7}{16}$	7.93	6.90	6.00	8.27	7.19	6.25	7.82	6.80
$\frac{1}{2}$	10.3	8.96	7.79	10.7	9.35	8.13	10.2	8.85
$\frac{9}{16}$	13.0	11.3	9.82	13.5	11.8	10.2	12.9	11.2
$\frac{5}{8}$	15.9	13.9	12.0	16.7	14.5	12.6	15.8	13.7
$\frac{3}{4}$	22.7	19.8	17.2	23.8	20.7	18.0	22.6	19.6
$\frac{7}{8}$	30.7	26.7	23.2	32.2	28.0	24.3	30.6	26.6
1	39.7	34.5	30.0	41.8	36.4	31.6	39.8	34.6
$1\frac{1}{8}$	49.8	43.3	37.7	52.6	45.7	39.8	50.1	43.5
$1\frac{1}{4}$	61.0	53.0	46.1	64.6	56.2	48.8	61.5	53.5
$1\frac{3}{8}$	73.1	63.6	55.3	77.7	67.5	58.8	74.1	64.5
$1\frac{1}{2}$	86.2	75.0	65.2	92.0	80.0	69.6	87.9	76.4
$1\frac{5}{8}$				107.0	93.4	81.2	103.0	89.3
$1\frac{3}{4}$				124.0	108.0	93.6	119.0	103.0
$1\frac{7}{8}$				141.0	123.0	107.0	136.0	118.0
2				160.0	139.0	121.0	154.0	134.0
$2\frac{1}{8}$				179.0	156.0		173.0	150.0
$2\frac{1}{4}$				200.0	174.0		193.0	168.0
$2\frac{1}{2}$				244.0	212.0		236.0	205.0
$2\frac{3}{4}$				292.0	254.0		284.0	247.0

TABLE AT 29 DESIGN DATA FOR BRAKES AND CLUTCHES

From mixed references, mostly *(18.8, 18.11, 18.13, 18.22)*. Unless otherwise indicated, values of f are for dry surfaces; if greasy or wet, they are much lower. If f at the lower end of the range (or below) is used in design, this in effect introduces a design factor. Use the lower pressure P given as the design maximum pressure where possible. The drum temperatures are the maximum values for steady operation. See manufacturers' catalogs for more detail.

MATERIAL	DRUM MAX. t, °F	f	P PSI (*max.*)	MAX. VEL. *fpm*
Metal on metal		0.2 to 0.25	150 (250)	
Wood on metal	150	0.2 to 0.25	50 (90)	
Leather on metal or wood	150	0.3 to 0.4	15 (40)	
Cork on iron		0.35	10 (15)	
Molded blocks	650		150	7500
Asbestos in rubber compound, compressed, on metal	400	0.3 to 0.4	75 (100)	
Asbestos in resin binder, molded, on metal	500	0.3 to 0.4	75 (100)	5000
in oil		$\approx$0.10	(600)	
Asbestos, flexible woven, on metal	300	0.35 to 0.45	50	
in oil		$\approx$0.12		
Sintered metal on cast iron	> 400	0.20 to 0.40	400	
in oil (as in automatic transmissions)		0.05 to 0.08		

KIND OF WELD AND STRESS	CODES, *ksi*
BUTTWELDS	
Tension	20(b) 8(f)
Compression	20(b) 18(c)
Shear	13(b)
Bending	
FILLET WELDS	
All	13.6(h)
Parallel	13.6(b) 12.4(c)
Transverse	

TABLE AT 30 DESIGN STRESSES FOR WELDED JOINTS

(a) Recommendations by Jennings.[19.1] (b) AWS Building Code. (c) AWS Bridge Code. (d) Lincoln Electric.[19.9] (e) Adapted from AWS Bridge Code and U.S. Steel data by Blodgett;[19.16] for a common class of structural steel or the equivalent; for $n_c = 2 \times 10^6$ cycles and $n_c = 10^5$ cycles; R = stress ratio = s_{min}/s_{max}, § 4.5. Maximum values for structural steel, butt welds, beads left on, say 18 ksi; for alloy steel, 54 ksi; for fillet welds, 12.5 and 37 ksi, respectively. The source does not mention stress concentration factors. (f) ASME Code, class 3, double-weld butt. (g) *Ultimate* strength of aluminum alloy welded with 1100 wire, argon shielded.[19.15] Use factor of safety. (h) AISC Building Code,[5.34] with E60XX electrodes.

SHIELDED WELDING		UNSHIELDED WELDING		FATIGUE DESIGN STRESSES(e)			
				Struct. Steel		*Q&T Alloy Steel*	
Steady	*Reversed*	*Steady*	*Reversed*	2×10^6	10^5	2×10^6	10^5
16(a) 11(g)	8(a)	13(a)	5(a)	$\frac{16}{1 - 0.8R}$	$\frac{18}{1 - 0.5R}$	$\frac{16.5}{1 - 0.8R}$	$\frac{31}{1 - 0.6R}$
18(a)	8(a)	15(a)	5(a)	$\frac{18}{1 - R}$	$\frac{18}{1 - 0.5R}$		
10(a) 15(d)	5(a)	8(a)	3(a)	$\frac{9}{1 - 0.5R}$	$\frac{13}{1 - 0.5R}$		
14(a) 11.3(g)	5(a)	11.3(a)	3(a)				
16(d) 12.7(g)				$\frac{7.2}{1 - 0.5R}$	$\frac{12.5}{1 - 0.5R}$	$\frac{9}{1 - 0.8R}$	$\frac{20.5}{1 - 0.6R}$

INDEX

C

D

E

G

H

I

J

N

O

P

Q

R

S

U

V

W

Y

Z